T0344967

Abstract Algebra

A Comprehensive Introduction

Through this book, upper undergraduate mathematics majors will master a challenging yet rewarding subject, and approach advanced studies in algebra, number theory and geometry with confidence. Groups, rings and fields are covered in depth with a strong emphasis on irreducible polynomials, a fresh approach to modules and linear algebra, a fresh take on Gröbner theory, and a group theoretic treatment of Rejewski's deciphering of the Enigma machine. It includes a detailed treatment of the basics on finite groups, including Sylow theory and the structure of finite abelian groups. Galois theory and its applications to polynomial equations and geometric constructions are treated in depth. Those interested in computations will appreciate the novel treatment of division algorithms. This rigorous text "gets to the point," focusing on concisely demonstrating the concept at hand, taking a "definitions first, examples next" approach. Exercises reinforce the main ideas of the text and encourage students' creativity.

John W. Lawrence is Professor Emeritus at the University of Waterloo. He was born in Ottawa, Canada, and received degrees from Carleton University and McGill University. After a year of postdoctoral work at the University of Chicago, he joined the Pure Mathematics Department of the University of Waterloo. He now lives with his wife Louise, in Thornhill Canada, where he continues his research in mathematics and probability.

Frank A. Zorzitto is Professor Emeritus at the University of Waterloo. After receiving his doctorate at Queen's University in Kingston, Canada, he served for forty years as a researcher in algebra and a professor in the Pure Mathematics Department at the University of Waterloo, and served as Department Chair for twelve years. In recognition of his commitment to the education of students, he received the University's Distinguished Teaching Award. Upon his retirement he was made an Honorary Member of the University. Currently he offers an online course to high school math teachers, based on his ebook *A Taste of Number Theory*. He continues to teach and to write in his retirement.

CAMBRIDGE MATHEMATICAL TEXTBOOKS

Cambridge Mathematical Textbooks is a program of undergraduate and beginning graduate-level textbooks for core courses, new courses, and interdisciplinary courses in pure and applied mathematics. These texts provide motivation with plenty of exercises of varying difficulty, interesting examples, modern applications, and unique approaches to the material.

A complete list of books in the series can be found at www.cambridge.org/mathematics
Recent titles include the following:

Chance, Strategy, and Choice: An Introduction to the Mathematics of Games and Elections, S. B. Smith
Set Theory: A First Course, D. W. Cunningham
Chaotic Dynamics: Fractals, Tilings, and Substitutions, G. R. Goodson
A Second Course in Linear Algebra, S. R. Garcia & R. A. Horn
Introduction to Experimental Mathematics, S. Eilers & R. Johansen
Exploring Mathematics: An Engaging Introduction to Proof, J. Meier & D. Smith
A First Course in Analysis, J. B. Conway
Introduction to Probability, D. F. Anderson, T. Seppäläinen & B. Valkó
Linear Algebra, E. S. Meckes & M. W. Meckes
A Short Course in Differential Topology, B. I. Dundas
Abstract Algebra with Applications, A. Terras
Complex Analysis, D. E. Marshall
Abstract Algebra: A Comprehensive Introduction, J. W. Lawrence & F. A. Zorzitto

Abstract Algebra

A Comprehensive Introduction

John W. Lawrence
University of Waterloo, Ontario

Frank A. Zorzitto
University of Waterloo, Ontario

Shaftesbury Road, Cambridge CB2 8EA, United Kingdom

One Liberty Plaza, 20th Floor, New York, NY 10006, USA

477 Williamstown Road, Port Melbourne, VIC 3207, Australia

314–321, 3rd Floor, Plot 3, Splendor Forum, Jasola District Centre, New Delhi – 110025, India

103 Penang Road, #05–06/07, Visioncrest Commercial, Singapore 238467

Cambridge University Press is part of Cambridge University Press & Assessment, a department of the University of Cambridge.

We share the University's mission to contribute to society through the pursuit of education, learning and research at the highest international levels of excellence.

www.cambridge.org
Information on this title: www.cambridge.org/9781108836654

DOI: 10.1017/9781108874328

First published 2021

A catalogue record for this publication is available from the British Library

ISBN 978-1-108-83665-4 Hardback

Additional resources for this publication at www.cambridge.org/abstract-algebra

Dedicated to our families...

Louise, Donna, Lisa, Joanna, Angela and Alex

Contents

Preface

Every practitioner of mathematics, pure or applied, could use some algebra. Algebra includes an adherence to precise definitions, a willingness to examine concrete problems through meaningful abstractions, and a thirst for lucid proofs based on the axiomatic method. Despite its challenges, algebra has for a long time now been a mainstay of the curriculum in mathematics programs. Not to mention that some people are actually drawn to its puzzles and surprises. Since Bartel Leendert van der Waerden's seminal book *Moderne Algebra* was published in 1930, a multitude of authors have offered their take on this subject. By means of this book we hope to make our contribution.

Our book is intended for senior undergraduate students of modern abstract algebra. Those more gifted in mathematics can well profit from it sooner. Beginning graduate students who need a refresher could find it useful as well. We expect that our reader comes with a bit of exposure to rigorous proofs, some familiarity with the common language of sets such as injective and surjective mappings, and some understanding of linear algebra including such notions as finite-dimensional vector spaces, non-singular matrices, and linear transformations.

In order for any mathematics book to be effective, it must be precise with the mathematics but also approachable to the student. We have tried to meet this goal by walking a fine line between proper mathematical formality and reader friendliness. We hope to have not written too much, but also not too little. There seem to be two points of view regarding the introduction of a concept. One is that the student should be gently exposed to motivating examples first. Another aims for efficiency, presenting the definitions and theorems at the outset. By and large we have taken a "definitions first – examples next" approach, thinking that too much effort at motivation can in its own way be distracting. Yet we have tried not to go overboard with this approach, and to offer motivating examples soon enough.

Obviously, the exercises are for learning. If the exercises in a book are too easy, the student will not learn enough, and quickly become bored. If they are too hard, discouragement can set in. We have decided to limit the number of exercises that are utterly routine. Yet once a student has digested a section, the bulk of the exercises that follow it should be of reasonable difficulty. Those exercises which might be considered hard are marked with an asterisk. In the end, the difficulty of the exercises will depend on the preparation, skill and stamina of the student.

Algebra now comprises several large areas and many special topics within them. This makes it challenging for authors to select which topics will meet the expectations of a broad segment of algebra instructors, whose task it is to pick a book for their students. We structured our book along the traditional lines of groups, rings, fields and Galois theory. For something extra we offer an in depth discussion of modules over principal ideal domains and of Gröbner bases, a subject which may be of interest to those who relish computing.

We open our book with a short discussion on the integers, primes and congruences, in order to ensure the student sees the ways of algebra within a truly basic example. These matters are treated in the spirit of a review, which could well be familiar to many a reader.

Our book more properly begins with Chapter 2 where we discuss groups, with an emphasis on finite groups, using a well trodden approach. The symmetric group gets a careful discussion. The fundamentals of Lagrange's theorem (including the failure of its converse), normal subgroups, homomorphisms, quotient groups, the first isomorphism theorem, internal and external products of groups, are covered along familiar lines. We close the chapter with a proof of the structure theorem for finite abelian groups. Since by its very nature this will demand a bit more of the student, we go at it with patience.

Chapter 3 focuses on group actions, orbits, stabilizers, some counting problems, the class equation, and the substantial Sylow theory. We introduce semi-direct products as a way to describe some groups. In anticipation of their use in Galois theory the chapter closes with the basics about solvable groups. Even though this is primarily a book in pure mathematics, we decided to close Chapter 3 with a discussion of the Enigma machine and how group theory was used to break the Enigma code. This is an interesting story of how abstract algebra influenced history.

Chapters 4 and 5 are about rings, with commutative domains as the primary focus. In Chapter 4 we cover the essential notions of units, zero divisors, ideals, integral domains, quotient rings, homomorphisms, the first isomorphism theorem, the correspondence theorem, maximal and prime ideals, and fraction fields. We also offer a somewhat more leisurely treatment of the notion of polynomials, which of course can be glossed over by those who feel it might be unnecessary. Chapter 5 is about primes, unique factorization, Euclidean domains, Gauss' lemma, irreducibility of polynomials, and the Hilbert basis theorem for Noetherian rings. We also introduce a souped up version of Eisenstein's criterion, which can be used in unexpected ways to compute degrees of field extensions.

In Chapter 6 we cover algebraic field extensions, splitting fields, separable polynomials, and the Galois group. Here the reader will find an unhurried treatment of the Galois correspondence. Some of the more important applications of Galois theory are offered in Chapter 7, including, of course, Galois' contribution on the solvability of polynomial equations by radicals. There is also a comprehensive treatment of ruler and compass constructions, which at one point uses a bit of Galois theory.

Chapter 8 deals with the structure theorem for finitely generated modules over principal ideal domains along with its application to the Jordan canonical form for matrices. Of course it also applies to finite (and even finitely generated) abelian groups, but that topic is dealt with in Chapter 2. A small bit of redundancy is useful because not every student will get to Chapter 8, and because students can benefit from seeing the same ideas more than once.

An expert may notice that our proofs of a few difficult points differ somewhat from the mainstream. We hope that the reader will like the way we have organized this challenging material.

Our treatment of Gröbner bases in Chapter 9 is more daring. We want to present the concept of a Gröbner basis in the setting of abstract algebra. For this reason we open up with a discussion of what we call a Gröbner domain. These can be seen as a generalization of Euclidean domains, which permits division algorithms in a larger context. Even though our chief example is the ring of polynomials in several variables over a field, we try to remain within the abstraction of Gröbner domains as much as possible.

Our closing Appendix A offers some informal set theory, including the equivalence of the axiom of choice and Zorn's lemma, a discussion of cardinality, and the use of these ideas to build the algebraic closure of any field. The preceding chapters do not make significant use of Appendix A. The proof that rings have maximal ideals is the main (and easy) application of Zorn's lemma. The material of Appendix A is offered for the student who might find it handy to have it here, and might appreciate our effort to make this abstract material approachable.

The book contains more than enough material for two semesters of instruction. We find it difficult to advise how much is to be covered in a given semester. This is because the length of a semester can differ across various jurisdictions, and also because classes can differ in their abilities and institutions can differ in their expectations. In our home university in a third year one semester course we typically would assume Chapter 1, and then cover much of Chapters 2, 3, 4 and 5 on groups and rings, possibly skipping a few themes such as semi-direct products, solvable groups, the Enigma machine, and irreducibility over Noetherian domains. This would be followed up in a second semester with an in depth discussion of field and Galois theory including a discussion of solvable groups. If the class is up to it, we might also treat Chapter 9 on Gröbner bases or possibly modules over principal ideal domains.

Here are some suggested ways to use our book for a one semester course:

- Chapters 1, 2, 3 for a course on groups
- Chapters 1, 4, 5, 6, 7 for a substantial course on commutative rings, fields and Galois theory
- Chapters 1, 4, 5, 8 for a course on commutative rings and modules
- Chapters 1, 4, 5, 9 for a course on rings and division algorithms.

Some instructors believe it is better to teach commutative rings before groups. Even though we have followed the common order of groups-rings-fields, an instructor who wishes to teach rings first should be able to jump into Chapter 4 from the get go. Our treatment of rings makes a small use of the basic facts about groups, but the very slight amount of group theory needed can readily be filled in by the instructor.

The book can also be used in algebra courses of a more general nature. We have tried to adopt a style that is student friendly by explaining the ideas and proofs in a detailed and self contained manner.

One irritant which we have tried to minimize is the somewhat distracting practice of referring to exercises or to other proofs within the proof of a result. Another distracting habit is to leave steps within a proof as an exercise. Of course this is deemed to encourage the student to work through some details. It can also be a reflection of an author's impatience. These features cannot be avoided entirely, but by proving virtually everything we claim at the moment we claim it, we think that reading our text will lead to a more seamless experience.

Our gratitude goes out to our Pure Mathematics Department at the University of Waterloo, which gave us the opportunity to teach courses in algebra and thereby come to appreciate what might be interesting, what might be important, and what might be fun. Our never ending streams of committed and often brilliant students taught us that algebra can be a pleasure to learn and to teach.

1 A Refresher on the Integers

A good place to start might be the set of ***integers***:

$$\mathbb{Z} = \{0, \pm 1, \pm 2, \pm 3, \dots\}.$$

We often encounter the notation $\mathbb{N}$ for the set of ***natural numbers***. There seems to be no consensus on whether to include 0 as a natural number. Following the advice of Paul Halmos (1916–2006) in his seminal book *Naive Set Theory*, first published in 1960, we shall include 0 as a natural number. Thus $\mathbb{N}$ stands for the set $\{0, 1, 2, 3, \dots\}$. We will on occasion use the somewhat less common notation $\mathbb{P}$ for the set $\{1, 2, 3, \dots\}$ of ***positive integers***. Often we shall prefer to say things like "let x be a positive integer," instead of "let $x \in \mathbb{P}$."

1.1 Euclidean Division and the Greatest Common Divisor

In the seventh book of Euclid we are told that any integer b, upon division by a positive integer a, yields a unique quotient q and a unique remainder r such that

$$b = aq + r \text{ and } 0 \leq r < a.$$

Indeed, q is the largest integer that is less than or equal to the fraction b/a, and r is simply $b - aq$. The procurement of q and r will be called ***Euclidean division***.

If the remainder $r = 0$, then $b = aq$. In this case we say that a ***divides*** b, and write $a \mid b$. For instance, $17 \mid 51$ since $51 = 17 \cdot 3$. Conversely, a bit of reflection shows that if $r \neq 0$, then $a \nmid b$. For example, the Euclidean division

$$-91 = 17 \cdot (-6) + 11 \text{ where } 0 < 11 < 17,$$

reveals that 17 does not divide -91. For visual clarity we may, on occasion just as above, use a dot to signify multiplication.

When $a \mid b$, we say that b is a ***multiple*** of a, and also that a is a ***divisor*** or ***factor*** of b. Here are a few simple things to note.

- If $a \mid b$ and $b \mid c$, then $a \mid c$.
- If $a \mid b$ and $a \mid c$ and x, y are any integers, then $a \mid bx + cy$.
- If $b \neq 0$ and $a \mid b$, then $|a| \leq |b|$.
- If $a \mid b$ and $b \mid a$, then $a = \pm b$.

The Greatest Common Divisor of Two Integers

If a, b are integers, an ***integer combination*** of a and b is any integer c built up as

$$c = ax + by \text{ for some integers } x, y.$$

For example, the equation $6 \cdot (-3) + 4 \cdot 7 = 10$ reveals that 10 is an integer combination of 6 and 4. But 11 is not an integer combination of 6 and 4, because integer combinations of 6 and 4 have to be even integers. We might note that the same integer combination c could be formed as an integer combination of a and b in more than one way. For example 10 also equals $6 \cdot 1 + 4 \cdot 1$.

If a, b are integers which are not both 0, then there do exist integer combinations of a and b which are positive. For instance $a \cdot a + b \cdot b = a^2 + b^2 > 0$. Hence there exist integers s, t which cause $as + bt$ to be minimal among the positive, integer combinations of a and b.

Now comes a noteworthy result.

Proposition 1.1 (Greatest common divisor). *Let a, b be integers, and suppose at least one of them is not 0. If an integer d satisfies any one of the following properties, then d satisfies them all.*

1. *d is the least among the positive, integer combinations of a and b.*
2. *d is a positive, integer combination of a and b, and d divides every integer combination of a and b.*
3. *d is a positive, common divisor of a and b, and every common divisor of a and b is a divisor of d.*
4. *d is the largest among the numbers that divide both a and b.*

Proof. We will prove that $1 \implies 2 \implies 3 \implies 4 \implies 1$.

$1 \implies 2$. Let $d = as+bt$ be that positive integer combination of a and b which is minimal, and let $c = ax+by$ represent any integer combination of a and b. By Euclidean division there exist integers q, r such that

$$c = dq + r \text{ where } 0 \leq r < d.$$

And so

$$r = c - dq = ax + by - (as + bt)q = a(x - sq) + b(y - tq),$$

which is evidently another integer combination of a and b. By the minimality of d, the non-negative remainder r cannot be positive. Thus $r = 0$, which gives $d \mid c$.

$2 \implies 3$. We have that $0 < d = ax + by$ for some integers x, y and d divides every integer combination of a and b. Thereby d is a common divisor of a and b, because a and b are themselves integer combinations of a and b. Also if c is another common divisor of a and b, then $c \mid ax + by$ too. So $c \mid d$.

$3 \implies 4$. By assumption every common divisor c of a and b is a divisor of d, and $d > 0$. Clearly then $d \geq c$.

$4 \implies 1$. By the proven implications $1 \implies 2 \implies 3 \implies 4$ the unique least, positive, integer combination of a and b is the largest integer that divides both a and b. To repeat, the largest integer that divides both a and b is the least, positive, integer combination of a and b. □

Definition 1.2. The number d that satisfies any one, and thereby all, of the conditions of Proposition 1.1 is called the ***greatest common divisor*** of a and b. We write that positive integer as

$$\gcd(a, b).$$

What should $\gcd(0, 0)$ be? Well, since the greatest common divisor should be an integer combination of a and b, it makes sense to say that $\gcd(0, 0) = 0$, because that is the only possible integer combination of 0 and 0. The choice $\gcd(0, 0) = 0$ is further justified by the fact that 0 is the only divisor of 0, which is also divisible by all other divisors of 0.

The Euclidean Algorithm

For pairs of small integers, their greatest common divisor can be seen by inspection. For instance, $\gcd(42, 30) = 6$. When the integers become large, there is an efficient technique for finding their greatest common divisor, based on repeated use of Euclidean division. Clearly,

$$\gcd(a, b) = \gcd(b, a) = \gcd(\pm a, \pm b), \ \gcd(0, b) = |b| \text{ and } \gcd(b, b) = |b|.$$

Thus, for the purpose of computing greatest common divisors of a and b, we need only consider the situation where $0 < a < b$.

If a, b are not both 0 and $b = aq + r$ for some integers q, r, one can see that the set of common divisors of b and a is the same as the set of common divisors of a and r. Therefore

$$\gcd(b, a) = \gcd(a, r).$$

The preceding remark points the way for Euclidean division to find $\gcd(b, a)$.

Say $0 < a < b$. Apply Euclidean division repeatedly as follows:

$$\begin{aligned} b &= aq_1 + r_1 & \quad 0 &< r_1 < a \\ a &= r_1 q_2 + r_2 & 0 &< r_2 < r_1 \\ r_1 &= r_2 q_3 + r_3 & 0 &< r_3 < r_2 \\ r_2 &= r_3 q_4 + r_4 & 0 &< r_4 < r_3, \end{aligned}$$

to obtain strictly decreasing remainders $r_1 > r_2 > r_3 > \cdots \geq 0$.

Sooner or later an integer remainder becomes 0. In other words, there must be an index n such that

$$\begin{aligned} r_{n-3} &= r_{n-2} q_{n-1} + r_{n-1} & \quad 0 &< r_{n-1} < r_{n-2} \\ r_{n-2} &= r_{n-1} q_n + r_n & 0 &< r_n < r_{n-1} \\ r_{n-1} &= r_n q_{n+1} + 0 & 0 &= r_{n+1}. \end{aligned}$$

From the remark preceding the above algorithm:

$$\begin{aligned} r_n &= \gcd(r_n, 0) = \gcd(r_{n-1}, r_n) = \gcd(r_{n-2}, r_{n-1}) = \cdots \\ &= \gcd(r_3, r_4) = \gcd(r_2, r_3) = \gcd(r_1, r_2) = \gcd(a, r_1) = \gcd(b, a). \end{aligned}$$

The last positive remainder r_n equals $\gcd(b,a)$. This famous process for obtaining greatest common divisors is called the ***Euclidean algorithm***.

For example, here comes $\gcd(8316, 4641)$:

$$\begin{aligned} 8316 &= 4641 \cdot 1 + 3675 \\ 4641 &= 3675 \cdot 1 + 966 \\ 3675 &= 966 \cdot 3 + 777 \\ 966 &= 777 \cdot 1 + 189 \\ 777 &= 189 \cdot 4 + 21 \\ 189 &= 21 \cdot 9 + 0. \end{aligned}$$

Thus, $\gcd(8316, 4641) = 21$.

Roughly, the number of steps in the Euclidean algorithm is no more than twice the base-two logarithm of b. Consequently, it comes as no surprise that machines can rapidly implement the Euclidean algorithm for enormous integers with hundreds of digits.

According to Proposition 1.1, $\gcd(a,b) = ax+by$ for some integers x, y. Inside the Euclidean algorithm lies the method of obtaining such x, y. It is a matter of backtracking up along the algorithm. For instance, in the preceding worked example:

$$\begin{aligned} \gcd(8316, 4641) &= 21 \\ &= 777 \cdot 1 - 189 \cdot 4 \\ &= 777 \cdot 1 - (966 - 777 \cdot 1) \cdot 4 \\ &= 777 \cdot 5 - 966 \cdot 4 \\ &= (3675 - 966 \cdot 3) \cdot 5 - 966 \cdot 4 \\ &= 3675 \cdot 5 - 966 \cdot 19 \\ &= 3675 \cdot 5 - (4641 - 3675 \cdot 1) \cdot 19 \\ &= 3675 \cdot 24 - 4641 \cdot 19 \\ &= (8316 - 4641 \cdot 1) \cdot 24 - 4641 \cdot 19 \\ &= 8316 \cdot 24 - 4641 \cdot 43. \end{aligned}$$

Thus, $\gcd(8316, 4641) = 8316x + 4641y$ where $x = 24$ and $y = -43$.

Calculations such as these can of course be done by machine when the integers in question are big.

Coprime Integers

Definition 1.3. If the greatest common divisor of integers a and b equals 1, we say that a, b are ***coprime***.

Evidently, two integers are coprime if and only if their only common divisors are ± 1. In accordance with Proposition 1.1, a, b are coprime precisely when

$$ax + by = 1$$

for some integers x, y. A couple of properties stand out for pairs of coprime integers.

Proposition 1.4. *If a, b, c are integers and a, b are coprime and $a \mid bc$, then $a \mid c$.*

Proof. We know that $ax + by = 1$ for some integers x, y. Then $acx + bcy = c$. Since $a \mid bc$ and evidently $a \mid ac$, it follows that $a \mid acx + bcy = c$. □

Proposition 1.5. *If a, b, c are integers and $a \mid c$ and $b \mid c$, then $ab \mid c$.*

Proof. We know that $ax + by = 1$ for some integers x, y. Then $acx + cby = c$. Since $b \mid c$, it is clear that $ab \mid ac$, and likewise $ab \mid cb$ because $a \mid c$. So, $ab \mid acx + cby = c$. □

EXERCISES

1. Use the Euclidean algorithm to find $\gcd(3150, 3003)$ and express this greatest common divisor as an integer combination of 3150 and 3003.
2. Using your favorite software, write a program to calculate the greatest common divisor of two integers by means of the Euclidean algorithm, and to express the greatest common divisor as an integer combination of the integers. By using your program, or otherwise, find $\gcd(2452548, 2943234)$ and express this greatest common divisor as an integer combination of the given integers.
3. These exercises can be done by using Proposition 1.1.
   (a) If a, b are non-zero integers and k is any integer, show that $\gcd(ka, kb) = k\gcd(a, b)$.
   (b) If a, b are non-zero integers and k is a positive integer, show that $\gcd(a + kb, b) = \gcd(a, b)$.
   (c) If $a \mid bc$, show that $a \mid b\gcd(a, c)$.
   (d) If a, b are coprime integers and $c \mid at$ and $c \mid bt$, show that $c \mid t$.
   (e) If a, b, c are integers with a, c coprime, prove that $\gcd(ab, c) = \gcd(b, c)$.
   (f) If a, b are each coprime with c, show that ab is coprime with c.
4. If a, b, c are integers with a, b non-zero, show that the equation $ax + by = c$ has integer solutions x, y if and only if $\gcd(a, b) \mid c$.
   Find an integer solution x, y to the equation $91x + 55y = 12$.
5. Show that 1 is the only complex number which satisfies both equations $x^{245} = 1$ and $x^{297} = 1$.
6. Let a, b be positive integers. Let $g = \gcd(a, b)$ and $\ell = ab/g$.
   (a) Explain very briefly why $a \mid \ell$ and $b \mid \ell$. Thus ℓ is a common multiple of a and b.

(b) It turns out that ℓ is the ***least common multiple*** of a and b. Prove this claim by showing that if a positive integer m is a common multiple of a and b, then $\ell \mid m$.

Hint. To get from $a \mid m, b \mid m$ to $\ell \mid m$, it suffices to show that $ab \mid mg$. Use Proposition 1.1.

Typically ℓ is written as $\operatorname{lcm}(a,b)$, and we have the identity

$$ab = \gcd(a,b)\operatorname{lcm}(a,b).$$

7. One might wonder about the efficiency of the Euclidean algorithm. Suppose a, b are integers such that $0 < a < b$, and that the Euclidean algorithm is applied to obtain $\gcd(a,b)$. By inspecting the algorithm one can see that each line of the algorithm consists of one Euclidean division, that there is a total of $n+1$ lines, and that $\gcd(a,b)$ appears on the nth line as r_n. We ask ourselves: how big could n get in terms of b?

(a) If $0 < a < b$ and $b = aq + r$ for a quotient q and remainder r with $0 \le r < a$, show that $q \ge 1$, and then $r < b/2$.

(b) Suppose that $r_1, r_2, r_3, \ldots, r_n$ are the positive strictly decreasing remainders which appear in the Euclidean algorithm used to obtain r_n as $\gcd(a,b)$.

If n is even, explain how the following inequalities emerge:

$$r_2 < \frac{b}{2}, \quad r_4 < \frac{b}{2^2}, \quad r_6 < \frac{b}{2^3}, \ldots, \quad r_n < \frac{b}{2^{n/2}}.$$

If n is odd, explain how the following inequalities emerge:

$$r_1 < \frac{b}{2}, \quad r_3 < \frac{b}{2^2}, \quad r_5 < \frac{b}{2^3}, \ldots, \quad r_n < \frac{b}{2^{(n+1)/2}} < \frac{b}{2^{n/2}}.$$

(c) If $\gcd(a,b)$ appears as r_n on the nth line of the Euclidean algorithm, use the fact $1 \le r_n$ to deduce that $2^{n/2} < b$, and then $n < 2\log_2(b)$.

Since line $n+1$ with a zero remainder is needed to terminate the Euclidean algorithm, we learn that the number of Euclidean divisions used in the Euclidean algorithm to compute $\gcd(a,b)$ is at most

$$1 + 2\log_2(b).$$

(d) If a, b are positive integers of size at most 987654321234567, show that the Euclidean algorithm will compute $\gcd(a,b)$ in no more than 100 lines.

If a, b are positive integers of size at most 2^{500}, show that the Euclidean algorithm will compute $\gcd(a,b)$ in at most 1000 lines.

Despite such enormous possibilities, that is not a lot of lines for a computer program to execute.

8. Suppose that a, b, c are integers and that the only factors common to all three are ± 1. Show that there exist integers x, y, z such that $ax + by + cz = 1$.

Generalize this to a statement about any number of integers $a_1, a_2, \ldots, a_n$.

1.2 Primes and Unique Factorization

Problems in $\mathbb{Z}$ invariably boil down to problems about primes.

Definition 1.6. An integer p is called ***prime*** when $p \neq 0, 1, -1$ and the only divisors of p are ± 1 and $\pm p$. An integer n other than $0, 1, -1$ and such that n has a divisor in addition to $\pm 1, \pm n$ is called **composite**.

Since p is prime whenever its negative is prime, attention is normally restricted to the positive primes, which, at a negligible loss of precision, tend to be called "prime" without the specification of positivity. Here are the primes up to 101:

2	3	5	7	11	13	17	19	23	29	31	37	41
43	47	53	59	61	67	71	73	79	83	89	97	101

With bigger integers it is no longer that easy to pick out primes. For example, a naive guess might be that 91 is prime, but $91 = 13 \cdot 7$.

Although we might feel confident that there are infinitely many primes, this is not obvious. But first comes a result which points to the importance of primes.

Proposition 1.7. *If n is an integer and $n \geq 2$, then n can be factored into primes.*

Proof. Use induction on n.

If $n = 2$, then n is factored as itself into primes. Suppose $2, 3, 4, \ldots, n-1$ can each be factored into primes. Now look at n. If n is prime, then n is a product of primes, namely itself. If n is composite, write $n = k\ell$, where $1 < k < n$ and $1 < \ell < n$. Since k, ℓ are among the integers $2, 3, 4, \ldots, n-1$, each of them factors into primes. That is,

$$k = p_1 \cdot p_2 \cdots p_r \text{ and } \ell = q_1 \cdot q_2 \cdots q_s, \text{ where the } p_j,\ q_j \text{ are primes.}$$

Then of course, the equations

$$n = k\ell = p_1 \cdots p_r \cdot q_1 \cdots q_s$$

give a factorization of n into primes. □

From the above comes a famous theorem already in Euclid's books.

Proposition 1.8. *There are infinitely many primes.*

Proof. Given any *finite* list of primes $p_1, p_2, \ldots, p_n$, here is how to come up with one more prime not on the list. Let

$$n = p_1 \cdot p_2 \cdots p_n + 1.$$

According to Proposition 1.7, n has a prime factor q. This q cannot be equal to any $p_1, \ldots, p_n$. Indeed, if q equalled some p_j, then

$$q \text{ would divide } n - p_1 \cdot p_2 \cdots p_n, \text{ which equals } 1.$$

Since no prime is a factor of 1, our q is a prime not equal to any of $p_1, p_2, \ldots, p_n$. This permits the build-up of new primes at will. □

Unique Factorization

The special thing about primes is that there is *only one way* to factor an integer into primes. Ambiguous factorings such as

$$24 = 6\cdot 4 = 8\cdot 3 = 1\cdot 1\cdot 12\cdot 2$$

do not occur when only primes are involved in the factors. Despite one's sense that such unique factorization must be true, this is not obvious.

It is worth noting that

an integer a is coprime with a prime p if and only if $p \nmid a$.

Indeed, if $p \mid a$, then $\gcd(p, a) = p \neq 1$. Conversely, if $d = \gcd(p, a)$ and $d \neq 1$, then the fact $d \mid p$ forces $d = p$, and then $d \mid a$.

The next result forms the cornerstone for the proof of unique factorization.

Proposition 1.9. *An integer p, other than $0, \pm 1$, is prime if and only if it has the property that whenever p divides a product ab, then p already divides a or b.*

Proof. Say p is prime and $p \mid ab$, and suppose $p \nmid a$. As observed, p and a are coprime. By Proposition 1.4 it follows that $p \mid b$.

For the converse, suppose that p divides a factor whenever p divides the product of two integers. Now let a be a factor of p. Thus $p = ab$ for some other integer b. Clearly $p \mid ab$, and thus $p \mid a$ or $p \mid b$. If $p \mid a$, then the fact $a \mid p$ yields $a = \pm p$. If $p \mid b$, write $b = qp$ for some integer q. Then $p = ab = aqp$. Cancel p to get $aq = 1$, and from that $a = \pm 1$. So, the only possible factors of p are $\pm p$ and ± 1. □

A strong case can be made that the alternative property in Proposition 1.9 ought to be taken as the *definition* of a prime number. Indeed, this is what makes the proof of unique factorization work.

Proposition 1.9 readily extends to the product of several integers. Namely, if a prime p divides the product $a_1 a_2 a_3 \cdots a_k$, then p already divides at least one of the a_j. The proof to follow, that factorization of integers into primes is unique, rests on the shoulders of Proposition 1.9.

Proposition 1.10. *If $p_1, p_2, \ldots, p_n$ and $q_1, q_2, \ldots, q_m$ are two lists of primes (positive) with repetitions allowed, and if*

$$p_1 \cdot p_2 \cdots p_n = q_1 \cdot q_2 \cdots q_m,$$

then, after a rearrangement of the q_j, the lists are identical. That is

$$m = n \text{ and } p_1 = q_1,\ p_2 = q_2,\ \ldots,\ p_n = q_n.$$

Proof. Clearly $p_1 \mid q_1 \cdot q_2 \cdots q_m$. By Proposition 1.9, p_1 divides some q_j. Rearrange the q_j, and say $p_1 \mid q_1$. Since q_1 is prime, $p_1 = 1$ or $p_1 = q_1$. The first option cannot hold because p_1 is prime. Thus $p_1 = q_1$, and then

$$p_1 \cdot p_2 \cdots p_n = p_1 \cdot q_2 \cdots q_m.$$

Cancel p_1 to get

$$p_2 \cdot p_3 \cdots p_n = q_2 \cdot q_3 \cdots q_m.$$

Repeat the argument, with the necessary rearrangement of the q_j, to get $p_2 = q_2$, and then

$$p_3 \cdot p_4 \cdots p_n = q_3 \cdot q_4 \cdots q_m.$$

Continuing in this fashion, after suitable rearrangement of the q_j, we end up with one of the following possibilities:

- $n < m$, and $p_1 = q_1, \ldots, p_n = q_n$, $1 = q_{m-n} \cdots q_m$
- $m < n$, and $p_1 = q_1, \ldots, p_m = q_m$, $p_{n-m} \cdots p_n = 1$
- $m = n$, and $p_1 = q_1, \ldots, p_n = q_n$.

The first two situations cannot happen, since only ± 1 can be factors of 1 and the p_j, q_j are not ± 1. So indeed, $m = n$ and all $p_j = q_j$, after some rearranging of the q_j. □

The Multiplicity of a Prime inside an Integer

It is customary to collect repeated primes in the unique factorization of a positive integer a as follows:

$$a = p_1^{e_1} \cdot p_2^{e_2} \cdots p_n^{e_n},$$

where p_j are now *distinct* primes and the exponents $e_j \geq 1$.

Definition 1.11. The unique number of times e_j in which a prime appears in the unique factorization of a non-zero integer a is called the ***multiplicity*** of p_j inside a. It is convenient to allow multiplicities to equal 0. A prime p has multiplicity 0 in a when $p \nmid a$.

For example, with

$$a = 2^3 \cdot 3^0 \cdot 5^1 \cdot 11^8 \cdot 29^0,$$

the primes 3 and 29 have multiplicity 0 in a, while the multiplicity of 11 in a is 8.

Proposition 1.10 is saying that for every positive integer a and every prime p, there is a unique non-negative multiplicity of p in a. Denote this multiplicity by $m_a(p)$. For example, $m_{320}(2) = 6$ because $320 = 2^6 \cdot 5^1$, while $m_{320}(7) = 0$.

For each positive a, the function $p \mapsto m_a(p)$ from the set of primes to the set of non-negative integers counts the number of times that each prime p appears in the unique factorization of a.

Here are a few observations about the multiplicity function. If a,b are positive integers, then

- $m_{ab}(p) = m_a(p) + m_b(p)$,
- $m_a(p) > 0$ for at most finitely primes p,
- $m_a(p) = m_b(p)$ for all primes p if and only if $a = b$,
- $m_a(p) = 0$ for all primes p if and only if $a = 1$.
- If m is a function defined on the set of primes with values in the set of non-negative integers, and if $m(p) > 0$ for at most a finite number of primes p, then there is a unique positive integer a such that $m(p) = m_a(p)$ for all primes p. Indeed, if $p_1, p_2, \ldots, p_n$ are the distinct primes at which $m(p_j) > 0$, the required a is given by

$$a = p_1^{m(p_1)} p_2^{m(p_2)} p_3^{m(p_3)} \cdots p_n^{m(p_n)}.$$

Divisibility and Unique Factorization

The notion of divisibility fits nicely into the language of multiplicities.

Proposition 1.12. *A positive integer a divides another positive integer b if and only if $m_a(p) \leq m_b(p)$ for every prime p.*

Proof. Suppose $a \mid b$. That is, $b = ac$ for some c. Then

$$m_b(p) = m_{ac}(p) = m_a(p) + m_c(p) \text{ for all primes } p.$$

Clearly $m_a(p) \leq m_b(p)$, because $m_c(p) \geq 0$.

Conversely, suppose $m_a(p) \leq m_b(p)$ for all primes p. For each such prime, let $m(p) = m_b(p) - m_a(p)$. Since $m(p) \geq 0$ for all primes p and $m(p) > 0$ for at most a finite number of primes, let c be the unique positive integer that has m as its multiplicity function, i.e. $m(p) = m_c(p)$. Then

$$m_{ac}(p) = m_a(p) + m_c(p) = m_a(p) + m(p) = m_a(p) + m_b(p) - m_a(p) = m_b(p)$$

for all primes p. It follows that $ac = b$, in other words $a \mid b$. □

Greatest Common Divisors in Terms of Multiplicity Functions

The greatest common divisor of two positive integers can be expressed in terms of the multiplicity function that arises from unique factorization.

Proposition 1.13. *If a,b are positive integers, then $\gcd(a,b)$ is the positive integer c whose multiplicity function is given by*

$$m_c(p) = \min\left(m_a(p), m_b(p)\right) \text{ for all primes } p.$$

Proof. By item 3 of Proposition 1.1, a positive integer c will equal $\gcd(a, b)$ if and only if

- $c \mid a$ and $c \mid b$, and
- whenever d is a positive integer such that $d \mid a$ and $d \mid b$, then $d \mid c$ as well.

The specified integer c is such that $m_c(p) \leq m_a(p)$ and $m_c(p) \leq m_b(p)$. By Proposition 1.12, $c \mid a$ and $c \mid b$.

Next, let d be any positive integer such that $d \mid a$ and $d \mid b$. Proposition 1.12 yields $m_d(p) \leq m_a(p)$ and $m_d(p) \leq m_b(p)$ for every prime p. So $m_d(p) \leq m_c(p)$ which, again by Proposition 1.12, gives $d \mid c$. Therefore, $c = \gcd(a, b)$. □

To illustrate Proposition 1.13:

$$\gcd(2^3 \cdot 3^0 \cdot 7^5 \cdot 13^4 \cdot 29^8, 2^5 \cdot 3^7 \cdot 7^3 \cdot 13^4 \cdot 29^0) = 2^3 \cdot 3^0 \cdot 7^3 \cdot 13^4 \cdot 29^0.$$

Coprimeness and Unique Factorization

Two positive integers a, b are coprime if and only if 1 is their greatest common divisor. This happens if and only if $\min(m_a(p), m_b(p)) = 0$ for all primes p, meaning that for every prime p either $m_a(p) = 0$ or $m_b(p) = 0$. In other words, a, b are coprime if and only if there is no prime p such that $p \mid a$ and $p \mid b$.

Why 1 Is Not a Prime

Some might argue that the integer 1 deserves to be a prime. After all, it cannot be factored down any further. However, if we allow 1 to be a prime, then unique factorization goes out the window. Indeed, we can factor the integer 1 from any integer a as much as we like:

$$a = 1 \cdot 1 \cdot 1 \cdot 1 \cdots 1 \cdot a.$$

True primes do not do that. The number of times they appear in the unique factorization of a is unique. Better that 1 should not be a prime.

EXERCISES

1. Show that integers a, b are coprime if and only if no prime divides them both.
2. If two positive integers s, t are such that $s + t$ is a prime, show that s, t are coprime.
3. (a) Show that an integer a is a perfect square if and only if the multiplicity $e_p(a)$ is even for all primes p.
   (b) If $a^2 \mid b^2$, show that $a \mid b$.
   (c) If a, b are coprime and their product ab is a perfect square, show that a and b are perfect squares. More generally show that if $ab = c^k$ for some integer c and some positive exponent k, then both $a = d^k$ and $b = e^k$ for some integers d, e.
4. If a positive integer a is such that $\sqrt{a} \in \mathbb{Q}$, show that $a = b^2$ for some integer b. In other words, show that square roots of integers that are not perfect squares must be irrational.

5. How many positive factors do 67375 and 70875 have in common?
6. If a is an integer such that $40 \mid a^2$, prove that $20 \mid a$.
7. (a) How many zeroes are at the end of the decimal expansion of 100! ?
   (b) Find $\gcd(100!, 3^{100})$ as a product of primes.
8. If r, n are integers with $0 \leq r \leq n$, the binomial coefficient $\binom{n}{r}$ is $\frac{n!}{r!(n-r)!}$. If n is prime and $1 \leq r \leq n-1$, show that n divides $\binom{n}{r}$.
9. Let a, b be positive integers. The *least common multiple* of a and b is the smallest integer that they both divide. For example, 6 and 4 both divide their product, which is 24, but that is not their least common multiple, which is 12. We denote this least common multiple by $\operatorname{lcm}(a, b)$. In the exercises of Section 1.1 we saw that

$$ab = \gcd(a, b) \operatorname{lcm}(a, b).$$

   (a) Write the formula for the multiplicity function of the least common multiple of two positive integers a, b in terms of the multiplicity functions of a and b.
   (b) Find $\operatorname{lcm}(100!, 3^{100})$ as a product of primes.
10. If a positive integer n is such that n divides $(n-1)!+1$, show that n must be prime.
    In due course we shall discover that the converse of this result is also true. For example, the prime 7 divides 721, which turns out to be $6!+1$.
11. If $\mathbb{P}$ is the set of positive integers, show that the function $f : \mathbb{P} \times \mathbb{P} \to \mathbb{P}$ defined by $f(m, n) = (2m-1)2^n$ is a bijection.

1.3 Congruences

Start with a positive integer n, and call it a ***modulus***.[1]

If a is any integer, Euclidean division gives unique integers q and r, where

$$a = nq + r \text{ and } 0 \leq r < n.$$

Definition 1.14. Two integers a, b are said to be ***congruent modulo*** n, and we write

$$a \equiv b \bmod n,$$

provided a, b have equal remainders, between 0 and $n-1$, upon Euclidean division by n.

When $n = 1$, the only possible remainder upon division by 1 is 0. In this trivial case, all pairs of integers are congruent. To make the story worthwhile, assume $n \geq 2$.

For instance, when the modulus is $n = 6$, the integers congruent to each other with a remainder of 2 are

$$\ldots, -10, -4, 2, 8, 14, 20, 26, \ldots$$

These are the integers that take the form $6q + 2$, where q is any integer.

[1] This comes from the Latin for "measure," to be used for comparing two integers.

With 6 as the modulus, the infinite set of integers $\mathbb{Z}$ gets partitioned into 6 disjoint parts in accordance with the 6 possible remainders $0, 1, 2, 3, 4, 5$:

- remainder of 0 $\{6q : q \in \mathbb{Z}\} = \{0, \pm 6, \pm 12, \ldots\}$
- remainder of 1 $\{6q + 1 : q \in \mathbb{Z}\} = \{\ldots, -11, -5, 1, 7, 13, \ldots\}$
- remainder of 2 $\{6q + 2, q \in \mathbb{Z}\} = \{\ldots, -10, -4, 2, 8, 14, \ldots\}$
- remainder of 3 $\{6q + 3 : q \in \mathbb{Z}\} = \{\ldots, -15, -9, -3, 3, 9, 15, \ldots\}$
- remainder of 4 $\{6q + 4 : q \in \mathbb{Z}\} = \{-14, -8, -2, 4, 10, 16, \ldots\}$
- and remainder of 5 $\{6q + 5 : q \in \mathbb{Z}\} = \{-13, -7, 5, 11, 17, \ldots\}$.

The same observation applies to any modulus n. Given a modulus n, the set of integers $\mathbb{Z}$ gets partitioned into n disjoint subsets according to the n possible remainders $0, 1, 2, \ldots, n-1$.

Clearly,

- if a is any integer, then $a \equiv a \bmod n$,
- if $a \equiv b \bmod n$, then $b \equiv a \bmod n$, and
- if $a \equiv b$ and $b \equiv c \bmod n$, then $a \equiv c \bmod n$.

Here is an alternative way to detect congruence.

Proposition 1.15. *Two integers a, b are congruent modulo n if and only if n divides $b - a$.*

Proof. If $a \equiv b \bmod n$, then $b = nq + r$ and $a = nt + r$, for some integers q, t and some common integer r. Then $b - a = n(q - t)$, whence $n \mid b - a$.

Conversely, suppose n divides $b - a$. By Euclidean division there are integers q, t, r, s such that

$$b = nq + r,\ 0 \leq r < n \text{ and } a = nt + s,\ 0 \leq s < n.$$

Then

$$r - s = (b - nq) - (a - nt) = b - a + n(t - q).$$

Since $n \mid b - a$ it follows that $n \mid r - s$. Together with the fact $|r - s| < n$, this forces $r - s = 0$, i.e. $r = s$. Therefore, $a \equiv b \bmod n$. □

In geometric language, two integers are congruent modulo n when the distance between them can be measured off in integer multiples of the modulus n.

Modular Arithmetic

Clearly $3 \equiv -6 \bmod 9$ and $11 \equiv 2 \bmod 9$. Now $3 + 11 = 14$ while $-6 + 2 = -4$, and notice that $14 \equiv -4 \bmod 9$. Furthermore, $3 \cdot 11 = 33$ while $-6 \cdot 2 = -12$. Again notice that $33 \equiv -12 \bmod 9$. These are not coincidences.

Proposition 1.16. *Let n be a modulus. If $a \equiv b \bmod n$ and $c \equiv d \bmod n$, then*

$$a \pm c \equiv b \pm d \text{ and } ac \equiv bd \bmod n.$$

Proof. By Proposition 1.15, the congruence $a+c \equiv b+d \bmod n$, comes down to showing that $n \mid (a+c)-(b+d)$. In other words, that $n \mid (a-b)-(c-d)$, which is obviously true.

The proof that $a-c \equiv b-d \bmod n$ imitates the one just done for addition.

In order to get $ac \equiv bd \bmod n$, observe that

$$ac - bd = ac - bc + bc - bd = (a-b)c + b(c-d),$$

which is an integer combination of $a-b$ and $c-d$. Since n divides both $a-b$ and $c-d$ it becomes clear that n divides $ac-bd$, meaning that $ac \equiv bd \bmod n$. □

By repeated use of Proposition 1.16 the following ***replacement principle*** emerges.

Proposition 1.17 (Replacement principle). *If $f(a_1, a_2, \ldots, a_k)$ is any integer formed by adding, subtracting, or multiplying integers $a_1, a_2, \ldots, a_k$, (i.e. $f(a_1, a_2, \ldots, a_k)$ is a polynomial built from the integers $a_1, a_2, \ldots, a_k$ and using only integer coefficients), and if*

$$a_1 \equiv b_1, a_2 \equiv b_2, \ldots, a_k \equiv b_k \bmod n,$$

then

$$f(a_1, a_2, \ldots, a_k) \equiv f(b_1, b_2, \ldots, b_k) \bmod n.$$

For instance, if

$$a_1 \equiv b_1, a_2 \equiv b_2 \text{ and } a_3 \equiv b_3 \bmod n,$$

then

$$(a_1^2 - 5a_2)a_3 \equiv (b_1^2 - 5b_2)b_3 \text{ and } a_1^{100}a_2^7 \equiv b_1^{100}b_2^7 \bmod n.$$

To illustrate the usefulness of the replacement principle, let us show that for every integer a, the fraction $(a^7-a)/7$ remains an integer. This comes down to showing that 7 divides a^7-a, which is equivalent to showing that $a^7 \equiv a \bmod 7$. Now, every integer a is congruent to its remainder r upon Euclidean division by 7. Thus, by Proposition 1.17, $a^7 \equiv r^7 \bmod 7$. So, all that is needed is to verify that $r^7 \equiv r \bmod 7$ for the only possible values of r. These possible values are $0, 1, 2, \ldots, 6$. This little task can be readily carried out. Indeed, modulo 7:

$$0^7 = 0 \equiv 0,\ 1^7 = 1 \equiv 1,\ 2^7 = 128 \equiv 2,\ 3^7 = 2187 \equiv 3,$$
$$4^7 = 16384 \equiv 4,\ 5^7 = 78125 \equiv 5,\ 6^7 = 279936 \equiv 6.$$

Additional shortcuts could have been taken in the above illustration of the replacement principle. For instance, since $6 \equiv -1 \bmod 7$, the congruence $6^7 \equiv 6 \bmod 7$ can be replaced by the much more obvious congruence $(-1)^7 \equiv -1 \equiv 6 \bmod 7$.

For yet another illustration of replacement, let us show that the sum of two squares is never congruent to 3 modulo 4. In other words, that the congruence

$$x^2 + y^2 \equiv 3 \bmod 4$$

has no integer solution x, y. Euclidean division gives that

$$x, y \equiv \text{ one of } 0, 1, 2, 3 \bmod 4.$$

By replacement,

$$x^2, y^2 \equiv \text{ one of } 0^2 \equiv 0, 1^2 \equiv 1, 2^2 \equiv 0, 3^2 \equiv 1 \bmod 4.$$

And then, by further replacement,

$$x^2 + y^2 \equiv \text{ one of } 0 + 0 = 0, 0 + 1 = 1, 1 + 0 = 1, 1 + 1 = 2 \bmod 4.$$

By inspection,

$$x^2 + y^2 \not\equiv 3 \bmod 4,$$

for all integers x, y. Having just seen that integers congruent to 3 modulo 4 are never the sum of two squares, one might ask a far more interesting question. Which integers can be expressed as the sum of two squares? This will be answered in due course.

Reduction Modulo *n*

The remainder of an integer a upon Euclidean division by n will be called the ***reduction*** of a modulo n. In other words, the reduction of a modulo n is the unique integer r between 0 and $n - 1$ such that $a \equiv r \bmod n$.

Reductions of not-too-large integers using not-too-large moduli, for example $2586 \equiv 3 \bmod 7$, can be done by hand or machine to implement Euclidean division. But replacement might afford a handier way to get the reduction. For instance, let us reduce the very large integer 306^{100} modulo 7, by repeated replacements modulo 7 as follows:

$$\begin{aligned} 306^{100} &\equiv 5^{100} && \text{since } 306 \equiv 5 \bmod 7 \\ &= (5^4)^{25} \\ &= 625^{25} \\ &\equiv 2^{25} && \text{since } 625 \equiv 2 \bmod 7 \\ &= (2^6)^4 \cdot 2 \\ &= 64^4 \cdot 2 \\ &\equiv 1^4 \cdot 2 && \text{since } 64 \equiv 1 \bmod 7 \\ &= 2. \end{aligned}$$

The reduction of 306^{100} modulo 7 is 2.

The Ring of Residues

The replacement principle reduces congruence problems about the infinite set of integers to a matter of finitely many cases. This important idea needs elaboration.

Definition 1.18. For any integer a, the ***congruence class***, or ***residue class***, or more briefly the ***residue*** of a modulo n is the set of integers which have the same reduction as a modulo n. This set will be designated by the box notation $[a]$.[2] The integer a is called a ***representative*** for its residue class.

In light of Proposition 1.15 this means that

$$[a] = \{b \in \mathbb{Z} : b \equiv a \bmod n\}.$$

Think of $[a]$ as *one thing* even though that one thing consists of an infinite set of numbers. For example, use the modulus $n = 9$ and $a = 15$. The reduction of 15 modulo 9 is 6. So, $[15]$ is the set of integers that reduce to 6 modulo 9. Alternatively, $[15]$ is the set of integers congruent to 15 modulo 9. That set is

$$[15] = \{\ldots, -21 - 12, -3, 6, 15, 24, 33, 42, \ldots\}.$$

A look at the definition of the box notation shows that

$$[a] = [b] \text{ if and only if } a \equiv b \bmod n.$$

To repeat, two integers represent the same residue if and only if they are congruent modulo n. Every integer is congruent to its remainder upon Euclidean division by n. The possible remainders are

$$0, 1, 2, \ldots, n - 1.$$

No two of these remainders are congruent to each other modulo n. Thus there are only a *finite number* of residues, in fact exactly n of them. This finite set of residues modulo n is denoted by

$$\mathbb{Z}_n.$$

For example, here are the residues of $\mathbb{Z}_5$:

$$[0] = \{0 + 5q : q \in \mathbb{Z}\},\ \ [1] = \{1 + 5q : q \in \mathbb{Z}\},\ \ [2] = \{2 + 5q : q \in \mathbb{Z}\},$$
$$[3] = \{3 + 5q : q \in \mathbb{Z}\},\ \ [4] = \{4 + 5q : q \in \mathbb{Z}\}.$$

The representatives $0, 1, 2, 3, 4$ chosen above are often preferred because they are the remainders that can result upon Euclidean division by 5. Yet the possibility of using other representatives needs to be kept in mind. For example, the five residues of $\mathbb{Z}_5$ can just as well be represented as

$$[0] = [10], \quad [1] = [6], \quad [2] = [-33], \quad [3] = [13], \quad [4] = [-1].$$

Proposition 1.16 when interpreted in the language of residues turns into the following statement.

[2] As we move along and it becomes more familiar, this slightly awkward notation will be used less, in favor of an unadorned a.

Proposition 1.19. *Suppose n is a modulus, and $[a], [b], [c], [d]$ are residues in $\mathbb{Z}_n$ represented by the integers a, b, c, d, respectively. If $[a] = [b]$ and $[c] = [d]$, then*

$$[a \pm c] = [b \pm d] \text{ and } [ac] = [bd].$$

Proposition 1.19 is an invitation to define the operations of addition, subtraction and multiplication on the finite set of residues $\mathbb{Z}_n$. Namely, define

$$[a] \pm [b] = [a \pm b] \text{ and } [a][b] = [ab].$$

According to Proposition 1.19 these definitions in $\mathbb{Z}_n$ do not depend on the chosen representatives for the residues $[a], [b]$. They are what we call ***well defined***.

Thus, from the integers a new, finite algebraic structure tolerating addition, subtraction and multiplication emerges. A set with the operations of addition, subtraction and multiplication, subject to suitable rules of calculation, is called a ***ring***. From the infinite ring of integers $\mathbb{Z}$ we have spawned the finite ring $\mathbb{Z}_n$. We shall come to visit $\mathbb{Z}_n$ on many occasions.

Inverses in $\mathbb{Z}_n$

If a is an integer coprime with n, then its residue $[a]$ in $\mathbb{Z}_n$ is such that $[a][s] = [1]$ for some other residue $[s]$ in $\mathbb{Z}_n$. Indeed, we assumed that $\gcd(a, n) = 1$. By Proposition 1.1 there exist integers s, t such that $as + nt = 1$. Since $[n] = [0]$ in $\mathbb{Z}_n$ we obtain

$$[1] = [as + nt] = [a][s] + [n][t] = [a][s].$$

In the parlance of rings $[s]$ is called an ***inverse element*** of $[a]$.

The Euclidean algorithm can be used to explicitly find an inverse for $[a]$. Rather than go into a general discussion, let us find an inverse for $[11]$ in $\mathbb{Z}_{350}$. The Euclidean algorithm gives $\gcd(350, 11)$ like so:

$$\begin{aligned} 350 &= 11 \cdot 31 + 9 \\ 11 &= 9 \cdot 1 + 2 \\ 9 &= 2 \cdot 4 + 1. \end{aligned}$$

This confirms that $1 = \gcd(350, 11)$. Working backwards in the above algorithm we see that

$$\begin{aligned} 1 &= 9 - 2 \cdot 4 \\ &= 9 - (11 - 9 \cdot 1) \cdot 4 \\ &= 9 \cdot 5 - 11 \cdot 4 \\ &= (350 - 11 \cdot 3) \cdot 5 - 11 \cdot 4 \\ &= 350 \cdot 5 - 11 \cdot 159. \end{aligned}$$

So, $[-159]$, which is the same as $[191]$, is an inverse for $[11]$ in $\mathbb{Z}_{350}$.

Note that if a is not coprime with n, then $[a]$ has no inverse in $\mathbb{Z}_n$. Indeed, suppose such $[a]$ has an inverse $[s]$. Then $[as] = [a][s] = [1]$, meaning that n divides $1 - as$, which says that $1 = as + nt$ for some t, thereby forcing a, n to be coprime.

The residues of $\mathbb{Z}_n$ which have an inverse in $\mathbb{Z}_n$ are called the ***units*** modulo n. For instance, the units of $\mathbb{Z}_{10}$ are $[1], [3], [7], [9]$, because $1, 3, 7, 9$ are coprime with 10. If the modulus is a prime p, then all integers $1, 2, \ldots, p-1$ are coprime with p, and so the $p-1$ residues represented by these integers are the units of $\mathbb{Z}_p$.

EXERCISES

1. Find the units of $\mathbb{Z}_{30}$, and for each unit find its inverse.
2. Use the replacement principle to find the last two digits in the decimal expansion of 59^{75}.
3. Show that a positive integer is congruent modulo 9 to the sum of its decimal digits. Deduce that 9 divides an integer a if and only if 9 divides the sum of its decimal digits.
4. If p is an odd prime and $x^2 \equiv 1 \bmod p$, show that $x \equiv \pm 1 \bmod p$. Show that the above still holds if the odd prime p is replaced by p^k, where k is any positive integer exponent.
   If $x^2 \equiv 1 \bmod 21$, can we conclude that $x \equiv \pm 1 \bmod 21$?
5. If a is a positive integer and $a \equiv 3 \bmod 4$, show that a has a prime divisor q such that $q \equiv 3 \bmod 4$.
   Use this fact to show that there exist infinitely many primes congruent to 3 modulo 4.
   Hint. If $p_1, p_2, \ldots, p_n$ is any finite list of primes all congruent to 3 modulo 4, let $a = 4p_1p_2 \cdots p_n + 3$. Show that a has a prime divisor q such that q is not any of the p_j and $q \equiv 3 \bmod 4$.
6. (a) If $[a], [b]$ are units of $\mathbb{Z}_n$, show that their product $[a][b]$ in $\mathbb{Z}_n$ is also a unit.
   (b) If $[a]$ is a unit of $\mathbb{Z}_n$ and if $[b], [c]$ in $\mathbb{Z}_n$ are such that $[a][b] = [a][c]$, prove that $[b] = [c]$. This is known as the cancellation property of units.
   (c) If $[a] \in \mathbb{Z}_n$ and $[a]$ has the cancellation property, show that $[a]$ is a unit of $\mathbb{Z}_n$.
7. Let $\mathbb{Z}_n^\star$ denote the set of units inside $\mathbb{Z}_n$, and suppose that

$$[a_1], [a_2], \ldots, [a_k]$$

   is a full listing of $\mathbb{Z}_n^\star$ without repetitions. We saw in the preceding exercise that any product of these units is again a unit. Also, units have the cancellation property.
   (a) If $[a]$ is any unit of $\mathbb{Z}_n^\star$, explain why the list

$$[a][a_1], [a][a_2], \ldots, [a][a_k]$$

   is merely a rearrangement of the original listing of $\mathbb{Z}_n^\star$.
   (b) Explain the following equality of products in $\mathbb{Z}_n^\star$:

$$([a][a_1])([a][a_2]) \cdots ([a][a_k]) = [a_1][a_2] \cdots [a_k].$$

   (c) Deduce from part (b) that $[a]^k = [1]$ in $\mathbb{Z}_n^\star$.

In terms of congruences we have just seen that if a is coprime with a positive integer n, and if k is the number of integers between 0 and $n-1$ that are coprime with n, then

$$a^k \equiv 1 \bmod n.$$

This is a famous result attributed to Leonhard Euler (1707–1783). We could call it Euler's theorem, but then it would be only one of a multitude.

If p is a prime, then the number of units in $\mathbb{Z}_p$, i.e. the number of elements in $\mathbb{Z}_p^\star$ is $p-1$. Indeed, in this case every integer from 1 to $p-1$ is coprime with p. Then Euler's theorem specializes to the statement that if p is a prime and a is an integer not divisible by p, then

$$a^{p-1} \equiv 1 \bmod p.$$

This famous result is called Fermat's Little Theorem in honor of its discoverer Pierre de Fermat (1607–1665). This is a result of stunning utility.

2 A First Look at Groups

The concept of an abstract group took some time to gel within mathematical culture. It began with problems on rigid motions in space as well as studies into the solutions of equations by means of what was then called the group of substitutions of the roots. Without a doubt the concept is here to stay. Pretty much along traditional lines, let us examine what a group is, look at a number of examples, and become familiar with the tools used to study them.

Definition 2.1. A non-empty set G is called a ***group*** when G is endowed with a binary operation subject to some possibly familiar conditions. This binary operation is a mapping

$$\star : G \times G \to G$$

which assigns to every pair (x, y), in the Cartesian product $G \times G$, an element $x \star y$ in G called the ***composite*** of x with y. This $x \star y$ is also known as the ***product*** of x with y in G.

The binary operation of composition must follow three laws.

- The ***associative law***:

$$(x \star y) \star z = x \star (y \star z) \text{ for every } x, y, z \text{ in } G.$$

- The ***neutral element*** requirement:

$$\text{there exists an element } e \text{ in } G \text{ such that } e \star x = x \star e = x \text{ for every } x \text{ in } G.$$

Such an e is called a neutral element of G.

- The ***inverse element*** requirement:

$$\text{every } x \text{ in } G \text{ has a } y \text{ in } G \text{ such that } x \star y \text{ and } y \star x \text{ are neutral elements.}$$

Such a y is called an ***inverse*** for x.

It is standard practice to suppress the symbol $\star$ for the composition of x and y and to simply write xy to mean the composite element $x \star y$.

2.1 Basic Properties and Examples of Groups

Uniqueness of Neutral Elements and Inverses

If e, f are neutral elements in a group G, then each acts neutrally when composed with the other to give

$$e = ef = f.$$

So, a group has a unique neutral element, which is typically denoted by the familiar symbol 1, and known as ***the identity element*** of G.

The symbol "1" will be called upon to shoulder a multitude of interpretations, and so its role in a given situation will need attention.

Inverses are also unique. Indeed, suppose that y, z are inverses for x. Thus $yx = 1$ and $xz = 1$, and then by the associative law:

$$y = y1 = y(xz) = (yx)z = 1z = z.$$

The unique inverse of an element x is typically denoted by x^{-1}.

A couple of things to notice are that

$$(x^{-1})^{-1} = x \text{ and } (xy)^{-1} = y^{-1}x^{-1},$$

for every x, y in a group. The first equation holds because x is the unique element that gives

$$x^{-1}x = xx^{-1} = 1.$$

The second equation follows from the associative law:

$$(xy)(y^{-1}x^{-1}) = ((xy)y^{-1})x^{-1} = (x(yy^{-1}))x^{-1} = (x1)x^{-1} = xx^{-1} = 1,$$

and likewise

$$(y^{-1}x^{-1})(xy) = 1.$$

The inverse of a composite is the composite of the inverses, *but taken in the reverse order*.

The possibilities for examples of groups are vast, but for starters take G to be the set of positive real numbers. The composition of two positive real numbers x, y is their familiar product xy. The real number 1 is its identity element. And for the inverse of x, take the reciprocal $x^{-1} = 1/x$.

Then there is the ***trivial group*** containing just one element 1. The binary operation in such a group is the only one possible, namely $1 \star 1 = 1$.

The Symmetric Group

For an important example, let L be any non-empty set, and let $\mathcal{S}(L)$ be the set of all bijections, i.e. maps that are one-to-one and onto,

$$\sigma : L \to L.$$

Such bijections are also known as ***permutations*** on L, because they rearrange the elements of L.

We compose two bijections $\sigma : L \to L$ and $\tau : L \to L$ to get a third bijection $\tau\sigma : L \to L$ given by

$$\tau\sigma : a \to \tau(\sigma(a)) \text{ for all } a \text{ in } L.$$

Note that for us the permutation appearing on the right acts first on the elements of L.

The composition of bijections is clearly associative. There is the identity bijection

$$1 : L \to L \text{ given by } 1 : a \mapsto a \text{ for all } a \text{ in } L.$$

Evidently $1\sigma = \sigma 1 = \sigma$ for every bijection σ. And every bijection $\sigma : L \to L$ has an inverse bijection $\sigma^{-1} : L \to L$ given by

$$\sigma^{-1}(a) = b \text{ if and only if } \sigma(b) = a.$$

A bit of reflection makes it clear that $\sigma\sigma^{-1} = \sigma^{-1}\sigma = 1$. And so, $\mathcal{S}(L)$ is a group.

Most important is the case where L is a finite set. Any finite set could be used, but the set L consisting of the first n positive integers $1, 2, 3, \ldots, n$ is the preferred prototype. These are treated merely as a set of symbols or "letters." A bijection

$$\sigma : \{1, 2, 3, \ldots, n\} \to \{1, 2, 3, \ldots, n\}$$

is then called a ***permutation on** n **letters***. The group of permutations on n letters is denoted by $\mathcal{S}_n$.

Every permutation σ on n letters is determined by a succession of n decisions. The value $\sigma(1)$ has n options. For $\sigma(2)$ there are $n - 1$ options, since the chosen $\sigma(1)$ is no longer available. Then $\sigma(3)$ is left with $n - 2$ options, and so on for $\sigma(4)$ with $n - 3$ options, down to $\sigma(n)$ with just one option. It follows that the number of permutations on n letters equals the product $n(n-1)(n-2)\cdots 3 \cdot 2 \cdot 1$, to be denoted by the factorial symbol $n!$. Thus, $\mathcal{S}_n$ is a finite group containing precisely $n!$ elements.

Abelian Groups

A property conspicuously absent from the axioms for a group G is the ***commutative*** law:

$$xy = yx \text{ for all } x, y \text{ in } G.$$

The group of positive real numbers under the usual multiplication satisfies the commutative law, but S_3 does not. For instance, take the permutations σ, τ on three letters given by

$$\sigma : 1 \mapsto 2 \mapsto 3 \mapsto 1 \text{ and } \tau : 1 \mapsto 2 \mapsto 1, 3 \mapsto 3.$$

Notice that

$$1 \overset{\tau}{\mapsto} 2 \overset{\sigma}{\mapsto} 3 \text{ while } 1 \overset{\sigma}{\mapsto} 2 \overset{\tau}{\mapsto} 1 \text{ and } 1 \neq 3.$$

Evidently $\sigma\tau$ and $\tau\sigma$ have a different effect on 1, and so commutativity fails in $\mathcal{S}_3$.

In honor of Neils Henrik Abel (1802–1829), groups that satisfy the commutative law are called ***abelian groups***.[1]

The non-abelian group $\mathcal{S}_3$ has 6 elements in it. It is a good exercise to see why all groups with fewer than 6 elements are abelian. Thus $\mathcal{S}_3$ is the smallest non-abelian group.

[1] Peculiarly the term "abelian" has come to be written with a lower case "a" in contrast with upper case terms such as Euclidean and Cartesian.

Additive Notation for Abelian Groups

If A is an abelian group, the composite of two elements a, b in A is at times denoted by the ***additive notation*** $a + b$, which is then called the ***sum*** of a and b. When additive notation is used, the neutral element is denoted by 0, and is naturally called ***zero***. The inverse of a is then denoted by $-a$, and is called the ***negative*** of a. The rules in an abelian group A using additive notation take on the familiar look:

- *associative law for addition*, $(a + b) + c = a + (b + c)$ for all a, b, c in A,
- the *neutrality* of zero, $0 + a = a + 0 = a$ for all a in A,
- the presence of *negatives*, $a + (-a) = -a + a = 0$ for all a in A,
- plus the *commutative law for addition*, $a + b = b + a$ for all a, b in A.

Clearly, the set of integers $\mathbb{Z}$ is an abelian group under the usual operation of addition. And so are the familiar sets of rational, real and complex numbers, whose notations, as a reminder, are as shown:

- the ***rational numbers*** $\mathbb{Q} = \{a/b : a, b \in \mathbb{Z} \text{ and } b \neq 0\}$,
- the ***real numbers*** $\mathbb{R}$, represented geometrically as the points on a straight line, and numerically as infinite decimal expansions,
- the ***complex numbers*** $\mathbb{C} = \{a + bi : a, b \in \mathbb{R}, i = \sqrt{-1}\}$, represented geometrically as the points of a plane.

Also $\mathbb{Z}_n$, the set of residues modulo a positive integer n, is an abelian group in which addition is defined by $[a] + [b] = [a + b]$. As noted in Chapter 1, this addition does not depend on the representatives a, b chosen for the respective residues $[a], [b]$. The group $\mathbb{Z}_n$ is a finite, abelian group containing exactly n elements.

For non-abelian groups, the multiplicative notation is used exclusively, while with abelian groups both multiplicative and additive notations can present themselves.

Unless otherwise specified we will use the multiplicative notation for groups.

Cancellation

All groups have the ***cancellation*** property. Namely, if $gh = gk$ for elements g, h, k in a group G, then $h = k$. Indeed,

$$h = 1h = (g^{-1}g)h = g^{-1}(gh) = g^{-1}(gk) = (g^{-1}g)k = 1k = k.$$

For abelian groups with additive notation, the cancellation property says that if a, b, c are in an abelian group A and if $a + b = a + c$, then $b = c$, which is hardly surprising.

The Group of Units Modulo n

One might ask if $\mathbb{Z}_n$ is an abelian group using the operation of multiplication given by $[a][b] = [ab]$ for any pair of residues $[a], [b]$. While the associative law holds, and there is an identity element $[1]$, some elements do not have an inverse. Most conspicuously, when $n \geq 2$, there is no inverse for $[0]$ since, regardless of $[b]$, the product $[0][b] = [0]$ instead of $[1]$.

If we try to repair the problem by taking just the set $\mathbb{Z}_n \setminus \{0\}$ where 0 is removed, another issue can emerge. Indeed, look at [3] and [2] in $\mathbb{Z}_6 \setminus \{0\}$, whose product $[3][2] = [6] = [0]$, which no longer sits in $\mathbb{Z}_6 \setminus \{0\}$.

However, there does reside an abelian group inside $\mathbb{Z}_n$, using multiplication. Recall that residue equality $[a] = [b]$ means that $b = a + nq$ for some integer q. So, if a, b represent the same residue, then $\gcd(a, n) = \gcd(b, n)$. In particular, a is coprime with n if and only if b is coprime with n. This prompts us to let

$$\mathbb{Z}_n^\star$$

denote the set of residues $[a]$ in $\mathbb{Z}_n$ for which a is coprime with n.

If a, b are each coprime with n, then so is their product ab. To see this note that since a, n have no prime factor in common and b, n have no prime factor in common, then ab, n have no prime factor in common. This means that if $[a], [b] \in \mathbb{Z}_n^\star$, then the product $[a][b] = [ab] \in \mathbb{Z}_n^\star$. Hence, $\mathbb{Z}_n^\star$ is endowed with a multiplication operation. The obvious identity element of $\mathbb{Z}_n^\star$ is $[1]$. And every element $[a]$ of $\mathbb{Z}_n^\star$ has an inverse. To see this last point, notice that there exist integers x, y such that $ax + ny = 1$, because a, n are coprime. From that,

$$[1] = [ax + ny] = [a][x] + [n][y] = [a][x] + [0][y] = [a][x] + [0] = [a][x].$$

An inverse for $[a]$ is this $[x]$. In practice, the Euclidean algorithm can be used to solve $ax + ny = 1$ for x, and thereby obtain the inverse $[x]$ of $[a]$ in $\mathbb{Z}_n^\star$. We have just proven that $\mathbb{Z}_n^\star$ is a finite group. Easily we also see that $\mathbb{Z}_n^\star$ is abelian.

Definition 2.2. The elements of the group $\mathbb{Z}_n^\star$ are very frequently called the ***units*** of $\mathbb{Z}_n$, and $\mathbb{Z}_n^\star$ is called the ***group of units modulo*** n. The number of elements in $\mathbb{Z}_n^\star$ (i.e. the number of units modulo n) is denoted by

$$\phi(n),$$

and the resulting function $\phi : \mathbb{P} \to \mathbb{P}$ is known as the ***Euler ϕ-function***.

We shall encounter the notion of units more generally in Chapter 4.

Since all residues in $\mathbb{Z}_n$ can be represented by the integers $0, 1, 2, \ldots, n-1$, the number of elements in $\mathbb{Z}_n^\star$, i.e. the value $\phi(n)$, equals the number of integers from $0, 1, 2, \ldots, n-1$ that are coprime with n.

For example, with 8 as the modulus, the units are $[1], [3], [5], [7]$. So $\mathbb{Z}_8^\star$ is an abelian group with 4 elements in it. Here is the composition table for the group $\mathbb{Z}_8^\star$. The box notations are omitted to avoid clutter.

$\star$	1	3	5	7
1	1	3	5	7
3	3	1	7	5
5	5	7	1	3
7	7	5	3	1

Taking 9 as the modulus, the units are [1], [2], [4], [5], [7], [8], constituting the group $\mathbb{Z}_9^\star$ with 6 elements in it. Its multiplication table is as follows:

$\star$	1	2	4	5	7	8
1	1	2	4	5	7	8
2	2	4	8	1	5	7
4	4	8	7	2	1	5
5	5	1	2	7	8	4
7	7	5	1	8	4	2
8	8	7	5	4	2	1

Both $\mathbb{Z}_9^\star$ and S_3 are groups containing 6 elements. However, S_3 is not abelian, while $\mathbb{Z}_9^\star$ is abelian.

The General Linear Groups

A major family of groups comes from linear algebra.

We are likely familiar with some examples of fields. A field is a set with the operations of addition, subtraction, multiplication, and division other than by the zero element, all subject to familiar rules. We shall, in due course, examine fields in depth. For now, let us simply note that the rational numbers $\mathbb{Q}$, the real numbers $\mathbb{R}$ and the complex numbers $\mathbb{C}$ are likely our first examples of fields.

Another significant example of a field is $\mathbb{Z}_p$, where p is a prime. We can certainly add, subtract and multiply residues modulo p, and every residue $[a]$ in $\mathbb{Z}_p$, except for $[a] = [0]$, is a unit. Indeed, $[a]$ is a unit if and only if a is coprime with p. Since p is prime, this happens if and only if $p \nmid a$, which in turn means that $[a] \neq [0]$.

Let F be one of the fields just mentioned (or any field, for those who are familiar with the concept). We may recall from linear algebra that the multiplication of $n \times n$ matrices with entries in F is associative. That is $(AB)C = A(BC)$ for all $n \times n$ matrices A, B, C whose entries are in F. There is the familiar identity matrix I, which satisfies $IA = AI = A$ for all matrices A. Also, every matrix A whose determinant is non-zero has an inverse matrix A^{-1}. Such matrices are called ***non-singular***, as well as ***invertible***. Since matrix multiplication is associative and the product of non-singular matrices is a non-singular matrix, the set of such matrices forms a group. This group, known as the ***general linear group*** over the field F, is denoted by

$$GL_n(F).$$

The Group $GL_2(\mathbb{Z}_5)$

Let us calculate the number of elements in the group $GL_2(\mathbb{Z}_5)$. Some familiarity with linear algebra is needed.

The number of 2×2 matrices $\begin{bmatrix} a & b \\ c & d \end{bmatrix}$ with entries in $\mathbb{Z}_5$ equals $5^4 = 625$, since each entry has five possible residues. However, for such a matrix to be in $GL_2(\mathbb{Z}_5)$, its determinant $ad - bc$

must not be the zero residue. This is the same as saying that the columns $\begin{pmatrix} a \\ c \end{pmatrix}, \begin{pmatrix} b \\ d \end{pmatrix}$ are linearly independent, which amounts to saying that

$$\begin{pmatrix} a \\ c \end{pmatrix} \neq \begin{pmatrix} 0 \\ 0 \end{pmatrix} \text{ and } \begin{pmatrix} b \\ d \end{pmatrix} \text{ is not a multiple of } \begin{pmatrix} a \\ c \end{pmatrix}.$$

Note that the 0 above stand for the zero residue. Hence, the number of possibilities for $\begin{pmatrix} a \\ c \end{pmatrix}$ is $5^2 - 1 = 24$. For each of these possibilities, the column $\begin{pmatrix} b \\ d \end{pmatrix}$ cannot be one of the five multiples of $\begin{pmatrix} a \\ c \end{pmatrix}$. Thus, for each possible choice of $\begin{pmatrix} a \\ c \end{pmatrix}$, there are $5^2 - 5 = 20$ choices for $\begin{pmatrix} b \\ d \end{pmatrix}$. This gives a combined total of $24 \cdot 20 = 480$ matrices in $GL_2(\mathbb{Z}_5)$.

Notice that $GL_2(\mathbb{Z}_5)$ is non-abelian. For instance, examine the following calculations:

$$\begin{bmatrix} 1 & 2 \\ 1 & 3 \end{bmatrix}\begin{bmatrix} 1 & 2 \\ 4 & 0 \end{bmatrix} = \begin{bmatrix} 4 & 2 \\ 3 & 2 \end{bmatrix} \neq \begin{bmatrix} 3 & 3 \\ 4 & 3 \end{bmatrix} = \begin{bmatrix} 1 & 2 \\ 4 & 0 \end{bmatrix}\begin{bmatrix} 1 & 2 \\ 1 & 3 \end{bmatrix}.$$

The entries of all matrices above are to be taken as residues modulo 5, with the box notation omitted to avoid clutter.

We have just built a non-abelian group of order 480. In general, $GL_n(\mathbb{Z}_p)$ is a rich source of finite, non-abelian groups. In due course we shall determine the number of elements in $GL_n(\mathbb{Z}_p)$ as a function of n and p.

The Power and Multiple Notations

The associative law, $(gh)k = g(hk)$, inside a group permits the triple product ghk to be written without any brackets at all. Likewise for any multiple product such as $ghk\ell m$, and so on. This allows for the introduction of the ***power notation***. If g is in a group G and n is any integer, let

$$g^n = \begin{cases} 1 & \text{if } n = 0 \\ ggg\cdots g & \text{taken } n \text{ times, if } n > 0 \\ g^{-1}g^{-1}g^{-1}\cdots g^{-1} & \text{taken } -n \text{ times, if } n < 0. \end{cases}$$

Note that the notation g^{-1} to mean the inverse of g and the very same notation g^{-1} for g taken to the -1 exponent, as above, yield exactly the same element. With the power notation, the common laws of exponents apply. Namely,

$$g^m g^n = g^n g^m = g^{m+n} \text{ and } (g^m)^n = g^{mn}$$

for all g in a group G and all integers m, n. The exponent law

$$g^m h^m = (gh)^m$$

applies if $gh = hg$ inside a group G. In particular, this law applies if G is abelian. However, if $gh \neq hg$, the latter law can well fail. For instance, $g^2h^2 = gghh$ while $(gh)^2 = ghgh$. Equality of these two expressions would mean $gghh = ghgh$, which after cancellation would yield $gh = hg$.

If A is an abelian group written additively, the ***multiple notation*** is the one to use. In this case, if $a \in A$ and n is any integer, let

$$na = \begin{cases} 0 & \text{if } n = 0 \\ a + a + a + \cdots + a & \text{taken } n \text{ times, if } n > 0 \\ (-a) + (-a) + (-a) + \cdots + (-a) & \text{taken } -n \text{ times, if } n < 0. \end{cases}$$

The rules for the multiple notation for an abelian group A, analogous to the laws of exponents, become

$$ma + na = (m + n)a, \quad n(ma) = (nm)a \quad \text{and} \quad na + nb = n(a + b)$$

for all a, b in A and all integers m, n.

2.1.1 The Order of a Group and of Its Elements

Definition 2.3. The number of elements in a group G is called the ***order***[2] of G.

If G is finite, the meaning of "order" is clear. For infinite groups we will have no occasion to make distinctions among the various so-called cardinalities, and we shall just say that they have infinite order. To avoid a surfeit of terminologies, we shall at times refer to the number of elements in any finite set as the ***order of the set***.

For every positive integer n, there is at least one group of order n, namely the abelian group $\mathbb{Z}_n$ with the operation of addition. There are also many groups of infinite order. To name one take the group of non-zero rational numbers under ordinary multiplication.

For many a positive integer n it can be an overwhelming project to discover all the multiplication tables for groups of order n.

Some groups contain elements g such that $g^m = 1$ for some positive exponent m. For instance, in the group $\mathbb{C}^\star$ of non-zero complex numbers under multiplication, the element $\omega = e^{2\pi i/7}$ satisfies $\omega^7 = 1$. In the group S_3, the permutation σ defined by

$$\sigma : 1 \mapsto 2 \mapsto 3 \mapsto 1$$

satisfies $\sigma^3 = 1$. In the group $GL_2(\mathbb{R})$ of invertible 2×2 matrices, the matrix $A = \begin{bmatrix} 0 & 1 \\ -1 & 0 \end{bmatrix}$ satisfies $A^4 = 1$.

Definition 2.4. An element g in any group G is said to have ***finite order*** when $g^m = 1$ for some positive integer m. If no such positive integer m exists, g is said to have ***infinite order***. If g is

[2] The use of this term has been around for much more than a century.

an element of finite order in a group G, the *smallest* positive integer m such that $g^m = 1$ is called the ***order*** of g.

We can observe right away that g has order 1 if and only if $g = 1$. The elements ω, σ, A in the groups mentioned above have orders $7, 3, 4$, respectively.

It may seem irksome that the term "order" has been assigned double duty to designate both the number of elements in a group, as well as the least power needed to bring an element down to 1. As we shall come to see, the two concepts of order are intertwined. In any case there will be little danger of confusion.

If the group A is abelian and written in additive notation, an element a has finite order when $ma = 0$ for some positive integer m, and the least such m is the order of a. For example, in $\mathbb{Z}_n$ under addition, the residue [1] has order n. The order of the residue [6] in $\mathbb{Z}_{15}$ under addition is 5.

A fairly notable theorem, to be shown eventually, says that if p is a prime, then the group of units $\mathbb{Z}_p^\star$ (under multiplication) contains an element whose order equals $p-1$, which coincides with the order of the group $\mathbb{Z}_p^\star$.

For now here are the fundamentals about the order of a group element.

Proposition 2.5. *Let g be any element in a group G.*

1. *The order of g is finite if and only if $g^k = g^\ell$ for some distinct integer exponents k, ℓ.*
2. *If the order of the group G is finite, then the order of g is also finite.*
3. *If n is the finite order of g, an exponent k causes $g^k = 1$ if and only if $n \mid k$.*
   Also, for any exponents ℓ, m, the equation $g^\ell = g^m$ holds if and only if $\ell \equiv m \bmod n$.
4. *If n is the finite order of g, then the finite list $1, g, \ldots, g^{n-1}$ never repeats, and includes all powers of g.*

Proof. 1. If $g^m = 1$ for some positive integer m, the duplication in powers of g can be illustrated by $g^m = g^0$. Conversely, suppose $g^k = g^\ell$ where $k < \ell$. Then $g^k = g^k g^{\ell-k}$. Cancel g^k to get $g^{\ell-k} = 1$, and thereby see that g has finite order.

2. If G is finite of order m, then the list of $m+1$ powers $1, g, g^2, \ldots, g^m$ must duplicate. By part 1, g has finite order.
3. If $n \mid k$, write $k = nm$ for some integer m. Then

$$g^k = g^{nm} = (g^n)^m = 1^m = 1.$$

Conversely, if $g^k = 1$, use Euclidean division to get integers q, r such that

$$k = nq + r \text{ where } 0 \leq r < n.$$

Then

$$1 = g^k = g^{nq+r} = (g^n)^q g^r = 1^q g^r = g^r.$$

Since n is the least positive integer such that $g^n = 1$ and since $0 \leq r < n$, it follows that $r = 0$, and thus $n \mid k$. The second claim of this item follows from the first because the equation $g^\ell = g^m$ amounts to saying that $g^{\ell-m} = 1$.

4. Suppose a repetition occurs in the list, say $g^i = g^j$ where $0 \leq i < j < n$. As seen in item 3, n divides $j - i$ which is impossible because $1 \leq j - i < n$.

   Furthermore, every integer k is congruent modulo n to its remainder r where $0 \leq r < n$. By item 3, $g^k = g^r$, which puts the arbitrary power g^k in the given short list. $\square$

Cyclic Groups

Definition 2.6. A group H consisting of powers of a single element g is called a ***cyclic group generated by the element*** g.

The laws of exponents reveal that every cyclic group is abelian.

Every group G contains within it groups that are cyclic. Simply take any element g from G and let $H = \{g^k : k \in \mathbb{Z}\}$, the set of integer powers of g. A quick inspection reveals that, using the group operation of G, the set H is a group as well, which by its very definition is cyclic. Indeed, $1 = g^0$ is a power of g. The product of two powers of g is a power of g, and the inverse of g^k is g^{-k} another power of g. Finally, the associative law holds in H because it holds in G.

According to item 4 of Proposition 2.5 an element g of finite order n will cause the group H to consist of the distinct powers $1, g, g^2, \ldots, g^{n-1}$. On the other hand if g has infinite order, the powers g^k as k runs over all integers never duplicate themselves, as item 1 of Proposition 2.5 reveals. Thus a reassuring observation ensues.

Proposition 2.7. *The order of a cyclic group equals the order of its generator.*

By way of illustration, take the set $\mathcal{T}$ of complex numbers z such that $|z| = 1$. The set $\mathcal{T}$ is an abelian group using multiplication of complex numbers, as can be readily checked. Geometrically, $\mathcal{T}$ is the circle consisting of all points in the complex plane having distance 1 to the origin. For this reason $\mathcal{T}$ is called the ***circle group***. For each positive integer n, the element $\omega = e^{2\pi i/n} \in \mathcal{T}$. The order of ω is n. According to Proposition 2.5 the powers $1, \omega, \omega^2, \ldots, \omega^{n-1}$ form a cyclic group of order n. These powers are the vertices of a regular polygon having n sides, as shown for $n = 10$.

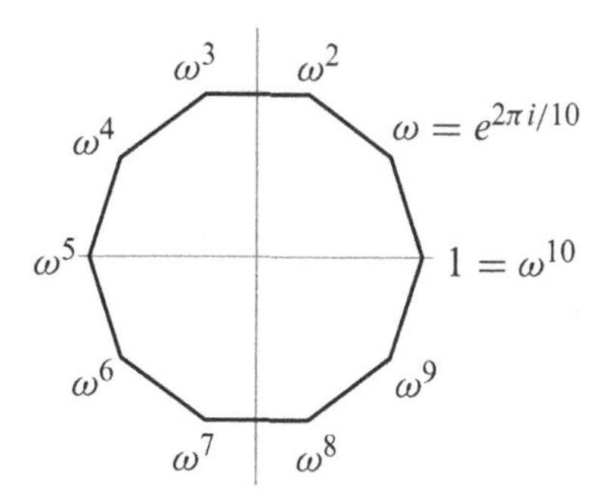

This diagram suggests the aptness of the term "cyclic."

The Order of the Product of Two Elements

The order of the product ab of two elements in a group G may have little connection to the individual orders of a and b. In fact, for any three integers, $r,s,t \geq 2$ there do exist groups with elements a,b of orders r,s, respectively, and such that ab has order t.

Yet, one useful circumstance at times presents itself.

Proposition 2.8. *If a,b are commuting elements in a group G and their respective orders m,n are coprime integers, then the order of ab is mn.*

Proof. Let ℓ be the order of ab. Since a,b commute, we have

$$(ab)^{mn} = a^{mn}b^{mn} = (a^m)^n(b^n)^m = 1^n1^m = 1.$$

By item 3 of Proposition 2.5, $\ell \mid mn$.

To get that $\ell = mn$, it suffices to have $mn \mid \ell$. Since m,n are coprime Proposition 1.5 makes it enough to show that $m \mid \ell$ and $n \mid \ell$. Because a,b commute we have $1 = (ab)^\ell = a^\ell b^\ell$, and thus $a^\ell = b^{-\ell}$. And then,

$$1 = 1^\ell = (a^m)^\ell = (a^\ell)^m = (b^{-\ell})^m = b^{-\ell m}.$$

By item 3 in Proposition 2.5, $n \mid -\ell m$. Since m,n are coprime, Proposition 1.4 reveals that $n \mid \ell$. (The minus sign is irrelevant.) For identical reasons $m \mid \ell$. □

The Order of a Power

Here is how to get the order of a power of a group element.

Proposition 2.9. *If g is an element of finite order n in a group G, then the order of a power g^k equals $n/\gcd(k,n)$.*

Proof. Let m be the order of g^k. So $g^{mk} = (g^k)^m = 1$. By item 2 in Proposition 2.5, $n \mid mk$. Then $n/\gcd(k,n)$ divides $mk/\gcd(k,n)$. Since the integers $n/\gcd(k,n)$ and $k/\gcd(k,n)$ are coprime, Proposition 1.4 reveals that $n/\gcd(k,n)$ divides m.

Also,

$$(g^k)^{n\gcd(k,n)} = (g^n)^{k/\gcd(k,n)} = (1)^{k/\gcd(k,n)} = 1.$$

Proposition 2.5, now applied to the order of g^k, reveals that m divides $n/\gcd(k,n)$. So $m = n/\gcd(k,n)$. □

Two noteworthy special cases emerge from Proposition 2.9.

Proposition 2.10. *If g has finite order n in a group G and k is a positive divisor of n, then the order of g^k equals n/k.*

Proof. In this case, $\gcd(k,n) = k$. □

Proposition 2.11. *If g has finite order n in a group G, a power g^k has order n if and only if the exponent k is coprime with n.*

Proof. The order $n/\gcd(k,n)$ of g^k is n if and only if $\gcd(k,n) = 1$. □

EXERCISES

1. Let G be the set of positive rational numbers. Let the binary operation $\star$ on G be given by $x \star y = xy/3$. Show that G is a group.
2. Verify that the set $\mathcal{T}$ of complex numbers z such that $|z| = 1$ is an abelian group under the operation of multiplication.
3. For each pair of real numbers (a,b) such that $a \neq 0$, let $f_{a,b} : \mathbb{R} \to \mathbb{R}$ be the function defined by $f_{a,b} : x \mapsto ax + b$. These are the functions whose graph is a straight line with non-zero slope. Verify that the set of such functions is a group under the operation of composition of functions.
4. In the abelian group $\mathbb{Z}_n$ using addition, find the formula for the order of a residue $[a]$ in terms of a and n.
5. Let G be a group of order 3 with distinct elements $1, a, b$. Show that $a^2 = b$ and $ab = 1$. Then write the composition table for G.
6. If G is a group and $a, b \in G$, show that ab and ba have equal order.
7. Let a, b be elements of a group G. Show that the equation $ab = x^2$ has a solution x in G if and only if the equation $yay = b$ has a solution y in G.
8. Show that the group of units $\mathbb{Z}_7^\star$ is a cyclic group, while $\mathbb{Z}_8^\star$ is not a cyclic group.
9. Let G be a non-empty set with an associative binary operation $(x,y) \mapsto xy$. Thus $x(yz) = (xy)z$ for all x, y, z in G.
   (a) If for every x, y in G there exist u, v in G such that $xu = y$ and $vx = y$, show that G must be a group.
   (b) If for every x, y in G there exists u in G such that $xu = y$, must G be a group? If, in addition, G is finite, must G be a group?
   (c) If for every x in G there exist e, u in G such that $xe = x$ and $xu = e$, must G be a group?
   (d) If for every x there exist e, u in G such that $xe = x$ and $ux = e$, must G be a group?
   (e) If G is finite and for every x, y, z in G such that $x \neq y$, it holds that $xz \neq yz$ and $zx \neq zy$, prove that G is a group.
   Must such G be a group, if the finiteness of G is not assumed?
10. Show that every element a in a group G of order n is such that $a^k = 1$ for some exponent $k \leq n$. In other words, the order of an element in a finite group is at most the order of the group.
11. If every element a in a group G is such that $a^2 = 1$, show that G is abelian.
12. Let G be a group of order 4 with distinct elements $1, a, b, c$. Show that either $1, a, a^2, a^3$ comprise all of G or else $a^2 = 1$. If $1, a, a^2, a^3$ comprise all of G, write the multiplication table for G. Repeat if $1, b, b^2, b^3$ comprise all of G, and do likewise if $1, c, c^2, c^3$ comprise all of G.
    If G is such that $a^2 = b^2 = c^2 = 1$, write the multiplication table for G.
13. Show that a group of order 5 must be abelian.

Hint. As we learn more about groups this result will become transparent, for now let G be non-abelian of order 5. Take a, b in G such that $ab \neq ba$. Deduce that the list $1, a, b, ab, ba$ comprises all of G, and then derive a contradiction.

Using the above and preceding exercises deduce that every group having fewer than six elements must be abelian.

14. If G is a finite group of even order, show that G has an element of order 2.

Hint. If no such element exists, then the inverse of every x in G, except for $x = 1$, will be distinct from x. Hence, the non-identity elements in G can be counted off in pairs.

In due course we will discover that in a finite group whose order is divisible by a prime p, there will always be an element of order p.

15. If p is a prime and n is a positive integer, show that the order of the finite group $GL_n(\mathbb{Z}_p)$ is

$$(p^n - 1)(p^n - p)(p^n - p^2) \cdots (p^n - p^{n-1}).$$

Hint. A matrix A belongs to $GL_n(\mathbb{Z}_p)$ if and only if its columns are linearly independent. This means that the first column is not zero, the second column is not a multiple of the first column, the third column is not a linear combination of the first two columns, and so on.

16. Some familiarity with the complex plane is needed for this exercise. Let $\alpha, \beta : \mathbb{C} \to \mathbb{C}$ be transformations of $\mathbb{C}$ defined by

$$\alpha : z \mapsto iz \text{ and } \beta : z \mapsto \overline{z},$$

where $i = \sqrt{-1}$ and $\overline{z}$ is the complex conjugate of z. Let 1 be the identity transformation of $\mathbb{C}$. Consider the eight transformations

$$1, \alpha, \alpha^2, \alpha^3, \beta, \alpha\beta, \alpha^2\beta, \alpha^3\beta.$$

Describe in geometric language what each transformation does to the complex plane $\mathbb{C}$, and thus observe that these eight transformations are distinct.

Explain why $\alpha^4 = 1, \beta^2 = 1$ and $\beta\alpha = \alpha^3\beta$, and use this information to show that these transformations constitute a non-abelian group of order 8 under the operation of composition of transformations.

Write the multiplication table for this group.

The group of this exercise is known as the ***dihedral group*** of order 8. When the transformations of this group are restricted to the square whose vertices are $1, i, -1, -i$ the square gets moved rigidly back onto itself. This is the group of rigid motions of the square.

17. Suppose that X is a non-empty set and let $\mathcal{P}(X)$ be the family of all subsets of X. If $A, B \in \mathcal{P}(X)$, their symmetric difference $A \star B$ is the subset defined by:

$$A \star B = (A \setminus B) \cup (B \setminus A).$$

This is the set of elements of A that are not in B together with the elements of B that are not in A.

Verify that $\mathcal{P}(X)$ is an abelian group whose identity element is the empty set $\emptyset$ and such that $A \star A = \emptyset$ for all A in $\mathcal{P}(X)$.

Hint. It may help to draw Venn diagrams to see what the associate law is saying.

18. Show that the matrices $A = \begin{bmatrix} 0 & -1 \\ 1 & 0 \end{bmatrix}$, $B = \begin{bmatrix} 0 & 1 \\ -1 & 1 \end{bmatrix}$, inside the group $GL_2(\mathbb{Q})$ of invertible 2×2 matrices over the rational numbers, have order 4 and 6, respectively, while their products AB and BA have infinite order.

When the group is not abelian, the order of a product can vary drastically from the orders of the factors.

19.* Suppose that a, b inside a group G are commuting elements of order m, n, respectively. Explain why it is that for any exponents k, ℓ, the order of $a^k b^\ell$ divides $\operatorname{lcm}(m,n)$. Then prove that there exist exponents k, ℓ such that the order of $a^k b^\ell$ equals $\operatorname{lcm}(m,n)$.

20. If a, b in a group G are such that $a^5 = 1$ and $aba^{-1} = b^2$, find the order of b.

21. If G is a group with an element a of order m and n is an integer coprime with m, show that there is an x in G such that $x^n = a$.

22.* If a, b in a group G are such that $ab^2 = b^3a$ and $ba^2 = a^3b$, show that $a = b = 1$.

2.2 The Symmetric Group

Definition 2.12. The group $\mathcal{S}_n$ of permutations on n letters, is known as the ***symmetric group***.

The order of $\mathcal{S}_n$ is $n!$. The name originated in the study of symmetric functions. A polynomial function $f(x_1, x_2, \ldots, x_n)$ in n variables, with the property that

$$f(x_1, x_2, \ldots, x_n) = f(x_{\sigma(1)}, x_{\sigma(2)}, \ldots, x_{\sigma(n)}),$$

for every permutation σ on n letters, is called a symmetric function. For instance, $f(x_1, x_2, x_3) = x_1^3 + x_2^3 + x_3^3$ is a symmetric function. A significant result in algebra is that if $g = x^n + a_{n-1}x^{n-1} + \cdots + a_1x + a_0$ is a polynomial in one variable with roots $\lambda_1, \lambda_2, \ldots, \lambda_n$, and if $f(\lambda_1, \lambda_2, \ldots, \lambda_n)$ is a polynomial function that is symmetric in the roots, then f is already a polynomial function of the coefficients a_j. For instance, if $g = x^2 + a_1x + a_0$ has roots λ_1, λ_2, then the symmetric expression $\lambda_1\lambda_2 + \lambda_1 + \lambda_2$ equals $a_0 - a_1$. These subtle matters will be explored in due course, but for now let us examine the symmetric group $\mathcal{S}_n$.

To keep things interesting *suppose throughout that* $n \geq 2$.

Permutation Notation

Let $L = \{1, 2, \ldots, n\}$ be the finite set of letters on which the group $\mathcal{S}_n$ acts. A permutation σ in $\mathcal{S}_n$ is readily described by the table

$$\begin{pmatrix} 1 & 2 & 3 & \cdots & n \\ \sigma(1) & \sigma(2) & \sigma(3) & \cdots & \sigma(n) \end{pmatrix},$$

which lists the values of σ beneath its possible inputs from L. For example,

$$\begin{pmatrix} 1 & 2 & 3 & 4 & 5 \\ 5 & 1 & 2 & 4 & 3 \end{pmatrix}$$

is the permutation in S_5 given by

$$1 \mapsto 5, 2 \mapsto 1, 3 \mapsto 2, 4 \mapsto 4, 5 \mapsto 3.$$

When permutations σ, τ are composed to give $\sigma\tau$, our convention is that *the permutation τ on the right acts first.*

With this understanding one can verify the following composition of two permutations in S_6:

$$\begin{pmatrix} 1 & 2 & 3 & 4 & 5 & 6 \\ 4 & 2 & 1 & 5 & 6 & 3 \end{pmatrix}\begin{pmatrix} 1 & 2 & 3 & 4 & 5 & 6 \\ 6 & 3 & 4 & 2 & 5 & 1 \end{pmatrix} = \begin{pmatrix} 1 & 2 & 3 & 4 & 5 & 6 \\ 3 & 1 & 5 & 2 & 6 & 4 \end{pmatrix}.$$

Here is an illustration of the inverse of a permutation:

$$\begin{pmatrix} 1 & 2 & 3 & 4 & 5 & 6 \\ 3 & 1 & 5 & 2 & 6 & 4 \end{pmatrix}^{-1} = \begin{pmatrix} 1 & 2 & 3 & 4 & 5 & 6 \\ 2 & 4 & 1 & 6 & 3 & 5 \end{pmatrix}.$$

2.2.1 Equivalence Relations and Partitions

Suppose A is a non-empty set. A ***relation*** on A is a rule for matching up pairs of elements from A. More precisely, a relation is a subset E of the Cartesian product $A \times A$. When a pair (x, y) lies in E, we declare x to be related to y.

For instance, taking the set $\mathbb{R}$ of real numbers, the subset E of $\mathbb{R}^2 = \mathbb{R} \times \mathbb{R}$, consisting of all points above the diagonal line "$y = x$," gives all points (x, y) where $x < y$. Thus the relation known as "less than" is defined.

We can write something like $x \sim y$ to denote that the pair $(x, y) \in E$. The relation $x \sim y$ is called an ***equivalence relation*** provided it comes with three likely familiar properties:

- ***reflexivity***, $x \sim x$ for all x in A,
- ***symmetry***, if $x \sim y$, then $y \sim x$,
- ***transitivity***, if $x \sim y$ and $y \sim z$, then $x \sim z$.

For instance, the relation $x \equiv y \bmod 7$ is an equivalence relation on the set of integers $\mathbb{Z}$. As another example of an equivalence relation, take any function $f : A \to B$. For each (x, y) in $A \times A$ declare that $x \sim y$ if and only if $f(x) = f(y)$.

A ***partition*** of a set A is a family of non-empty subsets of A, which are pairwise disjoint and whose union is all of A. For instance, here is a partition of the set $\{1, 2, 3, 4, 5, 6, 7, 8, 9, 10\}$ into 4 pairwise disjoint sets:

$$\{3, 5, 8\}, \ \{2, 4\}, \ \{10\}, \{9, 1, 7, 6\}.$$

Using some index set I, let $\{A_i\}_{i \in I}$ denote a family of subsets that form a partition of A. Declare $x \sim y$ if and only if x, y belong to the same A_i. This defines an equivalence relation on A. Partitions make instant equivalence relations.

Conversely, let us see that every equivalence relation $\sim$ makes a partition of A. For each x in A, define the ***equivalence class represented*** by x to be

$$[x] = \{y \in A : y \sim x\}.$$

The next result is quite simple and rather pedantic, but it deserves to be digested.

Proposition 2.13. *If $\sim$ is an equivalence relation on a non-empty set A, then $x \in [x]$ for all x in A. And for x, y in A, the following statements are equivalent:*

1. $y \sim x$
2. $[x] \cap [y] \neq \emptyset$
3. $[x] = [y]$.

Proof. The statement $x \in [x]$ is reflexivity in poor disguise.

Now suppose statement 1 that $y \sim x$. By symmetry $x \sim y$, and so $x \in [y]$, and since $x \in [x]$, these two equivalence classes do intersect. Thus 1 implies 2.

Next suppose 2 that $[x] \cap [y] \neq \emptyset$. Take a common element z in both $[x]$ and $[y]$. Let $u \in [x]$. Thus $u \sim x$. But $z \in [x]$ too, and so $z \sim x$, which by symmetry gives $x \sim z$. Then, by transitivity $u \sim z$. Since $z \in [y]$, we get $z \sim y$, and by transitivity again, $u \sim y$. Hence $u \in [y]$. We have just seen that $[x] \subseteq [y]$. A similar proof gives $[y] \subseteq [x]$, and so $[x] = [y]$. Thus 2 implies 3.

Finally, suppose $[x] = [y]$. Since $y \in [y] = [x]$, it is clear that $y \sim x$, whereby 3 implies 1, and the three statements are equivalent. □

Proposition 2.13 tells us that two elements from A are equivalent if and only if they represent the same equivalence class, and if two elements are not equivalent, they represent disjoint equivalence classes. Thus the equivalence classes of an equivalence relation form a partition of A. Let us apply these ideas to the study of a permutation.

2.2.2 The Orbits and Cycles of a Permutation

Orbits

If $\sigma \in \mathcal{S}_n$ and a, b are in the set of letters L being permuted, declare

$$b \sim a \text{ provided } b = \sigma^k(a) \text{ for some integer exponent } k.$$

This defines an equivalence relation on L. Indeed, the observation $a = \sigma^0(a)$ shows reflexivity. The fact that $b = \sigma^k(a)$ implies $a = \sigma^{-k}(b)$ shows symmetry. And if we had $b = \sigma^k(a)$ and $c = \sigma^\ell(b)$, substitution shows that

$$c = \sigma^k(\sigma^\ell(a)) = \sigma^{k+\ell}(a),$$

which shows that $\sim$ is transitive.

A partition of L thereby ensues. The disjoint equivalence classes that constitute the partition are called the ***orbits of the permutation*** σ.

Definition 2.14. If $a \in L$, the ***orbit*** of a under σ is the set

$$\mathcal{O}(a) = \{b \in L : b = \sigma^k(a) \text{ for some integer } k\}.$$

From Proposition 2.13 two letters represent the same orbit if and only if each letter is in the orbit of the other letter.

Even though an orbit is parametrized by the infinite set of integer exponents k, it remains a subset of the finite set L of n letters. For example, take the permutation $\sigma = \begin{pmatrix} 1 & 2 & 3 & 4 & 5 & 6 \\ 6 & 3 & 4 & 2 & 5 & 1 \end{pmatrix}$. The orbits of σ that partition the set of 6 letters are

$$\mathcal{O}(3) = \{3,4,2\},\ \mathcal{O}(1) = \{1,6\},\ \mathcal{O}(5) = \{5\}.$$

Since $\mathcal{S}_n$ is a finite group, every permutation σ in $\mathcal{S}_n$ has finite order. If ℓ is that order, then $\sigma^\ell(a) = 1(a) = a$ for every letter a. But for each such letter a, there could be a lesser positive exponent m such that $\sigma^m(a) = a$. For instance, the permutation $\sigma = \begin{pmatrix} 1 & 2 & 3 & 4 & 5 \\ 5 & 1 & 4 & 3 & 2 \end{pmatrix}$ has order 5, but $\sigma^3(2) = 2$. The smallest such exponent affords a helpful way to think about orbits.

Proposition 2.15. *If $\sigma \in \mathcal{S}_n$ and a is a letter, and m is the smallest positive integer such that $\sigma^m(a) = a$, then the orbit of a under σ consists of the list*

$$a, \sigma(a), \sigma^2(a), \ldots, \sigma^{m-1}(a).$$

The list never repeats a letter.

Proof. To see that the list $a, \sigma(a), \sigma^2(a), \ldots, \sigma^{m-1}(a)$ exhausts the orbit, take any integer k, and thereby the letter $\sigma^k(a)$ in $\mathcal{O}(a)$. By Euclidean division,

$$k = mq + r \text{ for some integers } q, r \text{ where } 0 \le r < m.$$

Hence,

$$\sigma^k(a) = \sigma^{mq+r}(a) = \sigma^r(\sigma^m)^q(a).$$

The power notation $(\sigma^m)^q$ represents the application of σ^m or its inverse a finite number of times, and since $\sigma^m(a) = a$, then so also is $(\sigma^m)^q(a) = a$. Thus,

$$\sigma^k(a) = \sigma^r(a),$$

and because $0 \le r < m$, the element $\sigma^k(a)$ falls into the desired short list.

Next, suppose that a repetition $\sigma^j(a) = \sigma^k(a)$ occurs in the short list. Here, $1 \le j < k < m$. Apply σ^{-j} to obtain $\sigma^{k-j}(a) = a$. Since $0 < k - j < m$, this contradicts the minimality of m. There is no repetition in the short list. □

Proposition 2.15 lets us think of an orbit of σ as a list of m distinct letters $a_1, a_2, a_3, \ldots, a_m$ on which σ acts as follows:

$$a_1 \mapsto a_2 \mapsto a_3 \mapsto \cdots \mapsto a_{m-1} \mapsto a_m \mapsto a_1.$$

The fact that repeated application of σ brings a_1 back to a_1 reinforces the "orbit" nomenclature.

The number of letters in $\mathcal{O}(a)$ is known as the ***length*** of the orbit of a. Proposition 2.15 shows that the length of $\mathcal{O}(a)$ equals the smallest positive integer m such that $\sigma^m(a) = a$.

Cycles

Definition 2.16. If m is a positive integer, a permutation τ on n letters is called a ***cycle of length*** m, or more briefly an m***-cycle***, whenever τ has an orbit of length m and all of its other orbits have length one.

Accordingly, a 1-cycle is a permutation for which all orbits have length one. Since an orbit $\mathcal{O}(a)$ of τ has length one if and only if $\tau(a) = a$, a 1-cycle is just the identity permutation 1, which fixes every letter. (Regrettably here, the notation "1" for the identity permutation, clashes with the symbol "1" representing the first of our n letters, as well as with "$m = 1$.")

More interesting is the case where the cycle τ has a unique orbit of length $m \geq 2$. Here there is a list of m distinct letters $a_1, a_2, \ldots, a_m$ in the unique orbit of length m. And the m-cycle τ is defined by

$$\tau : a_1 \mapsto a_2 \mapsto a_3 \mapsto \cdots \mapsto a_{m-1} \mapsto a_m \mapsto a_1$$

and also

$$\tau : b \mapsto b \text{ when } b \text{ is not any of the } a_j.$$

The cyclical behavior of τ reveals itself. The so-called ***cycle notation*** for an m-cycle is

$$\tau = (a_1, a_2, a_3, \ldots, a_m).$$

For example, in S_8 the cycle $(3, 7, 5, 2, 4)$ is the permutation given by

$$3 \mapsto 7 \mapsto 5 \mapsto 2 \mapsto 4 \mapsto 3 \text{ and } 1 \mapsto 1, 6 \mapsto 6, 8 \mapsto 8.$$

There is a bit of variability in the cycle notation. For instance, the notations $(3, 7, 5, 2, 4)$ and $(5, 2, 4, 3, 7)$ and $(4, 3, 7, 5, 2)$ specify the same cycle.

Factorization into Disjoint Cycles

Two permutations σ, τ on n letters, are ***disjoint*** provided there is no letter that is moved by both σ and τ. In other words,

$$\sigma(a) \neq a \text{ implies } \tau(a) = a.$$

As a special case, the identity permutation, which leaves every letter fixed, is disjoint from all permutations (and, paradoxically, even from itself).

In case σ, τ are cycles, how is their disjointness to be interpreted? In cycle notation, let

$$\sigma = (a_1, a_2, \ldots, a_\ell) \text{ and } \tau = (b_1, b_2, \ldots, b_m).$$

If $\ell = 1$, then σ is the identity permutation, which is disjoint from every permutation. Likewise for τ, if $m = 1$. But if $\ell, m \geq 2$, then the only letters moved by σ are the a_j, while the only letters moved by τ are the b_i. So, if $\ell, m \geq 2$, the cycles σ, τ are disjoint if and only if the a_j never overlap with the b_i. For example, in S_7 the 4-cycle $(6,7,3,1)$ is disjoint from the 2-cycle $(8,4)$, but the cycles $(6,7,3,1), (7,5,4)$ are not disjoint.

Even though the composition of permutations need not be commutative, commutativity holds if the permutations are disjoint.

Proposition 2.17. *If σ, τ are disjoint permutations on n letters, then $\sigma\tau = \tau\sigma$.*

Proof. Let L be the set of n letters that σ, τ permute, and let A be the subset of letters moved by σ, and B the subset of letters moved by τ. The disjointness of σ from τ implies that τ fixes the letters in A and σ fixes the letters in B.

If $a \in A$, then $\sigma(a) \in A$. For otherwise, we would have $\sigma(\sigma(a)) = \sigma(a)$, and thus $\sigma(a) = a$. Consequently, if $a \in A$, then τ fixes both a and $\sigma(a)$, and thus

$$\tau(\sigma(a)) = \sigma(a) = \sigma(\tau(a)).$$

Similarly,

$$\sigma(\tau(b)) = \tau(b) = \tau(\sigma(b)),$$

for every b in B. And if c is a letter neither in A nor in B, then obviously

$$\sigma(\tau(c)) = \sigma(c) = c = \tau(c) = \tau(\sigma(c)).$$

Since the effect of $\sigma\tau$ equals the effect of $\tau\sigma$ on every letter of L, it follows that $\tau\sigma = \sigma\tau$. □

Cycles are building blocks for the unique factorization of all permutations. A bit of terminology might help to explain this. If σ is a permutation and $\mathcal{O}$ is an orbit of σ of length m, then the restriction of σ to $\mathcal{O}$ is a bijection from $\mathcal{O}$ to $\mathcal{O}$, i.e. a permutation of $\mathcal{O}$. Let τ be the permutation that agrees with σ on the orbit $\mathcal{O}$ and fixes every letter not in $\mathcal{O}$.

Think of $\mathcal{O}$ as the list of distinct letters $a_1, a_2, \ldots, a_m$, where under the action of σ we have

$$\sigma : a_1 \mapsto a_2 \mapsto a_3 \mapsto \cdots \mapsto a_{m-1} \mapsto a_m \mapsto a_1.$$

On the letters outside of $\mathcal{O}$ the permutation σ does what it may. Since τ agrees with σ on $\mathcal{O}$ and fixes the letters not in $\mathcal{O}$, we see that τ is nothing but the cycle $\tau = (a_1, a_2, \ldots, a_m)$. Let us say that τ is the cycle ***spawned*** from the orbit $\mathcal{O}$. Obviously, the length of a cycle equals the length of the orbit that spawns it.

Proposition 2.18. *Every permutation on n letters is the product of disjoint, and thereby pairwise commuting, cycles.*

Proof. Let σ be our permutation on the set L of n letters, with pairwise disjoint orbits $\mathcal{O}_1, \mathcal{O}_2, \mathcal{O}_3, \ldots, \mathcal{O}_\ell$ that partition L. If m_j is the length of $\mathcal{O}_j$, let τ_j be the m_j-cycle spawned from $\mathcal{O}_j$. The cycles τ_j are pairwise disjoint because they are spawned by pairwise disjoint orbits. As a result the τ_j commute with each other.

Now check that $\sigma = \tau_1\tau_2\cdots\tau_\ell$. Take any letter a and let $\mathcal{O}_k$ be the orbit to which a belongs. To keep notations simple, say that $a \in \mathcal{O}_1$. Since the other cycles $\tau_2, \tau_3 \ldots, \tau_\ell$ fix a, so does their composite $\tau_2\tau_3\cdots\tau_\ell$. And then

$$\tau_1\tau_2\tau_3\cdots\tau_\ell(a) = \tau_1(a) = \sigma(a).$$

The last equality happens because τ_1 agrees with σ on $\mathcal{O}_1$. □

To actually find the factorization of a given permutation into disjoint cycles just track the letters along their orbits. For example, with the permutation

$$\sigma = \begin{pmatrix} 1 & 2 & 3 & 4 & 5 & 6 & 7 & 8 & 9 & 10 \\ 5 & 1 & 4 & 10 & 7 & 6 & 2 & 9 & 8 & 3 \end{pmatrix}$$

we see that

$$1 \mapsto 5 \mapsto 7 \mapsto 2 \mapsto 1 \text{ and } 3 \mapsto 4 \mapsto 10 \mapsto 3 \text{ and } 8 \mapsto 9 \mapsto 8 \text{ and } 6 \mapsto 6.$$

Our factorization into disjoint cycles is

$$\sigma = (1,5,7,2)(3,4,10)(8,9)(6).$$

The 1-cycle (6) is just the identity permutation, making its appearance redundant. Due to their redundancy, 1-cycles can be and usually are omitted from the factorization of a permutation. In our example, it is customary to write

$$\sigma = (1,5,7,2)(3,4,10)(8,9).$$

The order in which the disjoint cycles are written does not matter, since they commute with each other.

Uniqueness of the Cycle Decomposition

Given that cycles are spawned by the orbits of a permutation σ, and that orbits are uniquely determined by σ, there should be only one way to factor a permutation into disjoint cycles, taking into account the inconsequential addition or deletion of 1-cycles (which are just the identity) and the ordering of the commuting disjoint cycles. Let us prove this claim.

Proposition 2.19. *If $\sigma = \tau\rho$ is a factorization of a permutation σ into disjoint permutations τ, ρ, and τ is an m-cycle where $m \geq 2$, then the set of letters moved by τ is an orbit of σ, and that orbit spawns τ.*

Proof. In cycle notation let $\tau = (a_1, a_2, \ldots, a_m)$. Since $m \geq 2$, the letters moved by τ are precisely the a_j. By disjointness, ρ fixes the a_j, and so

$$\sigma(a_j) = \tau\rho(a_j) = \tau(a_j).$$

Under the action of τ we have

$$\tau : a_1 \mapsto a_2 \mapsto a_3 \mapsto \cdots \mapsto a_{m-1} \mapsto a_m \mapsto a_1.$$

And having just seen that τ agrees with σ on the a_j, the above action equally applies to σ. This reveals that the list of letters $a_1, a_2, \ldots, a_m$ is an orbit of σ, and that this orbit spawns τ. □

Proposition 2.20. *If $\sigma = \tau_1\tau_2 \cdots \tau_k$ is a factorization of a permutation σ into pairwise disjoint cycles τ_j that are not the identity, then σ has exactly k orbits of length at least two, and each such orbit spawns one of the cycles τ_j.*

Proof. By Proposition 2.19 applied to each cycle τ_j, each τ_j is spawned from an orbit of σ. Thus, σ has at least k orbits of length at least two.

All that is left is to check that every orbit of σ, having length at least two, spawns a τ_j, and not some cycle other than the τ_j. Well, if such an orbit existed, there would be a letter moved by σ but not by any τ_j, contradicting the fact that σ is the composite of the τ_j. □

Proposition 2.20 shows that the factorization of a permutation into disjoint cycles is uniquely determined by the orbits of σ, up to the order in which the commuting cycles are written.

The Orbit Structure

If σ is a permutation on the set of n letters L, let $\mathcal{O}_1, \mathcal{O}_2, \ldots, \mathcal{O}_k$ be the pairwise disjoint orbits of σ, which partition L. Let $\ell(\mathcal{O}_j)$ be the length of (i.e. the number of letters in) the orbit $\mathcal{O}_j$. By arranging the lengths in decreasing order, we come up with a k-tuple of positive integers

$$[\ell(\mathcal{O}_1), \ell(\mathcal{O}_2), \ell(\mathcal{O}_3), \ldots, \ell(\mathcal{O}_k)],$$

where

$$\ell(\mathcal{O}_1) \geq \ell(\mathcal{O}_2) \geq \ell(\mathcal{O}_3) \geq \cdots \geq \ell(\mathcal{O}_k).$$

And since the orbits partition L, we also have

$$\ell(\mathcal{O}_1) + \ell(\mathcal{O}_2) + \ell(\mathcal{O}_3) + \cdots + \ell(\mathcal{O}_k) = n.$$

Definition 2.21. The k-tuple of positive integers, written in descending order, that lists the lengths of the orbits of σ is called the ***orbit structure*** of σ.

The disjoint orbits of a permutation spawn the disjoint cycles that uniquely factor the permutation, and the length of each cycle is the length of the orbit that spawns it. Thus the orbit structure could just as readily be called the ***cycle structure*** of a permutation.

Definition 2.22. Any list of positive integers

$$[p_1, p_2, p_3, \dots, p_k]$$

such that

$$p_1 \geq p_2 \geq p_3 \geq \cdots \geq p_k \text{ and } p_1 + p_2 + p_3 + \cdots + p_k = n,$$

is called a ***partition of the integer*** n.

For instance, the integer 3 has 3 partitions:

$$[3], [2, 1], [1, 1, 1].$$

The integer 5 has 7 partitions:

$$[5], [4, 1], [3, 2], [3, 1, 1], [2, 2, 1], [2, 1, 1, 1], [1, 1, 1, 1, 1].$$

And the integer 6 has 11 partitions:

$$[6], [5, 1], [4, 2], [4, 1, 1], [3, 3], [3, 2, 1], [3, 1, 1, 1],$$
$$[2, 2, 2], [2, 2, 1, 1], [2, 1, 1, 1, 1], [1, 1, 1, 1, 1, 1].$$

It is a formidable combinatorial problem to count the number of partitions of n.

With a little imagination, it is easy to see that every partition of n is the orbit structure of some permutation on n letters. So, the number of orbit structures that permutations on n letters can have equals the number of partitions of n.

The Order of a Permutation

The orbit structure of a permutation reveals its order as an element of $\mathcal{S}_n$.

Proposition 2.23. *If σ, τ are disjoint permutations on n letters, having orders k, ℓ, respectively, then the order of their composite $\sigma\tau$ is the least common multiple of k and ℓ.*

Proof. If m is a positive integer, observe that

$$\sigma^m \tau^m = 1 \text{ if and only if } \sigma^m = \tau^m = 1.$$

One implication in this claim is clear. For the converse implication say $\sigma^m \neq 1$. Then some letter a is moved by σ^m, and thus by σ. By disjointness, a is fixed by τ, and hence by τ^m. Consequently,

$$\sigma^m \tau^m(a) = \sigma^m(a) \neq a,$$

which shows $\sigma^m \tau^m \neq 1$.

Since σ commutes with τ (because they are disjoint), $(\sigma\tau)^m = \sigma^m \tau^m$, and so

$$(\sigma\tau)^m = 1 \text{ if and only if } \sigma^m = \tau^m = 1.$$

The smallest exponent m to yield $(\sigma\tau)^m = 1$ is the smallest m to yield $\sigma^m = \tau^m = 1$. According to Proposition 2.5, such an m is the smallest positive integer divisible by both k and ℓ. In other words, m is the least common multiple of k and ℓ. □

As an aside, note that in case k, ℓ are coprime, Proposition 2.23 also follows from Proposition 2.8. However, it is not in general true that the order of a product of two commuting elements in a group is the least common multiple of the individual orders. For instance, the transposition $(1,2)$ has order 2. It commutes with itself. But the product $(1,2)(1,2)$ is the identity, whose order is 1 rather than 2.

Proposition 2.23 remains valid for the product of several pairwise disjoint permutations, with essentially the same proof as the one given. Alternatively one can give a simple argument by induction on the number of pairwise disjoint factors.

Next observe that the order of an m-cycle $\sigma = (a_1, a_2, \ldots, a_m)$ is m. Indeed, notice that

$$\sigma(a_1) = a_2, \sigma^2(a_1) = a_3, \ldots, \sigma^{m-1}(a_1) = a_{m-1}, \sigma^m(a_1) = a_1.$$

The lowest power of σ to fix a_1 is σ^m. The same applies to all a_j. Since σ fixes all other letters, $\sigma^m = 1$, and m is the least such exponent.

The orbit structure of a permutation is the list of the lengths of the disjoint cycles that factor the permutation. Since the order of a cycle is its length, Proposition 2.23 leads to the following result.

Proposition 2.24. *If* $[p_1, p_2, \ldots, p_k]$ *is the orbit structure of a permutation, then the order of the permutation is* $\operatorname{lcm}(p_1, p_2, \ldots, p_k)$.

For example, take the product of the disjoint cycles $\sigma = (1346)(297)(85)$ in $\mathcal{S}_{10}$. The orbit structure of σ is $[4,3,2,1]$. Its order is $\operatorname{lcm}(4,3,2,1) = 12$.

2.2.3 Transpositions and the Parity of a Permutation

Definition 2.25. A 2-cycle $\sigma = (a, b)$ acting on n letters is known as a ***transposition***.

The transposition σ maps a to b and b to a and leaves every other letter fixed. We know that every permutation on n letters is the composite of unique disjoint cycles. This will enable us to show that every such permutation is the composite of transpositions. Intuitively, this can be seen by realizing that any rearrangement of the letters $1, 2, \ldots, n$ can be achieved by switching them one pair at a time. But here is a proper proof.

Proposition 2.26. *Every permutation on n letters, with $n \geq 2$, is the composite of transpositions.*

Proof. Since every permutation is the composite of cycles, it suffices to show that every cycle is the composite of transpositions. Writing a cycle as $(a_1, a_2, a_3, \ldots, a_m)$, the desired fact is seen by looking closely at the equality

$$(a_1, a_2, a_3, \ldots, a_m) = (a_1, a_m)(a_1, a_{m-1}) \cdots (a_1, a_4)(a_1, a_3)(a_1, a_2).$$

In verifying this equality, keep in mind that the permutations on the right act first.

In case the cycle is the identity permutation 1, we can view it as the null product of transpositions, but if that seems puzzling, just note that the identity permutation equals $(1,2)(1,2)$. □

The factorization of a permutation into transpositions is far from unique. For instance,

$$(3,4)(3,1)(3,2) = (3,2,1,4) = (2,1)(2,4)(1,4)(1,2)(3,4).$$

What turns out to be unique is the ***parity*** of the number of transposition factors. If a permutation σ can be factored using an even number of transpositions, then any other factorization of σ into transpositions must also have an even number of transpositions. Let us proceed to show this somewhat intriguing result.

Permutations Acting on Functions

We start with polynomial functions $f(x_1, x_2, \ldots, x_n)$ in n variables with rational coefficients. For example, $x_1^2 + \frac{8}{3}x_2x_3 - 5x_1^3x_2x_3^4$ is such a function. We can take the domain and range of our functions to be the set of rational numbers $\mathbb{Q}$. Call this set of functions $\mathcal{F}$. Clearly, the sum of two functions in $\mathcal{F}$ and any rescaling of a function in $\mathcal{F}$ by a rational number multiple, remain functions in $\mathcal{F}$. In other words, $\mathcal{F}$ is a vector space over the field of scalars $\mathbb{Q}$. In due course, we will take a closer look at polynomials, but for now the above should be clear enough.

A permutation σ in S_n is given a new role as a transformation of the space $\mathcal{F}$. Namely, if $f(x_1, x_2, \ldots, x_n)$ is a polynomial function in $\mathcal{F}$ define a new polynomial function σf in $\mathcal{F}$ by permuting the variables according to σ. Namely,

$$\sigma f(x_1, x_2, \ldots, x_n) = f(x_{\sigma(1)}, x_{\sigma(2)}, \ldots, x_{\sigma(n)}).$$

For example, taking the function $f = x_1^2 + x_1x_2 + 6x_2^3x_3^5x_4$ and the cycle $\sigma = (4,1,2)$ in S_4, we obtain $\sigma f = x_2^2 + x_2x_4 + 6x_4^3x_3^5x_1$. A couple of things are easy to see.

- If $f \in \mathcal{F}$ and $\sigma, \tau \in S_n$, then $\sigma(\tau f) = (\sigma\tau)f$. In other words, a successive action on f by two (or more) permutations has the same effect as one action on f by the composite permutation.
- If $f \in \mathcal{F}$ and $\sigma \in S_n$ and $\lambda \in \mathbb{Q}$, then $\sigma(\lambda f) = \lambda(\sigma f)$. Also, if $f, g \in \mathcal{F}$ and $\sigma \in S_n$, then $\sigma(f + g)$. That is, that the action of σ on $\mathcal{F}$ is linear.

  In particular, $\sigma(-f) = -\sigma f$.

The following polynomial in $\mathcal{F}$ distinguishes something important about a permutation:

$$p = \prod_{i<j}(x_i - x_j).$$

This is the product of all differences $x_i - x_j$ where i,j are among the letters $1, 2, \ldots, n$ and $i < j$. For example, if $n = 4$, then

$$p = (x_1 - x_2)(x_1 - x_3)(x_1 - x_4)(x_2 - x_3)(x_2 - x_4)(x_3 - x_4).$$

One name for p is the ***Vandermonde polynomial***.

If σ is the cycle $(1, 3, 2, 4)$, then

$$\sigma p = (x_3 - x_4)(x_3 - x_2)(x_3 - x_1)(x_4 - x_2)(x_4 - x_1)(x_2 - x_1).$$

Notice, in the case of the 4-cycle, that all of the factors $x_i - x_j$ that are in p reappear with a possible change of sign in σp. Thus $\sigma p = \pm p$. Actually, in this case, $\sigma p = -p$.

The formula $\sigma p = \pm p$ happens in general. To see this, examine

$$\sigma p = \prod_{i<j} \left(x_{\sigma(i)} - x_{\sigma(j)} \right).$$

Each factor $x_i - x_j$ in p comes from taking a two-element subset i,j of the set of letters $L = \{1, 2, 3, \ldots, n\}$ and arranging it so that $i < j$. A permutation σ also makes a permutation of the family of two-element subsets of L. The two-element subset $\{i,j\}$ of L becomes the two-element subset $\{\sigma(i), \sigma(j)\}$. Thus σp is made up of the same factors as p is, except that some of the factors may have a change of sign. Thus $\sigma p = \pm p$.

The Parity of a Permutation

Definition 2.27. If $\sigma p = p$, we say that the permutation σ is ***even***, and if $\sigma p = -p$, we say that σ is ***odd***. We can also define the ***sign*** of a permutation σ as follows:

$$\operatorname{sgn}(\sigma) = \begin{cases} +1 & \text{if } \sigma p = p, \sigma \text{ is even} \\ -1 & \text{if } \sigma p = -p, \text{ i.e. } \sigma \text{ is odd.} \end{cases}$$

The value of $\operatorname{sgn} \sigma$, i.e. whether σ is even or odd, is known as the ***parity*** of σ.

The sign function comes with a noteworthy property.

Proposition 2.28. *If σ, τ are permutations on n letters, then*

$$\operatorname{sgn}(\sigma\tau) = \operatorname{sgn}(\sigma)\operatorname{sgn}(\tau).$$

Proof. Look at the four possible combined values of $\operatorname{sgn}(\sigma)$ and $\operatorname{sgn}(\tau)$. If $\operatorname{sgn}(\sigma) = \operatorname{sgn}(\tau) = +1$, then

$$(\sigma\tau)p = \sigma(\tau p) = \sigma p = p,$$

and so

$$\operatorname{sgn}(\sigma\tau) = +1 = \operatorname{sgn}(\sigma)\operatorname{sgn}(\tau).$$

If $\operatorname{sgn}(\sigma) = +1$ and $\operatorname{sgn}(\tau) = -1$, then

$$(\sigma\tau)p = \sigma(\tau p) = \sigma(-p) = -\sigma p = -p,$$

and so

$$\operatorname{sgn}(\sigma\tau) = -1 = \operatorname{sgn}(\sigma)\operatorname{sgn}(\tau).$$

And similarly for the two remaining cases. □

Since every permutation is a product of transpositions, Proposition 2.28 will reveal the parity of every permutation once the parity of a transposition is known.

To find the parity of σ by brute force, count the number of times that the factors $\left(x_{\sigma(i)} - x_{\sigma(j)}\right)$ appearing in σp are such that $\sigma(i) > \sigma(j)$. This will count the number of times that the factors $x_i - x_j$ in p reappear in σp with a change of sign. For every occurrence of

$$i < j \text{ together with } \sigma(i) > \sigma(j)$$

we say that σ implements an ***inversion***. The permutation σ is even if and only if σ implements an even number of inversions. For example, take

$$\sigma = \begin{pmatrix} 1 & 2 & 3 & 4 & 5 \\ 3 & 5 & 4 & 1 & 2 \end{pmatrix}.$$

By looking at the values of σ, we see that the inequality $1 < j$, for $j = 2, 3, 4, 5$, is inverted 2 times. The inequality $2 < j$ (for $j = 3, 4, 5$) is inverted 3 times. The inequality $3 < j$ (for $j = 4, 5$) is inverted 2 times, and $4 < j$ (for $j = 5$) is inverted 0 times. This gives a total of 7 inversions, which gives σ odd parity.

Fortunately, Proposition 2.28 can expedite matters.

Transpositions Have Odd Parity

The transposition $\sigma = (1, 2)$ is odd, as it implements just one inversion. Under the action of σ on the Vandermonde polynomial p, the factor $x_1 - x_2$ in p becomes $x_2 - x_1 = -(x_1 - x_2)$. The factors $x_1 - x_j$ where $2 < j$ become interchanged with the factors $x_2 - x_j$. The remaining factors do not change. Thus $\sigma p = -p$.

Proposition 2.29. *Every transposition is an odd permutation.*

Proof. The transposition $(1, 2)$ is odd.

If the transposition is of the type $(1, k)$ with $k > 2$, Proposition 2.28 applied to the factorization

$$(1, k) = (2, k)(1, 2)(2, k)$$

yields

$$\operatorname{sgn}(1, k) = \operatorname{sgn}(2, k)\operatorname{sgn}(1, 2)\operatorname{sgn}(2, k) = -(\operatorname{sgn}(2, k))^2 = -1.$$

If the transposition is of the type $(2, k)$ with $k > 2$, Proposition 2.28 applied to the factorization

$$(2, k) = (1, k)(1, 2)(1, k)$$

yields

$$\operatorname{sgn}(2,k) = \operatorname{sgn}(1,k)\operatorname{sgn}(1,2)\operatorname{sgn}(1,k) = -(\operatorname{sgn}(1,k))^2 = -1.$$

If the transposition is of the type (j,k) where j,k are distinct from $1,2$, Proposition 2.28 applied to the factorization

$$(j,k) = (1,j)(2,k)(1,2)(1,j)(2,k)$$

yields

$$\operatorname{sgn}(j,k) = -(\operatorname{sgn}(1,j))^2(\operatorname{sgn}(2,k))^2 = -1.$$

Thus all transpositions are odd. □

Alternatively, to see that a transposition (j,k) is odd, count directly the number of inversions that it implements. Say $j < k$. The inequalities

$$j < j+1, j < j+2, \ldots, j < k-1$$

get inverted by (j,k) into

$$k > j+1, k > j+2, \ldots, k > k-1.$$

And the inequalities

$$j+1 < k, j+2 < k, \ldots, k-1 < k$$

get inverted by (j,k) into

$$j+1 > j, j+2 > j, \ldots, k-1 > j.$$

This gives $2(k-j-1)$ inversions. There is one more inversion, namely, $j < k$ becomes $\sigma(j) = k > j = \sigma(k)$. That gives an odd number of inversions in total, and thus (j,k) is odd.

A Way to Detect Parity

Having seen that every permutation is the product of transpositions and that each transposition has odd parity, the parity of every permutation σ is now easily computed. Namely, factor $\sigma = \tau_1\tau_2\cdots\tau_m$ as a product of transpositions, in accordance with Proposition 2.26. By Proposition 2.28,

$$\operatorname{sgn}(\sigma) = \operatorname{sgn}(\tau_1)\operatorname{sgn}(\tau_2)\cdots\operatorname{sgn}(\tau_m) = (-1)^m.$$

The following becomes apparent.

Proposition 2.30. *A permutation σ is even if and only if it is the product of an even number of transpositions. All other factorizations of σ as a product of transpositions must also have an even number of transpositions. The analogous claims hold for odd permutations.*

The Parity of a Product of Disjoint Cycles

As noted in the proof of Proposition 2.26 the formula

$$(a_1, a_2, a_3, \ldots, a_m) = (a_1, a_m)(a_1, a_{m-1}) \cdots (a_1, a_4)(a_1, a_3)(a_1, a_2)$$

shows that every m-cycle is the composite of $m - 1$ transpositions. By Proposition 2.30, an m-cycle is odd if and only if m is even.

From this observation, the orbit structure of a permutation reveals its parity. For instance, suppose $\sigma \in S_{15}$ and the orbit structure of σ is $[6, 3, 3, 2, 1]$. This tells us that σ factors uniquely as the disjoint product of one 6-cycle, two 3-cycles and one 2-cycle. Applying Proposition 2.28, we get that

$$\operatorname{sgn}(\sigma) = (-1)(+1)(+1)(-1) = +1,$$

making σ even.

The Group $\mathcal{A}_n$

Denote the set of all even permutations by $\mathcal{A}_n$. The mapping $\sigma \mapsto (1,2)\sigma$ transforms even permutations into odd permutations, and vice versa. Thus there are exactly as many even permutations as odd permutations. The number of even permutations is $n!\,/2$. From Proposition 2.28 the product of even permutations is even, and clearly the identity permutation is even. Also, if σ is even, then so is σ^{-1}. This can be seen by looking at

$$1 = \operatorname{sgn}(1) = \operatorname{sgn}(\sigma\sigma^{-1}) = \operatorname{sgn}(\sigma)\operatorname{sgn}(\sigma^{-1}) = 1\operatorname{sgn}(\sigma^{-1}) = \operatorname{sgn}(\sigma^{-1}).$$

Thus, $\mathcal{A}_n$ is a group inside $\mathcal{S}_n$.

The important group $\mathcal{A}_n$ is known as the ***alternating*** group. This curious name goes at least as far back as the seminal monograph of Camille Jordan (1838–1922), published in 1870, about substitutions (a.k.a. permutations) and algebraic equations. In this monograph Jordan organizes and enhances the brilliant discoveries of Galois on the solvability of polynomial equations. Possibly, the term originates from the idea of alternating functions. These are the functions $f(x_1, x_2, \ldots, x_n)$ in n variables that simply change sign when two of the variables are transposed. The Vandermonde polynomial is such a function, and $x_1^2 - x_2^2$ is another one. When $\mathcal{A}_n$ acts on an alternating function by permuting its variables accordingly, the function does not change at all. Hence, the name "alternating."

Useful Generators for $\mathcal{A}_n$

As already noted, 3-cycles are even, and thus every product of 3-cycles is in $\mathcal{A}_n$. The converse to this remark is also true.

Proposition 2.31. *If $n \geq 3$ then every even permutation in $\mathcal{S}_n$ is a product of 3-cycles.*

Proof. An even permutation is a product of an even number of transpositions. By considering adjacent pairs of transpositions, the proof reduces to showing that for distinct letters a, b, c, d, the permutations of type $(a, b)(b, c)$ or $(a, b)(c, d)$ are products of 3-cycles. Well,

$$(a, b)(c, d) = (c, a, d)(a, b, c) \text{ and } (a, b)(b, c) = (a, b, c).$$ □

EXERCISES

1. Show that, in a finite group, the number of elements x such that $x^3 = 1$ is odd.
Hint. Let S be the set of x in a finite group G such that $x^3 = 1$ and $x \neq 1$. Show that S contains an even number of elements by showing that the relation on S given by

$$x \sim y \text{ provided that } y = x^2 \text{ or } y = x$$

is an equivalence relation, and examining the ensuing partition of S.

2. Find the order of $\begin{pmatrix} 1 & 2 & 3 & 4 & 5 & 6 & 7 & 8 & 9 & 10 \\ 5 & 1 & 7 & 10 & 2 & 6 & 3 & 8 & 9 & 4 \end{pmatrix}$. Is this permutation even or odd?

3. Is the cycle product $(1, 2, 3, 4, 5, 6)(1, 2, 4)(4, 5)$ even or odd? What is its order?

4. Count the number of permutations in $\mathcal{S}_n$ which fix at least one letter.

5. If the order of a permutation in $\mathcal{S}_n$ is odd, show that the permutation is even.

6. Show that every permutation in $\mathcal{S}_n$ can be factored as $(1, 2)\tau$ where τ is the product of 3-cycles.

7. If a permutation σ on n letters moves every letter and is a product of m disjoint cycles, show that σ is the product of $n - m$ transpositions.

8. An interlacing shuffle on a deck of $2n$ cards can be interpreted as the permutation

$$\begin{pmatrix} 1 & 2 & 3 & \cdots & n & n+1 & n+2 & \cdots & 2n-1 & 2n \\ 2 & 4 & 6 & \cdots & 2n & 1 & 3 & \cdots & 2n-3 & 2n-1 \end{pmatrix}.$$

On a deck of 52 cards what is the least number of interlacing shuffles that have to be repeated in order to restore the deck to its original configuration?
What if the deck has 50 cards?

9. How many orbit structures arise from the permutations of $\mathcal{S}_8$?

10. Find the orbit structures of all powers of the cycle $(1, 2, 3, \ldots, 8)$.

11. What is the maximum order that a permutation in $\mathcal{S}_7$ can have?

12. How many permutations of order 2 are there in $\mathcal{S}_7$? How many of order 2 in $\mathcal{S}_8$? Find a formula in terms of n for the number of permutations of order 2 in $\mathcal{S}_n$.

13. Let σ be the n-cycle $(1, 2, 3, \ldots, n)$ in $\mathcal{S}_n$ and k a positive exponent. We have noted that σ, along with all n-cycles, has order n.
(a) If k is not coprime with n, show that σ^k is no longer an n-cycle.
Hint. It can be checked directly, but Proposition 2.9 gives a quick solution.
(b) If k is coprime with n, Proposition 2.11 shows that σ^k has order n. Show that σ^k is actually an n-cycle.

14. If p is prime and σ in $\mathcal{S}_p$ has order p, show that σ is a p-cycle.
If n is not a power of a prime, find a permutation σ in $\mathcal{S}_n$ such that σ has order n but σ is not an n-cycle.

15. Explain why a permutation in $\mathcal{S}_n$ is even if and only if the number of even integers appearing in its orbit structure is even.

16. Let us call a permutation σ in $\mathcal{S}_n$ a perfect square when $\sigma = \tau^2$ for some permutation τ.
Show that perfect squares are even permutations.
Is every even permutation in $\mathcal{S}_n$ a perfect square?

17. Two permutations σ, τ in $\mathcal{S}_n$ are called ***conjugate*** when $\tau = \rho\sigma\rho^{-1}$ for some ρ in $\mathcal{S}_n$.

(a) Let $\sigma \sim \tau$ denote that σ, τ are conjugate. Show that this relation on the group $\mathcal{S}_n$ is an equivalence relation, and thereby partitions the group into disjoint classes.
These disjoint sets are called the ***conjugacy classes*** of $\mathcal{S}_n$.

(b) Let $(a_1, a_2, \ldots, a_k)$ be a k-cycle in $\mathcal{S}_n$ and let ρ be any permutation in $\mathcal{S}_n$. Show that the composite $\rho(a_1, a_2, \ldots, a_k)\rho^{-1}$ is the k-cycle $(\rho(a_1), \rho(a_2), \ldots, \rho(a_k))$. Thus the conjugate of a k-cycle is another k-cycle.

(c) Show that two cycles of equal length are conjugate.

(d) Show that conjugate permutations have the same orbit structure.

(e) If two permutations in $\mathcal{S}_n$ have the same orbit structure, show that they are conjugate.

We have just seen that two permutations are conjugate if and only if they have the same orbit structure.

18. (a) If $n \geq 3$, show that the 3-cycle $(1, 2, 3)$ can be written as the product of two n-cycles.

(b) Show that every 3-cycle in $\mathcal{S}_n$ is the product of two n-cycles.

(c) If n is odd, show that the n-cycles generate the alternating group $\mathcal{A}_n$.
If n is even, show that all of $\mathcal{S}_n$ is generated by its set of n-cycles.

19. Display the conjugacy classes of $\mathcal{S}_4$.
How many conjugacy classes does $\mathcal{S}_8$ have?

20. Show that the inverse of a permutation σ in $\mathcal{S}_n$ has the same orbit structure as σ.
Deduce that every permutation is conjugate to its inverse.

21. Permutations can be factored as the unique product of disjoint cycles and also as the product of transpositions. In this exercise we will discover that every permutation in $\mathcal{S}_n$ can be factored using just the transposition $(1, 2)$ and the n-cycle $(1, 2, \ldots, n)$.

(a) The transpositions $(1, 2), (2, 3), (3, 4), \ldots, (n-1, n)$ are called ***adjacent***. Here is a factorization of $(2, 6)$ into adjacent transpositions:

$$(2, 6) = (2, 3)(3, 4)(4, 5)(5, 6)(4, 5)(3, 4)(2, 3).$$

Show that every transposition can be factored using only adjacent transpositions.
Then show that every permutation is the product of adjacent transpositions.

(b) If $1 \leq j < n$, and σ is the n-cycle $(1, 2, \ldots, n)$, explain why the inverse of σ^j is σ^{n-j}. Then use a preceding exercise to show that $\sigma^j(1, 2)\sigma^{n-j} = (j+1, j+2)$ for $j = 0, 1, 2, \ldots, n-2$.

(c) Show that every permutation can be factored using only the transposition $(1,2)$ and the n-cycle $(1,2,\ldots,n)$.

22. For $n \geq 2$ prove that every permutation in $\mathcal{S}_n$ can be factored using only transpositions of the form $(1,k)$ where $2 \leq k \leq n$.

23. For each positive integer n find permutations σ,τ in $\mathcal{S}_n$ such that σ,τ each have order 2 while $\sigma\tau$ has order n.

This exercise reminds us that when two elements of a group do not commute, the order of their product can differ drastically from their original orders.

24. An element x in a group G is called ***central*** provided x commutes with every element of G. For sure 1 is central. The set of elements in G that are central is called the ***centre*** of G, and denoted by $Z(G)$.

If $n \geq 3$ and σ in $\mathcal{S}_n$ is not the identity permutation, show that there exists a permutation τ such that $\sigma\tau \neq \tau\sigma$, and thereby deduce that $Z(\mathcal{S}_n)$ contains only the identity permutation.

25. Find permutations σ,τ,ρ in $\mathcal{S}_5$ other than the identity permutation, and such that $\sigma\tau = \tau\sigma, \sigma\rho = \rho\sigma$ but $\rho\tau \neq \tau\rho$.

26. If σ is a permutation in $\mathcal{S}_n$, another permutation τ is said to ***centralize*** σ whenever $\sigma\tau = \tau\sigma$. Another way to say this is that $\tau\sigma\tau^{-1} = \sigma$. The set of all τ that centralize σ is called the ***centralizer*** of σ. For sure 1 always centralizes σ, as do all powers σ^k, as well as all permutations τ that are disjoint from σ. An interesting problem might be to compute explicitly all permutations that centralize a given σ. This exercise aims to answer this question when σ is a cycle in $\mathcal{S}_n$. A more complete answer is offered at the end of Section 3.8.4.

(a) Suppose that σ is the k-cycle $(1,2,3,\ldots,k)$ inside $\mathcal{S}_n$. Prove that the number of permutations τ which centralize σ equals $k \cdot (n-k)!$.

Does this formula change if we replace σ by any other k-cycle?

(b) How many permutations in $\mathcal{S}_{10}$ centralize the cycle $(1,2,3,4,5,6)$?

(c) If σ is our k-cycle, it should be clear that every permutation of the form $\sigma^j\rho$, where $0 \leq j < k$ and ρ is disjoint from σ, centralizes σ. If τ centralizes σ, prove that τ is of the above form.

(d) Find permutations σ,τ in $\mathcal{S}_4$ such that τ centralizes σ and yet τ is not of the form $\sigma^j\rho$, where ρ is disjoint from σ.

27. (a) How many permutations in $\mathcal{S}_5$ centralize the product of the disjoint cycles $(1,2,3)(4,5)$?

(b) How many permutations in $\mathcal{S}_8$ centralize $(1,2,3)(4,5)$?

(c) If the orbit structure of a permutation in $\mathcal{S}_n$ is $[n_1,n_2,\ldots,n_k]$, find a formula in terms of the orbit structure for the number of permutations in $\mathcal{S}_n$ that centralize σ.

2.3 Subgroups and Lagrange's Theorem

Definition 2.32. A subset H of a group G is called a ***subgroup*** of G provided H is a group in its own right using the composition from G. More precisely, we require that

- the identity $1 \in H$,
- if $g, h \in H$, then $gh \in H$, and
- if $g \in H$, then $g^{-1} \in H$.

The associative law holds automatically in H, since it holds already in G.

A Few Examples of Subgroups

- For any group G, the ***trivial group*** $\{1\}$, containing just the identity element, and the whole group G itself are obvious subgroups of G.
- For any positive integer n the set $n\mathbb{Z}$ of integers divisible by n is a subgroup of $\mathbb{Z}$, using the operation of addition.
- If $\mathbb{C}^\star$ is the abelian group of non-zero complex numbers under complex multiplication, the set of complex numbers $\mathcal{T}$ having absolute value 1, i.e. the complex numbers on the unit circle, is a subgroup of $\mathbb{C}^\star$, called the ***circle group***. In turn, for any positive integer n, the finite set of all nth roots of unity, i.e. the set

$$\mathcal{T}_n = \{1, e^{2\pi i/n}, e^{2(2\pi i)/n}, e^{3(2\pi i)/n}, \ldots, e^{(n-1)(2\pi i)/n}\},$$

  is a finite subgroup of $\mathcal{T}$.
- The alternating group $\mathcal{A}_n$ is a subgroup of the symmetric group $\mathcal{S}_n$.
- The group $\mathcal{A}_4$ has order 12. Here are its permutations in disjoint cycle notation:

$$1, (1,2)(3,4), (1,3)(2,4), (1,4)(2,3), (1,2,3), (1,3,2),$$
$$(1,2,4), (1,4,2), (1,4,3), (1,3,4), (2,3,4), (2,4,3).$$

  A routine verification shows that the set

$$H = \{1, (1,2)(3,4), (1,3)(2,4), (1,4)(2,3)\}$$

  is a subgroup of order 4 inside $\mathcal{A}_4$.
- The set of $n \times n$ matrices with entries in a field F and having determinant 1 is a subgroup of $GL_n(F)$. This subgroup, known as the ***special linear group***, is denoted by $SL_n(F)$.
- If G is any group, the ***centre*** of G is the subset

$$Z(G) = \{g \in G : gh = hg \text{ for all } h \text{ in } G\}.$$

  This is the set of elements that commute with all elements of G. Obviously $1 \in Z(G)$, and a small verification reveals that $Z(G)$ is a subgroup of G.

  Clearly $Z(G)$ is abelian. The group G is abelian if and only if $Z(G) = G$.

  For $n \geq 3$, one can check that the centre of $\mathcal{S}_n$ is the trivial group $\langle 1 \rangle$. An exercise in linear algebra reveals that for any field F, the centre of $GL_n(F)$ consists of all non-zero multiples of the identity matrix I.

Subgroups Generated by Subsets

An important way of obtaining subgroups of a given group G is by selecting a set of elements S from G. A ***word from*** S is any product

$$g_1 g_2 \cdots g_k,$$

where the g_j are either in S or inverses of elements from S. We also include the identity element 1 as the so-called *empty word*. A brief reflection reveals that the set H of all words from S is a subgroup of G. Clearly, any subgroup K containing S must contain all words from S, simply by what it means to be a subgroup. Since H is the smallest subgroup of G that contains S we say that H is the ***subgroup generated*** by S.

For example, Proposition 2.31 states that the set of 3-cycles in $\mathcal{S}_n$ generates the subgroup $\mathcal{A}_n$. An exercise also reveals that, for $n \geq 2$, $\mathcal{S}_n$ is generated by the transposition $(1,2)$ and the n-cycle $(1,2,3,\ldots,n)$.

In case the group G is *abelian* and S is a subset of G, a word $g_1g_2\cdots g_k$ from S can be rearranged (thanks to commutativity) to take the form $h_1^{e1}h_2^{e2}\cdots h_\ell^{e_\ell}$ where the h_j are distinct elements of S and the exponents e_j are integers. Just collect all appearances of g_1 or its inverse in the original product and call that h_1^{e1}. For the next g_i not already collected repeat the process to get h_2^{e2} and so on. If addition is used, a subset S generates G if and only if every element of G can be written in the linear combination form $e_1h_1 + e_2h_2 + \cdots + e_\ell h_\ell$ where the $h_j \in S$ and the e_j are integers. For example the abelian group $\mathbb{Z} \times \mathbb{Z} \times \mathbb{Z}$ under addition is generated by the set of elements $(1,0,0), (0,1,0), (0,0,1)$.

When a group G can be generated using a finite set we say that G is a ***finitely generated group***.

Cyclic Subgroups

If S is taken to be a singleton $\{g\}$, then the group generated by g is the set of all powers g^n where n is an integer. That group is denoted by $\langle g\rangle$, and is called the ***cyclic group*** generated by g. We have already encountered these in Section 2.1.1 where it was noted that the order of a cyclic group equals the order of its generator.

For example, given a positive integer n, the group $\mathcal{T}_n$, consisting of all nth roots of unity, is the cyclic group generated by the root $\zeta = e^{2\pi i/n}$. This is a subgroup of the group $\mathcal{T}$ consisting of all of all complex numbers having absolute value 1. The elements of $\mathcal{T}_n$ lie on the unit circle. As we go up along the powers of ζ we eventually return to ζ, and then repeat the cycle of powers.

Another cyclic group is $\mathbb{Z}$ using addition. Here, a generator is the integer 1, since every integer is a multiple of 1. Recall that, under addition, the multiple notation is used instead of the power notation.[3]

2.3.1 Partitions of a Group by Right Cosets of a Subgroup

Recall that two integers x, y are congruent modulo a positive integer n, provided n divides $x - y$. Another way to say this is that $x \equiv y \bmod n$ whenever $x - y$ belongs to the subgroup $n\mathbb{Z}$ of the group $\mathbb{Z}$ using addition.

The transport of this definition to any group G is an important idea. Since G is allowed to be non-abelian, multiplicative notation is called for.

[3] Admittedly, it may seem awkward to call a group such as $\mathbb{Z}$ "cyclic."

Definition 2.33. Let H be a subgroup of G. If $x,y \in G$, we say that x ***is congruent to*** y ***modulo*** H provided

$$xy^{-1} \in H.$$

Another way to say this is that

$$x = hy \text{ for some } h \text{ in } H.$$

It is also convenient to write

$$x \equiv y \bmod H,$$

to mean that x is congruent to y modulo H.

Proposition 2.34. *If H is a subgroup of a group G, then congruency modulo H is an equivalence relation on the set G.*

Proof. For each x in G, clearly $x = 1x$, and since $1 \in H$, the reflexivity requirement $x \equiv x \bmod H$ holds.

If $x \equiv y \bmod H$, this means $x = hy$ for some h in H. Then $y = h^{-1}x$, and since $h^{-1} \in H$, this says $y \equiv x \bmod H$. This verifies the symmetry of the congruence relation.

Congruency is also transitive. For if $x \equiv y \bmod H$ and $y \equiv z \bmod H$, this means that $x = hy$ and $y = kz$, for some h,k in H. Hence $x = hkz$, where hk is still in H, which means $x \equiv z \bmod H$. □

Clearly, $x \equiv y \bmod H$ if and only if x belongs to the set

$$Hy = \{hy : h \in H\}.$$

This is called the ***right coset*** of H represented by y. The right cosets are the equivalence classes for congruency modulo H. From Proposition 2.13, the family of right cosets forms a partition of the group G. Proposition 2.13, applied to the situation of congruency modulo H, can be interpreted to say

$$x \equiv y \bmod H \text{ if and only if } Hx = Hy.$$

And

$$x \not\equiv y \bmod H \text{ if and only if } Hx \cap Hy = \emptyset.$$

A word of caution! As G is not necessarily abelian, the statement $xy^{-1} \in H$ may not mean the same as saying that $y^{-1}x \in H$. The latter condition could well be looked into, but for now stick with $xy^{-1} \in H$ as the definition of congruence modulo H.

Examples of Partitions by Cosets

The partitioning of G by the right cosets of a subgroup H, despite its rather simple proof, has far reaching implications. But first a few examples.

- A familiar example comes from the cosets of the subgroup $n\mathbb{Z}$ of $\mathbb{Z}$, using addition. These cosets are the residues $[a]$ of the integers modulo n, which partition the set of integers.
- The group $\mathbb{C}^\star$ of non-zero complex numbers under multiplication is partitioned by the cosets of the subgroup $\mathcal{T}$ of complex numbers whose absolute value is 1, i.e. by the cosets of the unit circle. If $z \in \mathbb{C}^\star$, the coset

$$\mathcal{T}z = \{uz : |u| = 1\}$$

consists of all complex numbers whose distance to the origin equals $|z|$. Here the cosets of $\mathcal{T}$ inside $\mathbb{C}^\star$ are the concentric circles in $\mathbb{C}$ of positive radius all centred at the origin.
- Take the symmetric group S_3 of permutations on 3 letters, and let H be the subgroup of order 2, consisting of the identity 1 and the transposition $(1,2)$. Here are the three disjoint right cosets that result from the congruence modulo H:

$$\begin{aligned} H = H1 &= \{1, (1,2)\} = H(1,2) \\ H(1,3) &= \{(1,3), (1,3,2)\} = H(1,3,2) \\ H(2,3) &= \{(2,3), (1,2,3)\} = H(1,2,3). \end{aligned}$$

Notice how $1, (1,2)$ each represent the same coset of H (namely H itself), and $(1,3), (1,3,2)$ each represent the same coset of H, and $(2,3), (1,2,3)$ each represent the same coset of H. That is because

$$1 \equiv (1,2) \bmod H, (1,3) \equiv (1,3,2) \bmod H, \text{ and } (2,3) \equiv (1,2,3) \bmod H.$$

Just to check the second of these congruences:

$$(1,3)(1,3,2)^{-1} = (1,3)(1,2,3) = (1,2) \in H.$$

- Take the symmetric group $\mathcal{S}_n$ with the alternating group $\mathcal{A}_n$ as its subgroup. The cosets of $\mathcal{A}_n$ consist of $\mathcal{A}_n$ itself along with $\mathcal{A}_n(1,2)$, which is the set of all odd permutations. Here the partition of $\mathcal{S}_n$ is into two cosets.
- Take the general linear group $GL_n(\mathbb{Z}_7)$ consisting of the invertible $n \times n$ matrices with entries in the field $\mathbb{Z}_7$. This is the group of $n \times n$ matrices with a non-zero determinant. A small consideration shows that the set H consisting of all matrices whose determinant is 1 in the field $\mathbb{Z}_7$ is a subgroup of $GL_n(\mathbb{Z}_7)$. Two matrices A, B from $GL_n(\mathbb{Z}_7)$ belong to the same coset if and only if $\det(AB^{-1}) = 1$. But $\det(AB^{-1}) = \det(A)/\det(B)$. Thus A, B belong to the same coset of H if and only if $\det(A) = \det(B)$. From this it follows that H has 6 cosets, one for each of the six possible non-zero determinant values $1, 2, 3, 4, 5, 6$ in the field $\mathbb{Z}_7$.

Lagrange's Theorem and the Index of a Subgroup

The order of a finite group G is the number of elements in the group. A common notation for the order of G is the absolute value symbol

$$|G|.$$

If H is a subgroup of any group G, the number of right cosets of H is called the ***index*** of H in G. In case H has finitely many cosets, H is said to have ***finite index*** inside G, and ***infinite index*** otherwise. Obviously, subgroups of finite groups have finite index. Subgroups of an infinite group could also have finite index. For example, the subgroup $3\mathbb{Z}$ consisting of all integers divisible by 3 has index 3 inside the group of integers $\mathbb{Z}$ using addition. The common notation for the index of H in G is $[G : H]$.

The next result is paramount.

Proposition 2.35 (Lagrange's theorem). *If G is a finite group and H is a subgroup of G, then*

$$|G| = |H| \cdot [G : H].$$

In particular, the order of H divides the order of G.

Proof. The right cosets of H partition G. If Ha is such a coset, the mapping $h \mapsto ha$ is a bijection from H onto Ha, as can easily be verified. Thus all right cosets of H have the same order as H does. Since the finite group G is partitioned into sets with the same order as H, the order of H times the number of right cosets of H must equal the order of G. □

Proposition 2.35 is arguably the most used fact in group theory.

Lagrange's theorem tells us that for a finite group G, the order of each of its subgroups divides the order of G. This raises an interesting question. If G is a finite group of order n and d is a positive divisor of n, does G possess a subgroup of order d? We shall come to learn that, while this need not be true for all d, the answer is yes when d is a prime power.

Groups of Prime Order

Lagrange's theorem places a severe restriction on the order that a subgroup of a finite group can have.

For instance, suppose that the order of a group G is a prime p, such as, for example, the group $\mathbb{Z}_p$ under addition. By Lagrange, the only possible orders of subgroups of G are 1 and $|G|$. Hence the only subgroups of G are the trivial group, and the full group G. In particular, for any element x in G there is the cyclic subgroup $\langle x \rangle$ generated by x. This cyclic subgroup must either be trivial, when $x = 1$, or all of G when $x \neq 1$. Thus all groups of prime order are cyclic, and have no subgroups other than the two obvious ones.

The Order of Elements Inside Groups of Finite Order

When Lagrange's theorem is combined with Proposition 2.5 a widely used fact about groups emerges.

Proposition 2.36 (Lagrange's corollary). *If G is a finite group, then the order of every element x in G divides the order of G. Furthermore,*

$$x^{|G|} = 1.$$

Proof. The element x generates a cyclic group $\langle x \rangle$. By Proposition 2.7 the order of x as an element, say m, coincides with the order of the group $\langle x \rangle$ generated by x. By Lagrange, m divides $|G|$. Then item 3 of Proposition 2.5 gives $x^{|G|} = 1$. □

The Euler and Fermat Theorems

A very interesting application of Proposition 2.36 to the theory of integer congruences is attributed to Leonhard Euler (1707–1783). At first glance this result seems somewhat removed from group theory. Recall that for every positive integer n, the integer $\phi(n)$ is the number of integers from 0 to $n-1$ that are coprime with n. This is the same as the order of the group of units $\mathbb{Z}_n^\star$.

Proposition 2.37 (Euler's theorem). *If n is a positive integer and a is an integer coprime with n, then*

$$a^{\phi(n)} \equiv 1 \bmod n.$$

Proof. The residue $[a]$ lies in the group $\mathbb{Z}_n^\star$ of units modulo n. The order of this group is $\phi(n)$. By Proposition 2.36, $[a]^{\phi(n)} = [1]$. In the language of integer congruences, this says that $a^{\phi(n)} \equiv 1 \bmod n$. □

In case p is a prime, $\phi(p) = p-1$, and then Proposition 2.37 specializes to a famous result in the theory of numbers. Note that an integer a is coprime with a prime p if and only if $p \nmid a$.

Proposition 2.38 (Fermat's little theorem). *If p is a prime and a is an integer such that $p \nmid a$, then $a^{p-1} \equiv 1 \bmod p$.*

A Multiplicative Property of the Index

The index of a subgroup H inside a group G is the number of right cosets of H. This could be infinite, but the case where the index is finite is interesting enough. For instance, the infinite subgroup $n\mathbb{Z}$ of $\mathbb{Z}$ consisting of all multiples of n has finite index n. Suppose that K is a subgroup of H which in turn is a subgroup of G. Then of course K is a subgroup of G as well. It might be worthwhile to see how the indices $[G:H]$, $[H:K]$ and $[G:K]$ are related.

Proposition 2.39. *Suppose the group G has a subgroup H which in turn has a subgroup K. Say H has finite index in G and K has finite index in H. Then the index of K in G is finite, and*

$$[G:K] = [G:H][H:K].$$

Proof. If G is itself finite, a short proof emerges from Lagrange's theorem used three times. Indeed,

$$|K|[G:K] = |G| = |H|[G:H] = |K|[H:K][G:H],$$

and then cancel $|K|$.

But here is the proof which allows the order of G to be infinite. With $n = [G : H]$ take suitable elements $x_1, x_2, \ldots, x_n$ in G such that $Hx_1, Hx_2, \ldots, Hx_n$ are disjoint cosets which form a partition of G. Likewise with $m = [H : K]$ take suitable elements $y_1, y_2, \ldots, y_m$ in H such that the cosets $Ky_1, Ky_2, \ldots, Ky_m$ are disjoint and form a partition of H. We verify that the right cosets

$$Ky_ix_j, \text{ as } i \text{ runs from 1 to } m \text{ and } j \text{ runs from 1 to } n,$$

are disjoint and form a partition of G. Once that is done, the result follows from the observation that the number of Ky_ix_j is mn.

Now, if $x \in G$ it must be that $x \in Hx_j$ for some unique j. Then $xx_j^{-1} \in H$, whereby $xx_j^{-1} \in Ky_i$ for some unique i. Therefore $x \in Ky_ix_j$ for suitable j, i. This shows that the cosets Ky_ix_j exhaust all of G.

To see that the Ky_ix_j are disjoint for different pairs (i,j) suppose $Ky_ix_j \cap Ky_rx_s \neq \emptyset$ for some i, r between 1 and m and some j, s between 1 and n. This means there exist u, v in K such that

$$uy_ix_j = vy_rx_s.$$

Now uy_i, vy_r belong to H, which puts the above element in $Hx_j \cap Hx_s$. Since the right cosets of H are disjoint, it follows that $x_j = x_s$, i.e. $j = s$. Cancel to get $uy_i = vy_r$, which now lies in $Ky_i \cap Ky_r$. Because the right cosets of K are also disjoint it follows that $y_i = y_r$, i.e. $i = r$. The only way for Ky_ix_j to intersect Ky_rx_s is by having $(i, r) = (j, s)$, as was required to show. □

No Converse to Lagrange

The converse of Lagrange's theorem would say that if G is a group of order n and d is a divisor of n, then G has a subgroup of order d. As we shall come to learn this holds if G is abelian. It also holds if d is the power of a single prime. For now let us see that it does not hold in general.

Take the group $\mathcal{A}_4$ of even permutations on 4 letters. The order of $\mathcal{A}_4$ is 12. Yet $\mathcal{A}_4$ has no subgroup of order 6. Indeed, suppose H is a subgroup of order 6. Since the order of H is 6 and the order of $\mathcal{A}_4$ is 12, merely two right cosets of H will partition $\mathcal{A}_4$. Thus, for any permutation σ in $\mathcal{A}_4$ at least two of the cosets

$$H, H\sigma, H\sigma^2$$

must coincide. If $H = H\sigma$, then $\sigma \in H$. If $H = H\sigma^2$, then $\sigma^2 \in H$. And if $H\sigma = H\sigma^2$, then $\sigma = \sigma^{-1}\sigma^2 \in H$. We see that for all σ in $\mathcal{A}_4$ either σ or σ^2 lies in H, and since H is a subgroup it follows that $\sigma^2 \in H$ for all σ in $\mathcal{A}_4$. But $\mathcal{A}_4$ contains the eight 3-cycles

$$(1,2,3), (1,3,2), (1,2,4), (1,4,2), (1,3,4), (1,4,3), (2,3,4), (3,2,4),$$

and each of them is a perfect square, indeed $(a_1, a_2, a_3) = (a_1, a_3, a_2)^2$. The contradiction is that a group of order 6 already contains 8 elements.

2.3.2 Subgroups of Cyclic Groups

It is not always so simple to determine the subgroups of a given group. But when the group G is cyclic, let us see what those subgroups are.

Proposition 2.40. *Every subgroup of a cyclic group is cyclic.*

Proof. Let G be a cyclic group with generator g, and let H be a subgroup of G. If the only exponent k such that $g^k \in H$ is 0, then H is the trivial group, which is obviously cyclic. Otherwise H contains a power g^k where $k \neq 0$. By taking g^{-k} if need be, we can suppose that $k > 0$, and that k is the least positive exponent such that $g^k \in H$.

Since G is cyclic, the elements of H are powers of g. For any such power g^ℓ in H, Euclidean division gives integers q, r such that

$$\ell = kq + r \text{ where } 0 \leq r < k.$$

Then

$$g^r = g^{\ell - kq} = \left(g^\ell\right)\left(g^k\right)^{-q} \in H.$$

By the minimality of k it must be that $r = 0$, and thus

$$g^\ell = g^{kq+r} = g^{kq} = \left(g^k\right)^q.$$

This shows that every element of H is a power of g^k, making H cyclic with generator g^k. □

Proposition 2.41. *Let G be a cyclic group with generator g.*

If G is infinite, then the distinct subgroups of G are the cyclic groups $\langle g^m \rangle$ where m runs over the non-negative integers.

If G is finite of order n and d is a positive divisor of n, then G has exactly one subgroup of order d, namely the cyclic group $\langle g^{n/d} \rangle$. These are all the subgroups of G.

Proof. By Proposition 2.40 every subgroup of G, finite or infinite, is of the form $\langle g^m \rangle$ for some exponent m, and m can be made non-negative, if need be, by replacing g^m with its inverse g^{-m}.

Now if G is infinite and $\langle g^m \rangle = \langle g^k \rangle$ for some $m, k \geq 0$, we get an exponent ℓ such that $g^m = \left(g^k\right)^\ell = g^{k\ell}$. Since the order of G is infinite, so is the order of its generator g, due to Proposition 2.7. Then item 1 of Proposition 2.5 gives $m = k\ell$. So, $k \mid m$. By a similar argument $m \mid k$, and since m, k are non-negative, we get $m = k$.

Next suppose G is finite of order n. This n is also the order of the generator g. By Proposition 2.10 the element $g^{n/d}$ has order d, and by Proposition 2.7 $g^{n/d}$ generates a cyclic subgroup of order d. It remains to show that if H is a subgroup of order d, then $H = \langle g^{n/d} \rangle$. Well, the subgroup H is cyclic by Proposition 2.40. Since all elements of G are powers of g, a generator for H is g^k for some integer k. Again by Proposition 2.7 the order of g^k equals the order d of the cyclic group H that g^k generates. For this reason, or alternatively by Lagrange,

$$1 = \left(g^k\right)^d = g^{kd}.$$

By item 3 of Proposition 2.5, $kd = nr$ for some integer r. Then

$$g^k = g^{\frac{n}{d}r} = \left(g^{\frac{n}{d}}\right)^r.$$

This shows that g^k belongs to cyclic group $\langle g^{n/d}\rangle$. Consequently the group H generated by g^k is inside the group $\langle g^{n/d}\rangle$, and since both H and $\langle g^{n/d}\rangle$ have order d it follows that $H = \langle g^{n/d}\rangle$.

The last statement of our proposition comes out of Lagrange's theorem that the order of every subgroup divides the order of any finite group containing it. □

For example, the group of integers $\mathbb{Z}$ under addition is cyclic. Thus its subgroups are cyclic. Keeping additive notation in mind, the preceding result reveals that the subgroups of $\mathbb{Z}$ take the predictable form

$$n\mathbb{Z} = \{nk : k \in Z\}, \text{ where } n \geq 0.$$

For the finite cyclic group $\mathbb{Z}_{12}$ of residues modulo 12 under addition, we learn that there are 5 subgroups, which we can depict in the following so-called ***lattice diagram***.

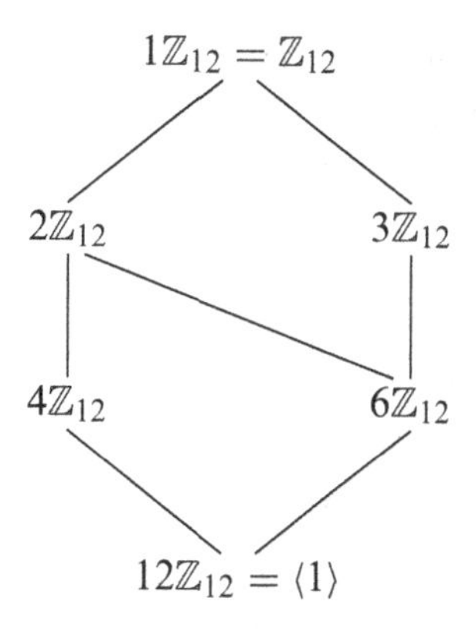

Keep in mind that in $\mathbb{Z}_{12}$ addition is being used.

The Number of Generators of a Finite Cyclic Group

Here is a small consequence of Proposition 2.11. Recall the Euler ϕ-function. For each positive integer n, the value $\phi(n)$ is the number of integers from 0 to $n-1$ that are coprime with n. This coincides with the order of the group $\mathbb{Z}_n^\star$ of residues which are units modulo n.

Proposition 2.42. *If G is a cyclic group of finite order n, then the number of generators of G equals $\phi(n)$.*

Proof. Let g generate G. Then G consists of the distinct powers $1, g, g^2, \ldots, g^{n-1}$. A power g^k from this list will generate G if and only if the order of g^k is n. By Proposition 2.11 this occurs if and only if k is coprime with n. So the number of such g^k is $\phi(n)$. □

For example, $\phi(30) = 8$, because the integers between 0 and 29 coprime with 30 are $1, 7, 11, 13, 17, 19, 23, 29$. Thus, every cyclic group of order 30 has 8 generators.

EXERCISES

1. Show that the intersection of any family of subgroups of a group G is a subgroup of G.
2. If G is a group with subgroups H, N such that G equals the union $H \cup N$, show that either $H \subseteq N = G$ or $N \subseteq H = G$.
3. If H, K are finite subgroups of a group G and the orders $|H|, |K|$ are coprime, show that the intersection $H \cap K$ is the trivial group $\langle 1 \rangle$.
4. If G is a group of finite order n and H, K are subgroups of order m, ℓ, respectively, show that $\mathrm{lcm}(m, \ell)$ divides n. Must G possess a subgroup whose order is $\mathrm{lcm}(m, \ell)$?
5. If G is a group and x, y in G are such that x, y and xy all have order 2, show that the elements $1, x, y, xy$ constitute a subgroup of G.
6. If G is a finite abelian group with at least two elements of order 2, show that 4 divides the order of G.
7. Obviously a finite group only has a finite number of subgroups. Prove that an infinite group has infinitely many subgroups.
8. If G is a group, containing a non-empty, finite subset H which is closed under the multiplication in G, show that H is a subgroup of G.
   Hint. If $x \in H$, the powers x^k, with $k \geq 1$, must overlap due to the finiteness of G.
9. Find a group G along with a subgroup H and elements x, y where $xy^{-1} \in H$ while $y^{-1}x \notin H$.
10. If a group G only has two subgroups, show that G is cyclic, and that the order of G is a prime.
11. If G is an abelian group generated by two elements a, b and such that $a^2 = b^3$, show that G is cyclic.
12. Show that none of the following groups can be generated by a finite subset:
    (a) the group $\mathbb{Q}$ of rational numbers under addition,
    (b) the group $\mathbb{Q}^\star$ of positive rational numbers under multiplication,
    (c) for each prime p the group G_p consisting of all complex numbers z whose order is a power of p, i.e. $z \in G_p$ if and only if there is a $k \geq 0$ such that $z^{p^k} = 1$.
    First verify that G_p is a subgroup of $\mathbb{C}^\star$.
13. Let H be the subgroup of S_5 generated by the cycles

$$\sigma = (1,2,3,4,5) \text{ and } \tau = (1,2,4,3).$$

    (a) Verify that $\tau\sigma = \sigma^2\tau$ and $\tau\sigma^{-1} = \sigma^3\tau$.
    (b) Explain why every element of H can be expressed in the form $\sigma^i\tau^j$ for some integers i,j.
    (c) Show that the order of H is at most 20.
    (d) Prove that the order of H is 20.

14. If $n \geq 3$, show that the centre of $\mathcal{S}_n$ is trivial.
If $n \geq 4$, show that the centre of $\mathcal{A}_n$ is trivial.

15. Show that the centre of $GL_n(\mathbb{R})$ consists of the non-zero scalar multiples of the identity matrix. Does it matter if the field $\mathbb{R}$ is replaced by another field such as $\mathbb{Q}, \mathbb{C}$, or $\mathbb{Z}_p$ where p is a prime?

16. If a, n are positive integers and ϕ is Euler's function, prove that $n \mid \phi(a^n - 1)$.
Hint. Let $m = a^n - 1$. Explain why the residue $[a]$ is an element in the group of units $\mathbb{Z}_m^\star$, then use Lagrange.

17. If p, q are primes such that q divides $2^p - 1$, show that p divides $q - 1$.
Hint. Work with the group $\mathbb{Z}_q^\star$ of units modulo q having order $p - 1$.

18. If a finite group G is not the trivial group, show that G has at least one element whose order is a prime.

19. Find a group G for which the subset $H = \{a^2 : a \in G\}$ is not a subgroup.

20. If m, n are positive integers, what is the index of the intersection of the subgroups $\mathbb{Z}m$ and $\mathbb{Z}n$ inside the group $\mathbb{Z}$, using addition?

21. If H, K are subgroups of a group G and both indices $[G : H]$ and $[G : K]$ are finite show that the intersection subgroup $H \cap K$ has finite index in G, and that the index is no more than the product $[G : H][G : K]$.

22. For $n > 2$, let $\mathbb{Z}_n^\star$ be the group of units modulo n using multiplication. Show that this group has an element of order 2, and deduce that the order of $\mathbb{Z}_n^\star$ is even.

23. This exercise requires some familiarity with linear algebra and determinants. In particular, keep in mind that for a pair of $n \times n$ matrices A, B, we have the property $\det(AB) = \det(A)\det(B)$.
Let $G = GL_2(\mathbb{Z}_5)$ be the group of 2×2 invertible matrices over the field of residues modulo 5.
(a) Find the order of G.
(b) Let H be the set of matrices A in G for which $\det A = 1$. (Here "1" stands for the residue of 1 modulo 5.) Show that H is a subgroup of G.
(c) Show that two matrices in G belong to the same right coset of H if and only if they have the same determinant.
(d) Find the index $[G : H]$ as well as the order of H.
(e) Answer the above questions for the group $GL_n(\mathbb{Z}_p)$, where p is a prime and n is a positive integer.

24. For a finite abelian group G, let H be the set of all elements x of G which equal their own inverse, i.e. such that $x^2 = 1$.
(a) Explain why H is a subgroup of G.
(b) Show that the product of all of the elements in H equals the product of all of the elements in G.
(c) Apply the preceding result to the group $\mathbb{Z}_p^\star$ of units modulo a prime p to prove that $(p - 1)! \equiv -1 \bmod p$. This result is known as Wilson's theorem.
(d) If $n \geq 2$ and $(n - 1)! \equiv -1 \bmod p$, show that n is prime.

25. If G is a cyclic group of order 36 with generator g, find its subgroups.
26. The group $\mathbb{Z}_n$, under addition, is cyclic. But for some choices of n, the group of units $\mathbb{Z}_n^\star$, under multiplication, is also cyclic. An interesting theorem says that this is always the case when n is a prime.

    Verify that $\mathbb{Z}_{29}^\star$ and $\mathbb{Z}_{41}^\star$ are cyclic. Find the number of generators of each group.
27. Suppose that the group G is cyclic and finite with generator g. Find the relationship between integers k and ℓ that makes g^k and g^ℓ generate the same subgroup of G.
28. What is the biggest order that a cyclic subgroup of $\mathcal{S}_6$ can have?
29. Show that the product group $\mathbb{Z} \times \mathbb{Z}$, using addition, is not a cyclic group.
30. If H is a cyclic group of order 3, show that the product group $\mathcal{A}_4 \times H$ has no subgroup of order 18.
31. Consider the pair of matrices

$$A = \begin{bmatrix} 1/2 & 3 \\ 1/4 & -1/2 \end{bmatrix}, \quad B = \begin{bmatrix} 1/8 & 15/2 \\ 1/8 & -1/4 \end{bmatrix}$$

inside the group of invertible 2×2 matrices over $\mathbb{Q}$.

(a) Verify that $A^2 = B^2 = I$, the identity matrix. Thus both A and B have order 2.

(b) Show that AB and BA have infinite order.

(c) * Let G be the group generated by A and B. Verify that $(AB)^{-1} = BA$ and use this to show that every element of G can be written in one of the forms

$$(AB)^n, (AB)^n A, (BA)^n, (BA)^n B,$$

where n is a non-negative exponent.

(d) * If H is the subgroup of G generated by AB, show that the index of H in G is 2.

(e) Show that the order of every element in G is either $1, 2$ or ∞.

32.* Let G be the group of 2×2 matrices with real entries and having determinant 1. Let S be the set of matrices of the form $\begin{bmatrix} 1 & x \\ 0 & 1 \end{bmatrix}$ together with matrices of the form $\begin{bmatrix} 1 & 0 \\ x & 1 \end{bmatrix}$ where $x \in \mathbb{R}$. Show that every matrix in G is a product of matrices from S, and thereby that S generates G.

2.4 Conjugation and Normal Subgroups

If x, y are in a group G and H is a subgroup, the congruence $x \equiv y \bmod H$ was defined to mean $xy^{-1} \in H$. This came down to saying that $Hx = Hy$. But why not use

$$y^{-1}x \in H?$$

This means that $x = yz$ for some z in H, or that x belongs to the set

$$yH = \{yz : z \in H\}.$$

The above sets are called the ***left cosets*** of H in G. Just as with right cosets, $xH = yH$ if and only if $y^{-1}x \in H$, and the left cosets partition G. Left cosets could serve equally well to prove Lagrange's theorem.

If G is abelian, then $Hy = yH$, obviously. But when G is not abelian this coincidence of left and right cosets might or might not happen.

For instance, take the alternating group $\mathcal{A}_n$ as a subgroup of the symmetric group $\mathcal{S}_n$. The right cosets of $\mathcal{A}_n$ consist of $\mathcal{A}_n$ itself, and $\mathcal{A}_n(1,2)$, which is the set of odd permutations. These coincide, respectively, with the left cosets $1\mathcal{A}_n$ and $(1,2)\mathcal{A}_n$.

On the other hand, take the group $\mathcal{S}_3$, whose order is 6, and the subgroup $H = \{1,(1,2)\}$ of order 2. Now, H will sprout 3 right cosets, which partition $\mathcal{S}_3$, and 3 left cosets, which also partition $\mathcal{S}_3$. We have already seen the right cosets of H, but here they are again:

$$H = \{1,(1,2)\},\ H(2,3) = \{(2,3),(1,2,3)\},\ H(1,3) = \{(1,3),(1,3,2)\}.$$

And here are the left cosets of H:

$$H = \{1,(1,2)\},\ (2,3)H = \{(2,3),(1,3,2)\},\ (1,3)H = \{(1,3),(1,2,3)\}.$$

The right cosets partition H, as do the left cosets, but they partition H differently! In this case $H(2,3) \neq (2,3)H$ and also $H(1,3) \neq (1,3)H$.

Definition 2.43. If a subgroup H of a group G satisfies

$$xH = Hx$$

for all x in G, the subgroup H is called ***normal***[4] in G.

For example $\mathcal{A}_n$ is normal in $\mathcal{S}_n$. If G is any group, its ***centre***

$$Z(g) = \{z \in G : zx = xz \text{ for all } x \text{ in } G\}$$

is a normal subgroup of G. This is obvious since every z in $Z(G)$ commutes with every x in G.

Conjugate Elements and Conjugate Subgroups

There is another way to think of normal subgroups, in terms of so-called conjugation.

Definition 2.44. Two elements x, y in a group G are ***conjugate*** provided

$$y = zxz^{-1} \text{ for some } z \text{ in } G.$$

It is easy to see that conjugacy is an equivalence relation on G. If $y = zxz^{-1}$ for some z in G, denote that fact by $x \sim y$. Using $z = 1$, the reflexivity $x \sim x$ is obvious. If $y = zxz^{-1}$, then

[4] This rather bland terminology is the one that has caught on, possibly because it suggests something to be desired of a subgroup. In the final letter of Evariste Galois (1811–1832), who more than anyone deserves to be called the founder of group theory, he refers to H giving rise to a "proper" partition of G when the partitions by right and left cosets coincide. In his famous treatise on permutations and algebraic equations, Camille Jordan refers to H as being "permutable" with the elements of G. Some have used the term "invariant subgroup" as well. In any case, the term "normal" has won out. It is an important concept.

$x = (z^{-1})y(z^{-1})^{-1}$, and thus $x \sim y$ implies the symmetric relation $y \sim x$. And if $y = zxz^{-1}$ and $u = vyv^{-1}$ for some z, v in G, then $u = v(zxz^{-1})v^{-1} = (vz)x(vz)^{-1}$. This proves the transitivity, that $x \sim y$ and $y \sim u$ imply $x \sim u$. The equivalence classes which emerge from the conjugacy relation, known as ***conjugacy classes***, will soon enough need deeper exploration.

The mapping

$$\varphi_z : G \to G \text{ defined by } \varphi_z(x) = zxz^{-1}$$

is called a ***conjugation*** by z. It is easy to see that φ_z is a bijection on G. Indeed, the equality $zxz^{-1} = zyz^{-1}$ implies $x = y$ by cancellation, which makes φ_z injective. While the equality $x = z(z^{-1}xz)z^{-1}$ shows that every x in G is the conjugate of $z^{-1}xz$, meaning that φ_z is surjective.

If H is a subgroup of G and $z \in G$, the image of H under the conjugation φ_z is the set

$$\varphi_x(H) = zHz^{-1} = \{zxz^{-1} : x \in H\}.$$

The image of H under φ_z remains a subgroup of G. To see that, notice the identities

$$\begin{aligned} 1 &= z1z^{-1} \\ (zxz^{-1})^{-1} &= zx^{-1}z^{-1} \\ zxyz^{-1} &= (zxz^{-1})(zyz^{-1}). \end{aligned}$$

The first equality shows zHz^{-1} contains the identity 1. The second equality shows zHz^{-1} is closed under inverses. The third equality shows zHz^{-1} is closed under composition.

The subgroup zHz^{-1} is called a ***conjugate subgroup*** of H. Here is the connection between conjugate subgroups and normality.

Proposition 2.45. *For a subgroup H of a group G, the following statements are equivalent.*

1. *$xH = Hx$ for all x in G. (This is the definition of normality.)*
2. *The partition of G by left cosets of H coincides with the partition of G by right cosets of H.*
3. *Every left coset of H is a subset of a right coset of H.*
4. *$xH \subseteq Hx$ for all x in G.*
5. *$xHx^{-1} \subseteq H$ for all x in G.*
6. *$xHx^{-1} = H$ for all x in G.*

Proof. Much of the proof is plain to see.

Item 2 is immediate from item 1, and item 3 is immediate from item 2.

Assuming item 3, each left coset xH is a subset of some right coset Hy. Since $x = x1 \in xH$, then $x \in Hy$, and by Proposition 2.34, $Hy = Hx$. Thus $xH \subseteq Hx$.

Item 4 implies item 5 routinely.

Assuming item 5, let $h \in H$. Then $x^{-1}hx \in H$, by item 5 applied to x^{-1}. And so, $h = x(x^{-1}hx)x^{-1} \in xHx^{-1}$. Thus, $xHx^{-1} = H$, which is item 6.

Finally, item 6 implies item 1 by a routine argument. Let $x \in G$. As h varies over H, the elements xhx^{-1} in xHx^{-1} pick up precisely all k in H. And so the elements xh in xH pick up precisely all kx in Hx. Thus $xH = Hx$, making H normal in G. $\square$

Subgroups of Index 2 Are Normal

Suppose that H is a subgroup of G of index 2. Thus H has two right cosets: H itself and Hy for some y not in H. Because of the partitioning of G into right cosets, Hy is the complementary set $G \setminus H$. Now, every left coset xH of H is either equal to H or disjoint from H, and in the latter case inside the complement $Hy = G \setminus H$ of H. This verifies that H satisfies item 3 of Proposition 2.45, and thereby that H is normal in G.

Why Normal Subgroups Matter

It might be reasonable to enquire as to why normal subgroups need to be considered. One of the breakthroughs in algebra was the discovery by Galois of the conditions for a polynomial equation $f(x) = 0$ to be solvable by radicals. Roughly, this means that the roots can be extracted by a succession of additions, subtractions, multiplications, divisions, and kth root operations applied to the coefficients. For instance, the roots of the equation $ax^2+bx+c=0$ are given by $(-b \pm \sqrt{b^2-4ac})/2a$. In order to answer yes to such a hard question for polynomials of all possible degrees, Galois showed that a group of permutations of the roots of the equation must be sufficiently endowed with normal subgroups. The details lie within that great development known as Galois theory.

Furthermore, we shall discover soon that normal subgroups are the essential tool for extending modular arithmetic to groups, and thereby enable new groups to be created.

2.4.1 Conjugates in $\mathcal{S}_n$

To define conjugacy in a group is one thing, but to detect it may be quite another. For instance, the conjugacy of two matrices in the group $GL_n(\mathbb{C})$ leads to an intriguing story about canonical forms. But let us address this question for the symmetric group $\mathcal{S}_n$.

Start with a mundane but crucial observation. If $\sigma, \tau \in \mathcal{S}_n$ and a is a letter, then

$$\tau\sigma\tau^{-1}(\tau(a)) = \tau(\sigma(a)).$$

In other words,

if σ maps a to b, then its conjugate permutation $\tau\sigma\tau^{-1}$ maps $\tau(a)$ to $\tau(b)$.

Consequently, if $\mathcal{O}(a)$ is the orbit of a letter a under σ, then the orbit of $\tau(a)$ under the conjugate $\tau\sigma\tau^{-1}$ is $\tau(\mathcal{O}(a))$. One way to see this is to let

$$\sigma : a = a_1 \mapsto a_2 \mapsto a_3 \mapsto \cdots \mapsto a_m \mapsto a_1 = a$$

describe the orbit of a under σ as in Proposition 2.15. From the mundane observation, the orbit of $\tau(a)$ under the conjugate $\tau\sigma\tau^{-1}$ is described by

$$\tau\sigma\tau^{-1} : \tau(a) = \tau(a_1) \mapsto \tau(a_2) \mapsto \tau(a_3) \mapsto \cdots \mapsto \tau(a_m) \mapsto \tau(a_1) = \tau(a),$$

which shows our claim. Another way to say this in the notation of cycles is that

$$\tau(a_1, a_2, \ldots, a_n)\tau^{-1} = (\tau(a_1), \tau(a_2), \ldots, \tau(a_n)).$$

Let

$$\mathcal{O}_1, \mathcal{O}_2, \ldots, \mathcal{O}_k$$

be the disjoint orbits for σ, which partition the set of letters L. They can be listed so that their lengths, say m_j, are decreasing (not necessarily strictly). Hence, the orbit structure of σ is $[m_1, m_2, \ldots, m_k]$. The images of these orbits under τ are

$$\tau(\mathcal{O}_1), \tau(\mathcal{O}_2), \ldots, \tau(\mathcal{O}_k).$$

Since τ is a permutation of L, the above list forms another partition of L. Also, each $\tau(\mathcal{O}_j)$ has exactly m_j elements. And as noted, each $\tau(\mathcal{O}_j)$ is an orbit of the conjugate $\tau\sigma\tau^{-1}$. This demonstrates that the orbit structure of $\tau\sigma\tau^{-1}$ coincides with the orbit structure of σ.

Conversely, suppose that ρ is a permutation whose orbit structure coincides with that of σ. This means that the disjoint orbits of ρ can be listed as

$$\mathcal{P}_1, \mathcal{P}_2, \ldots, \mathcal{P}_k$$

where each $\mathcal{P}_j$ has exactly m_j letters. Let

$$\rho : b_1 \mapsto b_2 \mapsto b_3 \mapsto \cdots \mapsto b_{m_j} \mapsto b_1$$

describe the orbit $\mathcal{P}_j$ using ρ, and let

$$\sigma : a_1 \mapsto a_2 \mapsto a_3 \mapsto \cdots \mapsto a_{m_j} \mapsto a_1$$

describe the matching orbit $\mathcal{O}_j$ using σ. Then define a permutation τ of L by mapping each orbit $\mathcal{O}_j$ onto the orbit $\mathcal{P}_j$ according to the rule

$$\tau(a_i) = b_i.$$

With such a choice of τ, the following holds for each b_i in the orbit $\mathcal{P}_j$:

$$\rho(b_i) = b_{i+1} = \tau(a_{i+1}) = \tau\sigma(a_i) = \tau\sigma\tau^{-1}(b_i).$$

In case $i = m_j$, the subscript $i+1 = m_j + 1$ should be interpreted as being 1. Since the orbits $\mathcal{P}_j$ cover L and since the b_i cover the $\mathcal{P}_j$, we have verified that

$$\rho = \tau\sigma\tau^{-1}.$$

And so a neat result emerges.

Proposition 2.46. *Two permutations in $\mathcal{S}_n$ are conjugate if and only if their orbit structures coincide.*

To illustrate Proposition 2.46, take these permutations on 9 letters:

$$\sigma = (8,5,3)(2,4,7)(1,6)(9)$$
$$\rho = (3,4,1)(7,8,9)(2,5)(5)$$

written in disjoint cycle notation. Their common orbit structure is $[3,3,2,1]$ as revealed by the disjoint cycles. Hence, these permutations are conjugate in $\mathcal{S}_9$. By the discussion preceding Proposition 2.46, a permutation τ that implements the conjugation of σ into ρ is given by matching up the entries in the top row of cycles for σ with those in the bottom row for ρ. That is, define τ by

$$8 \mapsto 3,\ 5 \mapsto 4,\ 3 \mapsto 1,\ 2 \mapsto 7,\ 4 \mapsto 8,\ 7 \mapsto 9,\ 1 \mapsto 2,\ 6 \mapsto 5,\ 9 \mapsto 5.$$

In cycle notation, $\tau = (1,2,7,9,5,4,8,3)$, which happens to be a full 9-cycle. Other choices of τ are also possible, because the two 3-cycles in σ or ρ could have been switched, and because there are variations in how a cycle can be designated.

2.4.2 Normal Subgroups of $\mathcal{A}_n$

Let us explore what the normal subgroups of the groups $\mathcal{A}_n$ of even permutations might be.

The group $\mathcal{A}_2$ is the trivial group $\langle 1 \rangle$, leaving nothing more to say. The group $\mathcal{A}_3 = \{1, (1,2,3), (1,3,2)\}$ which is a cyclic group of prime order 3, and thus has only itself and the trivial group as subgroups, and these are always normal.

The Subgroups of $\mathcal{A}_4$

The group $\mathcal{A}_4$, with 12 elements is slightly more interesting.

Here are its elements:

$$1, (1,2,3), (1,3,2), (1,2,4), (1,4,2), (1,3,4), (1,4,3),$$
$$(2,3,4), (3,2,4), (1,2)(3,4), (1,3)(2,4), (1,4)(2,3).$$

In addition to the identity, $\mathcal{A}_4$ has eight 3-cycles and three products of pairs of disjoint 2-cycles.

Lagrange tells us that every subgroup of $\mathcal{A}_4$ has order $1, 2, 3, 4, 6$ or 12. Let us identify the subgroups of $\mathcal{A}_4$, and decide which are normal.

- There are the obvious normal groups $\langle 1 \rangle$ and $\mathcal{A}_4$ of order 1 and 12, respectively.
- The subgroups of order 2 must be cyclic with generator of order 2. These are

$$\langle (1,2)(3,4) \rangle,\ \langle (1,3)(2,4) \rangle,\ \langle (1,4)(2,3) \rangle.$$

None of these subgroups is normal in $\mathcal{A}_4$. For instance, conjugate $(1,2)(3,4)$ by $(1,2,3)$ to get

$$(1,2,3)(1,2)(3,4)(2,1,3) = (2,3)(1,4) \notin \langle (1,2)(3,4) \rangle.$$

- The subgroups of order 3 must also be cyclic, and the generator of each of them must have order 3. This gives the four cyclic subgroups as shown

$$\langle(1,2,3)\rangle,\ \langle(1,2,4)\rangle,\ \langle(1,3,4)\rangle,\ \langle(2,3,4)\rangle.$$

The other four 3-cycles are the squares (as well as the inverses) of the above shown generators. None of these four subgroups is normal in $\mathcal{A}_4$. For instance, conjugate $(1,2,3)$ by $(2,3,4)$ to get

$$(2,3,4)(1,2,3)(3,2,4) = (1,3,4) \notin \langle(1,2,3)\rangle.$$

- Let K be the set of all products of pairs of disjoint transpositions. Namely,

$$K = \{1,\ (1,2)(3,4),\ (1,3)(2,4),\ (1,4)(2,3)\}.$$

A routine check confirms that K is a subgroup of $\mathcal{A}_4$. Furthermore K is normal in $\mathcal{A}_4$. Indeed, observe that 1 is the only conjugate of 1. And if σ is one of the products of pairs of disjoint transpositions, its cycle structure is $[2,2]$, which is the same for any conjugate of σ. Thus any conjugate of σ is another product of pairs of disjoint transpositions, and thus back inside K. Since K is conjugation invariant, Proposition 2.45 shows that K is normal in $\mathcal{A}_4$. Incidentally, this argument shows that K is normal in $\mathcal{S}_4$.

The cyclic subgroup $H = \langle(1,2)(3,4)\rangle$ is normal in K. This holds because K is abelian, but also because the index of H in K is 2. And yet we saw that H is not normal in $\mathcal{A}_4$. Thus it is possible to have the subgroup inclusions $H \subset K \subset G$, with H normal in K and K normal in G, but H not normal in G.

The only subgroup of $\mathcal{A}_4$ of order 4 is K. Indeed, let H be such a subgroup. All elements of H must have order $1, 2$ or 4. But in $\mathcal{A}_4$ there are no elements of order 4, and the 3-cycles in $\mathcal{A}_4$ have order 3, which prevents them from being in H. The only possibility is that $H = K$.

- Finally, $\mathcal{A}_4$ has no subgroups of order 6, as we already have proven to illustrate that the converse of Lagrange fails.

The Klein 4-Group

Before letting go of the subgroups of $\mathcal{A}_4$, let us look at the normal subgroup K of order 4 inside $\mathcal{A}_4$ a bit more. Let

$$\sigma = (1,2)(3,4), \quad \tau = (1,3)(2,4), \quad \rho = (1,4)(2,3).$$

The group K comprises σ, τ, ρ along with 1. Here is the composition table for K:

$\bullet$	1	σ	τ	ρ
1	1	σ	τ	ρ
σ	σ	1	ρ	τ
τ	τ	ρ	1	σ
ρ	ρ	τ	σ	1

One can observe that the multiplication table for K is identical, except for a replacement of symbols, to that of the group of units of $\mathbb{Z}_8^\star$, from Section 2.1.

Any group with this multiplication table is known as a ***Klein* 4-*group***, and is denoted by $\mathcal{V}_4$.

Here is another realization of the Klein 4-group as a group of matrices. Take the group G comprising the matrices

$$I = \begin{pmatrix} 1 & 0 \\ 0 & 1 \end{pmatrix}, \quad A = \begin{pmatrix} 1 & 0 \\ 0 & -1 \end{pmatrix}, \quad B = \begin{pmatrix} -1 & 0 \\ 0 & 1 \end{pmatrix}, \quad C = \begin{pmatrix} -1 & 0 \\ 0 & -1 \end{pmatrix}.$$

Its multiplication table is easy to write. In this way the Klein 4-group is realized as four linear transformations of $\mathbb{R}^2$. It is plain to see that

- I leaves $\mathbb{R}^2$ unmoved,
- A reflects $\mathbb{R}^2$ across the x-axis,
- B reflects $\mathbb{R}^2$ across the y-axis,
- C rotates $\mathbb{R}^2$ by $180°$.

The Simple Groups $\mathcal{A}_n$ for $n \geq 5$

We have enough tools to show that for $n \geq 5$ the groups $\mathcal{A}_n$ have no normal subgroups properly between the trivial group and all of $\mathcal{A}_n$.

Proposition 2.47. *If $n \geq 5$, then all 3-cycles are mutually conjugate in $\mathcal{A}_n$.*

Proof. It is not enough to say that, since all 3-cycles have identical orbit structure $[3, 1, 1, \ldots, 1]$, then they are mutually conjugate by Proposition 2.46. The conjugation must happen inside $\mathcal{A}_n$. So, starting with the 3-cycle $(1, 2, 3)$ and any other 3-cycle σ, there is a permutation τ in $\mathcal{S}_n$ such that

$$\sigma = \tau(1, 2, 3)\tau^{-1}.$$

If τ is even, the conjugation is in $\mathcal{A}_n$, and we are done. If τ is odd, replace τ by the even permutation $\tau(4, 5)$. Note that since $n \geq 5$, the transposition $(4, 5)$ is available. The inverse of $\tau(4, 5)$ is $(4, 5)\tau^{-1}$. By the disjointness of $(1, 2, 3)$ and $(4, 5)$ we still get

$$\tau(4, 5)(1, 2, 3)(4, 5)\tau^{-1} = \tau(4, 5)(4, 5)(1, 2, 3)\tau^{-1} = \tau(1, 2, 3)\tau^{-1} = \sigma.$$

Since any 3-cycle σ in $\mathcal{A}_n$ can be obtained from $(1, 2, 3)$ by a conjugation in $\mathcal{A}_n$, all 3-cycles are mutually conjugate in $\mathcal{A}_n$. □

And now for a result of some interest.

Proposition 2.48. *If $n \geq 5$, then the only normal subgroups of $\mathcal{A}_n$ are the trivial group and $\mathcal{A}_n$ itself.*

Proof. Let N be a normal subgroup of $\mathcal{A}_n$ other than the trivial group. The plan is to show that N contains a 3-cycle. Being normal, N will then contain all conjugates of the 3-cycle, which, by Proposition 2.47, will show that N contains all 3-cycles. Since, as seen in Proposition 2.31, the 3-cycles generate $\mathcal{A}_n$, the desired conclusion that $N = \mathcal{A}_n$ will follow.

By assumption, there is a σ in $\mathcal{A}_n$ other than the identity. For any even permutation ρ, the conjugate $\rho\sigma\rho^{-1} \in N$, and then the so-called ***commutator*** $\rho\sigma\rho^{-1}\sigma^{-1} \in N$. In order to obtain a 3-cycle from σ, a few cases need to be considered, and some virtuosity with commutators will be required.

The first possibility is that σ has a cycle of length at least 4 in its disjoint cycle decomposition. It does no harm to suppose $(1,2,3,4,\ldots,\ell)$ is that cycle and write $\sigma = (1,2,3,4,\ldots,\ell)\tau$, where τ is disjoint from the ℓ-cycle. The commutator of σ with $\rho = (1,2,3)$ simplifies as follows:

$$\begin{aligned}\rho\sigma\rho^{-1}\sigma^{-1} &= (1,2,3)(1,2,3,4,\ldots,\ell)\tau(2,1,3)\tau^{-1}(\ell,\ell-1,\ldots,4,3,2,1)\\ &= (1,2,3)(1,2,3,4,\ldots,\ell)(2,1,3)\tau\tau^{-1}(\ell,\ell-1,\ldots,4,3,2,1)\\ &= (1,2,3)(1,2,3,4,\ldots,\ell)(2,1,3)(\ell,\ell-1,\ldots,4,3,2,1)\\ &= (1,2,4) \in N.\end{aligned}$$

The second equality holds because τ and $(2,1,3)$ are disjoint, and the last equality holds by sheer calculation on all the letters, keeping in mind that $\ell \geq 4$. In this case, we have produced a 3-cycle in N.

The second possibility is that the length of the longest cycle in the disjoint cycle decomposition of σ is 3. Thus σ is a product of disjoint 3-cycles and maybe some 2-cycles. Since the inverse of every 2-cycle is itself and the square of every 3-cycle is another 3-cycle, σ^2 is a product of disjoint 3-cycles only. If σ^2 is a single 3-cycle, then N contains a 3-cycle as desired.

If not, then σ^2 is a product of two or more disjoint 3-cycles. (We might note, incidentally, that for $n = 5$, this current possibility cannot occur.) We can harmlessly take $(1,2,3)$ and $(4,5,6)$ as the first two of those disjoint factors and write $\sigma^2 = (1,2,3)(4,5,6)\tau$, where τ is disjoint from $(1,2,3)$ and $(4,5,6)$. The commutator of σ^2 with $\rho = (1,2,4)$ gives

$$\begin{aligned}\rho\sigma^2\rho^{-1}\sigma^{-2} &= (1,2,4)(1,2,3)(4,5,6)\tau(2,1,4)\tau^{-1}(5,4,6)(2,1,3)\\ &= (1,2,4)(1,2,3)(4,5,6)(2,1,4)\tau\tau^{-1}(5,4,6)(2,1,3)\\ &= (1,2,4)(1,2,3)(4,5,6)(2,1,4)(5,4,6)(2,1,3)\\ &= (1,2,5,3,4) \in N.\end{aligned}$$

Thus N contains a 5-cycle, and by the argument of the first possibility applied to this 5-cycle, it follows that N contains a 3-cycle.

The final possibility is that σ is the product of an even number of disjoint 2-cycles. Without harm we can suppose $\sigma = (1,2)(3,4)$ or possibly that $\sigma = (1,2)(3,4)(5,6)\tau$, where τ is disjoint from the preceding 2-cycles.

In the first case, take the commutator of σ with $\rho = (1,3)(2,5)$ to get:

$$\begin{aligned}\rho\sigma\rho^{-1}\sigma^{-1} &= (1,3)(2,5)(1,2)(3,4)(2,5)(1,3)(3,4)(1,2)\\ &= (1,2,4,5,3).\end{aligned}$$

This brings us back to the first possibility, whence N contains a 3-cycle.

In the second case, again take the commutator of σ with $\rho = (1,3)(2,5)$ to get:

$$\begin{aligned}\rho\sigma\rho^{-1}\sigma^{-1} &= (1,3)(2,5)(1,2)(3,4)(5,6)\tau(2,5)(1,3)\tau^{-1}(5,6)(3,4)(1,2)\\ &= (1,3)(2,5)(1,2)(3,4)(5,6)(2,5)(1,3)\tau\tau^{-1}(5,6)(3,4)(1,2)\\ &= (1,3)(2,5)(1,2)(3,4)(5,6)(2,5)(1,3)(5,6)(3,4)(1,2)\\ &= (1,6,3)(2,4,5) \in N.\end{aligned}$$

By the second possibility applied to this product of two 3-cycles, N contains a 3-cycle.

All possibilities are exhausted, and so $N = \mathcal{A}_n$. □

Definition 2.49. A group G having only itself and the trivial group as normal subgroups is called a ***simple group***.[5]

When $n \geq 5$, we have just seen that $\mathcal{A}_n$ is a simple group. As noted after the proof of Lagrange's theorem, all groups of prime order are simple, and these groups are cyclic and thereby abelian. In fact, groups of prime order have no subgroups whatsoever, except for the inevitable trivial group and the whole group. As we shall come to learn, the only simple abelian groups are those of prime order. An interesting fact, known to Galois already, is that the order of a simple non-abelian group must be at least 60, which happens to be the order of $\mathcal{A}_5$.

A truly deep discovery was the Feit–Thompson theorem. Walter Feit and John Thompson showed in 1962–1963 that every finite, simple, non-abelian group must have even order. This had already been conjectured by the renowned algebraist William Burnside in 1911. A monumental project covering the second half of the twentieth century has been to discover and tabulate all possible finite simple groups.

2.4.3 Normal Subgroups of the Quaternion Group

The more groups we witness, the better we can appreciate the theory.

Obviously, every subgroup of every abelian group is normal. But here is a non-abelian group of order eight in which every subgroup remains normal. Let $\mathbf{1}, \mathbf{i}, \mathbf{j}, \mathbf{k}$ stand for the following matrices inside the group $GL_2(\mathbb{C})$:

$$\mathbf{1} = \begin{bmatrix} 1 & 0 \\ 0 & 1 \end{bmatrix}, \quad \mathbf{i} = \begin{bmatrix} i & 0 \\ 0 & -i \end{bmatrix}, \quad \mathbf{j} = \begin{bmatrix} 0 & 1 \\ -1 & 0 \end{bmatrix}, \quad \mathbf{k} = \begin{bmatrix} 0 & i \\ i & 0 \end{bmatrix}.$$

By multiplying these matrices we see that

$$\mathbf{ij} = \mathbf{k},\ \mathbf{jk} = \mathbf{i},\ \mathbf{ki} = \mathbf{j},$$

while

$$\mathbf{ji} = -\mathbf{k},\ \mathbf{kj} = -\mathbf{i},\ \mathbf{ik} = -\mathbf{j},$$

and

$$\mathbf{i}^2 = \mathbf{j}^2 = \mathbf{k}^2 = -\mathbf{1}.$$

[5] This is an arguably simplistic terminology, since these groups are anything but simple in the everyday sense.

The above matrix multiplications are enough to reveal that the eight matrices

$$\pm\mathbf{1},\ \pm\mathbf{i}\pm\mathbf{j}\pm\mathbf{k},$$

form a non-abelian subgroup of $GL_2(\mathbb{C})$. This little group, called the ***quaternion group***, is denoted by $\mathcal{Q}$.

Obviously the trivial subgroup of $\mathcal{Q}$ as well as $\mathcal{Q}$ itself are normal. According to Lagrange the other subgroups have order 2 or 4. The subgroups of order 4 have index 2 in $\mathcal{Q}$, and are thereby normal. And if H is a subgroup of order 2, then H is generated by an element of order 2. The only such element is $-\mathbf{1}$. Thus $H = \{\pm\mathbf{1}\}$, which happens to be the centre $Z(\mathcal{Q})$ and is thereby normal in $\mathcal{Q}$.

EXERCISES

1. If N is a normal subgroup of G, and H is any subgroup of G, show that $N \cap H$ is normal in H.
2. If N is a finite subgroup of G and G has no other subgroups whose order equals the order of N, show that N is normal in G.
3. Let G be a group generated by subset S. Show that a subgroup H is normal if and only if $xHx^{-1} \subseteq H$ for all x in S.
4. If a, b are in a group, show that the products ab and ba are conjugates.
5. Let G be the group of 2×2 matrices with entries in $\mathbb{Q}$, that are both invertible and upper triangular, i.e. matrices of the form $\begin{bmatrix} a & b \\ 0 & c \end{bmatrix}$ where $ac \neq 0$. Show that the matrices of the form $\begin{bmatrix} 1 & b \\ 0 & 1 \end{bmatrix}$ constitute a normal subgroup of G.

   Is G itself a normal subgroup of the group $GL_2(\mathbb{Q})$ of all invertible 2×2 matrices?
6. If the order of a group G is odd and if x in G is conjugate to its inverse, show that $x = 1$.
7. If H, N are subgroups of G, their product set is

   $$HN = \{ab : a \in H, b \in N\}.$$

   Since G is not necessarily abelian the product set NH need not coincide with HN.
   (a) If N is a normal subgroup of G and H is any subgroup of G, prove that HN is a subgroup of G.
   (b) If both H, N are normal in G, show that HN is normal in G.
   (c) If neither H nor N is normal in G, is HN still a subgroup of G?
8. Find a group G with a subgroup K having a further subgroup H such that H is normal in K and K is normal in G, but H is not normal in G.
9. How many conjugacy classes are there in $\mathcal{S}_5$?
10. If $n \geq 5$, show that $\mathcal{A}_n$ has no subgroups of index 2.
11. Let $\mathcal{L}$ and $\mathcal{R}$ be the families of left and right cosets for a subgroup H of a group G. If H is normal, then $\mathcal{L}$ coincides with $\mathcal{R}$, but otherwise the family $\mathcal{L}$ will partition G differently than $\mathcal{R}$.

Verify that the mapping $\psi : \mathcal{L} \to \mathcal{R}$ given by $\psi(xH) = Hx^{-1}$ is well defined (i.e. does not depend on the choice of representative x) and that it is a bijection. This proves that H has exactly as many left cosets as it does right cosets. So, the notion of the index of H in G does not depend on which kinds of cosets are used.

12.* Find a group G with a subgroup H and an element x in G such that xHx^{-1} is a proper subgroup of H. Why does this not violate the implication of item 6 from item 5 in Proposition 2.45?

13. Show that the 3-cycle $(1, 2, 3)$ is not conjugate to its inverse $(2, 1, 3)$ in $\mathcal{A}_3$, and also that they are not conjugate in $\mathcal{A}_4$.

14. If $n \geq 5$, show that $\mathcal{S}_n$ has only three normal subgroups. What are they?

15. If G is a group and N is a subgroup of the centre $Z(G)$, prove that N is normal in G.

16. Suppose H is a normal subgroup of a group G and that the order of H is 2. Prove that H is a subgroup of the centre $Z(G)$.

17. If H is a normal subgroup of K and K is a normal subgroup of G, it need not be that H is normal in G. However, if in addition K is cyclic, prove that H is normal in G.

18. Suppose H, K are normal subgroups of a group G and $H \cap K = \langle 1 \rangle$. If $x \in H$ and $y \in K$, show that $xyx^{-1}y^{-1} = 1$ and thereby deduce that $xy = yx$.

19. If H is a subgroup of a group G, the ***normalizer*** of H in G is the set

$$\mathcal{N}(H) = \{x \in G : xHx^{-1} = H\}.$$

Show the following properties of $\mathcal{N}(H)$:

(a) $\mathcal{N}(H)$ is a subgroup of G,

(b) $H \subseteq \mathcal{N}(H)$,

(c) H is normal in $\mathcal{N}(H)$,

(d) if K is a subgroup of G such that $H \subseteq K$ and H is normal in K, then $K \subseteq \mathcal{N}(H)$,

(e) H is normal in G if and only if $\mathcal{N}(H) = G$.

Item (d) states that the normalizer of a subgroup H of G is the largest subgroup of G in which H is normal.

20. If H is a subgroup of G, show that the intersection of all conjugate groups $\bigcap_{x \in G} xHx^{-1}$ is a normal subgroup of G.

21.* Find ten normal subgroups of the Cartesian product $\mathcal{S}_3 \times \mathcal{S}_3$. Then prove that $\mathcal{S}_3 \times \mathcal{S}_3$ has precisely ten normal subgroups.

2.5 Homomorphisms

Inevitably, in the study of any category of objects, it becomes necessary to compare the objects. Tools are needed to make the comparison. For groups, that tool is the notion of ***group homomorphism***.[6]

[6] The term is a compound from the Greek "homo," meaning "same," and "morph," meaning "form." For instance, in biology, morphology is a study of living forms. A tadpole undergoes a metamorphosis when it becomes a frog, since its form has changed. Possibly, that is what has led to the current adoption of the root "morph" to suggest change. In mathematics, a homomorphism between two structures tells us that they follow the same form.

Definition 2.50. A mapping $\varphi : G \to H$ from a group G to a group H is called a ***group homomorphism*** provided

$$\varphi(xy) = \varphi(x)\varphi(y) \text{ for every } x, y \text{ in } G.$$

The homomorphism φ tells us that H follows the form of G. Right away, notice that

$$\varphi(1) = \varphi(1 \cdot 1) = \varphi(1) \cdot \varphi(1),$$

which, by cancelling, leads to

$$\varphi(1) = 1.$$

And then, for every x in G

$$1 = \varphi(1) = \varphi(xx^{-1}) = \varphi(x)\varphi(x^{-1}).$$

Thus, in addition to preserving the group composition, a homomorphism φ maps the identity in G to the identity in H, and

$$\varphi(x^{-1}) = \varphi(x)^{-1} \text{ for every } x \text{ in } G.$$

Examples of Homomorphisms

Most of the following examples are either obviously homomorphisms, or it is easy to check that they are.

1. The mapping that sends every x in a group G to 1 in a group H is called the ***trivial homomorphism***, and the mapping that sends every x in G back to x itself, is called the ***identity homomorphism*** on G.
2. If $\mathbb{Z}$ is the group of integers under addition and h is any element in any group H, the mapping $\varphi : \mathbb{Z} \to H$ given by $\varphi(n) = h^n$ is a homomorphism, as the law of exponents reveals.
3. If $\mathbb{R}$ is the group of real numbers under addition and $\mathcal{T}$ is the group of complex numbers on the unit circle, the mapping $\varphi : \mathbb{R} \to \mathcal{T}$ given by $\varphi(x) = e^{ix}$ is a homomorphism. This homomorphism is surjective but not injective.
4. If V, W are vector spaces, they are (by ignoring their scalar multiplication) groups under vector addition, and every linear transformation $T : V \to W$ is a homomorphism.
5. Let $\mathbb{Z}_n$ be the cyclic group of residues modulo n using addition, and let $\mathcal{T}_n$ be the group of nth roots of unity. If $\omega = e^{2\pi i/n}$, the cyclic group $\mathcal{T}_n$ is made up of

   $$1, \omega, \omega^2, \ldots, \omega^{n-1}.$$

   The mapping $\varphi : \mathbb{Z}_n \to \mathcal{T}_n$ given by $\varphi([k]) = \omega^k$, for every residue $[k]$ in $\mathbb{Z}_n$, is a group homomorphism, and it is also a bijection.
6. Let H be the group of order 2 consisting of ± 1 under multiplication. Proposition 2.28 says that the parity mapping $\text{sgn} : \mathcal{S}_n \to H$ is a homomorphism. This homomorphism is surjective, but far from being injective, since all even permutations get mapped to 1.

7. If $\mathbb{R}^\star$ is the group of non-zero real numbers under multiplication, our knowledge of linear algebra reminds us that the determinant mapping

$$\det : GL_n(\mathbb{R}) \to \mathbb{R}^\star$$

is a homomorphism. The above holds for any field F, not just for $\mathbb{R}$.

8. If x is in a group G, the conjugation mapping $\varphi_x : G \to G$, given by $\varphi_x : y \mapsto xyx^{-1}$ is a homomorphism. The identity

$$x(yz)x^{-1} = (xyx^{-1})(xzx^{-1})$$

makes this plain to see. This homomorphism is a bijection.

9. Here is a fancier homomorphism. The symmetric group $\mathcal{S}_4$ acts on the set of letters $L = \{1, 2, 3, 4\}$. There are three ways to split the set L into two equal parts. Here they are:

$$\begin{aligned} P_1 &= \{\{1,2\},\{3,4\}\} \\ P_2 &= \{\{1,3\},\{2,4\}\} \\ P_3 &= \{\{1,4\},\{2,3\}\}. \end{aligned}$$

Let E be the set $\{P_1, P_2, P_3\}$, consisting of these three partitions of L. Thinking of E as a set of letters, we get the group $\mathcal{S}(E)$ of permutations of E. It should be obvious that $\mathcal{S}(E)$ is just $\mathcal{S}_3$ in poor disguise.

Every permutation σ in $\mathcal{S}_4$ yields a permutation $\varphi(\sigma)$ in $\mathcal{S}(E)$ defined as follows:

$$\begin{aligned} \varphi(\sigma)(P_1) &= \{\{\sigma(1),\sigma(2)\},\{\sigma(3),\sigma(4)\}\} \\ \varphi(\sigma)(P_2) &= \{\{\sigma(1),\sigma(3)\},\{\sigma(2),\sigma(4)\}\} \\ \varphi(\sigma)(P_3) &= \{\{\sigma(1),\sigma(4)\},\{\sigma(2),\sigma(3)\}\}. \end{aligned}$$

For example, take the permutation $\sigma = (1, 4, 3)$ written as a cycle in $\mathcal{S}_4$. The permutation $\varphi(\sigma)$ in $\mathcal{S}(E)$ that results is given by:

$$\begin{aligned} \varphi(\sigma)(P_1) &= \{\{4,2\},\{1,3\}\} = P_2 \\ \varphi(\sigma)(P_2) &= \{\{4,1\},\{2,3\}\} = P_3 \\ \varphi(\sigma)(P_3) &= \{\{4,3\},\{2,1\}\} = P_1. \end{aligned}$$

Thus $\varphi(\sigma)$ is the cycle in $\mathcal{S}(E)$ described by

$$\varphi(\sigma) : P_1 \mapsto P_2 \mapsto P_3 \mapsto P_1.$$

A bit of checking reveals that the resulting mapping

$$\varphi : \mathcal{S}_4 \to \mathcal{S}(E)$$

is a group homomorphism.

Since φ is a mapping from a group of order 24 to a group of order 6, it cannot be injective. One could verify, however, that φ is surjective.

The Cayley Representation of a Group

Here is a significant homomorphism. Let G be a group and let $S(G)$ be the group of all bijections $\sigma : G \to G$. That is, think of G as a set of letters, and then $S(G)$ as the symmetric group on those letters. If the order of G is finite, say n, then the order of $S(G)$ is $n!$.

Take the mapping

$$\varphi : G \to S(G)$$

where, for every x in G, the value $\varphi(x)$ is the permutation $\varphi(x) : G \to G$ given by

$$\varphi(x)(y) = xy \text{ for every } y \text{ in } G.$$

To repeat, φ transforms G into a group of permutations of itself by letting $\varphi(x)$ be the permutation that sends every y in G into xy. Through this homomorphism G is said to ***act on itself by left multiplication***. However, we need to check a few things.

First we check that $\varphi(x)$ is a bijection on G. Well, suppose $\varphi(x)(y) = \varphi(x)(z)$ for some y, z in G. Thus $xy = xz$, and then $y = z$ by cancellation of x. This shows $\varphi(x)$ is injective. To see that $\varphi(x)$ is surjective, take z in G. Since $\varphi(x)(x^{-1}z) = x(x^{-1}z) = z$, surjectivity is checked too.

Next we check that $\varphi : G \to S(G)$ is a homomorphism. We want $\varphi(xy) = \varphi(x)\varphi(y)$ for every x, y in G. This comes down to checking that $\varphi(xy)(z) = \varphi(x)(\varphi(y)(z))$ for every z in G. Well, by the associative law,

$$\varphi(xy)(z) = (xy)z = x(yz) = \varphi(x)(\varphi(y)(z)).$$

Furthermore, φ is an injective homomorphism. Indeed, suppose $\varphi(x) = \varphi(y)$ for some x, y in G. This says that $xz = yz$ for all z in G. Put $z = 1$ to see that $x = y$.

The homomorphism φ is called the ***Cayley representation*** of G. Because of the Cayley representation every finite group can be viewed as a subgroup of some S_n. It also confirms that the subgroups of S_n are rife with as many complications as the whole theory of finite groups. No wonder the symmetric group is important.

2.5.1 The Kernel and Image of a Homomorphism

Definition 2.51. If $\varphi : G \to H$ is a group homomorphism, the ***kernel*** of φ is the set

$$\ker \varphi = \{x \in G : \varphi(x) = 1\}.$$

The ***image*** of φ is the set

$$\varphi(G) = \{\varphi(x) : x \in G\}.$$

At times $\varphi(G)$ is called the image of G under φ.

Here are some routine facts.

Proposition 2.52. *If $\varphi : G \to H$ is a group homomorphism, then $\varphi(G)$ is a subgroup of H and* $\ker(\varphi)$ *is a normal subgroup of G.*

Proof. Clearly $1 = \varphi(1) \in \varphi(G)$. If $x, y \in G$, the identity $\varphi(x)\varphi(y) = \varphi(xy)$ shows that $\varphi(G)$ is closed under composition in H. The identity $\varphi(x)^{-1} = \varphi(x^{-1})$ shows that $\varphi(G)$ is closed under inverses. Thus $\varphi(G)$ is a subgroup of H.

Turning to the kernel, we have $1 \in \ker\varphi$, since $\varphi(1) = 1$. If $x, y \in \ker\varphi$, the identity $\varphi(xy) = \varphi(x)\varphi(y) = 1 \cdot 1 = 1$ shows $xy \in \ker\varphi$. If $x \in \ker\varphi$, the identity $\varphi(x^{-1}) = \varphi(x)^{-1} = 1^{-1} = 1$ shows that $x^{-1} \in \ker\varphi$. Thus $\ker\varphi$ is a subgroup of G.

To see that $\ker\varphi$ is normal, take x in $\ker\varphi$ and y in G. Then

$$\varphi(yxy^{-1}) = \varphi(y)\varphi(x)\varphi(y^{-1}) = \varphi(y)1\varphi(y)^{-1} = 1,$$

which verifies that any conjugate of x in $\ker\varphi$ is again in $\ker\varphi$. By Proposition 2.45, $\ker\varphi$ is a normal subgroup of G. □

Proposition 2.53. *A group homomorphism $\varphi : G \to H$ is injective if and only if* $\ker\varphi$ *is the trivial group.*

Proof. If φ is injective, then only 1 can be mapped to 1 by φ, and so $\ker\varphi = \langle 1 \rangle$.

Conversely, suppose $\ker\varphi$ is trivial. Take x, y in G such that $\varphi(x) = \varphi(y)$. Then

$$1 = \varphi(x)\varphi(y)^{-1} = \varphi(x)\varphi(y^{-1}) = \varphi(xy^{-1}).$$

Since $\ker\varphi$ is trivial, $xy^{-1} = 1$, and thus $x = y$. □

2.5.2 Isomorphisms

Definition 2.54. A group homomorphism $\varphi : G \to H$ that is bijective is called an ***isomorphism***. When an isomorphism prevails between two groups G and H, the groups are said to be ***isomorphic***, and we write

$$G \cong H.$$

The inverse map of an isomorphism $\varphi : G \to H$ is the mapping $\psi : H \to G$ that sends every y in H back to the unique x in G such that $y = \varphi(x)$. Naturally we write ψ as φ^{-1}. This inverse map is just as much an isomorphism, which reconfigures the elements of H back to their original form in G. A very routine exercise shows that if a homomorphism $\varphi : G \to H$ comes with another homomorphism $\psi : H \to G$ such that $\varphi(\psi(y)) = y$ and $\psi(\varphi(x)) = x$ for every x in G and every y in H, then φ is an isomorphism with inverse ψ. At times it is convenient to establish that φ is an isomorphism by actually finding its inverse mapping.

If G and H are isomorphic groups, then they have identical properties as groups. So much so that it is natural (even desirable at times) to think of isomorphic groups as being the same group!

For example, take the cyclic group $\mathcal{T}_n$ of nth roots of unity generated by the element $\omega = e^{2\pi i/n}$, and take the group $\mathbb{Z}_n$ of residues modulo n under addition. The mapping $\varphi : \mathbb{Z}_n \to \mathcal{T}_n$ given by $\varphi : [k] \mapsto \omega^k$ is an isomorphism. Thus $\mathbb{Z}_n$ and $\mathcal{T}_n$ can be viewed as the same cyclic group of order n, brought to life in different forms.

For another example, take the group $\mathbb{Z}_8^\star$ of units modulo 8, which has four residues

$$[1], [3], [5], [7],$$

and the Klein 4-group $\mathcal{V}_4$, which is realized as the group of permutations

$$1, \quad \sigma = (1,2)(3,4), \quad \tau = (1,3)(2,4), \quad \rho = (1,4)(2,3).$$

A verification will show that a suitable isomorphism $\varphi : \mathbb{Z}_8^\star \to \mathcal{V}_4$ is given by

$$\varphi : [1] \mapsto 1, \quad [3] \mapsto \sigma, \quad [5] \mapsto \tau, \quad [7] \mapsto \rho.$$

As can be seen from Propositions 2.52 and 2.53, a homomorphism $\varphi : G \to H$ having a trivial kernel will yield an isomorphism $G \cong \varphi(G)$. For instance, the Cayley representation $\varphi : G \to S(G)$, by which each x in G becomes the permutation $\varphi(x)$ that moves each y in G to xy, has trivial kernel. Thus every group G is isomorphic to a subgroup of $S(G)$. If G is finite of order n, we can think of G as the set of letters $x_1, x_2, \ldots, x_n$. So, $S(G)$ is just $\mathcal{S}_n$. More precisely, $S(G)$ is isomorphic to $\mathcal{S}_n$. In summary, every finite group is isomorphic to a subgroup of $\mathcal{S}_n$ for some n.

Automorphisms

An isomorphism $\varphi : G \to G$ of a group to itself is called an ***automorphism*** of G. For instance, in the group of complex numbers $\mathbb{C}$ under addition, the conjugation map $a + ib \mapsto a - ib$ is a well known automorphism. In fact this map is also an automorphism of the group $\mathbb{C}^\star$ of non-zero complex numbers under multiplication, as well as an automorphism of the circle group $\mathcal{T}$, and an automorphism of the cyclic group $\mathcal{T}_n$ of nth roots of unity.

It is easy to check that the set $\mathrm{Aut}(G)$, consisting of all automorphisms of G, is a group in its own right using composition of mappings as the group operation. Naturally this group is called the ***automorphism group*** of G. A good problem for any given G is to describe $\mathrm{Aut}(G)$.

Inner Automorphisms

For each x in G, the conjugation map $\varphi_x : G \to G$ whereby $\varphi_x : y \mapsto xyx^{-1}$ is an important automorphism of G. By a routine exercise φ_x is an automorphism of G, with inverse mapping $\varphi_{x^{-1}}$. Automorphisms of G that arise from conjugation are called ***inner automorphisms***. Not only is each φ_x an automorphism of G, but the mapping

$$\varphi : G \to \mathrm{Aut}(G), \text{ given by } x \mapsto \varphi_x,$$

is itself a homomorphism. What needs checking is that $\varphi_x \circ \varphi_z = \varphi_{xz}$ for every x, z in G. This comes down to checking that

$$\varphi_x(\varphi_z(y)) = \varphi_{xz}(y)$$

for every y in G. In turn this requires us to check that

$$x(zyz^{-1})x^{-1} = (xz)y(xz)^{-1},$$

which is clear from the basic rules of group multiplication. As a consequence, the image of φ is a subgroup of $\mathrm{Aut}(G)$. This subgroup consists precisely of the inner automorphisms of G, and is denoted by $\mathrm{Inn}(G)$.

What is the kernel of this homomorphism $\varphi : G \to \mathrm{Aut}(G)$? Well, $x \in \ker\varphi$ if and only if φ_x is the identity automorphism on G. In other words, if and only if

$$xyx^{-1} = \varphi_x(y) = y,$$

for all y in G. So $x \in \ker\varphi$ if and only if $xy = yx$ for all y in G. This reveals that $\ker\varphi$ is the centre $Z(G)$.

When $Z(G)$ is trivial, such as for example when $G = S_n$, then φ is injective, and thereby every x in G gives a distinct inner automorphism φ_x. At the opposite extreme, if $Z(G) = G$, i.e. if G is abelian, then every φ_x is the identity automorphism.

2.5.3 Automorphisms of Cyclic Groups

Let us enquire about the automorphism groups of finite cyclic groups.

If G is an abelian group and k is an integer, then the ***power mapping***

$$\varphi_k : G \to G, \text{ given by } \varphi_k : x \mapsto x^k,$$

is a group homomorphism, by inspection. Note that when G is not abelian, φ_k need not be a homomorphism.

Proposition 2.55. *If G is an abelian group of finite order n, and k is an integer coprime with n, then the power mapping $\varphi_k : G \to G$ is an automorphism.*

Proof. Since k, n are coprime there exist integers ℓ, m such that $k\ell + nm = 1$. Then for every x in G:

$$\varphi_k\varphi_\ell(x) = \varphi_k(x^\ell) = (x^\ell)^k = x^{\ell k} = x^{(1-nm)} = x(x^n)^{-m} = x.$$

Similarly $\varphi_n(\varphi_\ell(x) = x$. Thus φ_ℓ is an inverse for φ_k, whereby φ_k is an automorphism. □

Proposition 2.56. *If G is a cyclic group of order n, then every automorphism of G is a power mapping φ_k for some k that is coprime with n.*

Proof. Let g be a generator for G and let $\varphi : G \to G$ be an automorphism. Since G is cyclic, we have $\varphi(g) = g^k$ for some integer exponent k. Since every x in G takes the form g^j for some j, the fact φ is a homomorphism gives

$$\varphi(x) = \varphi(g^j) = \varphi(g)^j = (g^k)^j = (g^j)^k = x^k.$$

In other words, $\varphi = \varphi_k$.

To check that k is coprime with n, note that for every x in G the elements x and $\varphi(x)$ have equal order, by a simple inspection using the fact φ is an automorphism. In particular the order of g^k equals the order of g. By Proposition 2.11, k must be coprime with the order of g, which is n, because g generates G. □

When G is cyclic and finite, we can now offer a concrete description of Aut(G) in terms of a group isomorphism.

Proposition 2.57. *If G is a cyclic group of order n, then its automorphism group is isomorphic to the group $\mathbb{Z}_n^\star$ of units modulo n.*

Proof. Let g generate G, whereby the order of g equals the order n of G. If $[k] = [\ell]$ as residues in $\mathbb{Z}_n^\star$, their integer representatives k and ℓ are coprime with n and $k \equiv \ell \bmod n$. Then $g^k = g^\ell$, by item 3 of Proposition 2.5. By Proposition 2.55 the power mappings $\varphi_k, \varphi_\ell : G \to G$ given by $x \mapsto x^k, x \mapsto x^\ell$, respectively, are automorphisms of G. As just noted they agree on the generator g. Hence $\varphi_k = \varphi_\ell$ as mappings on all of G.

Consequently the mapping

$$\Psi : \mathbb{Z}_m^\star \to \mathrm{Aut}(G) \text{ given by } [k] \mapsto \varphi_k$$

is well defined in the sense that it does not depend on the choice of representative k for the residue $[k]$. It is routine to check that Ψ is a group homomorphism. Indeed, for any integers k, ℓ and x in G we get

$$\varphi_k(\varphi_\ell(x)) = (x^\ell)^k = x^{k\ell} = \varphi_{k\ell},$$

which shows that $\varphi_k \circ \varphi_\ell = \varphi_{k\ell}$. For $[k], [\ell]$ in $\mathbb{Z}_n^\star$ this says that

$$\Psi([k][\ell]) = \Psi([k\ell]) = \varphi_{k\ell} = \varphi_k \circ \varphi_\ell = \Psi([k])\, \Psi([\ell]).$$

To see that Ψ is injective, let us check that its kernel is trivial. If $[k] \in \ker \Psi$, this means that φ_k is the identity automorphism, and thus that $g^k = g^1$. By item 3 of Proposition 2.5, $k \equiv 1 \bmod n$. In other words $[k] = [1]$.

Finally, Ψ is surjective, for that is what Proposition 2.56 says. □

If G is an infinite cyclic group, we leave it as an exercise to show that G has just two automorphisms: the identity map and the inverse map which sends every x to x^{-1}. If we think of the infinite cyclic group as the group of integers $\mathbb{Z}$ under addition, the automorphisms of $\mathbb{Z}$ are the identity and the negation map $x \mapsto -x$.

EXERCISES

1. If G is a group with binary operation given by $(x, y) \mapsto x \bullet y$, and a new operation, call it $\star$, on G is defined by $x \star y = y \bullet x$, show that G remains a group using the operation $\star$, and that G with the $\star$-operation is isomorphic to G with the original $\bullet$-operation.
2. If $\varphi : G \to H$ is a group homomorphism and x is an element of finite order in G, show that the order of $\varphi(x)$ divides the order of x.
3. For a group G the inverse mapping $\varphi : G \to G$ is given by $x \mapsto x^{-1}$. Prove that G is abelian if and only if the inverse mapping is an automorphism.
4. If G is a group of order 8 show that the mapping $\varphi : G \to G$ defined by $x \mapsto x^5$ is an automorphism.

   Hint. If G is not cyclic, what is the possible order of each element of G?

5. Find the automorphisms of the group $\mathbb{Z}_{12}$.
6. If $n \geq 3$ show that the group $\mathcal{S}_n$ is isomorphic to its group of inner automorphisms $\mathrm{Inn}(\mathcal{S}_n)$.
7. Show that an infinite cyclic group has only two automorphisms.
8. Explain why the number of generators of a finite cyclic group equals the number of automorphisms of the group.
9. Prove that the group $\mathrm{Inn}(G)$ of inner automorphisms of a group G is a normal subgroup of the full automorphism group $\mathrm{Aut}(G)$.
10. Show that a group with more than two elements has more than one automorphism.

    Hint. The somewhat more difficult case is that of a group G which is abelian and such that $x^2 = 1$ for every x in G. In additive notation this is a group G such that $2x = 0$ for every x in G. Show that such G is a vector space over the field $\mathbb{Z}_2$, and then use linear algebra.
11. Let $\varphi : \mathbb{Z} \times \mathbb{Z} \to \mathbb{Z}$ be the homomorphism defined by $\varphi : (x,y) \to 6x + 10y$. Here the groups use addition. Show that both the image and kernel of φ are cyclic groups, and find a generator for each of these groups.
12. If m,n are coprime integers and $\varphi : \mathbb{Z} \times \mathbb{Z} \to \mathbb{Z}$ is the group homomorphism defined by $(x,y) \mapsto mx + ny$ (using addition), show that $\ker\varphi$ is a cyclic subgroup of $\mathbb{Z} \times \mathbb{Z}$ and find its generator.
13. Suppose G is a finite group and that the mapping $\varphi : G \to G$ defined by $\varphi(x) = x^3$ is an automorphism. Prove that G is an abelian group.

    Hint. If $x,y \in G$, deduce that $x^2y^2 = (yx)^2$, and likewise $y^2x^2 = (xy)^2$. In the second identity, replace x by x^2 and y by y^2, and then show that $(xy)^4 = x^4y^4$ to get to $(xy)^3 = (yx)^3$. Now use the fact φ is injective.
14.* Let G be a group of odd order n (not assumed to be abelian). If the function $\varphi : G \to G$ given by $\varphi : x \mapsto x^5$ is a group automorphism, prove that n and 5 are coprime, and then prove that G is abelian.

2.6 Quotient Groups

The kernel N of a group homomorphism $\varphi : G \to H$ is normal in G. Conversely, as we shall now see, every normal subgroup of a group G is the kernel of some homomorphism $\varphi : G \to H$. The construction of a suitable H and a suitable φ will prove to be a worthwhile endeavor.

Suppose N is a normal subgroup of G. The family of right cosets of N coincides with the family of left cosets of N, and this family partitions G. Let

$$G/N$$

denote the family of cosets of N; left or right does not matter. We will now turn G/N into a group in its own right.

If Nx, Ny are cosets in G/N, compose them by the rule:

$$(Nx)(Ny) = Nxy.$$

In other words, the product of the cosets represented by x and y is the coset represented by xy. This definition is at first glance ambiguous because it depends on the representatives x, y chosen for their respective cosets. To overcome this, we must show that if $Nx = Ns$ and $Ny = Nt$, then $Nxy = Nst$. Given that $xs^{-1} \in N$ and $yt^{-1} \in N$, we want $(xy)(st)^{-1} \in N$. Well, since N is normal and $yt^{-1} \in N$, so is $x(yt^{-1})x^{-1} \in N$. And then

$$(xy)(st)^{-1} = xyt^{-1}s^{-1} = (xyt^{-1}x^{-1})xs^{-1} \in N.$$

The associative law for the composition in G/N is inherited directly from the same law in G. The identity element of G/N is the coset $N = N1$. The inverse of every coset Nx is Nx^{-1}. All of these claims can be seen by inspection.

Thus, when N is normal, G/N is a group!

The group operation in G/N is custom designed to ensure that the mapping

$$\varphi : G \to G/N, \text{ given by } x \mapsto Nx,$$

is a group homomorphism. Furthermore φ is clearly surjective.

Here is a summary of what we have just learned.

Proposition 2.58. *If N is a normal subgroup of a group G, then the set of cosets G/N is a group under the (well-defined) operation:*

$$(Nx)(Ny) = Nxy,$$

and the mapping

$$\varphi : G \to G/N \text{ given by } x \mapsto Nx$$

is a surjective group homomorphism with kernel N.

Definition 2.59. If N is a normal subgroup of G, the group of cosets G/N is called the ***quotient group*** of G ***modulo*** N. The mapping $\varphi : G \to G/N$ given by $x \mapsto Nx$ is called the ***canonical map*** or the ***canonical projection*** of G onto G/N.[7]

It should come as no surprise, for example, that the quotient group of $\mathbb{Z}$, under addition, modulo the subgroup $n\mathbb{Z}$ is the group $\mathbb{Z}_n$ of residues modulo n.

For another illustration of quotient groups, let $\mathcal{T}$ be the circle group inside the group of non-zero complex numbers $\mathbb{C}^\star$ under multiplication. The quotient group $\mathbb{C}^\star/\mathcal{T}$ turns out to be isomorphic to the group of $\mathbb{R}^+$ of positive real numbers under multiplication. Indeed, take the mapping

$$\varphi : \mathbb{R}^+ \to \mathbb{C}^\star/\mathcal{T} \text{ given by } x \to \mathcal{T}x.$$

This is an obvious homomorphism. Let us find its kernel. A positive real number x lies in $\ker\varphi$ if and only if $\mathcal{T}x = \mathcal{T}$. This is the same as saying $x \in \mathcal{T}$. Since x is a positive real this

[7] The term "canonical" suggests the map is standard, in the way some ancient scriptures are deemed to have set a standard.

means that $x = 1$. The kernel of φ is trivial, and so φ is an injection. To see that φ is surjective, take $\mathcal{T}z$ in $\mathbb{C}^\star/\mathcal{T}$. Since $z \neq 0$, its absolute value $|z| \in \mathbb{R}^+$. And then $\mathcal{T}|z| = \mathcal{T}z$, because $|z|z^{-1}$ has absolute value 1 and thereby belongs to $\mathcal{T}$.

Normal Subgroups Are Kernels

The kernel of the canonical map $\pi : G \to G/N$ consists of all x in G such that $Nx = N1$. That is $\ker \varphi = N$. The canonical map is a group homomorphism whose kernel is the original normal subgroup N. Our initial question about normal subgroups being kernels is answered.

Proposition 2.60. *A subgroup N of a group G is normal if and only if N is the kernel of some group homomorphism $\varphi : G \to H$.*

2.6.1 An Important Theorem of Cauchy

Lagrange told us that, in a finite group, the order of every subgroup divides the order of the parent group. On the other hand, it is possible to have a group G and a divisor m of $|G|$ while G has no subgroup of order m. For instance, the group $\mathcal{A}_4$ of order 12 has no subgroups of order 6. A remarkable result of Cauchy saves the day when m is a prime. We can use the quotient group construction to demonstrate Cauchy's theorem in case the group is abelian. The delicate argument is worthy of study.

Proposition 2.61 (Cauchy's theorem). *If G is a finite abelian group, and p is a prime dividing the order of G, then G contains an element (and thereby a subgroup) of order p.*

Proof. Suppose the result fails, and seek a contradiction.

In that case, there is a finite abelian group G whose order is as small as possible and for which the theorem fails. Such G has the following properties:

- p divides $|G|$,
- G contains no element of order p,
- every abelian group N, such that $|N| < |G|$ and p divides $|N|$, does contain an element of order p.

Our G cannot be cyclic, for otherwise Proposition 2.41 would ensure that G contained an element of order p. Thus any chosen non-identity element in G will generate a proper, non-trivial subgroup, call it H. If p divides $|H|$, then H must contain an element of order p, since $|H| < |G|$. That element of order p in H, is by default an element of order p in G, contrary to the choice of G. Thus p does not divide $|H|$.

Since G is abelian, H is normal, and so we can form the quotient group G/H. This quotient group is also abelian, quite obviously. By Lagrange,

$$|G| = |H||G/H|.$$

The prime p does not divide $|H|$, while p divides $|G|$, and so p divides $|G/H|$. Because H is not the trivial group, $|G/H| < |G|$. By the nature of G, the quotient group G/H contains an element of order p.

The elements of G/H are right cosets. So let g in G be such that Hg is an element of order p in G/H. Thus, inside the group G/H:

$$H \neq Hg \text{ while } H = (Hg)^p = H(g^p),$$

which tells us that

$$g \notin H \text{ while } g^p \in H.$$

If m is the order of H, Lagrange implies that

$$1 = (g^p)^m = (g^m)^p.$$

The order of g^m must then divide the prime p, and since G has no element of order p it follows that g^m has order 1, i.e. $g^m = 1$.

Since p does not divide m and p is prime, these integers are coprime. Hence there exist integers s, t such that $ps + mt = 1$. Then

$$g = g^1 = g^{ps+mt} = (g^p)^s(g^m)^t = (g^p)^s \in H.$$

We have a contradiction because $g \notin H$. □

2.6.2 The First Isomorphism Theorem

Two groups can seem different and yet be isomorphic. The next bit of abstraction can be a useful tool for establishing such isomorphisms.

Let us begin with a set theoretical preamble. Suppose $f : A \to B$ is a surjection from a set A onto a set B. For each b in B, form its ***pre-image***

$$f^{-1}(b) = \{a \in A : f(a) = b\}.$$

Clearly the family $\mathcal{P}$ of pre-images of every element in B forms a partition of A. Furthermore, the mapping $B \to \mathcal{P}$ given by $b \mapsto f^{-1}(b)$ is now a bijection. In other words, each pre-image $f^{-1}(b)$ (thought of as one thing) corresponds to the b in B that made the pre-image. So, the mapping

$$\bar{f} : \mathcal{P} \to B \text{ given by } f^{-1}(b) \mapsto b$$

is a bijection.

Transport this idea to the situation of a surjective group homomorphism

$$\varphi : G \to H.$$

Let $N = \ker \varphi$. As h goes over the elements of H, the pre-images $\varphi^{-1}(h)$ form a partition of G. What are these pre-images? For two elements x, y in G, the following statements are clearly equivalent:

$$\begin{aligned}
\varphi(x) &= \varphi(y) \\
\varphi(x)\varphi(y)^{-1} &= 1 \\
\varphi(xy^{-1}) &= 1 \\
xy^{-1} &\in N \\
Nx &= Ny.
\end{aligned}$$

Thus, two elements of G belong to the same pre-image under φ if and only if they belong to same right coset of N. The partition of G into pre-images under φ is precisely the family G/N of right cosets of N.

Since N is a kernel, it is normal in G. Thereby G/N becomes the quotient group. For every x in G, the coset Nx is the pre-image of $\varphi(x)$. As discussed in the preceding preamble with sets, there is the bijection

$$\overline{\varphi} : G/N \to H \text{ given by } Nx \mapsto \varphi(x).$$

This $\overline{\varphi}$ is an isomorphism simply by the way the quotient group is built. Indeed, for any two cosets Nx, Ny of N:

$$\overline{\varphi}(NxNy) = \overline{\varphi}(Nxy) = \varphi(xy) = \varphi(x)\varphi(y) = \overline{\varphi}(Nx)\overline{\varphi}(Ny).$$

Since every group homomorphism $\varphi : G \to H$ is surjective onto its image $\operatorname{im}\varphi$, the following noteworthy result emerges.

Proposition 2.62 (First isomorphism theorem). *If $\varphi : G \to H$ is a group homomorphism with kernel N, then the mapping*

$$\overline{\varphi} : G/N \to \varphi(G), \text{ defined by } Nx \mapsto \varphi(x),$$

is an isomorphism.

Briefly, every homomorphic image of a group is isomorphic to a quotient of that group.

Here is another way to think of the first isomorphism theorem. We have our homomorphism $\varphi : G \to H$ with kernel N. The canonical projection $\pi : G \to G/N$ sends x in G to its coset Nx. Then, for every x in G, we have

$$\varphi(x) = \overline{\varphi}(Nx) = \overline{\varphi}(\pi(x)).$$

The first isomorphism theorem says that

$$\varphi = \overline{\varphi} \circ \pi,$$

where $\overline{\varphi}$ is an isomorphism from G/N to $\operatorname{im}\varphi$. This factoring of φ through G/N is often pictured by a so-called ***commutative diagram*** as shown:

$$\begin{array}{ccc}
G & \xrightarrow{\ \varphi\ } & H \\
\downarrow{\scriptstyle \pi} & & \uparrow \\
G/H & \xrightarrow[\cong]{\ \overline{\varphi}\ } & \varphi(G)
\end{array}$$

The upward arrow in the diagram is the simple inclusion mapping of $\varphi(G)$ into H.

Orders of Images and Kernels

The first isomorphism theorem in conjunction with Lagrange gives a useful formula that relates the orders of the kernel and image of a homomorphism.

Proposition 2.63. *If $\varphi : G \to H$ is a homomorphism between groups and the order of G is finite, then*

$$|G| = |\ker \varphi||\varphi(G)|.$$

Proof. Since $G/\ker \varphi \cong \varphi(G)$, the order of $\varphi(G)$ equals the order of $G/\ker \varphi$, which is the index $[G : \ker \varphi]$. The conclusion now follows from Lagrange's theorem, Proposition 2.35. □

An Application to the Group $GL_n(\mathbb{Z}_p)$

The first isomorphism theorem can surface in unexpected situations. To illustrate, let p be a prime, and let $GL_n(\mathbb{Z}_p)$ be the group of $n \times n$ invertible matrices with entries in the finite field $\mathbb{Z}_p$. These are the matrices in $GL_n(\mathbb{Z}_p)$ whose determinant is non-zero. We might ask how many of the matrices in $GL_n(\mathbb{Z}_p)$ have determinant equal to 1? (Here the "1" stands for the residue $[1]$ in $\mathbb{Z}_p$.)

This problem seems daunting until we recall from linear algebra that the determinant function

$$\det : GL_n(\mathbb{Z}_p) \to \mathbb{Z}_p^\star,$$

into the group of non-zero elements of $\mathbb{Z}_p$, is a homomorphism. The order of the subgroup ker(det) is the desired quantity. One can readily check that det is surjective. By Proposition 2.63

$$|GL_n(\mathbb{Z}_p)| = |\ker(\det)||\mathbb{Z}_p^\star|.$$

Now $|\mathbb{Z}_p^\star| = p - 1$ and so,

$$|\ker(\det)| = \frac{|GL_n(\mathbb{Z}_p)|}{p-1}.$$

To get the order of the subgroup ker(det) it remains to figure out the order of the group $GL_n(\mathbb{Z}_p)$ of invertible matrices. A counting argument based on linear algebra will do that.

Let $A = [c_1, c_2, \ldots, c_n]$ be an $n \times n$ matrix with n-tuple columns c_j in $\mathbb{Z}_p^n$. Such A is invertible if and only if the c_j are linearly independent over the field $\mathbb{Z}_p$. This is the same as saying $c_1 \neq 0$, c_2 is not a $\mathbb{Z}_p$-multiple of c_1, c_3 is not a $\mathbb{Z}_p$-linear combination of c_1, c_2, and so on. Any invertible A can thus be determined by making a sequence of n suitable choices for its columns. For c_1 there are $p^n - 1$ possibilities, namely all but the zero column. For c_2 there are $p^n - p$ possibilities, namely all columns except for the p multiples of the non-zero c_1. For c_3 there are $p^n - p^2$ possibilities, namely all columns except for the p^2 columns that are linear combinations of the linearly independent c_1, c_2. Likewise c_4 has $p^n - p^3$ possibilities, and so on. The number of ways to build columns $c_1, c_2, \ldots, c_n$ such that A is invertible, and thereby the order of $GL_n(\mathbb{Z}_p)$, equals

$$(p^n - 1)(p^n - p)(p^n - p^2) \cdots (p^n - p^{n-1}).$$

Thus the number of $n \times n$ matrices over $\mathbb{Z}_p$ whose determinant is 1 equals

$$\frac{(p^n - 1)(p^n - p)(p^n - p^2) \cdots (p^n - p^{n-1})}{p - 1}.$$

With $p = 5$ and $n = 2$ there are $(5^2-1)(5^2-5)/4 = 120$ matrices over $\mathbb{Z}_5$ whose determinant equals 1.

Isomorphisms between Cyclic Groups

The first isomorphism theorem opens up elegant ways to show that two groups K, H are isomorphic. Find a group G and surjective homomorphisms $\varphi : G \to K$ and $\psi : G \to H$ such that $\ker \varphi = \ker \psi$. The first isomorphism theorem assures that both H and K are isomorphic to $G/\ker \varphi$, and thus to each other.

For example, any two cyclic groups of equal finite order are isomorphic. This is not surprising. For a clean proof, take a cyclic group H with generator h of order n. Recall, the order of H is the order of its generator h. Using the group of integers $\mathbb{Z}$, under addition, consider the surjective homomorphism $\varphi : \mathbb{Z} \to H$ defined by $\varphi(k) = h^k$. An integer k belongs to $\ker \varphi$ if and only if $h^k = 1$. By Proposition 2.5, $k \in \ker \varphi$ if and only if $n \mid k$. This shows that the kernel of φ is the group $n\mathbb{Z}$ of multiples of n. By the first isomorphism theorem $H \cong \mathbb{Z}/n\mathbb{Z} = \mathbb{Z}_n$.

All cyclic groups of order n are isomorphic to $\mathbb{Z}_n$, where the latter group uses addition.

Another Look at Inner Automorphisms

Each element x in a group G gives rise to the inner automorphism

$$\varphi_x : G \to G \text{ where } \varphi_x(y) = xyx^{-1} \text{ for all } y \text{ in } G,$$

and the mapping $\varphi : G \to \mathrm{Aut}(G)$ from G to the automorphism group of G, given by $\varphi : x \mapsto \varphi_x$ is itself a homomorphism whose image is the subgroup $\mathrm{Inn}(G)$ of all inner automorphisms of G and whose kernel is the centre $Z(G)$. By the first isomorphism theorem

$$G/Z(G) = G/\ker \varphi \cong \mathrm{Inn}(G).$$

We can expect the first isomorphism theorem to pop up with some regularity.

2.6.3 The Correspondence Theorem

Something else that is often encountered is the relationship between subgroups of a group and subgroups of its homomorphic images. If $\varphi : G \to H$ is a surjective group homomorphism, there is a natural bijection between subgroups of G containing $\ker \varphi$ and subgroups of H. The claims below are simple enough to check that the bulk of their proofs can be omitted.

Proposition 2.64 (Correspondence theorem). *Let $\varphi : G \to H$ be a surjective group homomorphism.*

- *If K is a subgroup of H, then its inverse image $\varphi^{-1}(K)$ is a subgroup of G, $\varphi^{-1}(K)$ contains* $\ker\varphi$, *and $\varphi(\varphi^{-1}(K)) = K$.*
  If K is normal in H, then $\varphi^{-1}(K)$ is normal in G.
- *If L is a subgroup of G containing* $\ker\varphi$, *then the direct image $\varphi(L)$ is a subgroup of H, and $L = \varphi^{-1}(\varphi(L))$.*
  If L is normal in G, then $\varphi(L)$ is normal in H.
- *In summary, the direct image mapping defined on the family of subgroups of G containing* $\ker\varphi$ *is a bijection onto the family of subgroups of H. The inverse of this bijection is the inverse image mapping from the family of subgroups of H to the family of subgroups of G containing* $\ker\varphi$. *This bijection restricted to the family of normal subgroups of G containing* $\ker\varphi$ *is onto the family of normal subgroups of H.*

Proof. Rather than check every detail we simply note that $\varphi(\varphi^{-1}(K)) = K$ because φ is surjective. This is also the reason for the normality of $\varphi(L)$ in H.

In the second item, the statement that $\varphi(L)$ is a subgroup of H clearly applies to all L not just to those containing $\ker\varphi$. Assuming L contains $\ker\varphi$, let us verify that $L = \varphi^{-1}(\varphi(L))$. From the definition of direct and inverse images it is clear that $L \subseteq \varphi^{-1}(\varphi(L))$. For the other inclusion suppose that $x \in \varphi^{-1}(\varphi(L))$. Then $\varphi(x) \in \varphi(L)$, which means that $\varphi(x) = \varphi(y)$ for some y in L. Then $\varphi(xy^{-1}) = 1$, and thereby $xy^{-1} \in \ker\varphi \subseteq L$. So, $x = (xy^{-1})y \in L$. □

The correspondence theorem can be applied to an arbitrary group homomorphism φ defined on G. Just let H be $\varphi(G)$. This theorem is frequently encountered in the case of a canonical projection $G \to G/N$ where N is a normal subgroup of G. Here it says that for every subgroup K of G/N, there is a unique subgroup L of G containing N and such that $K = L/N$. And K is normal if and only if L is normal.

By way of illustration, recall Proposition 2.41 which said that the subgroups of a cyclic group G order n correspond to the positive divisors of n. Let us use the correspondence theorem to see this once more. Take a generator g of G and using $\mathbb{Z}$ under addition define the surjective homomorphism $\varphi : \mathbb{Z} \to G$ by $k \mapsto g^k$. It can readily be seen that $\ker\varphi = \mathbb{Z}n$. By the correspondence theorem the subgroups of G correspond to the subgroups H of $\mathbb{Z}$ which contain the subgroup $\mathbb{Z}n$. Since every subgroup of the cyclic group $\mathbb{Z}$ is cyclic such H are of the form $\mathbb{Z}d$ for some integer d, which we can take to be positive. Clearly every multiple of n is a multiple of d if and only if $d \mid n$. In other words $\mathbb{Z}d \supseteq \mathbb{Z}n$ if and only if $d \mid n$.

EXERCISES

1. Complete the details in the proof of the correspondence theorem, Proposition 2.64.
2. Show that every group of order 4 is isomorphic either to the cyclic group $\mathbb{Z}_4$ under addition or to the Klein 4-group $\mathcal{V}_4$.
3. If H is a non-normal subgroup of a group G, show that there exist x, y, s, t in G such that $Hx = Hs, Hy = Ht$ and yet $Hxy \neq Hst$.

4. Show that the circle group $\mathcal{T}$ of complex numbers of absolute value 1 using multiplication is isomorphic to the quotient group $\mathbb{R}/\mathbb{Z}$ using addition.
5. If $\mathbb{R}^+$ is the group of positive real numbers under multiplication, $\mathbb{C}^\star$ is the group of non-zero complex numbers under multiplication with the circle group $\mathcal{T}$ as its subgroup, show that $\mathbb{C}^\star/\mathcal{T} \cong \mathbb{R}^+$.
6. Let $\varphi : G \to H$ be a group homomorphism, and let L be a subgroup of H. Prove the following order formula:

$$|\varphi^{-1}(L)| = |L||\ker \varphi|.$$

Hint. First dispose of the case where $\varphi^{-1}(L)$ is infinite, by showing that one of the factors on the right hand side must also be infinite. In case $\varphi^{-1}(L)$ is finite use Proposition 2.63.
7. If G is an abelian group and H consists of the elements in G having finite order, show that H is a subgroup of G and that the order of every element of G/H is either 1 or infinite.
8. Suppose G is an abelian group with a subgroup H. If a prime p divides the order of G/H, explain why there is an x in $G \setminus H$ such that $x^p \in H$.
9. Let G be an abelian group of finite order n, and let k be an integer. If the mapping $\varphi : G \to G$ given by $x \mapsto x^k$ is an automorphism, show that k is coprime with n. This result is the converse of Proposition 2.55.
10. The ***squares*** of a group G are the elements of the form x^2 as x runs over G. If H is a normal subgroup of G such that every element of H and of G/H is a square, show that every element of G is a square.
11. Let G be the abelian group $\mathbb{Z}\times\mathbb{Z}\times\mathbb{Z}$ using addition, and let H be the subgroup generated by $(1, 0, 0)$ and $(0, 3, 3)$. Show that $G/H \cong \mathbb{Z} \times \mathbb{Z}_3$ by finding a homomorphism $G \to \mathbb{Z} \times \mathbb{Z}_3$ whose kernel is H.
12. Let G be the product group $\mathbb{Z}_2 \times \mathbb{Z}_3 \times \mathbb{Z}_3$ using addition. If H is the cyclic subgroup generated by $(1, 1, 1)$ (where these 1s are the appropriate residues), show that $G/H \cong \mathbb{Z}_3$.
13. Show that S_4 has only one subgroup of order 12.
    Hint. If H is a subgroup of order 12, then H is normal and S_4/H has order 2.
14. (a) If a group G with centre Z is such that G/Z is cyclic show that G is abelian.
        Hint. Let g in G be such that the coset hg generates the quotient G/Z. This means that every x in G is of the form ag^k where $a \in Z$ and k is an integer exponent.
    (b) If the order of a group G is pq where p, q are distinct primes and G is not abelian, show that the centre of G must be the trivial group $\langle 1 \rangle$.
15. If G is an abelian group of order n and m is a positive divisor of n, show that G contains a subgroup of order m.
    Hint. This result is the converse of Lagrange's theorem for the case of abelian groups. Let p be a prime factor of m. Cauchy's theorem gives a subgroup H of order p. Apply an induction argument to the group G/H whose order n/p is divisible by m/p.
16. Let p be an odd prime and let $\mathbb{Z}_p^\star$ be the group of units modulo p. This group is of order $p-1$ with elements represented by the integers $1, 2, \ldots, p-1$.

(a) Observe that the squaring map $\varphi : \mathbb{Z}_p^\star \to \mathbb{Z}_p^\star$ given by $x \mapsto x^2$ is a homomorphism, and then find $\ker \varphi$.

(b) Let $H = \varphi(\mathbb{Z}_p^\star)$. The elements of H are the perfect squares of $\mathbb{Z}_p^\star$. They are also known as ***quadratic residues***. Show that the index of H in $\mathbb{Z}_p^\star$ is 2.

(c) If $a, b \in \mathbb{Z}_p^\star \setminus H$, show that $ab \in H$.

(d) If $p > 3$ show that at least one of $2, 3, 6$ represents a residue in H.

17. A finite group whose order is the power p^n of a single prime p is known as a ***p-group***.

If G is abelian and p is a prime, show that G is a p-group if and only if the order of every element in G is a power of p.

This result is also true for finite groups which are not abelian, but a bit more theory is needed to prove it.

18. If G is a p-group, n is a positive integer not divisible by p and a is any element of G, show that there exists x in G such that $x^n = a$.

19. Let G be an abelian group *using addition*. Such G is called ***divisible*** provided that for every y in G and every positive integer n there is an x in G such that $nx = y$.

(a) Show that each of the groups $\mathbb{Q}, \mathbb{R}, \mathbb{C}$, using addition, is divisible.

(b) Show that the only finite abelian group that is divisible is the trivial group containing just 0.

(c) If G is divisible and H is a subgroup of G, show that G/H is divisible.

Give an example of a divisible G with a non-divisible subgroup H.

(d) Show that the groups $\mathbb{Q}, \mathbb{R}, \mathbb{C}$ using addition do not have proper subgroups of finite index.

20. This exercise is about a peculiar infinite abelian group. Let p be a prime and let A be the set of all rational numbers which can be written in the form $\frac{a}{p^n}$ where $a \in \mathbb{Z}$ and $n \geq 0$. Let R be the set of all complex numbers z such that $z^{p^n} = 1$ for some $n \geq 0$.

(a) Verify that A is a subgroup of $\mathbb{Q}$ using addition, and that R is a subgroup of the circle group $\mathcal{T}$ (i.e. the complex numbers z such that $|z| = 1$) using multiplication.

(b) Show that the mapping $\varphi : A \to \mathcal{T}$ given by $x \mapsto e^{2\pi i x}$ is a homomorphism and that the image $\varphi(A) = R$.

(c) After observing that $\mathbb{Z}$ is a subgroup of A prove that $R \cong A/\mathbb{Z}$. Here R uses multiplication while $A/\mathbb{Z}$ uses addition.

(d) Show that $A/\mathbb{Z}$ is a divisible group, as discussed in the preceding exercise.

Hint. First notice that the order of each element in $A/\mathbb{Z}$ is a power of p. For y in $A/\mathbb{Z}$ and a positive integer n, solve the equation $nx = y$ in case $p \nmid n$ and in case n is a power of p.

(e) For each positive integer k show that the cyclic subgroup H_k of $A/\mathbb{Z}$ generated by $\frac{1}{p^k} + \mathbb{Z}$ has finite order p^k.

Show that, along with the trivial group and all of $A/\mathbb{Z}$, the H_k account for all subgroups of $A/\mathbb{Z}$.

Thus $A/\mathbb{Z}$ is an infinite abelian group such that all of its proper subgroups are finite, cyclic p-groups.

The common notation for the group $A/\mathbb{Z}$ is $\mathbb{Z}(p^\infty)$.

2.7 Products of Groups

Using Cartesian products we can easily build new groups from old ones, but it is even more interesting to see how some groups are isomorphic to such Cartesian products.

2.7.1 The External Product

If H, K are groups, their Cartesian product $H \times K$ is the set of ordered pairs (x, y), where $x \in H$ and $y \in K$. The operation of point-wise multiplication given by

$$(x, y)(s, t) = (xy, st), \text{ for every } (x, y), (s, t) \text{ in } H \times K,$$

creates a new group. The verification of associativity is trivial. It is equally easy to see that the identity of this group is $(1, 1)$, where each 1 is the identity in H or K as appropriate. The inverse of (x, y) is (x^{-1}, y^{-1}).

Definition 2.65. The group $H \times K$, formed in the above straightforward manner, is called the ***external product*** group of H and K.

It is equally easy and natural to form the external product of more than two groups.

The usefulness of external products comes from knowing when a given group G is isomorphic to an external product of two or more groups. That would provide a handy way to understand the group in terms of its simpler components. When a group G is isomorphic to an external product $H \times K$ of two other groups, we often refer to such an external product as a ***decomposition*** of G.

The Product of Cyclic Groups

In order to become acquainted with the external product, let us investigate it when the factors are cyclic groups.

Proposition 2.66. *If H, K are finite cyclic groups with generators h, k, respectively, and if the orders of H, K are coprime, then the product $H \times K$ is cyclic with generator (h, k). If the orders of H, K are not coprime, then the product $H \times K$ is not cyclic.*

Proof. Let m be the order of H, and n the order of K, which are also the respective orders of h and k. The orders of $(h, 1)$ and $(1, k)$ in $H \times K$ remain m and n, respectively. Also $(h, 1)$ commutes with $(1, k)$, and their product is $(h, 1)(1, k) = (h, k)$. By Proposition 2.8, the order of (h, k) is mn. Since mn is also the order of $H \times K$, the group $H \times K$ is cyclic with generator (h, k).

For the converse, suppose m, n are not coprime. In this case, the least common multiple ℓ of m and n is less than the product mn. For any (x, y) in $H \times K$ we obtain

$$(x, y)^{\ell} = (x^{\ell}, y^{\ell}) = (1, 1).$$

The second equality holds because both m and n divide ℓ. It is therefore impossible to have an element in $H \times K$ of order mn, and this prevents $H \times K$ from being cyclic. □

It is also the case that the product $H \times K$ of non-trivial cyclic groups is not cyclic when at least one of H, K is infinite. This can be left as an exercise.

The Chinese Remainder Theorem

A result from antiquity fits into the framework of external products of cyclic groups.

Proposition 2.67 (Chinese remainder theorem). *If m, n are coprime positive integers and a, b are any integers, then the simultaneous congruences*

$$x \equiv a \bmod m \quad \textit{and} \quad x \equiv b \bmod n$$

have an integer solution x. And if x, y are solutions, then $x \equiv y \bmod mn$.

Proof. It is a matter of interpreting Proposition 2.66. With m, n coprime, and using the cyclic groups $\mathbb{Z}_m, \mathbb{Z}_n$ under addition, the direct product $\mathbb{Z}_m \times \mathbb{Z}_n$ is once more a cyclic group under addition. Since the residues $[1]_m, [1]_n$ generate $\mathbb{Z}_m, \mathbb{Z}_n$, respectively, the pair $([1]_m, [1]_n)$ generates $\mathbb{Z}_m \times \mathbb{Z}_n$ as an abelian group using addition. So, for every pair of residues $([a]_m, [b]_n)$ in $\mathbb{Z}_m \times \mathbb{Z}_n$, there is an integer x such that

$$([x]_m, [x]_n) = x([1]_m, [1]_n) = ([a]_m, [b]_n).$$

In other words, for every pair of integers a, b, there is an integer x such that

$$x \equiv a \bmod m \text{ and } x \equiv b \bmod n.$$

Furthermore, Proposition 2.5, interpreted additively, reveals that if

$$x([1]_m, [1]_n) = ([a]_m, [b]_n) = y([1]_m, [1]_n),$$

then $x \equiv y \bmod mn$. □

Routine Practice with the Chinese Remainder Theorem

Two, or more, simultaneous congruences can be readily solved using suitable software. The main tool is the Euclidean algorithm applied to linear Diophantine equations. For those unfamiliar with this routine method, let us solve the three simultaneous congruences

$$x \equiv 3 \bmod 5, \ x \equiv 1 \bmod 9, \ x \equiv 6 \bmod 8.$$

The first congruence says that any solution x takes the form $x = 3 + 5t$ for some integer t. To also satisfy the second congruence, we must have $3 + 5t \equiv 1 \bmod 9$. This comes down to $5t \equiv 7 \bmod 9$. To get t we could solve the Diophantine equation $5t = 7 + 9u$ for t and u. But with our small integers we can see by inspection that $t \equiv 5 \bmod 9$ solves $5t \equiv 7 \bmod 9$. In other words, $t = 5 + 9s$ for some integer s.

Thus, the simultaneous solution of the first two congruences is given by

$$x = 3 + 5(5 + 9s) = 28 + 35s, \text{ where } s \in \mathbb{Z}.$$

We see that $x \equiv 28 \bmod 35$ is the solution of the first two congruences.

But x needs to satisfy the third congruence too. Consequently $28 + 35s \equiv 6 \bmod 8$. After some basic modular arithmetic this comes down to $3s \equiv 2 \bmod 8$, and then by inspection (or by solving a linear Diophantine equation in case of large integers), $s \equiv 6 \bmod 8$. That is $s = 6 + 8r$, where $r \in \mathbb{Z}$.

So,

$$x = 28 + 35(6 + 8r) = 238 + 280r, \text{ where } r \in \mathbb{Z}.$$

Our unique solution to the three congruences is $x \equiv 238 \bmod 280$. We should note that $280 = 5 \cdot 9 \cdot 8$. The fact that these three factors are pairwise coprime allowed the Chinese remainder theorem to be implemented.

Subgroups of External Products

If G, H are groups, the not so simple question of determining the subgroups of the product $G \times H$ arises. To that end the Chinese remainder theorem is useful when the orders of G, H are finite and coprime.

Proposition 2.68. *If G, H are finite groups and their orders are coprime, then all subgroups of $G \times H$ are of the form $A \times B$, where A is a subgroup of G and B is a subgroup of H.*

Proof. Clearly, if A is a subgroup of G and B is a subgroup of H, then $A \times B$ is a subgroup of $G \times H$. To see that these account for all subgroups of $G \times H$, suppose K is a subgroup of $G \times H$. Let A be the subgroup of G which is the image of the homomorphism $K \to G$ given by $(a, b) \mapsto a$. Likewise let B inside H be the image of the homomorphism $K \to H$ given by $(a, b) \mapsto b$. We will check that $K = A \times B$.

If $(a, b) \in K$, then $a \in A$ and $b \in B$ by definition of A and B. Thus $K \subseteq A \times B$. For the reverse inclusion it suffices to verify that $(a, 1) \in K$ for every a in A and that $(1, b) \in K$ for every b in B. Since $(a, b) = (a, 1)(1, b)$, it will then follow that $(a, b) \in K$ whenever $(a, b) \in A \times B$.

Now suppose $a \in A$. This means that $(a, c) \in K$ for some c in H. (Actually $c \in B$.) Let m be the order of a, and n the order of c. These m and n divide the orders of G and H, respectively, by Lagrange. Since the orders of G and H are coprime so are m and n coprime. By the Chinese remainder theorem, there is an integer r such that

$$r \equiv 1 \bmod m \text{ and } r \equiv 0 \bmod n.$$

From item 3 of Proposition 2.5

$$a^r = a^1 = a \text{ and } c^r = c^0 = 1.$$

Therefore

$$(a, 1) = (a^r, c^r) = (a, c)^r \in K.$$

Similarly $(1, b) \in K$, as desired. □

2.7.2 The Internal Product of Subgroups

Definition 2.69. If G is a group and H, K are subgroups, their ***internal product*** is the subset of G defined by:

$$HK = \{hk : h \in H \text{ and } k \in K\}.$$

The internal product need not be a subgroup, though at times it is. For example, take the group $\mathcal{S}_3$. Let H be the cyclic subgroup generated by $(1,2)$ and K the subgroup generated by $(1,3)$. Then HK consists of the four permutations

$$1, (1,2), (1,3), (1,2)(1,3) = (1,3,2),$$

and they do not form a subgroup. On the other hand, let L be the cyclic subgroup generated by $(1,2,3)$. Now with H as before, the product HL consists of the full group $\mathcal{S}_3$.

Here is the test for HK to be a subgroup of G.

Proposition 2.70. *Let G be a group and H, K subgroups of G. The internal product HK is a subgroup of G if and only if $HK = KH$. If either H or K is normal in G, then HK is a subgroup of G.*

Proof. The statement $HK = KH$ means that every product rs, where $r \in K$ and $s \in H$, equals a product uv, for some $u \in H$ and $v \in K$, and vice versa. It does not mean that $rs = sr$.

For all subgroups H and K, the set HK contains 1, because $1 = 1 \cdot 1$. Also, the identity $(hk)^{-1} = k^{-1}h^{-1}$ reveals that $y \in HK$ if and only if $y^{-1} \in KH$, regardless of assumptions about HK.

Now suppose that HK is a subgroup of G. If $x \in HK$, then $x^{-1} \in HK$. By the preceding observation $x = (x^{-1})^{-1} \in KH$. Thus $HK \subseteq KH$. On the other hand, if $x \in KH$, then $x^{-1} \in HK$, and since the latter is a subgroup, $x \in HK$ too. Hence $KH \subseteq HK$. Therefore $HK = KH$.

For the converse, suppose $HK = KH$. Now, if $x \in HK$, then $x^{-1} \in KH = HK$, which verifies that HK is closed under the taking of inverses. What is left is to see that HK is closed under composition. Let $h, \ell \in H$ and $k, m \in K$. So, $hk, \ell m \in HK$, and we want $hk\ell m \in HK$. Well, $k\ell \in KH$, and by assumption $k\ell = uv$ for some u in H and some v in K. Then

$$hklm = huvm = (hu)(vm) \in HK.$$

For the second statement in the proposition, suppose H is normal in G. If $h \in H$ and $k \in K$, then $hk = k(k^{-1}hk)$. The left side of this is an arbitrary element of HK. Since H is normal the right side is in KH. Thus $HK \subseteq KH$. Similarly $KH \subseteq HK$, and thus $HK = KH$. By the first statement of the proposition, HK is a subgroup of G. $\square$

The next item follows immediately from Proposition 2.70.

Proposition 2.71. *If H, K are subgroups of a group G, and if every element of H commutes with every element of K, then HK is a subgroup of G.*

Proposition 2.71 might seem more familiar when reduced to the situation of abelian groups written in additive notation.

Definition 2.72. If V is an abelian group, using addition notation, and H,K are subgroups, the internal product of H and K is written as $H+K$, and is called the ***sum*** of the subgroups H,K. That is

$$H+K = \{h+k : h \in H \text{ and } k \in K\}.$$

Perhaps this is most familiar when V is a vector space and H,K are subspaces. Since everything commutes here, we are in the situation of Proposition 2.71, and $H+K$ is indeed a subgroup of V.

A Mapping from the External to the Internal Product

What is the connection between the internal and external products of two subgroups? With every group G and subgroups H,K comes the following multiplication mapping

$$\mu : H \times K \to G \text{ given by } (h,k) \mapsto hk.$$

The image of μ is HK, but μ need not be a homomorphism. For instance, when HK is not a subgroup of G, it cannot be a homomorphism. It can also happen that μ is not a homomorphism even if HK is a subgroup. For example, we noted that $\mathcal{S}_3$ is the internal product of the cyclic groups H,K generated by $(1,2)$ and $(1,2,3)$, respectively. Thus $\mathcal{S}_3$ is the image under μ of $H \times K$. Since the latter group is abelian, while $\mathcal{S}_3$ is not abelian, μ cannot be a homomorphism.

When is μ a homomorphism?

Proposition 2.73. *Let G be a group with subgroups H,K. The mapping $\mu : H \times K \to G$, given by $(h,k) \mapsto hk$, is a homomorphism if and only if every element of H commutes with every element of K.*

Proof. Suppose the elements of H commute with those of K. For μ to be a homomorphism, we need to show that if $h,\ell \in H$ and $k,m \in K$, then

$$\mu((h,k)(\ell,m)) = \mu(h,k)\mu(\ell,m).$$

This is the same as showing that

$$\mu(h\ell,km) = \mu(h,k)\mu(\ell,m),$$

which boils down to

$$h\ell km = hk\ell m.$$

Since, by assumption, $\ell k = k\ell$, we get what we want.

Conversely, suppose μ is a homomorphism. Let $h \in H$ and $k \in K$. Obviously, inside the external product,

$$(h,1)(1,k) = (h,k) = (1,k)(h,1).$$

Apply the homomorphism μ to these equalities to obtain

$$hk = \mu(h,1)\mu(1,k) = \mu(h,k) = \mu(1,k)\mu(h,1)) = kh,$$

as desired. $\square$

Next let us enquire about $\ker\mu$. An element (h,k) from $H \times K$ is in $\ker\varphi$ if and only if $hk = 1$. This is equivalent to saying that $k = h^{-1} \in H \cap K$. Thus, $\ker\varphi$ is the subgroup consisting of all pairs (h, h^{-1}) where $h \in H \cap K$. The following is now apparent.

Proposition 2.74. *Let G be a group with subgroups H, K. The mapping $\mu : H \times K \to G$, given by $(h,k) \mapsto hk$, is an injective group homomorphism if and only if every element of H commutes with every element of K and $H \cap K = \langle 1 \rangle$. In that case, $H \times K \cong HK$.*

From this, a common way to establish an isomorphism between internal and external direct products emerges.

Proposition 2.75. *If G is a group and H, K are both normal subgroups with a trivial intersection, i.e. $H \cap K = \langle 1 \rangle$, then $H \times K \cong HK$.*

Proof. Py Proposition 2.74, it suffices to check that every element of H commutes with every element of K. Well, if $h \in H$ and $k \in K$, look at the so-called ***commutator*** $hkh^{-1}k^{-1}$. Since K is normal, the conjugate $hkh^{-1} \in K$, and thereby the commutator is in K. By the normality of H, this same commutator is likewise in H. Since this commutator is in $H \cap K$, it must be that $hkh^{-1}k^{-1} = 1$, and thus $hk = kh$. $\square$

If just one of H or K is normal in G, a simple exercise is to see that HK remains a subgroup of G. This condition, wherein just one of H or K is normal is intriguing, and will be looked at more deeply in the next chapter.

The Order of Any Internal Product

Say G is now finite with subgroups H, K. Proposition 2.73 said that if every element of H commutes with every element of K, then $\mu : H \times K \to G$ given by $(h,k) \mapsto hk$ is a homomorphism. In that case, by Proposition 2.62, $H \times K/\ker\varphi \cong HK$. And so by Lagrange,

$$|HK| = \frac{|H \times K|}{|\ker\varphi|}.$$

The order of the numerator obviously equals $|H||K|$, and by the observation preceding Proposition 2.74 the order of the denominator is seen to equal $|H \cap K|$. Thus $|HK| = |H||K|/|H \cap K|$. A direct argument, which follows, reveals that this formula prevails even if μ is not a homomorphism.

Proposition 2.76. *If G is a finite group and H, K are subgroups, then*

$$|HK| = \frac{|H||K|}{|H \cap K|}.$$

Proof. Use the multiplication mapping $\mu : H \times K \to G$ given by $(h,k) \mapsto hk$, whose image is HK. As x runs over HK, the inverse images $\mu^{-1}(x)$ form a partition $\mathcal{P}$ of $H \times K$. And the mapping $HK \to \mathcal{P}$, given by $x \mapsto \mu^{-1}(x)$, is a bijection.

To get the order of HK we can count the number of partitioning sets in $\mathcal{P}$. For that, it helps to first know the order of each of the partitioning sets. So, let A be one of those sets in $\mathcal{P}$, and let (h,k) be a representative for A. For every ℓ in $H \cap K$, we see that

$$\mu(h\ell, \ell^{-1}k) = h\ell\ell^{-1}k = hk = \mu(h,k).$$

As long as $\ell \in H \cap K$, the elements $(h\ell, \ell^{-1}k) \in A$, and the number of such elements equals $|H \cap K|$.

It turns out that every element in A takes the form $(h\ell, \ell^{-1}k)$ for some ℓ in $H \cap K$. Indeed, suppose $(r,s) \in A$. Then, $hk = rs$, which leads to $h^{-1}r = ks^{-1}$. It follows that $h^{-1}r \in H \cap K$. And then,

$$(r,s) = (hh^{-1}r, sk^{-1}k) = (hh^{-1}r, (ks^{-1})^{-1}k) = (hh^{-1}r, (h^{-1}r)^{-1}k) = (h\ell, \ell^{-1}k),$$

where $\ell = h^{-1}r \in H \cap K$.

We have just seen that the order of each set forming the partition $\mathcal{P}$ equals the order of $H \cap K$. Therefore, the number of sets in $\mathcal{P}$ equals $|H \times K|/|H \cap K|$. Since the number of sets in $\mathcal{P}$ equals $|HK|$ and since $|H \times K| = |H||K|$, the desired formula falls into place. □

A Remark on Groups of Order pq

One cannot predict which fragment of information might unveil further insights about groups. For instance, here is an application of Proposition 2.76 to groups whose order is the product of two distinct primes.

Proposition 2.77. *If p, q are distinct primes with $p < q$, and G is a group of order pq, then G has at most one subgroup of order q, and that subgroup is normal.*

Proof. Suppose that H, K are two subgroups of order q. If these subgroups are distinct, their intersection $H \cap K$ is a proper subgroup of them both, and its order must be a proper divisor of q, by Lagrange. Since q is prime, the order of $H \cap K$ is 1. By Proposition 2.76, the internal product HK has order q^2. But the maximum order that HK can have is that of G, which is pq. We come to the contradiction that $q^2 \leq pq$.

If H is that unique subgroup of order q, all conjugate subgroups of H have order q as well. By the uniqueness of subgroups of order q, all conjugates of H must be H. This means that H is normal. □

In the next chapter we will show that groups of order pq always have subgroups of order p and q, and we will determine all subgroups of order pq up to isomorphism.

EXERCISES

1. If G is a group and H, K are subgroups and H is normal in G, show that HK is a subgroup of G.
2. If H, K are cyclic non-trivial groups and H is infinite, show that $H \times K$ is not cyclic.
3. If G, H are groups with normal subgroups K, L, respectively, show that

$$(G \times H)/(K \times L) \cong G/H \times H/L.$$

   Generalize this fact to the product of more than two groups.
4. If the external product $H \times K$ of two finite groups is cyclic, show that H, K are finite cyclic groups whose orders are coprime.
5. Suppose H, K, L are normal subgroups of a group G such that

$$H \cap (KL) = K \cap (HL) = L \cap (HK) = \langle 1 \rangle$$

   and the set

$$HKL = \{abc : a \in H, b \in K, c \in L\}$$

   is all of G. Show that $G \cong H \times K \times L$.
   Show that the result still holds under the assumptions that $H \cap KL = K \cap L = \langle 1 \rangle$.
6. Show that the group $\mathbb{Q}^*$ of non-zero rational numbers under multiplication is isomorphic to the external product of two non-trivial subgroups.
   Then show that the group $\mathbb{Q}$ of all rational numbers under addition is not isomorphic to an external product of two non-trivial subgroups.
7. If A, B are finite groups of coprime orders and C is a subgroup of the direct product $A \times B$ having the same order as A, prove that $C = A \times \{1\}$.
8.* This exercise is about counting subgroups of certain external products.
   (a) Suppose that A is a finite, simple group and suppose that the order of the automorphism group of A is n. Show that the number of subgroups of $A \times A$ which are isomorphic to A equals $2 + n$.
   Hint. For every automorphism α of A, observe that $\varphi : A \to A \times A$, defined by $a \mapsto (a, \alpha(a))$, is an injective group homomorphism whose image is a subgroup of $A \times A$ isomorphic to A. Show that different automorphisms of A yield different images inside $A \times A$. This will account for n subgroups of $A \times A$, which are isomorphic to A.
   Conversely, if B is a subgroup of $A \times A$ and B is isomorphic to A, take an injective homomorphism $\varphi : A \to A \times A$ whose image is B. Observe that there exist homomorphisms $\alpha, \beta : A \to A$ such that $\varphi(x) = (\alpha(x), \beta(x))$ for all x in A. Use the fact that A is a simple group.
   (b) How many subgroups of $\mathbb{Z}_3 \times \mathbb{Z}_3$ are isomorphic to $\mathbb{Z}_3$?
   (c) How many subgroups of $\mathcal{A}_5 \times \mathcal{A}_5$ are isomorphic to $\mathcal{A}_5$?
9. Suppose that H, N are normal subgroups of a finite group G such that $G = HN$ and the orders of H and N are coprime. Show that if L is any subgroup of G, then $L = (H \cap L)(N \cap L)$.

10. This exercise is about a class of infinite groups known as the residually finite groups. Recall that if $\{H_j\}_{j \in J}$ is a family of non-empty sets indexed by a set J, the Cartesian product $\prod_{j \in J} H_j$ is the set of all functions $f : J \to \bigcup_{j \in J} H_j$ such that $f(j) \in H_j$. Such functions are called ***choice functions***, because each f chooses for each element in J an element $f(j)$ in H_j. The widely accepted ***axiom of choice*** states that such choice functions do exist. Often a choice function is written as a so-called J-tuple $(x_j)_{j \in J}$. For instance, when J is the set of $\mathbb{P}$ positive integers, the product $\prod_{j \in \mathbb{P}} H_j$ is the set of all sequences $(x_1, x_2, x_3, \ldots, x_j, \ldots)$ where $x_j \in H_j$.

An easy verification reveals that if the H_j are groups, the Cartesian product $\prod_{j \in J} H_j$ is also a group using the straightforward point-wise operation

$$(x_j)_{j \in J} (y_j)_{j \in J} = (x_j y_j)_{j \in J}.$$

Now let G be a group, possibly infinite.

(a) Prove that the following statements are equivalent.

- The intersection of all finite-index normal subgroups of G is the trivial group.
- For every a, b in G such that $a \neq b$, there is a finite group H and a homomorphism $\alpha : G \to H$ such that $\alpha(a) \neq \alpha(b)$.
- There exists a family of finite groups $\{H_j\}_{j \in J}$ and an injective homomorphism $\varphi : G \to \prod_{j \in J} H_j$.

A group satisfying any, and hence all, of these three conditions is called ***residually finite***.

(b) Prove that $\mathbb{Z}$, using addition, is residually finite.
Is $\mathbb{Q}$, using addition, residually finite?

11. (a) Suppose G, H are groups in which every element has finite order, and such that for every a in G and every b in H the orders of a and b are coprime. For example, two finite groups whose orders are coprime satisfy this condition. Here though, G, H could be infinite.

Prove that every subgroup of the product group $G \times H$ is of the form $K \times L$ where K is a subgroup of G and L is a subgroup of H.

(b) Suppose G, H are non-trivial groups such that the only subgroups of $G \times H$ are those of the form $K \times L$ where K is a subgroup of G and L is a subgroup of H. Prove that every element of G and of H has finite order and that for every a in G and every b in H, the orders of a and b are coprime. This is the converse of part (a).

2.8 Finite Abelian Groups

Although a comprehensive way to build all finite groups eludes us, this is not so if the groups are abelian. We have enough tools at our disposal to digest this satisfying nugget.

The finite abelian group theorem says that every such group is isomorphic to an external product of finitely many cyclic groups each of them having order equal to a prime power.

The list of prime powers arising in the cyclic factors involved is uniquely determined by the group, except for a possible rearrangement of the list. The proof of this result is intricate.

We have Cauchy's theorem, Proposition 2.61, at our disposal. Namely, if p is a prime that divides the order of an abelian group, then the group has an element of order p.

2.8.1 Groups of Prime Power Order

Definition 2.78. A group whose order is the power of a prime p is called a ***p-group***.

For examples of abelian p-groups take the external products

$$\mathbb{Z}_{p^{n_1}} \times \mathbb{Z}_{p^{n_2}} \times \cdots \times \mathbb{Z}_{p^{n_k}}$$

of cyclic groups $\mathbb{Z}_{p^j}$ using addition. The order of such a group is the prime power $p^{n_1+n_2+\cdots+n_k}$.

For a non-abelian example of a p-group take the subgroup H of $GL_n(\mathbb{Z}_p)$ consisting of the matrices of the form

$$\begin{bmatrix} 1 & x_1 & x_2 & x_3 \\ 0 & 1 & x_4 & x_5 \\ 0 & 0 & 1 & x_6 \\ 0 & 0 & 0 & 1 \end{bmatrix}, \text{ where the } x_j \in \mathbb{Z}_p.$$

Since these matrices are closed under multiplication and inverses, H is a group of order p^6. More examples can be had by taking larger upper triangular matrices of this type.

If G is a p-group with subgroup H, Lagrange forces H to be a p-group. If H is normal, the quotient G/H is once again a p-group.

Switch to Additive Notation

Here we discuss p-groups that are abelian.

Although it makes no difference mathematically, it might be more harmonious with our habits of thought if ***in this section we switch to additive notation*** when working with an abelian group. This will also make it more natural to appreciate the generalization of the upcoming ideas to the setting of modules in Chapter 8. Furthermore, the major result here is that every finite abelian group A is isomorphic to the direct product of the cyclic p-groups $\mathbb{Z}_{p^n}$, and the latter groups use addition. Thus, in this section the identity element of A will be written as 0, and multiple notation replaces power notation. If $x \in A$, the order of x, when finite, is the least positive integer n such that $nx = 0$.

If B, C are subgroups of an abelian group A, using addition, the internal product of B with C consists of all sums $x + y$ where $x \in B$ and $y \in C$. This product is now more aptly called the ***sum*** and denoted as $B + C$ (much like the sum of two subspaces of a vector space). Since A is abelian, the sum $B + C$ is a subgroup of A. If, in addition, $B \cap C$ is the trivial group $\langle 0 \rangle$, Proposition 2.74 ensures that $B + C \cong B \times C$, the external direct product of B and C.

By looking at the orders of the elements of a finite abelian group A we can detect if A is a p-group. This follows from Cauchy's theorem, Proposition 2.61.

Proposition 2.79. *A finite abelian group A is a p-group for some prime p if and only if the order of each of its elements is a power of p.*

Proof. If A is a p-group and $x \in A$, Lagrange forces the order of x to divide the order of A, which is a power of p. Thus the order of x is itself a power of p.

Conversely, suppose A is not a p-group. Thus another prime, say q, divides the order of A. By Cauchy's theorem, Proposition 2.61, A has an element of order q, which is surely not a power of p. □

The preceding observation holds as well for non-abelian groups, but that will have to wait until we have Cauchy's theorem for those groups.

Decomposing Abelian p-Groups into Cyclic Groups

Our first big objective is to show that if p is a prime and A is a finite abelian p-group, then A is isomorphic to a Cartesian product $\mathbb{Z}_{p^{e_1}} \times \mathbb{Z}_{p^{e_2}} \times \cdots \times \mathbb{Z}_{p^{e_k}}$ of cyclic p-groups $\mathbb{Z}_{p^{e_j}}$. With such an isomorphism we can arrange the cyclic p-groups so that the exponents e_j decrease, i.e. $e_1 \geq e_2 \geq \cdots \geq e_k$. The isomorphism would reveal that the first factor $\mathbb{Z}_{p^{e_1}}$ arises from an element whose order p^{e_1} is as big as possible. With this clue, we should focus on an element of A whose order is a maximum.

If $x \in A$, addition notation tells us that the cyclic subgroup generated by x consists of all integer multiples nx of x. For this reason we use the more symbolic $\mathbb{Z}x$, instead of $\langle x \rangle$, to denote the cyclic subgroup generated by x.

Here is a crucial result with an intricate proof.

Proposition 2.80. *Suppose p is a prime and A is a finite abelian p-group. For every element v in A of maximum order, there exists a subgroup B of A such that*

$$\mathbb{Z}v \cap B = \langle 0 \rangle \text{ and } \mathbb{Z}v + B = A.$$

Consequently, $A \cong \mathbb{Z}v \times B$.

Proof. Let B be a subgroup of A such that $\mathbb{Z}v \cap B$ is the trivial group $\langle 0 \rangle$ and B is maximal with respect to this property. The presence of such a B is assured because A is finite. So, any subgroup C of A which properly contains B must be such that $\mathbb{Z}v \cap C$ has a non-zero element in it. We will prove that $A = \mathbb{Z}v + B$.

Seeking a contradiction, suppose $\mathbb{Z}v + B \subsetneq A$, and take an element x from the complement $A \setminus (\mathbb{Z}v + B)$. Let p^e be the order of v, which is also the order of $\mathbb{Z}v$. This p^e is the maximum order that an element of A can have. Since A is a p-group, the order of every element of A is a power of p no higher than p^e. In particular $p^k x = 0 \in \mathbb{Z}v + B$ for some exponent k. Let k be the least positive exponent such that $p^k x \in \mathbb{Z}v + B$ and put $y = p^{k-1}x$. Clearly

$$y \notin \mathbb{Z}v + B \text{ while } py \in \mathbb{Z}v + B.$$

Let

$$py = mv + z \text{ where } m \in \mathbb{Z} \text{ and } z \in B.$$

If $p \nmid m$, then mv has the same order p^e as v, due to Proposition 2.11 interpreted in additive notation. Hence $p^{e-1}mv \neq 0$, and then

$$0 = p^e y = p^{e-1}py = p^{e-1}mv + p^{e-1}z.$$

The resulting equation $p^{e-1}mv = -p^{e-1}z$ shows that the intersection $\mathbb{Z}v \cap B$ is not the trivial group $\langle 0 \rangle$, contrary to the choice of B. Thus $p \mid m$, and we write $m = pk$ for some integer k. From $py = mv + z$ we then obtain $p(y - kv) = z \in B$. Clearly $y - kv \notin \mathbb{Z}v + B$, since kv lies in this subgroup and y does not lie in this subgroup. With $w = y - kv$ we have procured an element such that

$$w \notin \mathbb{Z}v + B \text{ while } pw \in B.$$

The subgroup $\mathbb{Z}w + B$ properly contains B, which then causes the intersection $\mathbb{Z}v \cap (\mathbb{Z}w + B)$ to be non-trivial, due to the maximal way B was chosen. Thus there exist u in B and an integer ℓ such that

$$0 \neq \ell w + u \in \mathbb{Z}v.$$

Now $p \nmid \ell$, for otherwise ℓw would lie in B, forcing the non-zero element $\ell w + u$ to lie in the zero intersection $\mathbb{Z}v \cap B$. Thus p and ℓ are coprime, whereby there exist integers s, t such that $s\ell + tp = 1$. Then

$$w = 1w = (s\ell + tp)w = s\ell w + tpw = s(\ell w + u) - su + tpw \in \mathbb{Z}v + B.$$

The inclusion above holds because $\ell w + u \in \mathbb{Z}v, u \in B$ and $pw \in B$.

We have arrived at a contradiction, namely that w simultaneously lies inside and outside of $\mathbb{Z}v + B$.

Having seen that $A = \mathbb{Z}v + B$, the conclusion that $A \cong \mathbb{Z}v \times B$ follows from Proposition 2.74 interpreted using addition notation. □

The door is now open to see that every finite abelian p-group is isomorphic to a direct product of cyclic groups.

Proposition 2.81. *If p is a prime and A is a finite abelian p-group, then A is isomorphic to an external product of cyclic groups. More explicitly, there exists a descending list of exponents $e_1 \geq e_2 \geq \cdots \geq e_k \geq 1$ such that*

$$A \cong \mathbb{Z}_{p^{e_1}} \times \mathbb{Z}_{p^{e_2}} \times \cdots \times \mathbb{Z}_{p^{e_k}}.$$

Proof. We might as well suppose that $A \neq \langle 0 \rangle$. Take an element v_1 in A of maximum order, say p^{e_1}. Proposition 2.80 yields a subgroup B_1 such that $A \cong \mathbb{Z}v_1 \times B_1$. If $B_1 = \langle 0 \rangle$, this means that A is isomorphic to the cyclic group $\mathbb{Z}v_1$.

If $B_1 \neq \langle 0 \rangle$, take an element v_2 in B_1 of maximum order among the elements of B_1. Say the order of v_2 is p^{e_2}. By the way v_1, v_2 are chosen it is clear that $e_1 \geq e_2$. Proposition 2.80

yields a subgroup B_2 inside B_1 such that $B_1 \cong \mathbb{Z}v_2 \times B_2$. Along with the fact $A \cong \mathbb{Z}v_1 \times B_1$ it is routine to see that $A \cong \mathbb{Z}v_1 \times \mathbb{Z}v_2 \times B_2$. If $B_2 = \langle 0 \rangle$, then $A \cong \mathbb{Z}v_1 \times \mathbb{Z}v_2$, a product of two cyclic groups.

If $B_2 \neq \langle 0 \rangle$, follow the developing pattern, and pick an element v_3 in B_2 of maximum order among the elements of B_2. Say the order of v_3 is p^{e_3}. By the choices of v_1, v_2, v_3 we can see that $e_1 \geq e_2 \geq e_3$. By Proposition 2.80 there is a subgroup B_3 of B_2 such that $B_2 \cong \mathbb{Z}v_3 \times B_3$. Then $A \cong \mathbb{Z}v_1 \times \mathbb{Z}v_2 \times \mathbb{Z}v_3 \times B_3$.

By iterating this argument, we pick up non-zero elements $v_1, v_2, \ldots, v_k$ of descending orders $p^{e_1}, p^{e_2}, \ldots, p^{e_k}$ and a subgroup B_k such that $A \cong \mathbb{Z}v_1 \times \mathbb{Z}v_2 \times \cdots \times \mathbb{Z}v_k \times B_k$. At some point B_k must be the trivial group $\langle 0 \rangle$, simply because A is finite. When that happens the desired isomorphism $A \cong \mathbb{Z}v_1 \times \mathbb{Z}v_2 \times \cdots \times \mathbb{Z}v_k$ is achieved.

Since all cyclic groups of equal order are isomorphic, each $\mathbb{Z}v_j \cong \mathbb{Z}_{p^{e_j}}$, and thus A is isomorphic to the direct product of these cyclic groups. □

A Counting Formula for Abelian p-Groups

If A is a finite abelian p-group, we just learned that there exists a descending list of positive integers $e_1 \geq e_2 \geq \cdots \geq e_k$ such that $A \cong \mathbb{Z}_{p^{e_1}} \times \mathbb{Z}_{p^{e_2}} \times \cdots \times \mathbb{Z}_{p^{e_k}}$. Could the same kind of isomorphism for A be achieved using a different descending list? Our next goal is to see that this descending list of e_j, or equivalently the list of powers p^{e_j}, is uniquely determined by A. Some fussy counting arguments will be used.

For every abelian p-group A and integer $m \geq 0$, the set

$$A_m = \{x \in A : p^m x = 0\}$$

is a subgroup of A, by inspection. If $A \cong \mathbb{Z}_{p^{e_1}} \times \mathbb{Z}_{p^{e_2}} \times \cdots \times \mathbb{Z}_{p^{e_k}}$, we wish to count the orders $|A_m|$ of this ascending chain of subgroups of A, in terms of the e_j at hand.

We first do this when the p-group is cyclic.

Proposition 2.82. *If p is a prime, and A is a cyclic group of order p^e for some exponent e and m is a non-negative integer, then the order of the subgroup $A_m = \{x \in A : p^m x = 0\}$ is $p^{\min(e,m)}$.*

Proof. We will show that the order of A_m is $\min(p^e, p^m)$, which is clearly the same as $p^{\min(e,m)}$.

Consider the case where $0 \leq m \leq e$. We want the order of A_m to be p^m. Since A is cyclic so is A_m, by Proposition 2.40. The order of A_m equals the order of any generator y of A_m. Since $p^m y = 0$, the order of such y is at most p^m. But in A there does exist an element y of order p^m. Just take a generator z of A and put $y = z^{e-m}$. This y lies in A_m and generates A_m because its order is maximal in the cyclic group A_m.

In case $m > e$, then all x in A satisfy $p^e x = 0$ by Lagrange, and thereby $p^m x = 0$ too. Now $A_m = A$ whose order is p^e. □

Next we count the order of A_m when an isomorphism of A with a product of cyclic groups is given.

Proposition 2.83. *Suppose A is a group such that*

$$A \cong \mathbb{Z}_{p^{e_1}} \times \mathbb{Z}_{p^{e_2}} \times \cdots \times \mathbb{Z}_{p^{e_k}}$$

for some prime p and some exponents $e_1, e_2, \ldots, e_k$. If m is a non-negative integer and A_m is the subgroup of all x such that $p^m x = 0$, then the order of A_m is

$$p^{\sum_{j=1}^{k} \min(e_j, m)}.$$

Proof. There is no harm is supposing that A actually is the product of the $\mathbb{Z}_{p^{e_j}}$. The desired formula is the same as the product

$$p^{\min(e_1,m)} \cdot p^{\min(e_2,m)} \cdots p^{\min(e_k,m)}.$$

If $x = (x_1, x_2, \ldots, x_k) \in A$, the equation $p^m x = 0$ holds if and only if all $p^m x_j = 0$. By Proposition 2.82 the number of such x_j in $\mathbb{Z}_{p^{e_j}}$ equals $p^{\min(e_j,m)}$. Then the number of possible k-tuples x must be the product of these $p^{\min(e_j,m)}$. □

Proposition 2.83 counts the order of each subgroup A_m in terms of a given isomorphism of A with a product of cyclic p-groups $\mathbb{Z}_{p^{e_j}}$. However, the order $|A_m|$ of each A_m is intrinsic to A, and does not depend on the cyclic p-groups $\mathbb{Z}_{p^{e_j}}$ used in the isomorphism with A. This means that the sequence of integers

$$\sum_{j=1}^{k} \min(e_j, m) \text{ for } m = 0, 1, 2, 3, \ldots$$

is uniquely determined by A, despite its apparent dependence on those e_j.

Here is the above observation stated more formally.

Proposition 2.84. *Let A be an abelian p-group for some prime p, and for each integer $m \geq 0$ let $A_m = \{x \in A : p^m x = 0\}$. If A is such that*

$$A \cong \mathbb{Z}_{p^{e_1}} \times \mathbb{Z}_{p^{e_2}} \times \cdots \times \mathbb{Z}_{p^{e_k}} \cong \mathbb{Z}_{p^{d_1}} \times \mathbb{Z}_{p^{d_2}} \times \cdots \times \mathbb{Z}_{p^{d_\ell}}$$

using positive exponents $e_1, e_2, \ldots, e_k$ and $d_1, d_2, \ldots, d_\ell$, then

$$\sum_{j=1}^{k} \min(e_j, m) = \log_p |A_m| = \sum_{j=1}^{\ell} \min(d_j, m),$$

for every $m \geq 0$.

Proof. Apply Proposition 2.83 first to the decomposition of A using the e_j and then reapply it using the decomposition using the d_j. □

The end goal is to see that not only is this sequence of sums $\sum_{j=1}^{k} \min(e_j, m)$ independent of the chosen isomorphism of A with a product of cyclic groups $\mathbb{Z}_{p^{e_j}}$, but also that the actual e_j are unambiguously determined by A.

The Histogram for a List of Integers

If $e_1, e_2, \ldots, e_k$ is a list of positive integers and m is a positive integer, let

$$h_m = \text{ the number of indices } j \text{ for which } e_j = m.$$

Let us call the resulting sequence of non-negative integers $h_1, h_2, \ldots, h_m, \ldots$ the ***histogram*** for the list of e_j. The sequence h_m eventually becomes 0. After a bit of reflection it should be evident that if the list of e_j is given so that $e_1 \geq e_2 \geq \cdots \geq e_k$, then the histogram sequence h_m completely determines the descending list of e_j.

It turns out that the histogram h_m is determined by the sequence of sums $\sum_{j=1}^{k} \min(e_j, m)$.

Proposition 2.85. *Suppose that $e_1, e_2, \ldots, e_k$ is a list of positive integers. For each integer $m \geq 0$ let $a_m = \sum_{j=1}^{k} \min(e_j, m)$, and for $m \geq 1$ let h_m be the histogram for the e_j. Then*

$$h_m = 2a_m - a_{m+1} - a_{m-1}.$$

Proof. For each subset J of the set of indices $\{1, 2, 3, \ldots, k\}$ let $|J|$ denote the number of elements in J. Examine the difference $a_{m+1} - a_m$ where $m \geq 0$. Well,

$$\begin{aligned} a_m &= \sum_{\{j: e_j > m\}} \min(e_j, m) + \sum_{\{j: e_j \leq m\}} \min(e_j, m) \\ &= \sum_{\{j: e_j > m\}} m + \sum_{\{j: e_j \leq m\}} e_j \\ &= |\{j : e_j > m\}| m + \sum_{\{j: e_j \leq m\}} e_j. \end{aligned}$$

In a similar vein

$$\begin{aligned} a_{m+1} &= \sum_{\{j: e_j \geq m+1\}} \min(e_j, m) + \sum_{\{j: e_j < m+1\}} \min(e_j, m) \\ &= \sum_{\{j: e_j \geq m+1\}} (m+1) + \sum_{\{j: e_j < m+1\}} e_j \\ &= |\{j : e_j > m\}| (m+1) + \sum_{\{j: e_j \leq m\}} e_j. \end{aligned}$$

Subtract to see that

$$a_{m+1} - a_m = |\{j : e_j > m\}|.$$

Replacing m by $m - 1$ we obtain

$$a_m - a_{m-1} = |\{j : e_j > m - 1\}|.$$

And then for $m \geq 1$ we get

$$2a_m - a_{m+1} - a_{m-1} = (a_m - a_{m-1}) - (a_{m+1} - a_m) = |\{j : e_j > m - 1\}| - |\{j : e_j > m\}|.$$

The final conclusion follows because

$$h_m = |\{j : e_j > m - 1\}| - |\{j : e_j > m\}|,$$

by inspection. □

Uniqueness of the Cyclic Decomposition of Finite Abelian p-Groups

The next significant result can be viewed as the other half of Proposition 2.81.

Proposition 2.86. *If p is a prime and A is a group such that*

$$A \cong \mathbb{Z}_{p^{e_1}} \times \mathbb{Z}_{p^{e_2}} \times \cdots \times \mathbb{Z}_{p^{e_k}} \cong \mathbb{Z}_{p^{d_1}} \times \mathbb{Z}_{p^{d_2}} \times \cdots \times \mathbb{Z}_{p^{d_\ell}},$$

where $e_1 \geq e_2 \geq \cdots \geq e_k \geq 1$ and $d_1 \geq d_2 \geq \cdots \geq d_\ell \geq 1$, then $k = \ell$ and all $e_j = d_j$.

Proof. For $m \geq 0$ let A_m be the subgroup consisting of all x in A such that $p^m x = 0$. According to Proposition 2.84,

$$\sum_{j=1}^{k} \min(e_j, m) = \log_p |A_m| = \sum_{j=1}^{\ell} \min(d_j, m).$$

For $m \geq 1$ let h_m be the histogram sequence for the descending e_j, and let k_m be the histogram sequence for the descending d_j. To get $k = \ell$ and all $e_j = d_j$ it suffices to see that $h_m = k_m$ for all m. Now Proposition 2.85, taken in light of the preceding equalities, gives

$$h_m = 2\log_p |A_m| - \log_p |A_{m+1}| - \log_p |A_{m-1}| = k_m,$$

as desired. □

The Invariant of a Finite Abelian p-Group

If p is a prime and A is a finite abelian p-group, Propositions 2.81 and 2.86 yield a unique integer k and a unique descending list of positive exponents $e_1 \geq e_2 \geq \cdots \geq e_k$ such that

$$A \cong \mathbb{Z}_{p^{e_1}} \times \mathbb{Z}_{p^{e_2}} \times \cdots \times \mathbb{Z}_{p^{e_k}}.$$

For brevity we can write the e_j as a k-tuple $[e_1, e_2, \ldots, e_k]$, where the box brackets are meant to indicate that the e_j decrease. We shall refer to such a decreasing list as the ***invariant*** of A. The prime powers p^{e_j} are traditionally known as the ***elementary divisors*** of A. If two finite abelian p-groups A, B have the same invariant, then $A \cong B$ obviously. From Proposition 2.86 it follows that if two finite abelian p-groups are isomorphic, then they have the same invariant.

To repeat briefly, every finite abelian p-group A determines a descending list of positive integers called its invariant. The invariant specifies the product of cyclic p-groups to which A is isomorphic. Two finite abelian p-groups are isomorphic if and only if they have the same invariant, which is why the term "invariant" is appropriate. The set of all possible invariants tabulates all possible finite abelian p-groups up to isomorphism. A satisfying result.

The Number of Abelian p-Groups of a Given Order

If p is a prime and n is a positive exponent, how many non-isomorphic abelian groups of order p^n are there? Such a group A must be isomorphic to $\mathbb{Z}_{p^{e_1}} \times \mathbb{Z}_{p^{e_2}} \times \cdots \times \mathbb{Z}_{p^{e_k}}$, where $[e_1, e_2, \ldots, e_k]$ is the invariant of A. If we insist that the order of A be p^n, then

$$p^n = p^{e_1} \cdot p^{e_2} \cdots p^{e_k} = p^{e_1+e_2+\cdots+e_k}$$

and thus $n = e_1 + e_2 + \cdots + e_k$. The question comes down to asking how many invariants are possible with the preceding constraint.

In discussing the orbit structure of a permutation in Section 2.2, we encountered the notion of a ***partition of a positive integer*** n. This is a decreasing list $[e_1, e_2, \ldots, e_k]$ of positive integers that add up to n.

The answer to our question comes in the language of partitions.

Proposition 2.87. *If p is a prime, then the number of non-isomorphic abelian groups of order p^n equals the number of partitions of n.*

For example, the integer 4 has the following five partitions

$$[4], [3, 1], [2, 2], [2, 1, 1], [1, 1, 1, 1].$$

Hence, regardless of the prime p, there are precisely five non-isomorphic abelian groups of order p^4. Here are those five groups

$$\mathbb{Z}_{p^4},\ \mathbb{Z}_{p^3} \times \mathbb{Z}_p,\ \mathbb{Z}_{p^2} \times \mathbb{Z}_{p^2},\ \mathbb{Z}_{p^2} \times \mathbb{Z}_p \times \mathbb{Z}_p \text{ and } \mathbb{Z}_p \times \mathbb{Z}_p \times \mathbb{Z}_p \times \mathbb{Z}_p.$$

For another example, the integer 6 has the following 11 partitions

$$[6],\ [5,1],\ [4,2],\ [4,1,1],\ [3,3],\ [3,2,1],\ [3,1,1,1],$$
$$[2,2,2],\ [2,2,1,1],\ [2,1,1,1,1],\ [1,1,1,1,1,1].$$

Thus there are precisely 11 non-isomorphic abelian groups of order p^6. These groups could be readily displayed. The prime p does not affect the count.

Tips on Finding the Invariant of a Finite Abelian p-Group

If a finite abelian group A of order p^n is put before us, a few observations may help find its invariant, particularly when the order of the group is moderate.

- If the order p^n of A is known, then the sought for invariant must be a partition of n, which brings us down to a finite search for the invariant.
- If an element in A of a maximum order p^e is detected, then the invariant $[e_1, e_2, \ldots, e_k]$ of A must be such that $e_1 = e$. That is because $A \cong \mathbb{Z}_{p^{e_1}} \times \mathbb{Z}_{p^{e_2}} \times \cdots \times \mathbb{Z}_{p^{e_k}}$, and in the latter group the highest order which an element attains is p^{e_1}.

- If the order of the subgroup A_1, as given in Proposition 2.83, is p^k, then k is the number of entries in the invariant. This coincides with the number of cyclic $\mathbb{Z}_{p^{e_j}}$ in the isomorphism with A.

For example, take the group of units $A = \mathbb{Z}_{30}^{\star}$ under multiplication. The order of this group is $\phi(30) = 8 = 2^3$, which makes A be a 2-group. Its only possible invariants are the partitions of 3. These are $[3], [2,1], [1,1,1]$. The elements of this group are the residues $\pm 1, \pm 7, \pm 11, \pm 13$ modulo 30. Looking at the residue 7, its order is one of 1,2,4 or 8. Now $7^2 = 49 \equiv -11$ mod 30. Then $7^4 \equiv 121 \equiv 1$ mod 30. The order of 7 is $4 = 2^2$. Likewise the order of -7 is 4. Since $(\pm 11)^2 = 121 \equiv 1$ mod 30 these two elements have order 2. The order of ± 13 is also 4 by checking. And trivially the orders of ± 1 are 1 and 2, respectively. Since the highest order that an element of our group can have is 2^2, we can be sure that $[2,1]$ is the invariant of our group. In other words $A \cong \mathbb{Z}_{2^2} \times \mathbb{Z}_2 = \mathbb{Z}_4 \times \mathbb{Z}_2$. Keep in mind that the given group on the left uses multiplication.

2.8.2 The Primary Decomposition

We turn our attention to arbitrary finite abelian groups, not just to those whose order is a prime power. We stick with addition notation. It turns out that most of the heavy lifting has already been done in obtaining the invariant of every finite p-group.

The Primary Components of a Finite Abelian Group

Definition 2.88. If A is an abelian group and p is a prime, let

$$A(p) = \{x \in A : p^k x = 0 \text{ for some exponent } k\}.$$

Such $A(p)$ will be called the ***p*-*primary component*** of A.

The k can vary with x. By item 3 of Proposition 2.5 the set $A(p)$ is the set of elements whose order is a power of p. Each $A(p)$ is a subgroup of A. Indeed, $0 \in A(p)$ obviously. Now suppose $x, y \in A(p)$. Thus $p^k x = 0 = p^\ell y$ for some exponents k, ℓ, and then $p^{\max(k,\ell)}(x \pm y) = 0$. This puts $x \pm y$ inside $A(p)$.

(As an aside we might note that when A is not abelian, the set $A(p)$, suitably defined using multiplicative notation, need not be a subgroup of A. For instance, in the group S_3 of permutations on 3 letters the set $S_3(2) = \{1, (1,2), (1,3), (2,3)\}$, and this cannot be a subgroup of S_3 because 4 does not divide 6.)

Furthermore if $A(p)$ is finite, Proposition 2.79 ensures that $|A(p)| = p^d$ for some $d \geq 0$. In other words, the p-primary component $A(p)$ is a p-group.

When A is finite, the orders of the primary components $A(p)$ are readily obtainable from the order of A.

Proposition 2.89. *For each finite abelian group A and each prime p, the order $|A(p)|$ of the subgroup $A(p)$ is the highest power of p that divides $|A|$.*

Proof. Let $|A| = p^e k$ where $e \geq 0$ and $p \nmid k$, and let $|A(p)| = p^d$ for some exponent d. By Lagrange p^d divides $p^e k$ and since $p \nmid k$, it must be that $d \leq e$. The goal is to see that $d = e$.

Well, Lagrange says that the order of the quotient group $A/A(p)$ is $p^e k / p^d = p^{e-d}k$. Now if $d < e$, Cauchy's theorem, Proposition 2.61, would provide an element $x + A(p)$ of order p in the quotient group. This would mean that the coset representative $x \notin A(p)$, while $px \in A(p)$. But then $p^{j+1}x = p^j(px) = 0$ for some exponent j, forcing the contradiction that $x \in A(p)$. □

It becomes evident from Proposition 2.89 that the subgroup $A(p)$ of a finite abelian group A is non-trivial if and only if p divides $|A|$.

For example, take the cyclic group $\mathbb{Z}_{45}$ of order 45, using addition, and take $p = 3$. The highest power of 3 that divides 45 is 9. Thus the 3-primary component $\mathbb{Z}_{45}(3)$ is a subgroup containing 9 of these 45 residues. With a little inspection here are the 9 elements of $\mathbb{Z}_{45}(3)$:

$$0, 5, 10, 15, 20, 25, 30, 35, 40.$$

Just to check: $3^2 \cdot 5 \equiv 0 \bmod 45$, and the same goes for the other multiples of 5.

Primary Decomposition of Finite Abelian Groups

The next result highlights the significance of the primary components. If y is an element in a finite abelian group, we shall be using the notation $\text{ord}(y)$ to denote its order.

Proposition 2.90 (Primary decomposition). *Let A be a finite abelian group of order n, and let $p_1^{e_1} p_2^{e_2} \cdots p_k^{e_k}$ be the unique factorization of n into powers of distinct primes p_j. If $A(p_j)$ is the p_j-primary component of A, then the mapping*

$$\varphi : A(p_1) \times A(p_2) \times \cdots \times A(p_k) \to A \text{ given by } (x_1, x_2, \ldots, x_k) \mapsto x_1 + x_2 + \cdots + x_k$$

is an isomorphism.

Proof. That φ is a homomorphism can be seen from a routine verification, or alternatively by appealing to Proposition 2.73 interpreted using addition. From Proposition 2.89 it follows that the order of the product group $A(p_1) \times A(p_2) \times \cdots \times A(p_k)$ is also n. Consequently, for φ to be an isomorphism it suffices to see that φ is injective. So we need to show that $\ker \varphi$ is trivial.

To that end let $(x_1, x_2, \ldots, x_k) \in \ker \varphi$. That is $x_1 + x_2 + \cdots + x_k = 0$. This equation can be rewritten as

$$\sum_{j=1}^{k} (0, \ldots, 0, x_j, 0, \ldots, 0) = (0, 0, \ldots, 0).$$

The order of each x_j, which is in the group $A(p_j)$, is a power of p_j. The order of its corresponding element $(0, \ldots, 0, x_j, 0, \ldots, 0)$ in the Cartesian product is the same power of p_j. As j varies, these powers of p_j are pairwise coprime. Then repeated use of Proposition 2.8, properly interpreted using addition, yields

$$1 = \operatorname{ord}(0, 0, \ldots, 0) = \prod_{j=1}^{k} \operatorname{ord}(0, \ldots, 0, x_j, 0, \ldots, 0).$$

Therefore every $\operatorname{ord}(0, \ldots, 0, x_j, 0, \ldots, 0) = 1$, and then every $x_j = 0$. Thus $\ker \varphi$ is trivial. □

It might be informative to see how the inverse isomorphism of φ in Proposition 2.90 can be calculated. For each prime divisor p_j of n put $d_j = n/p_j^{e_j}$. The greatest common divisor of all d_j combined is 1. Because of this, an exercise using greatest common divisors shows there exist integers t_j such that

$$d_1 t_1 + d_2 t_2 + \cdots + d_k t_k = 1.$$

Then for each x in A we have

$$x = (d_1 t_1 + d_2 t_2 + \cdots + d_k t_k)x = d_1 t_1 x + d_2 t_2 x + \cdots + d_k t_k x.$$

Lagrange gives

$$p_j^{e_j}(d_j t_j x) = t_j(d_j p_j^{e_j} x) = t_j(nx) = t_j 0 = 0,$$

for every p_j. Consequently $d_j t_j x \in A(p_j)$. The pre-image of x in A under the isomorphism φ is the k-tuple $(d_1 t_1 x, d_2 t_2 x, \ldots, d_k t_k x)$.

To illustrate the primary decomposition take $A = \mathbb{Z}_{45}$ of order $3^2 \cdot 5$. Here $A(3) = \{0, 5, 10, 15, 20, 25, 30, 35, 40\}$ and $A(5) = \{0, 9, 18, 27, 36\}$. These are cyclic groups of order 9 and 5, respectively. The isomorphism $A(3) \times A(5) \cong A$, given by $(x, y) \mapsto x + y$, says that every element in $\mathbb{Z}_{45}$ is the sum of a unique element of $A(9)$ and a unique element of $A(5)$. For instance, $1 \equiv 36 + 10 \bmod 45$ and $2 \equiv 27 + 20 \bmod 45$. We also notice that the cyclic group $\mathbb{Z}_{45}$ of order 45 is isomorphic to the product of cyclic groups of orders 9 and 5. Of course we could see that already from Proposition 2.66.

Uniqueness of the Primary Decomposition

We just saw that every finite abelian group A is isomorphic to the product of its p-primary components. Suppose now that A were isomorphic to another product of p-groups. It might be expected that these alternative factors are isomorphic to the p-primary components of A.

Proposition 2.91. *Suppose $B_1, B_2, \ldots, B_k$ are abelian groups of distinct prime power orders $p_1^{e_1}, p_2^{e_2}, \ldots, p_k^{e_k}$, respectively. If the group $A \cong B_1 \times B_2 \times \cdots \times B_k$, then the order of A equals $p_1^{e_1} p_2^{e_2} \cdots p_k^{e_k}$ and each $B_j \cong A(p_j)$.*

Proof. That the order of A equals the product of the $p_j^{e_j}$ is plain to see.

Let $\varphi : A \to B_1 \times \cdots \times B_k$ be the given isomorphism. For each j from 1 to k put

$$C_j = \{(x_1, \ldots, x_k) \in B_1 \times \cdots \times B_k : x_i = 0 \text{ when } i \neq j\}.$$

This C_j is the obvious isomorphic copy of B_j sitting inside the product. It will thereby suffice to see that $\varphi(A(p_j)) = C_j$. Since $A(p_j)$ and C_j each have $p_j^{e_j}$ elements and φ is injective, it merely suffices to check that $\varphi(A(p_j)) \subseteq C_j$.

An element x from the product $B_1 \times \cdots \times B_k$ lies in C_j if and only if $p^{e_j}x = (0, \ldots, 0)$. Indeed, the order of C_j is p^{e_j}, and thus by Lagrange $p^{e_j}x = (0, \ldots, 0)$ if $x \in C_j$. Conversely, suppose $x = (x_1, \ldots, x_k)$ and that $p^{e_j}x = (0, \ldots, 0)$. If $i \neq j$, we have that both $p_j^{e_j}x_i = 0$ and $p_i^{e_i}x_i = 0$. By Proposition 2.5, the order of x_i divides both $p_j^{e_j}$ and $p_i^{e_i}$. Since p_j and p_i are distinct primes, the order of x_i must be 1. That is, $x_i = 0$, and thereby $x \in C_j$.

Now if $y \in A(p_j)$, we have that $p_j^{e_j}y = 0$ and then

$$0 = \varphi(0) = \varphi(p_j^{e_j}y) = p_j^{e_j}\varphi(y).$$

By the preceding discussion $\varphi(y) \in C_j$, as desired. □

The only way to have a Cartesian product decomposition of A into p-groups, involving distinct primes p, is by using isomorphic copies of the p-primary components of A as in Proposition 2.90.

2.8.3 Putting the Bits Together

According to Proposition 2.90 every finite abelian group A is isomorphic to the product of the p-primary components $A(p)$, where p runs over the prime divisors of the order of A. Each of these p-primary components are themselves p-groups whose order is the highest power of p that divides the order of A, and so Proposition 2.81 applies to them. Taken together these form the first half of the ***structure theorem for finite abelian groups***.

Proposition 2.92 (Structure theorem for finite abelian groups). *Suppose A is a finite abelian group and that $p_1^{e_1} \cdot p_2^{e_2} \cdots p_k^{e_k}$ is the factorization of its order with distinct primes p_j and positive exponents e_j. For each prime power $p_j^{e_j}$, there is a partition $[e_{j1}, e_{j2}, \ldots, e_{j\ell_j}]$ of e_j such that*

$$\begin{aligned} A \cong & \left(\mathbb{Z}_{p_1^{e_{11}}} \times \mathbb{Z}_{p_1^{e_{12}}} \times \cdots \times \mathbb{Z}_{p_1^{e_{1\ell_1}}}\right) \\ & \times \left(\mathbb{Z}_{p_2^{e_{21}}} \times \mathbb{Z}_{p_2^{e_{22}}} \times \cdots \times \mathbb{Z}_{p_2^{e_{2\ell_2}}}\right) \times \cdots \times \left(\mathbb{Z}_{p_k^{e_{k1}}} \times \mathbb{Z}_{p_k^{e_{k2}}} \times \cdots \times \mathbb{Z}_{p_k^{e_{k\ell_k}}}\right). \end{aligned}$$

We refer to the isomorphism of Proposition 2.92 as a ***complete decomposition of A into cyclic p-groups***.

The second half of the structure theorem says that any change in the partitions $[e_{i1}, e_{i2}, \ldots, e_{i\ell_i}]$ which give a complete decomposition will lead to a group that is not isomorphic to A. This will follow from Propositions 2.86 and 2.91.

Proposition 2.93. *Every finite abelian group has precisely one complete decomposition into cyclic p-groups, except for a possible rearrangement of the primes that divide the order of the group.*

Proof. The proof entails a bit of bookkeeping.

Suppose that two complete decompositions of an abelian group A are given as shown:

$$A \cong \left(\mathbb{Z}_{p_1^{e_{11}}} \times \mathbb{Z}_{p_1^{e_{12}}} \times \cdots \times \mathbb{Z}_{p_1^{e_{1\ell_1}}}\right)$$
$$\times \left(\mathbb{Z}_{p_2^{e_{21}}} \times \mathbb{Z}_{p_2^{e_{22}}} \times \cdots \times \mathbb{Z}_{p_2^{e_{2\ell_2}}}\right) \times \cdots \times \left(\mathbb{Z}_{p_k^{e_{k1}}} \times \mathbb{Z}_{p_k^{e_{k2}}} \times \cdots \times \mathbb{Z}_{p_k^{e_{k\ell_k}}}\right)$$

and

$$A \cong \left(\mathbb{Z}_{q_1^{d_{11}}} \times \mathbb{Z}_{q_1^{d_{12}}} \times \cdots \times \mathbb{Z}_{q_1^{d_{1m_1}}}\right)$$
$$\times \left(\mathbb{Z}_{q_2^{d_{21}}} \times \mathbb{Z}_{q_2^{d_{22}}} \times \cdots \times \mathbb{Z}_{q_2^{d_{2m_2}}}\right) \times \cdots \times \left(\mathbb{Z}_{q_h^{d_{h1}}} \times \mathbb{Z}_{q_h^{d_{h2}}} \times \cdots \times \mathbb{Z}_{q_h^{d_{hm_h}}}\right).$$

From the first decomposition the order of A works out to be the product

$$p_1^{e_{11}+e_{12}+\cdots+e_{1\ell_1}} p_2^{e_{21}+e_{22}+\cdots+e_{2\ell_2}} \cdots p_k^{e_{k1}+e_{k2}+\cdots+e_{k\ell_k}}.$$

The alternative decomposition shows that the order of A equals

$$q_1^{d_{11}+d_{12}+\cdots+d_{1m_1}} q_2^{d_{21}+d_{22}+\cdots+d_{2m_2}} \cdots q_h^{d_{h1}+d_{h2}+\cdots+d_{hm_h}}.$$

By unique factorization, $h = k$ and $q_1 = p_1, q_2 = p_2, \ldots, q_k = p_k$, after a suitable rearrangement of the primes involved.

In the first complete decomposition of A the order of the ith factor

$$\mathbb{Z}_{p_i^{e_{i1}}} \times \mathbb{Z}_{p_i^{e_{i2}}} \times \cdots \times \mathbb{Z}_{p_i^{e_{i\ell_i}}}$$

is $p_i^{e_{i1}+e_{i2}+\cdots+e_{i\ell_i}}$. Then Proposition 2.91 reveals that the above ith factor is isomorphic to the p_i-primary component $A(p_i)$ of A. Likewise, keeping in mind that $q_i = p_i$, the ith factor

$$\mathbb{Z}_{p_i^{d_{i1}}} \times \mathbb{Z}_{p_i^{d_{i2}}} \times \cdots \times \mathbb{Z}_{p_i^{d_{im_i}}}$$

in the second complete decomposition of A is isomorphic to the same $A(p_i)$. Thus

$$\mathbb{Z}_{p_i^{e_{i1}}} \times \mathbb{Z}_{p_i^{e_{i2}}} \times \cdots \times \mathbb{Z}_{p_i^{e_{i\ell_i}}} \cong \mathbb{Z}_{p_i^{d_{i1}}} \times \mathbb{Z}_{p_i^{d_{i2}}} \times \cdots \times \mathbb{Z}_{p_i^{d_{im_i}}}.$$

Now Proposition 2.86 reveals that $\ell_i = m_i$ and that each $e_{ij} = d_{ij}$. Therefore the two complete decompositions of A are identical. □

The following should now be clear.

Proposition 2.94. *Two finite abelian groups are isomorphic if and only if they have the same complete decompositions into cyclic p-groups.*

The preceding result allows us to count the number of non-isomorphic abelian groups having a fixed order n.

Proposition 2.95. *If the unique factorization of an integer n into distinct primes is given by* $n = p_1^{e_1} p_2^{e_2} \cdots p_k^{e_k}$ *and* $\pi(e_j)$ *stands for the number of partitions of the positive exponent* e_j*, then the number of non-isomorphic abelian groups of order n equals the product*

$$\pi(e_1)\pi(e_2)\cdots\pi(e_k).$$

Proof. Count the complete decompositions into cyclic p-groups which produce groups having order n. Looking at such a complete decomposition in Proposition 2.92, the requirement that its order be n boils down to the equation

$$p_1^{e_1} p_2^{e_2} \cdots p_k^{e_k} = p_1^{e_{11}+e_{12}+\cdots+e_{1\ell_1}} \cdot p_2^{e_{21}+e_{22}+\cdots+e_{2\ell_2}} \cdots p_k^{e_{k1}+e_{k2}+\cdots+e_{k\ell_k}}.$$

This means that

$$e_j = e_{j1} + e_{j2} + \cdots + e_{j\ell_j}$$

for every j from 1 to k. In other words, a complete decomposition into cyclic p-groups gives a group of order n if and only if each $[e_{j1}, e_{j2}, \ldots, e_{j\ell_j}]$ is a partition of e_j. Thus each abelian group of order n corresponds to a list of partitions

$$[e_{11}, e_{12}, \ldots, e_{1\ell_1}], [e_{21}, e_{22}, \ldots, e_{2\ell_2}], \ldots, [e_{k1}, e_{k2}, \ldots, e_{k\ell_k}],$$

of $e_1, e_2, \ldots, e_k$, respectively. Clearly the number of such lists is the product of the $\pi(e_j)$. □

An Example

Let us count and build all non-isomorphic abelian groups of order 3969. Well, $3969 = 7^2 \cdot 3^4$. The exponent 2 has 2 partitions and the exponent 4 has 5 partitions. So there are precisely 10 non-isomorphic abelian groups of order 3969.

We know that such a group A is isomorphic to a product of groups $A(7) \times A(3)$ where the order of $A(7)$ is 7^2 and the order of $A(3)$ is 3^4. The possibilities for $A(7)$, up to isomorphism, are the following two groups

$$\mathbb{Z}_{49},\ \mathbb{Z}_7 \times \mathbb{Z}_7.$$

The possibilities for $A(3)$ are the following five groups

$$\mathbb{Z}_{81},\ \mathbb{Z}_{27} \times \mathbb{Z}_3,\ \mathbb{Z}_9 \times \mathbb{Z}_9,\ \mathbb{Z}_9 \times \mathbb{Z}_3 \times \mathbb{Z}_3, \mathbb{Z}_3 \times \mathbb{Z}_3 \times \mathbb{Z}_3 \times \mathbb{Z}_3.$$

The ten possible groups come from taking the products of each of the 7-groups with each of the 3-groups shown above. We might note that the primes 7 and 3 have nothing to do with the above counting. For any distinct primes p, q, the above argument reveals that the number of non-isomorphic abelian groups of order $p^2 q^4$ is 10.

EXERCISES

1. If A is an abelian p-group for some prime p, and q is an integer not divisible by p, show that for every a in G the equation $qx = a$ has a solution x in G.

2. How many non-isomorphic abelian groups of order $27, 35, 100, 288, 64827$, respectively, are there?
3. If p, q are distinct primes, count the non-isomorphic abelian groups of order p^5q^7.
4. If n is a product of distinct primes, show that every abelian group of order n is cyclic.
5. Show how to get all abelian groups of order $3^3 \cdot 5^4 \cdot 7$.
6. How many non-isomorphic abelian groups of order 16 have exactly 2 elements of order 2? How many with exactly 3 elements of order 2? How many with 4 such elements?
7. If A is an abelian p-group of order p^n, determined by a partition $[e_1, e_2, \ldots, e_k]$ of n, what is the maximum order attained by the elements of A?
8. If A is a non-trivial, finite, abelian p-group for some prime p, show that A is cyclic if and only if A has exactly one subgroup of order p.

   Show that the quaternion group discussed in Section 2.4.3 is a non-cyclic group with exactly one subgroup of order 2.
9. If A is a finite abelian group of order n and m divides n, use the structure theorem for finite abelian groups to prove that A contains a subgroup of order m.
10. Let A be a finite abelian group, and let x be an element of A such that $\text{ord}(x) \geq \text{ord}(y)$ for every y in A. Prove that $\text{ord}(x)$ is divisible by $\text{ord}(y)$ for every y in A.
11. Let $\mathbb{C}^\star$ be the group of non-zero complex numbers under multiplication. The dual of a group G is the set of all homomorphisms $\varphi : G \to \mathbb{C}^\star$. Let $\hat{G}$ be the set of such homomorphisms. If G is abelian using addition, an element φ of $\hat{G}$ is a mapping such that $\varphi(x + y) = \varphi(x)\varphi(y)$ for every x, y in G. For instance, if G is the group of real numbers under addition, the classical exponential function given by $x \mapsto e^x$ is in $\hat{G}$.

    If φ, ψ are in $\hat{G}$, define the product mapping $\varphi\psi : G \to \mathbb{C}^\star$ by the rule

$$\varphi\psi(x) = \varphi(x)\psi(x) \text{ for every } x \text{ in } G.$$

    A direct verification shows that under this multiplication $\hat{G}$ is an abelian group, whose identity element is the function which maps all of G to 1 in $\mathbb{C}^\star$. The group $\hat{G}$ is known as the *dual group* of G.

    This exercise will reveal that if G is a finite abelian group, then G is isomorphic to its dual $\hat{G}$. From here on assume that G is a finite abelian group using addition, and that $\hat{G}$ is its dual.

    (a) Show that $|\varphi(x)| = 1$ for every φ in $\hat{G}$ and every x in G. In other words, the range of every homomorphism in $\hat{G}$ is the circle group $\mathcal{T}$. (This holds even if G is not abelian.)

    (b) If G is cyclic, verify that $\hat{G}$ has the same order as G, and deduce from this that $\hat{G} \cong G$.

       Hint. If G has order n and generator a and $\varphi \in \hat{G}$, show that $\varphi(a)$ is an nth root of unity. This limits the possibilities for φ.

    (c) If $a \in G$ and a is not the zero element, show that $\varphi(a) \neq 1$ for some φ in $\hat{G}$. From this, show that if a, b are distinct elements of G, then $\varphi(a)$, $\varphi(b)$ are distinct for some φ in $\hat{G}$.

    (d)* Prove that $G \cong \hat{G}$, for every finite abelian group G.

    (e) If φ in $\hat{G}$ is not the identity element that maps all of G to 1 in $\mathcal{T}$, show that the sum $\sum_{a \in G} \varphi(a) = 0$.

12. Show that for every finite abelian group A there is a positive integer n such that A is isomorphic to a homomorphic image of the group $\mathbb{Z}_n^\star$.

13. Find the complete decomposition of the group $\mathbb{Z}_{24}^\star$ of units modulo 24 into cyclic p-groups. Repeat the exercise for $G = \mathbb{Z}_{19}^\star$ and for $\mathbb{Z}_{15}^\star$.

Hint. Calculate the orders of the elements of G.

14. Show that two finite abelian groups A, B are isomorphic if they have the same order n, and for every prime divisor p of n the p-primary components $A(p), B(p)$ are isomorphic.

15. Say two finite abelian p-groups A, B are such that for every n the number of elements in A having order p^n equals the number of elements in B having order p^n. Prove that A is isomorphic to B.

16. If A, B are finite abelian p-groups such that $A \times A \cong B \times B$, show that $A \cong B$.

Prove this claim without assuming that the abelian groups A, B are p-groups.

17.* Suppose that A is a finite abelian p-group for some prime p with $[e_1, e_2, \ldots, e_k]$ as its invariant, and that B is a subgroup of A having $[d_1, d_2, \ldots, d_\ell]$ as its invariant. Prove that $\ell \leq k$ and that $d_j \leq e_j$ for each j from 1 to ℓ.

3 Groups Acting on Sets

Groups become interesting when they do something, and that entails permutations of some set. Recall that if X is any set, then $\mathcal{S}(X)$ denotes the group of permutations of X (i.e. bijections $X \to X$) under the composition of mappings. For instance, if X is the set of n letters $\{1,2,\ldots,n\}$, then $\mathcal{S}(X)$ is the now familiar $\mathcal{S}_n$.

3.1 Definition and Some Illustrations of Actions

Definition 3.1. Let G be a group, X a non-empty set, and $\mathcal{S}(X)$ the group of permutations of X. A group homomorphism

$$\varphi : G \to \mathcal{S}(X)$$

is called an ***action homomorphism*** or more commonly an ***action*** of G on X. We also say that G ***acts*** on X.

The Dot Notation

The homomorphism φ allows every g in G to be viewed as the bijection

$$\varphi(g) : X \to X \text{ where } x \mapsto \varphi(g)(x).$$

It is convenient to think of g itself as the actual bijection of X. For every x in X and g in G, it is common to use the dot notation

$$g \cdot x \text{ to mean } \varphi(g)(x).$$

With this cleaner notation, the φ is understood to lie buried in the dot. The fact φ is a homomorphism means that

$$\varphi(gh)(x) = \varphi(g)(\varphi(h)(x)) \text{ for every } g, h \text{ in } G \text{ and every } x \text{ in } X,$$

which is the same as writing

$$g \cdot (h \cdot x) = (gh) \cdot x.$$

One might say that a group action is associative. There is no ambiguity in omitting brackets and writing $gh \cdot x$. The fact that $\varphi(1)$ is the identity permutation becomes the statement

$$1 \cdot x = x \text{ for every } x \text{ in } X.$$

Also,

$$\text{if } y = g \cdot x, \text{ then } x = g^{-1} \cdot y.$$

The elements of X are frequently called ***points***.

Recovering the Action Homomorphism from the Dot Notation

A homomorphism $\varphi : G \to \mathcal{S}(X)$ leads to the dot notation. Conversely, starting with a dot notation, the homomorphism φ can be recovered. Indeed, if some rule is specified that sends every pair (g, x) in the Cartesian product $G \times X$ to a point $g \cdot x$ in X, such a rule will give an action of G on X provided

$$g \cdot (h \cdot x) = (gh) \cdot x \text{ and } 1 \cdot x = x,$$

for every g, h in G and every x in X. From such notation, the homomorphism $\varphi : G \to \mathcal{S}(X)$ surfaces as

$$\varphi(g) : X \to X \text{ where } \varphi(g)(x) = g \cdot x \text{ for every } g \text{ in } G \text{ and } x \text{ in } X.$$

Clearly,

$$\varphi(gh)(x) = (gh) \cdot x = g \cdot (h \cdot x) = \varphi(g)(\varphi(h)(x))$$

for all g, h in G and all x in X, which means that $\varphi(gh) = \varphi(g)\varphi(h)$ as required of homomorphisms.

One little issue remains. How can we be sure that $\varphi(g)$ is actually in $\mathcal{S}(X)$? Well, that is what the requirement $1 \cdot x = x$ fulfills. Indeed, the mapping $\varphi(g) : X \to X$ is a bijection because its inverse mapping is $\varphi(g^{-1})$. To check that, let $x \in X$. Then

$$\varphi(g^{-1})(\varphi(g)(x)) = g^{-1} \cdot (g \cdot x) = (g^{-1}g) \cdot x = 1 \cdot x = x.$$

And similarly, $\varphi(g)(\varphi(g^{-1})(x)) = x$, which verifies that $\varphi(g)$ is a bijection on X.

Some versatility in flipping back and forth between the dot notation and the homomorphism that gives the action can be helpful.

Examples of Group Actions

- Every group G acts ***inertly*** on every set X by the rule

  $$g \cdot x = x \text{ for every } g \text{ in } G \text{ and every } x \text{ in } X.$$

  An inert action happens when the kernel of the action homomorphism $\varphi : G \to \mathcal{S}(X)$ is all of G.
- The group $G = GL_n(\mathbb{F})$ of invertible matrices, with entries in a field, acts on the column vector space $\mathbb{F}^n$ in the usual way of matrix multiplication. Namely, if $A = (a_{ij})$ is an invertible matrix and $v = (v_j)$ is in $\mathbb{F}^n$, the action of A on v is given by

$$Av = w, \text{ where } w \text{ is the column vector whose } i\text{th entry is } \sum_{j=1}^{n} a_{ij} v_j.$$

At times, as it is here, the dot is dropped.

- Every subgroup G of $\mathcal{S}_n$ acts on the set $\{1, 2, \ldots, n\}$ of n letters. Here, the homomorphism $\varphi : G \to \mathcal{S}_n$ that gives the action is the inclusion mapping $\sigma \mapsto \sigma$ for every σ in G. In dot notation the action looks like

$$\sigma \cdot x = \sigma(x), \text{ for every permutation } \sigma \text{ in } G \text{ and every letter } x.$$

- More generally, for every action of a group G on a set X, and any subgroup H of G, the subgroup H also acts on X. Just restrict the homomorphism that gives the action of G down to H.
- If G is any group, think of G also as a set of points. The Cayley representation given by

$$g \cdot x = gx \text{ for every } g \text{ in the group } G \text{ and every point } x \text{ in the set } G$$

is an action of G on itself. This has already been verified as an example of a homomorphism. We say that G ***acts on itself by left multiplication***.

Faithful and Unfaithful Actions

If the homomorphism $\varphi : G \to \mathcal{S}(X)$ that gives the action of a group G on a set X is injective, then G is isomorphic to its image inside $\mathcal{S}(X)$. With such a condition G may as well be a subgroup of the permutation group $\mathcal{S}(X)$, and we say that the group action is ***faithful***. With an action that is not faithful there will be a σ in G such that $\sigma \neq 1$ and $\sigma \cdot x = x$ for all x in X.

By way of illustration, take the Klein 4-group $\mathcal{V}_4$. An isomorphic copy consists of the matrices

$$I = \begin{pmatrix} 1 & 0 \\ 0 & 1 \end{pmatrix}, \quad A = \begin{pmatrix} 1 & 0 \\ 0 & -1 \end{pmatrix}, \quad B = \begin{pmatrix} -1 & 0 \\ 0 & 1 \end{pmatrix}, \quad C = \begin{pmatrix} -1 & 0 \\ 0 & -1 \end{pmatrix}.$$

Let X be the set comprising the following eight points in $\mathbb{R}^2$:

$$(1,0), (1,1), (0,1), (-1,1), (-1,0), (-1,-1), (0,-1), (1,-1),$$

sitting on a square around the origin. The group of matrices I, A, B, C acts on this set of eight points, by familiar matrix multiplication on the points written as columns. Since each of the three non-identity matrices moves at least one of the eight points in X, this is a faithful action.

However, $\mathcal{V}_4$ also acts on the set Y consisting of just the two points

$$(1,0), (-1,0).$$

Now, both I and A leave both $(1,0)$ and $(-1,0)$ fixed. The action of $\mathcal{V}_4$ on Y is not faithful.

The kernel of any action homomorphism $\varphi : G \to \mathcal{S}(X)$ is a normal subgroup of G, as all kernels are. If the action is not faithful, G must have a non-trivial normal subgroup.

If the unfaithful action is also not inert, that non-trivial, normal subgroup $\ker\varphi$ is also a proper subgroup of G. From these observations a potential method for showing that a group is not simple emerges. Here is a condition under which every group action of G on X must be unfaithful.

Proposition 3.2. *If a group G has order n and G acts on a set X having order m, and n does not divide $m!$, then the action is not faithful.*

Proof. This is yet another application of Lagrange. If the kernel of the homomorphism $\varphi: G \to \mathcal{S}(X)$ that implements the action were trivial, then G would be isomorphic to its image inside $\mathcal{S}(X)$. Thereby, the order of G would be a divisor of the order of $\mathcal{S}(X)$, which is $m!$, contradicting the assumption. Thus $\ker\varphi$ must be a non-trivial, normal subgroup of G. □

Proposition 3.2 offers a mechanism for discovering at times that a group is not simple, as in the ensuing situation.

3.1.1 A Group Action on Left Cosets

If G is a group and H is any subgroup, let G/H be the family of left cosets of H. Here G/H is just a set, not to be confused with a quotient group despite the suggestive notation. The group G acts on G/H as follows:

$$g \cdot xH = gxH \text{ for every } g, x \text{ in } G.$$

To possibly clarify the above, the coset xH, *viewed as a single point*, is sent by g to the coset gxH, again viewed as a single point. If $xH = yH$, then $gxH = gyH$, which ensures that our action of G on G/H does not depend on the chosen representatives of the left cosets. The action is well defined. The fact that it is a group action is readily verified. For if $g, h \in G$ and xH is a left coset, then

$$h \cdot (g \cdot xH) = h \cdot (gxH) = hgxH = (hg) \cdot xH \text{ and } 1 \cdot xH = (1x) \cdot H = xH.$$

Let us refer to such an action of G on G/H as a ***left coset action***. Here is something to notice about left coset actions.

Proposition 3.3. *Let H be a subgroup of a group G, and let $\varphi : G \to \mathcal{S}(G/H)$ be the left coset action of G on the set G/H of left cosets of H. Then $\ker\varphi$ is a normal subgroup of G that sits inside H, and $\ker\varphi$ contains all normal subgroups of G that sit inside H.*

Proof. The normality of $\ker\varphi$ is immediate because this subgroup of G is the kernel of a homomorphism. Next, if $x \in \ker\varphi$, then φ fixes all left cosets of H. That is $xyH = yH$ for all y in G. In particular, with $y = 1$ we get $xH = H$, and thus $x \in H$. This shows that $\ker\varphi \subseteq H$.

Finally, suppose N is a subgroup of H and that N is normal in G. To prove that $N \subseteq \ker\varphi$ we require that

$$xyH = yH \text{ for every } x \text{ in } N \text{ and every } y \text{ in } G.$$

This is the same as requiring that $y^{-1}xy \in H$. But $y^{-1}xy \in N$ because N is normal in G. The requirement then follows because $N \subseteq H$. □

Here is a potentially useful consequence of Propositions 3.2 and 3.3.

Proposition 3.4. *If G is a finite group of order n and H is a proper subgroup of index m where n does not divide $m!$, then H contains a proper, non-trivial normal subgroup of G. Consequently, G is not a simple group.*

Proof. Let φ be the left coset action of G on G/H. By Proposition 3.3 the normal subgroup $\ker\varphi$ sits inside the proper subgroup H, whereby $\ker\varphi$ is also a proper subgroup of G. It remains to see that $\ker\varphi$ is not the trivial group. This follows from Proposition 3.2 because the order of the set G/H being acted upon is m. □

Checking that Groups of Some Orders Are Not Simple

The assumption in Proposition 3.4 about the presence of a *proper* subgroup H whose index m in G is such that the order of G does not divide $m!$ might seem to be a bit much. Still, such a condition is met at times.

For instance, suppose G is a group of order 91. Note that $91 = 13 \cdot 7$. We shall come to learn, from a theorem of Cauchy, that G must have a subgroup H of order 13. The index of such H in G is 7. Observe that $91 \nmid 7!$, because $13 \nmid 7!$. By Proposition 3.4, H contains a non-trivial normal subgroup of G. Since the order of H is the prime 13, Lagrange forces that normal subgroup to be H itself. Thus we learn that every group of order 91 has a normal subgroup of order 13.

For another illustration suppose that G is a group of order 100, while H is a subgroup of G of order 25. A famous result due to Ludwig Sylow (1832–1918) shows that such an H must exist, but we are not yet ready to prove it. Now, the left coset action of G on G/H cannot be faithful since the index of H in G is 4 and 100 does not divide 4!. By Proposition 3.4, H contains a non-trivial subgroup N, which is normal in G. Since N is a non-trivial subgroup of H, the order of N is either 5 or 25, by Lagrange. Subject to Sylow's theorem, we have shown that every group of order 100 has a normal subgroup of order 5 or 25.

3.1.2 Groups of Order 6

Left coset actions provide a tool to show that every non-abelian group of order 6 is isomorphic to $\mathcal{S}_3$.

Suppose that G is such a group. The order of each element of G is either $1, 2, 3$ or 6, by Lagrange. If G had an element of order 6, then G would be cyclic and thereby abelian. Thus all elements of G, except for the identity element, have order 2 or 3. Since every element x of order 3 can be paired with its distinct inverse x^{-1}, the number of elements of order 3 is even. Thus one of the five non-identity elements of G has order 2, which then generates a subgroup H of order 2.

Let $\varphi : G \to \mathcal{S}(G/H)$ be the left coset action of G on G/H. According to Proposition 3.3 the kernel of φ is a subgroup of H that is normal in G. Clearly $\ker\varphi$ is either the trivial subgroup or $\ker\varphi = H$.

If $\ker\varphi$ is trivial, then φ is injective, which means that the image $\varphi(G)$ inside $\mathcal{S}(G/H)$ is isomorphic to G. By Lagrange, the order of the set G/H is 3, and so the order of $\mathcal{S}(G/H)$ is 6. It follows that $G \cong \varphi(G) = \mathcal{S}(G/H)$. Since the order of G/H is 3, it is patently clear that $\mathcal{S}(G/H)$ is isomorphic to $\mathcal{S}_3$. We have seen that if $\ker\varphi$ is trivial, then $G \cong \mathcal{S}_3$.

It remains to prove that $\ker\varphi$ cannot be H. Well, suppose to the contrary that $\ker\varphi = H$. In that case H is a normal subgroup of order 2 inside G. It follows that H is inside the centre of G. Indeed, if a is the generator of H and $x \in G$, the normality of H along with the fact H has order 2 shows that $xax^{-1} = a$. Then $xa = ax$, meaning that a, and thereby H, is central in G. Also the quotient group G/H is cyclic because it has prime order 3. The contradiction will be that G has to be an abelian group, in accordance with the following little result.

Proposition 3.5. *If H is a subgroup of the centre of a group G, then H is normal in G, and if G/H is cyclic, then G is abelian.*

Proof. Leaving the normality of H as a triviality, suppose G/H is cyclic with generator Hg for some g in G. If $x, y \in G$, then $Hx = Hg^k$ and $Hy = Hg^\ell$ for some integer exponents k, ℓ. This means that $x = ag^k, y = bg^\ell$ for some a, b in H, and thereby in the centre of G. Then

$$xy = ag^k bg^\ell = abg^k g^\ell = abg^{k+\ell} = bag^\ell g^k = bg^\ell ag^k = yx,$$

as desired. □

Non-abelian groups of order 6 are isomorphic to $\mathcal{S}_3$. From Proposition 2.92, abelian groups of order 6 are isomorphic to $\mathbb{Z}_2 \times \mathbb{Z}_3$, which by the Chinese remainder theorem is a cyclic group. The following tidbit emerges.

Proposition 3.6. *Every group of order 6 is either cyclic or isomorphic to $\mathcal{S}_3$.*

EXERCISES

1. Let G be a group of order $85 = 17 \cdot 5$. A theorem of Cauchy (yet to be proven) tells us that G contains a subgroup H of order 17. Show that H is a normal subgroup of G.
2. Suppose G is a group of order pm where p is prime and $p > m$. If H is a subgroup of order p, show that H must be normal in G.

   A theorem of Cauchy will soon reveal that such H do exist, and another more advanced result will reveal that there is only one such H. Thus groups of order pm with p prime and $p > m$ are not simple groups.
3. If G is a group of order 55 with a subgroup H of order 11, show that H must be normal.
4. If G is a group of order 10 show that G is isomorphic either to $\mathbb{Z}_2 \times \mathbb{Z}_5$ or to a subgroup of $\mathcal{S}_5$.

More generally if G is a group of order $2p$ where p is an odd prime, show that G is isomorphic either to $\mathbb{Z}_2 \times \mathbb{Z}_p$ or to a subgroup of $\mathcal{S}_p$.

Is $\mathbb{Z}_2 \times \mathbb{Z}_5$ itself isomorphic to a subgroup of $\mathcal{S}_5$?

3.2 Orbits and Stabilizers

If a group G acts on a set X, the relation on X given by

$$x \sim y \text{ provided } y = g \cdot x \text{ for some } g \text{ in } G$$

is an equivalence relation. Indeed, the equation $x = 1 \cdot x$ gives reflexivity. If $y = g \cdot x$, then $x = g^{-1} \cdot y$, which explains symmetry. And if $y = g \cdot x$ and $z = h \cdot y$, then $z = h \cdot (g \cdot x) = hg \cdot x$, which shows transitivity. By this relation, the set X is partitioned into disjoint equivalence classes. For each x in X its equivalence class is the set of all $g \cdot x$ as g runs over G.

Definition 3.7. If a group G acts on a set X and $x \in X$, the set

$$G \cdot x = \{g \cdot x : g \in G\}$$

inside X is known as the ***orbit*** of x under the action of G.

As observed, the orbits form a partition of X. Two orbits $G \cdot x$ and $G \cdot y$ are identical if and only if x belongs to the orbit of y, or vice versa.

We have already seen a special case of orbits. The group $\mathcal{S}_n$ acts on the set L of n letters. If $\sigma \in \mathcal{S}_n$ and $a \in L$, the orbit of a under σ is defined to be the set of letters $\sigma^k(a)$ taken over all integer exponents k. The cyclic group $G = \langle \sigma \rangle$ generated by σ acts on L. So, the orbit of a under σ is identical to the orbit of a under the action of this cyclic group.

Definition 3.8. When a group action of G on a set X satisfies $G \cdot x = X$ for some x in X, i.e. there is only one orbit, the action of G is said to be ***transitive*** on X. We also say that G ***acts transitively*** on X.

The term "transitive" as used here should not be confused with its use in the sense of equivalence relations.

For instance, the alternating group $\mathcal{A}_n$ acts transitively on the set L of n letters when $n \geq 3$. Just use an appropriate 3-cycle. If σ is a permutation on the set L of n letters, the cyclic group generated by σ acts transitively on L if and only if σ consists of one cycle of length n.

Orbits for an Action of the Klein 4-Group

As already noted the Klein 4-group $\mathcal{V}_4$ is the group of matrices

$$I = \begin{pmatrix} 1 & 0 \\ 0 & 1 \end{pmatrix}, \quad A = \begin{pmatrix} 1 & 0 \\ 0 & -1 \end{pmatrix}, \quad B = \begin{pmatrix} -1 & 0 \\ 0 & 1 \end{pmatrix}, \quad C = \begin{pmatrix} -1 & 0 \\ 0 & -1 \end{pmatrix},$$

which acts on the set X consisting of the eight points in $\mathbb{R}^2$:

$$(1,0), (1,1), (0,1), (-1,1), (-1,0), (-1,-1), (0,-1), (1,-1).$$

The action is according to matrix multiplication on the points written as columns. Since the order of $\mathcal{V}_4$ is 4, each orbit has at most 4 points in it. After a small calculation, we find that the orbit of $(1,0)$ is the set of points

$$(1,0), (0,1), (-1,0), (0,-1),$$

while the orbit of $(1,1)$ is the set

$$(1,1), (-1,1), (-1,-1), (1,-1).$$

This action partitions X into two disjoint orbits.

The Orbit-Stabilizer Relation

Even if a group action of G on a set X is faithful, it might still be that there exists a point x in X and an element g in G, other than the identity, such that $g \cdot x = x$ for that one x.

Definition 3.9. If $x \in X$, the set

$$G_x = \{g \in G : g \cdot x = x\}$$

is called the ***stabilizer*** of x under the action of G. The stabilizer of x is a subgroup of G, as can easily be verified.

For each $x \in X$ we ask how many points lie in the orbit $G \cdot x$. The stabilizer of x gives a connection. We might recall that the index of a subgroup can be defined as the number of right cosets or as the number of left cosets of the subgroup. Both definitions give the same result.

Proposition 3.10 (Orbit-stabilizer relation). *If a group G acts on a set X and $x \in X$, then there is a bijection between the orbit $G \cdot x$ and the family of left cosets of the stabilizer G_x inside G. In other words the order of $G \cdot x$ equals the index $[G : G_x]$ of the stabilizer G_x.*

Proof. The function

$$f : G \to G \cdot x, \text{ given by } g \mapsto g \cdot x \text{ for every } g \text{ in } G,$$

is a surjection from G onto the orbit of x. As p runs through the points in the orbit $G \cdot x$, the inverse images

$$f^{-1}(p) = \{g \in G : g \cdot x = p\}$$

form a partition, call it $\mathcal{P}$, of G. The inverse image mapping from $G \cdot x$ to $\mathcal{P}$ given by $p \mapsto f^{-1}(p)$ is a bijection (from a set of points to a set of sets). Thus, to count the number of points in the orbit of x, we can count the number of sets in the partition $\mathcal{P}$.

To count the partition $\mathcal{P}$, observe that group elements g, h both belong to the same set of $\mathcal{P}$, if and only if $g \cdot x = h \cdot x$. This is the same as saying that $h^{-1}g \cdot x = x$, and this means that

$h^{-1}g \in G_x$, the stabilizer of x. The latter comes down to saying that $gG_x = hG_x$. Thus the partition $\mathcal{P}$ is the family of left cosets of the stabilizer subgroup G_x. So, the orbit $G \cdot x$ has exactly as many points in it as the number of left cosets of G_x, which by definition is the index of G_x in G. □

If G is a *finite* group acting on a set X and $x \in X$, Lagrange says that $|G| = |G_x|[G : G_x]$. In the case of a finite group, Proposition 3.10 can be rephrased as follows.

Proposition 3.11. *If a finite group G acts on a set X and $x \in X$, then*

$$|G| = |G \cdot x|\ |G_x|.$$

In particular, the order of each orbit is a divisor of the order of the acting group.

Conjugate Stabilizers

A curiosity ensues. Suppose that a finite group G acts on a finite set X and that $x \in X$. The orbit $G \cdot x$ can be represented by any one of its points, not only by x. Suppose that $G \cdot x = G \cdot y$ for some other point y. From what we just saw

$$|G \cdot x|\ |G_x| = |G| = |G \cdot y|\ |G_y|.$$

It looks like, for x, y in the same orbit, the different stabilizers G_x and G_y have equal order. Here is a more informative way to see this.

Proposition 3.12. *If a group G acts on a set X and x, y are points in the same orbit, then their stabilizers are conjugate subgroups.*

Proof. Since x, y are in the same orbit, there is a g in G such that $y = g \cdot x$. A group element h lies in G_x, if and only if $x = h \cdot x$. Since $x = g^{-1} \cdot y$, the statement $x = h \cdot x$ becomes $g^{-1} \cdot y = hg^{-1} \cdot y$, which is the same as $y = ghg^{-1} \cdot y$. So, h stabilizes x if and only if ghg^{-1} stabilizes y. In other words $G_y = gG_xg^{-1}$, as required. □

Counting the Symmetries of a Cube

Let us indulge in a bit of geometrical imagination.

Picture a cube. Instead of labelling its eight vertices, let us label the six faces as $1, 2, 3, 4, 5, 6$, and let us do it so that $1, 2$ are opposite (top-bottom) faces, $3, 4$ are opposite (back-front) faces and $5, 6$ are opposite (left-right) faces, as shown.

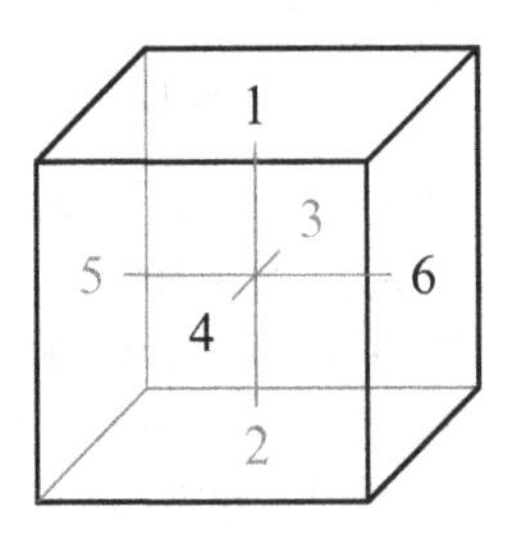

A ***symmetry*** of the cube is a rigid motion of the cube that brings the cube back onto itself. Such a symmetry must permute these six faces. Thereby the set G of symmetries lies inside the group $\mathcal{S}_6$ of order 720. Since the composite of two symmetries is a symmetry, we can see that G is a subgroup of $\mathcal{S}_6$. However, not every permutation in $\mathcal{S}_6$ comes from a symmetry. For instance, the transposition $(1,2)$ would require that two opposite faces be interchanged by a rigid motion without moving the other four faces. Our imagination tells us this can not be done. On the other hand the permutation $(1,2)(3,4)$ does come from the rotation of the cube through $180°$ about the axis joining the centres of faces 5 and 6. By Lagrange the order of G divides 720, but let us see how the orbit-stabilizer relation helps us find that order.

The group G acts transitively on the set of six faces. Indeed, we can easily picture a rotation though an axis joining two opposite face centres which moves any face i to any other face j. Thus the orbit $G \cdot 1$ of face 1 has order 6. What symmetries in G will stabilize face 1? We see only the 4 rotations through the axis joining the centre of face 1 to its opposite centre on face 2. These rotations are through the angles $0°, 90°, 180°$ and $270°$, and correspond to the identity permutation along with the permutations $(3,5)(4,6), (3,4)(5,6)$ and $(3,6)(4,5)$. Thus the stabilizer G_1 has order 4. According to Proposition 3.11 the order of G is $6 \cdot 4 = 24$.

To physically implement all 24 rigid motions of the cube requires some play with an actual cube.

3.2.1 An Action for the Dihedral Groups

This might be a good place to become acquainted with an interesting family of groups.

If n is a positive integer, the nth roots of 1 in the complex plane are the powers of $\omega = e^{2\pi i/n}$. Here is the set of those roots:

$$X = \{1, \omega, \omega^2, \ldots, \omega^{n-1}\}.$$

These powers form a cyclic group of course, but now just think of them as the vertices of a regular convex polygon with n equal sides. Here is the picture of the polygons for $n = 3, 4, 5, 6, 7$ and 8, commonly known as the triangle, square, pentagon, hexagon, heptagon and octagon, respectively.

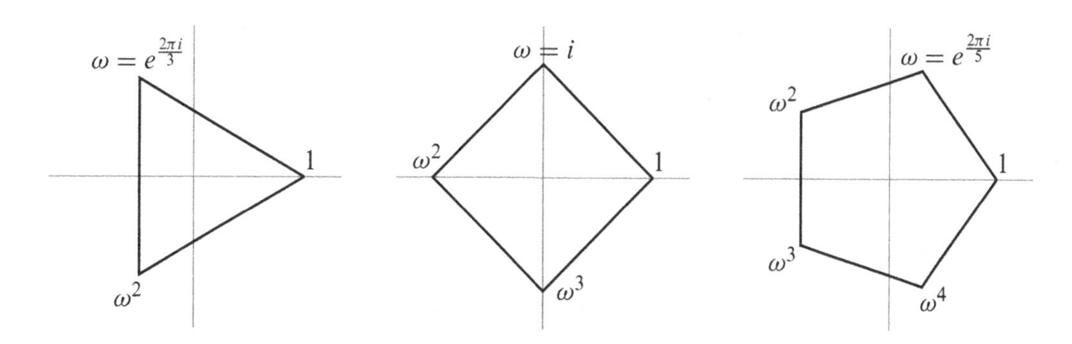

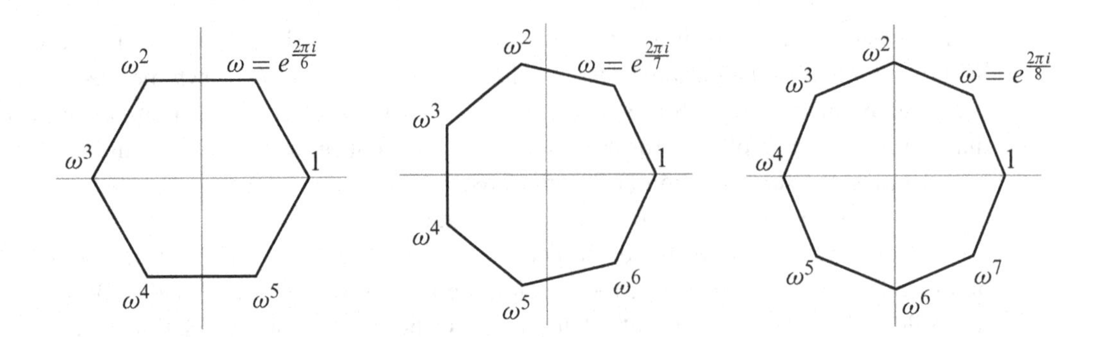

To specify their number of vertices and edges, such figures are often called ***n-gons***. A rigid motion in 3-space that superimposes the n-gon back on itself is called a ***symmetry*** of the figure. The composite of two symmetries is again a symmetry. There is the identity symmetry that leaves the polygon unmoved, and every symmetry can be followed up by another symmetry to restore the polygon to its original position. In short, the set of symmetries of the n-gon is a group. This group is known as a ***dihedral group*** to be denoted by $\mathcal{D}_n$.

It seems that there are $2n$ symmetries of the n-gon. We have the n counter-clockwise rotations through the angles $\theta, 2\theta, \ldots, (n-1)\theta$, and $n\theta$, where θ is the angle $\frac{2\pi}{n}$ (in radians). The last rotation brings the n-gon back to its original position, and thus it is the identity. In addition, the n-gon has n axes of symmetry about which we can carry out a reflection. If n is odd these are the axes from each vertex to the mid-point of its opposite side. If n is even there are $n/2$ axes joining opposite vertices and $n/2$ axes joining mid-points of opposite sides.

We need a way to do calculations within $\mathcal{D}_n$.

Every symmetry of the n-gon can be specified by a permutation of the set of vertices X. Thus, the group $\mathcal{D}_n$ becomes a subgroup of the permutation group $\mathcal{S}(X)$. We have an action of $\mathcal{D}_n$ on X. Of course, not every permutation of X gives a symmetry of the n-gon. For instance, using $n = 4$, the mapping that transposes 1 with i but leaves -1 and $-i$ unmoved cannot be a symmetry of the square.

There are two symmetries in $\mathcal{D}_n$ which generate all of $\mathcal{D}_n$. There is the rotation of the polygon counter-clockwise through the angle $\theta = 2\pi/n$. This is the permutation on X given by

$$\sigma : X \to X \text{ where } \omega^j \mapsto \omega^{j+1}.$$

In other words, σ is the mapping of the complex plane that multiplies every complex number by ω, restricted to the set X of vertices of our n-gon. Then there is the reflection of the n-gon across the x-axis. As a mapping of the complex plane, this reflection is given by taking the complex conjugate of each complex number. Restricted to the vertices of our n-gon this reflection is given by

$$\tau : X \to X \text{ where } \omega^j \mapsto \omega^{-j}.$$

The powers

$$\sigma, \sigma^2, \sigma^3, \ldots, \sigma^{n-1}, \sigma^n$$

give the rotations of the polygon though the angles

$$\theta, 2\theta, 3\theta, \ldots, (n-1)\theta, n\theta,$$

respectively. Clearly, $\sigma^n = 1$, the identity that leaves the n-gon unmoved. Obviously, $\tau^2 = 1$ as well.

How does the composite $\sigma\tau$ act? Notice that under any symmetry, the adjacent vertices 1 and ω go to adjacent vertices, and once their new positions are given, the placement of the other vertices is determined too. So, let us calculate

$$\sigma\tau(1) = \sigma(1) = \omega \text{ and } \sigma\tau(\omega) = \sigma(\omega^{-1}) = 1.$$

Since $\sigma\tau$ interchanges 1 and ω, it must be that $\sigma\tau$ is the reflection of the n-gon about the axis through the origin and the midpoint of the edge from 1 to ω.

What does $\sigma^2\tau$ do? Well,

$$\sigma^2\tau(1) = \sigma^2(1) = \omega^2 \text{ and } \sigma^2\tau(\omega) = \sigma^2(\omega^{-1}) = \omega.$$

Now 1 and ω go to ω^2 and ω, respectively. Hence, $\sigma^2\tau$ is the reflection of the n-gon about the axis through the origin and ω. In a similar spirit $\sigma^3\tau$ sends 1 to ω^3 and ω to ω^2, which is the reflection of the n-gon about the line through the origin and the midpoint of the edge joining ω and ω^2. In this vein, $\sigma^4\tau$ is the reflection about the axis through the origin and ω^2, and so on. Eventually, $\sigma^n\tau = \tau$, which is the original reflection across the x-axis.

Thus we see that the entire group $\mathcal{D}_n$, of order $2n$, consists of

$$n \text{ rotations } 1, \sigma, \sigma^2, \ldots, \sigma^{n-1} \text{ and } n \text{ reflections } \tau, \sigma\tau, \sigma^2\tau, \ldots, \sigma^{n-1}\tau.$$

In addition to the relations

$$\sigma^n = 1 \text{ and } \tau^2 = 1,$$

inside $\mathcal{D}_n$, there is one more worth noting. Namely,

$$\sigma\tau\sigma = \tau.$$

This can be readily seen by comparing the effects of the $\sigma\tau\sigma$ and τ on 1 and on ω. Alternatively just observe that $\sigma\tau\sigma(z) = \omega\overline{\omega z} = \bar{z} = \tau(z)$ for all complex numbers z. It is also clear geometrically, since $\sigma\tau$ is a reflection, which reveals the equivalent relation $\sigma\tau\sigma\tau = 1$. With the above three relations in hand, the multiplication table for $\mathcal{D}_n$ is fully determined. For instance, we can easily deduce that $\sigma^k\tau\sigma^k = \tau$ for all exponents k, and then

$$(\sigma^3\tau)(\sigma^2\tau) = \sigma\sigma^2\tau\sigma^2\tau = \sigma\tau\tau = \sigma,$$

and also

$$(\sigma^2\tau)\sigma^4 = \sigma^{n-2}\sigma^4\tau\sigma^4 = \sigma^{n-2}\tau.$$

In case it was not obvious already, the relation $\sigma\tau\sigma = \tau$ confirms that $\mathcal{D}_n$ is not abelian.

Since the dihedral group acts on the set X of vertices of the n-gon, we might enquire about the orbit-stabilizer relation. Clearly, this group acts transitively, having just one orbit.

Using the vertex ω as a representative for this orbit, Proposition 3.11 forces its stabilizer to have order 2. The identity symmetry 1 stabilizes ω. The other symmetry that stabilizes ω is the reflection $\sigma^2\tau$ across the axis through the origin and ω. If we take the vertex 1 as the representative of the orbit, then the two symmetries that stabilize 1 are the identity and τ.

We could also have used the orbit-stabilizer relation to confirm that $\mathcal{D}_n$ has order $2n$. Clearly the orbit of 1 under the action of $\mathcal{D}_n$ has order n and the stabilizer of 1 has order 2 (because the only rigid motions that fix 1 are the identity and the reflection τ across the x-axis). By the orbit-stabilizer relation, $\mathcal{D}_n$ has order $2n$.

Why Dihedral?

Maybe a remark on the name "dihedral" is warranted. When $\mathcal{D}_n$ acts as the group of symmetries of the polygon, the reflections, if carried out physically, would require that the polygon be flipped over in 3-space. Thus it makes a bit of sense to think of the polygon as a solid in 3-space having two faces: the front and the back face. These faces are joined at the edges which form the sides of the polygon. The faces just happen to be one and the same.

The most famous solids in 3-space are the five Platonic solids: the tetrahedron having four faces, the hexahedron (a.k.a. the cube) with six faces, the octahedron with eight faces, and the more complex dodecahedra and icosahedra with 12 and 20 faces, respectively. The suffix "hedron" is based on the Greek for "seat" or "base." The faces of our solids form the base on which they are built. In light of this, our polygon, with two faces coinciding with one another, has the more obscure name of *dihedron*. And its group of symmetries is, thereby, a dihedral group.

Finally, a caution that some authors denote $\mathcal{D}_n$ as $\mathcal{D}_{2n}$, referring to its own order $2n$ rather than the order n of the set of vertices on which it acts.

3.2.2 The Symmetries of a Tetrahedron

A regular tetrahedron is the Platonic solid with four vertices, four faces made up of congruent equilateral triangles and six edges as shown. The vertices are labelled by the letters $1, 2, 3, 4$.

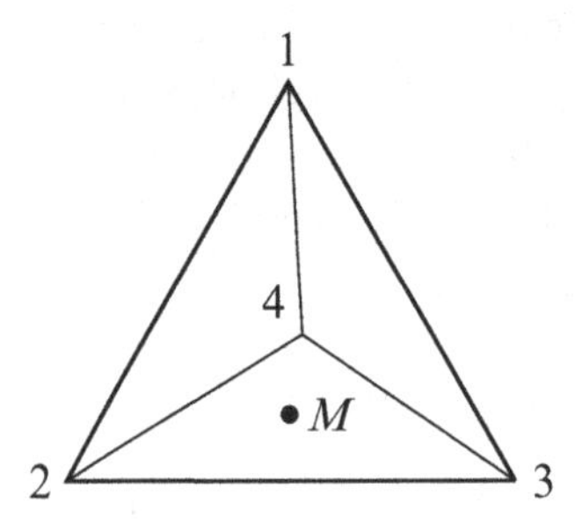

A symmetry of the tetrahedron is any permutation of the vertices that can be achieved as a rigid motion of the figure in 3-space which brings the figure back onto itself. Clearly the composite of two such symmetries is another symmetry. There is the identity symmetry, and every symmetry can be undone by another symmetry. The set of symmetries of the tetrahedron is a group, and it is a subgroup of $\mathcal{S}_4$. Call it $\mathcal{T}$.

In addition to the identity, we find in $\mathcal{T}$ two rotations about the axis through the vertex 1 and the mid-point M of its opposite face. These rotations are the cycles $(2,3,4)$ and its inverse $(4,3,2)$. Doing the same thing for the vertex 2 and its opposite face, we get the rotations $(1,3,4)$ and $(4,3,1)$. Likewise, for the vertex 3 we pick up $(1,2,4)$ and $(4,2,1)$, and for the vertex 4 we get $(1,2,3)$ and $(3,2,1)$.

Along with the identity we have discovered 9 symmetries of the tetrahedron, but Lagrange implies there must be more, since 9 does not divide the order of $\mathcal{S}_4$, which is 24. In fact, $|\mathcal{T}|$ must be either 12 or 24. Some permutations, for example $(1,2)$, do not give symmetries of the tetrahedron. Thus, $|\mathcal{T}| = 12$.

Another way to confirm that $|\mathcal{T}| = 12$ is to use the orbit-stabilizer relation. Obviously, $\mathcal{T}$ acts transitively on the vertices of the tetrahedron. There is only one orbit for the action of $\mathcal{T}$. Take the vertex 1 as a representative for this orbit $\mathcal{T} \cdot 1$. By inspection, the stabilizer $\mathcal{T}_1$ of 1 can only be the cyclic group consisting of $1, (2,3,4)$ and $(4,3,2)$. The orbit-stabilizer relation gives

$$|\mathcal{T}| = |\mathcal{T} \cdot 1||\mathcal{T}_1| = 4 \cdot 3 = 12.$$

Finally, we know that the alternating group $\mathcal{A}_4$ inside $\mathcal{S}_4$ has order 12 and this group is generated by the 3-cycles which we saw are also in $\mathcal{T}$. It follows that $\mathcal{T} = \mathcal{A}_4$. The missing three symmetries of the tetrahedron come from the permutations $(12)(34), (13)(24), (14)(23)$. The permutation $(12)(34)$, for instance, rotates the tetrahedron by $180°$ about the axis joining the mid-point of the side from vertex 1 to vertex 2 to the mid-point of the opposing side from vertex 3 to vertex 4. A bit of fiddling with an actual tetrahedron might be helpful to see this.

EXERCISES

1. Show that no two of the groups

$$\mathbb{Z}_8, \mathbb{Z}_4 \times \mathbb{Z}_2, \mathbb{Z}_2 \times \mathbb{Z}_2 \times \mathbb{Z}_2, \mathcal{D}_4 \text{ and } \mathcal{Q} \text{ (the quaternion group)}$$

are isomorphic.

2. Suppose n is a positive integer and that an integer r is such that $0 \le r \le n$. As we might well know, the binomial coefficient $\binom{n}{r}$ is defined to be the number of subsets of order r in a set of order n. There is a well known formula for this quantity, and the orbit-stabilizer relation, Proposition 3.11, can be used to find it. Let $\mathcal{X}$ be the family of all subsets of the set $\{1,2,\ldots,n\}$ of n letters. The group $\mathcal{S}_n$ acts not only on the latter set but also on $\mathcal{X}$. Indeed, if S is a subset of $\mathcal{X}$ and $\sigma \in \mathcal{S}_n$ put

$$\sigma \cdot S = \sigma(S), \text{ the image of } S \text{ under the bijection } \sigma.$$

(a) Verify that this is a group action of $\mathcal{S}_n$ on $\mathcal{X}$.

(b) If R is the subset $\{1,2,\ldots,r\}$ viewed as a point in $\mathcal{X}$, explain why the order of the orbit $\mathcal{S}_n \cdot R$ is $\binom{n}{r}$.

(c) Explain why a permutation σ in $\mathcal{S}_n$ stabilizes R if and only if σ is the disjoint product of a permutation of the set R with a permutation of the complementary set $\{r+1, r+2, \ldots, n\}$.

(d) What is the order of the stabilizer of R?

(e) Find the formula for $\binom{n}{r}$ based on the preceding calculations and the orbit-stabilizer relation.

3. (a) Show that the reflections σ and $\sigma\tau$ also generate $\mathcal{D}_4$.

(b) Show that the symmetries $1, \sigma^2, \tau, \sigma^2\tau$ form a subgroup of $\mathcal{D}_4$, and that this subgroup is isomorphic to the Klein 4-group $\mathcal{V}_4$.

(c) Find the centre of $\mathcal{D}_4$.

(d) Find all subgroups of the dihedral group $\mathcal{D}_4$ and determine which subgroups are normal.

4. Is $\mathcal{D}_3$ isomorphic to $\mathcal{S}_3$? Is $\mathcal{D}_6$ isomorphic to $\mathcal{A}_4$?

5. This set of exercises is about the normal subgroups of the dihedral groups $\mathcal{D}_n$. Recall that the generators σ, τ of $\mathcal{D}_n$ satisfy the following relations

$$\sigma^n = 1, \quad \tau^2 = 1, \quad \sigma\tau\sigma = \tau$$

and the group itself consists of the $2n$ distinct elements $\sigma^i\tau^j$ where $i = 0, 1, \ldots, n-1$ and $j = 0, 1$.

(a) Show that every subgroup H of the cyclic group $\langle\sigma\rangle$ generated by the rotation σ is normal in $\mathcal{D}_n$.

(b) Let K be the subset of $\mathcal{D}_n$ consisting of all symmetries which can be expressed in the form $\sigma^j\tau^i$ where j is even and i is any integer.

Show that K is a subgroup of $\mathcal{D}_n$. If n is odd show that $K = \mathcal{D}_n$. If n is even show that the index of K inside $\mathcal{D}_n$ is 2, and thereby deduce that K is a normal subgroup of $\mathcal{D}_n$.

(c) Let L be the subset of $\mathcal{D}_n$ consisting of all symmetries which can be put in the form $\sigma^j\tau^i$ where i,j are both even or both odd.

Show that L is a subgroup of $\mathcal{D}_n$. If n is odd show that $L = \mathcal{D}_n$. If n is even show that the index of L inside $\mathcal{D}_n$ is 2, and thereby deduce that L is a normal subgroup of $\mathcal{D}_n$.

(d) If N is a normal subgroup of $\mathcal{D}_n$ such that $\sigma^i\tau \in N$ for some even exponent i, show that $H \subseteq N$.

If N is such that $\sigma^j\tau \in N$ for some odd exponent i, show that $K \subseteq N$.

(e) Use the preceding information to draw the lattice of normal subgroups of $\mathcal{D}_n$ for each $n = 4, 5, 6, 8, 12$ and 27.

(f) Show that $\mathcal{D}_n$ is isomorphic to the Cartesian product of two proper subgroups of $\mathcal{D}_n$ if and only if $n = 2k$ where k is odd.

Hint. Use the preceding exercises along with Proposition 2.75.

6. If p is an odd prime and G is a group of order $2p$, show that G has a normal subgroup of order p.
7. Let G be any group of order p^2 where p is prime.
   (a) Show that G contains a subgroup of order p.
   (b) Show that every subgroup of G is normal.
   Hint. Only subgroups H other than the trivial group and G need be considered. Examine the left coset action of G on G/H.
   (c) Show that every proper subgroup of G lies inside the centre $Z(G)$.
   Hint. Take a subgroup H of G of order p. This H is normal and the quotient group G/H also has order p.
8. If a simple group G has a subgroup H of finite index n and $n > 1$, show that G is isomorphic to a subgroup of $\mathcal{S}_n$.

3.3 The Cauchy–Frobenius–Burnside Formula

Here is a combinatorial problem. The vertices of a square, call them $1, 2, 3, 4$, are each to be colored by one of three colors, say red, white or blue. Since each vertex has three possibilities for coloring it and since there are four vertices, the total number of vertex colorings is $3^4 = 81$. However, the square can be moved back onto itself by a symmetry from the dihedral group $\mathcal{D}_4$.

Using such a symmetry we may well be able match one coloring exactly with another one. For instance, the coloring below on the left can be matched to the coloring on the right by a reflection across the vertical axis.

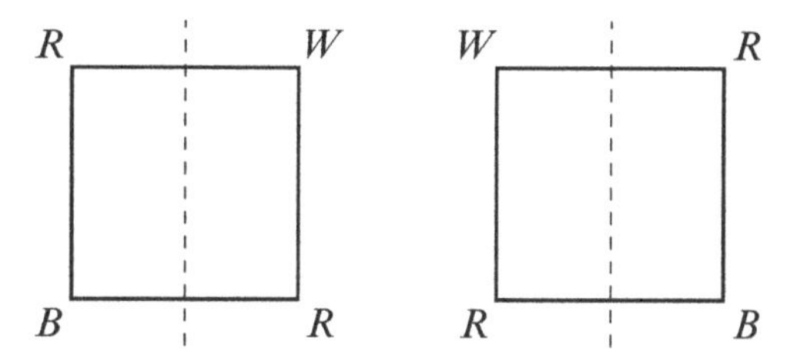

If a symmetry of the square can bring one coloring to coincide with another one, the two colorings are essentially the same. After agreeing to count all colorings that are essentially the same as one ***design***, how many different designs do we have? On first inspection this problem can seem elusive.

Burnside's Formula

The technique for solving such design problems was popularized by George Polya (1887–1985). The method is based on a formula commonly known as Burnside's lemma. However, in his 1897 book *Theory of Groups of Finite Order*, William Burnside (1852–1927) attributes the formula to Ferdinand Frobenius (1849–1917) who published the result in 1887. Yet the result goes even farther back to Augustin Louis Cauchy (1789–1857). At a loss for what to call it, let us settle for "Burnside's formula."

Suppose that a finite group G acts on a finite set X. For each x in X there is the stabilizer subgroup $G_x = \{g \in G : g \cdot x = x\}$. And for each group element g in G there is the dual ***fixed set*** of elements in X that g does not move:

$$X_g = \{x \in X : g \cdot x = x\}.$$

As usual, denote the orders of a finite set by the absolute value sign. Thus, $|X_g|$ is the number of points fixed by g, and $|G_x|$ is the number of elements that fix x.

The Burnside formula gives a way to count the number of orbits of this action.

Proposition 3.13 (Burnside's formula). *If a finite group G acts on a finite set X, then the number of orbits of the action equals*

$$\frac{1}{|G|}\sum_{g\in G}|X_g|.$$

Proof. The summation $\sum_{g\in G}|X_g|$ counts the number of pairs (g,x) that satisfy $g \cdot x = x$. Exactly the same thing is counted by the summation $\sum_{x\in X}|G_x|$. Thus

$$\sum_{g\in G}|X_g| = \sum_{x\in X}|G_x|.$$

Let $\mathcal{P}$ be the family of orbits under the action of G. This family is a partition of X. By collecting the second summation above first over each orbit S and then over the family of orbits $\mathcal{P}$, we see that

$$\sum_{x\in X}|G_x| = \sum_{S\in\mathcal{P}}\sum_{x\in S}|G_x|.$$

As x runs through an orbit S, the order $|G_x|$ of its stabilizer equals $|G|/|S|$, due to the orbit-stabilizer relation. Thus

$$\sum_{S\in\mathcal{P}}\sum_{x\in S}|G_x| = \sum_{S\in\mathcal{P}}\sum_{x\in S}|G|/|S| = |G|\sum_{S\in\mathcal{P}}\sum_{x\in S}1/|S| = |G|\sum_{S\in\mathcal{P}}1 = |G||\mathcal{P}|.$$

Putting all the equations together we obtain

$$|G||\mathcal{P}| = \sum_{g\in G}|X_g|,$$

which leads to the formula for $|\mathcal{P}|$ that we desire. □

Burnside's formula allows us to count the number of orbits of a group action by breaking up the problem into small bits.

3.3.1 A Counting Technique Based on Burnside

Let us illustrate the use of Burnside's formula in some delicate enumerations.

Building a Benzene Ring

A hypothetical compound is to be synthesized by attaching a molecule to each of the six vertices of a regular hexagon. Each vertex can receive any one of 4 molecules, called A, B, C, D. Call such a creation a ***configuration***. Since there are 6 vertices and each vertex accepts one of 4 possible molecules, the number of configurations is $4^6 = 4096$. To illustrate, here are three configurations.

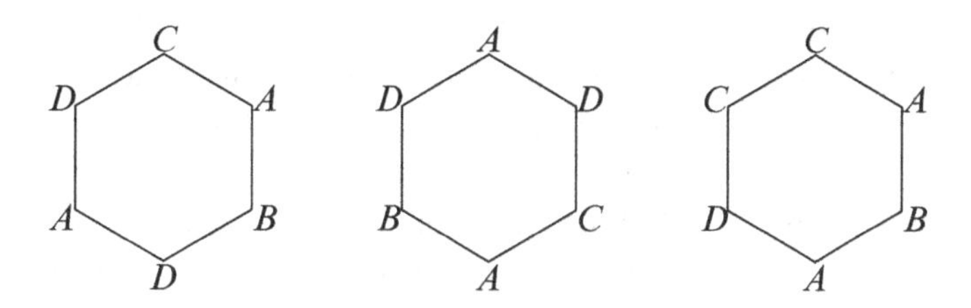

Let X be the set of these configurations. Two configurations make the same ***compound*** provided a symmetry of the hexagon superimposes one configuration on top of the other. For instance, the first two of the configurations shown make the same compound, because a clockwise rotation of the first hexagon through 120° superimposes one configuration upon the other.

The dihedral group $\mathcal{D}_6$ of symmetries of the hexagon, acts just as well on the set X of configurations. It is probably best to leave the formal definition of this action to the imagination. Two configurations make the same compound if and only if they are in the same orbit under the action of $\mathcal{D}_6$ on X. To count the number of distinct compounds, we need to count the number of orbits under the action of $\mathcal{D}_6$ on X. Burnside's formula is the tool to use.

For each of the 12 symmetries in $\mathcal{D}_6$, let us count the number of configurations that are fixed by that symmetry.

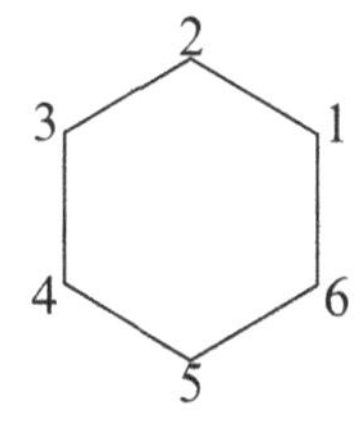

- The identity symmetry $\sigma = 1$ fixes all 4096 configurations.
- If σ is the reflection $(2, 6)(3, 5)$ across the axis joining vertex 1 to vertex 4, we are free to assign any one of A, B, C, D to the vertices $1, 2, 3, 4$. The requirement that σ fixes the configuration determines the assignment to the remaining vertices $5, 6$. Thus, there are 4^4

configurations fixed by this reflection. The similar reflections across the axes joining vertex 2 with vertex 5, and vertex 3 with vertex 6, each have 4^4 fixed configurations. All combined, these three reflections fix $4^4 + 4^4 + 4^4 = 768$ configurations.

- If σ is the reflection $(1,2)(3,6)(4,5)$ across the axis from the mid-point of the side from 1 to 2 to the mid-point of the side from 4 to 5, we are free to assign any one of A, B, C, D to each of the vertices $2, 3, 4$. The requirement that σ fixes the configuration determines the assignment to the vertices $4, 5, 6$. Thus, this reflection fixes 4^3 configurations. There are two similar reflections that each fix 4^3 configurations. From the three reflections, across axes joining the mid-points of opposite sides we pick up $4^3 + 4^3 + 4^3 = 192$ fixed configurations.
- If σ is the rotation $(1,2,3,4,5,6)$ of the hexagon through $60°$, we are free to assign any one of A, B, C, D to the vertex 1. After that the assignment to the other vertices is determined by the requirement that σ fixes the configuration. This rotation fixes just 4 configurations. Likewise, its inverse rotation $(6,5,4,3,2,1)$ fixes just 4 configurations. Together these two rotations fix 8 configurations.
- If σ is the rotation $(1,3,5)(2,4,6)$ through $120°$, we are free to assign any one of A, B, C, D to vertices 1 and 2. The requirement that a configuration be fixed by the rotation, determines the assignment to the remaining vertices. This rotation fixes 4^2 configurations. Its inverse rotation $(1,5,3)(2,6,4)$ also fixes 4^2 configurations. Combined, these two rotations fix $4^2 + 4^2 = 32$ configurations.
- The last remaining σ is the rotation $(1,4)(3,6)(2,5)$ through $180°$. After freely assigning any one of A, B, C, D to vertices $1, 2, 3$, the remaining assignment is determined by the requirement that this σ fixes the configuration. So, this rotation fixes $4^3 = 64$ configurations.

We have just seen that

$$\sum_{\sigma \in \mathcal{D}_6} X_\sigma = 4096 + 768 + 192 + 8 + 32 + 64 = 5160.$$

According to Burnside's formula, the total number of distinct compounds that can be formed is $5160/12 = 430$.

Painting a Tetrahedron

Let us use any one of n colors to paint each of the four faces of a regular tetrahedron, and call that a ***paint job***. Let X be the set of all possible paint jobs. There are n^4 of them. Two paint jobs give the same ***design*** when a symmetry of the tetrahedron superimposes one paint job onto the other one. We need to count the number of possible designs.

The group $\mathcal{T}$ of symmetries of the tetrahedron acts not only on its vertices $1, 2, 3, 4$, but also on the set X of paint jobs, in a natural way. To count the number of designs we need to count the number of orbits under the action of $\mathcal{T}$ on X. Burnside's formula is called for.

With the vertices of the tetrahedron labelled as $1, 2, 3, 4$, the group $\mathcal{T}$ coincides with the group $\mathcal{A}_4$ of even permutations on these letters.

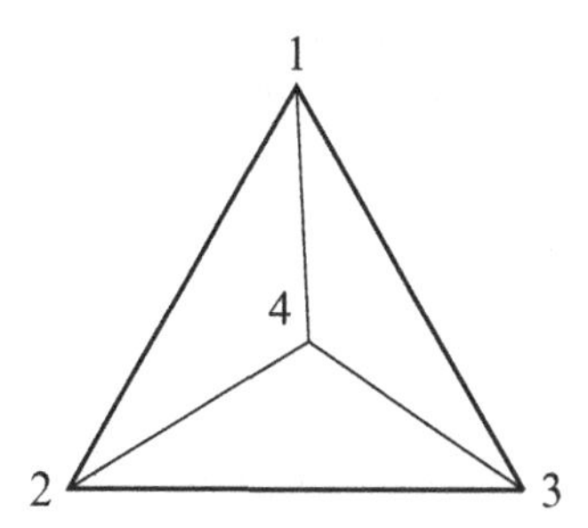

For each symmetry in $\mathcal{T}$ count the number of paint jobs that it fixes.

- If σ is the rotation $(2, 3, 4)$ that fixes the vertex 1, we can color the face opposite to 1 by any one of our n colors. Once we color another one of the four faces, the coloring of the remaining two faces is determined by this last coloring. Thus σ fixes n^2 paint jobs. Likewise the other seven rotations coming from the remaining 3-cycles in $\mathcal{T}$, each fix n^2 paint jobs. In all, the rotations coming from 3-cycles will fix $8n^2$ paint jobs.
- If σ is the (harder to visualize) rotation $(1, 2)(3, 4)$, then σ interchanges not only disjoint pairs of vertices, but also pairs of faces of the tetrahedron. Each face in an interchanged pair must have the same color in order to be fixed by σ. The number of paint jobs that are fixed equals n^2. The same argument applies to the rotations $(1, 3)(2, 4)$ and $(1, 4)(2, 3)$. In all, these three rotations fix $3n^2$ paint jobs.
- The remaining symmetry is the identity, which fixes all n^4 paint jobs.

The total number of paint jobs fixed by the symmetries in $\mathcal{T}$ is

$$8n^2 + 3n^2 + n^4 = n^4 + 11n^2.$$

Since $\mathcal{T}$ has 12 symmetries, Burnside's formula reveals that the tetrahedron allows for $\frac{n^4+11n^2}{12}$ designs.

Incidentally, it can be checked directly that $\frac{n^4+11n^2}{12}$ is an integer. Indeed, $n^4 \equiv n^2 \bmod 3$ and mod 4, by inspection. Thus $n^4 \equiv n^2 \bmod 12$, whereby $n^4 + 11n^2 \equiv 12n^2 \equiv 0 \bmod 12$.

Making a Necklace

Six white beads and four black beads are to be interspersed along a loop to make a necklace. How many different necklaces can be built?

To interpret this problem, take a decagon with vertices labelled $1, 2, 3, \ldots, 10$. To each vertex attach either a white bead W or a black bead B, using up all 10 beads. Here is an illustration of two such ***attachments***.

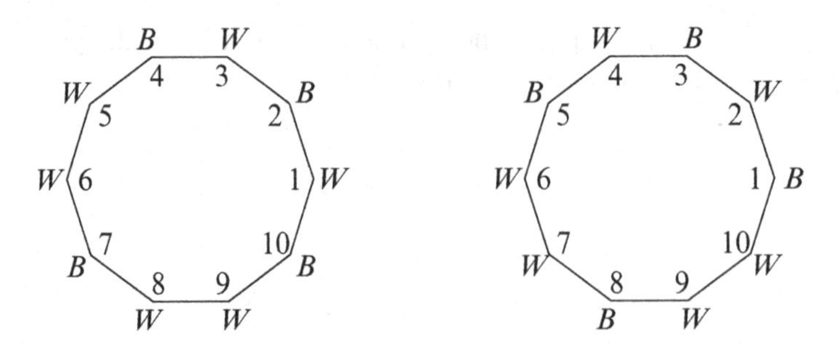

To count the number of ways of attaching beads, we need to choose where the four black beads should go. The white beads will go on the remaining empty vertices. The number of way to choose four vertices (to receive the black beads) from the 10 vertices is the binomial coefficient $\binom{10}{4} = 210$.

However, two bead attachments make the same necklace whenever a symmetry of the decagon brings one attachment to coincide with the other one. For instance, the attachments shown above make the same necklace because a counter-clockwise rotation of the first decagon through $1/10$ of a revolution will make its attachment coincide with that of the second decagon. Let X be the set consisting of the 210 bead attachments. By acting on the vertices $1, 2, \ldots, 10$, the dihedral group $\mathcal{D}_{10}$ also acts on X in a natural way. To count the number of distinct necklaces that can be made, we count the number of orbits under the action of $\mathcal{D}_{10}$ on X. Burnside's formula is called for.

We label the ten vertices of our decagon as shown in the preceding diagrams.

For each symmetry in $\mathcal{D}_{10}$, we need to count (using some imagination) the number of bead attachments that are fixed by that symmetry. Note that a symmetry of the decagon, written as the product of disjoint cycles on the letters $1, 2, \ldots, 10$, fixes a bead attachment if and only if the vertices in each of the disjoint cycles receive the same color of bead.

The group $\mathcal{D}_{10}$ is generated by the rotation $\sigma = (1, 2, \ldots, 10)$, through 1/10th of a revolution, and the reflection $\tau = (2, 10)(3, 9)(4, 8)(5, 7)$ across the axis joining vertex 1 to vertex 6. There are 20 symmetries to account for.

- The identity fixes all 210 attachments.
- The rotations $\sigma, \sigma^3, \sigma^7, \sigma^9$ are 10-cycles. As noted, any attachment of beads will require that all ten beads have the same color. For each of these four rotations the number of fixed attachments is 0. We do not have 10 beads of the same color.
- The rotations $\sigma^2, \sigma^4, \sigma^6, \sigma^8$ are each the product of two disjoint 5-cycles. To have a fixed attachment here we need at least five black beads or else ten white beads, which we do not have. The number of fixed attachments here is again 0.
- How about the rotation $\sigma^5 = (1, 6)(2, 7)(3, 8)(4, 9)(5, 10)$, though $1/2$ of a revolution? Here it looks like we can select any two of the five disjoint transpositions, and attach the four black beads to the vertices that get transposed. The other six vertices get white beads. Here, the number of fixed attachments is $\binom{5}{2} = 10$.

- Ten reflections remain to be accounted for. For the reflection τ, we can select any two of its four disjoint transpositions, assign the four black beads to their respective vertices, and give white beads to the remaining vertices. Such an attachment is fixed by τ. There are $\binom{4}{2} = 6$ ways to make such an attachment. We can also place a black bead on each of the vertices $1, 6$, and the other two black beads on one of the four pairs of vertices being transposed. This gives 4 more fixed attachments for a total of 10 fixed attachments under τ.

  There are four more reflections just like τ across axes joining opposite vertices. Each of them fixes 10 attachments. All combined, the reflections across axes that join opposite vertices will fix 50 attachments.
- There are five additional reflections in $\mathcal{D}_{10}$ across axes that join the mid-points of opposite sides, such as $\sigma\tau = (1,2)(3,10)(4,9)(5,8)(6,7)$. For this reflection, we can choose any two of its five disjoint transpositions, and assign black beads to the vertices they transpose, and white beads to the remaining vertices. In this way there are $\binom{5}{2} = 10$ ways of building attachments that are fixed by $\sigma\tau$.

  Taking all five reflections of this sort into account we get 50 ways to build attachments that are fixed by these reflections.

  All symmetries of $\mathcal{D}_{10}$ are accounted for.

The total number of fixed attachments under all symmetries of $\mathcal{D}_{10}$ equals

$$210 + 10 + 50 + 50 = 320.$$

By Burnside's formula the number of beads that can be built is $320/20 = 16$.

Evidently, counting problems such as these require patience and ingenuity. Yet, without Burnside's formula, they could well be overwhelming.

EXERCISES

1. Count the number of ways to color the four sides of a square using either the color red or the color blue. The same color can be used on all four sides.
2. Count the number of ways to string 3 black beads and 3 white beads on a necklace.
3. A configuration is to be formed by attaching any one of t letters to the vertices of a hexagon. Two configurations make the same object if a symmetry of the hexagon superimposes one configuration upon the other. Explain why the number of distinct objects equals

$$\frac{1}{12}(t^6 + 3t^4 + 4t^3 + 2t^2 + 2t).$$

4. Count the number of ways to color the edges of a regular tetrahedron using the colors red, white and blue.

5. A benzene molecule can be modelled as a regular hexagon with one of three radical compounds called A, B, C attached to the vertices of the hexagon. How many benzene molecules are there?
6. How many ways can the faces of a cube be colored using three colors?
7. If the edges of a square are colored using six colors and no color is used more than once, in how many ways can such a coloring be carried out?
8. A flag has 6 equal stripes each to be painted by one of three colors R, W, B. One coloring of the flag when viewed from the front coincides with another coloring when viewed from the back, we say the two flags are identical. For instance, using 6-tuples to denote the colorings, the flags

$$(R, W, B, B, R, W), \quad (W, R, B, B, W, R)$$

are identical. How many distinct flags can be colored?

What if 7 stripes and 3 colors are used?
9. Suppose that a finite group G acts transitively on a finite set X and that X has more than one element. Show that there is an element g in G having no fixed points in X, that is, such that $g \cdot x \neq x$ for all x in X.

3.4 The Class Equation and Its Implications

Every group G acts on itself by left multiplication. This is nothing but the left coset action of G on G/H where H is the trivial subgroup. But G comes with another action on itself.

Definition 3.14. If $g \in G$ and $x \in G$ (viewed as a point to be acted upon), let

$$g \cdot x = gxg^{-1}.$$

With this recipe, we shall say that G ***acts on itself by conjugation***.

That conjugation is a group action of G on G can be seen by inspection.

If $\varphi : G \to \mathcal{S}(G)$ is the homomorphism that implements the conjugation action, its kernel is the set:

$$\begin{aligned}
\ker\varphi &= \{g \in G : \varphi(g) \text{ is the identity map on } G\} \\
&= \{g \in G : \varphi(g)(x) = x \text{ for all } x \text{ in } G\} \\
&= \{g \in G : g \cdot x = gxg^{-1} = x \text{ for all } x \text{ in } G\} \\
&= \{g \in G : gx = xg \text{ for all } x \text{ in } G\}.
\end{aligned}$$

Evidently, the kernel of the conjugation action is the centre $Z(G)$ of the group.

3.4.1 The Class Equation

We ask about the orbits of the conjugation action.

Definition 3.15. If G acts on itself by conjugation and $x \in G$, then the orbit

$$G \cdot x = \{g \cdot x : g \in G\} = \{gxg^{-1} : g \in G\}$$

is called the ***conjugacy class*** of x.

For the group $\mathcal{S}_n$, Proposition 3.56 states that two permutations belong to the same conjugacy class if and only if the permutations have the same orbit structure. For instance, in $\mathcal{S}_5$ those permutations that are the disjoint product of a 3-cycle and a 2-cycle form one conjugacy class. And the set of all 4-cycles is another conjugacy class in $\mathcal{S}_5$.

The family of conjugacy classes forms a partition of G. If G is a finite group, Proposition 3.11 reveals that the order of each conjugacy class is a divisor of the order of G. Putting these observations together, we see that in the important case where G is finite:

the order $|G|$ is the sum of the orders of its conjugacy classes, and the orders of these classes are divisors of $|G|$.

When does a point x in G have just itself in its conjugacy class? This happens when $gxg^{-1} = x$ for all g in G, and this happens if and only if x belongs to the centre $Z(G)$. This observation enables us to repartition G using the family of subsets consisting of the centre $Z(G)$ and those conjugacy classes $G \cdot x$ that have order more than 1. We end up with a revealing result known as the ***class equation***.

Proposition 3.16 (Class equation). *If G is a finite group and if J is a set of representatives for the conjugacy classes that have order more than 1, then*

$$|G| = |Z(G)| + \sum_{x \in J} |G \cdot x|.$$

Each of the summands in the above equation is a divisor of $|G|$, and all $|G \cdot x| > 1$.

The Class Equation in Terms of Centralizers

For x in G, its stabilizer G_x under the conjugation action is the subgroup

$$G_x = \{g \in G : g \cdot x = gxg^{-1} = x\} = \{g \in G : gx = xg\}.$$

Because G_x is the subgroup consisting of those elements g that commute with x, its more common name is the ***centralizer*** of x, to be denoted by $C(x)$. The centralizer of x contains at least x itself as well as the centre $Z(G)$ of G, because the elements in the centre commute with every element in G. Also, $x \in Z(G)$ if and only if $C(x) = G$.

In light of the orbit-stabilizer relation, Proposition 3.10, the class equation, Proposition 3.16, has an alternative formulation.

Proposition 3.17 (Alternative class equation). *If G is a finite group and if J is a set of representatives for the conjugacy classes that have order more than 1, then*

$$|G| = |Z(G)| + \sum_{x \in J} [G : C(x)].$$

The Class Equation for $\mathcal{S}_4$

By way of illustration, let us write the class equation for the symmetric group $\mathcal{S}_4$. To do that we have to find the order of each conjugacy class. The orders of the conjugacy classes are determined by use of Proposition 2.46.

- The order of $\mathcal{S}_4$ is 24.
- The centre of $\mathcal{S}_4$ is trivial of order 1.
- The 2-cycles in $\mathcal{S}_4$ are

$$(1,2), (1,3), (1,4), (2,3), (2,4), (3,4),$$

  forming a conjugacy class of order 6.
- The products of disjoint pairs of 2-cycles are

$$(1,2)(3,4), (1,3)(2,4), (1,4)(2,3),$$

  forming a conjugacy class of order 3.
- The 4-cycles are

$$(1,2,3,4), (1,2,4,3), (1,3,2,4), (1,3,4,2), (1,4,2,3), (1,4,3,2),$$

  forming a class of order 6.
- The 3-cycles are

$$(1,2,3), (1,3,2), (1,2,4), (1,4,2), (1,3,4), (1,4,3), (2,3,4), (2,4,3),$$

  forming a class of order 8.

The class equation is the sum of these orders. Namely,

$$24 = 1 + 6 + 3 + 6 + 8.$$

As expected each summand is a divisor of 24.

The Class Equation for a Dihedral Group

Let us write the class equation for the dihedral group $\mathcal{D}_4$. Recall that $\mathcal{D}_4$ is the group of symmetries of the square whose vertices can be given by the points $1, i, -1, -i$ in the complex plane. The group $\mathcal{D}_4$ has order 8 and is generated by the rotation

$$\sigma \text{ given by } 1 \mapsto i \mapsto -1 \mapsto -i \mapsto 1,$$

and the reflection

$$\tau \text{ given by } 1 \mapsto 1, \; -1 \mapsto -1, i \mapsto -i \mapsto i.$$

The eight symmetries in $\mathcal{D}_4$ can be uniquely written as $\sigma^j\tau^i$ where $j = 0, 1, 2, 3$ and $i = 0, 1$. And the group is governed by the relations

$$\sigma^4 = 1, \; \tau^2 = 1, \; \sigma\tau\sigma = \tau.$$

By playing with these relations, let us determine the conjugacy classes of $\mathcal{D}_4$.

First, the centre $Z(\mathcal{D}_4)$ has to be non-trivial. Otherwise, the class equation would say that

$$8 = 1 + \text{a sum of even integers},$$

which cannot be.

From $\sigma\tau\sigma = \tau$, we get $\sigma^j\tau\sigma^j = \tau$ for all j, and then

$$\sigma^j\tau = \tau\sigma^{-j}.$$

In particular, $\sigma^2\tau = \tau\sigma^{-2}$. But $\sigma^{-2} = \sigma^2$, since $\sigma^4 = 1$. So, $\sigma^2\tau = \tau\sigma^2$. Since σ^2 commutes with the generators τ and σ of $\mathcal{D}_4$, it must lie in the centre of $\mathcal{D}_4$. We shall see from what follows that there is nothing else but 1 and σ^2 in $Z(\mathcal{D}_4)$.

What are the conjugates of σ? The relations in $\mathcal{D}_4$ reveal that $\tau\sigma\tau^{-1} = \tau\sigma\tau = \sigma^{-1} = \sigma^3$. Thus σ and σ^3 are conjugates. Since conjugate elements all have the same order and since σ, σ^3 are the only elements in $\mathcal{D}_4$ of order 4, these two rotations form a complete conjugacy class. They are also not in $Z(\mathcal{D}_4)$, because central elements only have themselves as conjugates.

What are the conjugates of τ? Well, the group relations yield $\sigma\tau\sigma^{-1} = \sigma^2\tau$. So, the reflections τ and $\sigma^2\tau$ are conjugate. This also implies they are not in $Z(\mathcal{D}_4)$. We shall see that these two reflections also form a complete conjugacy class.

But first check that $\sigma\tau$ and $\sigma^3\tau$ are conjugates. Well, by the group relations, $\tau(\sigma\tau)\tau^{-1} = \tau\sigma = \sigma^3\tau$. Hence these two elements are conjugate, and from this we see that they are not central either.

To finish the job, we need to be sure that $\tau, \sigma^2\tau$ form a separate conjugacy class from that of $\sigma\tau, \sigma^3\tau$. It suffices to check that τ and $\sigma\tau$ are not conjugates. Well, we could just conjugate τ by all eight elements of $\mathcal{D}_4$, and see that we never get $\sigma\tau$. Or, a bit more efficiently, notice for $j = 0, 1, 2, 3$ and $i = 0, 1$ that

$$\sigma^j\tau^i\tau(\sigma^j\tau^i)^{-1} = \sigma^j\tau^i\tau\tau^{-i}\sigma^{-j} = \sigma^j\tau\sigma^{-j} = \sigma^j\sigma^j\tau = \sigma^{2j}\tau,$$

which never equals $\sigma\tau$.

We have seen that $1, \sigma^2$ make up the centre of $\mathcal{D}_4$. And there are three conjugacy classes each of order 2. Namely,

$$\{\sigma, \sigma^3\},\ \{\tau, \sigma^2\tau\},\ \{\sigma\tau, \sigma^3\tau\}.$$

The class equation for $\mathcal{D}_4$ looks like:

$$8 = 2 + 2 + 2 + 2,$$

where the first "2" corresponds to the order of the centre.

At the risk of beating this problem to death, here is another way to see that τ is not conjugate to $\sigma\tau$. We can think of the vertices $1, i, -1, -i$ of the square as the letters $1, 2, 3, 4$, respectively. From this point of view, σ is the cycle $(1, 2, 3, 4)$, τ is the transposition $(2, 4)$, and $\mathcal{D}_4$ is the subgroup of $\mathcal{S}_4$ generated by σ and τ. Then

$$\sigma\tau = (1, 2, 3, 4)(2, 4) = (1, 2)(3, 4).$$

If τ were conjugate to $\sigma\tau$ in $\mathcal{D}_4$, they would remain conjugate in $\mathcal{S}_4$. However this cannot be, since τ and $\sigma\tau$ do not have the same orbit structure.

We might like to notice as well that the conjugacy classes of order 2 in $\mathcal{D}_4$ come with a nice geometrical symmetry. The rotations σ, σ^3 are those through $90°$. The reflections $\tau, \sigma^2\tau$ are across those axes joining opposite vertices of the square. The reflections $\sigma, \sigma^3\tau$ are across those axes joining the mid-points of opposite sides of the square.

The Class Equation and Groups of Order 124

When the order of a group is large and has a number of prime factors, the possibilities for the group can be mind boggling. Yet the class equation may still squeeze out fragments of information. To illustrate, let us show that groups of order 124 are never simple. Note that $124 = 31 \cdot 2 \cdot 2$.

If our group G, of order 124, is abelian, then G will contain an element whose order is greater than 1 and a proper divisor of 124. That element generates a proper, non-trivial, normal subgroup of G. If G is non-abelian and has a non-trivial centre $Z(G)$, then $Z(G)$ is a suitable normal subgroup.

That leaves the case where $Z(G)$ is trivial. Now the class equation, Proposition 3.16, implies that

$$124 = 1 + \text{a sum of possibly 2s, 4s, 31s and 62s.}$$

If only 31s and 62s appear in the sum, then $0 \equiv 1 \bmod 31$, which is impossible. Thus, a 2 or 4 appears in the class equation. Looking at the class equation, Proposition 3.17, we see that there is an x in G whose centralizer $C(x)$ has index 2 or 4. Clearly $124 \nmid 2!$ and $124 \nmid 4!$. By Proposition 3.4, this centralizer contains a non-trivial and normal subgroup of G. And so G is not simple.

3.4.2 The Class Equation and p-Groups

The point of the class equation is not so much to look at it but rather to shed some light on the group in question.

Definition 3.18. A group whose order is the power of a single prime p is called a ***p*-group**.

For instance $\mathcal{D}_4$, being of order 2^3 is a 2-group. We have already seen that the centre of $\mathcal{D}_4$ is non-trivial. The same fact holds for all p-groups.

Proposition 3.19. *For each prime p the centre of every p-group is non-trivial.*

Proof. If G is our group of order p^n, the class equation for G would look like

$$|G| = p^n = |Z(G)| + \text{a sum of positive powers of } p.$$

This implies that p divides $|Z(G)|$. Hence, $Z(G)$ is non-trivial. In fact its order is a positive power of p. □

Groups of Order p^2

Every group of prime order is cyclic and thus abelian. This can fail when the order is a higher prime power. For instance, $\mathcal{D}_4$ is a non-abelian group of order 2^3. Yet, groups whose order is the square of a prime remain abelian.

Proposition 3.20. *If p is a prime, then every group of order p^2 is abelian.*

Proof. The centre Z of our group G is non-trivial, due to Proposition 3.19. Thus the order of G/Z is either p or 1, by Lagrange. If the order is 1, then $G = Z$, an abelian group. If the order is p, then G/Z is cyclic because all groups of prime order are cyclic. Now G is abelian because of Proposition 3.5. □

We have Propositions 2.81 and 2.86 which tell us that abelian groups of order p^2 are isomorphic either to the cyclic group $\mathbb{Z}_{p^2}$ or to the Cartesian product $\mathbb{Z}_p \times \mathbb{Z}_p$. Evidently, groups of order p^2 are now completely classified.

Chains of Subgroups of a p-Group

For future consideration here is something deeper about subgroups of p-groups.

Proposition 3.21. *If G is a group of order p^m for some prime p and $m \geq 1$, then there exists a strictly decreasing chain of normal subgroups*

$$G = N_0 \supset N_1 \supset N_2 \supset \cdots \supset N_{m-1} \supset N_m = \langle 1 \rangle.$$

Each quotient group N_{j-1}/N_j has order p.

Proof. The result is obvious for groups of order p. Assuming the result for p-groups of order less than p^m, use this inductive assumption to get the result for G of order p^m.

The centre Z of G is non-trivial, by Proposition 3.19. Then Z contains a subgroup H of order p, by Cauchy's theorem, Proposition 2.61. Since H lies in the centre of G, this subgroup is normal in G. By Lagrange, the order of the quotient group G/H is p^{m-1}, and by the inductive hypothesis there is a properly decreasing chain of normal subgroups:

$$G/H = M_0 \supset M_1 \supset M_2 \cdots \supset M_{m-2} \supset M_{m-1} = \langle 1 \rangle,$$

such that each M_{j-1}/M_j has order p.

According to the correspondence theorem, Proposition 2.64, there is a decreasing chain of subgroups

$$G = N_0 \supset N_1 \supset N_2 \supset \cdots \supset N_{m-2} \supset N_{m-1} = H,$$

such that $M_j = N_j/H$ and these N_j are normal in G. The desired chain is completed by putting $N_m = \langle 1 \rangle$.

The fact that each quotient N_{j-1}/N_j has order p follows because the chain descending from G to $\langle 1 \rangle$ has m proper inclusions. By Lagrange each index $[N_{j-1} : N_j]$ is a positive power of p, say p^{t_j} where $t_j \geq 1$. Also by Lagrange we get

$$p^m = |G| = p^{t_1} p^{t_2} \cdots p^{t_m} = p^{t_1+t_2+\cdots+t_m}.$$

Thus $m = t_1 + t_2 + \cdots + t_m$ which means that all $t_j = 1$. □

Proposition 3.21 provides a significant family of what are known as ***solvable groups*** to be discussed in Section 3.7. Their significance arises from the classical problem of solving polynomial equations by means of formulas, which gave birth to Galois theory, a topic to be examined in Chapters 6 and 7.

A Converse to Lagrange for p-Groups

We might recall that the converse of Lagrange does not hold. The simplest illustration is the alternating group $\mathcal{A}_4$ of order 12 but having no subgroup of order 6. However, in the case of p-groups, Lagrange does have a converse.

Proposition 3.22. *If G is a finite p-group and d divides the order of G, then G contains a normal subgroup of order d.*

Proof. Since the order of G is p^m for some m, the divisor d equals p^k for some $k \leq m$. Examine the strictly decreasing chain of normal subgroups in Proposition 3.21, and exploit Lagrange's theorem. By Proposition 2.39 the order of G is the product of the indices $[N_{j-1} : N_j]$. That is

$$p^m = [N_0 : N_1] \cdot [N_1 : N_2] \cdot [N_2 : N_3] \cdots [N_{m-1} : N_m].$$

Since the inclusions in the chain are proper, each index must be p^r for some $r \geq 1$, and since there are m indices being multiplied, each $[N_{j-1} : N_j] = p$. From Lagrange we can see that the order of N_1 is p^{m-1}, the order of N_2 is p^{m-2}, the order of N_3 is p^{m-3} and so on. In particular, the order of the normal subgroup N_{m-k} is p^k, as desired. □

EXERCISES

1. Why is every group of order 49 isomorphic to one of two groups? What are those groups?
2. If Z is the centre of a group G, show that the order of the quotient group G/Z cannot be a prime.
3. Find the class equations for $\mathcal{D}_3, \mathcal{D}_5, \mathcal{A}_4$ and $\mathcal{Q}$ (the quaternion group).
4. How many conjugacy classes does $\mathcal{S}_6$ have?
5. A group of order 2 has just two conjugacy classes. Are there any other groups with only two conjugacy classes?
6. (a) How many n-cycles are there in $\mathcal{S}_n$?
   (b) Find the order of the centralizer of the cycle $(1, 2, 3, \ldots, n)$ in $\mathcal{S}_n$.
   (c) Exhibit all the permutations in $\mathcal{S}_n$ that commute with $(1, 2, 3, \ldots, n)$.
7. (a) How many 2-cycles are there in $\mathcal{S}_n$?
   (b) Find the order of the centralizer of the transposition $(1, 2)$ in $\mathcal{S}_n$.
   (c) Exhibit all the permutations in $\mathcal{S}_n$ that commute with $(1, 2)$.

8. Let p be a prime and let $GL_3(\mathbb{Z}_p)$ be the group of 3×3 invertible matrices with entries in the finite field $\mathbb{Z}_p$. If G is the subgroup consisting of those matrices having the value 1 on the main diagonal, verify that G is a group of order p^3 and such that $A^p = I$ for every A in G.

Since G is a p-group, its centre Z is non-trivial. Find Z.

9. The goal of this exercise is to describe all groups of order 8. Such groups are p-groups (where $p = 2$), and thereby come with a non-trivial centre. We can exploit this fact.

(a) Explain why every abelian group G of order 8 is isomorphic to one of three groups, and present those three groups.

(b) If G of order 8 is non-abelian, show that its centre Z has order 2, and that $x^2 \in Z$ for all x in G.

For the rest of this exercise suppose G of order 8 is non-abelian.

(c) Why must G have an element b of order 4?

Let H be the cyclic subgroup of order 4 generated by b. Either the elements of the complement $G \setminus H$ all have order 4, or there is an element in $G \setminus H$ of order 2.

(d) Let us see what happens if all elements of $G \setminus H$ have order 4.

(i) Using part (b) explain why, in addition to the identity element, G has one element z of order 2 and six elements of order 4. If x is an element of order 4, explain why $x^2 = z$.

(ii) If x is an element of order 4 show that $zx = x^{-1}$, also of order 4.

(iii) Let c be an element of order 4 other than b or zb, and let $d = bc$. Show that d is not one of $1, z, b, zb, c, zc$. Thus the elements of order 4 are b, zb, c, zc, d, zd.

(iv) Prove that

$$bc = d, cb = zd, cd = b, dc = zb, db = c, bd = zc.$$

The first of the above equations comes from the definition of d.

(v) Explain why $G \cong \mathcal{Q}$, the quaternion group from Section 2.4.3.

(e) Next let us see what happens when $G \setminus H$ has an element of order 2.

(i) In this case explain why $aHa = H$.

(ii) Show that $aba = b^{-1} = b^3$, and deduce that $bab = a$.

(iii) Explain why a, b generate G.

(iv) Using the relations $b^4 = 1, a^2 = 1, bab = a$, explain why $G \cong \mathcal{D}_4$.

We have learned that there are five non-isomorphic groups of order 8. Three are abelian, and two are the non-abelian groups $\mathcal{Q}$ and $\mathcal{D}_4$.

3.5 The Theorems of Cauchy and Sylow

We know that the converse of Lagrange does not hold. The group $\mathcal{A}_4$ comes to mind, as noted more than once. The trouble with 6 is that it takes more than one prime to factor it. If m is the power of a single prime and m divides the order of a group, subgroups of order m will exist, and quite a bit can be said about them. That is what the upcoming results are about. It takes a bit of slogging to work through the ensuing ideas, but the reward will be a more profound understanding of finite groups.

3.5.1 Cauchy's Theorem

We have seen in Proposition 2.61 that every abelian group, whose order is divisible by a prime, contains an element whose order equals that prime. The class equation lets us remove the requirement that the group be abelian. This was one of the first big results in the development of group theory.

Proposition 3.23 (Cauchy's theorem). *If G is a finite group and p is a prime that divides $|G|$, then G contains an element, and thereby a subgroup, of order p.*

Proof. Suppose the result is false, and seek a contradiction.

In that case there will be a group G that is as small as possible for which the theorem fails. More precisely, if the result is false, then there is a prime p and a finite group G such that

- p divides $|G|$,
- G has no elements of order p, and
- every group K, such that $|K| < |G|$ and p divides $|K|$, contains an element of order p.

The order of the centre $Z(G)$ cannot be divisible by p. Indeed, Cauchy's theorem holds for abelian groups. So, if p were to divide $|Z(G)|$, then $Z(G)$, and thereby G, would contain an element of order p.

Let J be a set of representatives for the conjugacy classes of order more than 1. If $x \in J$, then x is surely not in $Z(G)$, which means that the centralizer $C(x)$ is not all of G. Thus p cannot divide $|C(x)|$. For otherwise, the group $C(x)$, whose order is less than the order of G, would contain an element of order p. Then G would contain that same element of order p. Since p divides $|G|$ and does not divide $|C(x)|$, Lagrange forces the prime p to divide the index $[G : C(x)]$.

The class equation, Proposition 3.17, says $|G| = |Z(G)| + \sum_{x \in J}[G : C(x)]$. In this equation, every term is divisible by p except for $|Z(G)|$, and that is a contradiction. □

Using Cauchy to Identify p-Groups by Their Elements

As a simple application of Cauchy's theorem we can note that a p-group is detectable by looking at the order of its elements.

Proposition 3.24. *If p is a prime, a finite group G is a p-group if and only if the order of every element in G is a power of p.*

Proof. If G is a p-group, then Lagrange implies that the order of every element in G is a divisor of the pth power $|G|$. Hence, the order of every element in G is a pth power.

If G is not a p-group, then some other prime q divides $|G|$. Cauchy's theorem applied to q yields an element in G of order q, which is surely not a power of p. □

Groups of Order $2p$

Cauchy's theorem can be used to give an accounting of groups having order $2p$, where p is an odd prime.

Proposition 3.25. *If p is an odd prime and G is a group of order $2p$, then G is either cyclic or isomorphic to the dihedral group $\mathcal{D}_p$.*

Proof. By Cauchy's theorem, G contains a subgroup H of order 2 and a subgroup K of order p. The index of K in G is 2, which ensures K is normal in G. The order of the intersection $H \cap K$ divides both 2 and p. This forces the intersection to be trivial. By Proposition 2.76, the internal product HK has order $2p$. Hence $G = HK$.

Let a be a generator for H and b a generator for K. The order of b is p, and the order of a is 2, meaning that $a = a^{-1}$.

If a commutes with b, then the elements of H commute with those of K, and Proposition 2.74 leads to $H \times K \cong HK = G$. Since both H and K have prime order, they are both cyclic. Thus, $H \times K$, and thereby G, is cyclic by Proposition 2.66.

Suppose a, b do not commute. By the normality of K, the product aba (which is the same as aba^{-1}) is another generator of the cyclic group K. Hence, $aba = b^j$, where $1 \leq j \leq p-1$. Keeping in mind that $a^2 = 1$ we obtain

$$b = a^2ba^2 = a(aba)a = ab^ja = (aba)^j = (b^j)^j = b^{j^2}.$$

Seeing that $b^{j^2-1} = 1$ it follows that $p \mid (j^2-1) = (j-1)(j+1)$, from Proposition 2.5. Now if p were to divide $j-1$, the fact that $j-1 < p$ would imply $j-1 = 0$, and so $j = 1$. This would lead to $aba = b$, contrary to the fact a, b do not commute. Thus p divides $j+1$ and since $j+1 \leq p$, we must have that $j+1 = p$.

We have learned that if $ab \neq ba$, then $aba = b^{p-1} = b^{-1}$. That is $bab = a$.

Since the index of K in G is 2, the cosets K, aK form a partition of G. Using the fact b generates K, we see that the elements of G consist of all

$$b^j \text{ and } ab^j, \text{ where } j = 0, 1, \ldots, p-1.$$

The multiplication table for these elements is determined by the relations

$$a^2 = 1,\ b^p = 1,\ bab = a.$$

Every group of order $2p$ with generators a, b satisfying the above relations is isomorphic to $\mathcal{D}_p$, because these relations determine the multiplication table of such a group. □

If G is a group of order $2p$ generated by elements a, b and subject to the relations

$$a^2 = 1,\ b^p = 1,\ bab = a,$$

it might be interesting to see how G is isomorphic to $\mathcal{D}_p$ from a geometrical perspective.

The subgroup H of order 2 generated by a has p cosets:

$$H,\ bH,\ b^2H,\ b^3H, \ldots, b^{p-1}H.$$

There is a left coset action of G on this set of cosets, as discussed in Section 3.1.1. In particular the action of b on this set of cosets is given by

$$b \cdot H = bH,\ b \cdot bH = b^2H,\ \ldots, b \cdot b^{p-2}H = b^{p-1}H,\ b \cdot b^{p-1}H = b^pH = H.$$

For the action of a on the cosets of H, use the relation $ab = b^{-1}a = b^{p-1}a$. Thus,

$$a \cdot H = H,\ a \cdot bH = b^{p-1}H,\ a \cdot b^2H = b^{p-2}H,\ a \cdot b^3H = b^{p-3}H, \ldots .$$

Now use the cosets of H as labels for the vertices of a p-gon, as shown below for $p = 5$.

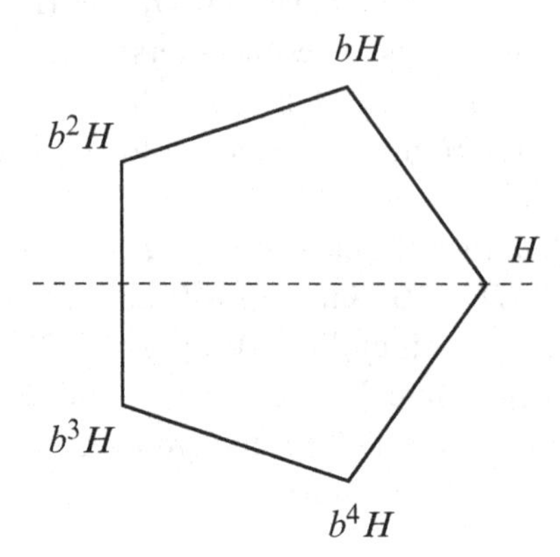

It is clear from this picture that b acts as the rotation of the p-gon through the angle $2\pi/p$, while a is the reflection of the p-gon through the horizontal axis passing through the vertex H. The generators of G act as the rotation and the reflection that generate $\mathcal{D}_p$. Thus G has a geometric incarnation as $\mathcal{D}_p$.

3.5.2 Sylow's Theorem

Ludwig Sylow (1832–1918) was a Norwegian mathematician who built upon Cauchy's theorem in a profound manner. His results now come with a variety of proofs. We offer a presentation of Sylow's theorem which uses yet another group action.

An Action on the Set of Conjugates of a Subgroup

If H is a subgroup of a group G and $x \in G$, the group

$$xHx^{-1} = \{xhx^{-1} : h \in H\}$$

is called a ***conjugate subgroup*** of H. Let $\mathcal{X}$ be the set of all conjugate subgroups of H. It is possible, for different x, y in G, to have $xHx^{-1} = yHy^{-1}$. For instance, when H is normal, all conjugate subgroups coincide, and the set $\mathcal{X}$ is made up of just one item, the subgroup H.

For a concrete example, take the group $\mathcal{S}_4$, and let H be the cyclic subgroup generated by $(1,2)$. The set $\mathcal{X}$ of conjugates of H consists of the 6 cyclic groups generated by each of the transpositions

$$(1,2), (1,3), (1,4), (2,3), (2,4), (3,4).$$

Thinking of the conjugates of H as points in $\mathcal{X}$, let G act on $\mathcal{X}$ according to

$$g \cdot K = gKg^{-1}, \text{ for every point } K \text{ in } \mathcal{X} \text{ and every } g \text{ in } G.$$

In other words, G conjugates the conjugate subgroups of H.

Although it is routine, let us check that if K is a conjugate subgroup of H and $g \in G$, then $g \cdot K = gKg^{-1}$ is a conjugate subgroup of H too. Well, write $K = xHx^{-1}$ for some x in G. Then

$$gKg^{-1} = gxHx^{-1}g^{-1} = gxH(gx)^{-1},$$

which is clearly a conjugate of H. We also need to confirm that we have a group action. If $g,f \in G$ and $K \in \mathcal{X}$, then

$$gf \cdot K = (gf)K(gf)^{-1} = gfKf^{-1}g^{-1} = g(fKf^{-1})g^{-1} = g \cdot (f \cdot K).$$

It is even more obvious that

$$1 \cdot K = 1K1^{-1} = K \text{ for all } K \text{ in } \mathcal{X}.$$

This verifies that the conjugation action of G on the set of conjugates of H is a bona fide group action.

The Normalizer

In the ensuing discussion H plays a dual role, as a subgroup of G, and as a point in $\mathcal{X}$ to be acted upon by G. Alertness as to the role of H will be helpful.

The orbit of the point H in $\mathcal{X}$ is

$$G \cdot H = \{gHg^{-1} : g \in G\}.$$

Clearly the orbit is all of $\mathcal{X}$. If G is finite, the orbit-stabilizer relation given in Proposition 3.11, applied to the orbit $\mathcal{X}$ of the point H, says that

$$|G| = |G \cdot H||G_H| = |\mathcal{X}||G_H|.$$

Here, G_H is the stabilizer of the point H. An element g in G stabilizes the point H if and only if

$$g \cdot H = gHg^{-1} = H.$$

This stabilizer subgroup G_H is more commonly known as the ***normalizer*** of the subgroup H in G. In keeping with tradition we denote the normalizer of H in G by $N_G(H)$.

The orbit-stabilizer relation for this conjugation action boils down to the following.

Proposition 3.26. *In a finite group G the number of conjugates of a subgroup H times the order of the normalizer of H equals the order of G.*

Clearly if $h \in H$, then $hHh^{-1} = H$, and therefore $H \subseteq N_G(H)$. The normalizer of a subgroup contains the subgroup. Another thing is that the subgroup H is normal in its normalizer. This is direct from the definition of the normalizer. Thus the normalizer of H in G is the biggest subgroup of G in which H remains a normal subgroup. For instance, if H is normal in G, then its normalizer is all of G.

An Illustration of a Normalizer in $\mathcal{S}_5$

For an example of a normalizer, take $\mathcal{S}_5$ and let H be the subgroup generated by the 3-cycle $(1,2,3)$. The conjugate subgroups of H are the cyclic groups generated by the conjugates of $(1,2,3)$. The conjugates of $(1,2,3)$ are all twenty of the 3-cycles that are in $\mathcal{S}_5$. These twenty 3-cycles generate exactly 10 subgroups of $\mathcal{S}_5$. (Keep in mind that the group generated by a 3-cycle includes the generator and its inverse.) Here are the 10 conjugate subgroups of H that make up the set $\mathcal{X}$ to undergo the conjugation action by $\mathcal{S}_5$:

$$\begin{aligned} H = K_1 &= \langle(1,2,3)\rangle, & K_2 &= \langle(1,2,4)\rangle, & K_3 &= \langle(1,3,4)\rangle, & K_4 &= \langle(1,3,5)\rangle, \\ K_5 &= \langle(1,4,5)\rangle, & K_6 &= \langle(1,4,5)\rangle, & K_7 &= \langle(2,3,4)\rangle, & K_8 &= \langle(2,3,5)\rangle, \\ K_9 &= \langle(2,4,5)\rangle, & K_{10} &= \langle(3,4,5)\rangle. \end{aligned}$$

According to Proposition 3.26, the normalizer of H will be a subgroup of $\mathcal{S}_5$ having order $120/10 = 12$. What are those 12 permutations that normalize H?

Well, the non-identity elements of H are the cycles $(1,2,3)$ and its inverse $(3,2,1)$. A permutation σ in $\mathcal{S}_5$ normalizes H if and only if

$$\sigma(1,2,3)\sigma^{-1} = (1,2,3) \text{ or } \sigma(1,2,3)\sigma^{-1} = (3,2,1).$$

As explained at the outset of Section 2.4.1, or by direct verification, we can see that $\sigma(1,2,3)\sigma^{-1}$ equals the cycle $(\sigma(1),\sigma(2),\sigma(3))$. Thus σ normalizes H if and only if the cycle $(\sigma(1),\sigma(2),\sigma(3))$ equals the cycle $(1,2,3)$ or the cycle $(3,2,1)$.

Noting that the cycle $(1,2,3)$ can also be represented as the 3-tuples $(2,3,1)$ and $(3,1,2)$, we see that $(\sigma(1),\sigma(2),\sigma(3))$ is the cycle $(1,2,3)$ if and only if

$$\begin{gathered} \sigma(1) = 1, \quad \sigma(2) = 2, \quad \sigma(3) = 3 \\ \text{or} \\ \sigma(1) = 2, \quad \sigma(2) = 3, \quad \sigma(3) = 1 \\ \text{or} \\ \sigma(1) = 3, \quad \sigma(2) = 1, \quad \sigma(3) = 2. \end{gathered}$$

The σ which satisfy the first of the above three conditions are the identity 1 as well as the transposition $(4,5)$. The σ which satisfy the second of the above three conditions are the cycle $(1,2,3)$ and the disjoint product $(1,2,3)(4,5)$. The σ which satisfy the third condition are the cycle $(1,3,2)$ and the disjoint product $(1,3,2)(4,5)$. Thus we pick up 6 of the 12 permutations in the normalizer of H. Namely,

$$1,\ (1,2,3),\ (1,3,2),\ (4,5),\ (1,2,3)(4,5),\ (1,3,2)(4,5).$$

The other 6 permutations in the normalizer come from the alternative requirement that the cycle $(\sigma(1),\sigma(2),\sigma(3))$ is the cycle $(3,2,1)$. By imitating the preceding analysis we can discover that the other 6 permutations in the normalizer of H are

$$(1,2),\ (1,3),\ (2,3),\ (1,2)(4,5),\ (1,3)(4,5),\ (2,3)(4,5).$$

All 12 elements of the normalizer of H have been discovered.

Maximal p-Subgroups

If G is a group and p is a prime, a subgroup H is called a ***p*-subgroup** when the order of H is a power of p. The trivial subgroup, having order p^0, is a p-subgroup. If p divides the order of G, Cauchy's theorem, Proposition 3.23, ensures that G contains a non-trivial p-subgroup, actually one of order p. A subgroup K is called a ***maximal p*-subgroup** when no p-subgroup of G properly contains K. Clearly, if G is a finite group and H is a p-subgroup, then H is contained in a maximal p-subgroup. In particular, since the trivial group is a p-subgroup, every finite group contains at least one maximal p-subgroup.

If p^r is the highest power of p that divides $|G|$, Lagrange implies that the biggest order that a maximal p-subgroup of G could have is p^r. A part of Sylow's upcoming theorem says that the order of every maximal p-subgroup of G is precisely p^r. Thereby, G always contains a subgroup of order p^r.

Note that if K is a maximal p-subgroup of G, and $x \in G$, then the conjugate group xKx^{-1} is another maximal p-subgroup of G. For if L were a p-subgroup of G that properly contained xKx^{-1}, then $x^{-1}Lx$ would be a p-subgroup that properly contained K. We can omit the checking.

Sylow's theorem comes with a variety of proofs. The proof which we offer hinges on an interplay among the maximal p-subgroups of G.

Proposition 3.27. *Let G be a finite group, p a prime, and K a maximal p-subgroup of G. If g is in the normalizer $N_G(K)$ and the order of g is a power of p, then $g \in K$.*

Proof. Let L be the cyclic subgroup of $N_G(K)$ generated by g. Since K is normal in $N_G(K)$ the product set KL is a subgroup of $N_G(K)$, according to Proposition 2.71. By Proposition 2.76 the order of KL is $|K||L|/|K \cap L|$. Thus KL is a p-group containing K. By the maximality of K, it follows that $KL = K$. Then $g \in L \subseteq KL = K$, as desired. □

Keep in mind that for any subgroup H of a group G, the normalizer $N_G(H)$ is designed so that H is normal in $N_G(H)$, and so the quotient $N_G(H)/H$ is a group.

Proposition 3.28. *If G is a finite group and K is a maximal p-subgroup for some prime p, then p does not divide the order of the group $N_G(K)/K$.*

Proof. Suppose to the contrary that p does divide $|N_G(K)/K|$. By Cauchy's theorem, Proposition 3.23, there is a coset Kg in $N_G(K)/K$ of order p. This means that the coset representative g lies in the complement $N_G(K) \setminus K$ while $g^p \in K$. Since K is a p-group, Lagrange forces the order of g^p to be a power of p. Thus the order of g is a power of p as well. By Proposition 3.27, $g \in K$, which is a contradiction. □

The next, more subtle, result captures the essence of Sylow's theorem.

Proposition 3.29. *If G is a finite group and p is a prime, then all maximal p-subgroups of G are conjugates and their number is congruent to* 1 *modulo p.*

Proof. Let H be a maximal p-subgroup of G and let $\mathcal{X}$ be the set of its conjugate subgroups. The set $\mathcal{X}$ consists of maximal p-subgroups. The task is to see that $\mathcal{X}$ accounts for all maximal p-subgroups of G, and that the number of subgroups in $\mathcal{X}$ is congruent to 1 modulo p.

Suppose F is any maximal p-subgroup. The subgroup F acts on $\mathcal{X}$ by conjugation. Explicitly, if $f \in F$ and $K \in \mathcal{X}$, take the action given by $f \cdot K = fKf^{-1}$. For each point K in $\mathcal{X}$, the orbit-stabilizer relation, Proposition 3.11, applied to the action of F on $\mathcal{X}$, reveals that the order $|F \cdot K|$ (i.e. the number of elements) of the orbit $F \cdot K$ divides the order of F, which is a power of p. Thus either

$$|F \cdot K| = 1, \text{ or } p \text{ divides } |F \cdot K|.$$

If $|F \cdot K| = 1$, we will prove that $F = K$. For that it suffices to see that $F \subseteq K$, because K is a p-subgroup and F is a maximal p-subgroup. The assumption that $|F \cdot K| = 1$ means that $fKf^{-1} = K$ for all f in F, which is the same as saying that F is a subgroup of the normalizer $N_G(K)$ of K. Since F is a p-group, Lagrange ensures that the order of every element of F is a power of p. By Proposition 3.27 applied to the maximal p-subgroup K, all elements of F are in K, i.e. $F \subseteq K$.

We have shown that if F is any maximal p-subgroup of G and K is any element of $\mathcal{X}$ (i.e. any conjugate of H) and $F \neq K$, then p divides the order of the orbit $F \cdot K$ under the conjugation action by F on $\mathcal{X}$.

If F is taken to be H (acting on its own set of conjugate subgroups $\mathcal{X}$ by conjugation) one of those orbits, namely $H \cdot H$ will consist of just the point H, while the order of each of the other orbits will be divisible by p. Keeping in mind that the orbits under the action of F (in this case H) on $\mathcal{X}$ partition the set $\mathcal{X}$, it follows that the number of elements in $\mathcal{X}$, i.e. the number of conjugates of H, is congruent to 1 modulo p. In particular, the number of elements of $\mathcal{X}$ is not divisible by p.

If F were taken outside of $\mathcal{X}$, then the order of every orbit inside $\mathcal{X}$ under the action of F would be divisible by p. Consequently, the number of elements in $\mathcal{X}$ would be divisible by p, which is not the case. This shows that there are no maximal p-subgroups outside of $\mathcal{X}$.

Everything has been checked. □

Sylow's Great Theorem

Proposition 3.29 has done the heavy lifting.

Proposition 3.30 (Sylow's theorem). *If p is a prime and G is a finite group of order $p^r m$ where $p \nmid m$, then*

1. *the maximal p-subgroups of G coincide with those of order p^r,*
2. *every maximal p-subgroup is conjugate to every other maximal p-subgroup,*
3. *if n_p is the number of maximal p-subgroups of G, then $n_p \equiv 1 \bmod p$,*
4. *n_p divides m.*

Proof. Regarding the first item, Lagrange makes it clear that subgroups of order p^r are maximal p-subgroups. On the other hand suppose H is a maximal p-subgroup of G. Let $\mathcal{X}$ be the family of those subgroups of G which are conjugate to H. The group G acts on $\mathcal{X}$ by conjugation, and the orbit of the point H under this action of G is all of $\mathcal{X}$. By the orbit-stabilizer relation, Proposition 3.11, applied to this action and then by Lagrange we get

$$p^r m = |G| = |\mathcal{X}| \cdot |N_G(H)| = |\mathcal{X}| \cdot |N_G(H)/H| \cdot |H|.$$

According to Proposition 3.29, p does not divide $|\mathcal{X}|$. Also p does not divide $|N_G(H)/H|$ by Proposition 3.28. Thus p^r divides $|H|$. But $|H|$ divides $p^r m$ by Lagrange, and since H is a p-subgroup its order divides p^r. Hence $|H| = p^r$.

The second and third items simply restate Proposition 3.29.

As to the fourth item, note that the equation $p^r m = |\mathcal{X}| \cdot |N_G(H)/H| \cdot |H|$ can now be rewritten as

$$p^r m = n_p \cdot |N_G(H)/H| \cdot p^r.$$

Cancel p^r to see that $n_p \,|\, m$. □

The maximal p-subgroups of G are called ***Sylow p-subgroups***. To repeat, here is the stunning result that we have just encountered.

- If p^r is the highest power of a prime p that divides the order of a finite group G, then G contains subgroups of order p^r. They are the Sylow p-subgroups.
- All Sylow p-subgroups of G are conjugate to each other.
- The number of Sylow p-subgroups of G is congruent to 1 modulo p and divides $|G|/p^r$.

Two trivial cases could be noted. If a prime p does not divide the order of a group G, then the only Sylow p-subgroup of G is the trivial group. If G itself is a p-group, then the only Sylow p-subgroup is G itself.

Normal Sylow Subgroups

Let H be a Sylow p-subgroup of G. The conjugates of H coincide with the Sylow p- subgroups of G. This reveals that H is normal if and only if H is the only Sylow p-subgroup of G. The constraints imposed by Sylow's theorem on the number n_p of Sylow p-subgroups can at times be used to establish that $n_p = 1$ and thereby that G has a normal Sylow p-subgroup.

Groups of Order 45

For a typical exercise on normal Sylow subgroups suppose G is a group of order 45, which equals $5 \cdot 3^2$. The number n_5 of Sylow 5-subgroups of G divides 9 and is congruent to 1 mod 5. By inspection the only possibility is that $n_5 = 1$. So, there is exactly one subgroup H of G having order 5, and H is normal.

Similarly the number n_3 of Sylow 3-subgroups of G divides 5 and is congruent to 1 mod 3. By inspection the only possibility is that $n_3 = 1$. So, there is exactly one subgroup K of order 9 inside G, and K is normal.

This information tells us a lot. The intersection group $H \cap K$ is trivial because its order divides both 5 and 9. According to Proposition 2.75 the internal product HK is a group isomorphic to $H \times K$. Since the order of $H \times K$ is $5 \cdot 9 = 45$, we also see that $HK = G$. The group H, being of prime order 5, is isomorphic to the cyclic group $\mathbb{Z}_5$. By Proposition 3.20 the group K of order 3^2 is abelian. The structure theorem for finite abelian groups, Proposition 2.92, implies that K is isomorphic to one of $\mathbb{Z}_9$ or $\mathbb{Z}_3 \times \mathbb{Z}_3$. Putting all these bits together we learn that every group of order 45 is isomorphic to one of the abelian groups: $\mathbb{Z}_5 \times \mathbb{Z}_9$, $\mathbb{Z}_5 \times \mathbb{Z}_3 \times \mathbb{Z}_3$. For an alternative representation of groups of order 45 the Chinese remainder theorem tells us that $\mathbb{Z}_5 \times \mathbb{Z}_9 \cong \mathbb{Z}_{45}$, while $\mathbb{Z}_5 \times \mathbb{Z}_3 \times \mathbb{Z}_3 \cong \mathbb{Z}_{15} \times \mathbb{Z}_3$.

Sylow's Theorem in the Abelian Case

In case G is a finite abelian group, all subgroups are normal. In that case for each prime p dividing the order of the group, there can only be one Sylow p-subgroup. That Sylow p-subgroup is nothing but the p-primary component of G discussed in Section 2.8.2.

The Sylow Subgroups of $\mathcal{S}_4$

For practice let us find the Sylow subgroups of $\mathcal{S}_4$. Since $|\mathcal{S}_4| = 24 = 2^3 \cdot 3$, the non-trivial Sylow subgroups are the Sylow 2-subgroups, which have order 8, and the Sylow 3-subgroups which have order 3.

If n_2 is the number of Sylow 2-subgroups of $\mathcal{S}_4$, we know that $n_2 \equiv 1 \bmod 2$ and $n_2 \,|\, 3$. The only numbers that do that are 1 and 3. Thus $\mathcal{S}_4$ has either one or three subgroups of order 8. Once we find one we can always get the others by conjugating the one we find.

In fishing for a subgroup of order 8, we might recall that the dihedral group $\mathcal{D}_4$ has order 8. This is the group of symmetries of the square.

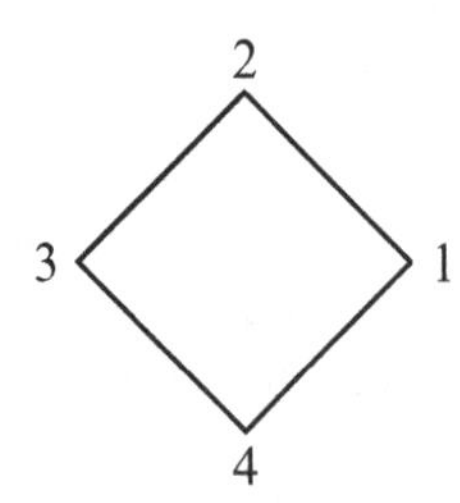

The cycle $\sigma = (1,2,3,4)$ gives the rotation of the square through 90°. The transposition $\tau = (2,4)$ gives the reflection of the square about the axis through vertices 1 and 3. The group $\mathcal{D}_4$ generated by σ and τ is a Sylow 2-subgroup of $\mathcal{S}_4$.

Could there be two more Sylow 2-subgroups in addition to $\mathcal{D}_4$? Well, we can conjugate $\mathcal{D}_4$ by various other permutations and see what we get.

Here is the group $\mathcal{D}_4$:

$$\begin{array}{llll} 1 & \sigma = (1,2,3,4) & \sigma^2 = (1,3)(2,4) & \sigma^3 = (4,3,2,1) \\ \tau = (2,4) & \sigma\tau = (1,2)(3,4) & \sigma^2\tau = (1,3) & \sigma^3\tau = (2,3)(1,4). \end{array}$$

Conjugate these permutations by $(1,2)$ to get the conjugate group $(1,2)\mathcal{D}_4(1,2)$:

$$\begin{array}{llll} 1 & (1,3,2,4) & (1,4)(2,3) & (2,4,3,1) \\ (1,4) & (1,2)(3,4) & (2,3) & (1,3)(2,4). \end{array}$$

We need to find one more Sylow 2-subgroup. When we conjugate $(1,2,3,4)$ by $(1,2,3)$, we get $(1,2,3)(1,2,3,4)(3,2,1) = (1,4,2,3)$, which is in neither of the preceding Sylow 2-subgroups. Thus, we can be sure that the conjugate group $(1,2,3)\mathcal{D}_4(3,2,1)$ will be our final Sylow 2-subgroup. Here it is:

$$\begin{array}{llll} 1 & (1,4,2,3) & (1,2)(3,4) & (3,2,4,1) \\ (3,4) & (1,4)(2,3) & (1,2) & (1,3)(2,4). \end{array}$$

In passing, note that the above three Sylow 2-subgroups have the Klein 4-group, made up of the permutations that are products of disjoint pairs of transpositions, as their intersection.

Finally, if n_3 is the number of Sylow 3-subgroups of $\mathcal{S}_4$, then $n_3 \equiv 1 \bmod 3$ and $n_3 \mid 8$. The possibilities for n_3 are 1 and 4. In fact, there are four Sylow 3-subgroups. Namely, the four cyclic groups generated by each of the 3-cycles:

$$(1,2,3),\ (1,2,4),\ (1,3,4),\ (2,3,4).$$

Some Groups of Order pq

Sylow's theorem can be applied in various circumstances. Let us now say something about groups whose order is the product of two primes.

Proposition 3.31. *Suppose p, q are primes such that $p < q$. If G is a group of order pq such that p does not divide $q - 1$, then G is cyclic.*

Proof. The number n_q of Sylow q-subgroups is congruent to $1 \bmod q$ and divides p. Since $p < q$, the only possible value of n_q is 1. Thus there is only one Sylow q-subgroup H of G, and such H is normal.

The number n_p of Sylow p-subgroups is congruent to $1 \bmod p$ and divides q. Thus $n_p = 1 + kp$ for some integer k, and $q = n_p\ell = (1 + kp)\ell$ for some integer ℓ. Since q is prime, we have either $1 + kp = q$ or $\ell = q$. The first possibility is ruled out by the assumption that p does not divide $q - 1$. Thus $\ell = q$ and $n_p = 1$. Accordingly there is only one Sylow p-subgroup, say K, and such K is normal.

Sylow's theorem has provided two normal subgroups: H of order p and K of order q. Their intersection is trivial because its order divides both p and q. By Proposition 2.75 the internal product HK is isomorphic to $H \times K$. Clearly the order of HK is pq, and so $HK = G$. Being of prime orders, H, K are cyclic. By Proposition 2.66, the product of cyclic groups of coprime order is again cyclic. Thus G is a cyclic group. $\square$

We have just learned that groups of orders such as $15 = 3 \cdot 5, 33 = 3 \cdot 11, 65 = 5 \cdot 13$ and $143 = 11 \cdot 13$ are cyclic.

By looking into the proof of the above result we might also find an alternative proof of Proposition 2.77.

The case where p does divide $q - 1$, such as with groups of order $55 = 5 \cdot 11$, is a different story. For instance, note that with a prime $q > 2$ the non-abelian dihedral groups $\mathcal{D}_q$ have order $2q$, where the prime 2 does divide $q - 1$. Groups of order pq where $p < q$ and p divides $q - 1$ will be examined in the next section.

The Existence of p-Subgroups of All Possible Orders

If a prime p divides the order of a group G, Cauchy's theorem, Proposition 3.23, says that G has a subgroup of order p. If p^n is the highest power of p dividing the order of G, Sylow says that G has a subgroup of order p^n. We can now check that for any power p^j dividing the order G, there exists a subgroup having order p^j.

Proposition 3.32. *If G is a group and a prime power p^j divides $|G|$, then G contains a subgroup of order p^j.*

Proof. Take a Sylow p-subgroup H of G. Since the order of H is the highest power p^n that divides $|G|$, we have $0 \leq j \leq n$, and p^j divides the order of the p-group H. From Proposition 3.22 regarding p-groups, H contains a subgroup of order p^j. Thereby so does G. $\square$

The Sylow theorem can be applied to glean information about the structure of some groups. Despite its powers, it does not always succeed in giving definitive results. Nevertheless, we can appreciate it when it does.

EXERCISES

1. For every prime p there is only one group, up to isomorphism, having order p. Find at least two non-primes that have such a property.
2. Find the normalizer in $\mathcal{S}_4$ of the subgroup generated by the cycle $(1, 2)$.
3. Find all Sylow 5-subgroups of $\mathcal{A}_5$.
4. Prove that every group of order 77 is cyclic.
5. Prove that every group of order 99 is isomorphic to one of two abelian groups. What are those abelian groups?
6. How many elements of order 5 does a group of order 20 contain?
7. For each prime p find a finite group with two Sylow p-subgroups having a non-trivial intersection.
8. If G is a group of order 28 and its Sylow 2-subgroup is normal, prove that G is abelian, and then describe all such G up to isomorphism.
9. How many Sylow 2-subgroups does the dihedral group $\mathcal{D}_6$ contain? Find these subgroups.

10. If H is a Sylow p-subgroup of a finite group G and $\varphi : G \to G$ is an automorphism of G, show that $\varphi(H)$ is a Sylow p-subgroup of G.

11. Let G be a finite group with a Sylow p-subgroup K for some prime p. Show that K is normal if and only if K contains all elements of G whose order is a pth power.

12. Let G be a finite group with a unique (and thereby normal) Sylow p-subgroup for some prime p. If H is a subgroup of G, show that H has a unique Sylow p-subgroup.

13. Prove that no group of order 40 is simple.

14. Prove that no group of order 132 is simple.

15. If p, q are distinct primes such that $p^3q \neq 24$, show that no group of order p^3q is simple.

16. Suppose n is a positive integer and p a prime factor of n. Also suppose n is such that the only divisor d of n that is congruent to 1 modulo p is $d = 1$. Prove there can be no simple group of order n.

Deduce that there are no simple groups having orders 64, 100, 124.

17. If a group G has order 175, show that G is isomorphic either to $\mathbb{Z}_{175}$ or to $\mathbb{Z}_{35} \times \mathbb{Z}_5$.

18. Suppose that G is a group of order $20449 = 11^2 \cdot 13^2$.

(a) Using Sylow's theorem show that G has only one Sylow 11-subgroup and only one Sylow 13-subgroup .

(b) If H, K are the unique Sylow subgroups of G having order $11^2, 13^2$ respectively, show that H, K are normal and that G is isomorphic to the exterior product $H \times K$.

(c) Explain why G is an abelian group.

d) Show that G is isomorphic to one of four groups and specify what those groups are.

19. Let H be a Sylow p-subgroup of a group G, let K be the normalizer of H, and let L be the normalizer of K. Prove that $K = L$.

Hint. Say x in G normalizes K, i.e. $xKx^{-1} = K$. To see that x must be in K, observe that H as well as xHx^{-1} are also Sylow p-subgroups of K. Apply Sylow's theorem to these subgroups of K.

20. Let G be a finite group having just one Sylow p-subgroup K for some prime p (which means that K is normal in G). If H is a subgroup of G, show that $H \cap K$ is a Sylow p-subgroup of H and that it is the only Sylow p-subgroup of H.

If K is not the only Sylow p-subgroup of G, show that there is a subgroup H of G such that $H \cap K$ is not a Sylow p-subgroup of H.

21. Suppose that G is a group of order 72.

(a) Show that the number of subgroups of G having order 9 is either 1 or 4.

(b) If G has 4 subgroups of order 9 and H is one of those 4 subgroups, show that the normalizer N of H has order 18 and thereby index 4 in G.

Hint. Explain why the four subgroups of order 9 are conjugates, and then use the orbit-stabilizer relation for the action of G by conjugation on this set of four subgroups.

(c) Using Proposition 3.4 deduce that if G has 4 subgroups of order 9, then G has a normal subgroup other than G and $\langle 1 \rangle$.

(d) Show that a group G of order 72 cannot be a simple group.

22. If G is a group of order 30, show that it has a normal subgroup of order 5 or a normal subgroup of order 3, and thereby deduce that such groups are never simple.

Hint. If there is more than one Sylow 5-subgroup and more than one Sylow 3-subgroup, show that G cannot have order 30.

23. Show that no group of order 56 is simple.

Hint. If G of order 56 were simple, how many Sylow 7-subgroups and Sylow 2-subgroups would G have? What does this imply regarding the order of G?

24. Show that every group of order $11^2 \cdot 13^2$ is abelian, and then show that there are only four such groups up to isomorphism.

25. Let p be a prime and let $GL_n(\mathbb{Z}_p)$ be the group of $n \times n$ invertible matrices with entries in the field $\mathbb{Z}_p$.

(a) Explain why the order of $GL_n(\mathbb{Z}_p)$ is

$$(p^n - 1)(p^n - p)(p^n - p^2) \cdots (p^n - p^{n-1}).$$

Hint. An $n \times n$ matrix A belongs to $GL_n(\mathbb{Z}_p)$ if and only if its columns are linearly independent. This comes down to saying that the first column of A is non-zero, the second column is not a multiple of the first, the third column is not a linear combination of the first two, and so on. How many non-zero first columns are there? How many second columns are not multiples of the first? How many third columns are not linear combinations of the first two, and so on?

(b) Let G be the set of all $n \times n$ upper triangular matrices over $\mathbb{Z}_p$ such that their entries on the main diagonal equal 1. Verify that G is a subgroup of $GL_n(\mathbb{Z}_p)$.

(c) Show that G is a Sylow p-subgroup of $GL_n(\mathbb{Z}_p)$.

26. Suppose G is a group of order 12. Prove that it has a normal Sylow 3-subgroup or a normal Sylow 4-subgroup.

Hint. If neither $n_3 = 1$ nor $n_2 = 1$, Sylow's theorem forces $n_3 = 4, n_2 = 3$. How many elements of order 3 does G have at least? How many elements of even order does G have at least?

27. Suppose G is a group of order pq^2, where p, q are distinct primes. Prove that it has a normal Sylow p-subgroup or a normal Sylow q-subgroup.

Hint. The harder case occurs when $p > q$. If $n_p \neq 1$, deduce that $p = q+1$, and thereby that $q = 2, p = 3$. Then use the preceding exercise.

28. If G is a finite group and H is a proper subgroup, show that the union of all conjugate subgroups xHx^{-1} is not all of G.

29.* If a group G has precisely 5 subgroups, including itself and the trivial group, show that G is isomorphic either to a cyclic group whose order is p^4 for some prime p, or to the Klein 4-group.

30. Let us explore the possibilities for the alternating group A_n to have subgroups of order $2n$.

(a) Suppose that n is a positive integer. If $\omega = e^{2\pi i n}$, the vertices of the regular n-gon can be taken as the powers $\omega^1, \omega^2, \ldots, \omega^{n-1}, \omega^n = 1$. The group of symmetries $\mathcal{D}_n$

of the n-gon permutes these vertices, and in this way $\mathcal{D}_n$ is a subgroup of order $2n$ inside $\mathcal{S}_n$.

Show that this $\mathcal{D}_n$ is a subgroup of $\mathcal{A}_n$ if and only if $n \equiv 1 \bmod 4$.

Deduce that if $n \equiv 1 \bmod 4$, then $\mathcal{A}_n$ contains a subgroup of order $2n$.

(b) * Suppose that p is prime and that $p \equiv 3 \bmod 4$. Prove that $\mathcal{A}_p$ contains no subgroup of order $2p$. This provides yet more illustrations that the converse of Lagrange fails, because for $p > 3$, the integer $2p$ divides the order $p!/2$ of $\mathcal{A}_p$.

Hint. Such a subgroup would have to be either cyclic or isomorphic to $\mathcal{D}_p$. Thus $\mathcal{A}_p$ would have to contain either an element of order $2p$ or else two elements σ and τ of order p and 2, respectively, and such that $\tau\sigma\tau = \sigma^{-1}$. Since p is prime, σ has to be a p-cycle. Show that a permutation α of order 2 such that $\alpha\sigma\alpha = \sigma^{-1}$ does exist in $\mathcal{S}_n \setminus \mathcal{A}_n$.

31.* If p is a prime, how many Sylow p-subgroups does $GL_3(\mathbb{Z}_p)$ have?

3.6 Semi-direct Products

As we saw in Proposition 2.70, if H, K are subgroups of a group G and every element of H commutes with every element of K, their internal product HK is a subgroup of G. Furthermore, if $H \cap K = \langle 1 \rangle$, then HK is isomorphic to the external product $H \times K$.

However, the internal product HK can still be a group without having every element of H commute with every element of K. By Proposition 2.70 this happens if one of H or K is normal in G. For instance, the alternating group $\mathcal{A}_n$ is normal in $\mathcal{S}_n$, being of index 2. So if K is the subgroup generated by the 2-cycle $(1, 2)$, their internal product $\mathcal{A}_nK$ is a subgroup of $\mathcal{S}_n$. In fact it is easy to see that $\mathcal{A}_nK = \mathcal{S}_n$. We want to shed some light on this situation using an important, if somewhat unusual, construction.

3.6.1 Automorphism Actions of One Group on Another

Definition 3.33. If H, K are groups, a homomorphism

$$\varphi : K \to \operatorname{Aut}(H)$$

is called an ***automorphism action*** of K on H.

This is an action of K on H, because automorphisms are bijections.

Here are a few illustrations of such actions.

- There is the ***trivial automorphism action*** which maps every y in a group K to the identity automorphism of a group H.
- Let H be a cyclic group of order 7, and K a cyclic group of order 3 with generator g. According to Proposition 2.56 there are precisely 6 automorphisms of H, and they take the form

$$\alpha_k : H \to H \text{ where } \alpha_k : x \mapsto x^k \text{ for } x \text{ in } H \text{ and } k = 1, 2, 3, 4, 5, 6.$$

Take the mapping $\varphi : K \to \text{Aut}(H)$ given by

$$1 \mapsto 1 \text{ (the identity map on } H), \ g \mapsto \alpha_2, \ g^2 \mapsto \alpha_4.$$

One can check that this φ is a homomorphism into $\text{Aut}(H)$ with image being the cyclic group generated by α_2.

- Suppose that F is a field, such as $\mathbb{Q}$ or $\mathbb{Z}_p$ for some prime p. The set of non-zero elements $F^\star$ is a group under multiplication and the field F is, by default, a group under addition. For every t in $F^\star$ the mapping $\varphi_t : F \to F$ defined by $x \mapsto tx$ is an automorphism of the abelian group F (using addition in F). Then the mapping $\varphi : F^\star \to \text{Aut}(F)$ given by $t \mapsto \varphi_t$ is an automorphism action of the group $F^\star$ (using multiplication) on F (using addition).
- For a general class of examples, let H be any normal subgroup of a group G and let K be any other subgroup of G. There is the action $\varphi : K \to \text{Aut}(H)$ which sends every y in K to the inner automorphism $\varphi_y : H \to H$ given by $\varphi_y : x \mapsto yxy^{-1}$. The normality of H ensures that φ_y is an automorphism of H.
- For any group H, let K be any subgroup of the automorphism group $\text{Aut}(H)$. The inclusion mapping $K \to \text{Aut}(H)$ is an obvious automorphism action.

Building a Group from an Automorphism Action

If $\varphi : K \to \text{Aut}(H)$ is an automorphism action and $y \in K$, denote the value of φ at y by the subscript notation φ_y. This might make the coming notations easier to keep track of. Next form the Cartesian product $H \times K$, just as a set. Upon $H \times K$ impose the following twisted operation, which we denote by a $\star$-notation. If $(x, y), (u, v) \in H \times K$ let

$$(x, y) \star (u, v) = (x\varphi_y(u), yv).$$

Definition 3.34. The Cartesian product $H \times K$, endowed with the $\star$-operation, is called the ***semi-direct product*** of H with K and is denoted by $H \rtimes_\varphi K$.

Proposition 3.35. *If H, K are groups and $\varphi : K \to \text{Aut}(H)$ is an automorphism action, then the semi-direct product $H \rtimes_\varphi K$ is a group. The identity of this group is $(1, 1)$, and for every (x, y) in $H \rtimes_\varphi K$ its inverse is $(\varphi_{y^{-1}}(x^{-1}), y^{-1})$.*

Proof. To check the associative law let $(x, y), (u, v), (s, t) \in H \rtimes_\varphi K$. Then

$$\begin{aligned}
(x, y) \star ((u, v) \star (s, t)) &= (x, y) \star (u\varphi_v(s), vt) \\
&= (x\varphi_y(u\varphi_v(s)), yvt) \\
&= (x\varphi_y(u)\varphi_y(\varphi_v(s)), yvt) \\
&= (x\varphi_y(u)\varphi_{yv}(s), yvt) \\
&= (x\varphi_y(u), yv) \star (s, t) \\
&= ((x, y) \star (u, v)) \star (s, t),
\end{aligned}$$

as desired.

The identity in $H \rtimes_\varphi K$ is $(1,1)$ because

$$(1,1) \star (u,v) = (1\varphi_1(u), 1v) = (u,v) \text{ and } (u,v) \star (1,1) = (u\varphi_v(1), v1) = (u,v),$$

for every (u,v) in $H \rtimes_\varphi K$. In the above, note that $\varphi_1(u) = u$ because φ_1 is the identity automorphism, and $\varphi_u(1) = 1$ because φ_u is an automorphism.

Checking for the inverse of (x,y) we have

$$\begin{aligned}(x,y) \star (\varphi_{y^{-1}}(x^{-1}), y^{-1}) &= (x\varphi_y(\varphi_{y^{-1}}(x^{-1})), yy^{-1}) \\ &= (x\varphi_{yy^{-1}}(x^{-1}), 1) \\ &= (x\varphi_1(x^{-1}), 1) \\ &= (xx^{-1}, 1) = (1,1),\end{aligned}$$

and the other way

$$\begin{aligned}(\varphi_{y^{-1}}(x^{-1}), y^{-1}) \star (x,y) &= (\varphi_{y^{-1}}(x^{-1})\varphi_{y^{-1}}(x), y^{-1}y) \\ &= (\varphi_{y^{-1}}(x^{-1}x), 1) = (\varphi_{y^{-1}}(1), 1) = (1,1).\end{aligned}$$

□

Properties of the Semi-direct Product

The following features of semi-direct products can readily be checked.

- If φ is the trivial automorphism action mapping all y in K to the identity automorphism on H, then the semi-direct product $H \rtimes_\varphi K$ collapses down to the usual direct product group $H \times K$.
- If φ is not the trivial homomorphism, then $H \rtimes_\varphi K$ is a non-abelian group, even though H, K could well be abelian. Indeed, there is a y in K such that φ_y is not the identity automorphism on H, which means there is an x in H such that $\varphi_y(x) \neq x$. And so

$$(x,1) \star (1,y) = (x\varphi_1(1), y) = (x,y) \neq (\varphi_y(x), y) = (1,y) \star (x,1).$$

- The mappings $H \to H \rtimes_\varphi K$ and $K \to H \rtimes_\varphi K$ given by $x \mapsto (x,1)$ and $y \mapsto (1,y)$, respectively, are injective homomorphisms. This is easy to check. Using these injections we take the liberty of identifying H with its isomorphic copy $H \times \langle 1 \rangle$, and K with its copy $\langle 1 \rangle \times K$, and declare that the semi-direct product contains H, K as subgroups.
- The subgroup H (when viewed as $H \times \langle 1 \rangle$) is normal in $H \rtimes_\varphi K$. One way to see this is to verify that the mapping $H \rtimes_\varphi K \to K$ given by $(x,y) \mapsto y$ is a homomorphism, that its kernel is H (when viewed as $H \times \langle 1 \rangle$), and then recall that kernels are normal.

  On the other hand, when the action φ is not the trivial homomorphism, the mapping $H \rtimes_\varphi K \to H$ given by $(x,y) \mapsto x$ will not be a homomorphism, and the kernel subgroup K (viewed as $\langle 1 \rangle \times K$) will not be normal.
- The intersection $H \cap K$ (actually $(H \times \langle 1 \rangle) \cap (\langle 1 \rangle \times K)$) is the trivial group.
- The conjugation action of K as a subgroup of $H \rtimes_\varphi K$ on H as a normal subgroup of $H \rtimes_\varphi K$ is the same as the original action φ of K on H. More precisely, for all x in H and y in K we have:

$$\begin{aligned}(1,y)\star(x,1)\star(1,y)^{-1} &= (1\varphi_y(x),y)\star(1,y^{-1})\\ &= (1\varphi_y(x)\varphi_y(1),yy^{-1}) = (\varphi_y(x),1).\end{aligned}$$

A Group of Order 12 as a Semi-direct Product

According to Proposition 2.92 all abelian groups of order 12 are isomorphic to $\mathbb{Z}_4 \times \mathbb{Z}_3$ or to $\mathbb{Z}_2 \times \mathbb{Z}_2 \times \mathbb{Z}_3$. We know of two non-abelian groups of order 12, namely the alternating group $\mathcal{A}_4$ and the dihedral group $\mathcal{D}_6$ of symmetries of the hexagon. These latter two groups are not isomorphic, because for example, $\mathcal{A}_4$ has no element of order 6 while $\mathcal{D}_6$ has such an element, namely the rotation of the hexagon through a 60° angle.

Semi-direct products can be used to construct a group of order 12, which is not isomorphic to any of the four preceding groups.

Other than the identity map, the only automorphism of $\mathbb{Z}_3$ is the negation map $x \mapsto -x$. This implies that there is only one non-trivial automorphism action $\varphi : \mathbb{Z}_4 \to \mathrm{Aut}(\mathbb{Z}_3)$. That action is such that φ_1, φ_3 give the negation automorphism while φ_0, φ_2 give the identity automorphism. For each representative $n = 0, 1, 2, 3$ of the residues of $\mathbb{Z}_4$ the automorphism φ_n can also be described by the rule $x \mapsto (-1)^n x$. Since $(-1)^n = (-1)^m$ whenever $n \equiv m \bmod 4$, the preceding calculation of $(-1)^n x$ does not depend on the integer chosen to represent a residue in $\mathbb{Z}_4$. The automorphism action φ permits the construction of the semi-direct product $\mathbb{Z}_3 \rtimes_\varphi \mathbb{Z}_4$. Keeping in mind that $\mathbb{Z}_3$ and $\mathbb{Z}_4$ use addition, the group operation in this semi-direct product is given by

$$(x,n)\star(y,m) = (x+(-1)^n y, n+m) \text{ where } x,y \in \mathbb{Z}_3 \text{ and } n,m \in \mathbb{Z}_4.$$

The identity element in the group is $(0,0)$.

The order of $\mathbb{Z}_3 \rtimes_\varphi \mathbb{Z}_4$ is 12. Since the automorphism action φ is not trivial, this group is not abelian. It is isomorphic to neither $\mathcal{A}_4$ nor $\mathcal{D}_6$. This can be seen by noting that the preceding two groups have no element of order 4 while our semi-direct product does. For instance, the higher powers of $(1,1)$ are:

$$\begin{aligned}(1,1)^2 &= (1,1)\star(1,1) = (0,2)\\ (1,1)^3 &= (1,1)\star(0,2) = (1,3)\\ (1,1)^4 &= (1,3)\star(1,1) = (0,0).\end{aligned}$$

An elaborate exercise using Sylow's theorem will show that every group of order 12 is isomorphic to one of

$$\mathbb{Z}_4 \times \mathbb{Z}_3, \mathbb{Z}_2 \times \mathbb{Z}_2 \times \mathbb{Z}_3, \mathcal{A}_4, \mathcal{D}_6, \mathbb{Z}_3 \rtimes_\varphi \mathbb{Z}_4.$$

3.6.2 Internal and Semi-direct Products

In addition to being a tool for the construction of new groups, semi-direct products can be helpful in describing old groups.

Proposition 3.36. *Suppose H, K are subgroups of a group G with H normal in G and $H \cap K = \langle 1 \rangle$. Then the internal product HK is isomorphic to a semi-direct product $H \rtimes_\varphi K$. The requisite automorphism action $\varphi : K \to \mathrm{Aut}(H)$ is the conjugation action of K on H that sends every y in K to the inner automorphism of H given by $\varphi_y : x \mapsto yxy^{-1}$.*

Proof. The desired isomorphism is given by the surjection

$$\theta : H \rtimes_\varphi K \to HK \text{ where } \theta : (x, y) \mapsto xy,$$

for every (x, y) in $H \rtimes_\varphi K$. To see that θ is a homomorphism we need

$$\theta((x, y) \star (u, v)) = \theta(x, y)\theta(u, v) \text{ for every } (x, y), (u, v) \text{ in } H \rtimes_\varphi K.$$

Well,

$$\begin{aligned}\theta((x, y) \star (u, v)) = \theta(x\varphi_y(u), yv) &= \theta(xyuy^{-1}, yv) \\ &= xyuy^{-1}yv = xyuv = \theta(x, y)\theta(u, v),\end{aligned}$$

as expected.

Also if $(x, y) \in \ker\theta$ then $xy = 1$, which leads to $y = x^{-1}$. Since $y \in K$ and $x^{-1} \in H$ and $H \cap K = \langle 1 \rangle$, it follows that $x = y = 1$. Thus $\ker\theta$ is trivial, which means that θ is injective, and we have our isomorphism. □

For instance, take the dihedral group $\mathcal{D}_n$ of symmetries of the regular n-gon, discussed in Section 3.2. If the vertices of the n-gon are labelled as the nth roots of unity $1, \omega, \ldots, \omega^{n-1}$ in $\mathbb{C}$, then $\mathcal{D}_n$ is the group generated by the permutations $\sigma : \omega^j \mapsto \omega^{j+1}$ and $\tau : \omega^j \mapsto \omega^{-j}$. Let $H = \langle \sigma \rangle$ and $K = \langle \tau \rangle$. The group H is of index 2 and thereby normal in $\mathcal{D}_n$. Also $\mathcal{D}_n = HK$ and $H \cap K = \langle 1 \rangle$. Thus, $\mathcal{D}_n \cong H \rtimes_\varphi K$ where φ is the conjugation action of K on H.

For another example, take the group $\mathcal{A}_4$ of even permutations on 4 letters. Let $H = \{1, (1,2)(3,4), (1,3)(2,4), (1,4)(2,3)\}$, a normal subgroup of order 4. Let K be the group of order 3 generated by the 3-cycle $(1,2,3)$. Clearly $H \cap K = \langle 1 \rangle$. By Proposition 2.76 the order of the subgroup HK is $4 \cdot 3 = 12$, which shows that $HK = \mathcal{A}_4$. Proposition 3.36 tells us that $\mathcal{A}_4 \cong H \rtimes_\varphi K$, where $\varphi : K \to \mathrm{Aut}\, H$ is the automorphism action given by $\varphi_t : x \mapsto txt^{-1}$ for every t in K and x in H.

3.6.3 Groups of Order pq – the Full Description

The semi-direct product construction with the help of the Cauchy and Sylow theorems enables us to identify all groups whose order is the product of two distinct primes. Here G is a group of order pq where p, q are primes and $p < q$. We have already seen that if $p \nmid q - 1$, then every such group is cyclic. We can now demonstrate this once more by means of semi-direct products, but also see what happens when p divides $q - 1$.

According to Cauchy's theorem, Proposition 3.23, G contains a subgroup K of order p, and a subgroup H of order q. These are also Sylow subgroups, and being of prime order, they are cyclic. By Sylow's theorem, Proposition 3.30, the number n_q of conjugates of H must be such that

$$n_q \,|\, p \text{ and } n_q \equiv 1 \bmod q.$$

By inspection it follows that $n_q = 1$, which tells us that H is a normal subgroup of G. We might note that the normality of H was also shown in Proposition 2.77.

The intersection $H \cap K$ is the trivial group since, by Lagrange, the order of the intersection is a divisor of both p and q. Proposition 2.76 reveals that the order of HK is pq, and thereby $HK = G$. Then by Proposition 3.36, $HK \cong H \rtimes_\varphi K$, for a suitable automorphism action $\varphi : K \to \mathrm{Aut}(H)$.

Knowing that G is isomorphic to a semi-direct product of H with K, all we need to do is account for all possible semi-direct products $H \rtimes_\varphi K$.

Since H is cyclic, Proposition 2.57 reveals that $\mathrm{Aut}(H)$ is isomorphic to the group of units $\mathbb{Z}_q^\star$ as discussed in Section 2.56. Thus the order of $\mathrm{Aut}(H)$ is $q - 1$.

Let $\varphi : K \to \mathrm{Aut}(H)$, where y in K goes to the automorphism φ_y, be any automorphism action of K on H. Since the order of y divides p by Lagrange, the order of its homomorphic image φ_y also divides p. But φ_y lies in $\mathrm{Aut}(H)$, which implies that its order divides $q-1$, again by Lagrange.

Thus, if $p \nmid q - 1$, the only possible order that φ_y can have is 1. In this case φ is the trivial action. Consequently, the only semi-direct product $H \rtimes_\varphi K$ is the direct product $H \times K$, and by Proposition 2.66, $H \times K$ is cyclic. For instance, every group of order 65 is cyclic, because $65 = 5 \cdot 13$ and $5 \nmid 12$.

We are down to the case where p divides $q - 1$.

If the action $\varphi : K \to \mathrm{Aut}(H)$ is trivial, then $H \rtimes_\varphi K$ is cyclic, as noted already. However, the condition $p \mid q - 1$ allows for non-trivial actions. To see that, let g be a generator for the cyclic group K of order p. Clearly, φ is non-trivial if and only if φ_g is not the identity automorphism. And this is the same as saying that φ_g has order p. Since $p \,|\, q - 1$, Cauchy's theorem, Proposition 3.23, assures that $\mathrm{Aut}(H)$ does contain an element of order p, say α. Then the action $\varphi : K \to \mathrm{Aut}(H)$ that sends each g^j in K to α^j is non-trivial. Thus, in the case where $p \,|\, q - 1$, there do exist non-abelian semi-direct products $H \rtimes_\varphi K$.

Let us now prove that if $\varphi, \psi : K \to \mathrm{Aut}(H)$ are two *non-trivial* automorphism actions, then $H \rtimes_\varphi K \cong H \rtimes_\psi K$. This will reveal that only two non-isomorphic semi-direct products are possible, the cyclic $H \times K$ and one that is non-abelian.

At this point we require a fact yet not proven, and begging the forbearance of the reader, we simply state that

for any group of prime order, its automorphism group is a cyclic group.

This is what Proposition 4.39 in Section 4.5.1 says.

As already noted, let g generate K. Since $\mathrm{Aut}(H)$ is cyclic of order $q - 1$ and $p \mid q - 1$, Proposition 2.41 reveals that $\mathrm{Aut}(H)$ has precisely one subgroup of order p, call it L. For the non-trivial action $\varphi : K \to \mathrm{Aut}(H)$ the automorphism φ_g must have order p, and thereby generate L. The powers φ_g^j where $j = 1, 2, \ldots, p-1$ (omitting $1 = \varphi_g^0$) are the other generators of L. Since ψ is non-trivial as well, the automorphism ψ_g is one of these other generators of L. Hence, $\psi_g = \varphi_g^m$ for some exponent m between 1 and $p-1$.

For any element $y = g^j$ in K we use the fact ψ and φ are automorphism actions to see that

$$\psi_y = \psi_{g^j} = \psi_g^j = (\varphi_g^m)^j = \varphi_{g^{mj}} = \varphi_{(g^j)^m} = \varphi_{y^m}.$$

Before writing our isomorphism between $H \rtimes_\varphi K$ and $H \rtimes_\psi K$, let us adopt the notations

$$(x,y) \star_\varphi (u,v) \text{ and } (x,y) \star_\psi (u,v)$$

for the respective operations in these semi-direct products. Now define

$$\theta : H \rtimes_\psi K \to H \rtimes_\varphi K \text{ by } (x,y) \mapsto (x, y^m).$$

The mapping $K \to K$ given by $y \mapsto y^m$ is an automorphism of the cyclic group K, since $p \nmid m$. Hence the above θ is a bijection. For θ to be an isomorphism it only needs to be a homomorphism, which we now check. If $(x,y),(u,v) \in H \rtimes_\psi K$, we obtain

$$\begin{aligned}
\theta((x,y) \star_\psi (u,v)) &= \theta(x\psi_y(u), yv) \\
&= (x\psi_y(u), (yv)^m) \\
&= (x\varphi_{y^m}(u), (yv)^m) && \text{since } \psi_y = \varphi_{y^m} \text{ from above} \\
&= (x\varphi_{y^m}(u), y^m v^m) && \text{since } K \text{ is abelian} \\
&= (x, y^m) \star_\varphi (u, v^m) \\
&= \theta(x,y) \star_\varphi \theta(u,v).
\end{aligned}$$

Our isomorphism is in place.

EXERCISES

1. Show that, up to isomorphism, there is only one group of order 91.
2. Let H, K be groups and form their semi-direct product $H \rtimes_\varphi K$ where φ is an automorphism action of K on H. Show that H, viewed as the subgroup $H \times \langle 1 \rangle$ of $H \rtimes_\varphi K$, is normal.
3. Explain why $\mathcal{S}_n \cong A_n \rtimes_\varphi K$, where K is the cyclic group generated by the 2-cycle $(1,2)$ and φ is a suitable automorphism action.
4. Let $\varphi : K \to \mathrm{Aut}(H)$ be an automorphism action of a group K on a group H leading to the semi-direct product $H \rtimes_\varphi K$. If the subgroup K (as identified with $\langle 1 \rangle \times K$) is normal in $H \rtimes_\varphi K$, prove that the action φ is trivial and thereby that the semi-direct product is the direct product group $H \times K$.

   Note. The action φ is trivial when $\varphi_y(x) = x$ for all x in H and all y in K.
5. If n is a positive integer, recall the Euler ϕ-function which gives $\phi(n)$ as the order of the group $\mathbb{Z}_n^\star$ of units modulo n. This exercise will show that if n is not coprime with $\phi(n)$, then there exist at least two non-isomorphic groups of order n. The failure of n to be

coprime with $\phi(n)$ can happen in one of two ways. The first way is that n has a repeated prime factor p. In this case the calculation of $\phi(n)$ shows that p still divides $\phi(n)$. The calculation of $\phi(n)$ shows that it could also happen when n has no repeated prime factor but there are two prime factors p, q such that p divides $q - 1$.

(a) If n has a repeated prime factor, find two non-isomorphic, abelian groups of order n.

(b) If n has no repeated prime factor and n is still not coprime with $\phi(n)$, construct a non-abelian group of order n, and thereby two non-isomorphic groups of order n.

Hint. Look at the construction of non-abelian groups of order pq.

6. (a) Let $\varphi : K \to \mathrm{Aut}(H)$ be an automorphism action of a group K on group H. If $\alpha : K \to K$ is an automorphism of K, show that $H \rtimes_\varphi K \cong H \rtimes_{\varphi \circ \alpha} K$.

Hint. The mapping $\theta : H \times K \to H \times K$ given by $(x, t) \mapsto (x, \alpha^{-1}(t))$ should do it.

(b) Suppose that $\varphi, \psi : \mathbb{Z}_2 \times \mathbb{Z}_2 \to \mathrm{Aut}(\mathbb{Z}_3)$ are two non-trivial automorphism actions of the four element group $\mathbb{Z}_2 \times \mathbb{Z}_2$ on the three element group $\mathbb{Z}_3$. Using part (a) show that

$$\mathbb{Z}_3 \rtimes_\varphi (\mathbb{Z}_2 \times \mathbb{Z}_2) \cong \mathbb{Z}_3 \rtimes_\psi (\mathbb{Z}_2 \times \mathbb{Z}_2).$$

For these groups the choice of non-trivial automorphism action does not affect the semi-direct product up to isomorphism.

Hint. Find all of the non-trivial automorphism actions for these groups.

7. For a positive integer n define

$$\varphi : \mathbb{Z}_2 \to \mathrm{Aut}(\mathbb{Z}_n) \text{ by } \varphi_0 = (x \mapsto x) \text{ and } \varphi_1 = (x \mapsto -x).$$

This automorphism action of the cyclic group $\mathbb{Z}_2$ on the cyclic group $\mathbb{Z}_n$ (both groups using addition) sends the residue 0 to the identity map on $\mathbb{Z}_n$ and the residue 1 to the negation automorphism on $\mathbb{Z}_n$.

Verify that φ is an automorphism action, and then show that the dihedral group $\mathcal{D}_n$ is isomorphic to the semi-direct product $\mathbb{Z}_n \rtimes_\varphi \mathbb{Z}_2$.

8.* This exercise reveals that there are precisely 5 non-isomorphic groups of order 12. The steps will involve group actions, Sylow theory and semi-direct products. Let G be a group of order 12. Let n_2 be the number of Sylow 2-subgroups, and let n_3 be the number of Sylow 3-subgroups.

(a) Show that $n_2 = 1$ or $n_2 = 3$, and that $n_3 = 1$ or $n_3 = 4$.

(b) Suppose $n_2 = n_3 = 1$.

Explain why G has just one subgroup of order 4 and one subgroup of order 3 and that these subgroups are normal.

If H is the unique subgroup of order 4 and K the unique subgroup of order 3, show that $G = HK \cong H \times K$.

Explain why $G \cong \mathbb{Z}_4 \times \mathbb{Z}_3$ or $G \cong \mathbb{Z}_2 \times \mathbb{Z}_2 \times \mathbb{Z}_3$.

(c) Show that the possibility $n_2 = 3, n_3 = 4$ cannot occur.

Hint. If G had four Sylow 3-subgroups, that would account for 8 elements of order 3 leaving just 4 elements with order not equal to 3.

(d) Suppose $n_2 = 1$ and $n_3 = 4$.

Let H be one of those four Sylow 3-subgroups. Since the order of H is 3, Lagrange says that H has four left cosets in G. So, let $\mathcal{X}$ be the set of those four left cosets of H, and let $\mathcal{S}(\mathcal{X})$ be the group of permutations of $\mathcal{X}$, which is nothing but $\mathcal{S}_4$. For each x in G the mapping $\varphi_x : \mathcal{X} \to \mathcal{X}$ given by $H \mapsto xH$ is a permutation of the four element set $\mathcal{X}$.

Verify that $\varphi : G \to \mathcal{S}(\mathcal{X})$ given by $x \mapsto \varphi_x$ is a homomorphism, i.e. a group action of G on $\mathcal{X}$.

Then prove φ is injective.

Hint. If $x \in \ker\varphi$ argue that $yxy^{-1} \in H$ for all y in G, and then that $x = 1$ because H is normal in G.

Deduce that in this case $G \cong \mathcal{A}_4$.

Hint. The image of G under φ has eight elements of order 3.

(e) Now suppose $n_2 = 3$ and $n_3 = 1$.

Let H be the unique, normal Sylow 3-subgroup of G and let K be one of the three Sylow 2-subgroups. The order of H is 3 and the order of K is 4.

Why is the internal product $HK = G$?

Explain why $G \cong H \rtimes_\varphi K$ for a suitable automorphism action φ of K on H.

(f) Continue with $n_2 = 3, n_3 = 1$ as in part (e).

The Sylow 2-subgroup K is either isomorphic to $\mathbb{Z}_4$ or isomorphic to $\mathbb{Z}_2 \times \mathbb{Z}_2$. If $K \cong \mathbb{Z}_4$, explain why $G \cong \mathbb{Z}_3 \rtimes_\varphi \mathbb{Z}_4$, where $\varphi : \mathbb{Z}_4 \to \mathrm{Aut}(\mathbb{Z}_3)$ is the unique non-trivial automorphism action of $\mathbb{Z}_4$ on $\mathbb{Z}_3$.

(g) Still with $n_2 = 3, n_3 = 1$, suppose that the Sylow 2-subgroup H is isomorphic to $\mathbb{Z}_2 \times \mathbb{Z}_2$. Explain why $G \cong \mathbb{Z}_3 \rtimes_\varphi (\mathbb{Z}_2 \times \mathbb{Z}_2)$ for some non-trivial automorphism action $\varphi : \mathbb{Z}_2 \times \mathbb{Z}_2 \to \mathrm{Aut}(\mathbb{Z}_3)$.

Show that the group $\mathcal{D}_6 \cong \mathbb{Z}_3 \rtimes_\psi (\mathbb{Z}_2 \times \mathbb{Z}_2)$ for some non-trivial automorphism action $\psi : \mathbb{Z}_2 \times \mathbb{Z}_2 \to \mathrm{Aut}(\mathbb{Z}_3)$.

Deduce that $G \cong \mathcal{D}_6$.

Thus we have an accounting of all groups of order 12.

9. In this exercise use automorphism actions along with Sylow's theorem and various other bits of group theory to prove that every group of order 255 is cyclic. Let G be such a group.

(a) Show that G contains a normal subgroup H of order 17.

(b) Explain why the order of the automorphism group $\mathrm{Aut}(H)$ is 16.

(c) Let $\varphi : G \to \mathrm{Aut}(H)$ be the automorphism action given by $\varphi_x(h) = xhx^{-1}$ for all x in G and all h in H. Show that $\ker\varphi = G$ by explaining why the order of $G/\ker\varphi$ divides both 255 and 16.

(d) Show that H is a subgroup of the centre Z of G.

(e) Deduce that the order of G/Z is one of $1, 3, 5, 15$, and then that G/Z is cyclic.

(f) Explain why G is abelian.

(g) Show that G is cyclic.

10. Let us check something that might feel obvious. Namely, if $H_1 \cong H_2$ and $K_1 \cong K_2$ are group isomorphisms, then the semi-direct product $H_1 \rtimes_\varphi K_1$ arising from some automorphism action φ of K_1 on H_1 should be isomorphic to the semi-direct product $H_2 \rtimes_\psi K_2$ arising from a suitable automorphism action ψ of K_2 on H_2. The key is to write the suitable action ψ. Here are the details.

Suppose that $\beta : H_1 \to H_2$ and $\alpha : K_2 \to K_1$ are group isomorphisms and that $\varphi : K_1 \to \text{Aut}(H_1)$ (where $t \mapsto \varphi_t$) is an automorphism action.

(a) Briefly explain why the conjugation mapping $\Gamma(\beta) : \text{Aut}(H_1) \to \text{Aut}(H_2)$ given by $\sigma \mapsto \beta \circ \sigma \circ \beta^{-1}$ is an isomorphism of the automorphism groups. In other words, the automorphism groups of two isomorphic groups are themselves isomorphic groups.

(b) The mapping $\Gamma(\beta) \circ \varphi \circ \alpha$ sends t in K_2 to $\Gamma(\beta)(\varphi_{\alpha(t)}) = \beta \circ \varphi_{\alpha(t)} \circ \beta^{-1}$, which is an automorphism of H_2. Let $\psi = \Gamma(\beta) \circ \varphi \circ \alpha$, and verify that ψ is an automorphism action of K_2 on H_2.

(c) Verify that the mapping $\theta : H_1 \rtimes_\varphi K_1 \to H_2 \rtimes_\psi K_2$ given by $(x, t) \mapsto (\beta(x), \alpha^{-1}(t))$ is an isomorphism.

11. This exercise illustrates the powers of the Sylow theorem in conjunction with semi-direct products to describe groups of order 75. Let G be such a group.

(a) Show that G has a normal subgroup H of order 25 and subgroup K of order 3.

(b) Prove that $G \cong H \rtimes_\varphi K$ for an automorphism action φ of K on H.

This shows that to describe groups of order 75 it suffices to consider the possible automorphism actions of K on H.

(c) Show that H is isomorphic either to the cyclic group $\mathbb{Z}_{25}$ or to the Cartesian product $\mathbb{Z}_5 \times \mathbb{Z}_5$. (The latter two groups use addition.)

(d) If $H \cong \mathbb{Z}_{25}$ show that the order of Aut(H) is 20, and thereby that every automorphism action $\varphi : K \to \text{Aut}(H)$ is trivial.

Deduce that if $H \cong \mathbb{Z}_{25}$, then $G \cong H \times K$ and that G is cyclic.

(e) We are down to the case where $H \cong \mathbb{Z}_5 \times \mathbb{Z}_5$. In this case why is G isomorphic to $(\mathbb{Z}_5 \times \mathbb{Z}_5) \rtimes_\varphi \mathbb{Z}_3$ for some automorphism action $\varphi : \mathbb{Z}_3 \to \text{Aut}(\mathbb{Z}_5 \times \mathbb{Z}_5)$?

It remains to examine such actions.

(f) Explain why $\text{Aut}(\mathbb{Z}_5 \times \mathbb{Z}_5)$ is isomorphic to the general linear group $GL_2(\mathbb{Z}_5)$.

(g) What is the order of $GL_2(\mathbb{Z}_5)$?

(h) Why does $GL_2(\mathbb{Z}_5)$ have a subgroup of order 3, and why are all subgroups of order 3 conjugates?

More explicitly, verify that the matrix $\begin{bmatrix} 4 & 2 \\ 2 & 0 \end{bmatrix}$ in $GL_2(\mathbb{Z}_5)$ generates a subgroup of order 3.

(i) Find a non-trivial automorphism action $\varphi : \mathbb{Z}_3 \to \text{Aut}(\mathbb{Z}_5 \times \mathbb{Z}_5)$, and thereby construct a non-abelian group of order 75.

(j) The last piece of the puzzle is to see that if $\varphi, \psi : \mathbb{Z}_3 \to \text{Aut}(\mathbb{Z}_5 \times \mathbb{Z}_5)$ are non-trivial automorphism actions, then

$$(\mathbb{Z}_5 \times \mathbb{Z}_5) \rtimes_\varphi \mathbb{Z}_3 \cong (\mathbb{Z}_5 \times \mathbb{Z}_5) \rtimes_\psi \mathbb{Z}_3.$$

In order to keep track of things, recall that $\mathbb{Z}_5 \times \mathbb{Z}_5$ and $\mathbb{Z}_3$ use addition. Let $\star_\varphi$ and $\star_\psi$ denote the group operations in $(\mathbb{Z}_5 \times \mathbb{Z}_5) \rtimes_\varphi \mathbb{Z}_3$ and $(\mathbb{Z}_5 \times \mathbb{Z}_5) \rtimes_\psi \mathbb{Z}_3$, respectively. If $x, y \in \mathbb{Z}_5 \times \mathbb{Z}_5$ and $s, t \in \mathbb{Z}_3$ we then have

$$(x,s) \star_\varphi (y,t) = (x + \varphi_s(y), s+t), \ (x,s) \star_\psi (y,t) = (x + \psi_s(y), s+t).$$

Check the following details.

- Why are the subgroups $\varphi(\mathbb{Z}_3), \psi(\mathbb{Z}_3)$ conjugate in $\mathrm{Aut}(\mathbb{Z}_5 \times \mathbb{Z}_5)$?
- Let β in $\mathrm{Aut}(\mathbb{Z}_5 \times \mathbb{Z}_5)$ be such that $\beta\varphi(\mathbb{Z}_3)\beta^{-1} = \psi(\mathbb{Z}_3)$. Show that $\psi_t = \beta \circ \varphi_t \circ \beta^{-1}$ or $\psi_t = \beta \circ \varphi_{2t} \circ \beta^{-1}$ for every t in $\mathbb{Z}_3$.
- Use the preceding exercise to deduce that the two semi-direct products are isomorphic.

(k) Show that there are only three non-isomorphic groups of order 75, two abelian and one non-abelian.

12. The purpose of this exercise is to create a finitely generated group having a subgroup that is not finitely generated. The first thing to do is digest the setup. Let

$$V = \{(x_j)_{j \in \mathbb{Z}} : x_j \in \mathbb{Z}_2 \text{ and all but finitely many } x_j = 0\}.$$

In other words V is the set of two-sided sequences of residues modulo 2 with only a finite number of non-zero entries, i.e. with the residue 1 appearing only a finite number of times. Under point-wise addition of sequences, V is a vector space over the field $\mathbb{Z}_2$ and thereby also an abelian group. Another way is to think of V as the set of functions $v : \mathbb{Z} \to \mathbb{Z}_2$ which take the residue 0 as its value except at finitely many inputs. For each integer n let $v_n : \mathbb{Z} \to \mathbb{Z}_2$ be the function defined by

$$v_n(m) = \begin{cases} 1 & \text{when } m = n \\ 0 & \text{when } m \neq n. \end{cases}$$

This is the two-sided sequence of residues modulo 2 having all entries 0 except for a 1 in the nth position. The abelian group V is generated by the set of functions v_n. In fact every element of V is the sum of finitely many uniquely determined v_j.

Now take the automorphism action $\varphi : \mathbb{Z} \to \mathrm{Aut}(V)$ defined by the rule

$$\varphi : n \mapsto \varphi_n = ((x_j)_{j \in \mathbb{Z}} \mapsto (x_{j-n})_{j\mathbb{Z}}).$$

Upon close inspection φ_n is seen to be the automorphism of V that shifts each generator v_j to v_{j+n}.

(a) Explain why the abelian group V is not finitely generated.

(b) Show that the semi-direct product $V \rtimes_\varphi \mathbb{Z}$ is finitely generated, in fact generated by two elements.

(c) Show that $V \rtimes_\varphi \mathbb{Z}$ has a subgroup which is not finitely generated.

3.7 Solvable Groups

Galois' great discovery was to connect the problem of solving for the roots of a polynomial to the structure of a group which acts as permutations of those roots. We are not yet ready for Galois theory, but this might be an opportune moment to take one step in preparing for it, by looking at a class of groups which plays a role in that theory.

Definition 3.37. If G is a group, a nested sequence of subgroups

$$G = N_0 \supset N_1 \supset N_2 \supset \cdots \supset N_{k-1} \supset N_k = \langle 1 \rangle$$

such that each N_j is a normal subgroup of N_{j-1} is called a ***subnormal chain*** or a ***subnormal series*** for G.

Note that each N_j is required to be normal merely as a subgroup of N_{j-1}, and need not be normal as a subgroup of G.[1]

For example, inside the alternating group $\mathcal{A}_4$ let H be the cyclic subgroup generated by the transpositions product $(1,2)(3,4)$ and let K be the abelian subgroup $\{1,(1,2)(3,4),(1,3)(2,4),(1,4)(2,3)\}$. The chain

$$\mathcal{A}_4 \supset K \supset H \supset \langle 1 \rangle$$

is subnormal, but H is not normal in $\mathcal{A}_4$. These observations were made already in Section 2.4 as part of the tabulation of subgroups of $\mathcal{A}_4$.

Definition 3.38. A group G is called ***solvable***[2] provided G has a subnormal chain for which the quotient groups N_{j-1}/N_j are abelian.

Our goal at this point is to build a bit of familiarity and skill in working with this somewhat subtle concept.

A Few Examples

- Every abelian group is solvable.
- The group $\mathcal{A}_4$ is solvable using the subnormal chain given above. Indeed, the quotients $\mathcal{A}_4/K, K/H$ and $H/\langle 1 \rangle$ have prime orders $3, 2$ and 2, respectively, and are therefore cyclic. Actually H can be dropped because K is already abelian.
- The group $\mathcal{S}_4$ is solvable. To see that take the abelian subgroup $K = \{1,(1,2)(3,4),(1,3)(2,4),(1,4)(2,3)\}$ which is normal in $\mathcal{A}_4$. The subnormal chain

$$\mathcal{S}_4 \supset \mathcal{A}_4 \supset K \supset \langle 1 \rangle$$

shows that $\mathcal{S}_4$ is solvable. Indeed, K is abelian, $\mathcal{A}_4/K$ is abelian because it has order 3, and the quotient $\mathcal{S}_4/\mathcal{A}_4$ is abelian since its order is 2.

[1] That is where the "sub" in "subnormal" comes from.

[2] Such strange terminology can only make sense after its connection to the solution of polynomial equations is established in Chapter 7.

- For each prime p, every finite p-group G is solvable. That follows from Proposition 3.21, which tells us that a group of order p^m comes with a strictly descending chain of normal subgroups:

$$G = N_0 \supset N_1 \supset N_2 \supset \cdots \supset N_{m-1} \supset N_m = \langle 1 \rangle.$$

  The multiplicative property of the index shows that p^m equals the product of the orders of the quotient groups N_j/N_{j+1}. Since there are m such quotient groups and since each quotient group is a p-group of order p^j for $j \geq 1$, it follows that each quotient has order exactly p. Thus each quotient is not only abelian but cyclic of prime order. Furthermore, the subgroups N_j are normal in N_{j-1} because they are normal in G.
- If H, K are abelian groups, then any semi-direct product $H \rtimes_\varphi K$ is solvable, due to the subnormal series $H \rtimes_\varphi K \supset H \supset \langle 1 \rangle$. Note that the semi-direct product was constructed to make H normal. Furthermore the mapping from $H \rtimes_\varphi K$ to K given by $(x, y) \mapsto y$ is a surjective homomorphism with kernel H (actually its isomorphic copy $H \times \langle 1 \rangle$). By the first isomorphism theorem, Proposition 2.62, $(H \rtimes_\varphi K)/H$ is isomorphic to the abelian group K.
- If p, q are distinct primes, groups of order pq are solvable because they are isomorphic to semi-direct products of cyclic, and thereby abelian, groups.
- To see some *non-solvable* groups take the alternating groups $\mathcal{A}_n$ with $n \geq 5$. In Section 2.4 we saw that this group is simple. Namely $\mathcal{A}_n$ has no normal subgroups except itself and the trivial group. If $\mathcal{A}_n$ were solvable, it would have to contain at least one proper normal subgroup N such that $\mathcal{A}_n/N$ is abelian. But the only normal candidate $\langle 1 \rangle$ will not work.
- The renowned Feit–Thompson theorem from 1963, which is far beyond our scope, states that every group of odd order is solvable.

3.7.1 The Second and Third Isomorphism Theorems

Two nuggets of a general nature are helpful in working with solvable groups.

The Second Isomorphism Theorem

Proposition 3.39 (Second isomorphism theorem). *If N is a normal subgroup of a group G and H is any subgroup of G, then the subgroup $N \cap H$ is normal in H and*

$$H/(N \cap H) \cong HN/N.$$

In particular, $H/(N \cap H)$ is isomorphic to a subgroup of G/N.

Proof. Recall that since N is normal in G, the internal product HN is a subgroup of G. The normality of $N \cap H$ in H is inherited from the normality of N in G. Then take the homomorphism

$$\psi : H \to G/N \text{ given by } x \mapsto xN \text{ for each } x \text{ in } H.$$

The kernel of this homomorphism is $N \cap H$, and its image is the subgroup HN/N of G/N. By the first isomorphism theorem, Proposition 2.62, $H/(N \cap H) \cong HN/N$. □

The Third Isomorphism Theorem

Proposition 3.40. *If $\varphi : G \to H$ is a group homomorphism and M is a normal subgroup of G containing* $\ker \varphi$, *then $\varphi(M)$ is normal in $\varphi(G)$ and*

$$G/M \cong \varphi(G)/\varphi(M).$$

Proof. The normality of $\varphi(M)$ in $\varphi(G)$ is easily checked. For every z in $\varphi(G)$ and x in M we want $z\varphi(x)z^{-1} \in \varphi(M)$. Since $z = \varphi(y)$ for some y in G and M is normal in G, we see that

$$z\varphi(x)z^{-1} = \varphi(y)\varphi(x)\varphi(y)^{-1} = \varphi(yxy^{-1}) \in \varphi(M).$$

Clearly the mapping

$$\psi : G \to \varphi(G)/\varphi(M) \text{ given by } x \mapsto \varphi(x)\varphi(M)$$

is a surjective homomorphism. The first isomorphism theorem, Proposition 2.62, will yield the desired result provided $\ker \psi = M$.

Well, if $x \in M$ then $\varphi(x) \in \varphi(M)$, which means that the coset $\varphi(x)\varphi(M)$ is the identity element of $\varphi(G)/\varphi(M)$. Thus $M \subseteq \ker \psi$. For the reverse inclusion suppose $x \in \ker \psi$. This means that $\varphi(x) \in \varphi(M)$. Then $\varphi(x) = \varphi(y)$ for some y in M. Hence $xy^{-1} \in \ker \varphi \subseteq M$, which leads to $x \in M$ because $y \in M$. □

Notice that the proofs of both the second and third isomorphism theorems hinge on the first isomorphism theorem, Proposition 2.62. Also note that Proposition 3.40 collapses down to the first isomorphism theorem when M is taken to be $\ker \varphi$.

It might be interesting to interpret the third isomorphism theorem in the situation where $H = G/K$ for some normal subgroup K of G and $\varphi : G \to G/K$ is the canonical projection whose kernel is K. If M is a normal subgroup of G containing K, its image $\varphi(M)$ is M/K. Then Proposition 3.40 says that

$$G/M \cong (G/K)/(M/K).$$

The quotient notation harmoniously lets us cancel the K. Another way to put it is that a quotient group of a quotient group of G is just another quotient group of G.

Here is a small follow-up on the third isomorphism theorem. In Proposition 3.40 the subgroup M was assumed to contain the kernel of the homomorphism φ. Our follow-up discusses the situation when a normal subgroup does not necessarily contain the kernel of φ.

Proposition 3.41. *If $\varphi : G \to H$ is a group homomorphism and N is any normal subgroup of G, then $\varphi(N)$ is normal in $\varphi(G)$ and $\varphi(G)/\varphi(N)$ is a homomorphic image of G/N.*

Proof. The normality of $\varphi(N)$ in H is routine to check as it was in Proposition 3.40.

Put $M = \varphi^{-1}(\varphi(N))$. Clearly M contains both N and $\ker \varphi$, and M is a normal subgroup of G by a routine verification. Clearly also $\varphi(M) = \varphi(N)$. By Proposition 3.40,

$$G/M \cong \varphi(G)/\varphi(M) = \varphi(G)/\varphi(N).$$

So it is enough to show that G/M is a homomorphic image of G/N. But this holds because $N \subseteq M$. Indeed, the mapping

$$\psi : G/N \to G/M \text{ given by } xN \mapsto xM$$

is well defined. For if $xN = xM$, then $y^{-1}x \in N \subseteq M$, whence $xM = yM$. This well-defined ψ is easily seen to be a surjective homomorphism. □

3.7.2 Subgroups and Quotients of Solvable Groups

A few straightforward results provide general ways to obtain solvable groups.

Proposition 3.42. *Every subgroup and every homomorphic image of a solvable group G is solvable.*

Proof. By assumption there is a chain

$$G = N_0 \supset N_1 \supset N_2 \supset \cdots \supset N_{k-1} \supset N_k = \langle 1 \rangle$$

such that each N_j is a normal subgroup of N_{j-1} and N_{j-1}/N_j is abelian.

If H is a subgroup of G, there results the chain of intersections

$$H = N_0 \cap H \supseteq N_1 \cap H \supseteq N_2 \cap H \supseteq \cdots \supseteq N_{k-1} \cap H \supseteq N_k \cap H = \langle 1 \rangle.$$

Some of the original strict inclusions might collapse down to equalities. By Proposition 3.39, applied to each normal subgroup N_j of N_{j-1}, the subgroup $N_j \cap H$ is normal in $N_{j-1} \cap H$ and $(N_{j-1} \cap H)/(N_j \cap H)$ is abelian because it is isomorphic to a subgroup of the abelian N_{j-1}/N_j. This proves that H is solvable. The fact that some of the inclusions above might collapse to equalities does no harm. Just throw away redundant $N_j \cap H$.

For the second claim, let $\varphi : G \to H$ be a surjective homomorphism. Apply φ to the given subnormal chain for G to obtain the chain

$$H = \varphi(N_0) \supseteq \varphi(N_1) \supseteq \varphi(N_2) \supseteq \cdots \supseteq \varphi(N_{k-1}) \supseteq \varphi(N_k) = \langle 1 \rangle.$$

By Proposition 3.41 applied to the restriction of φ to N_{j+1}, each $\varphi(N_j)$ is normal in $\varphi(N_{j-1})$. Furthermore the quotients $\varphi(N_{j-1})/\varphi(N_j)$ are abelian, being homomorphic images of the abelian N_{j-1}/N_j. Again, the possibility that some inclusions above might collapse to equalities does no harm. Thus H is solvable. □

It follows from Proposition 3.42 that every group containing a non-solvable subgroup is itself non-solvable. For instance, if $n \geq 5$, the symmetric group $\mathcal{S}_n$ is non-solvable since it contains the non-solvable $\mathcal{A}_n$.

Here is an even more useful result.

Proposition 3.43. *If $\varphi : G \to H$ is a surjective group homomorphism and both its image H and kernel K are solvable, then G is solvable.*

Proof. Both K and H have subnormal chains

$$K = M_0 \supset M_1 \supset M_2 \supset \cdots \supset M_{\ell-1} \supset M_\ell = \langle 1 \rangle$$

and

$$H = N_0 \supset N_1 \supset N_2 \supset \cdots \supset N_{k-1} \supset N_k = \langle 1 \rangle$$

where the adjacent quotients M_{j-1}/M_j and N_{j-1}/N_j are abelian. Apply inverse images of φ to the second chain above to obtain the chain

$$G = \varphi^{-1}(N_0) \supset \varphi^{-1}(N_1) \supset \varphi^{-1}(N_2) \supset \cdots \supset \varphi^{-1}(N_{k-1}) \supset \varphi^{-1}(N_k) = K.$$

The above inclusions are proper since φ is surjective, and each $\varphi^{-1}(N_j)$ is normal in $\varphi^{-1}(N_{j-1})$.

Once we see that the quotients $\varphi^{-1}(N_{j-1})/\varphi^{-1}(N_j)$ are abelian, the solvability of G comes from looking at the combined subnormal chain:

$$G = \varphi^{-1}(N_0) \supset \varphi^{-1}(N_1) \supset \cdots \supset \varphi^{-1}(N_k) = K \supset M_1 \supset \cdots \supset M_\ell = \langle 1 \rangle.$$

Since φ is surjective we can see that

$$\varphi(\varphi^{-1}(N_{j-1})) = N_{j-1} \text{ and } \varphi(\varphi^{-1}(N_j)) = N_j.$$

Apply Proposition 3.40 to the restriction of φ to $\varphi^{-1}(N_{j-1})$ using the normal subgroup $\varphi^{-1}(N_j)$ and the preceding observation to obtain

$$\varphi^{-1}(N_{j-1})/\varphi^{-1}(N_j) \cong N_{j-1}/N_j.$$

Since the groups N_{j-1}/N_j are abelian, the quotients of their inverse images are also abelian, as desired. $\square$

Typically, Proposition 3.43 is used as follows. A candidate group G is before us. We seek out a normal subgroup N which is solvable and such that the quotient group G/N is solvable. Using the canonical projection $G \to G/N$ having kernel N, it will follow that G is solvable.

Groups of Order 30 Are Solvable

Let us apply Proposition 3.43 along with some Sylow theory to show that every group G of order 30 is solvable.

Since $30 = 3 \cdot 10$, the number of Sylow 3-subgroups of G divides 10 and is congruent to 1 mod 3. So, the number of Sylow 3-subgroups is either 1 or 10. Similarly the number of Sylow 5-subgroups divides 6 and is congruent to 1 mod 5. So, the number of Sylow 5-subgroups is either 1 or 6. Furthermore every Sylow 3-subgroup has order 3 and every Sylow 5-subgroup has order 5. If there were 10 Sylow 3-subgroups, any two of them would intersect trivially (using Lagrange) and this would give rise to 20 elements of order 3 in G. Likewise if there were 6 Sylow 5-subgroups, any two of them would intersect trivially (using Lagrange again), which would give rise to 24 elements of order 5. Since $20 + 24 = 44 > 30$, it must be that there is either just one Sylow 3-subgroup or just one Sylow 5-subgroup.

Say just one Sylow 3-subgroup H exists. This H must be normal in G, since all of its conjugates, being Sylow 3-subgroups, equal H itself. The quotient G/H has order 10, by

Lagrange. Since 10 is a product of two distinct primes, G/H is solvable, as mentioned in the examples at the start of this section. By Proposition 3.43 our G is solvable. A similar argument shows that G is solvable in case just one Sylow 5-subgroup exists.

3.7.3 Improving the Subnormal Chains for Solvable Groups

Proposition 3.21 implies that not only is every finite p-group solvable, but it comes with a subnormal chain such that the index of each group in the chain relative to its preceding group is a prime. Thereby the adjacent quotients are cyclic groups of prime order. Let us verify now that this can be done for all finite solvable groups.

Proposition 3.44. *If G is a finite solvable group, then there exists a subnormal chain*

$$G = N_0 \supset N_1 \supset N_2 \supset \cdots \supset N_{k-1} \supset N_k = \langle 1 \rangle$$

such that each adjacent quotient N_{j-1}/N_j is cyclic of prime order.

Proof. As a first step we show this is so when G is abelian. In that case let N_1 be a maximal proper subgroup of G. The abelian group G/N_1 has prime order. For if not, there would be a non-trivial and proper subgroup H. The inverse image of H under the canonical projection $G \to G/N_1$ would then be a subgroup of G properly between N_1 and G, contrary to the maximality of N_1. Repeat this argument on N_1 to get a subgroup N_2 of N_1 such that N_1/N_2 has prime order. Continue in this way to get the desired chain, which has to terminate at the trivial group $\langle 1 \rangle$ because G is finite.

For the second step suppose G is a finite group with a normal subgroup H such that G/H is abelian. We will verify that there is a subnormal chain

$$G = L_0 \supset L_1 \supset L_2 \supset \cdots \supset L_n = H,$$

terminating at H and such that each quotient L_{j-1}/L_j has prime order. Well, by the discussion just held, applied to G/H, there is a subnormal chain

$$G/H = K_0 \supset K_1 \supset K_2 \supset \cdots \supset K_n = \langle 1 \rangle,$$

where each quotient K_{j-1}/K_j has prime order. Let $\varphi : G \to G/H$ be the canonical projection and put $L_j = \varphi^{-1}(K_j)$. This results in the chain

$$G = L_0 \supset L_1 \supset L_2 \supset \cdots \supset L_n = H.$$

Each inverse image L_j is normal in its predecessor because the abelian K_j are normal in their predecessors. Also, since φ is surjective, $\varphi(L_j) = K_j$ for all j. Apply the third isomorphism theorem, Proposition 3.40, to the restriction of φ to L_{j-1} using the normal subgroup L_j and the preceding observation to obtain

$$L_{j-1}/L_j \cong K_{j-1}/K_j,$$

which is a group of prime order.

For the final step let G be solvable with subnormal series

$$G = H_0 \supset H_1 \supset H_2 \supset \cdots \supset H_m = \langle 1 \rangle,$$

where each H_{j-1}/H_j is abelian. By the preceding second step applied to each abelian quotient H_{j-1}/H_j there is a chain starting at H_{j-1} and ending at H_j such that each adjacent quotient has prime order. To finish off, concatenate these chains in the obvious way to pick up the desired subnormal chain from G down to $\langle 1 \rangle$. □

3.7.4 Commutators and the Derived Series

In the proof of the simplicity of the groups $\mathcal{A}_n$ in Section 2.4 we encountered the commutator of two group elements. Commutators provide an alternative approach to solvability.

Definition 3.45. If x, y are in a group G, their ***commutator***, denoted by the box-bracket $[x, y]$, is the element

$$[x, y] = xyx^{-1}y^{-1}.$$

Clearly x, y commute if and only if $[x, y] = 1$. Thus, a group is abelian if and only if 1 is its only commutator.

If x, y, z are in a group, it is easy to verify that

$$[y, x] = [x, y]^{-1} \text{ and } z[x, y]z^{-1} = [zxz^{-1}, zyz^{-1}],$$

revealing that both inverses and conjugates of commutators are commutators. If $\varphi : G \to H$ is any homomorphism, it is also evident that

$$\varphi([x, y]) = [\varphi(x), \varphi(y)]$$

for any commutator $[x, y]$ in G. Thus, homomorphic images of commutators are commutators.

Commutators in $\mathcal{S}_n$

It is not so easy to recognize if an element in a group is a commutator. Indeed, given z in G, how does one decide if $z = xyx^{-1}y^{-1}$ for some x, y in G? Let us at least note, for $n \geq 3$, that every 3-cycle in $\mathcal{S}_n$ is a commutator. If $\sigma = (1, 2, 3)$ and $\tau = (1, 2)$, a simple calculation reveals that

$$[\sigma, \tau] = \sigma\tau\sigma^{-1}\tau^{-1} = (1, 2, 3)(1, 2)(3, 2, 1)(1, 2) = (1, 3, 2).$$

The 3-cycle $(1, 3, 2)$ is a commutator. According to Proposition 3.56, all 3-cycles are conjugate to each other in $\mathcal{S}_n$. And as we just saw, conjugates of commutators are again commutators. Thus all 3-cycles in $\mathcal{S}_n$ are commutators.

It might also be worth noting that in $\mathcal{S}_n$ all commutators are even permutations. Indeed, if $\operatorname{sgn} : \mathcal{S}_n \to \{\pm 1\}$ is the homomorphism that evaluates the parity of a permutation and $\sigma, \tau \in \mathcal{S}_n$, then

$$\operatorname{sgn}([\sigma, \tau]) = [\operatorname{sgn}(\sigma), \operatorname{sgn}(\tau)] = 1.$$

The last equality holds because the group $\{\pm 1\}$ is abelian.

Recall that all even permutations are products of 3-cycles. Since every 3-cycle is a commutator, every even permutation is a product of commutators. Because all commutators are even, the group generated by the set of commutators is the alternating group $\mathcal{A}_n$. Actually, $\mathcal{A}_n$ is not only generated by the commutators of $\mathcal{S}_n$, but the set of commutators already comprises all $\mathcal{A}_n$. To prove such a result requires considerably more effort.

The Commutator Subgroup and the Derived Series

While the inverse of every commutator is a commutator, it is not the case that the product of two commutators is always a commutator. And so, the set of commutators in a group G need not be a subgroup of G. At the end of the section we offer an example of a group of order 96 in which the set of commutators is not a subgroup.

Definition 3.46. The subgroup of a group G *generated* by the set of its commutators is called the ***commutator subgroup*** and also the ***derived subgroup*** of G, to be denoted by the accented G'.

Here are some key facts regarding commutator subgroups.

1. Since inverses of commutators are once again commutators, the group G' consists of all products of commutators.
2. If $\varphi : G \to H$ is a homomorphism, then $\varphi(G') \subseteq H'$. This is true because homomorphic images of commutators are commutators, and the same holds for any product of commutators. If φ is an automorphism, then $\varphi(G') = G'$.

   A subgroup H of G which is invariant under all automorphisms of G (i.e. $\varphi(H) = H$ for all automorphisms φ of G) is known as a ***characteristic subgroup***[3] of G. Being invariant under all automorphisms of G, characteristic subgroups remain invariant under inner automorphisms, and are thereby normal in G. Thus, G' is a normal subgroup of G.
3. Let $\varphi : G \to G/G'$ be the canonical projection. For any x, y in G we get

$$[\varphi(x), \varphi(y)] = \varphi([x, y]) = 1 \text{ in } G/G'.$$

   Here 1 is the identity of the quotient group G/G', and the above holds because $[x, y] \in G'$. Thus G/G' is an abelian group.

The commutator subgroup can also be identified from an abstract perspective.

Proposition 3.47. *If N is a normal subgroup of a group G and G/N is abelian, then $G' \subseteq N$. In other words, among the normal subgroups of G producing an abelian quotient, G' is the smallest.*

Proof. Let $\varphi : G \to G/N$ be the canonical projection. For every x, y in G we have

$$\varphi([x, y]) = [\varphi(x), \varphi(y)] = 1.$$

The last equality holds because G/N is abelian. Thus $[x, y] \in \ker\varphi = N$. Consequently $G' \subseteq N$, since the commutators that generate G' belong to N. □

[3] The choice of such a name in the literature for these kinds of subgroups remains a puzzle.

The Derived Series

What is done to G can be repeated on G'. Let $G^{(1)} = G'$. Then put

$$G^{(2)} = (G^{(1)})', G^{(3)} = (G^{(2)})',$$

and so on. There results a descending chain of subgroups:

$$G \supseteq G^{(1)} \supseteq G^{(2)} \supseteq G^{(3)} \supseteq \cdots \supseteq G^{(j)} \supseteq \cdots .$$

The inclusions will remain proper until at some point $G^{(j)} = G^{(j+1)}$. Should equality occur at some j, then the chain will stop descending properly and persist with equalities. This descending chain is known as the ***derived series*** for G.

Since $G^{(1)}$ is a characteristic subgroup of G, every automorphism φ of G restricts to an automorphism of $G^{(1)}$. And since $G^{(2)}$ is a characteristic subgroup of $G^{(1)}$, this φ in turn restricts to an automorphism of $G^{(2)}$. And so $G^{(2)}$ is again a characteristic subgroup of G. Thus the double commutator $G^{(2)}$ is normal in G. In a similar vein, all higher commutator subgroups $G^{(j)}$ are characteristic and thereby normal in G, not only in their preceding group $G^{(j-1)}$.

If the derived series terminates in the trivial group $G^{(k)} = \langle 1 \rangle$ for some k, then G is clearly solvable. Furthermore, not only is the derived series *sub*normal, but all subgroups $G^{(j)}$ in the series are normal in G. We now can check the converse statement, that the derived series of every solvable group terminates in the trivial group.

Proposition 3.48. *If G is a solvable group with a subnormal series*

$$G = N_0 \supset N_1 \supset N_2 \supset \cdots \supset N_{k-1} \supset N_k = \langle 1 \rangle$$

such that each N_j/N_{j+1} is abelian, and $G^{(j)}$ is the jth group in the derived series, then $G^{(j)} \subseteq N_j$. Consequently, the derived series of a solvable group terminates in the trivial group.

Proof. Use Proposition 3.47 repeatedly together with the routine fact that if H, K are subgroups of G and $H \subseteq K$, then $H' \subseteq K'$.

First $G^{(1)} \subseteq N_1$, since G/N_1 is abelian. Then $G^{(2)} \subseteq N_1' \subseteq N_2$, since N_1/N_2 is abelian. Then $G^{(3)} \subseteq N_2' \subseteq N_3$, since N_2/N_3 is abelian, and so on until we get to $G^{(k)} \subseteq N_k = \langle 1 \rangle$. □

From Proposition 3.48 we can see that the derived series of a solvable group is the shortest subnormal series that the group can have for which the adjacent quotients are abelian. The number of proper inclusions in the derived series is sometimes called the ***length*** of the group.

A Finite Group in which a Product of Commutators Is Not a Commutator

The derived subgroup G' of G is made up of all products of commutators. Although it is not so easy to check, many of the groups G that we know are such that their *set* of commutators is already the group G' with no need to take their products. In fact the smallest group containing a product of commutators which is not itself a commutator has order 96.

For curiosity's sake we now offer a finite group whose set of commutators is not closed under multiplication.

Let X be the set of all 5-tuples (a_1,a_2,a_3,a_4,a_5) where the entries a_j are 0 or 1. There are 32 such 5-tuples. Take L to be the set of functions $f : X \to \mathcal{S}_3$. There are 6^{32} such functions. The set L is a group using the point-wise multiplication of functions. That is, for $f,g \in L$, the product is the function fg which sends a 5-tuple a to $f(a)g(a)$. Here, the multiplication $f(a)g(a)$ is evaluated in the symmetric group $\mathcal{S}_3$. Alternatively, the group L can be seen as the product group of $\mathcal{S}_3$ taken 32 times. Since $\mathcal{S}_3$ is not abelian, neither is the very large group L.

In $\mathcal{S}_3$ let τ be the transposition $(1,2)$ and let σ be the 3-cycle $(1,2,3)$. In anticipation of future use, here is their commutator:

$$[\tau,\sigma] = \tau\sigma\tau^{-1}\sigma^{-1} = \sigma^2\sigma^{-1} = \sigma.$$

For each j from 1 to 5 let $f_j : X \to \mathcal{S}_3$ be the function in L defined by

$$f_j(a_1,a_2,a_3,a_4,a_5) = \tau^{a_j}.$$

In other words, f_j maps the 5-tuple (a_1,a_2,a_3,a_4,a_5) to the identity permutation if $a_j = 0$ and to τ if $a_j = 1$. For example, $f_1(1,1,0,0,0) = \tau$ while $f_4(1,1,0,0,0) = 1$ (the identity in $\mathcal{S}_3$). Take $g : X \to \mathcal{S}_3$ to be the constant function in L defined by $g(a) = \sigma$ for every a in X. For each j from 1 to 5 and each $(a_1,\ldots,a_5)$ in X, we have $f_j^2(a_1,\ldots,a_5) = (\tau^{a_j})^2 = 1$, and also $g^3(a_1,\ldots,a_5) = \sigma^3 = 1$. This tells us that the f_j have order 2 in L and g has order 3. It can also be checked that the f_j commute among themselves, but do not commute with g.

Now let G be the subgroup of L generated by all of f_1,f_2,f_3,f_4,f_5,g. Since g does not commute with the f_j, this is a non-abelian group. An element of G is a product $x_1x_2\cdots x_k$ where each x_i is an f_j or a g or an inverse of these. But since every f_j is its own inverse and the inverse of g is g^2, each x_i can be taken to be either an f_j or a g. We refer to such a product $x_1x_2\cdots x_k$ as a *word* in the f_j and g. Different words could represent the same element of G. Also note that if $f \in G$ and f is represented by the word $x_1x_2\cdots x_k$ and if a is a 5-tuple, then $f(a) = x_1(a)x_2(a)\cdots x_k(a)$, which is a word in τ and σ in $\mathcal{S}_3$.

It turns out that the product of commutators

$$[f_1,g][f_2,g][f_3,g][f_4,g][f_5,g]$$

inside G' is not a commutator.

To that end consider any commutator $[p,q]$ where $p,q \in G$. Take words $x_1x_2\cdots x_k$, $y_1y_2\cdots y_\ell$ in the f_j and g to represent (i.e. equal) p,q, respectively. For each j from 1 to 5 let

$$n_j = \begin{cases} 0 & \text{if an even number of } x_i \text{ equal } f_j \\ 1 & \text{if an odd number of } x_i \text{ equal } f_j \end{cases}$$

and

$$m_j = \begin{cases} 0 & \text{if an even number of } y_i \text{ equal } f_j \\ 1 & \text{if an odd number of } y_i \text{ equal } f_j. \end{cases}$$

The five pairs (n_j, m_j) cover four possibilities: $(0,0), (0,1), (1,0)$ and $(1,1)$. Hence there must be distinct i, j such that $(n_i, m_i) = (n_j, m_j)$. To keep notations simple say, without harm, that $(n_1, m_1) = (n_2, m_2)$. This tells us that the number of x_i which equal f_1 in the word for p has the same parity as the number of x_i which equal f_2 in the word for p. So, the total number of x_i which equal f_1 or f_2 in the word for p is even. Likewise, the total number of y_i which equal f_1 or f_2 in the word for q is even.

Evaluate p at the 5-tuple $a = (1,1,0,0,0)$ to get $p(a) = x_1(a)x_2(a)\cdots x_k(a)$, a word in τ and σ inside $\mathcal{S}_3$. An $x_i(a)$ will equal τ every time x_i is f_1 or f_2. The f_3, f_4, f_5 appearing in the word for p contribute only a 1 in the word for $p(a)$. Thus in the word for $p(a)$ there are an even number of τ appearing, the rest being σ.

Let us now check that a word inside $\mathcal{S}_3$ in τ and σ which has an even number of τ is already in the cyclic subgroup $\langle\sigma\rangle$. Well, let $\alpha_1\alpha_2\cdots\alpha_k$ be such a word. Its image under the canonical projection $\varphi : \mathcal{S}_3 \to \mathcal{S}_3/\langle\sigma\rangle$ is $\varphi(\alpha_1)\varphi(\alpha_2)\cdots\varphi(\alpha_k)$. If an $\alpha_i = \sigma$, then $\varphi(\alpha_i) = 1$ in the two element quotient group $\mathcal{S}_3/\langle\sigma\rangle$. Thus the image of the word is the product of an even number of $\varphi(\tau)$, which of course must be 1. Our word $\alpha_1\alpha_2\cdots\alpha_k$ lies in $\ker\varphi$ which is $\langle\sigma\rangle$.

We have just seen that $p(a)$ is in the abelian group $\langle\sigma\rangle$. Likewise, $q(a) \in \langle\sigma\rangle$. With this in hand we can evaluate the commutator $[p,q]$ at our 5-tuple a to get

$$[p,q](a) = [p(a), q(a)] = 1,$$

because $\langle\sigma\rangle$ is abelian. On the other hand, evaluate the original commutator product at our a to get

$$\begin{aligned}
[f_1,g][f_2,g][f_3,g][f_4,g][f_5,g](a) &= [f_1(a),g(a)][f_2(a),g(a)][f_3(a),g(a)][f_4(a),g(a)][f_5(a),g(a)] \\
&= [\tau,\sigma][\tau,\sigma][1,\sigma][1,\sigma][1,\sigma] \\
&= [\tau,\sigma]^2 = \sigma^2 \neq 1.
\end{aligned}$$

Since $[p,q]$ and our commutator product attain different values at a, there is no way they could be equal.

EXERCISES

1. Show that groups of order 490 and of order 1470 are solvable.
2. Why are simple non-abelian groups not solvable?
3. If H, K are finite solvable groups show that the semi-direct product $H \rtimes K$ is solvable.
4. If G is a group of order pq^k where p, q are distinct primes, k is a positive exponent, and $p < q$, show that G is solvable.

   Show that G remains solvable when $p > q$ and $k = 2$.
5. Find a group G with a normal subgroup N and an automorphism φ such that $\varphi(N) \nsubseteq N$. In other words, find a group with a normal subgroup that is not a characteristic subgroup.
6. If H is a subgroup of G, and $G' \subseteq H$, show that H is normal in G.
7. Find the commutator subgroup of $\mathcal{A}_4$. Then find the derived series of $\mathcal{A}_4$. What is the derived series of $\mathcal{S}_4$?

One way to approach this is to recall that the commutator subgroup is the smallest normal subgroup H such that $\mathcal{A}_4/H$ is abelian.

8. Show that all groups of order up to 59 are solvable.

   Note. Many cases are covered directly by the discussion in this section. Some use Sylow theory to get normal subgroups for which Proposition 3.43 might apply. For some cases, such as groups of order 48, Proposition 3.4 may help. It is necessary to stop at 59 because $\mathcal{A}_5$ has order 60, and this group is not solvable.
9. If the derived series of a group never stops descending, explain briefly why the group is infinite and non-abelian. Find an example of a group whose derived series never stops descending.
10. Is the even permutation $(1,2)(2,4)$ a commutator in $\mathcal{S}_4$? Is it a commutator in $\mathcal{A}_4$?
11. Find the derived series of the dihedral group $\mathcal{D}_4$.
12. If H is a Sylow p-subgroup of a finite group G and H is normal, why is H a characteristic subgroup of G?

3.8 Breaking the Enigma

We would like to digress a bit and outline, in some detail, a historically significant application of group theory. This was the breaking of an early version of the German Enigma encryption machine in the years before the outbreak of World War 2. This work had significant influence on the outcome of that war. The ideas originated from a brilliant Polish mathematician, Marian Rejewski (1905–1980). Soon after the outbreak of the War, Rejewski joined the code breaking group at Bletchley Park. For the history behind the breaking of the Enigma, we would recommend the book *Seizing the Enigma*, by David Kahn.

3.8.1 The Design of the Enigma Machine

In order to understand how the Enigma machine cipher was broken we need to examine first the machine itself.

The Enigma machine has six components: a keyboard, a lamp board on the top, a plug board on the front, a static disk inside the machine, a set of three rotors inside the machine, and a reflector also inside the machine.

- The keyboard consists of 26 typewriter keys each marked by a letter of the alphabet.
- The lamp board consists of 26 small electric lamps each marked with a letter of the alphabet.
- The plug board consists of 26 pairs of electrical sockets. Any two of these pairs of sockets can be connected together with an electric cord. Usually 12 sockets are paired off using 6 cords.
- Each of the 3 rotors is a thick disk with 26 electrical contacts on each side along the circumference of the disk. Inside the disk each electrical contact on one side is connected by

a wire to an electrical contact on the other side in a fairly random fashion. The 3 rotors are mounted on an axis so that the electrical contacts on a side of one rotor touch the electrical contacts of its adjoining rotor. Each of the 3 rotors rotates like a car odometer going through 26 positions. The right-most rotor rotates first going through the 26 positions. At that point the right-most rotor rotates through one additional position, and the middle rotor rotates through one position. When the middle rotor goes through 26 positions, the left-most rotor rotates by one position.

- At the left end of the axis there is the reflector consisting of a disk with 26 electrical contacts on one side. The electrical contacts are paired off inside the reflector by wires.
- At the right end of the axis is another disk with 26 contacts on one side. This static disk merely serves to connect the right-most rotor to the plug board.

When a key on the keyboard is pushed, the right-most rotor rotates one position. An electrical circuit is completed that runs from the keyboard, though the plug board, through the rotors, then through the reflector, back through the rotors and plug board, and finally to the lamp board, where the battery power lights one of the lamps. The letter corresponding to the lighted lamp is the encipherment of the letter pushed on the keyboard.

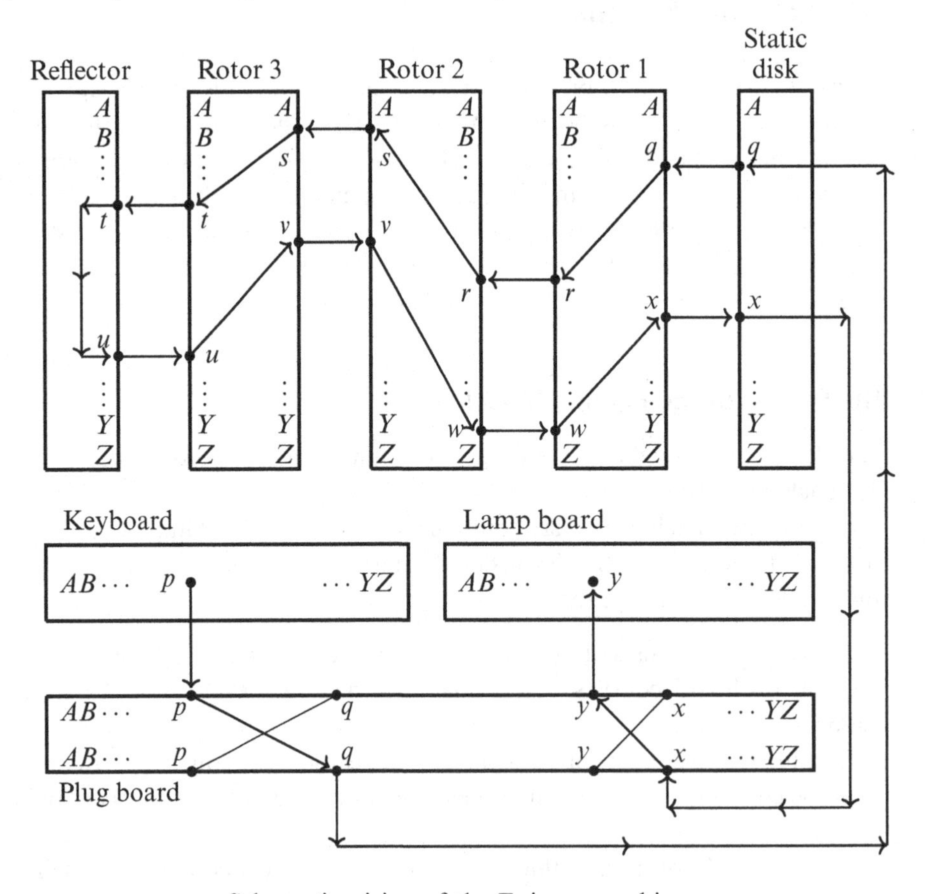

Schematic wiring of the Enigma machine.

The above diagram represents an electrical circuit, starting at an input p and ending at an output y. (We leave out the role of the necessary battery between p and y.) Here is how to read the diagram.

- The lower case letters $p, q, r, s, \ldots$ are to be treated as variables representing any of the 26 upper case letters $A, B, C, \ldots, Z$ realized as contact points in each of the boxes.
- An input letter p is typed into the keyboard and arrives at the plug board.
- The plug board will either transpose p with another letter q or it may leave p unchanged, depending on that day's plug board setting. We have shown p as being transposed with q.
- The letter q arrives at the static disk, and is passed on unchanged to the right side of rotor 1.
- The internal wiring of rotor 1 permutes q to a new letter r, and passes the letter r onto disk 2.
- The internal wiring of rotor 2 permutes r to a new letter s, and passes the letter s onto disk 3.
- The internal wiring of rotor 3 permutes s to a new letter t, and passes the letter t to the reflector.
- The reflector implements a transposition on t to get a new letter u, which is passed on back to rotor 3. Note that if the reflector had received u it would have passed back t.
- The wirings in rotors 3, 2, 1 permute u to new letters v, w, x, in turn.
- Rotor 1 passes letter x to the static disk and then on to the plug board which may or may not transpose x to a new letter y. We have shown x as being transposed with y.
- The final letter y appears on the lamp board as the encipherment of the original p.

Since a rotor moves each time a key is pressed, the encipherment of letters changes with each key press. Thus, for example, if the letter T is pushed twice, the letter L might light up on the lamp board after the first push, while the letter C might light up after the second push.

Even if a spy knows the general structure of the machine, they would not know how the rotors and the reflector are wired, or the starting position of the rotors when a message is sent.

This is what Rejewski used group theory to discover.

3.8.2 The Mathematical Representation of the Enigma Machine

At a given instant the Enigma machine represents a permutation on the 26 letters of the alphabet. How is this permutation made up? Each of the three rotors, as well as the reflector and the plug board contributes to the permutation.

Let

- δ be the permutation implemented by the plug board,
- ρ_1 the permutation implemented by (the right-most) rotor 1,
- ρ_2 the permutation implemented by (the middle) rotor 2,
- ρ_3 the permutation implemented by (the left-most) rotor 3, and
- ϵ the permutation implemented by the reflector.

As the wiring goes from the keyboard to the plug board, through the three rotors, then through the reflector, back through the rotors and the plug board, the permutation that the machine represents is given by

$$\tau = \delta^{-1}\rho_1^{-1}\rho_2^{-1}\rho_3^{-1}\epsilon\rho_3\rho_2\rho_1\delta.$$

This is the permutation ϵ implemented by the reflector conjugated by $\rho_3\rho_2\rho_1\delta$. The 26 contacts on the reflector are paired off using 13 wires. This means that ϵ is a product of 13 disjoint transpositions. By Proposition 3.56 the orbit structure of a permutation is preserved under conjugation. Thus the permutation represented by the Enigma machine at a given instant is the product of 13 disjoint transpositions. In particular, the order of the permutation is 2. Furthermore, since the machine always represents the product of 13 disjoint transpositions, no letter is fixed. That is, no letter is ever encrypted as itself.

If, in a given instance, the key B is pushed and if the letter Q lights up on the lamp board of an encrypting machine, then on an Enigma machine with the identical configuration a push of the letter Q on its keyboard will cause the letter B to light up on its lamp board. Thus, decryption on an Enigma machine receiving an encrypted message is done by typing in the encrypted message on its keyboard.

Of course, what made the Enigma machine difficult to break is that the permutation of the 26 letters represented by the machine changes each time a key is pressed. This is due to the movement of the rotors. If a key is pressed once and one rotor rotates through one notch, what is the permutation represented by the machine in its new configuration?

To answer this, notice that each of the three rotors is in contact with a disk to its left and one to its right (be it another rotor, the static disk or the reflector). For now, let the disk to the right of a rotor be called the *input disk*, and the disk to the left the *output disk*. In the middle lies the rotor with its 26 contacts to its right meeting the 26 contacts of the input disk and its 26 contacts to its left meeting the 26 contacts of the output disk. Label these contacts as $1, 2, \ldots, 26$. The labelled contacts of the input and output disk and the rotor are aligned as shown.

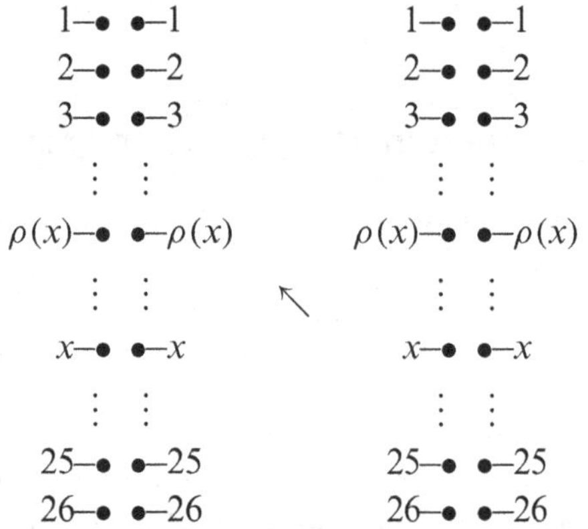

$$\begin{array}{ccccccc}
1 - \bullet & \bullet - 1 & & & 1 - \bullet & \bullet - 1 \\
2 - \bullet & \bullet - 2 & & & 2 - \bullet & \bullet - 2 \\
3 - \bullet & \bullet - 3 & & & 3 - \bullet & \bullet - 3 \\
\vdots & \vdots & & & \vdots & \vdots \\
\rho(x) - \bullet & \bullet - \rho(x) & & & \rho(x) - \bullet & \bullet - \rho(x) \\
\vdots & \vdots & & \nwarrow & \vdots & \vdots \\
x - \bullet & \bullet - x & & & x - \bullet & \bullet - x \\
\vdots & \vdots & & & \vdots & \vdots \\
25 - \bullet & \bullet - 25 & & & 25 - \bullet & \bullet - 25 \\
26 - \bullet & \bullet - 26 & & & 26 - \bullet & \bullet - 26
\end{array}$$

The left-most column of labelled contacts stands for the output disk. The two middle columns of contacts stand for the rotor with contacts on its left and right sides. The right-most column of contacts stands for the input disk. The internal wiring of the rotor matches each contact on the right side of the rotor to a contact on its left side. Let ρ represent the permutation on the letters $1, 2, \ldots, 26$ implemented by the internal wiring of the rotor in this initial position.

In this initial position contact x from the input disk touches contact x on the right of the rotor. The internal wiring of the rotor maps contact x to contact $\rho(x)$ on the left side of the rotor. Then contact $\rho(x)$ touches its namesake on the output disk. In this initial position the rotor implements the permutation ρ from the 26 letters (contacts) on the input disk to the 26 letters (contacts) on the output disk.

Now suppose that the rotor turns by one notch in the downward direction (in accordance with our diagram). The new configuration is as shown.

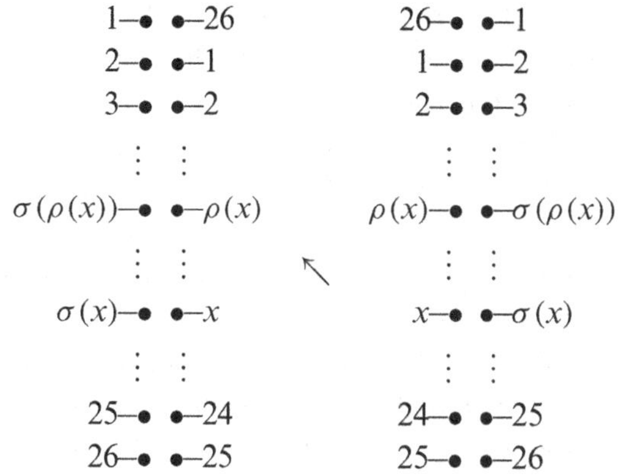

$$
\begin{array}{rl c rl}
1\text{—}\bullet & \bullet\text{—}26 & & 26\text{—}\bullet & \bullet\text{—}1 \\
2\text{—}\bullet & \bullet\text{—}1 & & 1\text{—}\bullet & \bullet\text{—}2 \\
3\text{—}\bullet & \bullet\text{—}2 & & 2\text{—}\bullet & \bullet\text{—}3 \\
\vdots & \vdots & & \vdots & \vdots \\
\sigma(\rho(x))\text{—}\bullet & \bullet\text{—}\rho(x) & & \rho(x)\text{—}\bullet & \bullet\text{—}\sigma(\rho(x)) \\
\vdots & \vdots & \nwarrow & \vdots & \vdots \\
\sigma(x)\text{—}\bullet & \bullet\text{—}x & & x\text{—}\bullet & \bullet\text{—}\sigma(x) \\
\vdots & \vdots & & \vdots & \vdots \\
25\text{—}\bullet & \bullet\text{—}24 & & 24\text{—}\bullet & \bullet\text{—}25 \\
26\text{—}\bullet & \bullet\text{—}25 & & 25\text{—}\bullet & \bullet\text{—}26
\end{array}
$$

If σ denotes the 26-cycle $(1, 2, \ldots, 26)$, we see now that contact $\sigma(x)$ on the input disk touches contact x on the right side of the rotor. As before, the internal wiring of the rotor sends x to contact $\rho(x)$ on the left side of the rotor. Since the rotor has rotated by one notch, contact $\rho(x)$ meets contact $\sigma(\rho(x))$ on the output disk. The net effect is that contact $\sigma(x)$ on the input disk is mapped to contact $\sigma(\rho(x))$ on the output disk. Because x is an arbitrary contact on the input disk, we can replace x by $\sigma^{-1}(x)$. We discover that contact $x = \sigma(\sigma^{-1}(x))$ on the input disk gets permuted to contact $\sigma(\rho(\sigma^{-1}(x)))$ on the output disk.

In summary, if ρ is the permutation from the input disk to the output disk implemented by the rotor between them, then the permutation implemented by the same rotor after it has rotated by one notch is

$$\sigma\rho\sigma^{-1}$$

where σ is the full cycle $(1, 2, \ldots, 26)$.

We can conclude from this that if $\tau_1 = \delta^{-1}\rho_1^{-1}\rho_2^{-1}\rho_3^{-1}\epsilon\rho_3\rho_2\rho_1\delta$ is the permutation implemented by the machine in an initial configuration, then the permutation implemented by the machine after the right-most rotor has turned one notch is given by

$$\tau_2 = \delta^{-1}(\sigma\rho_1^{-1}\sigma^{-1})\rho_2^{-1}\rho_3^{-1}\epsilon\rho_3\rho_2(\sigma\rho_1\sigma^{-1})\delta,$$

where σ is the cycle $(1, 2, 3, \ldots, 26)$. In general, if the right-most rotor turns through j notches, the middle rotor turns through k notches, and the left-most rotor turns through ℓ

notches, then the permutation implemented by the machine based on its initial configuration is given by

$$\delta^{-1}(\sigma^j\rho_1^{-1}\sigma^{-j})(\sigma^k\rho_2^{-1}\sigma^{-k})(\sigma^\ell\rho_3^{-1}\sigma^{-\ell})\epsilon(\sigma^\ell\rho_3\sigma^{-\ell})(\sigma^k\rho_2\sigma^{-k})(\sigma^j\rho_1\sigma^{-j})\delta.$$

This equals the conjugation of the reflector's ϵ by $(\sigma^\ell\rho_3\sigma^{-\ell})(\sigma^k\rho_2\sigma^{-k})(\sigma^j\rho_1\sigma^{-j})\delta$.

We will concentrate on the group-theoretic technique that Rejewski used (coupled with some good luck) to solve for the permutation ρ_1 inherent in the wiring of the first rotor.

3.8.3 How the Machine Was Used

There are four settings that can be changed manually in the Enigma machine before it is used. First, the order of the three rotors can be switched around. Second, the starting positions of each rotor can be reset (i.e. the starting selection of ρ_1, ρ_2, ρ_3). Third, the first and second rotors have a so-called "ring" setting that determines when the rotor to its left will rotate by one notch. Fourth, the settings of the plug board (i.e. δ) can be changed. For each day there existed in a secret log book a list of these particular settings which was common for all machines. The order of the rotors was changed about every three months, while the other settings were changed daily.

In order to send an encrypted message the operator looked up the daily settings for the machine. With these settings in place over all machines, the operator chose a sequence of three letters, from the keyboard, say xyz. (These lower case letters stand for any one of $A, B, \ldots, Z$.) This would mean that the recipient was to move the first rotor to position x, the second to position y and the third to position z. These would then be the rotor settings for the message to follow. Each message sent from any machine would thus have customized rotor settings. This would thwart the breaking of the code by the technique of frequency analysis, which could be useful if a large amount of information were sent with the same settings. This was a key reason why the Enigma machine was so difficult to break.

Say the permutation in effect for all machines at the start of the day was

$$\tau_0 = \delta^{-1}\rho_1^{-1}\rho_2^{-1}\rho_3^{-1}\epsilon\rho_3\rho_2\rho_1\delta.$$

Upon pushing the x key the first rotor turned by one notch. The new permutation in effect on the sending machine became

$$\tau_1 = \delta^{-1}(\sigma\rho_1^{-1}\sigma^{-1})\rho_2^{-1}\rho_3^{-1}\epsilon\rho_3\rho_2(\sigma\rho_1\sigma^{-1})\delta$$

where σ is the 26-cycle $(A, B, C, ..., Y, Z)$. (Here the use of capital letters or numbers $1, 2, \ldots, 26$ is immaterial.) On the lamp board the encryption of x, say $r = \tau_1(x)$, appeared.

Upon pushing the y key the first rotor turned by one more notch. The new permutation in effect on the sending machine became

$$\tau_2 = \delta^{-1}(\sigma^2\rho_1^{-2}\sigma^{-2})\rho_2^{-1}\rho_3^{-1}\epsilon\rho_3\rho_2(\sigma^2\rho_1\sigma^{-2})\delta.$$

On the lamp board the encryption of y, say $s = \tau_2(y)$, appeared.

Upon pushing the z key the first rotor again turned by one notch. The new permutation in effect on the sending machine became

$$\tau_3 = \delta^{-1}(\sigma^3 \rho_1^{-1} \sigma^{-3}) \rho_2^{-1} \rho_3^{-1} \epsilon \rho_3 \rho_2 (\sigma^3 \rho_1 \sigma^{-3}) \delta.$$

On the lamp board the encryption of z, say $t = \tau_3(z)$, appeared.

The letters rst were sent out openly by Morse code to a receiving machine, whose operator would then type these in to retrieve the original xyz. Note that the receiving machine was also set at τ_0 according to the daily settings. When the receiver typed in r, the first rotor on that machine turned to put the permutation τ_1 in effect. Then $\tau_1(r) = \tau_1(\tau_1(x)) = x$ was recovered on the lamp board. This is because all permutations represented by the machine have order 2. By typing in s the receiver got $\tau_2(s) = y$ on the lamp board, and by typing in t the receiver got $\tau_3(t) = z$ on the lamp board.

However, there was a major flaw in the protocol for transmitting the three rotor settings prior to each message. Namely, the transmission of xyz was **repeated**, in order to avoid transmission errors. The operator typed in $xyz\ xyz$. The resulting six turns of the first rotor implemented, in succession, the six permutations

$$\tau_j = \delta^{-1}(\sigma^j \rho_1^{-1} \sigma^{-j}) \rho_2^{-1} \rho_3^{-1} \epsilon \rho_3 \rho_2 (\sigma^j \rho_1 \sigma^{-j}) \delta \text{ where } j = 1, 2, 3, 4, 5, 6.$$

These permutations encrypted the input letters $xyz\ xyz$ as new letters, say

$$r = \tau_1(x), \quad s = \tau_2(y), \quad t = \tau_3(z), \quad u = \tau_4(x), \quad v = \tau_5(y), \quad w = \tau_6(z).$$

Rejewski knew that each τ_j was its own inverse (i.e. of order 2). Thus

$$\tau_4(\tau_1(r)) = \tau_4(x) = u, \quad \tau_5(\tau_2(s)) = \tau_2(y) = v, \quad \tau_6(\tau_3(t)) = \tau_6(z) = w.$$

A fragment of information was given away. Namely, the composite $\tau_4\tau_1$ permuted r to u, the composite $\tau_5\tau_2$ permuted s to v, and the composite $\tau_6\tau_3$ permuted z to w. After a good number of messages were sent for that day, with various letters r, s, t, u, v, w appearing out in the open, enough information was provided that the composite permutations

$$\tau_4\tau_1, \ \tau_5\tau_2, \ \tau_6\tau_3,$$

became fully known. So it became a matter of finding all possible factor permutations τ_j, from the above known products. This was predicated on the assumption that for these first six inputs, the other two rotors did not move.

Rejewski Factorization

Keep in mind that each of the above τ_j is the product of 13 disjoint transpositions, because they are all conjugates of the fixed product of transpositions ϵ built into the reflector. Rejewski's upcoming results about the product of two permutations which are themselves the product of 13 disjoint transpositions on 26 letters are informative. As it is no more difficult, we might as well work inside the group S_{2n} of permutations on $2n$ letters. A review of the orbit of a letter under the action of a permutation and the related factorization of a permutation into disjoint cycles is recommended at this point. This is discussed in Section 2.2.2.

In particular, recall that if σ is a permutation on a finite set L of letters and if a is a letter in L, the orbit of a under σ is the set of letters $\sigma^k(a)$ as k runs through all integers. Of course, since L is finite, this list of letters will repeat itself. We will refer to this as the ***σ-orbit*** of a. The set of letters L is partitioned by its family of σ-orbits. The number of letters in a σ-orbit is called the ***length*** of the orbit.

Proposition 3.49. *Suppose α, β are each the product of n disjoint transpositions on 2n letters, a is one of those letters, k is an integer, and put $\sigma = \alpha\beta$. Then*

1. $\sigma^k\alpha = \alpha\sigma^{-k}$ *and* $\beta\sigma^k = \sigma^{-k}\beta$,
2. *the permutations $\sigma^k\alpha$ and $\beta\sigma^k$ are each the product of n disjoint transpositions, and thereby move every letter,*
3. *the σ-orbits of a and $\beta(a)$ are disjoint,*
4. *the σ-orbits of $\beta(a)$ and $\alpha(a)$ coincide, and both β and α map the σ-orbit of a onto the common σ-orbit of $\beta(a)$ and $\alpha(a)$,*
5. *the lengths of the σ-orbits of a and of $\beta(a)$ are equal,*
6. *if b is a letter outside of the σ-orbits of a and of $\beta(a)$, then so is $\beta(b)$ outside of these σ-orbits.*

Proof. Note that α, β have order 2, and that they leave no letter fixed. Also note that the inverse of the product $\alpha\beta$ is $\beta\alpha$.

1. These identities come down to showing that

$$(\alpha\beta)^k\alpha = \alpha(\beta\alpha)^k \text{ and } \beta(\alpha\beta)^k = (\beta\alpha)^k\beta.$$

If $k \geq 0$, these identities can be seen by inspection. If $k < 0$, interchange α and β and replace k by the positive $-k$ in the first of these identities to get $(\beta\alpha)^{-k}\beta = \beta(\alpha\beta)^{-k}$. Multiply through by $(\beta\alpha)^k$ on the left and $(\alpha\beta)^k$ on the right to get $\beta(\alpha\beta)^k = (\beta\alpha)^k\beta$, which verifies the second identity for $k < 0$. The first identity for $k < 0$ is shown in a similar way.

2. If k is even, use the item just shown to obtain

$$\sigma^k\alpha = \sigma^{k/2}\sigma^{k/2}\alpha = \sigma^{k/2}\alpha\sigma^{-k/2}.$$

Thus $\sigma^k\alpha$ is a conjugate of α. By Proposition 3.56, $\sigma^k\alpha$ has the same orbit structure as α. In other words, $\sigma^k\alpha$ is a product of n disjoint transpositions.

If k is odd, write

$$\sigma^k\alpha = \sigma^{(k-1)/2}\alpha\beta\sigma^{(k-1)/2}\alpha = (\sigma^{(k-1)/2}\alpha)\beta(\alpha\sigma^{-(k-1)/2}).$$

Since the inverse of $\sigma^{(k-1)/2}\alpha$ is $\alpha\sigma^{-(k-1)/2}$ we see that $\sigma^k\alpha$ is a conjugate this time of β, and thereby a product of n disjoint transpositions.

Regarding $\beta\sigma^k$, replace it with its equal $(\beta\alpha)^k\beta$, and the result follows from the part just done after interchanging α with β.

3. Suppose to the contrary that the σ-orbits of a and $\beta(a)$ overlap. This means that $\sigma^m(a) = \beta(a)$ for some integer exponent m. Then $\beta\sigma^m(a) = a$, contradicting the fact that $\beta\sigma^m$ moves every letter.
4. That $\alpha(a)$ has the same σ-orbit as $\beta(a)$ follows trivially from the equality $\alpha(a) = \sigma(\beta(a))$ (which comes from the definition of σ).

   Now the σ-orbit of a consists of all $\sigma^m(a)$ as m runs through the integers. Then β maps this orbit to the set of all $\beta\sigma^m(a)$. Since

$$\beta\sigma^m(a) = \sigma^{-m}\beta(a)$$

   and since $-m$ runs through the integers as well, it becomes clear that β maps the σ-orbit of a onto the σ-orbit of $\beta(a)$.

   Regarding α we have

$$\alpha(\sigma^m(a)) = \sigma^{-m}(\alpha(a)) = \sigma^{-m}(\sigma\beta(a)) = \sigma^{1-m}(\beta(a)).$$

   This gives the recipe for how α maps the σ-orbit of a onto the σ-orbit of $\beta(a)$.
5. To see that these orbits have equal length, recall from Proposition 2.15 that the length of the orbit of a letter c under a permutation γ is the smallest positive integer m such that $\gamma^m(c) = c$. For every exponent m there is the (now well used) identity

$$\sigma^{-m}(\beta(a)) = \beta(\sigma^m(a)).$$

   Clearly the left side equals $\beta(a)$ if and only if $\sigma^m(a)$ equals a. This reveals that the σ^{-1}-orbit of $\beta(a)$ has the same length as the σ-orbit of a. But the orbits of any letter under inverse permutations coincide. Thus the length of the σ-orbit of $\beta(a)$ equals the length of the σ-orbit of a.
6. For this final item suppose $\beta(b)$ is in the σ-orbit of a or in the σ-orbit of $\beta(a)$. Thus $\beta(b) = \sigma^m(a)$ or $\beta(b) = \sigma^m(\beta(a))$ for some exponent m. Apply β to both sides of these equations and use the identities in item 1 to get

$$b = \beta\sigma^m(a) = \sigma^{-m}\beta(a) \text{ or } b = \beta\sigma^m(\beta(a)) = \sigma^{-m}(a).$$

   This puts b in the σ^{-1}-orbit of a or of $\beta(a)$. Since orbits under inverses do not change, the original assumption on b is contradicted. □

In order to exploit Proposition 3.49 some terminology is helpful.

Definition 3.50. If σ is a permutation on $2n$ letters, we say that σ has a ***Rejewski factorization*** when there exists a pair of permutations (α, β) such that $\sigma = \alpha\beta$ and α, β are themselves products of n disjoint transpositions. Such permutations α, β will be called ***Rejewski factors*** of σ.

Definition 3.51. If σ is a permutation on $2n$ letters, an ***equal length pairing*** of the σ-orbits is a partition of the family of σ-orbits into two element sets $\{A, B\}$ such that the orbits A, B have equal length.

A bit of reflection reveals that the orbits of σ have an equal length pairing if and only if for every positive integer ℓ, the number of orbits of length ℓ is even.

Proposition 3.52. *If a permutation σ on $2n$ letters has Rejewski factors α, β, then there exists a unique equal length pairing of its orbits:*

$$\{A_1, B_1\}, \{A_2, B_2\}, \ldots, \{A_m, B_m\},$$

such that for each pair of orbits $\{A_j, B_j\}$, the permutations α, β map A_j to B_j and B_j to A_j. Then the restrictions of α, β to $A_j \cup B_j$ are Rejewski factors of the restriction of σ to $A_j \cup B_j$.

Proof. Let a_1 be a letter and let A_1 be its orbit under σ. By items 3 and 5 of Proposition 3.49 the letter $\beta(a_1)$ determines a new orbit B_1 whose length equals that of A_1. The orbits A_1, B_1 of equal length, ℓ_1 say, are now paired off. Item 4 says that α, β map A_1 onto B_1. They also map B_1 onto A_1 because each is of order 2, and thus equals its own inverse.

The restriction of β to $A_1 \cup B_1$ is the product of ℓ_1 transpositions. Explicitly, we can list the orbit of a_1 under σ as

$$a_1, \sigma(a_1), \sigma^2(a_1), \ldots, \sigma^{\ell_1 - 1}(a_1).$$

Then the transpositions

$$(a_1, \beta(a_1)),\ (\sigma(a_1), \beta(\sigma(a_1))),\ (\sigma^2(a_1), \beta(\sigma^2(a_1))), \ldots, (\sigma^{\ell_1-1}(a_1), \beta(\sigma^{\ell_1-1}(a_1)))$$

are disjoint and their product is the restriction of β to $A_1 \cup B_1$. Similarly the restriction of α to $A_1 \cup B_1$ is the product ℓ_1 disjoint transpositions. The fact that σ restricted to $A_1 \cup B_1$ is the product of the restrictions of α and β to this union should be clear.

If $A_1 \cup B_1$ covers all $2n$ letters, we are done. Otherwise pick a_2 outside of $A_1 \cup B_1$. The orbit of a_2 under σ, say A_2, is disjoint from A_1 and B_1, and by items 3, 5 and 6 of Proposition 3.49 the orbit B_2 of $\beta(a_2)$ is disjoint from all of A_1, B_1, A_2 and has the same length as A_2. Thus new orbits A_2, B_2 of equal length are paired off. As argued in the case of A_1, B_1, both α and β map A_2 onto B_2 and B_2 onto A_2, and the restrictions of α, β to the union $A_2 \cup B_2$ are Rejewski factors of the restriction of σ to this union.

If the $A_1 \cup B_1 \cup A_2 \cup B_2$ covers all $2n$ letters, we are done. If not, take a_3 outside of $A_1 \cup B_1 \cup A_2 \cup B_2$. By repeating the preceding argument based on Proposition 3.49 a new pairing of orbits A_3, B_3, which satisfies all required conditions, emerges. Continue to create pairings in this way until the $2n$ letters are exhausted, and the equal length pairing is established.

The uniqueness of this pairing comes from the fact that every orbit A has to be paired with the orbit $\beta(A)$. □

Because of Proposition 3.52 we consider permutations σ on $2n$ letters which have an equal length pairing of their orbits. Our goal is not only to count the number of Rejewski factorizations of such σ, but also to give an algorithm that finds all Rejewski factorizations.

Proposition 3.53. *Suppose a permutation σ on $2n$ letters has an equal length pairing of its orbits:*

$$\{A_1, B_1\}, \{A_2, B_2\}, \ldots, \{A_m, B_m\}.$$

Also assume that each restriction $\sigma_{|A_j \cup B_j}$ has a Rejewski factorization (α_j, β_j) acting on the union $A_j \cup B_j$. Then the permutations α, β defined by

$$\alpha = \alpha_1 \alpha_2 \cdots \alpha_m, \quad \beta = \beta_1 \beta_2 \cdots \beta_m$$

give a Rejewski factorization of σ acting on all $2n$ letters.

Proof. We might note right away that each α_j, β_j can be extended to act on all $2n$ letters by declaring that they fix the letters outside of $A_j \cup B_j$. Thus it makes sense to multiply these α_j and these β_j. Since the set of $2n$ letters is the disjoint union of all $A_j \cup B_j$, it should now be clear that α, β are each products of n disjoint transpositions. Also $\sigma = \alpha\beta$, because $\sigma_{|A_j \cup B_j} = \alpha_j \beta_j$. □

Proposition 3.53 provides a way to build a Rejewski factorization of σ from partial Rejewski factorizations of σ restricted to the unions $A_j \cup B_j$ of the paired orbits of equal length. Proposition 3.52 says that all Rejewski factorizations of σ are built in this fashion. Thus we are led to consider permutations with just two orbits of equal length.

Proposition 3.54. *Suppose σ is a permutation on 2ℓ letters and having just two orbits A, B each of length ℓ. Then σ has precisely ℓ Rejewski factorizations.*

Proof. Keep in mind that any pair of Rejewski factors (α, β) of σ is such that both α and β map A onto B and B onto A, while σ maps A to A and B to B.

Let S be the set of all Rejewski factorizations (α, β) of σ. Choose any element a in A. Define the function $f : S \to B$ by the rule $(\alpha, \beta) \mapsto \beta(a)$. To see that σ has at most ℓ Rejewski factorizations, it suffices to show that f is injective.

Well, the set A, being the orbit of a under σ, can be listed as

$$a, \sigma(a), \sigma^2(a), \ldots, \sigma^{\ell-1}(a).$$

An inspection reveals that the transpositions

$$(a, \beta(a)),\ (\sigma(a), \beta\sigma(a)),\ (\sigma^2(a), \beta\sigma^2(a)), \ldots, (\sigma^{\ell-1}(a), \beta\sigma^{\ell-1}(a))$$

are disjoint and their product equals β. Now $\beta\sigma^i(a) = \sigma^{-i}\beta(a)$ for all exponents i, due to item 1 of Proposition 3.49. Thus we can write β as the following product of disjoint transpositions:

$$\beta = (a, \beta(a))\ (\sigma(a), \sigma^{-1}\beta(a))\ (\sigma^2(a), \sigma^{-2}\beta(a)) \cdots (\sigma^{\ell-1}(a), \sigma^{1-\ell}\beta(a)).$$

Should (γ, δ) be another pair of Rejewski factors of σ, we similarly obtain:

$$\delta = (a, \delta(a))\ (\sigma(a), \sigma^{-1}\delta(a))\ (\sigma^2(a), \sigma^{-2}\delta(a)) \cdots (\sigma^{\ell-1}(a), \sigma^{1-\ell}\delta(a)).$$

If we have that $\beta(a) = \delta(a)$, the above expressions for β and δ as products of ℓ disjoint transpositions reveal that $\beta = \delta$, and then $\alpha = \sigma\beta = \sigma\delta = \gamma$. Since $\beta(a) = \delta(a)$ implies that $(\alpha, \beta) = (\gamma, \delta)$ our f is injective, and there are at most ℓ Rejewski factorizations of σ.

Before we proceed to construct all Rejewski factors of σ it might be instructive to write the factorization of α into disjoint transpositions, along the lines of what we did for its companion factor β. Exactly as argued with β we see that

$$\alpha = (a,\alpha(a))\ (\sigma(a),\sigma^{-1}\alpha(a))\ (\sigma^2(a),\sigma^{-2}\alpha(a))\cdots(\sigma^{\ell-1}(a),\sigma^{1-\ell}\alpha(a)).$$

Since $\alpha = \sigma\beta$ we can replace $\alpha(a)$ by $\sigma\beta(a)$ to get

$$\alpha = (a,\sigma\beta(a))\ (\sigma(a),\beta(a))\ (\sigma^2(a),\sigma^{-1}\beta(a))\cdots(\sigma^{\ell-1}(a),\sigma^{2-\ell}\beta(a)).$$

We learn from the formulas for β and α as products of disjoint transpositions that they both depend on one letter a from A and one letter $b = \beta(a)$ from B. This informs us as to how to build the Rejewski factors of σ.

- Pick any element a from A and any element b from B. The set A is the orbit of a under σ, and so A can be listed as

$$a,\sigma(a),\sigma^2(a),\sigma^3(a),\ldots,\sigma^{\ell-1}(a).$$

  Likewise B can be listed as

$$b,\sigma(b),\sigma^2(b),\sigma^3(b),\ldots,\sigma^{\ell-1}(b).$$

- Define β to be the product of transpositions as shown:

$$\beta = (a,b)\ (\sigma(a),\sigma^{-1}(b))\ (\sigma^2(a),\sigma^{-2}(b))\cdots(\sigma^{\ell-1}(a),\sigma^{1-\ell}(b)).$$

- Define α to be the product of transpositions as shown:

$$\alpha = (a,\sigma(b))\ (\sigma(a),b)\ (\sigma^2(a),\sigma^{-1}(b))\cdots(\sigma^{\ell-1}(a),\sigma^{2-\ell}(b)).$$

The resulting pair (α,β) gives a Rejewski factorization of σ. Indeed, the first entries in the transpositions for α and β travel once through the orbit A and the second entries travel once through the disjoint orbit B. Thus the transpositions are disjoint. To see that $\sigma = \alpha\beta$, just inspect all possible inputs from the listings for A and B as elements of the form $\sigma^j(a)$ and $\sigma^j(b)$.

For a fixed a in A there are ℓ possible choices of b in B. Each choice of b gives a different permutation β, because β is defined to send a to b. Thus our method picks up the maximum possible number ℓ of Rejewski factorizations of σ. □

It might be informative to go over the preceding algorithm for constructing Rejewski factorizations, in the case where σ has just two orbits of length ℓ, using cycle notation. Choose an element a from A. The orbit of a is A which spawns the ℓ-cycle $(a_1,a_2,a_3,\ldots,a_\ell)$, where $a_1 = a$. There are ℓ ways to pick b from B. Having made a choice, the element b has B for its orbit which spawns the cycle $(b_1,b_2,b_3,\ldots,b_\ell)$, where $b_1 = b$. Then

$$\sigma = (a_1,a_2,a_3,\ldots,a_\ell)(b_1,b_2,b_3,\ldots,b_\ell).$$

Recall, this means that

$$\sigma : a_1 \mapsto a_2 \mapsto a_3 \mapsto \cdots \mapsto a_\ell \mapsto a_1 \text{ and } \sigma : b_1 \mapsto b_2 \mapsto b_3 \mapsto \cdots \mapsto b_\ell \mapsto b_1,$$

and the right-most cycle acts first.

With this interpretation the Rejewski factor β, defined in the proof of Proposition 3.54, comes out to be

$$\beta = (a_1, b_1)(a_2, b_\ell)(a_3, b_{\ell-1})(a_4, b_{\ell-2}) \cdots (a_{\ell-1}, b_3)(a_\ell, b_2),$$

and the factor α comes out to be

$$\alpha = (a_1, b_2)(a_2, b_1)(a_3, b_\ell)(a_4, b_{\ell-1}) \cdots (a_{\ell-1}, b_4)(a_\ell, b_3).$$

An easy inspection shows that $\sigma = \alpha\beta$. We get all Rejewski factorizations by going through all possible b in B, while not changing the a. By changing the a as well, only redundant Rejewski factorizations would emerge.

For an illustration of Proposition 3.54 we let $\sigma = (1,2,3)(4,5,6)$ and seek its Rejewski factorizations. There are three of them. If we take $a = 1, b = 4$, then σ is expressed as the original cycle product $(1,2,3)(4,5,6)$, and the Rejewski factors work out to be

$$\beta = (1,4)(2,6)(3,5), \quad \alpha = (1,5)(2,4)(3,6).$$

If we take $a = 1, b = 5$, then σ is expressed as $(1,2,3)(5,6,4)$, and the Rejewski factors work out to be

$$\beta = (1,5)(2,4)(3,6), \quad \alpha = (1,6)(2,5)(3,4).$$

If we take $a = 1, b = 6$, then σ is expressed as $(1,2,3)(6,4,5)$, and the Rejewski factors work out to be

$$\beta = (1,6)(2,5)(3,4), \quad \alpha = (1,4)(2,6)(3,5).$$

An inspection reveals that σ is indeed $\alpha\beta$ in each of the three cases.

Now we return to a general σ having an equal length pairing of its orbits.

Finding and Counting the Rejewski Factorizations

To see that a permutation on $2n$ letters has an equal length pairing of its orbits, simply verify that the number of orbits of every possible length is even. Given such a permutation, here is how to find its Rejewski factorizations.

- Construct an equal length pairing of the orbits

$$\{A_1, B_1\}, \{A_2, B_2\}, \ldots, \{A_m, B_m\}.$$

  This can often be done in a number of ways, and the different ways will lead to different factorizations.

- Find a Rejewski factorization (α_j, β_j) for each restriction $\sigma_{|A_j \cup B_j}$. Proposition 3.54 tells us that the number factorizations equals the length of the orbits A_j, B_j. It also gives the algorithm for finding them.
- Put $\alpha = \alpha_1\alpha_2 \cdots \alpha_m, \beta = \beta_1\beta_2 \cdots \beta_m$. Proposition 3.53 tells us that (α, β) is a Rejewski factorization of σ. Proposition 3.52 ensures that this process will get them all.

For every equal length pairing of the orbits of σ as above, the number of Rejewski factorizations coming out of that pairing equals the product $\prod_{j=1}^{m} \ell_j$. To count all Rejewski factorizations we need to multiply these products, one for every equal length pairing. A clean formula is given as follows.

Proposition 3.55. *Suppose that σ is a permutation on $2n$ letters with an equal length pairing of its orbits. For each positive integer ℓ, let $2m_\ell$ be the number of orbits of length ℓ. Then the number of Rejewski factorizations of σ equals*

$$\prod_{\ell=1}^{n} \frac{\ell^{m_\ell}(2m_\ell)!}{2^{m_\ell} m_\ell!}.$$

Proof. The length ℓ of any orbit is such that $1 \leq \ell \leq n$. This is because for every such orbit there is at least another one, whereby $2\ell \leq 2n$.

Take the orbits of common length ℓ and pair them off as shown:

$$\{A_1, B_1\}, \{A_2, B_2\}, \ldots, \{A_{m_\ell}, B_{m_\ell}\}.$$

First we count the number of such pairings.

To help cut through the clutter, let us count the number of ways to group the set of integers from 1 to $2m$ into pairs of integers. Well, the integer 1 can be paired with any one of the $2m - 1$ remaining integers. The next unpaired integer can be paired with any one of the $2m - 3$ remaining integers. The next unpaired integer can be paired with any one of the $2m - 5$ remaining integers. Continuing in this way we see that the number of ways in which the integers from 1 to $2m$ can be paired off is

$$(2m - 1)(2m - 3)(2m - 5) \cdots 5 \cdot 3 \cdot 1.$$

Multiply this product by

$$(2m)(2m - 2)(2m - 4) \cdots 4 \cdot 2,$$

and divide by its equal $2^m m!$ to see that the number of ways in which the integers from 1 to $2m$ can be paired off is $(2m)!/(2^m m!)$.

After replacing the set of integers from 1 to $2m$ by the set consisting of $2m_\ell$ orbits of length ℓ, we can see that the number of pairings of the family of orbits of length ℓ equals $(2m_\ell)!/(2^{m_\ell} m_\ell!)$.

For each j from 1 to m_ℓ, the number of Rejewski factorizations of the restriction $\sigma_{|A_j \cup B_j}$ equals ℓ, due to Proposition 3.54. Then the number of Rejewski factorizations of σ restricted to the union $\bigcup_{j=1}^{m_\ell}(A_j \cup B_j)$ equals $\dfrac{\ell^{m_\ell}(2m_\ell)!}{2^{m_\ell} m_\ell!}$.

A Rejewski factorization of σ is the product of the Rejewski factorizations of σ restricted to each union of orbits of equal length $\bigcup_{j=1}^{m_\ell}(A_j \cup B_j)$. This can be seen by imitating the arguments in Propositions 3.52 and 3.53. It follows that the total number of Rejewski factorizations of σ is the desired product. □

For instance, say that a permutation σ on 26 letters has 4 orbits of length 4 and 2 orbits of length 5. So $m_4 = 2, m_5 = 1$, and all other $m_\ell = 0$. The number of Rejewski factorizations of σ comes out to be $\left(\frac{4^2(2\cdot 2)!}{2^2\cdot 2!}\right)\left(\frac{5^1(2\cdot 1)!}{2^1\cdot 1!}\right) = 240$.

If σ has 6 orbits of length 3 and 4 orbits of length 2, then $m_3 = 3, m_2 = 2$ and all other $m_\ell = 0$. The number of Rejewski factorizations is $\left(\frac{3^3(2\cdot 3)!}{2^3\cdot 3!}\right)\left(\frac{2^2(2\cdot 2)!}{2^2\cdot 2!}\right) = 4860$.

If σ has just two orbits of order 13, then the number of Rejewski factorizations is 13, according to Proposition 3.54. Here $m_{13} = 1$ and all other $m_\ell = 0$. The formula of Proposition 3.55 gives $\frac{13^1(2\cdot 1)!}{2^1\cdot 1!} = 13$, as expected.

Back to the Enigma Story

Using the preceding results Rejewski determined that the first six permutations $\tau_1, \tau_2, \tau_3, \tau_4, \tau_5, \tau_6$ implemented by the Enigma machine on any given day came from a well determined and fairly restricted set of possibilities based on the orbit structures of the known products $\tau_4\tau_1, \tau_5\tau_2, \tau_6\tau_3$. They are what we have called the Rejewski factors of the known products. Rejewski had a very large number of assistants who rapidly did the requisite calculations. It turns out that the formula of Proposition 3.55, applied to the above three products, often gave a manageable number of Rejewski factors τ_j to contend with.

3.8.4 Finding the First Rotor Wiring

Now the challenge was to figure out the rotor wirings. In other words, to find the permutations $\delta, \rho_1, \rho_2, \rho_3, \epsilon$ introduced at the start of Section 3.8.2. This was a tall order with some guessing and luck needed. We concentrate on the contributions made by group theory in obtaining the wiring of the first rotor.

Thus for $j = 1, 2, 3, 4, 5, 6$ Rejewski knew that

$$\tau_j = \delta^{-1}(\sigma^j \rho_1^{-1} \sigma^{-j})(\rho_2^{-1}\rho_3^{-1}\epsilon\rho_3\rho_2)(\sigma^j\rho_1\sigma^{-j})\delta.$$

(This was also based on the reasonable guess that the other two rotors did not move on the first six inputs.) The six τ_j were narrowed down to come from a manageable list of possibilities. The σ is the known cycle $(1, 2, 3, \ldots, 26)$. The other five unknown permutations are buried deeply inside these six equations. By good fortune through the work of the spy, Hans-Thilo Schmidt,[4] the daily setting of the permutation δ coming from the plug board had been revealed to Rejewski for a two month period.

[4] See David Kahn's article "*The spy who most affected World War II*" in *Kahn on Codes*, Macmillan, 1983, pp. 76–88.

If we consider δ as known, the above equations can be re-expressed as

$$\sigma^{-j}\delta\tau_j\delta^{-1}\sigma^j = \rho_1^{-1}\sigma^{-j}(\rho_2^{-1}\rho_3^{-1}\epsilon\rho_3\rho_2)\sigma^j\rho_1.$$

Now the permutations $\sigma^{-j}\delta\tau_j\delta^{-1}\sigma^j$ are known to come from a small enough set of possibilities. For brevity let $\kappa_j = \sigma^{-j}\delta\tau_j\delta^{-1}\sigma^j$. Denote $\rho_2^{-1}\rho_3^{-1}\epsilon\rho_3\rho_2$ as a single unknown permutation η. With these notations the equations become

$$\kappa_j = \rho_1^{-1}\sigma^{-j}\eta\sigma^j\rho_1$$

for $j = 1,2,3,4,5,6$. Now there are two unknowns: ρ_1 and η. Next we manipulate these equations to isolate the unknown ρ_1 corresponding to the first rotor.

For j from 1 to 5 we obtain

$$\kappa_j\kappa_{j+1} = (\rho_1^{-1}\sigma^{-j}\eta\sigma^j\rho_1)(\rho_1^{-1}\sigma^{-j-1}\eta\sigma^{j+1}\rho_1) = \rho_1^{-1}\sigma^{-j}\eta\sigma^{-1}\eta\sigma^{j+1}\rho_1.$$

This leads to

$$\sigma^j\rho_1\kappa_j\kappa_{j+1}\rho_1^{-1}\sigma^{-j} = \eta\sigma^{-1}\eta\sigma.$$

Since the right side does not depend on j we can say that

$$\sigma^j\rho_1\kappa_j\kappa_{j+1}\rho_1^{-1}\sigma^{-j} = \sigma^{j+1}\rho_1\kappa_{j+1}\kappa_{j+2}\rho_1^{-1}\sigma^{-j-1}$$

for $j = 1,2,3,4$. This simplifies to

$$\kappa_j\kappa_{j+1} = (\rho_1^{-1}\sigma\rho_1)\kappa_{j+1}\kappa_{j+2}(\rho_1^{-1}\sigma\rho_1)^{-1}.$$

Up to a finite list of possibilities, the permutations $\kappa_j\kappa_{j+1}, \kappa_{j+1}\kappa_{j+2}$ are known. The last equation says that they are conjugate permutations. The task now is to find the possible conjugators $\rho_1^{-1}\sigma\rho_1$, and to count how many there are.

The Conjugacy Theorem

The ***orbit structure***[5] of a permutation in S_n is the list of its orbit lengths written in descending order. From Proposition 3.56, two permutations β, γ in S_n are conjugate if and only if their orbit structures coincide. We can see that β, γ have the same orbit structure if and only if for every integer ℓ from 1 to n, the number of β-orbits of length ℓ equals the number of γ-orbits of length ℓ.

Since the preceding pairs $\kappa_j\kappa_{j+1}, \kappa_{j+1}\kappa_{j+2}$, for $j = 1,2,3,4$, are conjugates, they have identical orbit structures. What we need are the possible conjugators $\rho_1^{-1}\sigma\rho_1$. The subsequent results address this question, which was already alluded to in Section 2.4.1.

Proposition 3.56. *If β, γ in S_n are conjugate permutations, then their orbit structures are identical. For every integer ℓ from 1 to n, let m_ℓ be the common number of β-orbits and γ-orbits of length ℓ. Then the number of permutations α, such that $\alpha\beta\alpha^{-1} = \gamma$, is at most*

[5] Various authors also refer to this as the cycle structure or the cycle type.

$$\prod_{\ell=1}^{n} \ell^{m_\ell}(m_\ell!).$$

Proof. Let α in S_n be such that $\alpha\beta\alpha^{-1} = \gamma$. For every integer k we also have $\alpha\beta^k\alpha^{-1} = \gamma^k$, and then $\alpha^k\beta = \gamma^k\alpha$. Thus,

$$\alpha\beta^k(a) = \gamma^k\alpha(a),$$

for every letter a and every integer k.

This reveals that the β-orbit of a is mapped by α to the γ-orbit of $\alpha(a)$. We also see that $\beta^k(a) = a$ if and only if $\gamma^k\alpha(a) = \alpha(a)$. This shows that the β-orbit of a has the same length as the γ-orbit of $\alpha(a)$. So, by taking direct images, α moves every β-orbit of length ℓ to a γ-orbit of length ℓ. Different β-orbits go to different γ-orbits under direct images, because α is a permutation. This is enough to see that β and γ have identical orbit structures. The rest of the proof is a matter of bookkeeping.

Let I be the set of integers ℓ between 1 and n for which β does possess an orbit of length ℓ. For each ℓ in I let

$$A_{\ell 1}, A_{\ell 2}, \ldots, A_{\ell m_\ell} \text{ and } B_{\ell 1}, B_{\ell 2}, \ldots, B_{\ell m_\ell}$$

be the respective lists of β-orbits and γ-orbits of length ℓ. The set of letters L is equal to the disjoint union

$$\bigcup_{\ell \in I} \left(\bigcup_{j=1}^{m_\ell} A_{\ell j} \right),$$

and likewise with the $B_{\ell j}$.

Fix ℓ in I and for each j from 1 to m_ℓ, select a letter a_j in the β-orbit $A_{\ell j}$. The equations $\alpha\beta^k(a_j) = \gamma^k\alpha(a_j)$ hold for every integer exponent k. This shows that the single value $\alpha(a_j)$ determines all values of the permutation α on the β-orbit $A_{\ell j}$. So, the restriction of α to the disjoint union $\bigcup_{j=1}^{m_\ell} A_{\ell,j}$ is completely determined by the m_ℓ-tuple $\big(\alpha(a_1), \alpha(a_2), \ldots, \alpha(a_{m_\ell})\big)$. As α varies over all permutations that conjugate β into γ, the number of restrictions of α to $\bigcup_{j=1}^{m_\ell} A_{\ell,j}$ is no more than the number of m_ℓ-tuples $\big(\alpha(a_1), \alpha(a_2), \ldots, \alpha(a_{m_\ell})\big)$.

Each $\alpha(a_j)$ belongs to the disjoint union $\bigcup_{j=1}^{m_\ell} B_{\ell,j}$. Since, under direct images, α maps different $A_{\ell j}$ to different $B_{\ell i}$, the $\alpha(a_j)$ are restricted so that no two of these letters belong to the same $B_{\ell i}$.

The entry $\alpha(a_1)$ can take at most ℓm_ℓ values, namely the number of letters in the disjoint union $\bigcup_{j=1}^{m_\ell} B_{\ell,j}$. The entry $\alpha(a_2)$ is excluded from taking values in the γ-orbit of $\alpha(a_1)$. Given the letter $\alpha(a_1)$, the letter $\alpha(a_2)$ can take at most $\ell(m_\ell - 1)$ values. In a similar vein, the letters $\alpha(a_1), \alpha(a_2)$ exclude $\alpha(a_3)$ from being in two of the orbits $B_{\ell i}$. Thus $\alpha(a_3)$ is restricted to at most $\ell(m_\ell - 2)$ values. Continue this counting argument to see that the number of possibilities for the m_ℓ-tuple $\big(\alpha(a_1), \alpha(a_2), \ldots, \alpha(a_{m_\ell})\big)$ is at most the product of the $\ell(m_\ell - j + 1)$. This product simplifies down to $\ell^{m_\ell}(m_\ell!)$.

As α varies over all conjugators from β to γ, we have just seen that the number of restrictions of α to the disjoint union $\bigcup_{j=1}^{m_\ell} A_{\ell,j}$ is at most $\ell^{m_\ell}(m_\ell!)$. Since the set L of all letters is the disjoint union $\bigcup_{\ell \in I}\left(\bigcup_{j=1}^{m_\ell} A_{\ell j}\right)$, the number of conjugators α is at most the product taken over all ℓ in I of the number of its restrictions to each $\bigcup_{j=1}^{m_\ell} A_{\ell j}$. Hence the number of conjugators is at most $\prod_{\ell \in I}^{n} \ell^{m_\ell}(m_\ell!)$. Finally, note that we can let ℓ run from 1 to n, instead of just through I, because when $m_\ell = 0$, the quantities $\ell^{m_\ell}(m_\ell!) = 1$. □

Given β, γ with identical orbit structures, we can now construct all possible conjugators from β to γ. The proof of the next result also provides the algorithm for doing so.

Proposition 3.57. *Suppose that β, γ are permutations in S_n with identical orbit structures. For every integer ℓ from 1 to n, let m_ℓ be the common number of β-orbits and γ-orbits of length ℓ. Then the number of permutations α such that $\gamma = \alpha\beta\alpha^{-1}$ equals the product*

$$\prod_{\ell=1}^{n} \ell^{m_\ell}(m_\ell!).$$

Proof. Because of Proposition 3.56 it suffices to construct at least the requisite number of conjugators α.

Let I be the set of all ℓ from 1 to n such that $m_\ell > 0$, i.e. such that β-orbits (and γ-orbits too) of length ℓ exist. For each ℓ in I enumerate the β-orbits of length ℓ:

$$A_{\ell 1}, A_{\ell 2}, \ldots, A_{\ell m_\ell}.$$

Then L is the disjoint union $\bigcup_{\ell \in I}\left(\bigcup_{j=1}^{m_\ell} A_{\ell j}\right)$. Select a letter $a_{\ell j}$ from each β-orbit $A_{\ell j}$. Then carry out the following steps, keeping track of the possible choices involved.

- Write a matching enumeration

$$B_{\ell 1}, B_{\ell 2}, \ldots, B_{\ell m_\ell}$$

  of the γ-orbits of length ℓ. The number of matching enumerations equals $m_\ell!$.
- For each ℓ in I and each matching enumeration $B_{\ell j}$ select a letter $b_{\ell,j}$ in $B_{\ell,j}$. There are ℓ possible choices for each $b_{\ell j}$, and consequently ℓ^{m_ℓ} choices for the m_ℓ-tuple $(b_{\ell 1}, \ldots, b_{\ell m_\ell})$.
- As ℓ varies over I and the $B_{\ell j}$ vary over the matching enumerations of the γ-orbits of length ℓ, the number of distinct m_ℓ-tuples of letters we can take equals $\prod_{\ell=1}^{n} \ell^{m_\ell}(m_\ell!)$. For each choice of m_ℓ-tuple of letters we now construct a distinct α which conjugates β to γ. This will produce the requisite number of conjugators α.
- The β-orbit $A_{\ell j}$ of $a_{\ell j}$ can be written explicitly as

$$A_{\ell j} = \{a_{\ell j}, \beta(a_{\ell j}), \beta^2(a_{\ell j}), \ldots, \beta^{\ell-1}(a_{\ell j})\}.$$

  Likewise the matching γ-orbit $B_{\ell j}$ of $b_{\ell j}$ is given by

$$B_{\ell j} = \{b_{\ell j}, \gamma(b_{\ell j}), \gamma^2(b_{\ell j}), \ldots, \gamma^{\ell-1}(b_{\ell j})\}.$$

- To specify a permutation α on L it suffices to give bijections

$$\alpha_{\ell j} : A_{\ell j} \to B_{\ell j},$$

for every ℓ in I and every $j = 1, \ldots, m_\ell$. Define these bijections by the rule

$$\alpha_{\ell j} : \beta^k(a_{\ell j}) \mapsto \gamma^k(b_{\ell j})$$

where $k = 0, 1, \ldots, \ell - 1$. The desired α is defined to agree with $\alpha_{\ell j}$ on each disjoint component $A_{\ell j}$.

It only remains to verify that α conjugates β to γ. For that it is convenient to notice that

$$\alpha_{\ell j}(\beta^h(a_{\ell j})) = \gamma^h(b_{\ell j})$$

for *all* integer exponents h, not just those from 0 to $\ell - 1$. For that we need Euclidean division along with the facts that β^ℓ is the identity on the orbits $A_{\ell j}$ and γ^ℓ is the identity of the orbits $B_{\ell j}$. Take integers q, k such that

$$h = \ell q + k \text{ and } 0 \le k < \ell.$$

Then in routine fashion:

$$\begin{aligned}\alpha_{\ell j}(\beta^h(a_{\ell j})) &= \alpha_{\ell j}((\beta^\ell)^q \beta^k(a_{\ell j})) = \alpha_{\ell j}(\beta^k(a_{\ell j}))\\ &= \gamma^k(b_{\ell j}) = (\gamma^\ell)^q \gamma^k(b_{\ell j}) = \gamma^h(b_{\ell j}).\end{aligned}$$

Each letter from L lies in a unique β-orbit $A_{\ell j}$, and can be written as $\beta^h(a_{\ell j})$ for some integer h. Using what we just checked above we get

$$\begin{aligned}\alpha\beta(\beta^h(a_{\ell j})) &= \alpha\beta^{h+1}(a_{\ell j}) = \alpha_{\ell j}\beta^{h+1}(a_{\ell j})\\ &= \gamma^{h+1}(b_{\ell j}) = \gamma\gamma^h(b_{\ell j}) = \gamma\alpha_{\ell j}\beta^h(a_{\ell j}) = \gamma\alpha(\beta^h(a_{\ell j})).\end{aligned}$$

Thus $\alpha\beta = \gamma\alpha$, and then $\alpha\beta\alpha^{-1} = \gamma$. □

We should observe that not only does Proposition 3.57 count the solutions α to $\gamma = \alpha\beta\alpha^{-1}$, but its proof offers an algorithm for finding the α. Also note that if we vary the choices of $a_{\ell j}$ in $A_{\ell j}$ or vary the ordering of the $A_{\ell j}$ no new conjugators α will be created, because of Proposition 3.56.

To illustrate Proposition 3.57 take the permutations

$$\beta = (1, 2, 3, 4)(5, 6, 7, 8)(9, 10) \text{ and } \gamma = (5, 8, 3, 2)(9, 1, 7, 10)(4, 6)$$

written as products of disjoint cycles. Each has 2 orbits of length 4 and 1 orbit of length 2. According to Proposition 3.57 there are

$$(4^2 \cdot 2!)(2^1 \cdot 1!) = 64$$

permutations α such that $\gamma = \alpha\beta\alpha^{-1}$. To build these, all we need to do is line up the orbits of β and γ and follow the directions in the proof of Proposition 3.57. Using the (visually) obvious alignment, one choice for α (in matrix notation) becomes

$$\begin{pmatrix} 1 & 2 & 3 & 4 & 5 & 6 & 7 & 8 & 9 & 10 \\ 5 & 8 & 3 & 2 & 9 & 1 & 7 & 10 & 4 & 6 \end{pmatrix}.$$

Another alignment of orbits is (visually) revealed by rewriting γ as

$$(7, 10, 9, 1)(2, 5, 8, 3)(6, 4).$$

From this another choice of α is

$$\begin{pmatrix} 1 & 2 & 3 & 4 & 5 & 6 & 7 & 8 & 9 & 10 \\ 7 & 10 & 9 & 1 & 2 & 5 & 8 & 3 & 6 & 4 \end{pmatrix}.$$

This can be repeated 62 more times.

For another illustration of our result, suppose β and γ permute 26 letters and the orbit structure of each is $[5, 5, 5, 3, 3, 2, 1, 1, 1]$. This means that each has three orbits of length 5, two orbits of length 3, one orbit of length 2, and three orbits of length 1. Then the number of α such that $\gamma = \alpha\beta\alpha^{-1}$ equals

$$(5^3 \cdot 3!)(3^2 \cdot 2!)(2^1 \cdot 1!)(1^3 \cdot 3!) = 162000.$$

This may seem like a lot, but the total number of permutations on 26 letters is 26!, which is approximately 10^{27}. By that measure the number of conjugators is not so big.

Counting the Centralizer and Conjugacy Class of a Permutation

Proposition 3.57 has a worthy specialization. If we take $\gamma = \beta$ in $\mathcal{S}_n$, our result finds and counts those α in $\mathcal{S}_n$ such that $\beta = \alpha\beta\alpha^{-1}$, i.e. such that $\beta\alpha = \alpha\beta$. The set of such α is the ***centralizer*** of β, discussed prior to Proposition 3.17. If for each ℓ the number of β-orbits is m_ℓ, Proposition 3.57 reveals that the order of the centralizer of β is $\prod_{\ell=1}^{n} \ell^{m_\ell}(m_\ell!)$.

Furthermore, letting $\mathcal{S}_n$ act on itself by conjugation, Proposition 3.11 (a variant of the orbit-stabilizer relation) reveals that the number of conjugates of β in $\mathcal{S}_n$ equals

$$\frac{n!}{\prod_{\ell=1}^{n} \ell^{m_\ell}(m_\ell!)}.$$

Back to the First Rotor

Rejewski knew that the products $\kappa_j\kappa_{j+1}$ for $j = 1, 2, 3, 4, 5$ had to come from a manageable set of possibilities arising from Rejewski factorizations of the revealed $\tau_4\tau_1, \tau_5\tau_2, \tau_6\tau_3$. For each possible $\kappa_j\kappa_{j+1}$ Rejewski could use his version of Proposition 3.57 to find another manageable set of possibilities for the conjugator $\rho_1^{-1}\sigma\rho_1$ from the $\kappa_{j+1}\kappa_{j+2}$ to $\kappa_j\kappa_{j+1}$. He actually got four manageable sets of possibilities for the conjugator $\rho_1^{-1}\sigma\rho_1$, one set for each j from 1 to 4.

By applying Proposition 3.57 once more the possible conjugators ρ_1 from the known 26-cycle σ to the known possible permutations $\rho_1^{-1}\sigma\rho_1$ were found.

Of course this work did not fish out the *unique* permutation ρ_1 coming from that day's setting of the first rotor. Rejewski's ingenious work inside $\mathcal{S}_{26}$ gave him a small enough set of possible permutations, and the knowledge that ρ_1 lurked among them. With his army of co-workers he found one such small set for each j for 1 to 4. By intersecting these four sets, the unique ρ_1 had a good chance of emerging.

To see why, consider the following non-rigorous heuristic. Imagine that we have four sets of permutations A, B, C, D, all subsets of $\mathcal{S}_{26}$ with at least one permutation in common. Say each set has about 10^6 seemingly random permutations. (Here we treat 10^6 as manageable.) Since $\mathcal{S}_{26}$ has almost 10^{27} permutations, the subsets A, B, C, D are tiny by comparison. It would be unlikely that their intersection would amount to more than the element they are known to have in common.

Epilogue

Needless to say, much more needed to be done to break the Enigma. There were the other two rotors to decipher. One lucky thing was that every few months the rotors were permuted, and thus every rotor got to be the first rotor sooner or later. There remained the reflector to solve and the puzzle of the daily settings when information from the spy was no longer available. A huge amount of computation was needed. For that various electromechanical devices were invented.

Rejewski did most of his work in the 1930s before World War 2 erupted. During this time and into the 1940s the Enigma machine and its method of use became more complex. For example, the U-boat Enigma later had four rotors instead of three. The work that Rejewski had started and taken to a magnificent level was to be continued at Bletchley Park, England. Among the many brilliant mathematicians who worked there were Alan Turing (1912–1954) and William Tutte (1917–2002).

Here are a couple of references for further reading.

- D. Kahn, *Seizing the Enigma*, Houghton-Mifflin (1991).
- W. Kozaczuk, *Enigma: How the German Machine Cipher was Broken and How it was Read by the Allies in World War Two*, edited and translated by Christopher Kasparek, University Publications of America (1979, 1984). (Includes several papers by Rejewski.)

EXERCISES

1. Find all Rejewski factorizations of the permutation $(1,2)(3,4)(5,6)(7,8)$.
2. If a permutation has 6 orbits of length 3 and 2 orbits of length 6 how many Rejewski factorizations does it have?
3. If a permutation has 2 orbits of length 10 and 2 orbits of length 3, how many Rejewski factorizations does it have?

4. If a permutation σ in S_{11} is the disjoint product of a 5-cycle and two 3-cycles, how many conjugates does σ have? How many α in S_{11} are such that $\sigma = \alpha\sigma\alpha^{-1}$? If τ is another disjoint product of a 5-cycle and two 3-cycles, how many α in S_{11} are such that $\beta = \alpha\sigma\alpha^{-1}$?
5. If n is odd, show that any $2n$-cycle σ can be expressed as the product $\alpha\beta\gamma$ where each α, β, γ is the product of n disjoint transpositions.

   Does this result hold if n is even?
6. Find the number of conjugates in S_{26} of the following permutation (written as a product of disjoint cycles):

$$(1,2,3,4)(5,6,7,8)(9,10,11,12)\ (13,14,15)(16,17,18)(19,20)(21,22)(23,24)(25)(26).$$

7. Every setting of the Enigma machine represents a product of 13 disjoint transpositions on 26 letters. How many permutations can the Enigma machine realize? In more mathematical terms, count the number of products of 13 disjoint transpositions on 26 letters.

4 Basics on Rings – Mostly Commutative

The great problems concerning the integers, such as the practical problem of factoring, the determination and distribution of primes, and the solution of equations all refer to the operations of addition and multiplication. For the study of integers as well as numerous other areas, it makes sense to bring these operations under the umbrella of a single concept, namely that of a ring. By working abstractly can we reap multiple rewards from a single effort.

4.1 Terminology and Examples

To begin with, a ***ring*** is an abelian group R under the operation of addition, but R also comes with a second operation of multiplication.

Definition 4.1. A non-empty set R is a ***ring*** when to every pair of elements (x, y) in $R \times R$, there are assigned two elements

$$x + y \text{ and } xy \text{ in } R,$$

in such a way that the following laws of operation apply.

- The ***associative law of addition***, $(x + y) + z = x + (y + z)$, holds for every x, y, z in R.
- The ***commutative law of addition***, $x + y = y + x$, holds for every x, y in R.
- There is a unique ***zero element*** 0 inside R with the neutral property that

$$0 + x = x + 0 = x \text{ for all } x \text{ in } R.$$

- For every x in R, there is a unique ***negative*** $-x$ in R such that $x + (-x) = (-x) + x = 0$.
- The ***associative law of multiplication***, $(xy)z = x(yz)$, holds for every x, y, z in R.
- There is a unique ***identity element*** 1 such that $1x = x1 = x$ for all x in R.
- The ***distributive laws***, $x(y + z) = xy + xz$ and $(x + y)z = xy + xz$, hold for every x, y, z in R.

Evidently, the first four of the above axioms are those of an abelian group under addition.

The *uniqueness* of the identity element 1 in the definition of a ring need not be assumed. For if there were elements $1, 1'$ such that $1x = x1 = 1'x = x1' = x$ for all x, the equation $1 = 11' = 1'$ would follow by taking $x = 1$ and then $x = 1'$.

The familiar sets $\mathbb{Z}, \mathbb{Q}, \mathbb{R}, \mathbb{C}$ of integer, rational, real and complex numbers, respectively, are rings under their usual operations of addition and multiplication. Also common are the rings $\mathbb{Z}_n$ of residues modulo a positive integer n, under the operations of modular arithmetic.

Among the laws of operation for a ring one seems conspicuously absent, namely

- the ***commutative law*** of multiplication $xy = yx$.

Naturally, rings in which the commutative law applies, such as the aforementioned rings of numbers and residues, are called ***commutative rings***.

Linear algebra provides a family of ***non-commutative*** rings. Take any commutative ring R, such as for example $\mathbb{Z}, \mathbb{Q}, \mathbb{R}, \mathbb{C}$ or $\mathbb{Z}_n$. The set of 2×2 matrices with entries in R is denoted by $M_2(R)$. This set is a ring under the usual addition and multiplication of matrices. Here, the zero matrix and the identity matrix play the roles of 0 and 1, respectively. As long as R is not the zero ring in which $1 = 0$, the matrix ring $M_2(R)$ is not commutative. For example, in the ring $M_2(R)$ we can readily check that

$$\begin{bmatrix} 1 & 1 \\ 0 & 0 \end{bmatrix} \begin{bmatrix} 1 & 0 \\ 1 & 0 \end{bmatrix} = \begin{bmatrix} 1+1 & 0 \\ 0 & 0 \end{bmatrix} \neq \begin{bmatrix} 1 & 1 \\ 1 & 1 \end{bmatrix} = \begin{bmatrix} 1 & 0 \\ 1 & 0 \end{bmatrix} \begin{bmatrix} 1 & 1 \\ 0 & 0 \end{bmatrix}.$$

More generally, for any positive integer n, there is the ring $M_n(R)$ of $n \times n$ matrices with entries in a commutative ring R. When $n \geq 2$, the rings $M_n(R)$ are non-commutative (unless $1 = 0$ in R).

Commutative rings will be the primary objects of our attentions. But there is no escaping the presence and importance of non-commutative rings in the development and applications of algebra. It is beneficial to have some facility in working with rings without the commutative law being assumed.

The zero and the identity elements inside a ring are not necessarily the same as the integers zero and one. Yet the common notations 0 and 1 are adopted. This efficient reuse of symbols makes for a cleaner exposition. No harm will result.

Some Non-rings

The set of positive integers $\mathbb{P} = \{1, 2, 3, \ldots\}$, is not a ring when operating with normal addition and multiplication. Indeed, $\mathbb{P}$ has no distinguished element that can play the role of 0. To make matters worse, its elements do not come with negatives.

The set of non-negative integers $\mathbb{N} = \{0, 1, 2, \ldots\}$ has the usual operations of addition and multiplication, and it comes with a suitable zero, but the elements in $\mathbb{N}$ other than 0 lack negatives inside $\mathbb{N}$.

The set of even integers $2\mathbb{Z} = \{0, \pm 2, \pm 4, \pm 6, \ldots\}$ comes with the usual operations of addition and multiplication, has a zero as well as negatives for every element. Since it fails to have the needed identity element, $2\mathbb{Z}$ is not a ring.

Some mathematicians do not require that rings have an identity element.[1] For them, $2\mathbb{Z}$ is considered to be a ring. This extra level of generality gives them access to some structures that are quite exotic, but the absence of an identity element comes with costs in technical minutiae that a beginner in algebra should not have to pay. Thus, for us, ***rings have a*** 1.

[1] Perhaps with tongue in cheek, such creatures might be called **rngs**.

Direct Products

For any rings R and S, the ***Cartesian product*** $R \times S$ is the set of all ordered pairs (r,s) where $r \in R$ and $s \in S$. To make $R \times S$ into a ring use the natural point-wise rules of addition and multiplication. Namely, if $(r_1,s_1),(r_2,s_2) \in R \times S$ put

$$(r_1,s_1) + (r_2,s_2) = (r_1 + r_2, s_1 + s_2) \text{ and } (r_1,s_1)(r_2,s_2) = (r_1r_2, s_1s_2).$$

For the zero and identity elements, take $(0,0)$ and $(1,1)$, respectively. For the negative of (r,s), take it to be $(-r,\,-s)$. Since the remaining ring axioms apply to each component R and S of the Cartesian product, they carry over to $R\times S$, which is typically called the ***direct product ring***.

In the same way we can build the direct product of more than two rings. Namely, take any number of rings $R_1, R_2, \ldots, R_n$. Use the Cartesian product $R_1 \times R_2 \times R_3 \times \cdots \times R_n$ consisting of n-tuples $(r_1, r_2, \ldots, r_n)$, where each $r_j \in R_j$. Add and multiply two such n-tuples by adding and multiplying their respective components. For the zero and identity elements in the product take the n-tuples $(0,0,\ldots,0)$ and $(1,1,\ldots,1)$, respectively. For the negative of such an n-tuple, negate its components.

Routine Consequences of the Axioms

A few unsurprising facts flow from the ring axioms. They are used regularly and without fuss.

Proposition 4.2. *Let R be any ring.*

1. *If $x,y,z \in R$ and $x + y = z + y$, then $x = z$, which is the familiar law of cancellation under addition.*
2. *If $x \in R$, then $-(-x) = x$.*
3. *If $x \in R$, then $0x = x0 = 0$, which is the old adage that anything times zero is zero.*
4. *If $x \in R$, then $(-1)x = x(-1) = -x$.*
5. *If $x,y \in R$, then $(-x)y = x(-y) = -(xy)$ and $(-x)(-y) = xy$.*

Proof. The required deductions are truly mundane.

1. This is merely the cancellation law for the abelian group R using addition.
2. This says that in a group, such as R under addition, the inverse of an inverse is the original element.
3. We have that $0 + 0x = 0x = (0+0)x = 0x + 0x$ using the additive neutrality of zero twice and the distributive law. Cancel $0x$ to get $0 = 0x$. A similar proof gives $x0 = 0$.
4. From the neutrality of 1, the distributive law, and item 3 we get

$$x + (-1)x = 1x + (-1)x = (1 + (-1))x = 0x = 0 = x + (-x).$$

   Cancel x to get $(-1)x = -x$. With a similar proof, $x(-1) = -x$.
5. Using item 4 along with the associative law we get

$$(-x)y = ((-1)x)y = (-1)(xy) = -(xy).$$

Similarly we can show that $x(-y) = -(xy)$. Finally, from these latest facts along with item 2 we also get $(-x)(-y) = -(x(-y)) = -(-(xy)) = xy$. □

Two Caveats

While using the ring axioms and the rules of Proposition 4.2, we should remain alert to the possible failure of the commutative law for multiplication. Another caveat is that, in a general ring, there is no cancellation law for multiplication. That is, $xy = zy$ need not imply $x = z$. Indeed, $x0 = z0$ for all possible x and z. Even if $y \neq 0$, cancellation can still fail. For instance, in the ring $M_2(\mathbb{Z})$ of 2×2 matrices with entries in $\mathbb{Z}$, we have the equation

$$\begin{bmatrix} 1 & -1 \\ -1 & 1 \end{bmatrix}\begin{bmatrix} 1 & 1 \\ 1 & 1 \end{bmatrix} = \begin{bmatrix} 2 & -2 \\ 0 & 0 \end{bmatrix}\begin{bmatrix} 1 & 1 \\ 1 & 1 \end{bmatrix}.$$

Here, cancellation of $\begin{bmatrix} 1 & 1 \\ 1 & 1 \end{bmatrix}$ would lead to the error: $\begin{bmatrix} 1 & -1 \\ -1 & 1 \end{bmatrix} = \begin{bmatrix} 2 & -2 \\ 0 & 0 \end{bmatrix}$.

In the commutative ring $\mathbb{Z}_6$ of residues modulo 6 cancellation of non-zero factors also fails. For instance, $[2][3] = [6] = [0] = [2][0]$ and yet $[2] \neq [0]$ in $\mathbb{Z}_6$.

The Zero Ring

The smallest ring of all is the ***zero ring*** $R = \{0\}$, consisting of just the element 0. In the zero ring, $1 = 0$. Conversely, in order to have the zero ring, just check that $1 = 0$. Indeed, the neutrality of 1 and item 3 of Proposition 4.2 tell us that $x = 1x = 0x = 0$ for all x in R. The zero ring is blatantly not interesting, but sometimes it pops up in connection with something else.

4.1.1 Compound Additions and Multiplications

From the associative laws we see that the use of brackets in the addition and multiplication of any number of ring elements is not necessary. In a ring, sums and products such as $x+y+z+w$ and $xyzw$ are perfectly well defined without resorting to brackets. More complicated expressions such as $xyx - zw + uvt$ are also well defined, with the usual understanding that multiplications get done before additions, and the understanding that the subtraction of an element such as zw indicates the addition of the negative of zw.

For a list of ring elements $x_1, x_2, \ldots, x_n$, we can use sigma notation to denote their sum. Thus $\sum_{j=1}^{n} x_j$ means $x_1 + x_2 + \cdots + x_n$. Then, for instance, the distributive and associative laws lead to expressions such as

$$\left(\sum_{i=1}^{m} y_i\right)\left(\sum_{j=1}^{n} x_j\right) = \sum_{i=1}^{m}\sum_{j=1}^{n} y_i x_j.$$

As with any abelian group under addition, repeated sums of x or its negative $-x$ with themselves make sense and we have the ***multiple notation*** for each x in R and each integer n:

$$nx = \begin{cases} x + x + \cdots + x, & \text{taken } n \text{ times when } n > 0 \\ 0 & \text{when } n = 0 \\ (-x) + (-x) + \cdots + (-x) & \text{taken } -n \text{ times when } n < 0. \end{cases}$$

In the multiple notation we need to remember that x is in the ring while n is an integer. It is also reassuring to note that $1x = x$ and $(-1)x = -x$ regardless of whether 1 is an integer or the identity of the ring. The usual distribution laws

$$n(x + y) = nx + ny \text{ and } (n + m)x = nx + mx$$

as well as the associative law

$$m(nx) = (mn)x$$

hold for all integers m, n and all x, y in R.

Likewise, for a positive integer n the repeated product $xx \cdots x$ taken n times is denoted by the ***power notation*** x^n. We also say that $x^0 = 1$. The power notation obeys the usual laws of exponents, which are:

$$(x^n)(x^m) = x^{n+m} \text{ and } (x^m)^n = x^{mn},$$

for non-negative integers m, n. However, we should be wary of writing $x^n y^n = (xy)^n$. For instance with $n = 2$, we have $x^2y^2 = xxyy$, while $(xy)^2 = xyxy$. If the ring is non-commutative, we cannot expect that $xxyy = xyxy$. On the other hand if $xy = yx$, for instance when the ring is commutative, then the law $x^n y^n = (xy)^n$ does apply.

As long as $xy = yx$ and n is a positive integer, the familiar ***binomial expansion*** prevails:

$$(x + y)^n = x^n + \binom{n}{1} x^{n-1}y^1 + \binom{n}{2} x^{n-2}y^2 + \cdots + \binom{n}{n-1} xy^{n-1} + y^n.$$

Henceforth, the various and common rules for compound additions and multiplications in a ring will be taken for granted.

4.1.2 Subrings

Many rings live inside other rings.

Definition 4.3. A non-empty subset S of a ring R is called a ***subring*** of R, provided S is closed under the operations of R. To be more precise:

- $0, 1 \in S$,
- $x, y \in S$ implies that $x + y, xy \in S$,
- $x \in S$ implies $-x \in S$.

Another way to say this is that S is a subring of R provided S is a subgroup of R under addition, $1 \in S$, and $xy \in S$ whenever $x, y \in S$. Once these ***closure*** conditions hold, it is clear that S is also a ring, because the axioms for the ring operations that hold in R continue to hold in S by default. In case S is a subring of R, the ring R is called an ***extension*** of S.

Examples of subrings abound. For instance, $\mathbb{Z}$ is a subring of $\mathbb{Q}$, while $\mathbb{C}$ is an extension of $\mathbb{R}$, which in turn is an extension of $\mathbb{Q}$. Inside the ring $\mathbb{C}$ there is the subring $\mathbb{Z}[i]$ consisting of all numbers $a + bi$ where $a, b \in \mathbb{Z}$ and $i = \sqrt{-1}$. Geometrically this consists of the lattice of points in the complex plane whose coordinates are integers. The elements of the ring $\mathbb{Z}[i]$ are known as ***Gaussian integers***. More generally, let $\sqrt{d}$ be a chosen square root of a given integer d. From this we obtain the subrings $\mathbb{Z}[\sqrt{d}]$ of $\mathbb{C}$ consisting of all complex numbers of the form $a + b\sqrt{d}$ where a, b are integers. This latter family of rings is a rich home for problems in the area of number theory.

For a non-commutative example, inside the ring $M_2(\mathbb{C})$ take the set $\mathcal{U}$ of all upper triangular matrices. These are the matrices $[a_{ij}]$ for which $a_{ij} = 0$ when $i > j$.

The Smallest Subring

Given a subring S of a ring R, the identity element 1 lies in S, and by its closure under negation so does -1 lie in S. Then the set P of all integer multiples $n1$ must belong to S. This P is nothing but the set of all sums of ± 1. A bit of reflection reveals that P is a subring. In fact P is the smallest subring inside any ring because, as we have just noted, it sits inside any subring S of R.

We can see from the above that the only subring of $\mathbb{Z}$ is $\mathbb{Z}$ itself. Likewise, the rings $\mathbb{Z}_n$ have only themselves as subrings.

EXERCISES

1. Let d be any fixed integer and $\sqrt{d}$ one of its square roots in the ring of complex numbers $\mathbb{C}$. Verify that the set $\mathbb{Z}[\sqrt{d}] = \{a + b\sqrt{d} : a, b \in \mathbb{Z}\}$ is a subring of $\mathbb{C}$.
2. (a) Let $\alpha = \frac{1}{2}\left(1 + \sqrt{-19}\right)$. Show that the set R of complex numbers of the form $a + b\alpha$, where $a, b \in \mathbb{Z}$, is a subring of $\mathbb{C}$.
   (b) Is the set R of real numbers of the form $\frac{1}{2}(a + b\sqrt{2})$ where $a, b \in \mathbb{Z}$, a subring of $\mathbb{R}$?
3. The ring $\mathbb{Q}$ has $\mathbb{Z}$ and $\mathbb{Q}$ as subrings. Show that $\mathbb{Q}$ has infinitely many more subrings.
4. If R is a finite ring and $x \in R$, prove that some finite sum $x + x + \cdots + x$ must equal 0.
5. Find all residues $[a]$ in $\mathbb{Z}_{10}$ that are invertible, i.e. for which there exist $[b]$ in $\mathbb{Z}_{10}$ such that $[a][b] = [1]$.
6. An element a in a ring R is called ***nilpotent*** when $a^m = 0$ for some positive integer n. If a is an integer, show that its residue in $\mathbb{Z}_n$ is nilpotent if and only if every prime dividing n also divides a.

   How many nilpotent elements does $\mathbb{Z}_{200}$ have?

7. Let A be any set, and let $\mathcal{P}(A)$ be the family of all subsets of A. Inside $\mathcal{P}(A)$ we can form the familiar unions $X \cup Y$ and intersections $X \cap Y$ of subsets X, Y of A. We also have the complement $A \backslash X$ of any subset X of A. For X, Y in $\mathcal{P}(A)$ define the operation of addition in $\mathcal{P}(A)$ according to

$$X + Y = (X \cup Y) \backslash (X \cap Y),$$

and the operation of multiplication according to

$$XY = X \cap Y.$$

Take the empty set $\emptyset$ as the zero and the full subset A as the identity element of $\mathcal{P}(A)$. For negation take the somewhat curious operation given by $-X = X$.

Verify that $\mathcal{P}(A)$ satisfies the eight axioms needed to make it a commutative ring.

8. A ***Boolean ring*** is a ring R such that $x^2 = x$ for all x in R.
   (a) Give examples of finite and of infinite Boolean rings.
   (b) Show that $x = -x$ for every x in a Boolean ring, and then show that every Boolean ring is commutative.
9. Find a commutative ring in which the equation $x^2 = 1$ has four solutions.

   Find a commutative ring in which this equation has infinitely many solutions.
10. Let R be the set of all complex numbers of the form $a + b\omega$, where $\omega = \sqrt[3]{2}$ and $a, b \in \mathbb{R}$. Is R a subring of $\mathbb{C}$?
11. (a) Suppose that R is a *finite* abelian group using additive notation. Show that it is possible to define a multiplication on R in such a way that R becomes a ring.
    (b) Let R be the abelian group which is the direct sum

$$\mathbb{Z}_2 \oplus \mathbb{Z}_3 \oplus \mathbb{Z}_4 \oplus \cdots \oplus \mathbb{Z}_n \oplus \cdots$$

where the $\mathbb{Z}_j$ are the cyclic groups of order j under addition. Note that each element of the direct sum is a sequence $(x_1, x_2, \ldots, x_n, \ldots)$ such that $x_j \in \mathbb{Z}_j$ and all but a finite number of x_j are zero. Using point-wise addition of sequences, R becomes an abelian group. Show that it is not possible to define multiplication on R in such a way that along with the given addition R becomes a ring.

12.* If $x^3 = x$ for all x in a ring R, prove that $6x = 0$ for all x in R, and then show that R is commutative.

   There are various ways to show this well known fact, all requiring some form of ingenuity with the ring operations.
13. Prove that every non-commutative ring has at least 8 elements in it. Then find a non-commutative ring with exactly 8 elements in it.

   Find a non-commutative ring containing precisely 16 elements.
14. Suppose that R is a commutative ring and that a, b are elements of R such that $au + bv = 1$ for some u, v in R. For example, this holds when a, b are coprime integers in $\mathbb{Z}$.

Show that there exist elements x, y in R such that $a^2x + by = 1$.

Use the above to prove that the equation $a^2x + b^2y = 1$ has a solution in R.

Explain why $a^4x + b^4y = 1$ has a solution in R, and then why $a^{2^k}x + b^{2^k}y = 1$ has a solution in R for all positive exponents k.

If m, n are arbitrary positive integers, show that $a^m x + b^n y = 1$ has a solution in R.

4.2 Units and Zero Divisors

The elements of a ring could use a bit of taxonomy.

4.2.1 The Group of Units

Definition 4.4. An element u in a ring R is called a ***unit*** provided every element in R is both a left and right multiple of u.[2] More precisely, for every x in $\mathbb{R}$, we have s, t in R such that

$$x = us = tu.$$

The ring identity 1 is a unit. Just take $s = t = x$.

Furthermore, 1 is a multiple of every unit u. Thus there exist s, t in R such that $1 = us = tu$. We might call s a right-inverse for u and t a left-inverse for u. But actually

$$s = 1s = tus = t1 = t,$$

which reveals that $s = t$. Thus every unit has a two-sided inverse, and, as the above equations reveal, that inverse is unique.

On the other hand, suppose that u in R has a two-sided inverse, say v. That is $uv = vu = 1$. Then every x in R is both a left and right multiple of u. Indeed,

$$x = 1x = uvx \text{ and } x = x1 = xvu.$$

The units are nothing but the elements u of R that possess a two-sided inverse.

As noted, the inverse of each unit u is unique, and we write it as u^{-1}. Furthermore, if u, v are units with respective inverses u^{-1}, v^{-1}, we have

$$uvv^{-1}u^{-1} = u1u^{-1} = uu^{-1} = 1 \text{ and } v^{-1}u^{-1}uv = v^{-1}1v = v^{-1}v = 1.$$

This reveals that a product of units is a unit. Taking this together with the associative law for multiplication, it follows that *the set of units is a group using the multiplication in R*. We shall adopt the ***star-notation*** $R^\star$ to denote the group of units in R.

Examples of Units

Upon encountering a ring, the question of identifying its group of units arises. For instance, in the rings $\mathbb{Q}$, $\mathbb{R}$ and $\mathbb{C}$ all elements except for 0 are units. In $\mathbb{Z}$ the only units are ± 1.

[2] In the sciences, a unit is an agreed upon fixed amount of some quantity, and every other amount is a multiple of that fixed amount. Possibly this observation gives credence to the use of the term in ring theory.

The units in the ring of $n \times n$ matrices $M_n(\mathbb{Q})$ are those matrices whose determinant is not zero. That is the general linear group $GL_n(\mathbb{Q})$ discussed in Section 2.1. This group of units is not abelian when $n \geq 2$.

Every ring has at least one unit, namely 1, while the ring $\mathbb{Z}_2$ has only the identity as a unit. In $\mathbb{Z}_6$ the residues [1] and [5] are units since $[1][1] = [1]$ and $[5][5] = [1]$, while $[2], [3], [4]$ are not units. More generally in the ring $\mathbb{Z}_n$ of residues modulo n, the group of units consists of all residues $[a]$ whose representative a is coprime with n. The order of the group $\mathbb{Z}_n^\star$ is given by the Euler function $\phi(n)$.

In a commutative ring the equality $uv = 1$ instantly gives $vu = 1$, but in some rings that are non-commutative it is possible to have elements u, v such that $uv = 1$ while $vu \neq 1$. Such elements u are not units. An example of such a ring is offered in the exercises at the end of this section.

Units in the Ring of Gaussian Integers

In the ring $\mathbb{Z}[i]$ of Gaussian integers there are the four obvious units $\pm 1, \pm i$, and no others.

Indeed, if $z = a + bi \in \mathbb{Z}[i]$, let $\phi(z) = a^2 + b^2$, the square of the usual norm of z. Clearly $\phi(z) \in \mathbb{Z}$, and by simple verification $\phi(zw) = \phi(z)\phi(w)$ when $z, w \in \mathbb{Z}[i]$. If $z = a + bi$ is a unit of $\mathbb{Z}[i]$ with inverse w, then $1 = \phi(1 + 0i) = \phi(zw) = \phi(z)\phi(w)$. Thus, $\phi(z)$ is a unit of $\mathbb{Z}$ with integer inverse $\phi(w)$, and $\phi(z)$ is positive. Since 1 is the only positive unit of $\mathbb{Z}$, we get $a^2 + b^2 = 1$. Because $a, b \in \mathbb{Z}$ it follows that $a = \pm 1, b = 0$, or $a = 0, b = \pm 1$. Thus z has to be one of $\pm 1, \pm i$.

Cancellation of Units

An element z in a ring R is called ***cancellable*** if the equality $zx = zy$ for x, y in R implies $x = y$, and the equality $xz = yz$ for x, y in R implies $x = y$.

Proposition 4.5. *Every unit in a ring R is cancellable.*

Proof. Let u be a unit and suppose that $ux = uy$ for some x, y in R. Taking v to be an inverse for u, we get

$$x = 1x = (vu)x = v(ux) = v(uy) = (vu)y = 1y = y.$$

A similar proof shows how $xu = yu$ implies that $x = y$. □

4.2.2 Zero Divisors

Units in a ring R are cancellable, as just noted. Some rings contain elements that are not units but remain cancellable. Most notably in $\mathbb{Z}$ we find that all integers a other than zero can be cancelled in an equation $ax = ay$ to get $x = y$, but only ± 1 are units. Let us examine the non-cancellable elements.

Proposition 4.6. *An element a in a ring R is not cancellable if and only if there is an accompanying non-zero x in R such that $ax = 0$ or $xa = 0$.*

Proof. If a is cancellable, then the equations $ax = 0$ or $xa = 0$ can be rewritten as $ax = a0$ or $xa = 0a$. Cancel the a to force $x = 0$.

If a is not cancellable, there must be some y, z in R such that $y \neq z$ while at the same time one of $ay = az$ or $ya = za$ holds. Thus either $a(y - z) = 0$ or $(y - z)a = 0$. In either case we get that $y - z$ is the non-zero x such that $ax = 0$ or $xa = 0$. □

Clearly 0 is not cancellable (except in the zero ring). The other non-cancellable elements get their own special name.

Definition 4.7. An element a in a ring R is called a ***zero divisor*** provided $a \neq 0$ and there is a non-zero x in R such that $ax = 0$ or $xa = 0$.

Despite the fact $0 = a0$ for any element a, keep in mind that both factors of 0 need to be non-zero before the term "zero divisor" is used.

Thus, the elements of a non-zero ring fall into four disjoint categories:

- the zero element 0,
- the group of units, which includes the identity element 1,
- the cancellable elements that are not units (if any),
- the zero divisors (if any).

By way of illustration, take the product ring $\mathbb{Z} \times \mathbb{Z}$. The cancellable elements consist of all integer pairs (a, b) where $a \neq 0$ and $b \neq 0$. Among these, the units are just the four pairs $(\pm 1, \pm 1)$. The zero divisors are the pairs of type $(a, 0)$ or $(0, a)$ where $a \neq 0$. Indeed, $(a, 0)(0, 1) = (0, 0)$ and $(0, a)(1, 0) = (0, 0)$.

For another example, let $\mathcal{U}$ be the subring of upper triangular matrices inside the ring $M_2(\mathbb{Z}_2)$ of 2×2 matrices whose entries are residues modulo 2. To lighten the clutter of notation, let 0 stand for the residue [0], and 1 for the residue [1]. The finite ring $\mathcal{U}$ is a ring made up of the following 8 matrices:

$$\begin{bmatrix} 0 & 0 \\ 0 & 0 \end{bmatrix}, \begin{bmatrix} 1 & 0 \\ 0 & 1 \end{bmatrix}, \begin{bmatrix} 0 & 1 \\ 0 & 0 \end{bmatrix}, \begin{bmatrix} 0 & 0 \\ 0 & 1 \end{bmatrix}, \begin{bmatrix} 1 & 0 \\ 0 & 0 \end{bmatrix}, \begin{bmatrix} 1 & 1 \\ 0 & 0 \end{bmatrix}, \begin{bmatrix} 0 & 1 \\ 0 & 1 \end{bmatrix}, \begin{bmatrix} 1 & 1 \\ 0 & 1 \end{bmatrix}.$$

In $\mathcal{U}$ the units are the identity $\begin{bmatrix} 1 & 0 \\ 0 & 1 \end{bmatrix}$ and $\begin{bmatrix} 1 & 1 \\ 0 & 1 \end{bmatrix}$ whose inverse is itself. The other 6 matrices are zero divisors. For instance, observe that $\begin{bmatrix} 0 & 1 \\ 0 & 1 \end{bmatrix}\begin{bmatrix} 0 & 1 \\ 0 & 0 \end{bmatrix} = \begin{bmatrix} 0 & 0 \\ 0 & 0 \end{bmatrix}$. Incidentally, notice that $\begin{bmatrix} 1 & 0 \\ 0 & 0 \end{bmatrix}\begin{bmatrix} 0 & 1 \\ 0 & 1 \end{bmatrix} = \begin{bmatrix} 0 & 1 \\ 0 & 0 \end{bmatrix}$, which shows that the ring $\mathcal{U}$ is non-commutative.

Cancellation in Finite Rings

The preceding example is a finite ring in which the only cancellable elements are the units. This is an interesting limitation that applies to all finite rings, as we shall now prove by exploiting the so-called ***pigeon-hole principle***. One way to state the pigeon-hole principle is that if F is a finite set, then every injective function $f : F \to F$ must also be surjective.

Proposition 4.8. *In a finite ring, every cancellable element is a unit. Thus a finite ring contains, in addition to zero, only units and zero divisors.*

Proof. Let a be a cancellable element in a finite ring F. We must come up with a^{-1}. Look at the function $f : F \to F$ given by $f(x) = ax$ for each x in F. If $f(x) = f(y)$, i.e. $ax = ay$ for some x, y in F, cancel the a to deduce $x = y$. By the pigeon-hole principle, f is surjective. Therefore there is some b in R such that $f(b) = 1$, or $ab = 1$.

So far we have an inverse b that works on the right side of a. By considering the function $g : F \to F$ given by $g(x) = xa$ for each x in F, and imitating the argument above, we pick up some c in F such $ca = 1$. Then, exploiting the associative law:

$$c = c1 = c(ab) = (ca)b = 1b = b.$$

Thus b is an inverse that works on both sides of a, as desired. □

For example, in the finite ring $\mathbb{Z}_{10}$ there are the units $[1], [3], [7], [9]$, the zero divisors $[2], [4], [5], [6], [8]$, and of course $[0]$.

A stunning and deep theorem due to J. H. M. Wedderburn (1882–1948) says that if all non-zero elements of a finite ring are units, then the ring must be commutative.

4.2.3 Integral Domains and Fields

Because of their significance in number theory and geometry, two types of rings will command our attention.

Definition 4.9. A ring R is called an ***integral domain***[3] provided:

$0 \neq 1$, R is commutative, and R has no zero divisors.

In other words, all non-zero elements in an integral domain R are cancellable. The ring R is called a ***field*** when

$0 \neq 1$, R is commutative, and all non-zero elements in R are units.

The classic examples of fields are $\mathbb{Q}$, $\mathbb{R}$ and $\mathbb{C}$, as well as the rings of residues $\mathbb{Z}_p$ modulo the primes p. Since units are cancellable, as seen in Proposition 1.2, every field is an integral domain. However, $\mathbb{Z}$ is an integral domain that is not a field, for in $\mathbb{Z}$ we find the cancellable integers ± 2, ± 3, $\pm 4, \ldots$ that are not units.

Every subring of an integral domain is another integral domain. Indeed, if the big ring has no zero divisors, then the subring will surely have none. In particular, every subring of a field is an integral domain. Thus, for example, the ring of Gaussian integers $\mathbb{Z}[i]$, being a subring of the field $\mathbb{C}$, is an integral domain. Furthermore, we shall come to learn that every integral domain is a subring of some field.

[3] A term originating in the works of E. Noether (1882–1935) and R. Dedekind (1831–1916).

Much awaits to be discovered about these structures, but for now here is an interesting nugget.

Proposition 4.10. *Every finite integral domain is a field.*

Proof. From Proposition 4.8, the only cancellable elements in a finite ring are the units. If that finite ring is also an integral domain, i.e. if all non-zero elements are cancellable, then all non-zero elements are units. That makes the domain a field. □

For example, when p is a prime integer the ring $\mathbb{Z}_p$ is an integral domain by simple verification. Thus this finite ring is a field. Actually in Section 1.3 we saw how to calculate the inverses of non-zero elements in this field.

EXERCISES

1. If R is a set satisfying all of the axioms of a ring except for the commutative law $x + y = y + x$, prove that this commutative law still must hold.
   Hint. Play with $(x + y)(1 + 1)$ using the axioms which are in effect.
2. Find the units of these rings: $\mathbb{Z}_{40}$, $\mathbb{Z}_{10} \times \mathbb{Z}_5$, $M_2(\mathbb{Z})$, $M_2(\mathbb{Z}_3)$.
3. Let R be the set of rational numbers which can be written in the form a/b where a, b are integers and b is odd. Show that R is a subring of $\mathbb{Q}$ and find the units of R.
4. This is a purely computational question. Decide if 71060 and 131859 are coprime, and if they are coprime find the inverse of the residue [71060] inside the ring $\mathbb{Z}_{131859}$.
5. Show that the set $R = \{a + b\sqrt{2} : a, b \in \mathbb{Q}\}$ is a field under the usual addition and multiplication of real numbers.
   If $i = \sqrt{-1}$ show that $R = \{a + bi : a, b \in \mathbb{Q}\}$ is a field under the usual addition and multiplication of complex numbers.
6. If a ring R has exactly three elements, must R be a field?
7. An element a in a ring R is called *nilpotent* when $a^n = 0$ for some positive integer n. If R is commutative and a is nilpotent, show that $1 - a$ is a unit.
   Hint. Factor $1 - a^n$.
   Show that the sum of a unit and a nilpotent element in a commutative ring is a unit in the ring.
8. In the non-commutative ring $M_n(\mathbb{R})$ of $n \times n$ matrices with entries in $\mathbb{R}$ the identity element is none other than the identity matrix I. Suppose that A, B are matrices in $M_n(\mathbb{R})$ and that $AB = I$. Using facts from linear algebra prove that $BA = I$ as well.
9. Using facts from linear algebra show that a non-zero matrix in the ring $M_n(\mathbb{R})$ is either a unit or a zero divisor.
10. This exercise requires some knowledge of infinite-dimensional vector spaces. Suppose V is a vector space. Let $\mathcal{L}(V)$ be the set of linear transformations $T : V \to V$. This set becomes a ring with operations as follows. The sum $S + T$ of two transformations S, T is given by the mapping $S + T : v \mapsto S(v) + T(v)$, where $v \in V$. The product ST

is given by the composite mapping $ST : v \mapsto S(T(v))$. The zero element and identity elements are given by the transformations $\mathbf{0} : v \mapsto 0$ and $I : v \mapsto v$. The negative $-T$ of a transformation T is given by $-T : v \mapsto -T(v)$. It is easy to check that $\mathcal{L}(V)$ is a ring using these operations.

Every linear transformation on V is determined by what it does to a basis of V. Now suppose that V has an infinite basis $v_1, v_2, \ldots, v_n, \ldots$. Let S be the transformation in $\mathcal{L}(V)$ given by $S : v_n \mapsto v_{n+1}$ for each basis vector v_n. Show that there is a transformation B in $\mathcal{L}(V)$ such that $BS = I$, but that there is no transformation C in $\mathcal{L}(V)$ such that $SC = I$. Thus you have a ring R and an element inside R possessing a left-inverse but not a right-inverse.

11. Show that, in a ring, the product of cancellable elements stays cancellable.

12. Consider the subring $\mathbb{Z}[\sqrt{-5}] = \{a + b\sqrt{-5} : a, b \in \mathbb{Z}\}$ of $\mathbb{C}$. Show that $a + b\sqrt{-5}$ is a unit of $\mathbb{Z}[\sqrt{-5}]$ if and only if $a^2 + 5b^2 = 1$ and then find all of its units.

13. Let $\sqrt{-9}$ be one of the square roots of -19 inside the field $\mathbb{C}$. And let $\alpha = \frac{1}{2}(1 + \sqrt{-19})$. Verify that $\alpha^2 = \alpha - 5$, and then show that the set $R = \{a + b\alpha : a, b \in \mathbb{Z}\}$ is a subring of $\mathbb{C}$.

14. Suppose d is an integer such that $d \equiv 1 \bmod 4$ and let $\alpha = \frac{1+\sqrt{d}}{2}$. Show that the set $R = \{a + b\alpha : a, b \in \mathbb{Z}\}$ is a subring of the ring of $\mathbb{C}$.

15. The set $C[0, 1]$ of continuous real valued functions, defined on the closed interval $[0, 1]$, is a ring when functions are added and multiplied point-wise. Identify the units and the zero divisors of the ring $C[0, 1]$. A bit of savvy with calculus is needed.

16. Show that there is no integral domain with exactly 6 elements.

17. Take the commutative ring $\mathbb{Z}[\sqrt{2}] = \{a + b\sqrt{2} : a, b \in \mathbb{Z}\}$, and let us explore what its group of units looks like.

(a) If $x = a + b\sqrt{2}$, let $N(x) = a^2 - 2b^2$. Verify that $N(xy) = N(x)N(y)$ for all x, y in $\mathbb{Z}[\sqrt{2}]$.

(b) Show that x is a unit of $\mathbb{Z}[\sqrt{2}]$ if and only if $N(x) = \pm 1$.

(c) Verify that the element $w = 1 + \sqrt{2}$ is a unit in this ring, along with all integer powers w^n as well as all negatives $-w^n$. Thus $\mathbb{Z}[\sqrt{2}]$ has infinitely many units.

(d) Show that there is no unit x in $\mathbb{Z}[\sqrt{2}]$ such that $1 < x < w$.

Hint. If $x = c + d\sqrt{2}$ were such a unit we would have

$$|c - d\sqrt{2}||c + d\sqrt{2}| = |c^2 - 2d^2| = |N(x)| = 1.$$

Now make an estimate on $|c - d\sqrt{2}|$.

(e) Suppose that x is a positive unit of $\mathbb{Z}[\sqrt{2}]$. Show that $w^n \leq x < w^{n+1}$ for some integer n. Deduce that $1 \leq w^{-n}x < 1 + \sqrt{2}$. Then deduce that $1 = w^{-n}x$. Conclude that all positive units are integer powers of w. Then deduce that all units are of type either w^n or $-w^n$.

18. Let F consist of all 2×2 matrices of the form $\begin{bmatrix} x & y \\ y & x+y \end{bmatrix}$, where $x, y \in \mathbb{Z}_2$.

How many elements does F contain? Verify that F is a subring of the ring $M_2(\mathbb{Z}_2)$. Show that F is a field.

19. Take the following matrices with entries in the field $\mathbb{R}$:

$$\mathbf{1} = \begin{bmatrix} 1 & 0 \\ 0 & 1 \end{bmatrix}, \quad \mathbf{i} = \begin{bmatrix} i & 0 \\ 0 & -i \end{bmatrix}, \quad \mathbf{j} = \begin{bmatrix} 0 & 1 \\ -1 & 0 \end{bmatrix}, \quad \mathbf{k} = \begin{bmatrix} 0 & i \\ i & 0 \end{bmatrix}.$$

We saw in Section 2.4.3 that they generate the so-called quaternion group $\mathcal{Q}$ using matrix multiplication. These matrices can be used as well to make what is known as the ***ring of real quaternions*** first discovered by the Irish mathematician William Rowan Hamilton (1805–1865). Let $\mathcal{H}$ be the set of all linear combinations

$$a\mathbf{1} + b\mathbf{i} + c\mathbf{j} + d\mathbf{k}, \text{ where } a, b, c, d \in \mathbb{R}.$$

(a) Verify that $\mathcal{H}$ is a subring of the ring $M_2(\mathbb{C})$ of 2×2 matrices over the field $\mathbb{C}$. Show that $\mathcal{H}$ is non-commutative.

(b) If $x = a\mathbf{1} + b\mathbf{i} + c\mathbf{j} + d\mathbf{k} \in \mathcal{H}$, show that $x \neq 0$ if and only if some of $a, b, c, d \neq 0$.

(c) If x as above is not zero and $e = a^2 + b^2 + c^2 + d^2$ show that x is a unit of $\mathcal{H}$ with inverse given by

$$\frac{1}{e}\left(a\mathbf{1} - b\mathbf{i} - c\mathbf{j} - d\mathbf{k}\right).$$

Thus $\mathcal{H}$ is a non-commutative ring in which all non-zero elements are units. Such structures are commonly known as ***division rings***.

(d) Show that the equation $x^2 + 1 = 0$ has infinitely many solutions in $\mathcal{H}$.

(e) Verify that the set R of linear combinations $a\mathbf{1} + b\mathbf{i} + c\mathbf{j} + d\mathbf{k}$ where $a, b, c, d \in \mathbb{Z}$ is a subring of $\mathcal{H}$.

(f) Identify the units of R.

4.3 Polynomials

Polynomials seem familiar. We factor them, find their roots, take their derivatives, sketch their graphs, and are comfortable in manipulating them. Despite this, it remains a sticky matter to properly define what is meant by a polynomial and to explain how they form rings. We might hope that the discussion to follow will not seem too pedantic, but if such be the case, there is not much harm in skimming through it.

A shortcut could be to think of a polynomial as a special kind of function. For instance, using the field $\mathbb{R}$ of real numbers, any list of numbers $r_0, r_1, r_2, \ldots, r_n$ gives rise to the function $f : \mathbb{R} \to \mathbb{R}$ given by $f(x) = r_0 + r_1x + r_2x^2 + \cdots + r_nx^n$ for every x in $\mathbb{R}$. Furthermore, any change in any r_j leads to a different function. Under the usual definition of point-wise function addition and multiplication the set of polynomial functions forms a ring. Informally, we speak of the polynomial function $f(x) = r_0 + r_1x + r_2x^2 + \cdots + r_nx^n$. However, such a point of view is too restrictive. For example, the polynomial function $f(x) = 3x^2 + 5x - 2$ can also be construed as the function $f : M_2(\mathbb{R}) \to M_2(\mathbb{R})$ that maps a matrix A to the matrix $3A^2 + 5A - 2I$. If we take polynomials to be functions, then we must ask a question lacking a good answer. Namely, what are polynomials functions of?

With no effort whatsoever we can replace the real field $\mathbb{R}$ by any commutative ring R, and imitate the function approach to obtain polynomial functions $f : R \to R$. Now, another problem emerges. Indeed, consider the two element ring $\mathbb{Z}_2$ of residues modulo 2 and look at the polynomial function $f(x) = x^2 + x$ on the ring $\mathbb{Z}_2$. If $0, 1$ are the residues in $\mathbb{Z}_2$, we see from modular arithmetic that $f(0) = 0^2 + 0 = 0$ and $f(1) = 1^2 + 1 = 0$. We run into the disconcerting situation of having a non-zero, quadratic polynomial function being the same as the zero function.

For the above reasons it seems wise to start from scratch and to define a polynomial to be something other than a function. Given a commutative ring R, one popular approach is to let x stand for a "formal symbol," also known by the equally foggy synonyms of "indeterminate" and "variable." Then we could define a polynomial to be any formal expression of the type $a_0+a_1x+a_2x^2+\cdots+a_nx^n$, where the $a_j \in R$ and the symbols in the exponents are non-negative integers. This too presents a difficulty. For instance, we would like to say that $3 + 4x + 5x^2 = 3+4x+5x^2+0x^3$. Yet, if polynomials are formal expressions, then the above two polynomials, having a formally different appearance, should not be equal. In rebuttal to this objection one hears that such polynomials are equal because they only differ by the zero term $0x^3$. However, we do not yet have a ring that allows us to speak of the zero polynomial, nor of differences. In practice we tend to ignore these seemingly petty nuances, declare that polynomials form a ring, and proceed to work with them.

Nevertheless, an important a concept as that of polynomials deserves a proper accounting. We may recall from linear algebra that a vector is simply an element of a vector space. Thus, in linear algebra what needs to be defined is the notion of a vector space. With a similar attitude in mind, let us say that a polynomial is an element of a polynomial ring, and proceed to define polynomial rings.

4.3.1 The Definition of Polynomial Rings

Start with a ***commutative*** ring R that is a subring of another commutative ring T. If x is an element chosen from T, we can speak about the smallest subring of T that includes x and all of R. If A is the smallest subring of T containing x and R, then the closure laws for subrings demand that A contain every element of the form $r_0+r_1x+r_2x^2+\cdots+r_nx^n$, where the $r_j \in R$ and n runs through all non-negative integers. On the other hand, it is easy to inspect that the set of all such elements satisfies the closure requirements to be a subring of T, and contains R and x. Thus we know what the elements of A look like. Rings such as A are worthy of a definition.

Definition 4.11. If A is a commutative ring with a subring R and $x \in A$, we say that A is ***generated*** by x over R provided every element in A can be expressed in the form

$$r_0 + r_1x + r_2x^2 + \cdots + r_nx^n,$$

where $r_j \in R$ and n is any non-negative integer. The notation for a commutative ring that is generated by some x over R is $R[x]$.

It is tempting to say that the commutative rings $R[x]$ are polynomial rings, but before rushing into such terminology let us consider an example. Let A be the smallest subring of $\mathbb{R}$ containing both $\mathbb{Z}$ and $\sqrt{2}$. It is straightforward to see that A consists of all numbers of the form $r_0+r_1\sqrt{2}$ where $r_0, r_1 \in \mathbb{Z}$, and therefore that $A = \mathbb{Z}[\sqrt{2}]$. In this example we do not need the higher powers of $\sqrt{2}$ to account for all of A. This stems from the fact that $\sqrt{2}^2 = 2$. There is a collapse in the powers of $\sqrt{2}$ that prevents $\mathbb{Z}[\sqrt{2}]$ from making the grade as a polynomial ring over $\mathbb{Z}$.

In order to consider generators that avoid such collapses we adopt a definition that mimics the notion of linear independence inside vector spaces.

If R is a subring of a commutative ring A and $x \in A$, we say that the powers of x are ***R-linearly independent*** provided the equation

$$r_0 + r_1x + r_2x^2 + \cdots + r_nx^n = 0,$$

using r_j in R and a non-negative integer n, forces all r_j to be 0.

In the example preceding this definition we saw that the powers of $x = \sqrt{2}$ are not $\mathbb{Z}$-linearly independent because $2 - x^2 = 0$. On the other hand, Charles Hermite (1822–1901) proved the remarkable and deep result that the powers of $x = e$, the base for natural logarithms, are $\mathbb{Q}$-linearly independent. Soon after, Ferdinand von Lindemann (1852–1939) showed that the powers of $x = \pi$ are $\mathbb{Q}$-linearly independent, and thereby settled an ancient geometry problem on the impossibility of squaring the circle.

We come to an important definition.

Definition 4.12. A *commutative* ring A containing a subring R is called a ***polynomial ring over R*** provided A is generated over R by some x in A whose powers are R-linearly independent. The elements of a polynomial ring $A = R[x]$ are called ***polynomials*** in x ***over*** R.

Given a polynomial ring $R[x]$ we are likely already familiar with the ensuing assortment of terminologies and observations.

Anatomy of a Polynomial

Suppose that $R[x]$ is a polynomial ring over a commutative ring R.

The generating element x is sometimes called an ***indeterminate over*** R. It is also called a ***variable*** as a reminder of our habit of thinking of polynomials as functions. Note that the exact nature of x is immaterial. All that matters is for the powers of x to be R-linearly independent.

The R-polynomials in x all take the form $f = r_0+r_1x+r_2x^2+\cdots+r_nx^n$. If $f = 0$, then the R-linear independence of powers of x forces all r_j to be 0, and we might as well not bother writing any of the r_jx^j. If f is not the zero element of $R[x]$, then some $r_j \neq 0$, and by dropping some 0 that appear as $0x^i$ we can suppose that

$$f = r_0 + r_1x + r_2x^2 + \cdots + r_nx^n \text{ where } r_n \neq 0.$$

Let us agree to call the above the ***standard form*** of the non-zero polynomial f. Of course there is the equally good, alternative form written in descending powers of x, that we will freely adopt. In addition to its standard form, the same polynomial can assume other appearances. For instance, in the polynomial ring $\mathbb{Z}[x]$ (presuming it exists) we have

$$\begin{aligned} -5 - 7x + 6x^2 &= 0x^3 + 6x^2 - 7x - 5 = 6x^2 - 7x - 5 \\ &= -5 + 6x^2 - 7x = (6x - 7)x - 5 = (3x - 5)(2x + 1). \end{aligned}$$

However, as we might suspect, a polynomial can have only one standard form. At the risk of being somewhat overbearing, let us show this important fact.

Proposition 4.13. *Every non-zero polynomial in a polynomial ring has just one standard form.*

Proof. In the polynomial ring $R[x]$ let f have standard forms

$$f = r_0 + r_1x + r_2x^2 + \cdots + r_nx^n \text{ and } f = s_0 + s_1x + s_2x^2 + \cdots + s_mx^m.$$

Thus all $r_j \in R, r_n \neq 0$, and all $s_i \in R, s_m \neq 0$.

If $n \neq m$, say $n > m$, then subtract f from itself to obtain

$$0 = f - f = (r_0 - s_0) + (r_1 - s_1)x + \cdots + (r_m - s_m)x^m + \cdots + r_nx^n.$$

Since $r_n \neq 0$, this contradicts the R-linear independence of the powers of x. Thus $n = m$.

If $r_j \neq s_j$ for some $j = 0, 1, 2, \ldots, n$, again subtract f from itself to get

$$0 = f - f = (r_0 - s_0) + (r_1 - s_1)x + \cdots + (r_j - s_j)x^j + \cdots (r_n - s_n)x^n.$$

Now since $r_j - s_j \neq 0$, we again contradict the R-linear independence of powers of x. Thus $r_j = s_j$ for all j, which proves that every polynomial only has one standard form. □

In contrast with Proposition 4.13, take the ring $\mathbb{Z}[\sqrt{2}]$, where $x = \sqrt{2}$ does not have $\mathbb{Z}$-linearly independent powers. Here we have $3 = 1 + 0x + x^2 = -1 - 2x^2 + 2x^4$, which are three distinct standard representations for the same element.

Now we can lay down the familiar jargon that goes with polynomials.

Suppose that f is a non-zero polynomial in a polynomial ring $R[x]$ written in its unique standard form

$$f = r_0 + r_1x + r_2x^2 + \cdots + r_nx^n, \text{ where } r_n \neq 0.$$

- The element r_j in R is called the ***coefficient*** of x^j. The ring R is also known as the ***ring of coefficients*** of $R[x]$.
- The non-zero r_n is known as the ***leading coefficient*** of f.
- If the leading coefficient of f is 1, the polynomial is called ***monic***.
- The highest non-negative integer n appearing in the standard representation is called the ***degree*** of f, and we write $n = \deg f$.

- The polynomials $r_j x^j$ are called the ***terms*** of f. They are alternatively known as the ***monomials*** of f. Proposition 4.13 reveals that two polynomials are equal if and only if they have the same non-zero terms. The term $r_n x^n$ is called the ***leading term*** of f, while the term r_0 is known as the ***constant term*** of f. Those terms of f which are not zero are frequently referred to as the ***terms appearing in*** f. According to this jargon the zero polynomial is the one in which no terms appear.
- The elements of the ring R are called ***constant polynomials***. The constant polynomials consist of the polynomials of degree 0 along with the zero polynomial, which is the zero element in the polynomial ring $R[x]$. The polynomials of degree 0 have the standard representation $f = r_0$ where $r_0 \neq 0$. They certainly belong to R, but they leave out the constant 0 polynomial. The zero polynomial has no leading coefficient, and thereby does not have a degree.
- Polynomials of degree $1, 2, 3, 4$ and 5 are often called ***linear***, ***quadratic***, ***cubic***, ***quartic*** and ***quintic***, respectively.

The Existence of Polynomial Rings

We now show that every commutative ring is the ring of coefficients of a polynomial ring.

Proposition 4.14. *For every commutative ring R there exists a polynomial ring with R as its ring of coefficients.*

Proof. We have to construct a commutative ring A that is generated over R by some x whose powers are R-linearly independent. Our expectation is that the elements of A should look like $r_0 + r_1 x + r_2 x^2 + \cdots + r_n x^n$ where x is something in A, the $r_j \in R$, and the n covers all non-negative integers. This suggests that, in order to capture the notion of coefficients, we should be looking at sequences from R.

Let A be the set of all infinite sequences

$$(r_0, r_1, r_2, \ldots, r_j, \ldots)$$

where the $r_j \in R$ and all but finitely many r_j are 0. The operations that follow will give A the structure of a commutative ring.

- For the zero element in A take the sequence $0 = (0, 0, 0, \ldots, 0)$.
- For the identity element of A take the sequence $1 = (1, 0, 0, \ldots, 0, \ldots)$.
- Let the addition operation in A be given by adding the corresponding entries of sequences in A point-wise as shown:

$$(r_0, r_1, r_2, r_3, \ldots) + (s_0, s_1, s_2, s_3, \ldots) = (r_0 + s_0, r_1 + s_1, r_2 + s_2, r_3 + s_3, \ldots).$$

- To negate a sequence just negate each of its entries as indicated:

$$-(r_0, r_1, r_2, r_3, \ldots) = (-r_0, -r_1, -r_2, -r_3, \ldots).$$

- To multiply two sequences in A perform the process sometimes known as convolution. Namely:

$$(r_0, r_1, r_2, r_3, \dots)(s_0, s_1, s_2, s_3, \dots)$$
$$= (r_0 s_0, r_0 s_1 + r_1 s_0, r_0 s_2 + r_1 s_1 + r_2 s_0, r_0 s_3 + r_1 s_2 + r_2 s_1 + r_3 s_0, \dots).$$

Regarding the addition it is clear that since the r_j and s_j are zero eventually, so are the $r_j + s_j$ zero eventually. Thus the sum of two sequences in A is again in A. This is also obvious for the negation of a sequence. As to the product of two sequences, note that the entry in the nth position is the summation $\sum_{i+j=n} r_i s_j$. If r_k is the last non-zero entry in the sequence $(r_0, r_1, r_2, r_3, \dots)$ and s_ℓ is the last non-zero entry in $(s_0, s_1, s_2, s_3, \dots)$, then once $n > k+\ell$ every term in the summation $\sum_{i+j=n} r_i s_j$ vanishes because either $i > k$ or $j > \ell$. Consequently the summation $\sum_{i+j=n} r_i s_j = 0$ when $n > k+\ell$, which confirms that the product of two sequences in A is again in A.

A bit of checking reveals that the given operations on A satisfy the ring axioms including the commutativity of multiplication. For instance, the axioms involving addition, zero and subtraction hold because they are inherited from R at each entry of our sequences. The commutative law of multiplication in A follows from the symmetry between the r_i and s_j in the formula $\sum_{i+j=n} r_i s_j$ for the nth entry of the product sequence, along with the fact R itself is commutative. The neutrality of 1 is easy to see.

Since it is the messiest to check, let us verify the associative law for multiplication. We resort to sigma notation. Suppose

$$r = (r_0, r_1, r_2, r_3, \dots), \ s = (s_0, s_1, s_2, s_3, \dots), \ t = (t_0, t_1, t_2, t_3, \dots) \in A.$$

The entry in the kth position for the product rs is $\sum_{i+j=k} r_i s_j$. Then the entry in the nth position for the product $(rs)t$ works out to be

$$\sum_{k+\ell=n} \left(\sum_{i+j=k} r_i s_j \right) t_\ell = \sum_{i+j+\ell=n} r_i s_j t_\ell.$$

To see the above equality note that in order to partition n as a sum $i + j + \ell$ of three non-negative integers, we can first partition n as the sum $k + \ell$ of two non-negative integers and then partition k as the sum $i + j$ of two non-negative integers. Of course, the fact R is a commutative ring was used implicitly. By a similar argument the nth entry of $r(st)$ also works out to the right side of the above equation. Thus $(rs)t = r(st)$.

A similar routine verifies the distributive law.

Notice that R itself is a subring of A after we make the identification $r = (r, 0, 0, 0, \dots)$ for each r in R. Now put

$$x = (0, 1, 0, 0, \dots, 0, \dots).$$

By direct calculation of the convolution product we obtain that

$$x^2 = (0, 0, 1, 0, 0, \dots), x^3 = (0, 0, 0, 1, 0, 0, \dots), \dots, x^n = (0, 0, \dots, 0, 1, 0 \dots),$$

where the 1 appears in the nth slot of the sequence (starting with $n=0$). Since for any $r = (r, 0, 0, \ldots)$ in the subring R of A we also have that

$$rx^n = (0, 0, \ldots, 0, r, 0, \ldots)$$

with r in the nth slot, we come to the important revelation that every element in A can be written in the form

$$r_0 + r_1 x + r_2 x^2 + \cdots + r_n x^n \text{ where } r_j \in R.$$

Thus A is generated by x over R.

It remains to see that the powers of x are R-linearly independent. Well, notice that in A the equation

$$r_0 + r_1 x + r_2 x^2 + \cdots + r_n x^n = 0,$$

where $r_j \in R$ and n is a non-negative integer, is the same as the equation

$$(r_0, r_1, r_2, \ldots, r_n, \ldots) = (0, 0, 0, \ldots, 0, \ldots).$$

Thus all $r_j = 0$ in the presence of such an equation, and our polynomial ring over R has been constructed. □

There is little need to remember the synthetic construction of polynomial rings given in Proposition 4.14. The point is that every ring is the ring of coefficients of a polynomial ring. There is also no need to worry that one polynomial ring $R[x]$ is somehow different than another one $R[t]$. Proposition 4.13 makes them all essentially the same, or as we could say, "isomorphic."

In order to liberate the letters x, y, t for other purposes, henceforth we shall prefer (with occasional exceptions) to use capital letters, such as X, Y, Z, to denote our indeterminates.

4.3.2 Properties of the Degree

We remind ourselves that the zero polynomial does not have a degree. Also observe that if a commutative ring R is not an integral domain, then neither is the polynomial ring $R[X]$, because the zero divisors in R persist in the extension $R[X]$. Additional zero divisors in $R[X]$ can also crop up.

The next result looks at the degree of sums and products of polynomials.

Proposition 4.15. *Let f, g be non-zero polynomials in a polynomial ring $R[X]$.*

1. *If f and g have unequal degrees, then $f + g \neq 0$ and $\deg(f + g) = \max\{\deg f, \deg g\}$.*
2. *If $\deg f = \deg g$ and $f + g \neq 0$, then $\deg(f + g) \leq \deg f$.*
3. *If $fg \neq 0$, then $\deg(fg) \leq \deg f + \deg g$.*
4. *If R is an integral domain, then $fg \neq 0$, whereby $R[x]$ is also an integral domain. Furthermore $\deg(fg) = \deg f + \deg g$.*

Proof. Let $f = r_0 + r_1X + \cdots + r_nX^n$ and $g = s_0 + s_1X + \cdots + s_mX^m$ be the standard forms of f and g. Thus $r_n \neq 0, s_m \neq 0$ and $n = \deg f, m = \deg g$.

Regarding item 1, if $n > m$ say, then the standard form of $f + g$ is

$$f + g = (r_0 + s_0) + (r_1 + s_1)X + (r_2 + s_2)X^2 + \cdots + (r_m + s_m)X^m + \cdots + r_nX^n.$$

Clearly $f + g \neq 0$ since $r_n \neq 0$, and just as clearly $\deg(f + g) = n = \max\{\deg f, \deg g\}$.

For item 2 we have $n = m$ and so

$$f + g = (r_0 + s_0) + (r_1 + s_1)X + (r_2 + s_2)X^2 + \cdots + (r_n + s_n)X^n.$$

If $r_n + s_n \neq 0$, then the above is the standard form of $f + g$ and $\deg(f + g) = n = \deg f$. If, on the other hand, $r_n + s_n = 0$, then the standard form of $f + g$ requires that the monomial $(r_n + s_n)X^n = 0X^n$ be dropped and possibly some lower monomials too. Hence in this case $\deg(f + g) < n = \deg f$. In either case $\deg(f + g) \leq \deg f$.

For item 3 multiply f and g in accordance with the operational rules for commutative rings to get:

$$fg = r_0s_0 + (r_0s_1 + r_1s_0)X + \cdots + \left(\sum_{i+j=k} r_is_j\right)X^k + \cdots + r_ns_mX^{n+m}.$$

If $r_ns_m \neq 0$, then the above is the standard form of fg and $\deg(fg) = n + m = \deg f + \deg g$. In case that R might not be an integral domain, we could have $r_ns_m = 0$. Then some zero monomials have to be dropped in the form of fg above, including $0X^{n+m}$, before it becomes standard. Then $\deg(fg) < n + m = \deg f + \deg g$. In any case $\deg(fg) \leq \deg f + \deg g$.

Item 4 is covered within the proof of item 3 above. This is because in an integral domain R the product r_ns_m of the leading coefficients is never 0. □

To illustrate that the assumption $fg \neq 0$ in item 3 of Proposition 4.15 is not redundant, consider the polynomial ring $\mathbb{Z}_4[X]$. With residues in $\mathbb{Z}_4$ denoted by $0, 1, 2, 3$, notice that $2+2X$ is a non-zero polynomial, and yet, after we remember that $2 \cdot 2 = 0$ in $\mathbb{Z}_4$, we get $(2 + 2X)(2 + 2X) = 0$.

Units in Polynomial Rings

Let us enquire about the units of $R[X]$. The units of R remain units in the extension ring $R[X]$. Depending on the ring R, additional units might emerge. For example, consider the polynomial ring $\mathbb{Z}_4[X]$. Denote the residues of $\mathbb{Z}_4$ by $0, 1, 2, 3$. Since $2 + 2 = 2 \cdot 2 = 0$ in $\mathbb{Z}_4$ we notice that $(1 + 2X)(1 + 2X) = 1$. Hence $1 + 2X$ is a unit in $\mathbb{Z}_4[X]$ that is not one of the two original units $1, 3$ in the subring $\mathbb{Z}_4$ of constant polynomials.

In the case of integral domains such extra units do not crop up.

Proposition 4.16. *If R is an integral domain, then the units of $R[X]$ are the constant polynomials that are already units in R.*

Proof. If f is a unit of $R[X]$ with inverse g, then part 4 of Proposition 4.15 along with $fg = 1$ implies that $0 = \deg 1 = \deg f + \deg g$. Thus $\deg f = \deg g = 0$, which forces f, g to be constant polynomials back in R. Thus f, g are units of R. □

4.3.3 Polynomials in Several Variables

In this discussion the only sane way to present things is by means of sigma-notations.

Suppose that $R[X]$ is a polynomial ring over a commutative ring R. Using another indeterminate, say Y, we can form the polynomial ring $R[X][Y]$. A polynomial f in $R[X][Y]$ takes the form $f = \sum_j f_j Y^j$, where the sum is taken over a finite number of indices j until the degree of f is attained, and the $f_j \in R[X]$. Being in $R[X]$, each $f_j = \sum_i a_{ij} X^i$ where the sum is taken over finitely many i until the degree of f_j is reached. Thus the polynomials of $R[X][Y]$ take the form

$$f = \sum_j \left(\sum_i a_{ij} X^i \right) Y^j = \sum_{i,j} a_{ij} X^i Y^j,$$

where the pairs of indices (i,j) range over a finite set. The second equation follows from the laws of operation in the ring $R[X][Y]$. By interchanging the roles of X and Y, it should thus be clear that the ring $R[Y][X]$ coincides with the ring $R[X][Y]$. This common ring is denoted by $R[X, Y]$ and is known as the ***polynomial ring in two variables*** X and Y. It is a routine matter to verify that $\sum_{i,j} a_{ij} X^i Y^j = \sum_{i,j} b_{ij} X^i Y^j$ if and only if $a_{ij} = b_{ij}$ for all pairs of indices (i,j). For this reason the above expansion is known as the ***standard form*** of a polynomial in two variables.

More generally, a ***polynomial in*** n ***variables*** $X_1, X_2, \ldots, X_n$ takes the standard form

$$\sum_{i_1, i_2, \ldots, i_n} a_{i_1 i_2 \cdots i_n} X_1^{i_1} X_2^{i_2} \cdots X_n^{i_n},$$

where the indices i_j are non-negative integers, the n-tuples $(i_1, i_2, \ldots, i_n)$ run over a finite set, and the coefficients $a_{i_1 i_2 \cdots i_n} \in R$. Two standard forms are equal if and only if their coefficients that correspond to each product $X_1^{i_1} X_2^{i_2} \cdots X_n^{i_n}$ are equal. Sums and products are brought to standard form by simply using the laws of a commutative ring.

Things get much more complicated with polynomials in more than one variable. For instance, using the field $\mathbb{R}$ of real numbers we might recall that the polynomial equation $f(X) = 0$ determines a finite number of roots, while the polynomial equation in two variables $f(X, Y) = 0$ typically defines a curve in 2-space, which is something quite a bit more intricate. Chapter 9 is dedicated to the study of polynomials in several variables.

4.3.4 The Ring of Formal Power Series

The more rings we know, the better we can appreciate their peculiarities. Here is a family of commutative rings that are cousins to the polynomial rings.

Definition 4.17. For any commutative ring R and indeterminate symbol X, a ***formal power series*** in X with coefficients in R is any expression f of the form

$$f = a_0 + a_1X + a_2X^2 + \cdots + a_kX^k + \cdots = \sum_{k=0}^{\infty} a_kX^k.$$

As with polynomials a_j is the ***coefficient*** of X^j, the a_jX^j are the ***monomials*** of f, and those a_jX^j in which $a_j \neq 0$ are the ***terms*** of f. Two formal series are equal if and only if they look identical. That is, $\sum_{k=0}^{\infty} a_kX^k = \sum_{k=0}^{\infty} b_kX^k$ if and only if $a_j = b_j$ for all j.

Unlike polynomials, power series allow for infinitely many of their coefficients to be non-zero. Curiously, power series are easier to define than polynomials, because the issue of making identifications such as $1 + 2X = 1 + 2X + 0X^2$, prior to having a ring to work with, does not arise. The set of all formal power series is denoted by the double-bracket notation $R[[X]]$. To activate $R[[X]]$ into a commutative ring, we define the ring operations as follows.

- As zero element in $R[[X]]$ take the series $0 = 0 + 0X + 0X^2 + 0X^3 + \cdots$.
- For the identity take $1 = 1 + 0X + 0X^2 + \cdots + 0X^n + \cdots$.
- For two series $f = a_0 + a_1X + a_2X^2 + \cdots,\ g = b_0 + b_1X + b_2X^2 + \cdots$, let their sum be given by

$$f + g = (a_0 + b_0) + (a_1 + b_1)X + (a_2 + b_2)X^2 + \cdots = \sum_{k=0}^{\infty}(a_k + b_k)X^k.$$

- For the product of the above two series take the convolution

$$\begin{aligned} fg &= (a_0b_0) + (a_1b_0 + a_0b_1)X + (a_2b_0 + a_1b_1 + a_0b_2)X^2 + \cdots \\ &= \sum_{k=0}^{\infty}(a_kb_0 + a_{k-1}b_1 + \cdots + a_1b_{k-1} + a_0b_k)X^k = \sum_{k=0}^{\infty}\left(\sum_{i+j=k} a_ib_j\right)X^k. \end{aligned}$$

- For the negative of the series f, take it to be $-f = -a_0 - a_1X - a_2X^2 + \cdots$.

These natural operations make $R[[X]]$ into a commutative ring. We omit the routine verifications of the ring axioms.

By thinking of a polynomial $a_0 + a_1X + a_2X^2 + \cdots + a_nX^n$ as the formal power series $a_0 + a_1X + a_2X^2 + \cdots + a_nX^n + 0X^{n+1} + \cdots + 0X^k + \cdots$, it becomes evident that the ring $R[X]$ of polynomials is a subring of the power series ring $R[[X]]$. Yet oddities can happen with power series that polynomials avoid. For instance, notice the classic geometric series identity in $R[[X]]$:

$$(1 - X)(1 + X + X^2 + X^3 + \cdots) = 1 + 0X + 0X^2 + 0X^3 + \cdots = 1.$$

Thus we find that the polynomial $1 - X$, when viewed as a formal power series in $\mathbb{Q}[[X]]$, becomes a unit. A good exercise, left to the reader, is to determine the units of $R[[X]]$.

EXERCISES

1. Let R be a commutative ring and let $a \in R$. If a is cancellable in R, does a remain cancellable as an element of the polynomial ring $R[X]$?
2. If R is a commutative ring, what are the units of the polynomial ring $R[X]$ and of the polynomial ring in two variables $R[X, Y]$?
3. Find a commutative ring R and polynomials f, g in $R[X]$ such that $\deg(fg) < \deg f + \deg g$.
4. Show that a polynomial ring over any commutative ring R cannot be a field.
5. If R is an integral domain, show that the power series ring $R[[X]]$ is also an integral domain.
6. Find an infinite commutative ring R and a non-zero polynomial f such that $f : R \to R$ is the zero function.
7. If R is a commutative ring show that an element a in R causes $1 - aX$ to be a unit in the polynomial ring $R[X]$ if and only if a is nilpotent in R, i.e. if and only if $a^n = 0$ for some positive exponent n.
8. If R is a commutative ring, show that the units of the power series ring $R[[X]]$ are those series $\sum_{j=0}^{\infty} a_j X^j$ for which a_0 is a unit of R.
9. If K is a field, show that every non-zero formal power series f in $K[[X]]$ can be written in the form $X^n g$, where n is a non-negative integer and g is a unit of $K[[X]]$.
10. Let $\mathbb{Z}_2[X]$ be the polynomial ring in the variable X with coefficients in the finite field $\mathbb{Z}_2$. Let $R = \mathbb{Z}_2[X^2, X^3 + X]$. This is the set of all sums of the form $\sum a_{ij}(X^2)^i(X^3 + X)^j$ where $a_{ij} \in \mathbb{Z}_2$ and i, j are non-negative integers.

    Show that R is a proper subring of $\mathbb{Z}_2[X]$.
11. * The polynomial ring $\mathbb{Z}[X]$ is a subring of $\mathbb{Q}[X]$ in the obvious way. If f in $\mathbb{Z}[X]$ is the square of a polynomial in $\mathbb{Q}[X]$, show that f is already the square of a polynomial in $\mathbb{Z}[X]$.
12. * If R is a commutative ring and some polynomial f in $R[X]$ is a zero divisor, show that $af = 0$ for some polynomial a of degree 0, that is, for some non-zero a in R.
13. * If $f(X, Y)$ is a non-zero polynomial in $\mathbb{Z}[X, Y]$, show that the equation $f(2^n, 3^n) = 0$ has a solution for at most finitely many positive integers n.

    Hint. Write $f(X, Y) = \lambda X^a Y^b + \sum \lambda_j X^{a_j} Y^{b_j}$, where λ, λ_j are non-zero integers, a, b, a_j, b_j are non-negative exponents, and $2^a 3^b > 2^{a_j} 3^{b_j}$ for all j.

4.4 Homomorphisms and Ideals

Every trade needs its tools. For rings (as it was with groups) and the problems that fall within their scope, none is more fundamental than the notion of ring homomorphism. In the early part of this section, the rings involved are allowed to be non-commutative, but our focus will soon enough drift towards commutative rings.

4.4.1 Ring Homomorphisms

Definition 4.18. A ***ring homomorphism*** from a ring R to a ring A is a mapping $\varphi : R \to A$ such that

$$\varphi(x+y) = \varphi(x) + \varphi(y)$$
$$\varphi(xy) = \varphi(x)\varphi(y)$$
$$\varphi(1) = 1$$

for every x, y in R.

A ring homomorphism φ is, in particular, a homomorphism between the underlying groups R and A using the addition operations. Thus, as with all group homomorphisms,

$$\varphi(0) = 0,\ \varphi(-x) = -\varphi(x) \text{ and } \varphi(x-y) = \varphi(x) - \varphi(y) \text{ for all } x, y \text{ in } R.$$

While $\varphi(0) = 0$ automatically, the requirement that $\varphi(1) = 1$ does not come for free from the fact φ also preserves the multiplication. For instance, take the ring $\mathbb{Z} \times \mathbb{Z}$ where addition and multiplication are done coordinate by coordinate. The mapping $\psi : \mathbb{Z} \times \mathbb{Z} \to \mathbb{Z} \times \mathbb{Z}$ where $\psi(x, y) = (x, 0)$ preserves the addition and multiplication of $\mathbb{Z} \times \mathbb{Z}$ but not the identity element, which is $(1, 1)$ in this case. In keeping with our requirement that rings have a 1, as well as their subrings, consistency demands that homomorphisms map the identity of the domain ring to the identity of the target ring. Because of this requirement the above ψ is not a homomorphism. For an even simpler example, take the zero map that sends every x in a ring R to 0 in a ring A. This clearly respects the addition and multiplication but, unless A is the zero ring, the zero map will not send 1 to 1.

The fact that $\varphi(1) = 1$ brings a small but useful consequence.

Proposition 4.19. *If $\varphi : R \to S$ is a ring homomorphism and x is a unit of R with inverse y, then $\varphi(x)$ is a unit of S with inverse $\varphi(y)$.*

Proof. Apply φ to the equation $xy = 1$ to get $1 = \varphi(1) = \varphi(xy) = \varphi(x)\varphi(y)$. Similarly for the equation $yx = 1$. □

If $R^\star, S^\star$ denote the groups of units in R, S, respectively, we have just observed that the restriction of φ to $R^\star$ is a group homomorphism into $S^\star$.

Definition 4.20. If $\varphi : R \to A$ is a ring homomorphism, the ***image of R under*** φ is the set of values

$$\{\varphi(R) = \{\varphi(x) : x \in R\} \text{ inside } A\}.$$

The image is a subring of A, by simple verification. This set is sometimes called the *image of the homomorphism* φ.

Here are some examples of homomorphisms.

- If R is a subring of a ring A, then the ***inclusion map*** $\varphi : R \to A$, given by $\varphi(r) = r$ for each r in R, is a homomorphism. Obviously its image inside A is R. If $R = A$, this homomorphism is none other than the ***identity map*** on A.

- The mapping $\mathbb{C} \to \mathbb{C}$ given by $z \mapsto \bar{z}$, where $\bar{z}$ is the complex conjugate of the complex number z, is a well known homomorphism. The image of this homomorphism is all of $\mathbb{C}$.
- For each positive integer n, the mapping $\varphi : \mathbb{Z} \to \mathbb{Z}_n$ given by $\varphi(a) = [a]$, where $[a]$ is the residue of a modulo n, is a homomorphism. The ring operations in $\mathbb{Z}_n$ are expressly designed to make the process of taking residues into a homomorphism. Here the image is all of $\mathbb{Z}_n$.

At times there are no homomorphisms from one ring to another. For instance, it is impossible to have a homomorphism $\varphi : \mathbb{Z}_2 \to \mathbb{Z}$. Indeed, if φ were such a homomorphism, we would get

$$0 = \varphi([0]) = \varphi([1] + [1]) = \varphi([1]) + \varphi([1]) = 1 + 1 = 2,$$

an obvious contradiction.

The Homomorphisms on $\mathbb{Z}$

If A is any ring, what are the homomorphisms $\varphi : \mathbb{Z} \to A$? For the moment let $0_A, 1_A$ denote the zero and the identity of A in order to distinguish them from the zero 0 and the identity 1 of $\mathbb{Z}$. We have seen that $\varphi(0) = 0_A$. The requirement that $\varphi(1) = 1_A$ along with the fact φ respects addition, forces $\varphi(2) = \varphi(1+1) = \varphi(1)+\varphi(1) = 1_A+1_A$. Likewise, for any positive integer n we must have $\varphi(n) = 1_A + 1_A + \cdots + 1_A$ taken n times. Also $\varphi(-1) = -1_A$ as we have seen, and then, for each negative integer n, we must get $\varphi(n) = -1_A - 1_A - \cdots - 1_A$ taken $-n$ times. Thus there can be at most one homomorphism from $\mathbb{Z}$ to any ring A, and it must be given by the ***multiple notation***:

$$\varphi(n) = n1_A = \begin{cases} 1_A + \cdots + 1_A & \text{taken } n \text{ times when } n \text{ is positive} \\ 0_A & \text{for } n = 0 \\ -1_A - \cdots - 1_A & \text{taken } -n \text{ times when } n \text{ is negative.} \end{cases}$$

It is easy to confirm that such φ is a homomorphism, and as we just argued, this is the only ring homomorphism from $\mathbb{Z}$ to a ring A.

Substitution Maps

Polynomial rings provide a truly useful source of homomorphisms.

Definition 4.21. Take a polynomial ring $R[X]$ over a commutative ring R. Let A be any commutative ring extension of R, and pick any element a in A. The ***substitution map***

$$\varphi_a : R[X] \to A$$

is defined as follows. If $f = r_0 + r_1X + r_2X^2 + \cdots + r_nX^n$ is a non-zero polynomial in $R[X]$ written in its unique standard form with $r_n \neq 0$, put

$$\varphi_a(f) = r_0 + r_1a + r_2a^2 + \cdots + r_na^n.$$

Also put $\varphi_a(0) = 0$.

It may seem evident that the substitution map is a homomorphism. However, this is only because we tend to think of polynomials as functions, but once we recall that polynomials are not functions a priori, a proof that the substitution map is a homomorphism might be called for, no matter how pedantic it may seem.

One thing we can observe is that both $R[X]$ and A contain R as a subring and that $\varphi_a(r) = r$ for every r in R. When a mapping between two sets, such as φ_a, leaves the elements of a common subset R unchanged we say that the mapping ***fixes*** the subset R. We also say that R is ***fixed*** by the mapping.

Proposition 4.22. *If A is a commutative ring extension of a ring R and a is an element of A, then the substitution map $\varphi_a : R[X] \to A$ is a ring homomorphism that fixes R. Conversely, every homomorphism $\varphi : R[X] \to A$ that fixes R is a substitution map for some a in A.*

Proof. If a polynomial is written as $f = r_0 + r_1X + \cdots + r_nX^n$ where the coefficient r_n could be 0, in which case f is not exactly in standard form, it remains true that $\varphi_a(f) = r_0 + r_1a + \cdots + r_na^n$, by inspection.

With this observation it is routine to see that φ_a is a homomorphism. Indeed, write two polynomials as $f = \sum_{j=0}^{n} r_jX^j$ and $g = \sum_{j=0}^{n} s_jX^j$. We can ignore standard forms and arrange for them to both end with the same X^n. Then $f + g = \sum_{j=0}^{n}(r_j + s_j)X^j$, and then

$$\varphi_a(f+g) = \sum_{j=0}^{n}(r_j + s_j)a^j = \sum_{j=0}^{n} r_ja^j + \sum_{j=0}^{n} s_ja^j = \varphi_a(f) + \varphi_a(g).$$

Also $fg = \sum_{k=0}^{2n}\left(\sum_{i+j=k} r_is_j\right)X^k$, and using the laws for commutative rings we get

$$\varphi(fg) = \sum_{k=0}^{2n}\left(\sum_{i+j=k} r_is_j\right)a^k = \left(\sum_{j=0}^{n} r_ja^j\right)\left(\sum_{j=0}^{n} s_ja^j\right) = \varphi_a(f)\varphi_a(g).$$

It is quite evident that φ_a fixes R, and in particular that $\varphi_a(1) = 1$.

Regarding the converse, let $\varphi : R[X] \to A$ be a homomorphism that fixes R. For each polynomial $f = r_0 + r_1X + r_2X^2 + \cdots + r_nX^n$, the preceding assumptions give:

$$\begin{aligned}\varphi(f) &= \varphi(r_0) + \varphi(r_1)\varphi(X) + \varphi(r_2)\varphi(X)^2 + \cdots + \varphi(r_n)\varphi(X)^n \\ &= r_0 + r_1\varphi(X) + r_2\varphi(X)^2 + \cdots + r_n\varphi(X)^n.\end{aligned}$$

With $a = \varphi(X)$ we can see that φ is the substitution map φ_a. □

Clearly the image of a substitution map $\varphi_a : R[X] \to A$ is the subring $R[a]$ generated by a over R. However, we should not construe the image $R[a]$ to be another polynomial ring because the powers of a need not be R-linearly independent.

From Proposition 4.22 it follows that no matter how a polynomial is presented, its image under φ_a is obtained by substituting a for X. For instance, in $\mathbb{Z}[X]$ if $f = 3X(X^4-1)(X^2+7)$, then $\varphi_a(f) = 3a(a^4 - 1)(a^2 + 7)$. This is because $\varphi_a(X) = a$ and φ_a is a homomorphism that respects the ring operations. It might be suggested that we should define the substitution map φ_a directly by simply plugging in a for X no matter the appearance of the polynomial f.

The trouble with that idea is that it raises the matter of checking if such a substitution gives the same value for different appearances of the same polynomial f. Thus our seemingly laborious approach was unavoidable. Having made our apology we can also write polynomials f as $f(X)$, and then speak of the very important ***substitution map***:

$$\varphi_a : f(X) \mapsto f(a).$$

When using this map we often speak informally and say things like: "put $X = a$."

For an example, let $T : V \to V$ be a linear transformation on a vector space V over a field K. The set $K[T]$ consisting of all linear operators of the form $r_0 I + r_1 T + r_2 T^2 + \cdots + r_n T^n$, where $r_j \in K$, is a commutative ring using the (possibly) familiar addition and composition of linear operators. This ring contains the field K disguised as the scalar operators rI where $r \in K$. We then have the substitution map $\varphi_T : K[X] \to K[T]$ given by

$$f = \sum_j r_j X^j \mapsto f(T) = \sum_j r_j T^j.$$

The substitution $f(X) \mapsto f(T)$ from polynomials to rings of operators is a significant instrument for studying linear operators, leading to notable results like the Cayley–Hamilton theorem.

Polynomials as Functions

Suppose A is a commutative ring extension of a ring R. For every a in A the substitution map $\varphi_a : R[X] \to A$ is given by $\varphi_a(f) = f(a)$. If we now fix f and let the a vary through A, this formula gives meaning to the concept of a ***polynomial function***. This is the function $A \to A$ defined by $a \mapsto f(a)$. It is common practice to confound the notation for this function with the notation for the polynomial that spawned it, namely f or sometimes $f(X)$. For example, Fermat's theorem that $a^p \equiv a$ modulo a prime p reveals that the distinct polynomials $X^p + X^3 - X$ and $X^3 - X$ in $\mathbb{Z}_p[X]$ give the same function $\mathbb{Z}_p \to \mathbb{Z}_p$. Even though for some rings A, different polynomials can yield the same polynomial function, this notation should cause no harm.

Substitution in Several Variables

Suppose that R is a commutative subring of a commutative ring A and that $a, b \in A$. If $R[X, Y]$ is the polynomial ring in two variables over R, one would expect that there is a unique homomorphism $\psi : R[X, Y] \to A$ which fixes R and such that $\psi(X) = a, \psi(Y) = b$. If $f = \sum_{ij} r_{ij} X^i Y^j$ is a polynomial in $R[X, Y]$, such a homomorphism would have to be defined by

$$\psi(f) = \sum_{ij} a_{ij} a^i b^j.$$

The common notation for $\psi(f)$ is $f(a, b)$, where a replaces X and b replaces Y. One way to see that this formula $f \mapsto f(a, b)$ does define a homomorphism is to observe that ψ is the composite of two substitution maps on polynomial rings in one variable.

The ring $R[X, Y]$ can be viewed as the ring $R[X][Y]$ of polynomials in Y with coefficients in $R[X]$. In a natural way $R[X] \subseteq A[X]$ and $b \in A[X]$. As already seen there is the substitution map $\varphi_b : R[X][Y] \to A[X]$ which maps $f(X, Y)$ to $f(X, b)$. There is also the substitution map $\varphi_a : A[X] \to A$ which maps $g(X)$ to $g(a)$. The composite of these two substitution maps is given by

$$f \mapsto f(X, b) \mapsto f(a, b),$$

which is the desired $\psi(f)$.

Similarly if $R[X_1, X_2, \ldots, X_n]$ is a polynomial ring in several variables and $a_1, a_2, \ldots, a_n$ are elements in a commutative ring A containing R as a subring, there is the unique multiple substitution map $\psi : R[X_1, X_2, \ldots, X_n] \to A$ which sends each X_j to a_j and fixes R.

We shall come to appreciate that, when dealing with polynomial rings, substitution maps are ubiquitous.

4.4.2 The Kernel

Definition 4.23. If $\varphi : R \to A$ is a homomorphism between two rings, the ***kernel*** of φ is the subset of R given by

$$\ker \varphi = \{a \in R : \varphi(a) = 0\}.$$

For instance, the kernel of the homomorphism $\varphi : \mathbb{Z} \to \mathbb{Z}_n$ that maps each integer a to its residue $[a]$ modulo n is the set $n\mathbb{Z}$ of integers that are divisible by n. For another example, take the polynomial ring $\mathbb{Q}[X]$ over $\mathbb{Q}$. The kernel of the substitution map $\mathbb{Q}[X] \to \mathbb{Q}$ given by $f(X) \mapsto f(1)$ consists of all polynomials f that vanish at 1. Clearly all polynomials of the form $(X - 1)g$, where $g \in \mathbb{Q}[X]$, belong to the kernel of this map. We shall soon come to see that there are no other polynomials in this kernel.

Every kernel comes with three fundamental properties.

- $0 \in \ker \varphi$, since $\varphi(0) = 0$.
- If $a, b \in \ker \varphi$, then $a + b \in \ker \varphi$, as has been already checked for abelian groups.
- If $a \in \ker \varphi$ and $r \in R$, then $ra \in \ker \varphi$ and $ar \in \ker \varphi$. To see this, observe that $\varphi(ra) = \varphi(r)\varphi(a) = \varphi(r)0 = 0$, and likewise for $\varphi(ar)$.

Note that if $1 \in \ker \varphi$, then the homomorphism requirement $\varphi(1) = 1$ leads to the conclusion $1 = 0$ in A, forcing A to be the zero ring and φ to be the zero map that sends all of R to 0. Except for this rarity, $1 \notin \ker \varphi$. Thus $\ker \varphi$ is not usually a subring of R, quite unlike the image $\varphi(R)$, which is a subring of A.

When $\ker \varphi$ happens to contain only 0, we say that $\ker \varphi$ is ***trivial***.

Proposition 4.24. *A ring homomorphism $\varphi : R \to A$ is injective if and only if $\ker \varphi$ is trivial.*

Proof. This is a consequence of Proposition 2.53 applied to the underlying abelian groups R and A using addition. □

Homomorphisms which are injective are sometimes called ***embeddings***. As just observed, embeddings are detected by inspecting their kernel.

Isomorphisms

Definition 4.25. If a homomorphism $\varphi : R \to A$ is both injective and surjective, we call it an ***isomorphism*** of rings. When two rings R, A have an isomorphism between them, we write $R \cong A$ and call them ***isomorphic***.

At times isomorphic rings R and A can be treated as if they were identical. For example, suppose that $R[X]$ and $R[Y]$ are polynomial rings over R, possibly obtained by entirely different methods. Regardless, there is the isomorphism $\varphi : R[X] \to R[Y]$ that maps a polynomial f in $R[X]$ with standard representation $r_0 + r_1X + r_2X^2 + \cdots + r_nX^n$ to the polynomial $\varphi(f)$ in $R[Y]$ with standard representation $r_0 + r_1Y + r_2Y^2 + \cdots + r_nY^n$. This is what it means to say that all polynomial rings over R are essentially the same.

If $R = A$, whereby the homomorphism is from a ring back into itself, then φ is called an ***endomorphism***, and if, in addition, φ is injective and surjective, then φ is called an ***automorphism***. For example, the conjugation map $z \to \bar{z}$ on $\mathbb{C}$ is an automorphism.

The Sticky Business of Identification

If we have an embedding $\varphi : R \to A$ between two rings, then the image $\varphi(R)$ is a subring of A that is isomorphic to R. It is thereby *extremely convenient* to think of R and $\varphi(R)$ as being identical, and to say that R itself becomes a subring of A via the embedding φ. Although technically improper, ring identifications via embeddings will avoid great amounts of needless writing.

For instance, we all agree that $\mathbb{Z}$ is a subring of the field of fractions $\mathbb{Q}$. However, technically $\mathbb{Z}$ is only embedded into $\mathbb{Q}$ by the homomorphism that maps each integer a into the fraction $a/1$. Yet we all can appreciate the waste of time in such hair-splitting. Another example is the embedding $\mathbb{R} \to M_n(\mathbb{R})$ that maps each real number r to the scalar matrix rI, where I is the identity matrix. Again, technically the number r is not the same as the matrix rI, but after identification it makes sense to say that $\mathbb{R}$ lives inside $M_n(\mathbb{R})$ as the subring of scalar matrices. Then there is the example of the ring $C(\mathbb{R})$ of continuous functions $f : \mathbb{R} \to \mathbb{R}$ under the usual operations of point-wise addition and multiplication. The embedding of $\mathbb{R}$ into $C(\mathbb{R})$ sends every real number c to the constant function taking constant value c. Technically, constant functions are not the same as real numbers, but it makes good sense to say that $\mathbb{R}$ is identified with the subring of $C(\mathbb{R})$ consisting of the constant functions.

We should seize opportunities to make such identifications, for the sake of removing clutter and making room to concentrate on more meaningful problems.

4.4.3 Ideals

In 1876 Richard Dedekind (1831–1916) put forth the notion of an ideal in his famous book on number theory, wherein he rescued the failure of unique factorization in some rings of algebraic numbers by replacing the notion of a ring element by something more "idealized." Since its inception the concept has more than proven its worth.

Definition 4.26. A subset J of a ring R is called an ***ideal*** of R provided

- $0 \in J$,
- if $x, y \in J$, then $x + y \in J$, and
- if $x \in J$, $y \in R$, then $yx \in J$ and $xy \in J$.

Clearly the set containing only 0 is an ideal of R, known as the ***zero or trivial ideal***. The entire ring R is an ideal as well. As noted prior to Proposition 4.24, the kernel of every ring homomorphism $\varphi : R \to A$ is an ideal of R.

It might help to call the third property of an ideal its ***absorption property***. If a is in an ideal J, the absorption property tells us that since $-a = (-1)a$, then $-a \in J$. Thus an ideal is an abelian subgroup of R (using the addition within R) with the absorption property.

In case the ring R is commutative, the absorption property does not need to be stated as a two-sided condition, but for non-commutative rings, both conditions need to hold in order have an ideal. For example, take the non-commutative ring $M_2(\mathbb{Q})$. Let $\mathcal{J}$ be the set of matrices of the type $\begin{bmatrix} a & 0 \\ b & 0 \end{bmatrix}$ where $a, b \in \mathbb{Q}$. It is easy to verify that if $A \in \mathcal{J}$ then $BA \in \mathcal{J}$ for all B in $M_2(\mathbb{Q})$. Clearly $\mathcal{J}$ is closed under addition and the matrix 0 lies in $\mathcal{J}$. Thus $\mathcal{J}$ is almost an ideal, except for the fact that $AB \notin \mathcal{J}$ for some A in $\mathcal{J}$ and some B in $M_2(\mathbb{Q})$. In the setting of non-commutative rings, such structures as these are called ***left ideals***.

Although we shall primarily be concerned with commutative rings, here is an example of an ideal in a non-commutative ring. For any field K, let R be the subring of $M_n(K)$ consisting of all upper triangular matrices. When $n \geq 2$ the ring R is non-commutative. Let $\mathcal{J}$ be the set of strictly upper triangular matrices, which have only 0 on the main diagonal. Such $\mathcal{J}$ is an ideal of R, but we omit the routine checking.

Proper Ideals Are Not Subrings

Except for R itself, ideals are not subrings of R.[4] Indeed, if an ideal J is a subring of R, then $1 \in J$, and after that, since every element a in R can be written as $a1$, the absorption property forces a to be in J. The ideals J, which are not all of R, are called the ***proper*** ideals of R.

Here is a little test for an ideal J to be all of R.

Proposition 4.27. *An ideal in a ring is equal to the entire ring if and only if the ideal contains a unit of the ring.*

Proof. Let J be an ideal of a ring R. If $J = R$, then J contains the unit 1. Conversely if J contains a unit u, then the absorption property shows that J contains $au^{-1}u$ for any a in R. Since $a = au^{-1}u$, the ideal J contains all a in R. □

[4] Mathematicians who allow rings to not have an identity element might disagree with this claim.

A commutative ring R is a field if and only if 0 is its only non-unit. Thus in a field all ideals except for the zero ideal contain units and thereby must be all of R.

Proposition 4.28. *A commutative ring R is a field if and only if its only ideals are the zero ideal and the full ring R.*

Finitely Generated and Principal Ideals

If J is an ideal in a ring R, and $a_1, \ldots, a_n \in J$, while $x_1, \ldots, x_n \in R$, the properties of J, applied repeatedly, show that the sum $x_1a_1 + \cdots + x_na_n \in J$. Such a sum will be called an ***R-linear combination*** of $a_1, \ldots, a_n$. Ideals are closed under the formation of R-linear combinations of its elements. Furthermore, if the ring R is commutative, R-linear combinations make ideals. Pick any elements $a_1, \ldots, a_n$ from a commutative ring R. An inspection reveals that the set

$$\{x_1a_1 + \cdots + x_na_n : x_1, \ldots, x_n \in R\},$$

consisting of all R-linear combinations of $a_1, \ldots, a_n$, is an ideal of R. The elements $a_1, \ldots, a_n$ are called ***generators***, and sometimes a ***basis***, of the ideal. The ideal in a commutative ring R generated by elements $a_1, \ldots, a_n$ is denoted by the wedge bracket notation:

$$\langle a_1, a_2, \ldots, a_n \rangle,$$

and sometimes the more informative notation

$$Ra_1 + Ra_2 + \cdots + Ra_n.$$

Definition 4.29. An ideal J that can be generated by a finite number of its elements is called a finitely generated ideal. If $J = \langle a \rangle$, generated by just one element, the ideal is called a principal ideal. In a principal ideal all elements are multiples of a single element.

Non-uniqueness of Ideal Generators

Different sets can generate the same ideal in a commutative ring R. If $a_1, \ldots, a_n \in R$ and $b_1, \ldots, b_m \in R$, the ideals $\langle a_1, \ldots, a_n \rangle$ and $\langle b_1, \ldots, b_m \rangle$ coincide if and only if every a_i is an R-linear combination of the b_j and every b_j is an R-linear combination of the a_i. In $\mathbb{Z}$ for example, $\langle 6, 15 \rangle = \langle 3 \rangle$. Indeed, both 6 and 15 are multiples of 3, while 3 is a $\mathbb{Z}$-linear combination of 6 and 15. Namely, $3 = -2 \cdot 6 + 1 \cdot 15$.

Given two lists of elements $a_1, \ldots, a_n$ and $b_1, \ldots, b_m$ in a commutative ring R, it is not so easy to decide in practice whether or not $\langle a_1, \ldots, a_n \rangle = \langle b_1, \ldots, b_m \rangle$. This is an issue we examine in Chapter 9. For now here is a small observation regarding principal ideals.

Proposition 4.30. *Two elements a, b in an integral domain R generate the same principal ideal if and only if $b = ua$ for some unit u.*

Proof. If $b = ua$ for some unit u, then $a = u^{-1}b$. The first equation shows that multiples of b are multiples of a, while the second equation shows that multiples of a are multiples of b. Thus $\langle a \rangle = \langle b \rangle$.

Conversely, if these ideals are equal, then $b = ua$ and $a = vb$ for some u, v in R. Thus $b = u(vb) = (uv)b$. If $b = 0$, then so is $a = 0$, whereby $b = 1a$. If $b \neq 0$ cancel b to get $uv = 1$, which verifies that u is a unit with inverse v. □

A Non-principal Ideal

Some commutative rings have ideals that are not principal. For example, the ideal $\langle 2, X \rangle$ in the polynomial ring $\mathbb{Z}[X]$, presented here with two generators, is not principal. Suppose to the contrary that $\langle 2, X \rangle = \langle f \rangle$ for a single polynomial f in $\mathbb{Z}[X]$. Then

$$2 = fg,\ X = fh,\ f = 2k + X\ell,\ \text{for some polynomials } g, h, k, \ell \text{ in } \mathbb{Z}[X].$$

The first equation above implies that $\deg f = \deg g = 0$, meaning that $f, g \in \mathbb{Z}$. Thus $f = \pm 1$ or $f = \pm 2$. If $f = \pm 2$, the second equation above yields $X = \pm 2h$. This is impossible since the leading coefficient of X is 1, not an even number. The alternative is $f = \pm 1$. Then the third equation gives $\pm 1 = 2k + X\ell$. The substitution map, by which $X \mapsto 0$, applied to this equation yields $\pm 1 = 2k + 0\ell$, an impossibility since ± 1 are odd integers. Thus $\langle 2, X \rangle$ is not a principal ideal in $\mathbb{Z}[X]$.

A Non-finitely Generated Ideal

For an example of a commutative ring with an ideal that is not finitely generated, let R be the ring of all functions $f : \mathbb{N} \to \mathbb{Z}$ with the operations of point-wise addition and multiplication. The set

$$J = \{f \in R : f(n) = 0 \text{ for all but finitely many } n \text{ in } \mathbb{N}\}$$

is an ideal, by inspection. Let $f_1, \ldots, f_k$ be any finite list of functions in J. Since each $f_j(n) = 0$ for all n beyond some n_j, there is an m in $\mathbb{N}$ beyond all n_j at which all f_j vanish. Then every R-linear combination of $f_1, \ldots, f_k$ vanishes at this m. Clearly there exist functions in J that do not vanish at this m. Such functions are not R-linear combinations of the f_j. Thus, no finite list $f_1, \ldots, f_k$ can generate J.

EXERCISES

1. If $\varphi : R \to A$ is a ring homomorphism and u is a unit of R, show that $\varphi(u)$ is a unit of A with inverse $\varphi(u^{-1})$.
2. If K is a field, show that every homomorphism from K to a non-zero ring A is injective. This little result is worth remembering.
3. (a) Show that there are no homomorphisms between $\mathbb{Z}_2$ and $\mathbb{Z}_3$, in either direction.
   (b) Find all homomorphisms $\mathbb{Z}_6 \to \mathbb{Z}_3$ and $\mathbb{Z}_3 \to \mathbb{Z}_6$.

(c) Describe all pairs of positive integers (m, n) for which there is a ring homomorphism $\mathbb{Z}_m \to \mathbb{Z}_n$.

4. Let

$$\mathbb{Q}\left[\sqrt{2}\right] = \{a + b\sqrt{2} : a, b \in \mathbb{Q}\} \text{ and } \mathbb{Q}\left[\sqrt{3}\right] = \{a + b\sqrt{3} : a, b \in \mathbb{Q}\}.$$

Verify that these sets are subrings of $\mathbb{R}$, and that they are fields. Then show there is no ring homomorphism between them.

5. Briefly explain why no two of the rings

$$\mathbb{Z}, \mathbb{Q}, \mathbb{R}, \mathbb{C}, \mathbb{Z}_n, \mathbb{Z}[i], \mathbb{Q}[\sqrt{2}]$$

are isomorphic. Note. Rings can fail to be isomorphic for a variety of reasons.

6. Show that the only homomorphisms $\mathbb{Z} \to \mathbb{Z}$ and $\mathbb{Q} \to \mathbb{Q}$ are the respective identity maps.

7. For this exercise some dexterity with matrix operations will help.

(a) If K is a field, show that the ring $M_n(K)$ of $n \times n$ matrices over K has only two ideals.

(b) If A is a ring (other than the zero ring), prove that every homomorphism $\varphi : M_n(K) \to A$ is an embedding.

8. Let R be a commutative ring. An element e inside R is called ***idempotent*** provided $e^2 = e$.

(a) Find the idempotents of the product rings $\mathbb{Z} \times \mathbb{Z}$ and $\mathbb{Z}_4 \times \mathbb{Z}_6$.

(b) Let e be an idempotent in a commutative ring R. If $e \neq 1$, show that 1 is not in the ideal $Re = \{re : r \in R\}$, and thereby that Re is not a subring of R. Yet, verify that Re is a ring in its own right.

(c) If e is an idempotent of a commutative ring R, verify that $1 - e$ is an idempotent of R, and then show that $R \cong Re \times R(1 - e)$.

9. Find all ring homomorphisms $\varphi : \mathbb{R} \to \mathbb{R}$.

Hint. First show that a homomorphism leaves every integer and then every rational number unchanged. Then show that non-negative real numbers go to non-negative reals, using the fact these are the squares of other real numbers. Then show that each homomorphism is increasing on $\mathbb{R}$. Then decide what the increasing functions on R that fix $\mathbb{Q}$ have to be.

10. Show that every non-zero ideal in the ring $\mathbb{Z}[i]$ of Gaussian integers must contain a positive integer.

11. If $K[X, Y]$ is the ring of polynomials in two variables over a field K, show that the ideal $\langle X, Y \rangle$ is not principal.

12. Let I, J be ideals in a commutative ring R.

(a) Show that the set of all sums $a + b$ where $a \in I, b \in J$ is an ideal of R, and that this ideal contains both I and J. This ideal is usually denoted by $I + J$ and is called the ***sum ideal*** of I and J.

(b) Show that the set of all finite sums $\sum_{ij} a_i b_j$ where $a_i \in I$ and $b_j \in J$ is an ideal of R, and that this ideal is contained in $I \cap J$. This ideal is denoted by IJ and is called the ***product ideal*** of I and J.

(c) Show by an example that the set of all products ab where $a \in I, b \in J$ need not be an ideal of R.

13. If $\varphi : R \to T$ is a homomorphism between integral domains R and T, and I, J are ideals of R, show that $\varphi(I \cap J) \subseteq \varphi(I) \cap \varphi(J)$. Find an example for which the above inclusion is proper.

4.5 Ideals in $\mathbb{Z}$ and in Polynomial Rings

Regarding the ring of integers we might note that its ideals coincide with its subgroups under addition. Indeed, if J is a subgroup of Z with x in J and n in $\mathbb{Z}$, then the product nx is the sum of finitely many copies of x or its negative. Thereby $nx \in J$, which gives J the absorption property. The next result along with its proof are nothing but a variation of Proposition 2.40, which stated that all subgroups of cyclic groups are cyclic.

Proposition 4.31. *The ideals of $\mathbb{Z}$ are principal.*

Proof. Suppose J is an ideal of $\mathbb{Z}$. If $J = \langle 0 \rangle$, then J is principal, obviously. If $J \neq \langle 0 \rangle$, there must be a non-zero integer in J. Multiply that integer by -1, if need be, to see that a positive integer lies in J. Among the positive integers that lie in J let a be the least one. Clearly $\langle a \rangle \subseteq J$. If b is any integer in J, use Euclidean division to write

$$b = aq + r \text{ where } q, r \in \mathbb{Z} \text{ and } 0 \leq r < a.$$

Since J is an ideal, $r = b - aq \in J$, because r is a $\mathbb{Z}$-linear combination of b and a. From the minimality of a among the positive integers in J deduce that $r = 0$. Hence $b = aq \in \langle a \rangle$, which shows $J = \langle a \rangle$. □

The Greatest Common Divisor of Two Integers and the Ideals of $\mathbb{Z}$

Suppose that a non-zero ideal J of $\mathbb{Z}$ is presented using two generators as $J = \langle a, b \rangle$. Since this ideal is principal we have that $\langle a, b \rangle = \langle c \rangle$ for some other integer c. By changing the sign of c if need be, we can take $c > 0$. Thus

$$a = cs, b = ct \text{ and } c = ax + by \text{ for some integers } s, t, x, y.$$

If d is another integer that divides both a and b, then surely $d|c$. We see that our single generator c of the ideal $\langle a, b \rangle$ is the greatest common divisor of a and b, which was discussed just after Proposition 1.1. In particular, recall that a and b are coprime when their greatest common divisor is 1. In the language of ideals this amounts to saying that $\langle a, b \rangle = \langle 1 \rangle = \mathbb{Z}$.

Euclidean Division with Polynomials

With an approach that imitates the proof for $\mathbb{Z}$, let us show that if K is a field, then all ideals of a polynomial ring $K[X]$ are principal. In the case of $\mathbb{Z}$, Euclidean division was key to the proof. In $K[X]$ an analogous (and probably well known) Euclidean division is available. We can even do a bit better, and present our Euclidean division in a way that encompasses some polynomials over commutative rings.

Proposition 4.32 (Euclidean division with polynomials). *If R is a commutative ring and g is a non-zero polynomial in $R[X]$ such that its leading coefficient is a unit of R, then for every f in $R[X]$ there exist* unique *polynomials q, r in $R[X]$ so that*

$$f = gq + r \text{ and } \deg r < \deg g \text{ or } r = 0.$$

In particular, if R is a field, then the above holds for all non-zero g.

Proof. If $f = gq$ for some polynomial q in $R[X]$, we are done by using $r = 0$. Otherwise, let q in $R[X]$ be chosen such that the degree of $f - gq$ is as low as possible. By putting $r = f - gq$ we obtain $f = gq + r$, but it remains to see that $\deg r < \deg g$.

Well, if $\deg r \geq \deg g$, let $n = \deg r - \deg g$. Let a be the leading coefficient of g (which is assumed to be a unit) and b the leading coefficient of r. The polynomial $h = r - ba^{-1}X^n g$ has degree less than $\deg r$. But since

$$h = f - gq - ba^{-1}X^n g = f - g\left(q - ba^{-1}X^n\right),$$

this h is a polynomial whose degree is lower than the minimal $\deg(f - gq)$. Thus $\deg r < \deg g$.

Regarding the uniqueness, suppose q_1, r_1 are also in $R[X]$ and such that

$$f = gq_1 + r_1 \text{ and } \deg r_1 < \deg g \text{ or } r_1 = 0.$$

Then $gq + r = gq_1 + r_1$ which gives $g(q - q_1) = r_1 - r$. If $q \neq q_1$, then

$$\deg(r_1 - r) = \deg(g(q - q_1)) = \deg g + \deg(q - q_1).$$

This is because the leading coefficient of g is a unit. So, if b is the leading coefficient of $q - q_1$ and a is the leading coefficient of g, then the leading coefficient of $g(q - q_1)$ remains the non-zero ab. A contradiction results because $\deg g + \deg(q - q_1) \geq \deg g > \deg(r_1 - r)$. Hence $q = q_1$ and $r = r_1$.

The final statement about Euclidean division when R is a field is obvious, because all non-zero elements in a field are units. □

The polynomials q and r appearing in the Euclidean division are known as the ***quotient*** and ***remainder***, respectively, when g is divided into f.

A special class of polynomials g in $R[X]$ to which Euclidean division applies are those which are monic. For example with the monic $g = X^3 - 2$ and the polynomial $f = 5X^5 - 4X^3 + X - 1$ in $\mathbb{Z}[X]$ we have

$$5X^5 - 4X^3 + X - 1 = (X^3 - 2)(5X^2 - 4) + 10X^2 + X - 9.$$

We can leave the actual algorithm which found $q = 5X^2 - 4$ and $r = 10X^2 + X - 9$ to our high school recollections, or else defer it to Chapter 9.

The Ideals of $K[X]$

By imitating the situation with the ring of integers, Proposition 4.32 can be exploited to describe the ideals of $K[X]$.

Proposition 4.33. *In the ring $K[X]$ of polynomials over a field K every ideal is principal. If J is a non-zero ideal of $K[X]$ and g in J is a non-zero polynomial of minimal degree among the non-zero polynomials of J, then g generates J. Any two generators of J are constant multiples of each other.*

Proof. The zero ideal is obviously principal.

Suppose now that J is a non-zero ideal and that g is a polynomial in J of minimal degree among the non-zero polynomials of J. If f is any polynomial in J, Proposition 4.32 applied to g and f, provides a polynomial q such that

$$f = gq \text{ or } \deg(f - gq) < \deg g.$$

Since J is an ideal, $f - gq \in J$, and then the minimality of $\deg g$ among the non-zero polynomials of J rules out the second option above. Thus $f = gq$, which shows that J is principal with generator g.

Next suppose that h is another generator of J. Then $g = hk$ and $h = g\ell$ for some polynomials k, ℓ in $K[X]$. This gives $g = g\ell k$ and then $1 = \ell k$ by cancelling g. Since the only units of $K[X]$ are the non-zero constant polynomials, we see that g, h are constant multiples of each other. □

If g is a generator of a non-zero ideal J in $K[X]$ and a is the leading coefficient of g, then the polynomial $a^{-1}g$ is monic (i.e. with 1 as its leading coefficient) and generates the same ideal J. Clearly there can only be one monic generator of J. It is common practice to select the ***unique monic generator*** of the ideal J.

Roots of Polynomials

Knowing that ideals in $K[X]$ are principal puts us in a position to demonstrate something familiar but important about the number of roots of a polynomial with coefficients in a field.

Definition 4.34. Given a commutative ring extension A of a ring R, and a polynomial f in $R[X]$, a ***root*** of f is any element a in A such that $f(a) = 0$.

The first thing we need is the ***factor theorem***. Given a polynomial ring $R[X]$, we say that a polynomial g ***divides*** a polynomial f in $R[X]$ provided $f = gh$ for some polynomial h in $R[X]$. Another way to put it is that g divides f if and only if f lies in the principal ideal $\langle g \rangle$.

Proposition 4.35 (Factor theorem). *Let K be a field, f a non-zero polynomial in $K[X]$, and let a be an element of K. The polynomial $X - a$ divides f in $K[X]$ if and only if a is a root of f.*

Proof. Let $\varphi_a : K[X] \to K$ be the substitution map given by $h \mapsto h(a)$. Since the polynomial $X - a \in \ker \varphi_a$, the ideal $\ker \varphi_a$ is not the zero ideal. By Proposition 4.33 this kernel is a principal ideal generated by a polynomial of minimal degree inside it. Well, $X - a$ is such a polynomial, and thus $\ker \varphi_a = \langle X - a \rangle$.

Clearly a is a root of f if and only if $f \in \ker \varphi_a$, which holds if and only if $f \in \langle X - a \rangle$. That is, if and only if $X - a$ divides f. □

Proposition 4.36. *If K is a field, then the number of roots of a non-zero polynomial in $K[X]$ is at most the degree of the polynomial.*

Proof. Looking for a proof by induction, note that polynomials of degree 0 have 0 roots in K. Suppose that the result holds for all polynomials of degree less than n, and take f in $K[X]$ of degree n. If f has no root in K, then surely $0 \leq n$. On the other hand if f has a root a in K, the factor theorem, Proposition 4.35, gives

$$f = (X - a)g \text{ for some } g \text{ in } K[X].$$

Clearly $\deg g = n - 1$, which by assumption can have at most $n - 1$ roots. Now if b in K is any root of f, substitute $X = b$ to get that

$$0 = f(b) = (b - a)g(b).$$

Evidently either $b = a$ or b is a root of g. So, f has at most n roots in K. □

For commutative rings R which are not integral domains, the number of roots of a polynomial in $R[X]$ can exceed its degree. In such R there exist non-zero a, b such that $ab = 0$. Then the polynomial aX in $R[X]$ has degree 1 and yet comes with at least two roots: 0 and b.

4.5.1 Finite Groups of Units in a Field

While it remains fresh on our minds, it might be worthwhile to exploit the preceding degree bound to say something surprising about finite groups of units in a field.

Proposition 4.37. *If G is a finite subgroup of the group $F^\star$ of units of a field F, then G is a cyclic group.*

Proof. Such an abelian group G is isomorphic to a product of p-groups for various distinct primes p, according to the primary decomposition theorem, Proposition 2.90. If each of these p-groups is cyclic, then so is G cyclic, by Proposition 2.66. So we may as well suppose G is a p-group where p is a prime. The order of every element of G is a power of p. Let p^k be the maximum order that occurs among the elements of G. Since every element a in G satisfies $a^{p^j} = 1$ for some $j \leq k$, it follows that $a^{p^k} = (a^{p^j})^{p^{k-j}} = 1^{p^{k-j}} = 1$.

Thus every a in G is a root of the polynomial $X^{p^k} - 1$ in $F[X]$. By Proposition 4.36, the order of G is at most p^k. But G contains an element of order p^k. That element has to be a generator of G. □

By way of illustration, take the set G of all complex numbers that are nth roots of unity, i.e. $G = \{z : z^n = 1\}$. This is also the set of roots of the polynomial $X^n - 1$ in $\mathbb{C}[X]$. Under multiplication G is a group of order n, and G is cyclic with generator $\zeta = e^{2\pi i/n}$. A generator for this group is called a ***primitive nth root of unity***.

A noteworthy special case emerges from Proposition 4.37.

Proposition 4.38. *The group of non-zero elements of every finite field is a cyclic group.*

For example, for every prime p, the group of non-zero elements $\mathbb{Z}_p^\star$ inside the finite field $\mathbb{Z}_p$ is cyclic. In the language of congruences modulo p this result says that there exists an integer a such that the powers

$$a, a^2, a^3, \ldots, a^{p-2}, a^{p-1},$$

after reduction mod p, constitute a permutation of the integers $1, 2, \ldots, p-1$. This interesting and unexpected result has found uses in current techniques of cryptography.

In light of Proposition 2.57 another corollary emerges.

Proposition 4.39. *The automorphism group of a group of prime order q is a cyclic group of order $q-1$.*

Proof. By Proposition 2.57 such an automorphism group is isomorphic to the group of units $\mathbb{Z}_q^\star$ of the finite field $\mathbb{Z}_q$, and we just proved that such groups of units are cyclic. □

The above result was used in Section 3.6.3 to provide the complete classification of groups of order pq, where p, q are distinct primes. We might glean from this that algebra (and much of mathematics for that matter) cannot truly be compartmentalized into subdisciplines.

EXERCISES

1. Let K be an infinite field and suppose that a polynomial f in $K[X]$ is such that $f(a) = 0$ for all a in K. In other words f gives the zero function on K. Prove that f is the zero polynomial. If f, g are polynomials which agree as functions on K, i.e. f, g are such that $f(a) = g(a)$ for all a in K, show that $f = g$ as polynomials in $K[X]$.

   Show that the above claims do not hold when K is a finite field.
2. If $\varphi : R \to A$ is a homomorphism between two commutative rings, verify that the mapping $\psi : R[X] \to A[X]$ given by

   $$r_0 + r_1X + r_2X^2 + \cdots + r_nX^n \mapsto \varphi(r_0) + \varphi(r_1)X + \varphi(r_2)X^2 + \cdots + \varphi(r_n)X^n$$

   is a homomorphism between their polynomial rings.

   In particular, for each positive integer m, the mapping $\varphi : \mathbb{Z}[X] \to \mathbb{Z}_m[X]$ defined by

   $$a_0 + a_1X + a_2X^2 + \cdots + a_nX^n \mapsto [a_0] + [a_1]X + [a_2]X^2 + \cdots + [a_n]X^n$$

   is a homomorphism. This mapping is known as the process of ***reducing the coefficients modulo*** m. If $f \in \mathbb{Z}[X]$, the polynomial $\varphi(f)$ is also known as the ***reduction of f modulo*** m.
3. Find a polynomial f with integer coefficients, having no root in $\mathbb{Q}$, but such that for every prime p its reduction modulo p does have a root in $\mathbb{Z}_p$.

Hint. For $p > 3$ an exercise from Section 2.6 involving quadratic residues can help get one. If p is 2 or 3 find a such a polynomial by trying a few possibilities. Multiply these polynomials to get one that works for all p.

4. Let $g = X^2 + X + 1$ and $f = X^4 + 2X^3 + 5X^2 + X + 3$ viewed as polynomials in $\mathbb{Z}_n[X]$ where n is a positive integer. Find all n such that g divides f in $\mathbb{Z}_n[X]$.

5. (a) If $\varphi : A \to B$ is a surjective homomorphism between commutative rings and all ideals of A are principal, prove that all ideals of B are principal.

(b) Prove that in any homomorphic image of a polynomial ring $K[X]$, where K is a field, all ideals are principal.

6. Find generators for the kernel of each of the following substitution maps:

(a) $\mathbb{Z}[X] \to \mathbb{C}$ where $f(X) \mapsto f(\sqrt[3]{2})$,

(b) $R[X] \to \mathbb{C}$ where $f(X) \mapsto f(1 + 2i)$,

(c) $\mathbb{Z}[X] \to \mathbb{R}$ where $f(X) \mapsto f(\sqrt{3} + 1)$,

(d) $\mathbb{Z}[X] \to \mathbb{C}$ where $f(X) \mapsto f(\sqrt{2} + \sqrt{3})$.

7. If R is a commutative ring with polynomial ring $R[X]$, describe all possible homomorphisms $\varphi : R[X] \to R[X]$ that fix R. Which such homomorphisms are injective? Which are surjective? Which are automorphisms?

8. Take the polynomial ring $\mathbb{Z}[X]$ and form the product ring $\mathbb{Z}[X] \times \mathbb{Z}[X]$. Let

$$A = \{(f, a) : f \in \mathbb{Z}[X], a \in Z, f(0) = a\}.$$

Verify that A is a subring of $\mathbb{Z}[X] \times \mathbb{Z}[X]$. Show that A is isomorphic to $\mathbb{Z}[X]$ and thereby an integral domain, while some elements from A are zero divisors in $\mathbb{Z}[X] \times \mathbb{Z}[X]$.

9. This multi-step exercise explores some of the interplay between a polynomial and the function that arises from it. It will help to be familiar with the notion of a vector space and its dimension.

We start with a field K. Every polynomial f in $K[X]$ gives a function from K to K according to $a \mapsto f(a)$ for each a in K.

Take any list $a_0, a_1, a_2, \ldots, a_n$ of $n + 1$ distinct elements from K. Say $n \geq 1$, and then examine closely the following polynomials in $K[X]$:

$$\begin{aligned}
p_0 &= (X - a_1)(X - a_2)(X - a_3)\ldots(X - a_n)\\
p_1 &= (X - a_0)(X - a_2)(X - a_3)\ldots(X - a_n)\\
p_2 &= (X - a_0)(X - a_1)(X - a_3)\ldots(X - a_n)\\
&\vdots\\
p_n &= (X - a_0)(X - a_1)(X - a_2)\ldots(X - a_{n-1}).
\end{aligned}$$

Here, each p_k is the product of all the linear polynomials $X - a_j$ but omitting $X - a_k$. These p_k are monic polynomials of degree n.

(a) Explain briefly why $p_k(a_k) \neq 0$ but $p_k(a_j) = 0$ when $j \neq k$.

(b) The set V of polynomials in $K[X]$ of degree at most n (along with the zero polynomial) is a vector space of dimension $n+1$ since the monomials $1, X, X^2, \ldots, X^n$ form a basis of V. Using part (a) show that the p_k are linearly independent and thereby form a basis of V.

(c) From part (b) we have that every polynomial in $K[X]$ of degree at most n is a linear combination of the p_k. Use this fact to show that if f is a polynomial in V, i.e. if $\deg f \le n$, and if f as a function vanishes at all a_j, then f is the zero polynomial. This proves in a somewhat non-standard way that, for a polynomial over a field K, the number of a in K such that $f(a) = 0$, i.e. the number of roots, is bounded by its degree.

(d) Suppose that the field K is infinite, and that a polynomial f in $K[X]$ gives the zero function on K. Prove that $f = 0$ as a zero polynomial in $K[X]$. Use this to show that if K is infinite, then distinct polynomials in $K[X]$ give distinct functions on K.

This reveals that when the ring of coefficients is an infinite field, it is harmless to identify a polynomial with the function that it makes. However, keep in mind that for all finite rings R, as well as some infinite ones, different polynomials can give the same function on R.

(e) Given any function $g : K \to K$ find a suitable K-linear combination f of the polynomials $p_0, p_1, \ldots, p_n$ such that f agrees with g at the distinct points $a_0, a_1, \ldots, a_n$. This achieves the so-called Lagrange interpolation, whereby any function g on a field, such as $\mathbb{R}$, has a polynomial of degree at most n that agrees with g at a specified list of $n + 1$ points.

Does this result hold if K is merely an integral domain?

(f) If K is a finite field having q elements, use part (e) to show that every function $g : K \to K$ is a polynomial function.

Does this hold if K is a finite integral domain?

10. If p is a prime, count the number of distinct polynomial functions $\mathbb{Z}_p \to \mathbb{Z}_p$.

11. Let J be an ideal in a commutative ring R which is in turn a subring of a commutative ring S. Let SJ denote the set of all finite sums $\sum s_k j_k$ where $s_k \in S$ and $j_k \in J$. Verify that SJ is an ideal inside S, and that any ideal of S containing J must contain SJ. If J is principal in R with generator x, show that SJ is principal in S also with generator x.

12. Let $\varphi : \mathbb{Z}[X] \to \mathbb{C}$ be the substitution map given by $f \mapsto f(\sqrt{2})$. Clearly $X^2 - 2 \in \ker\varphi$. Show that if $f \in \ker\varphi$, then $X^2 - 2$ divides f in $\mathbb{Z}[X]$, and thereby deduce that $\ker\varphi = \langle X^2 - 2\rangle$.

Hint. Use Proposition 4.32.

13. This exercise requires some knowledge of mathematical analysis. Let $C[0, 1]$ be the ring of continuous functions on the closed unit interval $[0, 1]$. Addition and multiplication of functions are to be done point-wise. Pick $p \in [0, 1]$ and let M_p be the set of functions f in $C[0, 1]$ such that $f(p) = 0$.

Show that M_p is an ideal of $C[0, 1]$, but that it is not finitely generated.

Prove that there are no ideals J such that $M_p \subsetneq J \subsetneq C[0, 1]$.

14. For each commutative ring R a polynomial f in $R[X]$ also represents a function $R \to R$. Show that those polynomials which represent the zero function on R form an ideal J of $R[X]$. If p is a prime integer explain why all ideals of $\mathbb{Z}_p[X]$ are principal, and then find a generator for the ideal J in this case.

15. Find the ideals of the ring $K[[X]]$ of power series over a field K.

16. Let R be the set of all polynomials in $K[X]$ for which the coefficient of X is 0. Verify that R is a subring of $K[X]$. Show that the ideal $\langle X^2, X^3 \rangle$ in R is not principal.

17. Let R be the subring of $\mathbb{Q}$ consisting of all fractions of the form a/b, where a, b are integers and b is odd. Find the units of R. Show that all ideals of R are principal by finding them all.

18.* If R is a commutative ring with elements a, b such that $\langle a \rangle = \langle b \rangle$, then a, b are multiples of each other. If R is a domain, then a, b must be unit multiples of each other. That is, $b = ua$ and $a = u^{-1}b$ where u is a unit.

Find a commutative ring (obviously not a domain) with elements a, b that are multiples of each other but not unit multiples of each other.

4.6 Quotient Rings and the Isomorphism Theorem

Start with any ring R, not necessarily commutative, and any ideal J of R. Using its addition operation, R is an abelian group and J is a subgroup. If $a \in R$, the ***coset of a modulo J*** is the subset of R given by:

$$a + J = \{a + x : x \in J\}.$$

As with all groups, the element a that determines the coset $a + J$ is called a ***representative*** of the coset. Two elements a, b of R represent the same coset of J if and only if $b - a \in J$. The cosets form a partition of the abelian group R. In accordance with Definition 2.59, the sets of this partition form the quotient group R/J inherited from the addition operation of R. In the abelian group R/J the addition is given by

$$(a + J) + (b + J) = (a + b) + J \text{ for all } a, b \text{ in } J.$$

As well, there is the canonical projection

$$\varphi : R \to R/J \text{ given by } a \mapsto a + J \text{ for every } a \text{ in } J.$$

The canonical projection is a group homomorphism between two abelian groups using the operation of addition, and its kernel is J.

The absorption property of the ideal J affords a way to make R/J into a ring.

Proposition 4.40. *If J is an ideal in a ring R, and a, b, c, d in R are such that*

$$a + J = c + J \text{ and } b + J = d + J,$$

then

$$ab + J = cd + J.$$

Consequently, the multiplication in R/J given by

$$(a + J)(b + J) = ab + J$$

does not depend on the representatives chosen for the cosets $a + J, b + J$.

Under this multiplication the abelian group R/J is a ring.
The canonical projection $\varphi : R \to R/J$ given by $a \mapsto a + J$ is a surjective ring homomorphism, and its kernel is J.

Proof. We have that $d - b, c - a \in J$. From the absorption property of the ideal J it follows that

$$cd - ab = cd - cb + cb - ab = c(d - b) + (c - a)b \in J.$$

Thus $ab + J = cd + J$.

Since the multiplication and addition in R/J are inherited from their respective operations in R, all of the requirements for R/J to be a ring automatically flow from the fact R is a ring. For instance, the equations

$$(1 + J)(a + J) = 1a + J = a + J = a1 + J = (a + J)(1 + J)$$

reveal that $1 + J$ is the identity element of R/J.

Furthermore, the addition and multiplication in R/J are defined to ensure that the canonical projection $\varphi : R \to R/J$ is a ring homomorphism.

Finally, φ is surjective and its kernel is J because J is a subgroup of R using the addition in R, whereby Definition 2.59 applies. □

Definition 4.41. The ring R/J is called the ***quotient ring*** of R ***modulo*** J.

As noted in Section 4.4.2 the kernel of every ring homomorphism is an ideal. From Proposition 4.40 the converse is evident.

Proposition 4.42. *A subset of a ring is an ideal if and only if the subset is the kernel of a ring homomorphism.*

For a familiar example of quotient rings take the principal ideal $\mathbb{Z}n$ in $\mathbb{Z}$ generated by a positive integer n. The quotient ring $\mathbb{Z}/\mathbb{Z}n$ is precisely the ring $\mathbb{Z}_n$ of residues modulo n.

The ring R/J is the zero ring, if and only if it consists of just one coset, which happens if and only if J is the full ideal R itself. At the other extreme, if J is the trivial ideal $\langle 0 \rangle$, then each coset $x + J$ contains only the element x itself. In this case the canonical projection $\varphi : R \to R/J$ is injective and thereby an isomorphism, which means that R/J is just replicating R. The interesting quotient rings arise when J is non-trivial and proper.

4.6.1 The First Isomorphism Theorem for Rings

In the case of rings the first isomorphism theorem amounts to a small adaptation of its namesake for groups, Proposition 2.62.

Proposition 4.43 (First isomorphism theorem). *If $\varphi : R \to A$ is a surjective ring homomorphism with kernel J, then the mapping*

$$\psi : R/J \to A, \text{ given by } x + J \mapsto \varphi(x),$$

is a well-defined isomorphism of rings.

Proof. By default φ is a homomorphism of the underlying abelian groups R and A under the addition operation. The first isomorphism theorem, Proposition 2.62, applied to these groups says that the mapping ψ is a well-defined group isomorphism using the addition operations on R/J and A. All that remains is to observe that ψ is also an isomorphism of rings. But that is plain to see because

$$\psi((x+J)(y+J)) = \psi(xy+J) = \varphi(xy) = \varphi(x)\varphi(y) = \psi(x+J)\psi(y+J),$$

for every x, y in R, and

$$\psi(1+J) = \varphi(1) = 1.$$

The last 1 just above is in A. □

Every homomorphism $\varphi : R \to A$, surjective or not, remains surjective onto the image $\varphi(R)$ inside A. Thus we pick up the isomorphism

$$R/\ker\varphi \cong \varphi(R).$$

In other words, the image of a homomorphism is isomorphic to its domain modulo its kernel.

The Characteristic of a Ring

For any ring R, there is only one homomorphism $\varphi : \mathbb{Z} \to R$, given by $n \mapsto n1$, in accordance with the multiple notation.

The image $\varphi(\mathbb{Z})$ consists of all finite sums of ± 1, and is thereby the smallest subring of R. Since the ideals of $\mathbb{Z}$ are principal, there is an integer n such that $\ker\varphi = \mathbb{Z}n$, and we can take n to be non-negative. Different non-negative integers generate different principal ideals.

Definition 4.44. The unique non-negative integer n generating $\ker\varphi$ is known as the ***characteristic*** of the ring.

If $n = 0$, then φ is an embedding of $\mathbb{Z}$ into R, due to Proposition 4.24. We then identify the image ring $\varphi(\mathbb{Z})$ with its isomorphic copy $\mathbb{Z}$, and deem that every ring of characteristic 0 contains the ring of integers $\mathbb{Z}$.

If $n > 0$, the first isomorphism theorem, Proposition 4.43, applied to the ideal $\mathbb{Z}n$ in $\mathbb{Z}$, tells us that the mapping

$$\psi : \mathbb{Z}_n = \mathbb{Z}/\mathbb{Z}n \to \varphi(\mathbb{Z}), \text{ given by } x + \mathbb{Z}n \mapsto \varphi(x),$$

is an isomorphism. We then make the identification of $\varphi(\mathbb{Z})$ with $\mathbb{Z}_n$, and deem that R contains $\mathbb{Z}_n$ as its smallest subring.

Let us highlight this observation.

Proposition 4.45. *Every ring is either an extension of the ring of integers $\mathbb{Z}$ in the characteristic zero case, or the ring $\mathbb{Z}_n$ in the case of positive characteristic n.*

If R is an integral domain with positive characteristic n, then its smallest subring $\mathbb{Z}_n$ is also an integral domain. This forces n to be prime. For otherwise, we have $n = rs$ using factors

r, s greater than 1 and less than n, and then the equation $[r][s] = [rs] = [n] = [0]$ shows that $[r], [s]$ are zero divisors.

Proposition 4.46. *The characteristic of every integral domain is either 0 or a prime integer p, and thereby a domain contains either the ring of integers $\mathbb{Z}$ or the finite field $\mathbb{Z}_p$.*

Note that the characteristic 1 only happens when $1 = 0$ in R, i.e. only when R is the zero ring.

The Complex Numbers as a Quotient Ring

The substitution map $\varphi_i : \mathbb{R}[X] \to \mathbb{C}$ given by $f(X) \mapsto f(i)$ is surjective. Indeed, if $a, b \in \mathbb{R}$, then $\varphi_i(a + bX) = a + bi$, which covers all complex numbers. According to Proposition 4.33, $\ker \varphi_i$ is a principal ideal generated by a polynomial in $\ker \varphi_i$ of minimal degree. Well, $X^2 + 1$ is such a polynomial, since $\varphi_i(X^2 + 1) = i^2 + 1 = 0$ and since no polynomial of lower degree, linear or constant, lies in $\ker \varphi_i$. By the first isomorphism theorem, Proposition 4.43,

$$\mathbb{C} \cong \mathbb{R}[X]/\langle X^2 + 1\rangle.$$

Because of this isomorphism it would be entirely acceptable to *define* the complex numbers as the quotient ring $\mathbb{R}[X]/\langle X^2 + 1\rangle$.

Another Illustration of the First Isomorphism Theorem

Let K be any field, and let $\varphi : K[X, Y] \to K[X]$ be the substitution map given by

$$f(X, Y) \mapsto f(X^2, X^3).$$

For instance $\varphi(Y^2 - X^3) = (X^3)^2 - (X^2)^3 = 0$, which shows that $Y^2 - X^3 \in \ker \varphi$.

In fact $\ker \varphi = \langle Y^2 - X^3\rangle$. To see this let $f(X, Y) \in \ker \varphi$. Think of $f(X, Y)$ and $Y^2 - X^3$ as polynomials in Y with coefficients in the polynomial ring $K[X]$. In other words, observe that $K[X, Y] = K[X][Y]$. Since $Y^2 - X^3$ is monic in Y, Euclidean division as in Proposition 4.32 yields polynomials $q(X, Y), r(X, Y)$ in $K[X][Y]$ such that

$$f(X, Y) = (Y^2 - X^3)q(X, Y) + r(X, Y),$$

where $r(X, Y)$ is either 0 or a polynomial of degree less than 2 in Y. So $r = r_1(X)Y + r_2(X)$ for some $r_1(X), r_2(X)$ in $K[X]$. Apply the substitution map φ to get

$$0 = f(X^2, X^3) = r_1(X^2)X^3 + r_2(X^2).$$

The polynomial $r_1(X^2)X^3$ is either 0 or all of its monomials are of odd degree in X. While the polynomial $r_2(X^2)$ is either 0 or all of its monomials are of even degree in X. Because of this the preceding equation implies that both $r_1(X^2)X^3$ and $r_2(X^2)$ are 0. Then $r_1(X) = r_2(X) = 0$, and $f(X, Y) = (Y^2 - X^3)q(X, Y)$. This means that $f(X, Y) \in \langle Y^2 - X^3\rangle$.

Now let us find the image $K[X^2, X^3]$ of φ. For each $f = \sum_{i,j\geq 0} r_{ij}X^iY^j \in K[X, Y]$ we get

$$\varphi(f) = \sum_{i,j\geq 0} r_{ij}(X^2)^i(X^3)^j = \sum_{i,j\geq 0} r_{ij}X^{2i+3j}.$$

For each integer $n \geq 2$ there exist non-negative integers i,j such that $n = 2i + 3j$, but not for $n = 1$. Thus the image of φ consists of all polynomials in $K[X]$ for which the coefficient of X is 0. That is all polynomials of the form $r_0 + r_2X^2 + r_3X^3 + \cdots + r_nX^n$. Then, by the first isomorphism theorem, this ring is isomorphic to $K[X, Y]/\langle Y^2 - X^3 \rangle$.

4.6.2 Computing the Euler Function

For each positive integer n the group of units in the finite ring $\mathbb{Z}_n$ was introduced in Section 2.1. A residue $[a]$ is a unit in $\mathbb{Z}_n$ if and only if its representative a is coprime with n.

Definition 4.47. The number of units in the ring $\mathbb{Z}_n$ is denoted by $\phi(n)$. The resulting function ϕ, defined on the set of positive integers, is called the ***Euler function***.

Every residue in $\mathbb{Z}_n$ has precisely one representative among the list of integers $0, 1, 2, \ldots, n-1$. Thus, $\phi(n)$ equals the number of integers in this list which are coprime with n. Let us see how the first isomorphism theorem can be used to develop a method for calculating $\phi(n)$.

Note that 0 is coprime with 1, and thereby $\phi(1) = 1$. Suppose p is a prime and that $n = p^e$ for some positive integer exponent e. The elements in the list $0, 1, 2, \ldots, p^e - 1$ that are not coprime with p^e are the multiples of p. These multiples are $0p, 1p, 2p, \ldots, (p^{e-1} - 1)p$. There are p^{e-1} of them. Thus

$$\phi(p^e) = p^e - p^{e-1}.$$

We are left with having to compute $\phi(n)$ when n is not a prime power.

The Chinese Remainder Theorem for Rings

It will be useful to have a variant of Proposition 2.66 in the setting of rings. The first isomorphism theorem gets used again.

Proposition 4.48 (Chinese remainder theorem). *If r, s are coprime positive integers, then $\mathbb{Z}_{rs} \cong \mathbb{Z}_r \times \mathbb{Z}_s$.*

Proof. For each integer x let $[x]_n$ denote its residue in the ring $\mathbb{Z}_n$. Define the ring homomorphism

$$\varphi : \mathbb{Z} \to \mathbb{Z}_r \times \mathbb{Z}_s \text{ by } x \mapsto ([x]_r, [x]_s).$$

By inspection $x \in \ker\varphi$ if and only if both r and s divide x. Since r, s are coprime, $x \in \ker\varphi$ if and only if the product rs divides x. This says that $\ker\varphi$ is the principal ideal $\mathbb{Z}rs$.

By the first isomorphism theorem, Proposition 4.43, the mapping

$$\psi : \mathbb{Z}_{rs} = \mathbb{Z}/\mathbb{Z}rs \to \mathbb{Z}_r \times \mathbb{Z}_s \text{ given by } [x]_{rs} \mapsto ([x]_r, [x]_s)$$

is an isomorphism onto the subring $\varphi(\mathbb{Z})$ of $\mathbb{Z}_r \times \mathbb{Z}_s$. Since ψ is injective and each of $\mathbb{Z}_{rs}$ and $\mathbb{Z}_r \times \mathbb{Z}_s$ has precisely rs elements, the subring $\varphi(\mathbb{Z}) = \mathbb{Z}_r \times \mathbb{Z}_s$. So, $\psi : \mathbb{Z}_{rs} \cong \mathbb{Z}_r \times \mathbb{Z}_s$. □

The Chinese remainder theorem has an obvious extension to the product of any number of pairwise coprime integers, which we can leave as an exercise

Now under any ring isomorphism, units correspond to units. In particular the isomorphism ψ of Proposition 4.48 maps the group of units of $\mathbb{Z}_{rs}$ onto the group of units of $\mathbb{Z}_r \times \mathbb{Z}_s$. A bit of reflection shows that a pair $([a]_r, [b]_s)$ is a unit of $\mathbb{Z}_r \times \mathbb{Z}_s$ if and only if $[a]_r$ is a unit of $\mathbb{Z}_r$ and $[b]_s$ is a unit of $\mathbb{Z}_s$. These remarks show that ψ restricts to the following isomorphism of groups:

$$\psi : \mathbb{Z}_{rs}^\star \cong \mathbb{Z}_r^\star \times \mathbb{Z}_s^\star.$$

The multiplicative nature of the Euler function now reveals itself. Namely, if r, s are coprime, then

$$\phi(rs) = \phi(r)\phi(s).$$

In particular, if we have the factorization of an integer n into powers of distinct primes, then we can readily calculate $\phi(a)$. For example, instead of counting on our fingers, we can say that

$$\phi(637) = \phi(13 \cdot 7^2) = \phi(13)\phi(7^2) = 12 \cdot (7^2 - 7) = 12 \cdot 42 = 504.$$

4.6.3 The Correspondence Theorem

It is useful to have some connection between the ideals in the domain of a homomorphism and the ideals in its image. The ***correspondence theorem*** provides that link. The statement and proof are not much different than those of its namesake for groups, Proposition 2.64. The essence is that if φ is a ring homomorphism, then there is a natural bijection between the set of ideals inside the image of φ and the set of ideals in the domain of φ which contain $\ker \varphi$. There are numerous details, mostly trivial, to check.

Proposition 4.49 (Correspondence theorem). *Let $\varphi : R \to A$ be a surjective ring homomorphism. (Here R, A need not be commutative.) Let $\mathcal{J}$ be the family of ideals of R containing* $\ker \varphi$, *and $\mathcal{L}$ the family of ideals of A.*

1. *If $J \in \mathcal{J}$, then $\varphi(J) \in \mathcal{L}$.*
2. *If $L \in \mathcal{L}$, then $\varphi^{-1}(L) \in \mathcal{J}$.*
3. *If $J \in \mathcal{J}$, then $\varphi^{-1}(\varphi(J)) = J$.*
4. *If $L \in \mathcal{L}$, then $\varphi(\varphi^{-1}(L)) = L$.*
5. *The mapping $\mathcal{J} \to \mathcal{L}$ defined by $J \mapsto \varphi(J)$ is a bijection whose inverse is the mapping $\mathcal{L} \to \mathcal{J}$ defined by $L \mapsto \varphi^{-1}(L)$.*
6. *If $J \in \mathcal{J}$, then $R/J \cong A/\varphi(J)$.*

Proof. 1. Obviously $\varphi(J)$ is an abelian group under addition. The other thing to check is the absorption property of $\varphi(J)$. Suppose $r \in J$, and thereby that $\varphi(r) \in \varphi(J)$. Since φ is surjective, every a in A equals $\varphi(s)$ for some s in R. Then $a\varphi(r) = \varphi(s)\varphi(r) = \varphi(sr) \in \varphi(J)$ because $sr \in J$. Similarly, $\varphi(r)a \in J$.

2. The fact that $\varphi^{-1}(L)$ is an ideal is trivial to check. Also $\varphi^{-1}(L)$ does contain $\ker\varphi$ because $0 \in L$ and $\varphi^{-1}(0) = \ker\varphi$.
3. By the definition of inverse images, $J \subseteq \varphi^{-1}(\varphi(J))$. For the reverse inclusion let $r \in \varphi^{-1}(\varphi(J))$. Then $\varphi(r) \in \varphi(J)$. That is, $\varphi(r) = \varphi(s)$ for some s in J. Hence $r - s \in \ker\varphi$, which by assumption is inside J. Since $r - s \in J$ and $s \in J$, we see that $r \in J$.
4. This holds routinely for all surjective functions on sets.
5. This summarizes what items 1 to 4 are saying.
6. Take the mapping

$$\psi : R \to A/\varphi(J) \text{ defined by } r \mapsto \varphi(r) + \varphi(J).$$

Being the composite of φ followed by the canonical projection $A \to A/\varphi(J)$, this ψ is a ring homomorphism. Since φ is surjective, so is ψ surjective. An element r lies in $\ker\psi$ if and only if $\varphi(r) \in \varphi(J)$, in other words if and only if $r \in \varphi^{-1}(\varphi(J))$. By item 3 just proven, this amounts to saying that $r \in J$. So, $\ker\psi = J$. Then $R/J \cong A/\varphi(J)$ by the first isomorphism theorem 4.43. □

If N is an ideal of R, the correspondence theorem applied to the canonical projection $R \to R/N$ says that every ideal of R/N is of the form J/N for a unique ideal J containing N. All ideals of R/N are accounted for in this way. Furthermore $R/J \cong (R/N)/(J/N)$.

An Illustration of the Correspondence Theorem

The isomorphism in item 6 of the correspondence theorem can be useful in detecting the nature of some quotient rings. Here is a scenario which could occur.

Say R is a commutative ring and $a, b \in R$. What can Proposition 4.49 reveal about the quotient ring $R/\langle a,b\rangle$? Well, the canonical projection $\varphi : R \to R/\langle b\rangle$ is a surjective homomorphism, and its kernel is the principal ideal $\langle b\rangle$. The ideal $\langle a,b\rangle$ contains $\langle b\rangle$. By a simple verification the image $\varphi(\langle a,b\rangle)$ is the ideal $\langle\varphi(a),\varphi(b)\rangle$ inside $R/\langle b\rangle$. Since $\varphi(b) = 0$ the image $\varphi(\langle a,b\rangle)$ is the principal ideal $\langle\varphi(a)\rangle$. According to the correspondence theorem

$$R/\langle a,b\rangle \cong \left(R/\langle b\rangle\right)/\langle\varphi(a)\rangle.$$

The right side is a succession of quotients of two principal ideals, which might more readily be deciphered.

For example, let us describe the quotient ring $\mathbb{Z}[X]/\langle X^2+1, X-2\rangle$. Start with the substitution map $\varphi : \mathbb{Z}[X] \to \mathbb{Z}$ given by $f \mapsto f(2)$. This is a surjection and its kernel is the ideal $\langle X-2\rangle$. Indeed, $\langle X-2\rangle \subseteq \ker\varphi$ because $X-2$ has a root at 2. On the other hand let $f \in \ker\varphi$. Since $X-2$ is monic, Proposition 4.32 gives polynomials q, r in $\mathbb{Z}[X]$ such that $f = (X-2)q + r$ and r is constant. From $f(2) = 0$ it follows that $r = 0$, and thus $f \in \langle X-2\rangle$.

The correspondence theorem gives a bijection between the ideals of $\mathbb{Z}[X]$ containing $\langle X-2\rangle$ and the ideals of $\mathbb{Z}$. Now $\langle X^2+1, X-2\rangle$ contains $\langle X-2\rangle$. By item 6 of the correspondence theorem

$$\mathbb{Z}[X]/\langle X^2+1, X-2\rangle \cong \mathbb{Z}/\varphi(\langle X^2+1, X-2\rangle).$$

In $\mathbb{Z}$ the ideal

$$\varphi(\langle X^2+1, X-2\rangle) = \langle \varphi(X^2+1), \varphi(X-2)\rangle = \langle 2^2+1, 2-2\rangle = \langle 5\rangle.$$

Thus

$$\mathbb{Z}[X]/\langle X^2+1, X-2\rangle \cong \mathbb{Z}/\langle 5\rangle = \mathbb{Z}_5.$$

The original quotient is isomorphic to the finite field $\mathbb{Z}_5$.

Let us see what happens if we carry out the process in a different order. Take the substitution map $\psi : \mathbb{Z}[X] \twoheadrightarrow \mathbb{Z}[i]$ given by $f \mapsto f(i)$. This is a surjective homomorphism and its kernel is $\langle X^2+1\rangle$. Indeed, $X^2+1 \in \ker\psi$ because i is a root of X^2+1. On the other hand let $f \in \ker\psi$. Since X^2+1 is monic, Euclidean division gives polynomials q, r in $\mathbb{Z}[X]$ such that $f = (X^2+1)q + r$ and r is of the form $a + bX$ for some a, b in $\mathbb{Z}$. Since $f(2) = 0$ we get $r(2) = a + bi = 0$. By the nature of complex numbers $a = b = 0$, and thus $r = 0$. Having seen that $f = (X^2+1)q$ it follows that $\ker\psi = \langle X^2+1\rangle$. By item 6 of the correspondence theorem

$$\begin{aligned}\mathbb{Z}[X]/\langle X^2+1, X-2\rangle &\cong \mathbb{Z}[i]/\psi(\langle X^2+1, X-2\rangle)\\ &= \mathbb{Z}[i]/\langle \psi(X^2+1), \psi(X-2)\rangle = \mathbb{Z}[i]/\langle i-2\rangle.\end{aligned}$$

Putting all of this together we come up with the less than obvious discovery that

$$\mathbb{Z}[i]/\langle i-2\rangle \cong \mathbb{Z}_5.$$

EXERCISES

1. Find the value of $\phi(6615)$.
2. If $n = p_1^{e_1} p_2^{e_2} \cdots p_k^{e_k}$ is the unique factorization of n into powers of distinct primes, find the formula for $\phi(n)$ in terms of the p_j and e_j.
3. Prove that $\mathbb{Q}[X]/\langle x^2-2\rangle\rangle \cong \mathbb{Q}[\sqrt{2}]$.
4. Show that $\mathbb{Z}[X]/(X^2+1) \cong \mathbb{Z}[i]$.
5. (a) Find a root of the polynomial $X^2 - 2X + 2$ inside the field $\mathbb{C}$.
   Hint. $X^2 - 2X + 2 = (X-1)^2 + 1$.
   (b) Use the first isomorphism theorem to prove $\mathbb{R}[X]/\langle X^2 - 2X + 2\rangle \cong \mathbb{C}$.
   (c) Show that $\mathbb{R}[X]/\langle X^2 + X + 1\rangle \cong \mathbb{C}$.
6. Suppose that $n_1, n_2, \ldots, n_k$ are pairwise coprime positive integers, and let $n = n_1 n_2 \cdots n_k$. Prove the ring isomorphism

$$\mathbb{Z}_n \cong \mathbb{Z}_{n_1} \times \mathbb{Z}_{n_2} \times \cdots \times \mathbb{Z}_{n_k}.$$

7. Use the correspondence theorem to show that in $\mathbb{Q}[X]$ there are precisely two ideals containing $\langle X^2+1\rangle$.

   Show that in $\mathbb{Q}[X]$ there are precisely four ideals containing $\langle X^2-1\rangle$, and find those ideals.

8. If a, b are coprime integers, prove that

$$\mathbb{Z}[i]/\langle a+bi\rangle \cong \mathbb{Z}/\langle a^2+b^2\rangle.$$

Hint. Use the correspondence theorem.

9. If $n = p^e$ is a prime power, then

$$\begin{aligned} n &= (p^e - p^{e-1}) + (p^{e-1} - p^{e-2}) + \cdots + (p^1 - p^0) + 1 \\ &= \phi(p^e) + \phi(p^{e-1}) + \cdots + \phi(p) + \phi(1). \end{aligned}$$

Evidently this sum is taken over all positive divisors of $n = p^e$. So, if n is a prime power, then

$$n = \sum_{d|n} \phi(d),$$

where the sum is taken over all positive divisors of n. Show that this formula holds for every positive integer n.

10. If R is a commutative ring and its characteristic is a prime p, show that the mapping $\varphi : R \to R$ defined by $x \mapsto x^p$ is a ring homomorphism.
11. Let $\varphi : R \to A$ be a surjective homomorphism between commutative rings, and let J be an ideal containing $\ker\varphi$. If $r_1, r_2, \ldots, r_n$ generate J, verify that $\varphi(r_1), \varphi(r_2), \ldots, \varphi(r_n)$ generate the ideal $\varphi(J)$.
12. Let $K[X]$ be a polynomial ring over a field K and let R be the subring of $K[X]$ consisting of those polynomials for which the coefficient of X is 0. We saw for instance that $R \cong K[X, Y]/\langle Y^2 - X^3\rangle$. Find a non-principal ideal in R.
13.* Let $\mathbb{Z}[X, Y]$ be the ring of polynomials in two variables with integer coefficients. Prove that the ideal $\langle 2, X, Y\rangle$ cannot be generated with just two polynomials.
14. Suppose that

$$\varphi : A \to B \text{ and } \psi : A \to C$$

are surjective ring homomorphisms. Prove that

$$B/\varphi(\ker\psi) \cong A/(\ker\varphi + \ker\psi) \cong C/\psi(\ker\varphi).$$

4.7 Maximal and Prime Ideals

A central problem in algebra has been to solve polynomial equations. This quest culminated with the work of Evariste Galois (1881–1832), who showed that this comes down to a study of symmetries within fields. In Chapters 6 and 7, we shall undertake a detailed study of fields. For now, let us see how to construct fields by means of quotient rings.

4.7.1 Maximal Ideals and the Construction of Fields

Given a commutative ring R and an ideal M in R, what must M be like in order to ensure that the quotient ring R/M is a field? The answer lies in the following concept.

Definition 4.50. An ideal M in a ring R is ***maximal*** provided

- M is a proper ideal of R, i.e. $M \subsetneq R$, and
- there are no other proper ideals between M and R, i.e. if N is an ideal of R such that $M \subseteq N \subseteq R$, then $N = M$ or $N = R$.

Here is why this definition matters.

Proposition 4.51. *An ideal M in a commutative ring R is maximal if and only if the quotient ring R/M is a field.*

Proof. Suppose that M is a maximal ideal. Since M is proper, the quotient R/M is not the zero ring. So, $1 \neq 0$ in R/M. To be a field every non-zero coset $a + M$ in R/M needs to have an inverse in R/M. By definition of the quotient ring R/M, we must come up with some b in R such that $1 - ab \in M$. This will ensure that $1 + M = (a + M)(b + M)$.

The assumption $a + M \neq 0 + M$ means that $a \notin M$. Let

$$N = \{x + ab : x \in M \text{ and } b \in R\}.$$

By inspection N is an ideal of R containing M. Since $a \in N \setminus M$, the ideal N properly contains M. Thus $N = R$ due to the maximality of M. In particular, $1 \in N$, and so there exist some x in M and some b in R which give $x + ab = 1$. Such b in R causes $1 - ab \in M$, as desired.

For the converse, suppose that R/M is a field and N is an ideal of R such that $M \subsetneq N \subseteq R$. We need to have $N = R$. Since N properly contains M there is some a in $N \setminus M$. Then the coset $a + M$ in R/M is non-zero. Since R/M is a field there must be a coset $b + M$ such that $(a+M)(b+M) = 1+M$. In other words, $1-ab \in M$ for some b in R. Therefore $1-ab$ is in the larger ideal N as well. Since $a \in N$, absorption gives $ab \in N$, and then $1 = (1 - ab) + ab \in N$. Hence $N = R$. □

For an illustration of Proposition 4.51 consider the (possibly overused) ring $\mathbb{Z}_n = \mathbb{Z}/\mathbb{Z}n$ of residues modulo n. A small consideration shows that $\mathbb{Z}_n$ is an integral domain if and only if n is prime. By Proposition 4.10 finite integral domains coincide with fields. So, by Proposition 4.51, the ideal $\mathbb{Z}n$ in $\mathbb{Z}$ is maximal if and only if n is a prime. A small exercise will also confirm this directly.

For another illustration, take the polynomial ring $\mathbb{Z}[X]$. Define the map $\varphi : \mathbb{Z}[X] \to \mathbb{Z}_2$ by $\varphi : f \to [f(0)]$. This φ is a homomorphism, because it is the composite of the substitution map $f \mapsto f(0)$ and the canonical projection $\mathbb{Z} \to \mathbb{Z}_2$. The image of φ is $\mathbb{Z}_2$. Also $f \in \ker \varphi$ if and only if $f(0)$ is an even integer, in other words, if and only if the constant term of f is an even integer. The polynomials f in $\mathbb{Z}[X]$ whose constant term is even constitute the ideal $\langle 2, X\rangle$. By the first isomorphism theorem, Proposition 4.43, $\mathbb{Z}[X]/\langle 2, X\rangle \cong \mathbb{Z}_2$. Since $\mathbb{Z}_2$ is a field, the ideal $\langle 2, X\rangle$ is maximal in $\mathbb{Z}[X]$.

Proposition 4.51 when applied to the special case of the zero ideal tells us that R, when viewed as $R/\langle 0\rangle$, is a field if and only if $\langle 0\rangle$ is a maximal ideal. This comes down to saying that the only other ideal of R is R itself, which is what Proposition 4.28 already told us.

4.7.2 Existence of Maximal Ideals

While we may readily be able to produce maximal ideals in some rings, there remains the theoretical matter of knowing if every ring (other than the zero ring) possesses a maximal ideal. To that end the axiom of choice and its equivalent Zorn's lemma, as discussed in Appendix A, is needed.

Definition 4.52. One version of the ***axiom of choice*** says that if S is a non-empty set and $\mathcal{P}$ is the family of all *non-empty* subsets of S, then there is a function $f : \mathcal{P} \to S$ such that $f(A) \in A$ for each A in $\mathcal{P}$.

The function f "chooses" an element from each non-empty subset of S. Thereby f is called a ***choice function*** for the set S. For example, taking $\mathbb{N}$ to be the set of natural numbers, a good choice function $f : \mathcal{P} \to \mathbb{N}$ comes from letting $f(A)$ be the least element inside A, for every non-empty subset A of $\mathbb{N}$. For sets in general, such an easy specification of f is not available. By accepting the axiom of choice, its equivalent ***Zorn's lemma*** can be used.

In order to state Zorn's lemma a few concepts need to be set up.

Definition 4.53. A ***partial order*** on a non-empty set $\mathcal{F}$ is a relation on pairs from $\mathcal{F}$, denoted by the inequality symbol $\leq$, and having the following properties.

- *Reflexivity*: $A \leq A$ for all A in $\mathcal{F}$.
- *Transitivity*: if $A \leq B$ and $B \leq C$, then $A \leq C$.
- *Anti-symmetry*: if $A \leq B$ and $B \leq A$, then $A = B$.

The set $\mathcal{F}$ is then called a ***partially ordered set***.

In a partially ordered set $\mathcal{F}$ there could be pairs A, B in $\mathcal{F}$ which are not comparable by the relation $\leq$.

A subset $\mathcal{K}$ of a partially ordered set $\mathcal{F}$ is called a ***chain*** in $\mathcal{F}$, provided that for any two elements A, B in $\mathcal{K}$, either $A \leq B$ or $B \leq A$. A subset $\mathcal{K}$ of a partially ordered set $\mathcal{F}$ is ***bounded above*** provided there is an element B in $\mathcal{F}$ such that $A \leq B$ for all A in $\mathcal{K}$. An element M in $\mathcal{F}$ is called ***maximal*** provided the only A in $\mathcal{F}$ that satisfies $M \leq A$ is M itself.

For an example of a partially ordered set, take any set S and let $\mathcal{F}$ be any family of subsets of S. On $\mathcal{F}$ take the relation $\leq$ to mean set inclusion. That is, for A, B in $\mathcal{F}$ let $A \leq B$ indicate that $A \subseteq B$. The three requirements for this to be a partial order are easy to see.

Of current interest take a ring R, and let $\mathcal{F}$ be the family of all proper ideals of R partially ordered by inclusion. The maximal elements of $\mathcal{F}$ are the maximal ideals of R. For instance with $R = \mathbb{Z}$, let $\mathcal{F}$ be the set of proper ideals of $\mathbb{Z}$ partially ordered by inclusion. Then the family $\mathcal{K}$ of ideals of the form $\mathbb{Z}2^n$, where $n = 1, 2, \ldots$ is a chain. This chain is bounded above by the ideal $\mathbb{Z}2$.

Zorn's lemma was formulated originally in 1922 by Kazimierz Kuratowski (1896–1980), and later independently in 1935 by Max Zorn (1906–1992), not to mention the related work of Ernst Zermelo back in 1904. For some reason, the name "Zorn's lemma" has caught on, in glaring neglect of the other discoverers. Nonetheless it is a very handy result.

Here is what ***Zorn's lemma*** says:

If $\mathcal{F}$ is a partially ordered set in which every chain is bounded above, then $\mathcal{F}$ contains a maximal element.

In Appendix A we prove the equivalence between the axiom of choice and Zorn's lemma. Taking this as given for now, we come to the application of Zorn's lemma in ring theory.

Proposition 4.54. *Every proper ideal in a non-zero ring is contained in a maximal ideal.*

Proof. If R is a ring containing a proper ideal J, consider the family $\mathcal{F}$ of all proper ideals in R that contain J, and partially order $\mathcal{F}$ by inclusion. The union of ideals is generally not an ideal, but if $\mathcal{K}$ is a chain in $\mathcal{F}$, then a routine verification reveals that the union $L = \bigcup_{K \in \mathcal{K}} K$ is an ideal, and $J \subseteq L$. In addition, $1 \notin K$ for any ideal K in $\mathcal{K}$ since all K in $\mathcal{F}$ are proper. Therefore $1 \notin L$, which ensures that $L \in \mathcal{F}$.

Since every chain of ideals in $\mathcal{F}$ is bounded above by an ideal in $\mathcal{F}$, Zorn's lemma produces an ideal that is maximal in $\mathcal{F}$, which is precisely a maximal ideal of R containing J. □

4.7.3 Prime Ideals and Integral Domains

We might also consider those ideals P in a commutative ring R that cause R/P to be an integral domain.

Definition 4.55. An ideal P in a commutative ring R is a ***prime ideal*** provided

- P is a proper ideal of R, i.e. $P \subsetneq R$, and
- whenever a, b in R satisfy $ab \in P$, it must be that a or b is already in P.

For instance, the zero ideal $\langle 0 \rangle$ is prime in R if and only if R is an integral domain. For another example, in $\mathbb{Z}[X]$ the principal ideal $\langle X \rangle$ is prime. To see this suppose that $fg \in \langle X \rangle$ for some polynomials f, g in $\mathbb{Z}[X]$. Thus $fg = Xh$ for some polynomial h. Substitute $X = 0$ to obtain $f(0)g(0) = 0h(0) = 0$ in $\mathbb{Z}$. Thus $f(0) = 0$ or $g(0) = 0$. Say, $f(0) = 0$. This informs us that the constant term of f is 0, and by simple inspection we can see that $f = Xh$ for some polynomial h in $\mathbb{Z}[X]$. Thus $f \in \langle X \rangle$.

In the ring of integers $\mathbb{Z}$, where all ideals are principal, a non-zero ideal $\mathbb{Z}n$ is prime if and only if its generator n is a prime integer. This is the essence of Proposition 1.9. This suggests why the term "prime ideal" is used.

With commutative rings, prime ideals are defined precisely to ensure their quotient rings are integral domains.

Proposition 4.56. *An ideal P in a commutative ring R causes R/P to be an integral domain if and only if P is prime.*

Proof. The quotient R/P is an integral domain provided $0 + P \neq 1 + P$, and provided the only way to get $(a + P)(b + P) = 0 + P$ is by having $a + P = 0 + P$ or $b + P = 0 + P$.

The first of these conditions means that $1 \notin P$, which is the same as saying that P is a proper ideal of R. The second of these conditions means that the only way to get $ab \in P$ is by having $a \in P$ or $b \in P$.

Evidently the integral domain requirements rephrase the prime ideal requirements. □

Since all fields are integral domains, an immediate corollary emerges from Propositions 4.51 and 4.56.

Proposition 4.57. *Every maximal ideal in a commutative ring is a prime ideal.*

The converse of Proposition 4.57 does not hold. For instance, the ideal $\langle X \rangle$ in $\mathbb{Z}[X]$ is prime, and yet it properly sits inside the maximal ideal $\langle 2, X \rangle$ mentioned after Proposition 4.51. For an even simpler example, let R be an integral domain which is not a field. In this case the zero ideal $\langle 0 \rangle$ is prime and properly contained in every other ideal. Since R is not a field some of these other ideals are proper.

4.7.4 Building Fields from Polynomial Rings

If K is a field, then every ideal of the polynomial ring $K[X]$ is principal. If $\langle f \rangle$ is such an ideal and it happens to be maximal, Proposition 4.51 informs us that $K[X]/\langle f \rangle$ is a field. So, what does it take for the polynomial f to generate a maximal ideal?

Irreducible Polynomials

Definition 4.58. A polynomial f in $K[X]$ is called ***irreducible*** provided f is not a constant polynomial and the only way to have a factorization $f = gh$ using polynomials g, h in $K[X]$ is by having one of g or h be a constant polynomial. We shall also say that f is ***reducible*** when f is not constant and there do exist non-constant polynomials g, h such that $f = gh$.

Note that the opposing terms reducible and irreducible are applied only to polynomials of positive degree, leaving the constant polynomials in a class by themselves.

Linear polynomials are irreducible. Indeed, if $f = gh$ and f is linear, then one of g or h must be constant because either g or h has to have degree 0. If a polynomial f has degree more than 1 and f has a root in K, the factor theorem, Proposition 4.35, shows that f has a linear factor in $K[X]$. Thus irreducible polynomials of degree more than 1 never have a root in K. Naturally, the question arises: how to decide if a general polynomial is irreducible? We will offer some answers in the next chapter. In the meantime, here is an irreducibility test for polynomials of degree 2 or 3.

Proposition 4.59. *For any field K, a polynomial f of degree 2 or 3 is irreducible in $K[X]$ if and only if f has no roots in K.*

Proof. If f has a root in K, then f is reducible with a linear factor, as already noted. Conversely suppose f is reducible, and say $f = gh$ for some non-constant polynomials g, h. If f has degree 2, then both g and h must have degree 1, and if f has degree 3, then one of g or h has degree 1. In either case f has a linear factor of the form $aX + b$, which then causes f to have $-b/a$ as a root. □

For instance, $X^2 + 1$ has no root in $\mathbb{R}$, meaning that $X^2 + 1$ is irreducible in $\mathbb{R}[X]$. However $X^2 + 1$ is reducible in $\mathbb{C}[X]$ because $\pm i$ are its roots.

Here is how irreducible polynomials make fields.

Proposition 4.60. *Let $K[X]$ be the ring of polynomials over a field K. A polynomial f in $K[X]$ is such that $K[X]/\langle f \rangle$ is a field if and only if f is irreducible.*

Proof. According to Proposition 4.51 this comes down to showing that $\langle f \rangle$ is a maximal ideal of $K[X]$ if and only if f is irreducible.

Suppose that $\langle f \rangle$ is maximal. Now f cannot be the zero polynomial because in $K[X]$ the zero ideal is not maximal. Also f cannot be a non-zero constant, because non-zero constants are units of $K[X]$. Units generate all of $K[X]$, which is not a maximal ideal because maximal ideals are proper. Thus the degree of f is at least 1. Now, suppose g, h are polynomials in $K[X]$ such that $f = gh$. Assuming that g is not constant, let us show that h is constant. Clearly

$$\langle f \rangle \subseteq \langle g \rangle \subseteq K[X].$$

Since g is not constant, the ideal $\langle g \rangle$ is proper in $K[X]$. Then the maximality of $\langle f \rangle$ forces $\langle f \rangle = \langle g \rangle$. In particular, $g = fk$ for some polynomial k. The fact that $f = gh$ leads to $f = fkh$. Cancel f to get $1 = kh$, whereby h is a unit, i.e. a non-zero constant. This proves f is irreducible.

Conversely, suppose f is irreducible. To see that $\langle f \rangle$ is maximal, first note that $\langle f \rangle$ is a non-zero and proper ideal. This is because the irreducibility of f assumes $\langle f \rangle$ is not constant. Next suppose that J is an ideal of $K[X]$ such that

$$\langle f \rangle \subseteq J \subseteq K[X].$$

To get the maximality of $\langle f \rangle$ we can suppose $J \subsetneq K[X]$ and show that $J = \langle f \rangle$. By Proposition 4.33 every ideal of $K[X]$ is principal, and so $J = \langle g \rangle$ for some polynomial g. The proper inclusion $\langle g \rangle \subsetneq K[X]$ shows that g is not a constant polynomial. The inclusion $\langle f \rangle \subseteq \langle g \rangle$ means that $f = gh$ for some polynomial h. Since f is irreducible and g is not constant, h must be constant, and clearly non-zero. Thus $g = fh^{-1}$, which puts g inside $\langle f \rangle$, and yields the desired equality $\langle f \rangle = \langle g \rangle$. □

For example, take the polynomial $X^3 + X^2 + X + 2$ in $\mathbb{Z}_3[X]$. (Here 2 is a residue in $\mathbb{Z}_3$.) None of the possibilities $0, 1, 2 \pmod 3$ are roots, by simple substitution. Being cubic, this polynomial is irreducible and therefore the quotient $\mathbb{Z}_3[X]/\langle X^3 + X^2 + X + 2 \rangle$ is a field. It would be a digression at this time to show that the number of elements in this field is 27.

An examination of the proof of Proposition 4.60 would show that it carries through to all integral domains whose ideals are principal. The discussion of unique factorization in Chapter 5 will provide the details, as well as a comprehensive study of irreducible polynomials.

EXERCISES

1. Show that the ideal $\langle 2, X\rangle$ in $\mathbb{Z}[X]$ consists of all polynomials whose constant term is even.

   Show that the ideal $\langle 2, X\rangle$ is maximal in $\mathbb{Z}[X]$ directly from the definition of maximal ideals.
2. Let $\mathbb{Q}[X, Y]$ be the ring of polynomials in two variables with coefficients in $\mathbb{Q}$. Show that the ideal $\langle X, Y\rangle$ is maximal.
3. Give a direct proof, based solely on definitions, that every maximal ideal in a commutative ring is prime, without appealing to the fact every field is an integral domain.
4. Show that in a finite, non-zero commutative ring, every prime ideal is maximal.

   Hint. What if the finite ring is an integral domain?
5. If every ideal in an integral domain R is principal, prove that every non-zero prime ideal in R is maximal.

   Hint. Study the proof of Proposition 4.60.
6. This exercise requires some knowledge of real analysis. Suppose that $C[0, 1]$ is the set of all real valued continuous functions $f : [0, 1] \to \mathbb{R}$.

   (a) Verify that, using point-wise addition and multiplication of functions, $C[0, 1]$ is a commutative ring. This is a truly routine item.
   (b) Is $C[0, 1]$ an integral domain?
   (c) What are the units of $C[0, 1]$?
   (d) If p is a point in the interval $[0, 1]$, show that the set $J_p = \{f \in C[0, 1] : f(p) = 0\}$ is a maximal ideal of $C[0, 1]$.
   (e) If M is a maximal ideal of $C[0, 1]$, prove that $M = J_p$ for some p in $[0, 1]$.
   (f) Is J_p a principal ideal?

7. Show that $\mathbb{R}[X]/\langle X^2 - 2\rangle \cong \mathbb{R} \times \mathbb{R}$.
8. If $a, b \in \mathbb{Q}$ and $a \neq b$, show that $\mathbb{Q}[X]\langle (X - a)(x - b)\rangle \cong \mathbb{Q} \times \mathbb{Q}$.
9. Let R be the set of matrices of the form $\begin{bmatrix} a & b \\ 0 & a \end{bmatrix}$ where $a, b \in \mathbb{Q}$. Verify that R is a commutative subring of $M_2(\mathbb{Q})$. Then prove $R \cong \mathbb{Q}[X]/\langle X^2\rangle$.
10. Show that $\mathbb{Q}[X]/\langle X^2\rangle \cong \mathbb{Q}[X]/\langle (X - 1)^2\rangle$.
11. Show that $\mathbb{Z}_5[X]/\langle X^3 + X + 1\rangle$ is a field.
12. Let R be the set of all eventually constant sequences $(x_1, x_2, \ldots, x_n, \ldots)$ of real numbers. This means that for each such sequence there is an index m (depending on the sequence) such that $x_m = x_{m+1} = x_{m+2} = \cdots$. Verify that, under the usual operations of point-wise addition and multiplication, R is a subring of the ring of all sequences of real numbers. Then find the maximal ideals of R.

4.8 Fractions

Much of high school mathematics engages fractions and their abundant applications. A fraction $\frac{a}{b}$ is made up of a pair of integers a, b such that $b \neq 0$, but the same fraction can have different appearances. For instance, $\frac{-2}{3} = \frac{6}{-9}$. In fact $\frac{a}{b} = \frac{c}{d}$ precisely when $ad = bc$. Fractions can be added and multiplied so that $\mathbb{Q}$ becomes a ring extension of $\mathbb{Z}$. The common rules for fraction addition and multiplication are given by:

$$\frac{a}{b} + \frac{c}{d} = \frac{ad + bc}{bd} \quad \text{and} \quad \frac{a}{b}\frac{c}{d} = \frac{ac}{bd}.$$

This fraction building process works for commutative rings in general. From the abstract construction of fractions we will learn that every integral domain is a subring of some field. Since it comes at little extra cost we will keep the process fairly general. The overall picture should be quite transparent, but there will be a number of mundane details to check.

4.8.1 Localizations at Denominator Sets

The addition and multiplication rules for fractions in $\mathbb{Q}$ seem to require that the product of two denominators should also be a denominator and that such a product had better not be 0. Furthermore, in order to have every integer a be a fraction we must insist that 1 be an eligible denominator. That gives us the right to say $a = \frac{a}{1}$. These are the key ingredients for a set of denominators.

Definition 4.61. A subset D inside a non-zero commutative ring R will be called a ***denominator set*** provided

- $1 \in D$ while $0 \notin D$,
- D contains no zero divisors, and
- for all $a, b \in D$, the product $ab \in D$.

Here are some examples of denominator sets. The verifications are straightforward.

- The set of all non-zero elements in any integral domain R, such as $\mathbb{Z}$ or any polynomial ring $K[X]$ over a field K.
- The singleton $\{1\}$ in any non-zero commutative ring R.
- The set of all cancellable elements in any commutative ring R.
- The set $\{1, 2, 2^2, \ldots, 2^n, \ldots\}$ inside $\mathbb{Z}$.
- The set $\{a \in \mathbb{Z} : 2 \nmid a\}$ inside $\mathbb{Z}$.
- The complement $D = R \setminus P$ of any prime ideal P in an integral domain R.

To verify the last example, note that $1 \in D, 0 \notin D$ because $0 \in P, 1 \notin P$. Since R is an integral domain, there are no zero divisors in D. To see that D is closed under multiplication, suppose to the contrary that $a, b \in D$, that is $a, b \notin P$. Since the ideal P is prime, $ab \notin P$ either, that is $ab \in D$.

Localization at a Denominator Set

Suppose that R is a commutative ring with a denominator set D.

Definition 4.62. A ***localization***[5] of R at D is any commutative ring extension A of R such that

- the elements of D are units of A, and
- every element in A takes the form rs^{-1}, for some r in R and some s in D.

By way of example, the field $\mathbb{Q}$ of rational numbers is a localization of $\mathbb{Z}$ at the set D of non-zero integers. For another example, take the denominator set $D = \{1, 2, 2^2, \ldots, 2^n, \ldots\}$ inside $\mathbb{Z}$. Now a localization of $\mathbb{Z}$ at this D is the ring A of all rational numbers that are of the form $a/2^n$ for some integer a and some non-negative exponent n. Note that every element 2^n in D is a unit in A with inverse $1/2^n$. Furthermore, every element of A has the form rs^{-1} where $r, s \in \mathbb{Z}$ and $s \in D$. Thus A localizes $\mathbb{Z}$ at D.

We might also note that if the elements of D are already units in R, then R itself is a suitable localization of R at D.

If A is a localization of a ring R at a denominator set D and rs^{-1}, with r in R and s in D, is a typical element of A, we could take the liberty of writing this element in the ***fraction notation*** as shown:

$$rs^{-1} = \frac{r}{s} = r/s.$$

The fraction notation helps to validate the term "denominator set." In any localization, the familiar fraction rules hold. Indeed, for $r, t \in R$ and s, u in D, we have:

- $\frac{r}{s} = \frac{t}{u}$ if and only if $rs^{-1} = tu^{-1}$ if and only if $ru = ts$
- $r = r1^{-1} = \frac{r}{1}$
- $\frac{r}{s} + \frac{t}{u} = rs^{-1} + tu^{-1} = ruu^{-1}s^{-1} + tss^{-1}u^{-1}$
  $= (ru + ts)u^{-1}s^{-1} = (ru + ts)(su)^{-1} = \frac{ru + ts}{su}$
- $\frac{r}{s}\frac{t}{u} = rs^{-1}tu^{-1} = rts^{-1}u^{-1} = rt(su)^{-1} = \frac{rt}{su}.$

For the above rules to hold the commutativity of R is essential.

Fields of Fractions

A special localization needs to be singled out.

Proposition 4.63. *A localization of an integral domain at its denominator set of all non-zero elements is a field.*

[5] This name has its origins in algebraic geometry, a subject which, along with number theory, has been the driving force in the development of commutative algebra.

Proof. Let A be a localization of an integral domain R at the denominator set of all non-zero elements of R. The non-zero elements of R are units in A, and every element of A takes the form rs^{-1} for some r, s in R where $s \neq 0$. Now if $rs^{-1} \neq 0$, then $r \neq 0$, which means that r is a unit in A, whereby the element sr^{-1} exists in A. By commutativity in A we get $(rs^{-1})(sr^{-1}) = 1$. Since every non-zero element of A has an inverse, A is a field. □

Definition 4.64. A localization of a domain R at its set of non-zero elements is known as a ***field of fractions*** of R.

Of course it must be verified that every integral domain does possess a field of fractions. This will be done shortly.

4.8.2 Uniqueness of Localizations

A commutative ring can possess at most one localization at a given denominator set. This fact will emerge from the following ***homomorphism extension property*** of localizations.

Proposition 4.65 (Homomorphism extension). *Suppose that A is a localization of a commutative ring R at some denominator set D. If $\varphi : R \to B$ is a homomorphism into a commutative ring B such that every u in D gets mapped to a unit $\varphi(u)$ in B, then φ has a unique homomorphism extension $\psi : A \to B$.*

Proof. The elements of D are units in A, and every element of A takes the form rs^{-1} where $r \in R$ and $s \in D$.

If $\psi : A \to B$ is such an extension, then

$$\psi(rs^{-1}) = \psi(r)\psi(s^{-1}) = \psi(r)\psi(s)^{-1} = \varphi(r)\varphi(s)^{-1}, \text{ for all } r \text{ in } R \text{ and } s \text{ in } D.$$

This proves that φ has at most one extension to A, and it also indicates how such an extension must be defined.

So, we need to verify that the formula $\psi(rs^{-1}) = \varphi(r)\varphi(s)^{-1}$ yields a well-defined homomorphism $\psi : A \to B$. To see that this formula is well defined, suppose $r, t \in A$, $s, u \in D$ and $rs^{-1} = tu^{-1}$ in A. Then $ru = ts$ in R, whereby

$$\varphi(r)\varphi(u) = \varphi(ru) = \varphi(ts) = \varphi(t)\varphi(s),$$

which gives $\varphi(r)\varphi(s)^{-1} = \varphi(t)\varphi(u)^{-1}$. Thus the formula for ψ is a properly defined mapping from A to B.

The verification that ψ gives a homomorphism is routine but tedious. With regards to addition, let $r, t \in R$ and $s, u \in D$. Then

$$\begin{aligned}\psi(rs^{-1} + tu^{-1}) &= \psi(rs^{-1}uu^{-1} + tu^{-1}ss^{-1}) \\ &= \psi\left((ru + ts)(us)^{-1}\right) \\ &= \varphi(ru + ts)\varphi(us)^{-1}\end{aligned}$$

$$= (\varphi(r)\varphi(u) + \varphi(t)\varphi(s))\,\varphi(u)^{-1}\varphi(s)^{-1}$$
$$= \varphi(r)\varphi(s)^{-1} + \varphi(t)\varphi(u)^{-1}$$
$$= \psi(rs^{-1}) + \psi(tu^{-1}).$$

We omit the easier verifications that $\psi\left((rs^{-1})(tu^{-1})\right) = \psi(rs^{-1})\psi(tu^{-1})$ and $\psi(1) = 1$.

Finally, for each r in R we get

$$\psi(r) = \psi(r1^{-1}) = \varphi(r)\varphi(1)^{-1} = \varphi(r)1^{-1} = \varphi(r)1 = \varphi(r).$$

Thus ψ extends φ. □

Our desired uniqueness result ensues.

Proposition 4.66. *If A and B are localizations of a commutative ring R at a denominator set D, then $A \cong B$.*

Proof. By the definition of localization the inclusion maps $R \to A, R \to B$ map the elements of D to units in A, B respectively. Then Proposition 4.65 says there exist unique homomorphisms

$$\psi : A \to B \text{ and } \gamma : B \to A,$$

such that $\psi(s) = \gamma(s) = s$ for all s in D. Now we obtain two ring homomorphisms $A \to A$ which extend the inclusion map $R \to A$. They are $\gamma \circ \psi$ and the identity map $A \to A$. By the uniqueness part of Proposition 4.65 it must be that $\gamma \circ \psi$ is the identity map on A. Similarly $\psi \circ \gamma$ is the identity map on B. The desired isomorphism from A to B is ψ, with inverse γ. □

A special case of Proposition 4.66 says that every integral domain has at most one field of fractions, up to isomorphism.

4.8.3 Existence of Localizations

The preceding considerations would be rather moot if localizations did not exist. We need to construct them for all commutative rings R and all denominator sets D, using some form of synthesis.

Begin with the Cartesian product $R \times D$ of pairs (r, s), where $r \in R$ and $s \in D$, and impose an equivalence relation upon the product as follows. Given two pairs (r, s) and (t, u) in $R \times D$, declare that

$$(r, s) \sim (t, u) \text{ if and only if } ru = ts.$$

In order for this relation to partition the set $R \times D$ into disjoint blocks of related elements, we need to verify that $\sim$ is an equivalence. It is trivial to see that $\sim$ is both reflexive and symmetric. For this relation to be transitive we require that D have no zero divisors. Suppose that $(r, s) \sim (t, u)$ and that $(t, u) \sim (v, w)$. Then $ru = ts$ and $tw = vu$. To get $(r, s) \sim (v, w)$ we need $rw = vs$. From the given equations and the commutativity of R we get

$$urw = ruw = tsw = tws = vus = uvs.$$

Now u is cancellable because $u \in D$, which has no zero divisors. Hence $rw = vs$, which shows that $\sim$ is transitive.

The resulting family of equivalence classes that partition $R \times D$ will be denoted by R_D. Each equivalence class will be tagged by any one of its representatives (r, s) that are in the class, and be denoted by the fraction-like notation $\frac{r}{s}$, where $r \in R$ and $s \in D$. That is, for each (r, s) in $R \times D$ let

$$\frac{r}{s} = \{(t, u) \in R \times D : ru = ts\}.$$

The classes in R_D will be called ***fractions with denominators*** in D. The thought that a set of ordered pairs could be denoted as a fraction may seem unorthodox. However we should appreciate, for example in $\mathbb{Q}$, that the fraction $\frac{1}{3}$ is really the totality of expressions $\frac{2}{6}, \frac{-5}{-15}, \frac{7}{21}$, etc. Our fraction notation is good because, as we have just seen, $\frac{r}{s} = \frac{t}{u}$ if and only if $(r, s) \sim (t, u)$, which is the same as saying $ru = ts$.

Here are the predictable operations that make R_D into a commutative ring.

- For addition let $\frac{r}{s} + \frac{t}{u} = \frac{ru + ts}{su}$.
- For multiplication let $\frac{r}{s}\frac{t}{u} = \frac{rt}{su}$.
- For the zero and identity elements take $\frac{0}{1}$ and $\frac{1}{1}$, respectively.
- For negation let $-\frac{r}{s} = \frac{-r}{s}$.

The operations of addition, multiplication, and negation seem to depend on the representatives chosen for the fractions under consideration. But as might be expected, these operations are well defined.

Proposition 4.67. *If* $\frac{r}{s} = \frac{r'}{s'}$ *and* $\frac{t}{u} = \frac{t'}{u'}$, *then*

$$\frac{ru + ts}{su} = \frac{r'u' + t's'}{s'u'}, \quad \frac{rt}{su} = \frac{r't'}{s'u'}, \text{ and } \frac{-r}{s} = \frac{-r'}{s'}.$$

Proof. According to the definition of fraction equality coming from the equivalence $\sim$, we are given that

$$rs' = r's \text{ and } tu' = t'u.$$

The first of the desired equations comes down to showing that

$$(ru + ts)s'u' = (r'u' + t's')su.$$

By making liberal use of the commutativity of R along with the given equations, we get

$$\begin{aligned}(ru+ts)s'u' &= rus'u' + tss'u' \\ &= rs'uu' + tu'ss' \\ &= r'su'u + t'uss' \\ &= r'u'su + t's'su \\ &= (r'u' + t's')su.\end{aligned}$$

The second desired equation comes down to $rts'u' = r't'su$. This is the same as having $rs'tu' = r'st'u$, which is immediate by multiplying the given equalities.

The final desired equation comes down to $(-r)s' = (-r')s$, which is obvious from $rs' = r's$. □

Having seen that addition, multiplication, and negation in R_D do not depend on the representatives for the fractions involved, we can now undertake the verification that R_D is a commutative ring. Let us refer to the given operations inside R_D as the fraction operations.

Proposition 4.68. *Given a denominator set D in a commutative ring R, the set of fractions R_D with its fraction operations, is a commutative ring.*

Proof. We have to verify the axioms that define a commutative ring. Begin with the associative law of addition. If $\frac{r}{s}, \frac{t}{u}, \frac{v}{w} \in R_D$, we get

$$\begin{aligned}\left(\frac{r}{s}+\frac{t}{u}\right)+\frac{v}{w} &= \frac{ru+ts}{su}+\frac{v}{w} = \frac{(ru+ts)w+vsu}{suw} = \frac{ruw+tsw+vsu}{suw} \\ &= \frac{ruw+(tw+vu)s}{suw} = \frac{r}{s}+\frac{tw+vu}{uw} = \frac{r}{s}+\left(\frac{t}{u}+\frac{v}{w}\right), \text{ as needed.}\end{aligned}$$

To check the neutrality of $\frac{0}{1}$, for every $\frac{r}{s}$ in R_D observe that

$$\frac{r}{s}+\frac{0}{1} = \frac{r1+0s}{s1} = \frac{r}{s}.$$

To verify the law of negation for $\frac{r}{s}$ note that

$$\frac{r}{s}+\frac{-r}{s} = \frac{rs+(-r)s}{s^2} = \frac{0}{s^2} = \frac{0}{1}.$$

The last equality above is justified because $0 \cdot 1 = 0 \cdot s^2$.

We omit the verification of the remaining axioms, as they follow along the same lines. □

The construction of localizations is virtually complete.

Proposition 4.69. *For any commutative ring R and any denominator set D, the ring R_D is a localization of R at D.*

Proof. The first thing is to notice that R becomes a subring of R_D by means of ring identification. Indeed, the mapping $\alpha : R \to R_D$ given by $r \mapsto \frac{r}{1}$ is a homomorphism, as can be easily checked. It is also an embedding. To see that, let $r \in \ker\alpha$. Then $\frac{r}{1} = \frac{0}{1}$, which gives $r = r \cdot 1 = 0 \cdot 1 = 0$. Having a trivial kernel, α is an embedding. This allows for the identification $r = \alpha(r) = r/1$ which causes R to be a subring of R_D.

Finally, R_D is a localization of R, since every s in D has its inverse $\frac{1}{s}$ in R_D, and since every fraction $\frac{r}{s}$ in R_D is the same as rs^{-1}. □

If R is an integral domain and D is the set of non-zero elements of R, the construction of the localization R_D together with Proposition 4.63 tells us the following.

Proposition 4.70. *Every integral domain is a subring of a field, namely its field of fractions.*

The field of fractions of $\mathbb{Z}$ is none other than the field $\mathbb{Q}$ of rational numbers. Another example comes from the integral domain of polynomials $K[X]$, where K is any field. The fraction field of $K[X]$ is known as the field of ***rational functions***, and is denoted (using rounded brackets) by $K(X)$. Thus an element of $K(X)$ is a fraction $\frac{f}{g}$ where $f, g \in K[X]$ and $g \neq 0$, with the usual understanding that $\frac{f}{g} = \frac{h}{k}$ if and only if $fk = gh$ in $K[X]$. Strictly speaking, the fractions in $K(X)$ are not, a priori, functions. After all, polynomials are not a priori functions either, but the name "functions" seems to be entrenched.

Laurent Polynomials

For a different example of localization, take the polynomial ring $K[X]$ over a field K. For a denominator set in $K[X]$, take $D = \{1, X, X^2, \ldots, X^n, \ldots\}$, the non-negative powers of X. In this example the localization of $K[X]$ at D consists of all rational functions of the type $\frac{f}{X^j}$, where f is a polynomial in $K[X]$ and j is a non-negative integer. By expanding f in its standard representation we can see that this localization $K[X]_D$ consists of all rational functions that can be put in the form

$$\sum_{j=m}^{n} a_j X^j = a_m X^m + a_{m+1} X^{m+1} + \cdots + a_n X^n,$$

where m, n are *arbitrary* integers, $m \leq n$ and the $a_j \in K$. The rational functions of this form are known as ***Laurent polynomials***. For instance, $3X^{-2} - 5X^{-1} + 6 + X$ is a Laurent polynomial. If both a_m and a_n are non-zero, let us say that the Laurent polynomial presented as above is in *canonical form*. Even though it looks pretty clear, it should be noted that two Laurent polynomials in canonical form are equal if and only if the corresponding coefficients of the X^j are equal. Indeed, suppose that

$$a_m X^m + a_{m+1} X^{m+1} + \cdots + a_n X^n = b_k X^k + b_{k+1} X^{k+1} + \cdots + b_\ell X^\ell,$$

where $m \le n, k \le \ell$ and a_m, a_n, b_k, b_ℓ are non-zero. For argument's sake, say that $m \le k$. Multiply through by X^{-m} to get

$$a_m + \cdots + a_n X^{n-m} = b_k X^{k-m} + \cdots + b_\ell X^{l-m}.$$

This is an equality of polynomials in $K[X]$. From the K-linear independence of powers of X, it follows that $\ell - m = n - m,\ k - m = 0$ and thereby that $\ell = n$ and $k = m$, and also that $a_j = b_j$ for all j from m to n.

EXERCISES

1. If $R = \mathbb{Z} \times \mathbb{Z}$ and D is the denominator set of all (a, b) in $\mathbb{Z} \times \mathbb{Z}$ where $a \neq 0$ and $b \neq 0$, show that $\mathbb{Q} \times \mathbb{Q}$ is a localization of $\mathbb{Z} \times \mathbb{Z}$ at D.
2. If a, b are in an integral domain R and m, n are coprime positive integers such that $a^n = b^n$ and $a^m = b^m$, show that $a = b$.
3. Let D be a denominator set in a commutative ring R, and let R_D be the localization of R at D. The elements of R_D are the fraction $\frac{a}{b}$ where $a \in R$ and $b \in D$. If a is also in D and u is a unit of R, then $\frac{ua}{b}$ is a unit of R_D with inverse $\frac{u^{-1}b}{a}$. This exercise explores the possible converse of the preceding statement.
   (a) Inside $\mathbb{Z}$ let D be the denominator set consisting of all integers 4^n where $n \ge 0$. Find an integer a such that $\pm a \notin D$ such that $\frac{a}{4^n}$ is a unit in R_D.
   (b) A denominator set D in a commutative ring R is called ***saturated*** provided for every a in D and every c in R such that $c \mid a$, there is a unit of R such that $uc \in D$. That is, all factors of elements in D are unit multiples of elements in D.
   Show that the set of powers 4^n in $\mathbb{Z}$ is not saturated.
   If P is a prime ideal of R and D is the complement $R \setminus D$, show that D is saturated.
   If p is a prime element of R and D is the set of powers p^n where $n \ge 0$, show that D is saturated.
   (c) Suppose D is a saturated denominator set in a commutative ring R, and that a in R and b in D is such that $\frac{a}{b}$ is a unit in R_D. Show that there is a unit u in R such that $ua \in D$.
4. Let $K[X, Y]$ be the polynomial ring in two variables X, Y over a field K, and let $\langle XY - 1 \rangle$ be the ideal generated by $XY - 1$. Using the first isomorphism theorem show that $K[X, Y]/\langle XY - 1 \rangle$ is isomorphic to the ring of Laurent polynomials in the variable X.
5. Let R be a commutative ring. Prove that R is an integral domain if and only if the number of roots in R of every non-zero polynomial f in $R[X]$ is at most $\deg f$.

6. Let R be a commutative ring in which every ideal is principal, for example $\mathbb{Z}$ or the polynomial ring $\mathbb{Q}[X]$. If A is a localization of R at some denominator set D, prove that every ideal of A is also principal.

   Hint. If J is an ideal of A, show that a generator of $R \cap J$ as an ideal of R is also a generator of J as an ideal of A.
7. Let K be a field, $K[X]$ the ring of polynomials over K, and $K(X)$ the field of fractions of $K[X]$, also known as the field of rational functions over K. Prove that there is no rational function $\frac{f(X)}{g(X)}$ whose square is X.
8. Suppose P is a maximal ideal in a commutative ring R and take the denominator set $D = R \setminus P$. Show that the localization of R at D has just one maximal ideal. That maximal ideal consists of the set of non-units in the localization.

5 Primes and Unique Factorization

An integer p is prime when its only factors are $\pm p$ and ± 1. According to Proposition 1.9, if p is prime and a, b are integers such that $p \mid ab$, then $p \mid a$ or $p \mid b$. Every non-zero integer a can be factored as $a = p_1 p_2 \cdots p_n$, for some primes $p_1, p_2, \ldots, p_n$. The only thing that can vary in this factorization is the order in which the prime factors p_j appear, and possibly their sign.

This chapter deals with the notions of primeness and unique factorization in the general setting of integral domains.

Unless otherwise indicated, the rings in this chapter are integral domains.

5.1 Primes, Irreducibles and Factoring

Definition 5.1. If a, b are elements in an integral domain A, we say that a ***divides*** b, or that a is a ***factor*** of b, when $b = ac$ for some element c in A. When this happens, write $a \mid b$ in A.

Irreducibles

The units of A divide every element of A. Elements b that can be factored only by having a unit as a factor deserve attention.

Definition 5.2. An element p in an integral domain A is called ***irreducible***, provided p is non-zero and not a unit, and is such that

$$\text{if } p = ac \text{ for some } a, c \text{ in } A, \text{ then } a \text{ or } c \text{ is a unit.}$$

If a non-zero, non-unit element b can be factored as $b = ac$ using non-units a, c we shall say that b is ***reducible***.

The zero element and the units of A are in a class by themselves, neither irreducible nor reducible. Thus, for instance, in a field there are no reducible elements and no irreducible elements. Taking some liberties with grammar we often refer to the irreducible elements as the ***irreducibles***, just as prime integers are often called primes.

The usual primes $\pm 2, \pm 3, \pm 5, \pm 7, \ldots$ are the irreducibles of $\mathbb{Z}$. In other rings the irreducibles might not be so easy to detect, and even with $\mathbb{Z}$ this is a daunting computation for very large integers. In Section 4.7.4 we saw that in the polynomial ring $K[X]$ over a field K, the irreducibles are the non-constant polynomials which cannot be factored except by use of a constant polynomial as a factor. We saw that all linear polynomials are irreducible, while polynomials of degree 2 or 3 are irreducible in $K[X]$ if and only if they have no root in K.

One version of the fundamental theorem of algebra (which we take as given for now) says that over $\mathbb{C}$ the irreducible polynomials can only be linear. The problem of finding irreducible polynomials of higher degree over various fields, $\mathbb{Q}$ in particular, will deserve our attention.

An alternative way to think of irreducible elements comes from the next item.

Proposition 5.3. *An element p in an integral domain A is irreducible if and only if p is non-zero and not a unit, and is such that*

$$\textit{if } p = ac \textit{ for some } a, c \textit{ in } A, \textit{ then } p \mid a \textit{ or } p \mid c.$$

Proof. Suppose p has this property. To show irreducibility say $p = ac$ for some a, c in A. Now $p \mid a$ or $p \mid c$. Say $a = pd$ for some d in A. From that, $p = pdc$. Cancel p to get $1 = dc$, whence c is a unit.

Conversely, if p is irreducible and $p = ac$ for some a, c in A, then one of a or c is a unit. Say c is a unit. Thus $a = pc^{-1}$, whence $p \mid a$. □

Primes

Proposition 5.3 might be compared to the following concept.

Definition 5.4. An element p in an integral domain A is a called ***prime element***, or more briefly a ***prime***, provided p is non-zero and not a unit, and is such that

$$\text{if } p \mid ac \text{ for some } a, c \text{ in } A, \text{ then } p \mid a \text{ or } p \mid c.$$

An element p in A is prime if and only if the principal ideal $\langle p \rangle$ generated by p is a prime ideal. This reassuring coincidence can be seen by inspection. In light of Proposition 4.56, p is prime if and only if the quotient ring $A/\langle p \rangle$ is an integral domain. Proposition 5.3 also reveals the following.

Proposition 5.5. *Every prime in an integral domain is irreducible.*

Proof. Suppose p is prime in an integral domain A, and let $p = ac$ for some a, c in A. Obviously p divides ac, and being prime, p must divide either a or c. In light of Proposition 5.3 p is irreducible. □

In Section 1.2 the term "prime" was used to define the irreducibles of $\mathbb{Z}$, and (fortunately) Proposition 1.9 reveals that in the case of $\mathbb{Z}$ the irreducibles and the primes are one and the same. According to tradition, the term "prime" is predominantly used when dealing with integers.

Proposition 4.60 tells us that if K is a field and if p is an irreducible polynomial in $K[X]$, then the ideal $\langle p \rangle$ is maximal and thereby a prime ideal in $K[X]$. Thus in $K[X]$ too the irreducibles coincide with the primes. In contrast to the case of integers, tradition dictates that the term "irreducible" be used when dealing with polynomial rings $K[X]$.[1]

[1] Mathematical language should be consistent, but on occasion human traditions get in the way.

To dispel unjustified inferences based on $\mathbb{Z}$ and on $K[X]$, let us look at an integral domain in which some irreducibles are not prime.

Irreducibles Need Not Be Prime

Consider the ring $\mathbb{Z}[\sqrt{-5}]$ consisting of all numbers of the form $a + b\sqrt{-5}$ where a, b are integers. This is a subring of $\mathbb{C}$, and thereby an integral domain.

In this ring the element 2, viewed as $2 + 0\sqrt{-5}$, is irreducible but not prime. To prove this claim we need the so-called norm function

$$N : \mathbb{Z}[\sqrt{-5}] \to \mathbb{Z} \text{ defined by } N(a + b\sqrt{-5}) = a^2 + 5b^2.$$

A simple check reveals that $N(xy) = N(x)N(y)$ for all x, y in $\mathbb{Z}[\sqrt{-5}]$. Consequently, if $u \mid v$ in $\mathbb{Z}[\sqrt{-5}]$, then $N(u) \mid N(v)$ in $\mathbb{Z}$.

Let us verify that an element u in $\mathbb{Z}[\sqrt{-5}]$ is a unit if and only if $N(u) = 1$. Well, if u is a unit in $\mathbb{Z}[\sqrt{-5}]$, this means that u divides 1 in $\mathbb{Z}[\sqrt{-5}]$. Hence $N(u)$ divides $N(1) = 1$ in $\mathbb{Z}$, and then $N(u) = 1$ because $N(u)$ is a non-negative integer. On the other hand if $N(u) = 1$, write $u = a + b\sqrt{-5}$ for some a, b in $\mathbb{Z}$. Given that $N(u) = a^2 + 5b^2 = 1$, it becomes clear that $a = \pm 1$ and $b = 0$. That is, $u = \pm 1$, which are clearly units of $\mathbb{Z}[\sqrt{-5}]$. The above argument reveals, in passing, that the units of $\mathbb{Z}[\sqrt{-5}]$ are ± 1.

Now, to see that 2 is irreducible in $\mathbb{Z}[\sqrt{-5}]$, suppose $2 = xy$ with x a non-unit in $\mathbb{Z}[\sqrt{-5}]$. Then

$$4 = N(2) = N(xy) = N(x)N(y),$$

while the integer $N(x) \neq 1$. Since $a^2 + 5b^2$ is never 2 when a, b are integers, we see that $N(x) \neq 2$. Thus $N(x) = 4$. This implies that $N(y) = 1$, whereby y is a unit in $\mathbb{Z}[\sqrt{-5}]$.

To see that 2 is not prime in $\mathbb{Z}[\sqrt{-5}]$, observe the equation

$$2 \cdot 3 = (1 + \sqrt{-5}) \cdot (1 - \sqrt{-5}),$$

and note that 2 does not divide either of $1 \pm \sqrt{-5}$. For if 2 divides one of $1 \pm \sqrt{-5}$, we would get that $N(2)$ divides one of $N(1 \pm \sqrt{-5})$. That would say that 4 divides 6 in $\mathbb{Z}$.

One can similarly check that the elements 3 and $1 \pm \sqrt{-5}$ are irreducible but not prime in $\mathbb{Z}[\sqrt{-5}]$.

A Prime in $\mathbb{Z}[\sqrt{-5}]$

While we are on the ring $\mathbb{Z}[\sqrt{-5}]$, one may ask whether it does contain any actual primes. By means of the correspondence theorem, Proposition 4.49, we can demonstrate that the integer 11 remains a prime in $\mathbb{Z}[\sqrt{-5}]$.

For each integer a let $[a]$ denote its residue in $\mathbb{Z}_{11}$. Look at the map $\varphi : \mathbb{Z}[X] \to \mathbb{Z}_{11}[X]$ that reduces the coefficients of polynomials modulo 11. Namely $\varphi : \sum a_j X^j \mapsto \sum [a_j] X^j$. Such φ is a surjective homomorphism and its kernel is the principal ideal $\langle 11 \rangle$ taken inside $\mathbb{Z}[X]$.

The ideal $\langle 11, X^2+5\rangle$ inside $\mathbb{Z}[X]$ contains the kernel $\langle 11\rangle$. By the correspondence theorem, Proposition 4.49:

$$\begin{aligned}\mathbb{Z}[X]/\langle 11, X^2+5\rangle &\cong \mathbb{Z}_{11}[X]/\varphi(\langle 11, X^2+5\rangle)\\ &= \mathbb{Z}_{11}[X]/\langle \varphi(11), \varphi(X^2+5)\rangle = \mathbb{Z}_{11}[X]/\langle X^2+[5]\rangle.\end{aligned}$$

Now examine the substitution map $\psi : \mathbb{Z}[X] \to \mathbb{Z}[\sqrt{-5}]$ given by $f \mapsto f(\sqrt{-5})$. Clearly ψ is surjective and $X^2+5 \in \ker\psi$. On the other hand suppose $f \in \ker\psi$. That is, suppose $f(\sqrt{-5}) = 0$. Since X^2+5 is monic in $\mathbb{Z}[X]$, Proposition 4.32 gives polynomials q, r in $\mathbb{Z}[X]$ such that

$$f = (X^2+5)q + r \text{ where } r = aX + b \text{ for some integers } a, b.$$

Apply ψ to this equation, i.e. put $X = \sqrt{-5}$, to obtain $a + b\sqrt{-5} = 0$. It follows that $a = b = 0$, for otherwise $\sqrt{-5}$ would be a rational number. Thus $r = 0$ and X^2+5 divides f in $\mathbb{Z}[X]$. This proves that $\ker\psi = \langle X^2+5\rangle$. Now the ideal $\langle 11, X^2+5\rangle$ contains the kernel $\langle X^2+5\rangle$. Once more by the correspondence theorem, Proposition 4.49:

$$\begin{aligned}\mathbb{Z}[X]/\langle 11, X^2+5\rangle &\cong \mathbb{Z}[\sqrt{-5}]/\psi(\langle 11, X^2+5\rangle)\\ &= \mathbb{Z}[\sqrt{-5}]/\langle \psi(11), \psi(X^2+5)\rangle = \mathbb{Z}[\sqrt{-5}]/\langle 11\rangle.\end{aligned}$$

We might note here that ψ maps the integer 11 to itself as an element of $\mathbb{Z}[\sqrt{-5}]$.

The preceding isomorphisms reveal that

$$\mathbb{Z}[\sqrt{-5}]/\langle 11\rangle \cong \mathbb{Z}_{11}[X]/\langle X^2+[5]\rangle.$$

In the field $\mathbb{Z}_{11}$ the polynomial $X^2+[5]$ has no root, by testing all the possibilities. (In the parlance of number theory, -5 is not a quadratic residue mod 11.) Since this polynomial is quadratic, it must be irreducible in $\mathbb{Z}_{11}[X]$ according to Proposition 4.59. By Proposition 4.51 $\mathbb{Z}_{11}[X]/\langle X^2+[5]\rangle$ is a field. Hence $\mathbb{Z}[\sqrt{-5}]/\langle 11\rangle$ is a field, and therefore an integral domain. By Proposition 4.56 the ideal $\langle 11\rangle$ is prime in $\mathbb{Z}[\sqrt{-5}]$, and so the generator 11 of this ideal is prime in $\mathbb{Z}[\sqrt{-5}]$.

Associates

Suppose that an element a in our integral domain A is factored as $a = pq$, where p, q are irreducible. If u is any unit of A, then $a = (pu)(qu^{-1})$, and pu, qu^{-1} remain irreducible. By replacing irreducible factors with unit multiples, alternative factorizations of the elements in A are created trivially. To better cope with this ambiguity in factoring, a little terminology will come in handy.

Definition 5.6. Two elements a, b in an integral domain A are said to be ***associates*** in A provided $b = au$ for some unit u of A.

The relation of being associate is an equivalence relation on an integral domain A. Thereby A gets partitioned into disjoint subsets where each subset contains the elements that are associates of each other. For example, in $\mathbb{Q}[X]$ the associates of X^2+1 are the polynomials aX^2+a where $a \in \mathbb{Q}$ and $a \neq 0$. In $\mathbb{Z}$ the associates of 5 are ± 5. In the ring $\mathbb{Z}[i]$ of Gaussian integers, where the units are $\pm 1, \pm i$, the associates of $2+i$ are $2+i, -2-i, -1+2i, i-2i$.

A brief consideration reveals that the only associate of 0 is 0, associates of units are units and all units are associates of each other, associates of irreducible elements remain irreducible, and associates of primes remain prime.

Another item to observe is that a, b are associates if and only if a, b generate the same principal ideal in A.

As noted prior to Definition 5.6, a factorization of an element into irreducibles can be altered trivially by suitably replacing the factors with associates. However, the alternative factorizations of 6 in the ring $\mathbb{Z}[\sqrt{-5}]$ as $6 = 2 \cdot 3$ and $6 = (1+\sqrt{-5}) \cdot (1-\sqrt{-5})$ go beyond this trivial replacement of factors by associates. Thus 6 has a factorization into irreducibles in $\mathbb{Z}[\sqrt{-5}]$ that is truly non-unique. What goes wrong is that the irreducible factors $2, 3, 1+\sqrt{-5}, 1-\sqrt{-5}$ are not prime.

The next uniqueness result highlights the value of using primes in the factorization of an element.

Uniqueness of Factorizations Using Primes

Proposition 5.7. *If $p_1, p_2, \ldots, p_n$ and $q_1, q_2, \ldots, q_m$ are primes in an integral domain A and*

$$p_1 p_2 \cdots p_n = q_1 q_2 \cdots q_m,$$

then $n = m$ and, after a possible relabelling of the primes, each p_j is an associate of q_j.

Proof. The proof is pretty much the same as that of Proposition 1.10.

Say that $n \leq m$, and use induction on n, starting with $n = 0$.

For $n = 0$, the assumed factorization equation collapses to $1 = q_1 q_2 \cdots q_m$. Now m must be zero as well, for otherwise a non-unit q_j would divide 1. In this case there are no primes on either side to worry about.

Suppose that for some n and m, with $n \leq m$, we have primes $p_1, p_2, \ldots, p_n$ and $q_1, q_2, \ldots, q_m$ that give

$$p_1 p_2 \cdots p_{n-1} p_n = q_1 q_2 \cdots q_{m-1} q_m.$$

Evidently the prime p_n divides the product $q_1 q_2 \cdots q_{m-1} q_m$. Hence p_n divides one of the factors q_j, and by relabelling the primes we might as well say that $p_n | q_m$. Say $q_m = u p_n$ for some u in A. Being prime, q_m is also irreducible, and since p_n is not a unit, u must be a unit. Thus p_n and q_m (after possible relabelling of primes) are associates.

We now have $p_1 p_2 \cdots p_{n-1} p_n = q_1 q_2 \cdots q_{m-1} u p_n$. Cancel p_n on the far right to get

$$p_1 p_2 \cdots p_{n-1} = q_1 q_2 \cdots (q_{m-1} u).$$

Apply the inductive hypothesis to this last equation involving products of primes (noting that $q_{m-1} u$ remains a prime), to conclude that $n - 1 = m - 1$, whence $n = m$. The inductive hypothesis also yields, after a relabelling of primes, that $p_1,p_1,\ldots,p_{n-2},p_{n-1}$ are associates of $q_1,q_2,\ldots,q_{n-2},q_{n-1}u$, respectively. Since u is a unit, we also see that p_{n-1} is an associate of q_{n-1}. Given that p_n and q_n were already shown to be associates, the result on uniqueness of factorization follows. □

While the factorization of elements into primes is unique except for their unavoidable replacement by associates, there is nothing in Proposition 5.7 guaranteeing that integral domains are endowed with primes nor that factorizations of ring elements into primes are in fact possible. We must consider these matters.

Unique Factorization Domains

The next definition is motivated by Proposition 5.7.

Definition 5.8. An integral domain A is called a ***unique factorization domain*** provided every non-zero, non-unit element in A can be expressed as a product of primes.

Because of Proposition 5.7 the factorization of every non-zero, non-unit element in such a domain will be unique, except for the unavoidable replacement of prime factors by associates and the order in which the factors are written. The commonly used acronym for such a ring is ***UFD***.

As seen in Proposition 1.10, $\mathbb{Z}$ is a unique factorization domain. Before getting to more examples of unique factorization domains one nugget about them can be picked up.

Proposition 5.9. *If A is a unique factorization domain, then every irreducible in A is prime.*

Proof. Let p be our irreducible. Since A admits prime factorizations, we have $p = p_1p_2\cdots p_n$ for some primes p_j. The irreducibility of p forces all but one of the p_j to be units. Since primes are not units, there is only one p_j to begin with. Thus $p = p_1$, a prime. □

EXERCISES

1. Show that two elements a, b in an integral domain A are associates if and only if they generate the same principal ideal.
2. Show that an element p, in an integral domain A, is prime if and only if the ideal $\langle p\rangle$ that it generates is a prime ideal.
3. Find the Gaussian integers $a + bi$ that are associates of their conjugate $a - bi$.
4. Show that in the ring $\mathbb{Z}[\sqrt{-5}]$ the elements, 3 and $1 \pm \sqrt{-5}$ are irreducible but not prime. Deduce that $\mathbb{Z}[\sqrt{-5}]$ is not a unique factorization domain.
5. Let K be any field and let A be the set of polynomials of the form

$$a_0 + a_2X^2 + a_3X^3 + \cdots + a_nX^n,$$

where the coefficient of X^1 is 0. It is routine to see that A is a subring of $K[X]$. Verify that the polynomials X^2 and X^3 are irreducible in A but not prime. Find a polynomial f in

A which has a factorization using two irreducible factors and another factorization using three irreducible factors. Is A a unique factorization domain?

6. (a) Is the integer 13 a prime in $\mathbb{Z}[\sqrt{-5}]$? What about 7?
   (b) Show that an integer prime p remains a prime in $\mathbb{Z}[\sqrt{-5}]$ if and only if the residue of -5 is a not perfect square in the field $\mathbb{Z}_p$.
7. Show that a prime integer p remains a prime in the larger ring of Gaussian integers $\mathbb{Z}[i]$ if and only if the polynomial $X^2 + 1$ has no root in the finite field $\mathbb{Z}_p$. In other words, if and only if -1 is not a quadratic residue modulo p.

   Hint. The kernel of the substitution map $\varphi : \mathbb{Z}[X] \to \mathbb{Z}[i]$ given by $f \mapsto f(i)$ is $\langle X^2+1\rangle$. The kernel of the map $\psi : \mathbb{Z}[X] \to \mathbb{Z}_p[X]$, which reduces modulo p the integer coefficients of every polynomial, is the ideal $\langle p\rangle$ in $\mathbb{Z}[X]$. The ideal $\langle X^2 + 1, p\rangle$ contains both of these kernels. Use the correspondence theorem in two different ways.
8. If A is a unique factorization domain and P is a non-zero prime ideal of A, show that P contains a prime element of A.

5.2 Principal Ideal and Noetherian Domains

Definition 5.10. An integral domain in which all ideals are principal is called a ***principal ideal domain***.

The common acronym for such a domain is ***PID***. We have seen in Propositions 4.31 and 4.33 that $\mathbb{Z}$, as well as all polynomial rings $K[X]$ over any field K, are principal ideal domains. On the other hand, we saw in Section 4.4.3 that the ring $\mathbb{Z}[X]$ is not a principal ideal domain. Our purpose here is to demonstrate that every PID is a UFD. This opens up a mechanism for finding unique factorization domains.

In a Principal Ideal Domain Irreducibles Are Prime

In Proposition 4.60 we saw that if K is a field and f is an irreducible polynomial in $K[X]$, then the principal ideal $\langle f\rangle$ is maximal. Since maximal ideals are prime ideals, it follows that in $K[X]$ every irreducible polynomial is a prime. The proof of Proposition 4.60 made use of the fact all ideals of $K[X]$ are principal. We can imitate that proof for all principal ideal domains.

Proposition 5.11. *For each element p in a principal ideal domain A, the following properties are equivalent:*

- *p is irreducible,*
- *the ideal $\langle p\rangle$ is maximal,*
- *p is prime.*

Proof. If p is irreducible, then p is not a unit, and thus $\langle p\rangle \subsetneq A$. To see that the ideal $\langle p\rangle$ is maximal let J be an ideal of A such that $\langle p\rangle \subseteq J \subseteq A$. Given that all ideals of A are principal,

$J = \langle u \rangle$ for some u in A. Then $p = uv$ for some v in A. Since p is irreducible, u or v must be a unit. If v is a unit, then u and p are associates, which implies that $\langle p \rangle = \langle u \rangle$. If u is a unit, then $J = \langle u \rangle = A$ by Proposition 4.27. This proves $\langle p \rangle$ is a maximal ideal.

Next suppose $\langle p \rangle$ is maximal. Then $\langle p \rangle$ is a prime ideal by Proposition 4.57, and so its generator p is a prime.

Finally, if p is a prime, then p is irreducible due to Proposition 5.5. This completes the circle of implications, making these statements equivalent. □

We have seen that $\mathbb{Z}$ as well as polynomial rings $K[X]$ over fields K are principal ideal domains. In principal ideal domains we are entitled to use the terms "irreducible" and "prime" interchangeably. In the ring of integers the term "prime" seems to be favored, while in polynomial rings the term "irreducible" predominates.

In due course we shall encounter more principal ideal domains.

5.2.1 Noetherian Domains

We wish to prove that every principal ideal domain is a unique factorization domain. In light of Proposition 5.11, all that is required is to prove that every non-zero, non-unit element in a principal ideal domain can be factored as a product of irreducibles.

To that end, an idea developed by the great Emmy Noether (1882–1935) plays a role.

Definition 5.12. A commutative ring A is called ***Noetherian*** provided every increasing chain of ideals

$$J_1 \subseteq J_2 \subseteq J_3 \subseteq \cdots \subseteq J_n \subseteq J_{n+1} \subseteq \cdots$$

must be such that $J_k = J_\ell$ for all indices k, ℓ beyond some index n.

Briefly, we say that in a Noetherian ring every increasing chain of ideals ***terminates***.

Proposition 5.13. *A commutative ring A is Noetherian if and only if every ideal in A is finitely generated. In particular, all principal ideal domains are Noetherian.*

Proof. Suppose A contains an ideal J that is not finitely generated. Pick some a_1 in J. Clearly $\langle a_1 \rangle \subsetneq J$. Then pick a_2 in the complement $J \setminus \langle a_1 \rangle$. Again clearly $\langle a_1 \rangle \subsetneq \langle a_1, a_2 \rangle \subsetneq J$. Now pick a_3 in the complement $J \setminus \langle a_1, a_2 \rangle$. Clearly once more $\langle a_1 \rangle \subsetneq \langle a_1, a_2 \rangle \subsetneq \langle a_1, a_2, a_3 \rangle \subsetneq J$. Proceed in this fashion to construct a strictly increasing chain of ideals that never terminates. Thus A is not Noetherian.

Conversely, suppose that every ideal of A is finitely generated, and let

$$J_1 \subseteq J_2 \subseteq J_3 \subseteq \cdots \subseteq J_n \subseteq \cdots$$

be an increasing chain of ideals in A. It is easy to verify that, since this is a chain, the union $J = \bigcup J_n$ of all the ideals in the chain is also an ideal of A. By assumption, J has a finite set of generators $a_1, a_2, \ldots, a_k$. Being in the union J, each a_i belongs to some J_n, and since these ideals form a chain, all of the finitely many a_i belong to one common J_n. Thus

$$J = \langle a_1, a_2, \ldots, a_k \rangle \subseteq J_n \subseteq J_{n+1} \subseteq J_{n+2} \subseteq \cdots \subseteq J.$$

The chain terminates at the index n, if not sooner, and so A is Noetherian. □

Proposition 5.14. *Every non-zero, non-unit element in a Noetherian integral domain has a factorization into irreducibles.*

Proof. Let a be our non-zero, non-unit element in a Noetherian domain A. We suppose that a is not a product of irreducibles, and seek to contradict the fact A is Noetherian.

To begin with a itself must be reducible, whereby $a = a_1 b_1$, with neither a_1 nor b_1 a unit. Furthermore, if we could factor both a_1 and b_1 into irreducibles, then the totality of those factors would yield a factorization of a into irreducibles. Thus one of a_1 or b_1 is not a product of irreducibles. Let us say that a_1 is not a product of irreducibles.

Since b_1 is not a unit, the factorization $a = a_1 b_1$ implies the proper inclusion of ideals: $\langle a \rangle \subsetneq \langle a_1 \rangle$. For if we had equality $\langle a \rangle = \langle a_1 \rangle$, then $a_1 = ac$ for some c in A, and then we would get $a = a_1 b_1 = acb_1$. Cancelling a, we would get that b_1 is a unit, which is not so.

The assumption that a is not a product of irreducibles has led to a proper containment of principal ideals $\langle a \rangle \subsetneq \langle a_1 \rangle$, where a_1 is also a non-unit and not a product of irreducibles. This fact can now be applied to a_1, and then repeatedly to pick up a chain of ideals in A that does not terminate. □

The fact that elements of Noetherian domains factor into irreducibles does not imply they factor into primes, and we shall see examples of that. However by combining Propositions 5.11, 5.13 and 5.14 we do come to an interesting result.

Proposition 5.15. *Every non-zero, non-unit element in a principal ideal domain has a factorization into primes. In other words, every principal ideal domain is a unique factorization domain.*

For the ring $\mathbb{Z}$ the above says that every integer other than $0, \pm 1$ is a product of primes and the primes involved are unique, which we saw in Propositions 1.7 and 1.10. When Proposition 5.15 is applied to the polynomial rings $K[X]$ over any field K, the result says that any polynomial of degree at least one is a product of irreducible polynomials. The polynomial factors involved are unique, including the number of times they repeat, except for a possible replacement by associates, that is by non-zero constant multiples.

The value of Proposition 5.15 increases with more examples of principal ideal domains. The feature that made $\mathbb{Z}$ and $K[X]$ be principal ideal domains is that they possess a division algorithm. We explore that notion in the next section.

EXERCISES

1. Show that every homomorphic image of a Noetherian ring is another Noetherian ring. Is every subring of a Noetherian ring again a Noetherian ring?
2. If A is an integral domain but not a field, show that the polynomial ring $A[X]$ has ideals that are not principal.
3.* Let $K[X, Y]$ be the ring of polynomials in two variables over a field K.
   (a) Why is this a Noetherian ring?

(b) Show that the ideal generated by X, Y is not principal.
(c) Show that the ideal generated by X^2, XY, Y^2 cannot be generated by two elements.
(d) For each positive integer n, show that the ideal generated by the $n+1$ monomials of degree n cannot be generated by n elements.

Hint. If f is in such an ideal, every monomial appearing in f must belong to the ideal. Thus every monomial in f has degree at least n.

4. Suppose that A is an integral domain in which every non-zero, non-unit element has a unique factorization into irreducibles. In more detail, suppose that for every non-zero, non-unit element a in A there are irreducible elements $p_1, p_2, \ldots, p_n$ such that $a = p_1 p_2 \cdots p_n$. Also suppose that every other factorization of a as a product $q_1 q_2 \cdots q_m$ using irreducibles q_i must be such that $n = m$ and every p_j is an associate of q_j after a possible relabelling of the irreducibles.

Prove that in such A every irreducible is prime, and deduce that A is a unique factorization domain.

5. If A is a Noetherian integral domain containing a prime p, prove that the intersection $\bigcap_{n=0}^{\infty} \langle p^n \rangle = \langle 0 \rangle$.

6. Here is an interesting example of an integral domain with an assortment of features. Let $A = \mathbb{Z} + X\mathbb{C}[X]$. This is the ring of polynomials f with complex coefficients, but such that the constant term $f(0) \in \mathbb{Z}$.

(a) Verify that A is a subring of $\mathbb{C}[X]$.
(b) Show that the units of A consist of ± 1.

Hint. If u in A is a unit, then u is also a unit of $\mathbb{C}[X]$ and $u(0)$ is a unit of $\mathbb{Z}$.

(c) If p is a prime in $\mathbb{Z}$, show that the constant polynomial p remains a prime in A.

Hint. If p divides fg in A, then p divides $f(0)g(0)$ in $\mathbb{Z}$. Deduce that $p \mid f$ or $p \mid g$ in A.

(d) For any integer prime p show that $\langle X \rangle \subsetneq \langle p^{-1}X \rangle$ as ideals of A. Build on this to show that A is not Noetherian.
(e) For any positive exponent n and any integer prime p show that $p^n \mid X$ in A. Deduce that $\bigcap_{n=0}^{\infty} \langle p^n \rangle \neq \langle 0 \rangle$.

Since p is also prime in A, the above behavior could be unexpected.

(f) Show that the set $X\mathbb{C}[X]$ of elements h in A such that $h(0) = 0$ is a prime ideal of A which contains no irreducible elements, and hence no prime elements.

If $g \in X\mathbb{C}[X]$ and $h = h_1 h_1 \cdots h_n$ is a factorization of g into irreducible elements of A, show that some $h_j \in X\mathbb{C}[X]$. Deduce that no element of $X\mathbb{C}[X]$ has a factorization into irreducibles.

(g) If g is an irreducible element of A, show that either g is constant and irreducible as an element of $\mathbb{Z}$ or that $g = \pm 1 + aX$ where a is a non-zero complex number.

Hint. If g is constant and $g = uv$ for some u, v in A, then u, v must be constant and thereby integers. If g is non-constant show that $g(0) \neq 0$. If $g(0) \neq \pm 1$ find a way to factor g using non-units. If $\deg g > 1$, use the assumed factorization of g into linear factors in $\mathbb{C}[X]$ to get a factorization of g into linear factors in $A[X]$.

(h) If g is an irreducible element in A, show that g is prime in A.

(i) Show that every g in A such that $g(0) \neq 0$, except for the units ± 1, has a factorization into primes of A, and that such factorization is unique up to a change of sign in the factors.

5.3 Euclidean Domains

As noted in Chapter 1, Euclid has taught us that if a, b are integers and $a > 0$ then there exist integers q, r such that

$$a = bq + r \text{ and } 0 \leq r < a.$$

If $a < 0$, the above Euclidean division applies to the positive $-a$, to give integers q, r such that

$$b = -aq + r \text{ and } 0 \leq r < -a.$$

After noting that $(-a)q = a(-q)$, we see that for all $a \neq 0$, there exist integers q, r such that

$$b = aq + r \text{ and } 0 \leq r < |a|.$$

A similar Euclidean division applies to polynomials over a field K. Proposition 4.32 told us that if g, f are polynomials in $K[X]$ and $g \neq 0$, then there exist polynomials q, r such that

$$f = gq + r \text{ and } \deg r < \deg g \text{ or } r = 0.$$

Euclidean division in both $\mathbb{Z}$ and $K[X]$ was key to proving these rings were principal ideal domains. It seems worthwhile to consider an abstraction of this notion by which more principal ideal domains, and thereby unique factorization domains, can be found.

Euclidean Functions

Definition 5.16. A ***Euclidean function*** on an integral domain A is a mapping $\phi : A \setminus \{0\} \to \mathbb{N}$ with the property that for every pair of elements a, b in A where $a \neq 0$, there exist elements q, r in A such that

$$b = aq + r \text{ and } \phi(r) < \phi(a) \text{ or } r = 0.$$

An integral domain that admits a Euclidean function is called a ***Euclidean domain***. The elements q, r are called ***quotients*** and ***remainders***, respectively, for division of b by a.

There is no need to insist on the uniqueness of q and r. Another way to say this is that a Euclidean function $\phi : A \setminus \{0\} \to \mathbb{N}$ is such that if $a, b \in A$ and $a \neq 0$ and $a \nmid b$, then there is a q in A such that

$$\phi(b - aq) < \phi(a).$$

For the ring of integers $\mathbb{Z}$, the absolute value function is Euclidean and for polynomial rings $K[X]$ over a field K the degree function is Euclidean. The next result tells us that in our search for principal ideal domains we could start by looking for Euclidean functions.

Proposition 5.17. *Every Euclidean domain is a principal ideal domain and thereby a unique factorization domain.*

Proof. The proof is a blatant imitation of the proofs of Propositions 4.31 and 4.33. Let J be an ideal in a domain A with Euclidean function ϕ. If J is the zero ideal, then J is clearly principal. Otherwise let a be a non-zero element in J for which $\phi(a)$ is minimal. If b is any element in J we have q, r in A such that

$$b = aq + r \text{ and } \phi(r) < \phi(a).$$

Since $r = b - aq$ and a, b are in J, then so is r in J. By the minimality of $\phi(a)$, taken over all non-zero elements in J, it follows that $r = 0$. Hence $b = aq$, which proves that $J = \langle a \rangle$. □

A Euclidean Function for the Gaussian Integers

The function

$$\phi : \mathbb{C} \to \mathbb{R} \text{ given by } a + ib \mapsto a^2 + b^2 \text{ for all } a, b \text{ in } \mathbb{R}$$

is such that

- $\phi(zw) = \phi(z)\phi(w)$ for all z, w in $\mathbb{C}$,
- $\phi(z) = 0$ if and only if $z = 0$,
- $\phi(z) \in \mathbb{N}$ when z is a Gaussian integer in $\mathbb{Z}[i]$, and
- for every z in $\mathbb{C}$, there is a q in $\mathbb{Z}[i]$ such that $\phi(z - q) < \frac{1}{2}$.

The last item can be seen by noting that the Gaussian integers form the lattice of points in the complex plane with integer coordinates, and thus every complex number is at a distance at most $1/\sqrt{2}$ from a Gaussian integer.

The restriction of ϕ to $\mathbb{Z}[i]\setminus\{0\}$ is a Euclidean function. Indeed, the image of the restriction is inside $\mathbb{N}$. Next take x, y in $\mathbb{Z}[i]$ and such that $x \neq 0$. As noted above there is a q in $\mathbb{Z}[i]$ such that

$$\phi\left(\frac{y}{x} - q\right) < \frac{1}{2}.$$

Since $\phi(x) > 0$ we obtain:

$$\phi(y - xq) = \phi\left(\left(\frac{y}{x} - q\right)x\right) = \phi\left(\frac{y}{x} - q\right)\phi(x) \leq \frac{1}{2}\phi(x) < \phi(x).$$

Evidently this q along with $r = y - xq$ provide a suitable quotient and remainder for division of y by x.

Since the ring of Gaussian integers has a Euclidean function, this ring is a principal ideal domain, and thereby also a unique factorization domain. In the section to follow we will identify the primes of $\mathbb{Z}[i]$, and in the process answer a classical question about integers that are sums of squares.

By way of contrast, the ring $\mathbb{Z}[\sqrt{-5}]$, which we have considered, does not possess a Euclidean function. For if it did, then Proposition 5.17 would say that this ring was a principal ideal domain. However, in $\mathbb{Z}[\sqrt{-5}]$ we had found irreducibles that were not primes, and because of this, Proposition 5.11 shows that $\mathbb{Z}[\sqrt{-5}]$ is not a principal ideal domain.

EXERCISES

1. If K is a field show that the integral domain $K[X]$ has infinitely many irreducibles.

Hint. One way is to imitate the proof that there are infinitely many primes in $\mathbb{Z}$.

If the field K is finite show that $K[X]$ contains irreducible polynomials of arbitrarily high degree. In due course we shall come to see that for every possible degree n, there exists an irreducible in $K[X]$ of degree n.

2. If K is a field, show that the function ϕ, given by $\phi(a) = 1$ when $a \neq 0$, is a Euclidean function.

3. Let $\mathbb{Z}[\sqrt{-2}]$ be the integral domain consisting of all complex numbers of the form $a + b\sqrt{-2}$ where a, b are integers. Show that $\phi\left(a + b\sqrt{-2}\right) = a^2 + 2b^2$ defines a Euclidean function ϕ on $\mathbb{Z}[\sqrt{-2}]$.

4. Let K be a field, and $K[[X]]$ the ring of formal power series in the variable X with coefficients in K.

(a) If $f = a_0 + a_1X + a_2X^2 + \cdots + a_nX^n + \cdots$ is a non-zero series in $K[[X]]$, let $\phi(f)$ be the smallest index n such that $a_n \neq 0$. Show that ϕ is a Euclidean function on $K[[X]]$.

(b) Identify the units of $K[[X]]$.

(c) Identify the primes of $K[[X]]$, and indicate which primes are associates of each other.

(d) If $f \in K[X]$ and f is not a unit, describe the unique factorization of f into primes.

5. Let $\phi : A \setminus \{0\} \to \mathbb{N}$ be a Euclidean function on an integral domain A. Prove that a non-zero element a of A is such that $\phi(a)$ is minimal in the set $\phi(A \setminus \{0\})$, if and only if a is a unit of A.

6. Let $\mathbb{Z}[\sqrt{-5}]$ be the domain of all numbers of the form $a + b\sqrt{-5}$ where a,b are integers. We have seen that this ring is not a principal ideal domain. Prove that the ideal $J = \langle 2, 1 + \sqrt{-5}\rangle$ inside $\mathbb{Z}[\sqrt{-5}]$ is not principal.

7. Is a homomorphic image of a unique factorization domain another unique factorization domain? Is a subring of a unique factorization domain another unique factorization domain?

8. The set $\mathbb{Q}[\sqrt{2}]$ consists of all real numbers of the form $a + b\sqrt{2}$ where $a, b \in \mathbb{Q}$ is a field inside $\mathbb{R}$. Inside $\mathbb{Q}[\sqrt{2}]$ there is the subring $\mathbb{Z}[\sqrt{2}]$ consisting of all $a + b\sqrt{2}$ such that $a, b \in \mathbb{Z}$. Let $\phi : \mathbb{Q}[\sqrt{2}] \to \mathbb{Q}$ be the function given by $a + b\sqrt{2} \mapsto |a^2 - 2b^2|$. The restriction of ϕ to $\mathbb{Z}[\sqrt{2}] \setminus \{0\}$ maps the subring $\mathbb{Z}[\sqrt{2}]$ into the non-negative integers $\mathbb{N}$. This exercise is to show that ϕ is a Euclidean function for $\mathbb{Z}[\sqrt{2}]$.

(a) Clearly $\phi(x) \geq 0$ for all x in $\mathbb{Q}[\sqrt{2}]$.
Verify that $\phi(x) = 0$ if and only if $x = 0$.

(b) Verify that $\phi(xy) = \phi(x)\phi(y)$ for all x, y in $\mathbb{Q}[\sqrt{2}]$.

(c) Let $x, y \in \mathbb{Z}[\sqrt{2}]$, and such that $x \neq 0$. The quotient $\frac{y}{x}$ lies in $\mathbb{Q}[\sqrt{2}]$, and thus takes the form $r + s\sqrt{2}$ where $r, s \in \mathbb{Q}$. Clearly there exist integers m, n such that $|r - m| \leq \frac{1}{2}$ and $|s - n| \leq \frac{1}{2}$. The element $q = m + n\sqrt{2}$ is in $\mathbb{Z}[\sqrt{2}]$.

Show that $\phi(y - xq) < \phi(x)$, and deduce that ϕ is a Euclidean function for $\mathbb{Z}[\sqrt{2}]$.

5.4 Gaussian Primes and Sums of Squares

Some integers are the sum of squares of two integers. For example, $73 = 7^2 + 3^2$ and $29 = 2^2 + 5^2$. Others, such as $3, 7, 19$, are not. Besides checking the myriad of possibilities for a given integer, is there a way to tell? This problem is closely related to finding the primes in the ring of Gaussian integers. Indeed, the identity $a^2 + b^2 = (a + ib)(a - ib)$ reveals that the sum of squares problem is a factoring problem in $\mathbb{Z}[i]$.

As already shown, the function $\phi : \mathbb{Z}[i] \to \mathbb{N}$ given by $\phi(a + bi) = a^2 + b^2$, for every Gaussian integer $a+bi$, is a Euclidean function. For convenience let us call ϕ the ***norm function*** on $\mathbb{Z}[i]$. Note that $\phi(z) = z\overline{z}$ for every Gaussian integer z. Clearly, an integer is a sum of two squares if and only if the integer is equal to $\phi(z)$ for some Gaussian integer z.

As seen at the start of Section 4.2.1, a Gaussian integer u is a unit if and only if $\phi(u) = 1$, and this occurs if and only if u is one of ± 1, $\pm i$. Since $\mathbb{Z}[i]$ is a principal ideal domain, the primes and irreducibles of $\mathbb{Z}[i]$ coincide, and every non-zero, non-unit of $\mathbb{Z}[i]$ has a factorization into primes. Furthermore, the primes z involved in the factorization must be unique except for their possible replacement by associates $\pm z$, $\pm iz$.

A Gaussian integer z is prime if and only if its complex conjugate $\overline{z}$ is prime. This is because the conjugation mapping $\mathbb{Z}[i] \to \mathbb{Z}[i]$ given by $z \mapsto \overline{z}$ is a ring automorphism.

Some primes in $\mathbb{Z}$ fail to remain primes in $\mathbb{Z}[i]$. For example, 5 is a prime in $\mathbb{Z}$, but not in $\mathbb{Z}[i]$, because $5 = (2 + i)(2 - i)$. We shall refer to primes in $\mathbb{Z}$ as ***rational primes***, and to primes in $\mathbb{Z}[i]$ as ***Gaussian primes***. For instance, 5 is a rational prime but not a Gaussian prime.

Proposition 5.18. *A Gaussian integer z is a Gaussian prime if and only if:*

- *its norm $\phi(z)$ is a rational prime, or*
- *one of its associates $\pm z$, $\pm iz$ is a positive, rational prime that cannot be expressed as a sum of two squares.*

Proof. Suppose $\phi(z)$ is a rational prime. To see that z is a Gaussian prime, it suffices to check that z is irreducible in $\mathbb{Z}[i]$. To that end suppose $z = xy$ for some Gaussian integers x, y. Then $\phi(z) = \phi(xy) = \phi(x)\phi(y)$. Since $\phi(z)$ is a rational prime, one of $\phi(x)$ or $\phi(y)$ is ± 1. Hence one of x or y is a unit in $\mathbb{Z}[i]$. Thus z is irreducible.

Next suppose that one of the associates $\pm z$, $\pm iz$ is a positive rational prime that is not a sum of two squares. Without harm say z itself is a positive integer and z is not a sum of two squares. Now suppose $z = xy$ for some Gaussian integers x, y. Thus $z^2 = \phi(z) = \phi(x)\phi(y)$. Since z is a positive, rational prime, unique factorization in $\mathbb{Z}$ implies that $\phi(x)$ is one of $1, z$ or z^2. The possibility that $\phi(x) = z$ is ruled out by the fact z is not a sum of two squares. Thus $\phi(x)$ equals 1 or z^2. In the first case x is a unit in $\mathbb{Z}[i]$. In the second case $\phi(y) = 1$, and y is a unit in $\mathbb{Z}[i]$. Thus z is irreducible in $\mathbb{Z}[i]$, and thereby a Gaussian prime.

For the converse, we suppose that z is a Gaussian prime but that $\phi(z)$ is not a rational prime, and prove that one of $\pm z$, $\pm\, iz$ is a positive rational prime which is not the sum of two squares.

Note that $\phi(z) > 1$, because the Gaussian prime z is non-zero and not a unit. Thus $\phi(z)$ is a product of some rational primes. Since $\phi(z) = z\overline{z}$ and z is a Gaussian prime, z divides one of the rational prime factors p of $z\overline{z}$. Therefore

$$z\overline{z} = pt \text{ and } p = zu$$

for some positive integer t and some Gaussian integer u. Hence $z\overline{z} = zut$, and then $\overline{z} = ut$.

Because $\overline{z}$ is a Gaussian prime, one of u or t is a unit in $\mathbb{Z}[i]$. If the positive integer t is a unit, it must be that $t = 1$, and then $\phi(z) = z\overline{z} = p$, contrary to the assumption that $\phi(z)$ is not a rational prime. Hence u is a unit, and therefore $z = u^{-1}p$, a unit multiple of a rational prime.

It remains to see that p is not the sum of two squares. If p were such a sum, we would have that $z = u^{-1}p = u^{-1}w\overline{w}$ for some Gaussian integer w, which is an impossible factorization because z is a Gaussian prime. □

Proposition 5.18 leads to an inevitable question. How do we detect if a prime (or any integer for that matter) is the sum of two squares?

Primes that are Sums of Squares

Proposition 5.18 tells us how to build all Gaussian primes from rational primes. Start with a positive, rational prime p. If p is a sum of two squares, say $p = a^2 + b^2 = (a+ib)(a-ib)$, then the Gaussian integers $a \pm bi$ and their four respective associates are Gaussian primes. If p is not a sum of two squares, then p itself and its associates are Gaussian primes. We are hereby bound to ask what makes a positive prime integer be a sum of two squares. The answer lies in a theorem of Pierre Fermat. Since $2 = 1^2 + 1^2$, we need only consider odd primes.

Proposition 5.19. *An odd, positive, prime integer p is a sum of two squares if and only if $p \equiv 1 \bmod 4$.*

Proof. Suppose $p = a^2 + b^2$ for some integers a, b. The squares a^2, b^2 are each congruent mod 4 to 0 or 1. Thus, p is congruent mod 4 to one of the sums

$$0 = 0 + 0, \quad 1 = 1 + 0, \quad 1 = 0 + 1, \quad 2 = 1 + 1.$$

In particular, $p \not\equiv 3 \bmod 4$. Since p is odd, $p \not\equiv 0$ and $p \not\equiv 2 \bmod 4$ either. Hence $p \equiv 1 \bmod 4$.

For the converse, suppose $p \equiv 1 \bmod 4$. This means that $\frac{p-1}{4}$ is an integer. We exploit the finite field $\mathbb{Z}_p$. The polynomial $X^{(p-1)/2} - 1$ over $\mathbb{Z}_p$ has at most $\frac{p-1}{2}$ roots, while $\mathbb{Z}_p$ has $p - 1$ non-zero elements. So there is a non-zero element c in $\mathbb{Z}_p$ which is not a root of this polynomial. Put

$$a = c^{(p-1)/4} \text{ and } b = a^2 = c^{(p-1)/2}.$$

According to Fermat's little theorem, Proposition 2.38, $b^2 = c^{p-1} = 1$ in $\mathbb{Z}_p$. Since the polynomial $X^2 - 1$ has ± 1 as its only roots, $b = \pm 1$ in $\mathbb{Z}_p$. By the choice of c it must be that $b = -1$, and thus $a^2 = -1$. The element a in $\mathbb{Z}_p$ is such that $a^2 + 1 = 0$.

We have just shown that if $p \equiv 1 \bmod 4$, then there is an integer x such that $p \mid x^2 + 1$. Within the Gaussian integers $p \mid (x + i)(x - i)$. Evidently $p \nmid x + i$ and $p \nmid x - i$, whereby p is not a Gaussian prime. According to Proposition 5.18, our positive, rational prime p is a sum of two squares. □

Having seen what it takes for a prime to be a sum of two squares, we might be curious about the situation with an arbitrary, positive integer. For that, we need to discuss the number of times that a prime appears in the factorization of elements in unique factorization domains.

Multiplicity of Primes in Unique Factorizations and Sums of Squares

If z is a non-zero element in a unique factorization domain A and p is a prime in A, there is a unique non-negative exponent m such that

$$p^m \mid z \text{ while } p^{m+1} \nmid z.$$

Such m, which is precisely the number of times that p appears in the unique factorization of z, is called the ***multiplicity*** of p in z. Only a finite number of p actually appear in the factorization of z. These have positive multiplicity, and the other primes have multiplicity zero. In particular, z is a unit if and only if all primes have zero multiplicity in z. All associates of p have the same multiplicity in z.

Proposition 5.20. *A positive integer n is a sum of two squares if and only if all primes p, such that $p \equiv 3 \bmod 4$, have even multiplicity in n.*

Proof. If integers a, b are each sums of two squares, then so is their product ab. Indeed, with the norm function ϕ, write $a = \phi(x)$ and $b = \phi(y)$ for some Gaussian integers x, y. Then $ab = \phi(x)\phi(y) = \phi(xy)$, which is a sum of two squares.

Suppose all primes which are congruent to 3 modulo 4 have even multiplicity in n. Then $n = s^2 t$ for some positive integers s, t, where the prime factors of t are either equal to 2 or are congruent to 1 modulo 4. By Proposition 5.19 the prime factors of t are sums of two squares. Also $s^2 = s^2 + 0^2$, a sum of squares. Thus n is a product of integers which are sums of two squares. As noted above, n is itself a sum of two squares.

The converse hinges on the fact $\mathbb{Z}[i]$ is a unique factorization domain, along with the observation that an ordinary integer c divides a Gaussian integer $a + ib$ if and only if c divides both a and b in $\mathbb{Z}$. Thus c divides $a + ib$ if and only if c divides $a - ib$.

Suppose that $n = a^2 + b^2$ and that p is a prime such that $p \equiv 3 \bmod 4$. By Propositions 5.19 and 5.18 in that order, p is a Gaussian prime. From the preceding observation, the multiplicity of the Gaussian prime p in $a + ib$ equals its multiplicity in $a - ib$. Since $n = (a + ib)(a - ib)$, the multiplicity of p in the factorization of n into Gaussian primes is even. That even multiplicity of p is the same multiplicity that p has in the factorization of n into ordinary integers. That is because an ordinary integer divides another ordinary integer in $\mathbb{Z}[i]$ if and only if it does so already in $\mathbb{Z}$. □

A Way to Factor a Gaussian Integer

Even though a practical algorithm for factoring huge integers within $\mathbb{Z}$ remains out of reach, here is a way to factor a Gaussian integer z subject to having the factorization of its norm $\phi(z)$ in $\mathbb{Z}$.

Clearly, every non-zero, non-unit Gaussian integer can be written in the form $x(a+bi)$ where x is a positive rational integer and a, b are coprime integers. Just take out the greatest common divisor of the real and imaginary parts of z. Thus the factorization of a Gaussian integer reduces to factoring rational integers into Gaussian integers and factoring Gaussian integers of the form $a+ib$ where a, b are coprime.

To factor a positive integer x into Gaussian primes, first write $x = p_1 p_2 \cdots p_k$ where the p_j are positive primes in $\mathbb{Z}$. If any $p_j \equiv 3 \bmod 4$, such p_j is not the sum of two squares, and by Proposition 5.18, such p_j is already a Gaussian prime. Those $p_j \equiv 1 \bmod 4$ or $p_j = 2$ are such that

$$p_j = c^2 + d^2 = (c+id)(c-id)$$

for some integers c, d, because of Proposition 5.19. By Proposition 5.18 the factors $c \pm id$ are Gaussian primes. Thus all of the prime factors of x inside $\mathbb{Z}[i]$ get picked up.

For example,

$$6125 = 5^3 \cdot 7^2 = ((2+i)(2-i))^3 \cdot 7^2 = (2+i)^3 \cdot (2-i)^3 \cdot 7^2.$$

Next consider a non-unit $z = a+ib$ where a, b are coprime.

Say the unique factorization of z into Gaussian primes, up to associates, is given by $z = w_1 w_2 \cdots w_k$. All of the Gaussian primes w_j are such that their norms $\phi(w_j)$ are primes in $\mathbb{Z}$. Otherwise, in accordance with Proposition 5.18, some w_j would be a unit multiple of a rational prime p. That p would obviously divide z, and thereby divide both a and b. This would be contrary to the fact a, b are coprime.

Thus, the equation $\phi(z) = \phi(w_1)\phi(w_2)\cdots\phi(w_k)$ gives the unique factorization of $\phi(z)$ into rational primes. This shows that if p is a rational prime factor of $\phi(z)$, then there exists a prime factor w of z such that $\phi(w) = p$.

How do we find such a Gaussian prime factor w of z given that we have a rational prime factor p of $\phi(z)$? Well, being the norm of a Gaussian prime w, the rational prime p is a sum of two squares. Suppose we found integers c, d (for instance by checking the possibilities) such that

$$p = c^2 + d^2 = (c+id)(c-id).$$

Both $c \pm id$ are Gaussian primes by Proposition 5.18. We also have $p = w\overline{w}$. From these two factorizations of p in the unique factorization domain $\mathbb{Z}[i]$, it follows that w is an associate of $c+id$ or of $c-id$. In particular, one of the Gaussian primes $c+id$ or $c-id$ divides z. We can now divide both of these into z as complex numbers to see which quotient gives a Gaussian integer. In this way we can peel off the Gaussian prime factors of z.

For example, take $z = 55 + 63i$. By inspection 55 and 63 are coprime.

The prime factorization of the norm $\phi(z)$ is given by

$$\phi(z) = 55^2 + 63^2 = 6694 = 2 \cdot 13 \cdot 269.$$

Since 2 divides $\phi(z)$ and $2 = (1+i)(1-i)$, one of the Gaussian primes $1 \pm i$ divides z. After testing both divisions, we learn that $z = (1-i)(-4+59i)$.

Now work on $-4+59i$. Its norm is $\phi(-4+59i) = 3497 = 13 \cdot 269$. Since the rational prime 13 divides $-4+59i$ and $13 = 2^2 + 3^2 = (2+3i)(2-3i)$, one of the Gaussian primes $2 \pm 3i$ divides $-4+59i$. After dividing we learn that $-4+59i = (2+3i)(13+10i)$.

It remains to factor $13+10i$. Calculating its norm we get $\phi(13+10i) = 269$, which is a rational prime. Therefore $13+10i$ is a Gaussian prime. The factorization of z into Gaussian primes is given by

$$z = (1-i)(2+3i)(13+10i),$$

up to a suitable replacement of the factors by associates.

We should observe that our ability to factor in $\mathbb{Z}[i]$ in practice hinges on our ability to factor in $\mathbb{Z}$ as well as to express primes which are congruent to 1 modulo 4 as sums of squares.

EXERCISES

1. Factor the Gaussian integers 273 and $8 - 11i$ into Gaussian primes.
2. Is 637 a sum of two squares? What about 4459?
3. Suppose that p is a prime and that $p = a^2 + b^2$ for some integers a, b. Show that a, b are uniquely determined by p, except for a possible change of sign and possible interchange of a and b.
4. Show that the function $\epsilon : \mathbb{Z}[i] \to \mathbb{N}$ given by $\epsilon(a+ib) = |a| + |b|$ is not a Euclidean function on the ring of Gaussian integers.
5. (a) If x is a Gaussian integer that is the sum of squares of finitely many other Gaussian integers, prove that x is also the sum of squares of three Gaussian integers.
   (b) Prove that $2+2i$ is a sum of three squares in $\mathbb{Z}[i]$, but not a sum of two squares.
6. Find an integral domain B and a subring A with an element p which is prime in A but not in B. Then find examples where $p \in A$, p is prime in B but p is not prime in A.

5.5 Greatest Common Divisors

For two integers a, b, we know that their greatest common divisor, denoted by $\gcd(a, b)$, is a common factor of a and b that is divisible by all other common factors. Except for a possible change of sign, $\gcd(a, b)$ is uniquely determined by a and b. In Section 1.1 we outlined the Euclidean algorithm that computes $\gcd(a, b)$.

This concept extends to any finite set of elements in any integral domain.

Definition 5.21. Let A be an integral domain, and let $a_1, a_2, \ldots, a_n$ be a list of elements in A. An element d in A is called a ***greatest common divisor*** of $a_1, a_2, \ldots, a_n$ whenever

- $d \mid a_i$ for all i, and
- $c \in A$ and $c \mid a_i$ for all i, then $c \mid d$.

Notice that 0 is a greatest common divisor if and only if all a_i are 0. We can see that any two greatest common divisors of $a_1, a_2, \ldots, a_n$ are mutual factors of each other, and are thereby associates. Conversely, any associate of a greatest common divisor of $a_1, a_2, \ldots, a_n$ is again a greatest common divisor. In particular, the identity element 1 is a greatest common divisor of $a_1, a_2, \ldots, a_n$ if and only if the only common factors of these elements are the units of A.

Definition 5.22. We say that $a_1, a_2, \ldots, a_n$ are ***coprime*** when they have 1 as a greatest common divisor.

In the language of ideals, the first condition for being a greatest common divisor states that the principal ideal $\langle d \rangle$ contains the ideal $J = \langle a_1, a_2, \ldots, a_n \rangle$ generated by the a_i. The second condition states that any other principal ideal $\langle c \rangle$ containing J already contains the ideal $\langle d \rangle$. Thus, $a_1, a_2, \ldots, a_n$ have a greatest common divisor if and only if the ideal $\langle a_1, a_2, \ldots, a_n \rangle$ is contained in a principal ideal that, in turn, sits inside all other principal ideals containing $\langle a_1, a_2, \ldots, a_n \rangle$. This leads to a simple fact.

Proposition 5.23. *Every finite set of elements in a principal ideal domain has a greatest common divisor.*

Proof. If $a_1, a_2, \ldots, a_n$ are those elements, the ideal $\langle a_1, a_2, \ldots, a_n \rangle$ is principal, and clearly minimal among all ideals containing it. □

A Domain without Greatest Common Divisors

In some domains greatest common divisors fail to exist.

For example, let A be the subring of $\mathbb{Q}[X]$ consisting of those polynomials of the form

$$a_0 + a_2X^2 + a_3X^3 + \cdots + a_nX^n,$$

where the coefficient of X is 0. In $\mathbb{Q}[X]$ the polynomials X^5 and X^6 have X^5 as their greatest common divisor. Yet, in A they have no greatest common divisor.

To see this, suppose that g is a greatest common divisor of X^5 and X^6 in A. Since g divides X^5 in A, it must be that g divides X^5 in $\mathbb{Q}[X]$. Thus g is one of $1, X, X^2, X^3, X^4$ or X^5, or any non-zero constant multiple of these. Certainly $g \neq X$ because $X \notin A$. We can rule out X^5 because $X^5 \nmid X^6$ in A, due to the fact $X \notin A$. Likewise, we rule out X^4 because $X^4 \nmid X^5$ in A. If g were equal to X^3, then the fact $X^2 \mid X^5$ and $X^2 \mid X^6$ in A would force $X^2 \mid X^3$ in A, which is again prohibited by the fact that $X \notin A$. If g equals 1 or X^2, then the fact that $X^3 \mid X^5$ and

$X^3 \mid X^6$ in A forces $X^3 \mid g$, which is not so because the degree of g is too small. Since none of the above powers of X is a greatest common divisor of X^5 and X^6, neither are any of their constant multiples. Thus X^5 and X^6 have no greatest common divisor in the domain A.

The Greatest Common Divisor Algorithm for Euclidean Domains

A replica of the GCD-algorithm, outlined for the ring of integers in Section 1.1, will find greatest common divisors in any Euclidean domain. Here is the GCD-algorithm in the case of two elements a, b with $a \neq 0$ in any domain A with Euclidean function ϕ.

Use the Euclidean function as much as possible to pick up quotients $q_1, q_2, \ldots$ and non-zero remainders $r_1, r_2, \ldots$ in A such that

$$\begin{aligned} b &= aq_1 + r_1 & \qquad \phi(r_1) &< \phi(a) \\ a &= r_1q_2 + r_2 & \phi(r_2) &< \phi(r_1) \\ r_1 &= r_2q_3 + r_3 & \phi(r_3) &< \phi(r_2) \\ r_2 &= r_3q_4 + r_4 & \phi(r_4) &< \phi(r_3) \\ &\vdots & &\vdots \end{aligned}$$

The $\phi(r_j)$ cannot strictly decrease indefinitely, because they are non-negative integers. At some point we must arrive at:

$$\begin{aligned} r_{k-2} &= r_{k-1}q_k + r_k & \qquad \phi(r_k) &< \phi(r_{k-1}) \\ r_{k-1} &= r_kq_{k+1} + 0. \end{aligned}$$

The first of our equations reveals that the common divisors of b, a coincide with the common divisors of a, r_1. Likewise, the second equation gives that the common divisors of a, r_1 coincide with the common divisors of r_1, r_2. Therefore the pair b, a has the same common divisors as the pair r_1, r_2. Repeating this observation down the equations of the GCD-algorithm we discover that the pair b, a has the same common divisors as the pair $r_k, 0$. It follows that a greatest common divisor for b, a is a greatest common divisor for $r_k, 0$. By inspection r_k is a greatest common divisor of this latter pair, and thus r_k is a greatest common divisor of b, a.

We recall that the ideals of Euclidean domains are principal, and in the language of ideals, a greatest common divisor of b, a is any single generator for the ideal $\langle b, a \rangle$. Such a generator d is an A-linear combination of b and a. That is $d = ax + by$ for some x, y in A. The GCD-algorithm shows us how to find such x and y, as follows.

The second last equation of the algorithm reveals that r_k is an A-linear combination of r_{k-1} and r_{k-2}. The equation prior to that reveals that r_{k-1} is an A-linear combination of r_{k-2} and r_{k-3}. By substitution r_k becomes an A-linear combination of r_{k-2} and r_{k-3}. Proceed with this argument up along the equations of the GCD-algorithm until it emerges that r_k is an A-linear combination of b and a.

An Example for Polynomials

Take the polynomials

$$f(X) = X^4 + X^3 - X^2 + 1 \text{ and } g(X) = X^3 + X^2 + X + 1.$$

Let us compute a greatest common divisor for these two polynomials by using the GCD-algorithm in the Euclidean domain $\mathbb{Q}[X]$. Thus we have

$$X^4 + X^3 - X^2 + 1 = (X^3 + X^2 + X + 1)(X) + (-2X^2 - X + 1)$$
$$X^3 + X^2 + X + 1 = (-2X^2 - X + 1)\left(-\frac{1}{2}X - \frac{1}{4}\right) + \left(\frac{5}{4}X + \frac{5}{4}\right)$$
$$-2X^2 - X + 1 = \left(\frac{5}{4}X + \frac{5}{4}\right)\left(-\frac{8}{5}X + \frac{4}{5}\right) + 0.$$

A greatest common divisor for the given polynomials $f(X)$ and $g(X)$ is the last non-zero remainder, $\frac{5}{4}X + \frac{5}{4}$. This could be replaced by its associate $X + 1$, which is more pleasing to the eye.

Since the greatest common divisor $\frac{5}{4}X + \frac{5}{4}$ is a single generator for the ideal in $\mathbb{Q}[X]$ generated by $f(X)$ and $g(X)$, we can express this greatest common divisor as a $\mathbb{Q}[X]$-linear combination of $f(X)$ and $g(X)$. This is done in practice by working up in the GCD-algorithm above. The middle equation above gives

$$\frac{5}{4}X + \frac{5}{4} = g(X) - (-2X^2 - X + 1)\left(-\frac{1}{2}X - \frac{1}{4}\right).$$

The first equation gives

$$-2X^2 - X + 1 = f(X) - g(X)X.$$

Substitution of the latter expression for $-2X^2 - X + 1$ into the equation above it gives

$$\frac{5}{4}X + \frac{5}{4} = g(X) - (f(X) - g(X)X)\left(-\frac{1}{2}X - \frac{1}{4}\right).$$

Thus

$$\frac{5}{4}X + \frac{5}{4} = f(X)\left(\frac{1}{2}X + \frac{1}{4}\right) + g(X)\left(-\frac{1}{2}X^2 - \frac{1}{4}X + 1\right),$$

which expresses our greatest common divisor as a $\mathbb{Q}[X]$-linear combination of $f(X)$ and $g(X)$.

Multiplicity Functions in Unique Factorization Domains and Greatest Common Divisors

In a unique factorization domain A every finite list of elements has a greatest common divisor, obtainable from unique factorization in A. To see how it comes about, the use of multiplicity functions might be helpful.

Suppose one prime from each equivalence class of associated primes in A has been selected, and let $\mathcal{P}$ be the collection of these selected primes. For instance, in the case of

$\mathbb{Z}$, we normally take $\mathcal{P}$ to be the set of positive primes. In the case of $K[X]$, where K is a field, we normally take $\mathcal{P}$ to be the set of irreducible polynomials which are monic. In the domain $\mathbb{Z}[i]$ of Gaussian integers, each prime z has one associate in each of the four quadrants of the complex plane. So we could take $\mathcal{P}$ to be the set of Gaussian primes that lie in the first quadrant, including the x-axis and excluding the y-axis. We shall refer to such $\mathcal{P}$ as a ***representative set of primes*** for A.

Unique factorization says that every non-zero, non-unit element a in A can be factored in the form

$$a = up_1^{m_1} p_2^{m_2} \cdots p_k^{m_k}$$

where u is a unit, the p_j are distinct primes taken from $\mathcal{P}$, and the m_j are positive exponents *uniquely* determined by a. Typically m_j is called the ***multiplicity*** of the prime p_j in a.

For each non-zero a in A and every prime p in $\mathcal{P}$, let

$$m_a(p) = m_j \text{ if } p \text{ is one of the primes } p_j \text{ appearing in the unique factorization of } a.$$

Otherwise, let

$$m_a(p) = 0.$$

For each non-zero element a, the resulting function

$$m_a : \mathcal{P} \to \mathbb{N}$$

is called the ***multiplicity function*** of a on the representative set of primes $\mathcal{P}$.

A bit of reflection reveals the following properties of multiplicity functions. If a, b are non-zero elements of A, then

- $m_a(p) > 0$ only at the finitely many p in $\mathcal{P}$ that divide a,
- a is a unit if and only if $m_a(p) = 0$ for all primes p in $\mathcal{P}$,
- $m_{ab}(p) = m_a(p) + m_b(p)$ for all p in $\mathcal{P}$,
- $a \mid b$ in A if and only if $m_a(p) \leq m_b(p)$ for all p in $\mathcal{P}$,
- $a = p^{m_a(p)}c$ for some c such that $p \nmid c$,
- a, b are associates if and only if $m_a(p) = m_b(p)$ for all p in $\mathcal{P}$.
- If $m : \mathcal{P} \to \mathbb{N}$ is any function which takes at most finitely many non-zero values, then there exists a non-zero c in A such that $m = m_c$. Namely, let $p_1, p_2, \ldots, p_k$ be the finitely many p_j in $\mathcal{P}$ for which $m(p_j) > 0$. After that take c to be the product

  $$p_1^{m(p_1)} p_2^{m(p_2)} \cdots p_k^{m(p_k)}.$$

  For instance, if m is the zero function, then $c = 1$.

Using the preceding observations here is how to construct greatest common divisors in unique factorization domains.

Proposition 5.24. *Let A be a unique factorization domain and let $\mathcal{P}$ be a representative set of primes for A. Let $a_1, a_2, \ldots, a_n$ be a list of elements of A, not all of them 0. Define $m : \mathcal{P} \to \mathbb{N}$ according to*

$$m(p) = \min\{m_{a_j}(p) : a_j \neq 0\}, \textit{ for all } p \textit{ in } \mathcal{P}.$$

If d in A is constructed such that its multiplicity function $m_d = m$, then d is a greatest common divisor of $a_1, a_2, \ldots, a_n$.

Proof. To see that d divides all non-zero a_j, observe that $m_d(p) \leq m_{a_j}(p)$ for all p in $\mathcal{P}$. Clearly d divides all a_j which are 0. Next suppose that c in A is such that c divides all a_j. Thus $m_c(p) \leq m_{a_j}(p)$ for all non-zero a_j and all p in $\mathcal{P}$. It follows that $m_c(p) \leq m(p) = m_d(p)$, whereby $c \mid d$. □

Although a greatest common divisor of $a_1, \ldots, a_n$ is uniquely determined only up to associates, a rule for selecting one is usually adopted, and with an innocuous abuse of precision we often write a chosen greatest common divisor as

$$\gcd(a_1, a_2, \ldots, a_n).$$

For instance, Proposition 5.24 tells us that in $\mathbb{Z}$:

$$\gcd(2^4 3^0 5^6 7^1, 2^2 3^8 5^3 7^0, 2^4 3^2 5^3 7^6) = 2^2 3^0 5^3 7^0 = 500.$$

Coprimeness in Unique Factorization Domains

Elements $a_1, a_2, \ldots, a_n$ in an integral domain A are ***coprime*** when 1 is a suitable greatest common divisor. If A is a unique factorization domain with a representative set of primes $\mathcal{P}$, this is saying that the function $m : \mathcal{P} \to \mathbb{N}$ defined in Proposition 5.24 is the zero function. Therefore the elements $a_1, a_2, \ldots, a_n$ are coprime when there is no prime p in $\mathcal{P}$ that divides all a_j. Since every prime of A is associated to a prime of $\mathcal{P}$, we see that the a_j are coprime if and only if no prime of A divides all a_j.

Here are a few observations that harken back to Propositions 1.4 and 1.5.

Proposition 5.25. *Let a, b, c lie in a unique factorization domain A with a representative set of primes $\mathcal{P}$.*

- *If a, b are coprime and $a \mid bc$, then $a \mid c$.*
- *If a, b are coprime and $a \mid c$ and $b \mid c$, then $ab \mid c$.*
- *If $d = \gcd(a, b)$ and r, s in A are such that $a = dr, b = ds$, then r, s are coprime.*

Proof. As to the first item, since a, b are coprime, there is no p in $\mathcal{P}$ for which both $m_a(p) > 0$ and $m_b(p) > 0$. In particular if $m_b(p) > 0$, then $m_a(p) = 0 \leq m_c(p)$. On the other hand if $m_b(p) = 0$, then

$$m_a(p) \leq m_{bc}(p) = m_b(p) + m_c(p) = m_c(p).$$

In any case $m_a(p) \leq m_c(p)$, meaning that $a \mid c$.

Since $m_{ab}(p) = m_a(p) + m_b(p)$, the second item comes down to showing that

$$m_a(p) + m_b(p) \leq m_c(p) \text{ for all } p \text{ in } \mathcal{P}.$$

Since a, b are coprime, one of $m_a(p)$ or $m_b(p)$ is 0. Thus the inequality holds because $m_a(p) \leq m_c(p)$ and $m_b(p) \leq m_c(p)$.

With the third item, note that $\gcd(a, b)$ is defined by the multiplicity function $m(p) = \min(m_a(p), m_b(p))$. For all p in $\mathcal{P}$ we have that

$$m_a(p) = m_d(p) + m_r(p) \text{ and } m_b(p) = m_d(p) + m_s(p).$$

If r, s were not coprime there would exist a prime p such that $m_r(p) > 0$ and $m_s(p) > 0$. Consequently, at this p we would get:

$$m_a(p) > m_d(p) \text{ and } m_b(p) > m_d(p).$$

But two numbers cannot both exceed their minimum. □

EXERCISES

1. Using the norm function on the ring of Gaussian integers, carry out the Euclidean algorithm to find a greatest common divisor of $2 + 8i$ and $19 - 5i$. Then obtain such a greatest common divisor by finding a factorization of these Gaussian integers into Gaussian primes.
2. Let A be a unique factorization domain with a representative set of primes $\mathcal{P}$. Show that a non-zero element a in A is a unit multiple of a perfect square if and only if its multiplicity function $m_a : \mathcal{P} \to \mathbb{N}$ takes on only even integer values.

   Describe the unit multiples of perfect cubes in terms of multiplicity functions. Generalize to perfect nth powers.
3. Use the definition of greatest common divisors to show that any $\gcd(a, b, c)$ and any $\gcd(a, \gcd(b, c))$ must be associates. Thus, the problem of finding greatest common divisors for more than two elements can be handled recursively working with just two elements at a time.
4. Let a, b, c, d be in an integral domain A. Show that a greatest common divisor for the pair a, b is also a greatest common divisor for the pair c, d if and only if the set of common divisors of a, b coincides with the set of common divisors of c, d.
5. Two elements a, b in an integral domain A are called *coprime* provided they are not units of A and the only common factors of a and b are units of A.
   (a) If $ax + by = 1$ for some x, y in A, prove that a, b are coprime.
   (b) In a principal ideal domain A show that a, b in A are coprime, if and only if $ax + by = 1$ for some x, y in A.
   (c) If A is a unique factorization domain, explain how a, b are coprime if and only if none of the primes appearing in a prime factorization of a are associated to any of the primes appearing in a prime factorization of b.
   (d) If a, b are coprime in a unique factorization domain A, and the product $ab = c^2$ for some c in A, show that $a = ut^2$ for some t, u in A where u is a unit.
   (e) Let $\mathbb{Q}(X, Y)$ be the field of rational functions in two variables X, Y and having rational coefficients. Let A be the subset $\mathbb{Q}(X, Y)$ consisting of all rational functions

of the form $\dfrac{f(X,Y)}{(X^2+Y^2)^k}$ where k runs over the non-negative integers, and f is a polynomial in $\mathbb{Q}[X,Y]$ for which the coefficients of all X^n and all Y^n are 0.

Show that A is a subring of $\mathbb{Q}(X,Y)$.

Show that $X^2s + Y^2t = 1$ for some s,t in A, and hence deduce that X^2 and Y^2 are coprime.

Show that X^2Y^2 is a perfect square in A.

Show that X^2 is not a unit multiple of a perfect square in A.

Now we see that it matters, in the preceding item, for A to be a unique factorization domain.

6. Suppose that A is a unique factorization domain, and a,b,c in A are such that $c = ab$ and a,b are coprime.
   (a) If c is a perfect square in A, show that a,b are perfect squares in A.
   (b) In the ring $\mathbb{Z}[\sqrt{-5}]$ find elements a,b,c such that the only common factors of a and b are ± 1, $c = ab$, c is a perfect square, but a,b are not perfect squares.
7. Here we exploit the fact that the ring of polynomials $K[X]$ over a field K is Euclidean to develop the partial fraction expansion of a rational function. In $K[X]$ the greatest common divisor for a pair of polynomials g,h, not both 0, is the unique monic generator of the ideal $\langle g,h\rangle$. Two such polynomials are coprime when the above ideal is all of $K[X]$.
   (a) If $f,g,h \in K[X]$ and g,h are coprime, show that

$$\frac{f}{gh} = \frac{p}{g} + \frac{q}{h}$$

   for some polynomials p,q.
   (b) If $f,g \in K[X]$ and $g \neq 0$, show that

$$\frac{f}{g} = q + \frac{r}{g},$$

   for some polynomials q,r such that either $r = 0$ or $\deg r < \deg g$.

   Use induction to show that for any positive exponent n:

$$\frac{f}{g^n} = q + \frac{r_1}{g} + \frac{r_2}{g^2} + \cdots + \frac{r_n}{g^n},$$

   where q and r_j are polynomials such that $\deg r_j < \deg g$ or $r_j = 0$.
   (c) In the polynomial ring $\mathbb{C}[X]$ over the complex numbers, the only irreducible polynomials in $\mathbb{C}[X]$ are the linear polynomials. If $f,g \in \mathbb{C}[X]$ and g is monic, use the unique factorization in $\mathbb{C}[X]$ to show that

$$\frac{f}{g} = \frac{f_1}{(X-\theta_1)^{n_1}} + \frac{f_2}{(X-\theta_2)^{n_2}} + \cdots + \frac{f_k}{(X-\theta_k)^{n_k}},$$

   where the $f_j \in \mathbb{C}[X]$, $\theta_1, \theta_2, \ldots, \theta_k$ are the roots of g and the exponents $n_j \geq 1$.

(d) Show that every rational function $\frac{f}{g}$ in $\mathbb{C}(X)$ can be expressed as the sum $p+q$ where p is a polynomial and q is a linear combination of functions of the form $\frac{1}{(X-\theta)^n}$ where $\theta \in \mathbb{C}$ and $n \geq 1$.

Such an expression is known as the ***partial fraction expansion*** of $\frac{f}{g}$.

(e) For this part some familiarity with infinite-dimensional vector spaces is required. Show that the set of functions

$$X^j \text{ where } j \geq 0, \text{ and } \quad \frac{1}{(X-\theta)^k} \text{ where } \theta \in \mathbb{C} \text{ and } k \geq 1$$

forms a basis for $\mathbb{C}(X)$ viewed as a vector space over $\mathbb{C}$.

8. A ***least common multiple*** for a pair of non-zero elements a, b in an integral domain A, is an element c that is divisible by a and b and divides all elements that are divisible by a and b. We explore the question of their existence and their relationship to greatest common divisors.

(a) Show that a, b in a domain A have a least common multiple if and only if the intersection $\langle a\rangle \cap \langle b\rangle$ is a principal ideal. In that case show that a generator for the intersection is a least common multiple of a and b.

(b) Let a, b be non-zero elements in a domain A and let c be a least common multiple of a, b. Since ab is divisible by both a and b it follows that $ab = cd$ for some element d in A. Show that d is a greatest common divisor of a and b.

Hint. From the fact $a \mid c$ deduce that $d \mid b$, and similarly that $d \mid a$. If e is such that $a = et$ and $b = es$ for some t, s, show that $bt = as$, and deduce that c divides this common element.

(c) Suppose that a, b are non-zero elements in a unique factorization domain A. Find a formula in terms of multiplicity functions to obtain a least common multiple of a and b. If $\operatorname{lcm}(a, b)$ denotes such a least common multiple, verify that

$$ab = \gcd(a, b) \operatorname{lcm}(a, b).$$

(d) Let K be a field and let A be the subring of $K[X]$ consisting of all polynomials for which the coefficient of X is 0. Show that in A the pair X^2, X^3 has 1 as its greatest common divisor, but does not have a least common multiple.

Curiously the presence of least common multiples ensures that of greatest common divisors, but not conversely.

9. (a) Suppose that in an integral domain A every pair of elements has a greatest common divisor. Show that every irreducible in A is also prime.

Hint. Let p be an irreducible that divides a product ab. Let $d = \gcd(ap, ab)$. Explain why $p \mid d$, $d = ae$ for some e in a, $ae \mid ap$, and e divides the irreducible p.

(b) In an integral domain containing an irreducible which is not prime, show how to find a pair of elements without a greatest common divisor.

(c) Let A be subring of $\mathbb{Q}[X]$ consisting of polynomials of the form $a_0 + a_2X^2 + a_3X^3 + \cdots + a_nX^n$ where the coefficient of X is 0. Explain why X^2 is irreducible but not prime. Then use part (b) to specify two elements in A having no greatest common divisor.

(d) In the domain $\mathbb{Z}[\sqrt{-5}]$ recall that the element 2 is irreducible but not prime. Find two elements in $\mathbb{Z}[\sqrt{-5}]$ which do not have a greatest common divisor.

5.6 Polynomials over Unique Factorization Domains

One goal of this section is to see that if a domain A has unique factorization into primes, then so does its polynomial ring $A[X]$. Thus for instance, the ring $\mathbb{Z}[X]$ will be a unique factorization domain, even though not all ideals of $\mathbb{Z}[X]$ are principal.

5.6.1 Gauss' Lemma

We begin with a discussion of factorization in $A[X]$ and its relationship to the polynomials of $K[X]$, where K is the field of fractions of A.

Primes of Degree Zero in Polynomial Rings

A prime in a subring of a given domain may lose its primeness as an element of the larger ring. For example, the integer 2 is prime in $\mathbb{Z}$, but in the larger ring of Gaussian integers it even loses its irreducibility, since 2 factors as $(1 + i)(1 - i)$. In the ring $\mathbb{Z}[\sqrt{-5}]$ the integer 2 keeps its irreducibility, but still loses its primeness because 2 does not divide either of $1 \pm \sqrt{-5}$ but does divide their product which is 6. Such a collapse of primeness does not occur when the larger ring is a ring of polynomials.

Even though they are most useful when the domains have unique prime factorization, the next two results hold for all integral domains,

Proposition 5.26. *If A is an integral domain and p is a prime in A, then p remains a prime in $A[X]$.*

Proof. With the quotient ring $A/\langle p\rangle$ comes the canonical projection $A \to A/\langle p\rangle$ whereby a in A gets mapped to its residue $\overline{a} = a + \langle p\rangle$ in $A/\langle p\rangle$. This gives rise to the mapping $\varphi : A[X] \to A/\langle p\rangle[X]$ defined by

$$\varphi(a_nX^n + a_{n-1}X^{n-1} + \cdots + a_1X + a_0) = \overline{a_n}X^n + \overline{a_{n-1}}X^{n-1} + \cdots + \overline{a_1}X + \overline{a_0},$$

for every polynomial $a_nX^n + a_{n-1}X^{n-1} + \cdots + a_1X + a_0$ in $A[X]$. Sometimes we say that φ reduces the coefficients of the polynomial modulo p. A simple check shows that φ is a ring homomorphism.

Since p is a prime in A, the principal ideal $\langle p\rangle$ is a prime ideal, and the quotient ring $A/\langle p\rangle$ is an integral domain. Therefore, the polynomial ring $A/\langle p\rangle[X]$, with coefficients in $A/\langle p\rangle$, is an integral domain.

Now suppose that $f, g \in A[X]$ and that $p \mid fg$ in $A[X]$. This says that p divides every coefficient of fg, and therefore

$$0 = \varphi(fg) = \varphi(f)\varphi(g) \text{ in } A/\langle p \rangle[X].$$

Since $A/\langle p \rangle[X]$ is an integral domain, either $\varphi(f) = 0$ or $\varphi(g) = 0$ in $A/\langle p \rangle[X]$. Hence, p divides all coefficients of f, or p divides all coefficients of g. That is, $p \mid f$ or $p \mid g$ in $A[X]$. □

Primitive Polynomials

Definition 5.27. If A is an integral domain, a non-zero polynomial f in $A[X]$ is called ***primitive*** provided there is no prime in A which divides all the coefficients of f.

Another way to say this is that there is no prime p in A such that $p \mid f$ in $A[X]$. For example, in $\mathbb{Z}[X]$ the polynomial $6X^3 - 9X^2 + 10X + 15$ is primitive. The primitive polynomials of degree 0 are the units of A.

Here is a somewhat important consequence of Proposition 5.26.

Proposition 5.28. *If A is an integral domain and f, g are primitive polynomials in $A[X]$, then the product fg remains primitive.*

Proof. If fg is not primitive, then some prime p in A divides all coefficients of fg. That is, $p \mid fg$ in $A[X]$. According to Proposition 5.26, $p \mid f$ or $p \mid g$, which implies that one of f or g is not primitive. □

The Content of Polynomials over Unique Factorization Domains

At this point let A be a unique factorization domain, with polynomial ring $A[X]$. For instance, the ring $\mathbb{Z}[X]$, the ring $K[Y][X]$ of polynomials in the variables Y, X over any field K, and the ring $\mathbb{Z}[i][X]$ of polynomials whose coefficients are Gaussian integers. We suppose that a representative set of primes $\mathcal{P}$ has been chosen for A. Each non-zero a in A has its multiplicity function $m_a : \mathcal{P} \to \mathbb{N}$.

Observe that an element a in A divides a polynomial f in $A[X]$ if and only if a divides every coefficient of f.

Proposition 5.29. *If a, b are in a unique factorization domain A and f is a primitive polynomial in $A[X]$ such that a divides bf in $A[X]$, then a already divides b in A.*

Proof. Using a contrapositive argument suppose that $a \nmid b$. Then there is a prime p in $\mathcal{P}$ such that $m_a(p) > m_b(p)$. This means there exist c, d in A and a non-negative exponent k such that

$$b = p^k c, p \nmid c \text{ and } a = p^{k+1} d.$$

We are given that $bf = ag$ for some g in $A[X]$. Thus $p^k cf = p^{k+1} dg$. Cancel p^k to get $cf = pdg$. Thus p divides all coefficients of cf. Since $p \nmid c$, the prime p divides all coefficients of f. Thus f is not primitive. □

If $f = a_0 + a_1X + a_2X^2 + \cdots + a_nX^n$ is a non-zero polynomial, Proposition 5.24 gives a recipe to obtain $\gcd(a_0, a_1, \ldots, a_n)$ in A. Namely, let

$$m(p) = \min\{m_{a_j}(p) : a_j \neq 0\}$$

for all p in $\mathcal{P}$, and define

$$\gcd(a_0, a_1, \ldots, a_n) = \prod p^{m(p)}$$

where the product is taken over the finitely many primes p where $m(p) > 0$.

Definition 5.30. If $f = a_0 + a_1X + a_2X^2 + \cdots + a_nX^n$ is a polynomial over a unique factorization domain A, the ***content*** of f is $\gcd(a_0, a_1, \ldots, a_n)$, and is denoted by $\kappa(f)$.

Using the positive primes in $\mathbb{Z}$ for example,

$$\kappa(6X^3 - 10X^2 + 22X - 4) = 2 \text{ and } \kappa(6X^3 - 70X^2 + 21X - 4) = 1.$$

Proposition 5.31. *Let A be a unique factorization domain with a representative set of primes $\mathcal{P}$ used to calculate the content $\kappa(f)$ of non-zero polynomials f. If f is such a polynomial, then $f = \kappa(f)g$ where g is a primitive polynomial.*

Proof. By the definition of content, $f = \kappa(f)g$ for some g in $A[X]$. If g were not primitive, there would be a prime p such that $p \mid g$. Thus $f = \kappa(f)ph$ for some h in $A[X]$. Since $\kappa(f)p$ divides every coefficient of f and $\kappa(f)$ is a greatest common divisor of those coefficients, it follows that $\kappa(f)p \mid \kappa(f)$. Thereby $p \mid 1$, which is a contradiction. □

We remind ourselves that every unique factorization domain A sits inside its field of fractions K, as discussed in Section 4.8.

Proposition 5.32 (Gauss' lemma). *Let A be a unique factorization domain with fraction field K, and let $A[X], K[X]$ be their respective polynomial rings.*

1. *If $f, g \in A[X]$ with f primitive and $f \mid g$ as polynomials in $K[X]$, then already $f \mid g$ in $A[X]$.*
2. *If $f \in A[X]$ and $f = gh$ for some polynomials g, h in $K[X]$, then there exist non-zero scalars r, s in the field K such that rg, sh are in $A[X]$ and $f = (rg)(sh)$.*
3. *If $f \in A[X]$ and $f = gh$ for some monic polynomials g, h in $K[X]$, then $g, h \in A[X]$.*

Proof. Regarding item 1, say $g = fh$ for some h in $K[X]$. We require that $h \in A[X]$. The coefficients of h are fractions a/b where $a, b \in A$. Let d be the product of all denominators appearing in all coefficients of h. Then $dh \in A[X]$, and $dg = fdh$.

Write $dh = \kappa(dh)k$, where k is a primitive polynomial in $A[X]$. Consequently, $dg = \kappa(dh)fk$. Since f, k are primitive so is fk according to Proposition 5.28. Then $d \mid \kappa(dh)$ in A, by Proposition 5.29. Let $\kappa(dh) = db$ for some b in A, and cancel d from the equation $dg = dbfk$ to obtain $g = fbk$. This polynomial bk is just h, but we now see that it lies in $A[X]$.

As to item 2, let a be the product of the denominators appearing in the coefficients of g. Clearly $ag \in A[X]$. According to Proposition 5.31 the polynomial $\frac{a}{\kappa(ag)}g \in A[X]$ and it is primitive. Then

$$f = \left(\frac{a}{\kappa(ag)}g\right)\left(\frac{\kappa(ag)}{a}h\right).$$

Since the factor $\frac{a}{\kappa(ag)}g$ is primitive and $f \in A[X]$ it follows from our first item above that the second factor $\frac{\kappa(ag)}{a}h$ is automatically in $A[X]$. Thus we see our alternative factorization in $A[X]$.

Item 3 follows readily from the just proven item 2. Take scalars r, s in K such that $f = (rg)(sh)$ and $rg, sh \in A[X]$. Since g, h are monic so is f monic, and the leading coefficients of rg, sh are r, s, respectively. This puts r, s in A. Also $sr = 1$. Thus $g = s(rg) \in A[X]$, and likewise $h \in A[X]$. □

Another way to state item 1 of Gauss' lemma is that if $f, g \in A[X]$ and f is primitive and the rational function g/f is a polynomial in $K[X]$, then g/f is already a polynomial in $A[X]$.

5.6.2 The Primes of $A[X]$ and Its Unique Factorization

If A is a unique factorization domain, we now offer some insight into the primes of $A[X]$. Note that the units of $A[X]$ are simply the units of A. Let K be the field of fractions of A. Recall that a polynomial in $K[X]$ is reducible precisely when it is the product of two polynomials of positive degree. Those polynomials of positive degree which cannot be so factored are called irreducible. Since $K[X]$ is a Euclidean domain, the primes of $K[X]$ are its irreducible polynomials.

Proposition 5.33. *Let A be a unique factorization domain with fraction field K.*

A constant polynomial in $A[X]$ is prime as an element of $A[X]$ if and only if it is already prime as an element of A.

A polynomial in $A[X]$ of positive degree is prime in $A[X]$ if and only if it is primitive, and as a polynomial in $K[X]$ it is irreducible.

Proof. First consider a constant polynomial p, i.e. $p \in A$. If p is prime as an element of A, Proposition 5.26 says that p is prime in $A[X]$. Conversely, suppose p is prime in $A[X]$. To see that p is prime in A suppose $p \mid rs$ where $r, s \in A$. By default $p \mid rs$ in the polynomial ring $A[X]$ wherein p is prime. Thus p divides one of r or s in $A[X]$. Say $r = pq$ for some q in $A[X]$. Since $\deg p = \deg r = 0$, it must be that $\deg q = 0$. That is, $q \in A$, whereby p is prime in A.

For the rest suppose f in $A[X]$ is a polynomial of positive degree.

Say f is primitive, and irreducible as a polynomial in $K[X]$. To see that f is prime in $A[X]$ let g, h in $A[X]$ be such that $f \mid gh$ in $A[X]$. By default $f \mid gh$ inside $K[X]$. In the Euclidean domain $K[X]$, the irreducible f is prime, and thus f divides one of g or h in $K[X]$. Say $f \mid g$ in $K[X]$. Now since $f, g \in A[X]$ and f is primitive, the first part of Gauss' lemma, Proposition 5.32, says that $f \mid g$ in $A[X]$.

For the converse we suppose that f is either not primitive in $A[X]$ or is reducible as a polynomial of $K[X]$, and show that f is not prime in $A[X]$.

If f is not primitive, then some prime p of A divides f. Write $f = pg$ where $g \in A[X]$. Neither p nor g is a unit of $A[X]$, and with this proper factorization of f, it follows that f is not prime in $A[X]$.

If f is reducible in $K[X]$, there exist polynomials g, h in $K[X]$ of positive degree such that $f = gh$. By the second part of Gauss' lemma these g, h can be rescaled by elements of K to obtain a factorization of f in $A[X]$ into polynomials of positive degree. This proper factorization of f in $A[X]$ ensures that f is not prime. □

Unique Factorization in $A[X]$

Now to the unique factorization of polynomials over unique factorization domains.

Proposition 5.34. *If A is a unique factorization domain, then so is the polynomial ring $A[X]$.*

Proof. Let f be a non-zero, non-unit of $A[X]$.

Take the content out of f, and write $f = \kappa(f)g$ where g is primitive. Factor $\kappa(f) = p_1 p_2 \cdots p_n$ where the p_j are primes in A. The p_j are also primes in $A[X]$, according to Proposition 5.33.

After that, let K be the fraction field of A, and factor g as $g = g_1 g_2 \cdots g_m$, where the g_j are irreducible in $F[X]$. This is possible (at least in principle) because $K[X]$ is Euclidean, and thereby a unique factorization domain. Use the second part of Gauss' lemma, Proposition 5.32, repeatedly if need be, to pick up non-zero r_j in K such that $r_j g_j \in A[X]$ and $g = (r_1 g_1)(r_2 g_2) \cdots (r_m g_m)$. These $r_j g_j$ remain irreducible in $K[X]$. They are also primitive in $A[X]$. The latter is true because their product g is primitive. By Proposition 5.33, these $r_j g_j$ are primes in $A[X]$.

Accordingly, the factorization of f into primes in $A[X]$ is given by:

$$f = p_1 p_2 \cdots p_n (r_1 g_1)(r_2 g_2) \cdots (r_m g_m).$$ □

By way of a reminder, the uniqueness of the factorization in Proposition 5.34 came from Proposition 5.7. For a tiny example of prime factorization in $\mathbb{Z}[X]$, we have

$$2X^2 - 8 = 2(X + 2)(X - 2).$$

Proposition 5.34 can be applied any number of times to get that polynomial rings in more than one variable are unique factorization domains. For instance, we just learned that $\mathbb{Z}[X]$ is a unique factorization domain. Letting Y be another indeterminate we get that the polynomial ring in two variables $\mathbb{Z}[X, Y] = \mathbb{Z}[X][Y]$ has unique factorization. By repeating this construction indefinitely, a multitude of unique factorization domains can be constructed.

Proposition 5.35. *If A is a unique factorization domain, then so is the polynomial ring $A[X_1, X_2, \ldots, X_n]$ in any number of variables.*

Proposition 5.33 makes it worthwhile to ask whether a polynomial in $A[X]$ is irreducible in $K[X]$. In the next section we turn to this matter.

EXERCISES

1. This problem reveals a practical test for a polynomial to have a linear factor of multiplicity greater than one.

   Let $K[X]$ be a polynomial ring with coefficients over a field. If

$$f = a_nX^n + a_{n-1}x^{n-1} + \cdots + a_2X^2 + a_1X + a_0 \in K[X],$$

   its formal derivative is the polynomial

$$f' = na_nX^{n-1} + (n-1)a_{n-1}X^{n-2} + \cdots + 2a_2x + a_1.$$

   The coefficient ja_j of X^{j-1} is to be interpreted as the sum of a_j with itself, taken j times inside K. Obviously, if the field is $\mathbb{R}$, this formal derivative is the derivative that is familiar to us all.

   (a) Verify the usual derivative rules for addition and multiplication, namely that

$$(f+g)' = f' + g' \text{ and } (fg)' = f'g + fg'.$$

   (b) If $a \in K$ and $(X-a)^2 \,|\, f$ in $K[X]$, show that $(X-a) \,|\, f'$.
   (c) If $a \in K$ and $X - a \,|\, f$ but $(X-a)^2 \nmid f$, show that $(X-a) \nmid f'$.
   Hint. Use the factor theorem, Proposition 4.35.

   Thus, a linear factor $X - a$, appearing in the unique factorization of f, has multiplicity one if and only if $X - a$ is not a factor of f' or, equivalently, if and only if $f'(a) \neq 0$.
2. If K is a field, the ideals of the polynomial ring $K[X]$ are principal. Prove the converse of this result. Namely, if K is an integral domain and the ideals of the polynomial ring $K[X]$ are principal, show that K is a field.
3. Let A be an integral domain with fraction field F. If g is a non-constant polynomial in $A[X]$ and g is prime, show that g is irreducible in $F[X]$.
4. If A is an integral domain such that its polynomial ring $A[X]$ is a unique factorization domain, prove that A is a unique factorization domain.
5. If A is a unique factorization domain with fraction field K and f is a primitive polynomial in $A[X]$, Gauss' lemma says that if $f \in A[X]$ and f has two non-constant factors in $K[X]$, then f already has two non-constant factors in $A[X]$. When A is not a unique factorization domain, this result can fail as in the following example.

   Let $A = \mathbb{Z}[2i] = \{a + b2i : a, b \in \mathbb{Z}\}$. Explain why the fraction field of A is $\mathbb{Q}[i] = \{u + vi : u, v \in \mathbb{Q}\}$. Show that $X^2 + 1$ is reducible in $K[X]$, but has no linear factor in $A[X]$. Deduce that A is not a unique factorization domain. Find a non-unit element in A which does not have a factorization into primes of A.

5.7 Irreducible Polynomials

Now comes the problem of identifying the irreducible polynomials over a field F. The more irreducibles we have, the more fields we can construct by means of Proposition 4.60.

For starters, all linear polynomials in $F[X]$ are irreducible, and we take it as given for now that in the polynomial ring $\mathbb{C}[X]$ over the complex numbers there are no more irreducible polynomials. Polynomials in $F[X]$ of degree 2 or 3 are irreducible if and only if they have no root in F, as seen in Proposition 4.59.

A common situation is to have a unique factorization domain A with fraction field F. If a non-constant polynomial in $F[X]$ is irreducible, then any one of its non-zero constant multiples is irreducible. Thus, when we examine the irreducibility of a given f in $F[X]$ we can always rescale f by the product of the denominators of the coefficients of f. In this way we can suppose f is in $A[X]$, without harm. If need be we can also divide f by its content, and thereby ensure that f is both in $A[X]$ and primitive. Every time such f is irreducible in $F[X]$, Proposition 5.33 will also yield that f is a prime in $A[X]$.

We shall focus on three sufficiency tests for irreducibility: the rational roots test, the mod-p test, and Eisenstein's criterion.

The Irreducibility of $X^4 + 1$ in $\mathbb{Q}[X]$

Just to have it on hand, let us first verify the irreducibility of $X^4 + 1$ in $\mathbb{Q}[X]$. Clearly this polynomial has no rational roots, and thus it has no linear factors. If it were reducible, then it would be the product of two quadratics in $\mathbb{Q}[X]$. By Gauss' lemma, $X^4 + 1$ could then be factored into quadratics in $\mathbb{Z}[X]$, and those quadratic factors could be made monic. Say we have

$$X^4 + 1 = (X^2 + aX + b)(X^2 + cX + d),$$

where $a, b, c, d \in \mathbb{Z}$. Multiply out the right side and equate coefficients to get

$$a + c = 0,\ b + d + ac = 0,\ bc + ad = 0,\ bd = 1.$$

Then $c = -a$, and either $b = d = 1$ or $b = d = -1$. Substitute these possibilities into the second equation above to get $a^2 = \pm 2$, which the integer a cannot satisfy. Thus $X^4 + 1$ is irreducible in $\mathbb{Q}[X]$.

Incidentally, the equations $c = -a, b = d = 1, a^2 = 2$ show us that in the larger polynomial ring $\mathbb{R}[X]$ we have

$$X^4 + 1 = (X^2 + \sqrt{2}X + 1)(X^2 - \sqrt{2}X + 1).$$

These quadratics have no root in $\mathbb{R}$ because $X^4 + 1$ has no root in $\mathbb{R}$. In $\mathbb{R}[X]$ they give the unique factorization of $X^4 + 1$ into irreducibles.

The alternative equations $c = -a, b = d = -1, a^2 = -2$ reveal the factorization

$$X^4 + 1 = (X^2 + \sqrt{2}iX - 1)(X^2 - \sqrt{2}iX - 1).$$

This factorization is in $\mathbb{C}[X]$. But now the quadratic factors are reducible into linear factors in $\mathbb{C}[X]$. The possibly well known quadratic formula could be used to find those linear factors. In any case the linear factors come from the four complex roots of $X^4 + 1$. The roots are $e^{\pi i/4}, e^{3\pi i/4}, e^{5\pi i/4}, e^{7\pi i/4}$.

5.7.1 The Rational Root Test

A polynomial of degree 2 or 3 is irreducible in $F[X]$ if and only if it has no roots in F, which raises the question of how to decide that a polynomial in $F[X]$ has no root in F.

The next result applies to any field F that is the field of fractions of a unique factorization domain A. If r, s in A are used to form the fraction r/s in F, the third item of Proposition 5.25 allows us to assume that r, s are coprime in A. Just factor out $\gcd(r, s)$ from the numerator and denominator and cancel it. A fraction r/s such that r, s are coprime is said to be in ***reduced form***.

Proposition 5.36 (Rational root test). *Suppose A is a unique factorization domain with fraction field F, and let $f = a_nX^n + a_{n-1}x^{n-1} + \cdots + a_1X + a_0$ be a polynomial in $A[X]$. If a fraction r/s with r, s coprime in A is a root of f, then $r \mid a_0$ and $s \mid a_n$ in A.*

Proof. Clear denominators in the equation $f(r/s) = 0$ to obtain:

$$a_nr^n + a_{n-1}r^{n-1}s + \cdots + a_1rs^{n-1} + a_0s^n = 0.$$

By inspection $s \mid a_nr^n$ and $r \mid a_0s^n$. Since r, s are coprime, Proposition 5.25 yields that $s \mid a_n$. Likewise $r \mid a_0$. □

The utility of Proposition 5.36 is that the roots r/s of a polynomial come from a highly restricted set of possibilities. After reducing the fraction r/s into lowest terms, the r is restricted to the factors of the constant coefficient of f, while the s is restricted to the factors of the leading coefficient of f.

Proposition 5.36 is useful to decide on the irreducibility of polynomials of degree 3 in $\mathbb{Q}[X]$. By clearing denominators of the coefficients, such a cubic polynomial f lies in $\mathbb{Z}[X]$ with no harm done. Being cubic, f is irreducible in $\mathbb{Q}[X]$ if and only if f has no rational root. Then Proposition 5.36 can help decide.

For example, the polynomial $3X^3-2$ is irreducible in $\mathbb{Q}[X]$, because it has no rational roots. Indeed, suppose r, s are coprime integers such that $s > 0$ and r/s is a root of $3X^3 - 2$. By the rational roots test $s \mid 3$ and $r \mid (-2)$. Thus s equals one of $1, 3$, while r equals one of $1, -1, 2, -2$. The possibilities for r/s are

$$1, -1, 2, -2, 1/3, -1/3, 2/3, -2/3.$$

However, none of these fractions r/s ever gives a root of $3X^3 - 2$, as can be seen by substitution.

We might also note that $3X^3-2$ is a primitive polynomial in $\mathbb{Z}[X]$, and since it is irreducible in $\mathbb{Q}[X]$, Proposition 5.33 assures that $3X^3 - 2$ is a prime in $\mathbb{Z}[X]$.

The Rational Root Test Applied to Fields of Rational Functions

The rational root test can at times be used to establish the irreducibility of polynomials in more than one variable. Before considering examples, we digress briefly to look at the ***degree of a rational function***. Suppose $K[X]$ is a polynomial ring over a field K and that $r, s \in K[X]$, both non-zero. The degree of the rational function r/s is defined by:

$$\deg(r/s) = \deg r - \deg s.$$

This definition does not depend on the representation of the rational function. For if we had $r/s = t/u$ for some polynomials r, s, t, u, then $ru = ts$. Thus

$$\deg r + \deg u = \deg(ru) = \deg(ts) = \deg t + \deg s,$$

and then $\deg r - \deg s = \deg t - \deg u$.

This integer valued degree function inherits two key properties of the degree function on polynomials.

- If f, g are non-zero rational functions, then $\deg(fg) = \deg f + \deg g$. This can be seen by inspecting our definition of degree and using the additivity of the degree on polynomials.
- If f, g are non-zero rational functions and $\deg f < \deg g$, then $f + g$ is non-zero and

$$\deg(f + g) = \deg g.$$

To see this, let $f = r/s, g = t/u$, where r, s, t, u are polynomials. Then

$$f + g = (ru + ts)/su \text{ and } \deg(f + g) = \deg(ru + ts) - \deg(su).$$

We are given that

$$\deg r - \deg s = \deg f < \deg g = \deg t - \deg u.$$

Hence

$$\deg(ru) = \deg r + \deg u < \deg t + \deg s = \deg(ts).$$

This ensures that $ru + ts \neq 0$ and so $f + g \neq 0$. Then

$$\begin{aligned}\deg(f + g) &= \deg(ru + ts) - \deg(su) = \deg(ts) - \deg(su)\\ &= \deg t + \deg s - (\deg t + \deg u) = \deg t - \deg u = \deg g.\end{aligned}$$

We can use the rational root test along with our degree function to present a (somewhat specialized) family of irreducible polynomials in two variables. It will be useful to view polynomials in $K[X, Y]$ as polynomials in Y with coefficients in $K[X]$.

Proposition 5.37. *Suppose K is a field and $f(X, Y)$ a polynomial in $K[X, Y]$ of the form*

$$f(X, Y) = a(X)Y^3 + b(X)Y^2 + c(X),$$

where $a(X), b(X), c(X)$ are polynomials in X of degree 3, 4, 3 respectively, have no irreducible factor in common, and such that $a(X)$ is irreducible and $c(X)$ has no repeated linear factor. Then such $f(X, Y)$ is irreducible in $K[X, Y]$.

Proof. The polynomial $f(X, Y)$ is a cubic in Y over the unique factorization domain $K[X]$. The assumption that $a(X), b(X), c(X)$ have no common irreducible factor in $K[X]$ is stating that $f(X, Y)$ is primitive as a polynomial in $K[X][Y]$. By Proposition 5.33 $f(X, Y)$ will be irreducible (i.e. prime) in $K[X][Y]$ provided it is irreducible as a polynomial in Y over the fraction field $K(X)$ of rational functions in X. Since $f(X, Y)$ is cubic in Y over the field $K(X)$, it suffices to check that $f(X, Y)$ has no root in $K(X)$.

The rational root test can show that there is no rational function $r(X)/s(X)$ such that $f(X, r(X)/s(X)) = 0$. Suppose on the contrary that $r(X)/s(X)$ is such a function. We take the polynomials $r(X), s(X)$ to be coprime in $K[X]$ and take $s(X)$ to be monic. By the rational root test $s(X)$ divides $a(X)$ and $r(X)$ divides $c(X)$. Since $a(X)$ is cubic and irreducible, $\deg s(X)$ is either 0 or 3, and since $c(X)$ is cubic, $\deg r(X)$ is one of $0, 1, 2, 3$. By doing the requisite subtractions $\deg(r(X)/s(X))$ must be one of $0, 1, 2, 3, -3, -2, -1$. We also note that $\deg(r(X)/s(X)) = 1$ precisely when the monic $s(X) = 1$ and $r(X)$ is linear.

In summary, if $g(X) \in K(X)$ is such that $f(X, g(X)) = 0$, then $\deg g(X)$ is one of $0, 1, 2, 3, -3, -2, -1$. Furthermore if $\deg g(X) = 1$, then $g(X)$ is a linear polynomial in $K[X]$.

The following degrees chart, computed using the properties of the degree function, shows that, except for the case $\deg g(X) = 1$, it never happens that $f(X, g(X)) = 0$ when $g(X)$ has one of the above degrees.

$\deg g(X)$	0	1	2	3	-3	-2	-1
$\deg a(X)g(X)^3$	3	6	9	12	-6	-3	0
$\deg b(X)g(X)^2$	4	6	8	10	-2	0	2
$\deg c(X)$	3	3	3	3	3	3	3
$\deg f(X, g(X))$	4	?	9	12	3	3	3

In case $g(X)$ is a linear polynomial in $K[X]$, the equation

$$a(X)g(X)^3 + b(X)g(X)^2 + c(X) = 0$$

implies that $c(X)$ has a repeated linear factor, contrary to assumption. Therefore such $g(X)$ cannot satisfy $f(X, g(X)) = 0$ either.

There is no $g(X)$ in $K(X)$ such that $f(X, g(X)) = 0$. □

For example, the polynomials

$$(3x^3 - 2)Y^3 + X^4Y^2 + (X - 1)(X^2 + 1), \quad (X^3 + X + 1)Y^3 + (X^4 + X^2)Y^2 + (X^3 + 2)$$

are irreducible in $\mathbb{Q}[X, Y]$. Furthermore, as polynomials in Y with coefficients in $\mathbb{Z}[X]$, they are primitive, and being irreducible as polynomials over the fraction field $\mathbb{Q}[X]$ of $\mathbb{Z}[X]$, they are prime polynomials in $\mathbb{Z}[X, Y]$, due to Proposition 5.33.

5.7.2 Using a Natural Extension of Homomorphisms

If $\varphi : A \to B$ is a homomorphism between commutative rings A, B, the map $\psi : A[X] \to B[X]$ defined by

$$\psi : a_0 + a_1X + a_2X^2 + \cdots + a_nX^n \mapsto \varphi(a_0) + \varphi(a_1)X + \varphi(a_2)X^2 + \cdots + \varphi(a_n)X^n$$

becomes a ring homomorphism between the respective polynomial rings. This is a matter of simple verification. Evidently ψ agrees with φ on the constant polynomials.

Definition 5.38. For each homomorphism $\varphi : A \to B$ between commutative rings A, B, the homomorphism $\psi : A[X] \to B[X]$ which agrees with φ on A and sends X in $A[X]$ to X in $B[X]$ will be called the ***natural extension*** of φ to the polynomial rings.

Since the natural extension of φ to the polynomial rings agrees with φ on the constant polynomials, it does no harm to use the same symbol φ for the natural extension of φ. This little inaccuracy is at times worth it for the convenience it affords.

For example, if $\varphi : \mathbb{Z} \to \mathbb{Z}_2$ is the canonical projection of $\mathbb{Z}$ onto the quotient ring $\mathbb{Z}_2$ of residues modulo 2, then the image of $X^3 - 4X^2 + 3X - 1$ under the natural extension of φ is $X^3 + X + 1$ in $\mathbb{Z}_2[X]$.

Here is a simple consequence of the natural extension being a homomorphism.

Proposition 5.39. *Let A be a unique factorization domain with fraction field F, and let $\varphi : A \to B$ be a ring homomorphism with natural extension $\varphi : A[X] \to B[X]$. If a polynomial f in $A[X]$ is monic of positive degree n and $\varphi(f)$ has no factor in $B[X]$ of positive degree less than n, then f is irreducible in $F[X]$ (and prime in $A[X]$).*

Proof. Suppose on the contrary that f is reducible in $F[X]$. That is, $f = gh$ for some polynomials g, h in $F[X]$ of positive degree less than n. By Gauss' lemma we can take g, h to be in $A[X]$. Then $\varphi(f) = \varphi(g)\varphi(h)$, and $\varphi(f)$ stays monic of degree n. If $\varphi(g)$ in $B[X]$ had degree n, then g would have degree n, which is not the case. Thus $\varphi(g)$ has degree less than n, and likewise for $\varphi(h)$. In that case $\varphi(g)$ and $\varphi(h)$ would be factors of $\varphi(f)$ of positive degree less than n, contrary to the assumption on $\varphi(f)$. Also f is primitive in $A[X]$ because it is monic, and by Proposition 5.33 it is prime in $A[X]$. □

For an illustration of the above result, take the monic polynomial $X^3 - 3X - 1$ in $\mathbb{Z}[X]$. Use the canonical projection $\varphi : \mathbb{Z} \to \mathbb{Z}_2$ from the integers to their residues modulo 2. The image of our polynomial under the natural extension of φ is $X^3 + X + 1$. This cubic is irreducible in $\mathbb{Z}_2[X]$ because it has no roots in the field $\mathbb{Z}_2$. By Proposition 5.39, $X^3 - 3X - 1$ is irreducible in $\mathbb{Q}[X]$.

For another example, take the polynomial $X^4 + Y^3X^3 - Y^2X^3 + Y^4$ in the ring $\mathbb{Q}[Y, X]$ with two variables X, Y. When $\mathbb{Q}[Y, X]$ is viewed as the ring $\mathbb{Q}[Y][X]$ of polynomials in X with coefficients in $\mathbb{Q}[Y]$ our given polynomial is monic in the variable X. The substitution map $\mathbb{Q}[Y] \to \mathbb{Q}$ that sends polynomials $f(Y)$ to $f(1)$ has a natural extension $\varphi : \mathbb{Q}[Y][X] \to \mathbb{Q}[X]$ given by $f(Y, X) \mapsto f(1, X)$. The image of our given polynomial under the natural extension is $X^4 + 1$ in $\mathbb{Q}[X]$. As we have seen $X^4 + 1$ is irreducible in $\mathbb{Q}[X]$. By Proposition 5.39, the given polynomial is irreducible in $\mathbb{Q}[Y, X]$, actually in $Q(Y)[X]$.

5.7.3 Eisenstein's Criterion

A sufficiency test for irreducibility, discovered by Ferdinand Eisenstein (1823–1852), has proven its worth.

Our proof of Eisenstein's criterion uses a small observation. If B is an integral domain with polynomial ring $B[X]$, then the factors of a non-zero monomial bX^n can only be other such monomials. This might seem obvious, until we notice that when B is not a domain, such as, for example, a ring of characteristic 4, anomalies like $X^2 = (X + 2)(X + 2)$ do crop up.

To check our observation, suppose to the contrary that $bX^n = gh$ where $g, h \in B[X]$ and g has more than one term appearing in its standard representation. Let cX^k be the non-zero term with the lowest power of X appearing in the standard representation of g. Thus $0 \leq k < \deg g$. Let dX^ℓ be the lowest term in the standard representation of h. Since $f = gh$, the lowest term in the standard representation of f is $cdX^{k+\ell}$. Since $k + \ell < \deg g + \deg h = \deg f$ and $cd \neq 0$ in our domain B, the polynomial f could not have been a monomial.

Proposition 5.40 (Eisenstein's criterion). *Let A be a unique factorization domain with fraction field F and polynomial ring $A[X]$. If $f = a_0 + a_1X + \cdots + a_nX^n \in A[X]$ and p is a prime in A such that*

$$p \mid a_j \text{ for all } j = 0, \ldots, n-1,\ p \nmid a_n \text{ and } p^2 \nmid a_0,$$

then f is irreducible in $F[X]$.

Proof. Let $\varphi : A \to A/\langle p\rangle$ be the canonical projection onto the quotient ring, and (using the same letter) let $\varphi : A[X] \to A/\langle p\rangle[X]$ be its natural extension to the polynomial rings.

Now suppose to the contrary that f is reducible in $F[X]$, and write $f = gh$ where g, h are polynomials in $F[X]$ of positive degree. Since A has unique factorization, Gauss' lemma, Proposition 5.32, says there is no harm in supposing that $g, h \in A[X]$. Then we have

$$\varphi(f) = \varphi(gh) = \varphi(g)\varphi(h).$$

By our assumptions $\varphi(f) = \varphi(a)X^n$, where a is the leading coefficient of f, and its residue $\varphi(a)$ in $A/\langle p\rangle$ is not 0. As noted prior to this proof both $\varphi(g)$ and $\varphi(h)$ are monomials. Furthermore, since

$$\deg g + \deg h = \deg f = \deg \varphi(f) = \deg \varphi(g) + \deg \varphi(h),$$

it follows that $\deg \varphi(g) = \deg g$ and $\deg \varphi(h) = \deg h$, and these degrees are at least 1. Then the constant terms of $\varphi(g)$ and $\varphi(h)$ must be 0. This means that the constant terms of g, h are both divisible by p. If c, d are these constant terms, respectively, the constant term a_0 of f is their product cd, which is divisible by p^2, contrary to the assumed properties of f. □

For an illustration of Eisenstein's criterion, take the polynomial $f = X^n - p$ in $\mathbb{Z}[X]$, where p is any prime integer. The conditions of Eisenstein's criterion are met using the prime p, and thus f is irreducible in $\mathbb{Q}[X]$. If $n \geq 2$, then f has no linear factor and thus no root in $\mathbb{Q}[X]$. This shows that any $\sqrt[n]{p}$ is an irrational number. Furthermore, the example reveals that $\mathbb{Q}[X]$ contains irreducible polynomials of arbitrarily high degree. This example holds just as well if the ring $\mathbb{Z}$ is replaced by any unique factorization domain A and p is any prime in A.

An Application of Eisenstein to Some Cyclotomic Polynomials

Definition 5.41. For a prime integer p, the pth ***cyclotomic polynomial*** is defined to be

$$f(X) = \frac{X^p - 1}{X - 1} = X^{p-1} + X^{p-2} + \cdots + X + 1.$$

The roots of this polynomial are the complex pth roots of unity $e^{2\pi ik/p}$, where k runs from 1 to $p-1$. To see that $f(X)$ is irreducible in $\mathbb{Q}[X]$, a direct application of Eisenstein's criterion to $f(X)$ does not seem at hand. To get around this, replace $f(X)$ by $g(X) = f(X+1)$. That is, use the substitution map $\varphi : \mathbb{Z}[X] \to \mathbb{Z}[X]$, where $h(X) \mapsto h(X+1)$. Since φ is a ring isomorphism, $f(X)$ is irreducible if and only if $g(X)$ is irreducible.

Proposition 4.65 applies to φ, and yields the extension homomorphism $\psi : K(X) \to K(X)$ given by $h(X)/k(X) \mapsto h(X+1)/k(X+1)$. This extension, in conjunction with the binomial theorem, gives:

$$g(X) = \varphi(f(X)) = \psi\left(\frac{X^p - 1}{X - 1}\right) = \frac{(X+1)^p - 1}{X + 1 - 1}$$

$$= X^{p-1} + pX^{p-2} + \cdots + \binom{p}{r} X^{p-r-1} + \cdots + p.$$

The coefficients $\binom{p}{r}$ in this monic polynomial are divisible by p, and the last coefficient p is not divisible by p^2. By Eisenstein's criterion $g(X)$ is irreducible, and therefore $f(X)$ is too.

We should mention that there exist cyclotomic polynomials for all positive integers. A look at those will come as part of our study of Galois theory.

Using Eisenstein on Polynomials with More than One Variable

Eisenstein is commonly applied to polynomials with integer coefficients. Still, there are other unique factorization domains over which it can be exploited.

To illustrate, let us show that for every n the polynomials $Y^n + X^2 + 1$ are irreducible in the polynomial ring $\mathbb{C}[X, Y]$. View $\mathbb{C}[X, Y]$ as the ring $\mathbb{C}[X][Y]$ of polynomials in Y with coefficients in the Euclidean ring $\mathbb{C}[X]$. The coefficient $X^2 + 1$ factors into irreducibles in $\mathbb{C}[X]$ as $(X+i)(X-i)$. Since $Y^n + X^2 + 1 = Y^3 + (X+i)(X-i)$, Eisenstein's criterion can be applied using the prime $X + i$ to see that $Y^n + X^2 + 1$ is irreducible, in $\mathbb{C}(X)[Y]$.

EXERCISES

1. For integers n, r where $0 \le r \le n$, the binomial coefficient $\binom{n}{r}$ is defined to be $\frac{n!}{r!(n-r)!}$. Explain why this coefficient is an integer, and if $0 < r < n$ and n is prime, show that n divides $\binom{n}{r}$. This fact was used to show that the cyclotomic polynomials for prime integers are irreducible in $\mathbb{Q}[X]$.
2. Find all irreducible quadratics, cubics and quartics in $\mathbb{Z}_2[X]$.

   Hint. For the quadratics and cubics, eliminate those polynomials which have a root in $\mathbb{Z}_2$. Then eliminate the quartics which have a root. Finally, eliminate the quartics which are the product of irreducible quadratics.
3. If p is prime, factor the polynomial $X^p - X$ into irreducible polynomials in $\mathbb{Z}_p[X]$.
4. Show that $8X^3 - 6X - 1$ is irreducible in $\mathbb{Q}[X]$.
5. Prove that $X^4 - 5X^2 + 6X + 1$ is irreducible in $\mathbb{Q}[X]$.
6. If A is a unique factorization domain and f is a monic polynomial in $A[X]$ with a root α in the fraction field F of A, show that $\alpha \in A$.
7. Let n be an integer which n is not a perfect square. The ring $\mathbb{Z}[\sqrt{n}]$ consists of all complex numbers $a + b\sqrt{n}$ where $a, b \in \mathbb{Z}$. If $n \equiv 1 \bmod 4$ show that $\mathbb{Z}[\sqrt{n}]$ is not a unique factorization domain.

   Hint. If $\mathbb{Z}[\sqrt{n}]$ were a unique factorization domain with fraction field F, then every monic polynomial in $\mathbb{Z}[\sqrt{n}]$ would have no roots in $F \setminus \mathbb{Z}[\sqrt{n}]$. Find a monic polynomial for which $\frac{1}{2} + \frac{1}{2}\sqrt{n}$ is a root.
8. Use the natural extension of the canonical projection of $\mathbb{Z}$ onto $\mathbb{Z}_2$ to show that X^4+X+1 is irreducible in $\mathbb{Q}[X]$.
9. Show that $X^5 + 2X + 4$ is irreducible modulo 3, and thus irreducible in $\mathbb{Q}[X]$.
10. Factor $3X^3 - 2X^2 + 5X + 2$ into irreducibles in $\mathbb{Q}[X]$.
11. Show that $15X^4 - 10X^2 + 9X + 21$ is irreducible in $\mathbb{Q}[X]$, by showing it is irreducible modulo 2.
12. Show that the polynomial $X^4 + X^3Y + X^2Y^2 + XY^3 + Y^4$ is irreducible in $\mathbb{Q}[X, Y]$.
13. Show that $6X^5 + 14X^2 - 21X + 56$ is irreducible in $\mathbb{Q}[X]$.
14. If n is an integer such that $X^3 + nX + 2$ is reducible in $\mathbb{Q}[X]$, show that $n = 1, -3$ or -5.
15. If n is not prime, show that $X^{n-1} + X^{n-2} + \cdots + X^2 + X + 1$ is reducible in $\mathbb{Z}[X]$.
16. Let p be a positive prime and n a positive odd integer. Prove that $X^n - p^2$ is irreducible in $\mathbb{Q}[X]$.

    Hint. Suppose $f(X)g(X) = X^n - p^2$ for some non-constant polynomials f, g in $\mathbb{Q}[X]$. Then $f(X^2)g(X^2) = X^{2n} - p^2 = (X^n - p)(X^n + p)$. Use the fact $\mathbb{Q}[X]$ has unique factorization into irreducibles along with the fact the left factors have even degree while the right factors have odd degree.
17. Show that $iX^4 - 3X^3 + (6 + 3i)X^2 - 6i$ is irreducible in $\mathbb{Z}[i][X]$.
18. Show that $X^3 + 12X^2 + 18X + 6$ is irreducible in $\mathbb{Z}[i][X]$.

19. If n is an odd positive integer, prove that $X^n + 4$ is irreducible in $\mathbb{Q}[X]$.
Hint. Suppose $X^n + 4 = f(X)g(X)$ for some polynomials f, g of positive degree, then $X^{2n} + 4 = g(X^2)f(X^2)$. Factor the left side into irreducibles in $\mathbb{Q}[i][X]$ and use unique factorization in $\mathbb{Q}[i][X]$.

20. If p is an odd prime, show that $X^n - p$ is irreducible in $\mathbb{Z}[i][X]$ for every positive n.

21. Use Eisenstein's criterion to show that $f(X) = X^4 + 1$ is irreducible in $\mathbb{Q}[X]$.
Hint. Apply the automorphism $\mathbb{Q}[X] \to \mathbb{Q}[X]$ given by $g(X) \mapsto g(X + 1)$.

22. Show that $Y^2 + X^2 - 1$ is irreducible in $\mathbb{Q}[X, Y]$.

23. If K is any field and n is a positive integer, prove that $Y^n - X$ is irreducible in the ring $K(X)[Y]$ of polynomials in Y with coefficients in the field $K(X)$. Is $Y^n - X$ a prime in $\mathbb{Q}[X, Y]$?

24. The ring $\mathbb{Z}[\sqrt{2}]$ is a Euclidean domain and its fraction field is $\mathbb{Q}[\sqrt{2}]$ consisting of all numbers of the form $r + s\sqrt{2}$ where $r, s \in \mathbb{Q}$. Show that $X^n - \sqrt{2}$ is irreducible in $\mathbb{Q}[\sqrt{2}][X]$.

25. The polynomial $f = X^4 + 1$ is irreducible in $\mathbb{Q}[X]$. Prove that f is reducible in $\mathbb{Z}_p[X]$ for all primes p.
Hint. Follow these steps.
First dispose of the case $p = 2$.
If p is odd, and there exists an a in Z_p such that $a^2 = -1$, factor $X^4 + 1$ in $\mathbb{Z}_p[X]$.
If there is no a in $\mathbb{Z}_p$ such that $a^2 = -1$, show that either 2 or -2 in $\mathbb{Z}_p$ equals a^2 for some b in $\mathbb{Z}_p$. Then show that $X + bX + 1$ or $X^2 + bX - 1$, respectively, divides $X^4 + 1$.

26. Prove that the polynomial $f = (X - 1)(X - 2)(X - 3) \cdots (X - n) - 1$ is irreducible in $\mathbb{Q}[X]$ for all positive n.
Hint. If f is reducible then, by Gauss' lemma, f has a proper factor g in $\mathbb{Z}[X]$, and without loss of generality $1 \le \deg g \le n/2$. Show that $g^2 - 1$ has the same distinct roots $1, 2, \ldots, n$ as $f + 1$, and thereby deduce that $f + 2 = g^2$, a perfect square. Since perfect squares take only non-negative values on $\mathbb{Q}$, deduce that the even integer $n < 6$. Then eliminate the cases $n = 2, 4$.

27. Prove that the polynomial $f = (X - 1)(X - 2)(X - 3) \cdots (X - n) + 1$ is irreducible for all positive n, except for $n = 4$.

5.8 Polynomials over Noetherian Rings

One might enquire if the assumption that A has unique factorization is essential for Eisenstein's criterion to work. It turns out that the result prevails even for Noetherian domains. Recall from Section 5.2.1 that a commutative ring A is Noetherian when every ascending chain of ideals eventually ceases to properly increase. By Proposition 5.13 this is the same as saying that every ideal in A is finitely generated.

5.8.1 Hilbert's Basis Theorem: a Source of Noetherian Rings

Principal ideal domains are Noetherian, but a way to obtain more examples of Noetherian rings would be satisfying. In that regard, nothing surpasses a famous and important result due to David Hilbert (1862–1943).

Proposition 5.42 (Hilbert's basis theorem). *If A is a commutative Noetherian ring, then so is its polynomial ring $A[X]$.*

Proof. Suppose that some ideal J of $A[X]$ is not finitely generated, and seek a contradiction.

As J is not finitely generated, J is not the zero ideal, and there is a polynomial f_1 in J of minimal degree. Also $\langle f_1\rangle \subsetneq J$. In the non-empty set $J\setminus\langle f_1\rangle$ there is a polynomial f_2 of minimal degree. Still $\langle f_1,f_2\rangle \subsetneq J$, and so in the non-empty set $J \setminus \langle f_1,f_2\rangle$ there is a polynomial f_3 of minimal degree. Again $\langle f_1,f_2,f_3\rangle \subsetneq J$, which yields an f_4 in $J \setminus \langle f_1,f_2,f_3\rangle$ of minimal degree. Continue in this fashion to pick up a sequence of polynomials $f_1,f_2,\dots,f_n,f_{n+1},\dots$ such that f_{n+1} is of minimal degree among the polynomials in $J\setminus\langle f_1,f_2,\dots,f_n\rangle$. Clearly $\deg f_n \leq \deg f_{n+1}$ for all n.

Let a_n be the leading coefficient of f_n. Inside the Noetherian ring A, the ascending chain of ideals

$$\langle a_1\rangle \subseteq \langle a_1,a_2\rangle \subseteq \langle a_1,a_2,a_3\rangle \subseteq \cdots \subseteq \langle a_1,a_2,\dots,a_n\rangle \subseteq \cdots$$

must terminate. This means that $a_{m+1} \in \langle a_1,a_2,\dots,a_m\rangle$ for some m. And thus,

$$a_{m+1} = u_1a_1 + u_2a_2 + \cdots + u_ma_m \text{ for some } u_j \text{ in } A.$$

Now the leading coefficient of the polynomial

$$g = \sum_{j=1}^{m} u_j X^{\deg f_{m+1} - \deg f_j} f_j$$

is a_{m+1}. The degree of g equals $\deg f_{m+1}$, and g lies in the ideal $\langle f_1,f_2,\dots,f_m\rangle$. Consequently $f_{m+1} - g$ is a polynomial in $J \setminus \langle f_1,f_2,\dots,f_m\rangle$, and its degree is less than $\deg f_{m+1}$, in contradiction to the minimality of $\deg f_{m+1}$. □

Since $\mathbb{Z}$ is Noetherian, so is the polynomial ring $\mathbb{Z}[X]$. For any field K, the principal ideal domain $K[X]$ is Noetherian. Thus the polynomial ring in two variables $K[X, Y]$ is Noetherian since it coincides with the ring $K[X][Y]$ of polynomials in Y over $K[X]$. Inductively we see that the polynomial ring $K[X_1, X_2, \dots, X_n]$ in any number of indeterminates is Noetherian. For identical reasons, $\mathbb{Z}[X_1, \dots, X_n]$ is Noetherian too.

The next straightforward result produces a wealth of Noetherian rings.

Proposition 5.43. *Every homomorphic image of a commutative Noetherian ring is Noetherian.*

Proof. Let A be a commutative Noetherian ring, and let $\varphi : A \to B$ be a surjective ring homomorphism. Suppose J is an ideal of B. Its inverse image $\varphi^{-1}(J)$ is an ideal of A. Since A

is Noetherian, $\varphi^{-1}(J)$ is finitely generated, say by $a_1, \ldots, a_n$. If $b \in B$, then $b = \varphi(a)$ for some a in $\varphi^{-1}(J)$, because φ is surjective. Such $a = r_1a_1 + \cdots + r_na_n$ for some r_j in A. Apply the homomorphism φ to get $b = \varphi(a) = \varphi(r_1)\varphi(a_1) + \cdots + \varphi(r_n)\varphi(a_n)$, which proves that J is finitely generated by $\varphi(a_1), \ldots, \varphi(a_n)$. □

For example, with any finite list of complex numbers $\alpha_1, \alpha_2, \ldots, \alpha_n$, consider the substitution map

$$\varphi : \mathbb{Z}[X_1, X_2, \ldots, X_n] \to \mathbb{C} \text{ given by } f(X_1, X_2, \ldots, X_n) \mapsto f(\alpha_1, \alpha_2, \ldots, \alpha_n).$$

The image of φ is the subring $\mathbb{Z}[\alpha_1, \alpha_2, \ldots, \alpha_n]$ of $\mathbb{C}$, consisting of all polynomial expressions in the α_j having integer coefficients. This is the smallest subring of $\mathbb{C}$ that contains all of the α_j. As just seen, all such rings are Noetherian. We knew that the ring $\mathbb{Z}[i]$ of Gaussian integers is Noetherian, since all of its ideals are principal. Now we know that rings such as $\mathbb{Z}[\sqrt{2}], \mathbb{Z}[\sqrt{-5}], \mathbb{Z}[\sqrt{2}, \sqrt{-5}, i]$ are Noetherian too.

A Noetherian Domain that Is Not a Unique Factorization Domain

The domain $\mathbb{Z}[\sqrt{-5}]$ is Noetherian, as a consequence of Hilbert's theorem. According to Proposition 5.14 its non-zero, non-unit elements can be factored into irreducibles. But as seen in Section 5.1, this domain has irreducibles that are not prime. By Proposition 5.9, $\mathbb{Z}[\sqrt{-5}]$ is an example of a Noetherian domain that is not a unique factorization domain.

A Unique Factorization Domain that Is Not Noetherian

Here is a unique factorization domain that is not Noetherian.

Let A be the ring $\mathbb{Z}[X_1, X_2, \ldots, X_n, \ldots]$ of polynomials in infinitely many indeterminates X_j. The ring A can be viewed as the union of the ascending chain of subrings:

$$\mathbb{Z} \subset \mathbb{Z}[X_1] \subset \mathbb{Z}[X_1, X_2] \subset \cdots \subset \mathbb{Z}[X_1, X_2, \ldots, X_n] \subset \cdots.$$

To add or multiply two polynomials in A, carry out the operations in one of the subrings $\mathbb{Z}[X_1, X_2, \ldots, X_n]$ in the chain to which they belong. The result of the operations does not depend on which subring containing the polynomials is used.

Every ring $\mathbb{Z}[X_1, X_2, \ldots, X_n]$ is a unique factorization domain due to Proposition 5.35, and a Noetherian ring by repeated application of the Hilbert basis theorem. Their union A retains unique factorization, but is no longer Noetherian.

To see that A retains unique factorization, note that in each of the unique factorization domains $\mathbb{Z}[X_1, X_2, \ldots, X_n]$, primes and irreducibles coincide. By Proposition 5.26, every prime g in $\mathbb{Z}[X_1, X_2, \ldots, X_n]$ remains so in the larger domain $\mathbb{Z}[X_1, X_2, \ldots, X_n][X_{n+1}]$, which is nothing but $\mathbb{Z}[X_1, X_2, \ldots, X_n, X_{n+1}]$. By repeating the argument, g remains prime in every ring $\mathbb{Z}[X_1, X_2, \ldots, X_m]$ where $m \geq n$.

Such g stays prime in A as well. Indeed, suppose g divides a product fh where $f, h \in A$. Clearly $g \mid fh$ also in one of the subrings $\mathbb{Z}[X_1, \ldots, X_m]$, where $m \geq n$. In this larger subring g remains prime. Thus $g \mid f$ or $g \mid h$ in this larger subring, and so $g \mid f$ or $g \mid h$ in A.

For A to be a unique factorization domain we need to factor every non-zero, non-unit polynomial f in A into primes in A. Well, every such f lies in some $\mathbb{Z}[X_1, \dots, X_n]$. In this unique factorization domain, $f = h_1 h_2 \cdots h_k$, where the h_j are primes in $\mathbb{Z}[X_1, \dots, X_n]$. These h_j remain primes in A, as noted. Thus A is a unique factorization domain.

To see that A is not Noetherian, we show that the inclusions in the ascending chain of ideals

$$\langle X_1 \rangle \subseteq \langle X_1, X_2 \rangle \subseteq \langle X_1, X_2, X_3 \rangle \subseteq \cdots \subseteq \langle X_1, X_2, \dots, X_n \rangle \subseteq \cdots .$$

are proper. If, to the contrary, some inclusion were an equality, then some X_{n+1} would be in the ideal $\langle X_1, X_2, \dots, X_n \rangle$. That is

$$X_{n+1} = X_1 f_1 + X_2 f_2 + \cdots + X_n f_n, \text{ for some } f_j \text{ in } A.$$

There is a subring $\mathbb{Z}[X_1, X_2, \dots, X_m]$ that contains all f_j as well as X_{n+1}. Then the substitution map $\varphi : \mathbb{Z}[X_1, X_2, \dots, X_m] \to \mathbb{Z}$ which maps X_{n+1} to 1 and all other X_j to 0, applied to the preceding expression for X_{n+1}, would lead to the contradiction $1 = 0$.

The above argument carries through for any unique factorization domain A. The polynomial ring $A[X_1, X_2, \dots, X_n, \dots]$ in infinitely many variables X_j is a unique factorization domain which is not Noetherian.

5.8.2 Eisenstein's Criterion for Noetherian Domains

It might be less well known that Eisenstein's criterion applies equally well to Noetherian domains. We offer a proof that uses the result for unique factorization domains.

The Localization of a Ring at a Prime Element

The proof of Eisenstein for Noetherian domains hinges on the localization of an integral domain at a prime. Suppose that A is an integral domain with field of fractions F, and that p is a prime in A. Let A_p be the set of all fractions in F that can be written in the form

$$\frac{a}{b} \text{ where } a, b \in A \text{ and } p \nmid b.$$

By simple verification A_p is a subring of F. In fact, the set of elements b in A where $p \nmid b$ is a denominator set D as discussed in Section 4.8. The ring A_p is nothing but the ***localization*** of A at this D. We shall refer to A_p as the ***localization of*** A ***at the prime*** p. Since every a in A can be viewed as the fraction $\frac{a}{1}$, the domain A is a subring of A_p.

Here are the basic features of the localization of A at p.

- An element $\frac{a}{b}$, where $p \nmid b$, is a unit of A_p if and only if $p \nmid a$. The inverse of such $\frac{a}{b}$ is $\frac{b}{a}$.
- The prime p divides $\frac{a}{b}$ in A_p if and only if $p \mid a$ in A. One direction of the proof is immediate. For the other direction, say $p \mid \frac{a}{b}$ in A_p. Then $\frac{a}{b} = p\frac{c}{d}$ where $\frac{c}{d} \in A_p$ and $p \nmid d$. Hence, $ad = pcb$ in A, and then $p \mid a$ in A, because p is prime and $p \nmid d$.

In particular, if $a \in A$ and $p \mid a$ in A_p, then $p \mid a$ in A. Also, in light of the preceding remark, the units of A_p are the elements not divisible by p.

- The element p remains a prime in A_p. Indeed, suppose $p \mid \frac{a}{b}\frac{c}{d}$ where $\frac{a}{b}, \frac{c}{d}$ are in A_p. As noted $p \mid ac$ in A, and since p is prime, $p \mid a$ or $p \mid c$. Thus $p \mid \frac{a}{b}$ or $p \mid \frac{c}{d}$.
- In fact p is the only prime in A_p, other than its associates pu, where u runs over the units of A_p. Indeed, suppose q is a prime in A_p. Say $q = \frac{a}{b}$ where $a, b \in A$ and $p \nmid b$. Since q is not a unit, $p \mid a$, meaning that $p \mid q$ in A_p. Write $q = pu$ for some u in A_p. Since the prime q is irreducible and p is not a unit, u must be a unit in A_p. Thus q is an associate of p.

  It follows from this observation that any prime factorization of a non-zero, non-unit element of A_p could only take the form $p^k u$, where k is a positive exponent and u is a unit of A_p.

Proposition 5.44. *If A is a Noetherian domain, then its localization A_p at a prime p is both a Noetherian and a unique factorization domain.*

Proof. For the Noetherian part, let J be an ideal of A_p. Clearly $J \cap A$ is an ideal of A, which by assumption is finitely generated. Say $J \cap A = \langle a_1, \ldots, a_n \rangle$ where the $a_j \in A$. Now if $x = \frac{a}{b} \in J$, we see that $bx \in J \cap A$ and so $bx = a_1 t_1 + \cdots + a_n t_n$ for some $t_j \in A$. Then $x = a_1 \frac{t_1}{b} + \cdots + a_n \frac{t_n}{b}$, which shows that the finitely many a_j also generate J as an ideal of A_p.

To see that A_p is a unique factorization domain, it suffices to see that if x is a non-zero, non-unit element of A_p, then there is a positive exponent k such that $p^k \mid x$ while $p^{k+1} \nmid x$. For then we have $x = p^k u$ where u, not being divisible by p, is a unit.

Well, if the above does not occur, there is an x such that $p^j \mid x$ for all positive exponents j. Write $x = p^j y_j$ where $y_j \in A_p$. The equations $p^j y_j = x = p^{j+1} y_{j+1}$ lead to $y_j = p y_{j+1}$. So there is the ascending chain of ideals

$$\langle y_1 \rangle \subseteq \langle y_2 \rangle \subseteq \cdots \subseteq \langle y_j \rangle \subseteq \langle y_{j+1} \rangle \subseteq \cdots$$

in the Noetherian ring A_p. For some j we must have $\langle y_j \rangle = \langle y_{j+1} \rangle$, which means that $y_{j+1} = t y_j$ for some t in A_p. And then $y_{j+1} = t p y_{j+1}$, which gives $1 = tp$. This is a contradiction since p is not a unit of A_p. □

An ingredient in the above proof is that for any prime q and any non-zero x in a Noetherian ring, there is an exponent k such that $q^k \mid x$ while $q^{k+1} \nmid x$. This is the same as saying that the intersection $\bigcap_{j=1}^{\infty} \langle q^j \rangle$ is the zero ideal. Those with a yearning for algebraic pathologies may wish to seek out an integral domain A with a prime p and a non-zero element x such that every power p^k divides x.

Eisenstein in the Noetherian Case

Proposition 5.45. *Suppose A is a Noetherian domain with fraction field F, and that p is a prime in A. If $f = a_n X^n + a_{n-1} X^{n-1} + \cdots + a_0 \in A[X]$ and is such that*

$$p \text{ divides all } a_0, \ldots, a_{n-1}, \ p \nmid a_n \text{ and } p^2 \nmid a_0,$$

then f is irreducible in $F[X]$.

Proof. Since $A \subset A_p$, the polynomial f also lies in $A_p[X]$. By Proposition 5.44, A_p is a unique factorization domain. The element p remains a prime in A_p. Clearly p divides all of $a_0, \ldots, a_{n-1}$ also in A_p. If p divided a_n or p^2 divided a_0 in A_p, the same would occur also in A, contrary to assumption. Thus f satisfies the conditions of Eisenstein's criterion, Proposition 5.40, with respect to the unique factorization domain A_p. By Proposition 5.40, f is irreducible over the fraction field of A_p, which is still F. □

For example, we saw in Section 5.1 that the integer 11 remains a prime as an element of the ring $\mathbb{Z}[\sqrt{-5}]$ of numbers of the form $a + b\sqrt{-5}$ where $a, b \in \mathbb{Z}$. The fraction field of this domain is $\mathbb{Q}[\sqrt{-5}]$ made up of all $a + b\sqrt{-5}$ where $a, b \in \mathbb{Q}$. Evidently the latter is a field wherein the inverse of every non-zero $a + b\sqrt{-5}$ is $\frac{a}{a^2+5b^2} - \frac{b}{a^2+5b^2}\sqrt{-5}$. As noted in Section 5.1 the domain $\mathbb{Z}[\sqrt{-5}]$ does not have unique factorization. It is Noetherian however, being a homomorphic image of the Noetherian ring $\mathbb{Z}[X]$, via the substitution map that sends X to $\sqrt{-5}$. By Proposition 5.45 the polynomial $X^n - 11$ is irreducible as a polynomial in $\mathbb{Q}[\sqrt{-5}][X]$, regardless of the exponent n. Such information is useful because, in accordance with Proposition 4.60, we can now build another field: $\mathbb{Q}[\sqrt{-5}][X]/\langle X^n - 11\rangle$.

Eisenstein's Criterion and Gauss' Lemma Can Fail in Some Domains

Having seen that Eisenstein's criterion, Proposition 5.40, applies to unique factorization domains and also to Noetherian domains, it might be worth noting that it does not apply to all integral domains. Here is a neat counter-example shown to us by the young Austrian mathematician Daniel Smertnig. Let

$$A = \mathbb{Z} + Y\mathbb{C}[Y].$$

In other words, A is the subring of the polynomial ring $\mathbb{C}[Y]$ consisting of those complex polynomials $f(Y)$ such that $f(0) \in \mathbb{Z}$. It is easy to see that A is a subring of $\mathbb{C}[Y]$.

The fraction field F of A is $\mathbb{C}(Y)$, the field of all rational functions in Y. Indeed, any rational function $f(Y)/g(Y)$ can be rewritten as $Yf(Y)/Yg(Y)$ where $Yf(Y), Yg(Y)$ are now in A.

The units of A are ± 1. Indeed, suppose $u(Y)$ is a unit of A. Substitute $Y = 0$ to see that $u(0)$ is a unit, now in $\mathbb{Z}$. That is, $u(0) = \pm 1$. Since $u(Y)$ also remains a unit in the larger polynomial ring $\mathbb{C}[Y]$, it must be that $u(Y)$ is a constant polynomial. Hence, $u(Y) = \pm 1$.

Each prime p in $\mathbb{Z}$ remains a prime in A. To see this suppose $f(Y), g(Y) \in A$ and that $p \mid f(Y)g(Y)$. Substitute $Y = 0$ to see that p divides $f(0)g(0)$, now in $\mathbb{Z}$. Then $p \mid f(0)$ or $p \mid g(0)$, because p is prime in $\mathbb{Z}$. Say $f(0) = pq$ for some integer q. This means there is a polynomial $h(Y)$ in $\mathbb{C}[Y]$ such that

$$f(Y) = pq + Yh(Y) = pq + pYp^{-1}h(Y) = p(q + Yp^{-1}h(Y)).$$

This reveals that p divides $f(Y)$ in A, and therefore that p is prime in A.

Now, for any integer prime p and any positive integer n, take the polynomial

$$X^n - p \text{ in the new variable } X.$$

This polynomial sits in $\mathbb{C}[X]$, and thereby also in $\mathbb{C}[Y][X]$. Actually our polynomial is in $A[X]$, because the constant term $-p$ is an integer. Since p is prime in A, this polynomial satisfies the assumptions of Eisenstein's criterion, Proposition 5.40.

However, polynomials over $\mathbb{C}$ factor into linear factors (take that as given). Our $X^n - p$ is a product of linear factors in $\mathbb{C}[X]$ and hence a product of linear factors in $F[X]$, where $F = \mathbb{C}(Y)$ is the fraction field of A. Thus $X^n - p$, being reducible in $F[X]$, fails the conclusion of Eisenstein's criterion.

The polynomial $X^n - p$ also illustrates that Gauss' lemma, Proposition 5.32, need not hold over all integral domains. As we have seen $X^n - p$ factors in $F[X]$. However, it is irreducible in $A[X]$. One way to see that is by noting that $X^n - p$ is monic in $A[X]$. The homomorphism $\varphi : A \to \mathbb{Z}$ given by $f(X) \mapsto f(0)$ has its natural extension to $\varphi : A[X] \to \mathbb{Z}[X]$. This extension maps $X^n - p$ in $A[X]$ to $X^n - p$ in $\mathbb{Z}[X]$. By Eisenstein and by Gauss' lemma applied in $\mathbb{Z}[X]$, the polynomial $X^n - p$ is irreducible in $\mathbb{Z}[X]$. So, by Proposition 5.39, $X^n - p$ is irreducible in $A[X]$. Gauss' lemma fails in the ring $A[X]$.

5.8.3 Primes in Rings Coming from Algebraic Integers

The usefulness of Proposition 5.45 hinges on our capacity to discover primes in Noetherian domains. A prime integer could well lose its primeness in a larger ring. For instance, in the ring of Gaussian integers, 5 is no longer prime because $5 = (2 + i)(2 - i)$, thereby losing its irreducibility. In $\mathbb{Z}[\sqrt{-5}]$, which is not a unique factorization domain, the integer 3 remains irreducible but loses its primeness, because 3 divides $6 = (1 + \sqrt{-5})(1 - \sqrt{-5})$ without dividing either of these factors.

For each prime integer p we now explore what it takes for it to survive as a prime inside rings of the form $\mathbb{Z}[\alpha]$ for suitable complex numbers α. This will create conditions for us to apply Eisenstein's criterion on these Noetherian domains.

Algebraic Integers

Start with a monic, irreducible polynomial $g = X^n + a_{n-1}X^{n-1} + \cdots + a_1X + a_0$ in $\mathbb{Z}[X]$ and take a root α in $\mathbb{C}$ of g. Such α are known as ***algebraic integers***. By Gauss' lemma, Proposition 5.32, g remains irreducible as a polynomial in $\mathbb{Q}[X]$. There is the substitution map

$$\varphi : \mathbb{Z}[X] \to \mathbb{C} \text{ given by } f \mapsto f(\alpha).$$

The image of φ is denoted by $\mathbb{Z}[\alpha]$. Being a homomorphic image of the Noetherian ring $\mathbb{Z}[X]$, the image $\mathbb{Z}[\alpha]$ is likewise Noetherian. The elements of $\mathbb{Z}[\alpha]$ are polynomial expressions in α with integer coefficients. Since

$$\alpha^n = -a_{n-1}\alpha^{n-1} - \cdots - a_1\alpha - a_0,$$

every element of $\mathbb{Z}[\alpha]$ can be reduced to the form

$$b_0 + b_1\alpha + \cdots + b_{n-1}\alpha^{n-1}, \text{ where } b_j \in \mathbb{Z}.$$

We typically say that $\mathbb{Z}[\alpha]$ is the ***ring generated*** by α, because it is the smallest subring of $\mathbb{C}$ which contains α. For example, a root of $X^2 + 1$ generates the Gaussian integers $\mathbb{Z}[i]$, while a root of $X^2 + 5$ generates $\mathbb{Z}[\sqrt{-5}]$.

Every such $\mathbb{Z}[\alpha]$ contains $\mathbb{Z}$ as a subring. Our interest is to explore the conditions for a prime p in $\mathbb{Z}$ to remain a prime in $\mathbb{Z}[\alpha]$. This will increase the scope of Eisenstein's criterion for Noetherian rings. First comes an observation on the kernel of φ.

Proposition 5.46. *If g in $\mathbb{Z}[X]$ is an irreducible, monic polynomial with a root α in $\mathbb{C}$, then the kernel of the substitution map $\varphi : \mathbb{Z}[X] \to \mathbb{C}$ given by $f \mapsto f(\alpha)$ is the principal ideal generated by g.*

Proof. Clearly $g\mathbb{Z}[X] \subseteq \ker\varphi$.

For the reverse inclusion suppose $f \in \mathbb{Z}[X]$ and that $f \in \ker\varphi$, i.e. $f(\alpha) = 0$. Clearly f and g belong to the ideal $J = \{h \in \mathbb{Q}[X] : h(\alpha) = 0\}$ inside the larger ring $\mathbb{Q}[X]$. Since $\mathbb{Q}$ is a field, this ideal is principal. Say $J = k\mathbb{Q}[X]$ for some generator k, which we can take to be monic. Then $g = k\ell$ for some ℓ in $\mathbb{Q}[X]$. The polynomial g remains irreducible in $\mathbb{Q}[X]$ by Gauss' lemma. Hence ℓ is a constant in $\mathbb{Q}$, and since both g and k are monic, $\ell = 1$ and $k = g$. Therefore, $f \in J = g\mathbb{Q}[X]$. We have that $f, g \in \mathbb{Z}[X]$, g divides f inside $\mathbb{Q}[X]$ and, being monic, g is primitive. By Gauss' lemma, $g \mid f$ in $\mathbb{Z}[X]$. That is $f \in g\mathbb{Z}[X]$. □

The Survival of Primes

Definition 5.47. For every prime integer p there is the canonical projection $\psi : \mathbb{Z} \to \mathbb{Z}_p$, and its resulting natural extension:

$$\psi : \mathbb{Z}[X] \to \mathbb{Z}_p[X] \text{ defined by } \sum a_j X^j \mapsto \sum \psi(a_j) X^j.$$

For each polynomial f in $\mathbb{Z}[X]$, the polynomial $\psi(f)$ with coefficients reduced modulo p is known as the ***reduction of f modulo p***. If $\psi(f)$ in $\mathbb{Z}_p[X]$ is irreducible, we say that f is ***irreducible modulo***[2] p.

Proposition 5.48. *Let g be any irreducible, monic polynomial in $\mathbb{Z}[X]$ with a root α in $\mathbb{C}$, and let p in $\mathbb{Z}$ be any prime. If $\psi(g)$ is the reduction of g modulo p, then*

$$\mathbb{Z}[\alpha]/\langle p\rangle \cong \mathbb{Z}_p[X]/\langle \psi(g)\rangle.$$

Consequently, p remains a prime in $\mathbb{Z}[\alpha]$ if and only if g is irreducible modulo p.

Proof. Apply the correspondence theorem, Proposition 4.49, to two surjective ring homomorphisms. The substitution map:

$$\varphi : \mathbb{Z}[X] \to \mathbb{Z}[\alpha] \text{ where } f \mapsto f(\alpha)$$

[2] With mild regret that the cognate terms "reduction" and "irreducible" appear in the same definition, and yet have diverse connotations.

and the natural extension of the canonical projection

$$\psi : \mathbb{Z}[X] \to \mathbb{Z}_p[X] \text{ where } \psi(f) \text{ is the reduction of } f \text{ modulo } p.$$

By Proposition 5.46 $\ker \varphi = \langle g \rangle$, and clearly $\ker \psi = \langle p \rangle$ as an ideal of $\mathbb{Z}[X]$. The ideal $\langle g, p \rangle$ contains both $\langle g \rangle$ and $\langle p \rangle$.

By the correspondence theorem, Proposition 4.49, applied to φ:

$$\mathbb{Z}[X]/\langle g,p \rangle \cong \mathbb{Z}[\alpha]/\varphi(\langle g,p \rangle) = \mathbb{Z}[\alpha]/\langle \varphi(g), \varphi(p) \rangle = \mathbb{Z}[\alpha]/\langle p \rangle.$$

Note that $\varphi(g) = g(\alpha) = 0$ and $\varphi(p) = p$ since the constant p evaluated at α is p.

Likewise by the correspondence theorem, Proposition 4.49, applied to ψ:

$$\mathbb{Z}[X]/\langle g,p \rangle \cong \mathbb{Z}_p[X]/\psi(\langle g,p \rangle) = \mathbb{Z}_p[X]/\langle \psi(g), \psi(p) \rangle = \mathbb{Z}_p[X]/\langle \psi(g) \rangle.$$

Obviously here $\psi(p) = 0$.

From the preceding pair of isomorphisms it follows immediately that

$$\mathbb{Z}[\alpha]/\langle p \rangle \cong \mathbb{Z}_p[X]/\langle \psi(g) \rangle.$$

One of these quotient rings is a domain if and only if the other quotient ring is a domain. Thus p is a prime in $\mathbb{Z}[\alpha]$ if and only if $\psi(g)$ is a prime in $\mathbb{Z}_p[X]$. Since $\mathbb{Z}_p$ is a field, the primes of $\mathbb{Z}_p[X]$ are the irreducible polynomials, and so the proof is complete. □

Here is an embellishment of part of Proposition 5.48.

Proposition 5.49. *Let p be a prime in $\mathbb{Z}$ and g a monic polynomial in $\mathbb{Z}[X]$ with a complex root α. If g is irreducible modulo p, then $\mathbb{Z}[\alpha]/\langle p \rangle$ is a field and p remains a prime in $\mathbb{Z}[\alpha]$.*

Proof. Proposition 5.39 applies to the natural extension $\mathbb{Z}[X] \to \mathbb{Z}_p[X]$ of the canonical projection $\mathbb{Z} \to \mathbb{Z}_p$. Consequently g is already irreducible in $\mathbb{Z}[X]$. Then Proposition 5.48 applies to reveal that p is prime in $\mathbb{Z}[\alpha]$. Furthermore, since the reduction $\psi(g)$ of g modulo p is irreducible, the quotient $\mathbb{Z}_p[X]/\langle \psi(g) \rangle$ is a field, due to Proposition 4.60. By Proposition 5.48, $\mathbb{Z}[\alpha]/\langle p \rangle$ is a field. □

Using Eisenstein in Conjunction with Algebraic Integers

Here is an illustration of Proposition 5.49 used in conjunction with Eisenstein's criterion, Proposition 5.45, for Noetherian rings.

Let $g = X^3 - 2$ in $\mathbb{Z}[X]$. This cubic polynomial is irreducible modulo 7 because $X^3 - [2]$ has no roots in $\mathbb{Z}_7$, by inspection. According to Proposition 5.48, the prime 7 remains so in the Noetherian ring $\mathbb{Z}[\sqrt[3]{2}]$. Now for any positive exponent n, Eisenstein's criterion for Noetherian rings shows that the polynomial $X^n - 7$ is irreducible in the polynomial ring $K[X]$, where K is the field of fractions of $\mathbb{Z}[\sqrt[3]{2}]$.

What is that field of fractions K? There is the substitution map $\varphi : \mathbb{Q}[X] \to \mathbb{C}$ given by $f \mapsto f(\sqrt[3]{2})$. The image of φ is $\mathbb{Q}[\sqrt[3]{2}]$ consisting of all $\mathbb{Q}$-linear combinations of powers $\sqrt[3]{2}$.

The polynomial g generates the principal ideal $\ker\varphi$. Indeed, if h is the monic generator of $\ker\varphi$, then h divides g in $\mathbb{Q}[X]$. Since g is irreducible in $\mathbb{Q}[X]$ and monic, g must be h. By the first isomorphism theorem, $\mathbb{Q}[X]/\langle g\rangle \cong \mathbb{Q}[\sqrt[3]{2}]$. But $\mathbb{Q}[X]/\langle g\rangle$ is a field. Hence $\mathbb{Q}[\sqrt[3]{2}]$ is a field. This field contains $\mathbb{Z}[\sqrt[3]{2}]$, and this field is clearly contained in the field of fractions of $\mathbb{Z}[\sqrt[3]{2}]$. Consequently $\mathbb{Q}[\sqrt[3]{2}]$ is the fraction field of $\mathbb{Z}[\sqrt[3]{2}]$.

We have learned that, regardless of n, the polynomial $X^n - 7$ is irreducible in $\mathbb{Q}[\sqrt[3]{2}][X]$. In the next chapter we shall be examining algebraic field extensions. An example of such an extension will be the field $\mathbb{Q}[\sqrt[3]{2}, \sqrt[n]{7}]$ of all polynomial expressions in $\sqrt[3]{2}$ and $\sqrt[n]{7}$. A significant problem will be to determine a quantity called the ***degree*** of such an extension. Our result with Eisenstein over Noetherian rings will be helpful.

5.8.4 A Principal Ideal Domain that Is Not Euclidean

Recalling from Section 5.3, a function ϕ from an integral domain A to the set of non-negative integers $\mathbb{N}$ is called Euclidean provided that for every pair of elements a, b in A where $a \neq 0$ and $a \nmid b$, there is an element q in A such that

$$\phi(b - aq) < \phi(a).$$

With $r = b - aq$ this can be restated as $b = aq + r$ where $\phi(r) < \phi(a)$. Domains with a Euclidean function are called ***Euclidean domains***. The ideals of a Euclidean domain are principal, according to Proposition 5.17. This raises the challenge of finding a non-Euclidean domain whose ideals remain principal.

A popular example of such a domain is $\mathbb{Z}[\alpha]$ where $\alpha = \frac{1+\sqrt{-19}}{2}$, which is a root of the polynomial $X^2 - X + 5$. For those who might find it interesting, we offer a quite different example of a principal ideal domain that is not Euclidean. The example is rather intricate.

Let $T = \mathbb{Z}_2[X_1, Y_1, X_2, Y_2, \ldots, X_i, Y_i, \ldots]$ be the ring of polynomials in infinitely many variables X_i, Y_i, with coefficients in the two element field $\mathbb{Z}_2$. The units of T are the constant polynomials that are units of $\mathbb{Z}_2$, which happen to be just the identity element 1. As demonstrated at the end of Section 5.8.1, T is a unique factorization domain, and thereby the irreducible polynomials of T are the primes of T. Every non-constant polynomial has a factorization into irreducible polynomials, which is unique up to associates. Since T has but one unit, each irreducible polynomial has only itself as an associate. Thus the factorization of each non-constant polynomial is absolutely unique. As shown in Section 5.5 every pair of non-zero polynomials f, g in T has a greatest common divisor $\gcd(f, g)$. This is a common divisor of f and g which is divisible by all other common divisors of f and g. To get the $\gcd(f, g)$ multiply the irreducible polynomials that divide both f and g, repeating each irreducible as many times as it divides both f and g. If $\gcd(f, g) = 1$, we say that f, g are coprime. This means that f, g have no common irreducible factor.

Further terminology will facilitate working with T. A ***monomial*** in T is a product of finitely many X_i and Y_j. This can be written as:

$$X_{i_1}^{d_1} X_{i_2}^{d_2} \cdots X_{i_k}^{d_k} Y_{j_1}^{e_1} Y_{j_2}^{e_2} \cdots Y_{j_\ell}^{e_\ell},$$

where the X_{i_r} and Y_{j_r} are distinct variables and the integer exponents d_r, e_r are non-negative. We say that d_r is the ***multiplicity*** of X_{i_r} in such a monomial. Likewise e_r is the multiplicity of Y_{j_r}. A variable X_i ***appears in a monomial*** if its multiplicity in that monomial is positive. Likewise for the Y_j. The only monomial in which all variables have zero multiplicity is the constant 1. This is also the only unit of T. Since the coefficients of every polynomial f in T can only be 0 or 1, the elements of T are finite sums of monomials. The unique monomials which add up to give f are called the ***monomials of*** f.

Since coefficients come from $\mathbb{Z}_2$, the identity

$$(f+g)^2 = f^2 + 2fg + g^2 = f^2 + g^2$$

prevails for all f, g in T. Consequently, if $f = m_1 + m_2 + \cdots + m_k$, where the m_j are the monomials of f, then $f^2 = m_1^2 + m_2^2 + \cdots + m_k^2$. So all variables appearing in the monomials of f^2 have even multiplicity in those monomials.

Let us say that f is ***even in the*** Y provided that each Y_j appears in the monomials of f only with even multiplicity. If f, g in T are both even in the Y, then so is their product fg. Now let

$$D = \{f \in T : f \neq 0 \text{ and every irreducible factor of } f \text{ is even in the } Y\}.$$

Clearly $1 \in D$ and D is closed under multiplication. Thus D is a denominator set as discussed in Section 4.8. In addition D is what is known as a ***saturated denominator set***. This means that if $g \in D$ and $f \in T$ and f divides g, then $f \in D$ as well. This is because the irreducible factors of f are also irreducible factors of g, which are even in the Y. Furthermore, every polynomial in D is even in the Y, because its irreducible factors are even in the Y.

Obviously every polynomial in which only X_j appear is in D. For more intricate examples of irreducible polynomials in D, suppose g, h are coprime in T and even in the Y, and suppose X_n is a variable that does not appear in any of the monomials of g or h, then the polynomial $f = g + hX_n$ is both even in the Y and irreducible, and thereby in D. That f is even in the Y should be easy to visualize. Regarding the irreducibility, it helps to interpret T as the polynomial ring $S[X_n]$ in the single variable X_n where the coefficients come from the subring S of polynomials in T whose monomials never have an appearance of X_n. In this light, we can speak of the degree $\deg k$ of a polynomial k in T as its degree in the variable X_n. So our f is a degree one polynomial in X_n with coefficients in S. Suppose $f = g + hX_n = uv$ for some u, v in T. Since $\deg f = 1$, it must be that one of u or v has degree 0 while the other has degree 1 in the variable X_n. Say $u \in S$ and $v = w + zX_n$ where $w, z \in S$. Then $g + hX_n = u(w + zX_n) = uw + uzX_n$. Since uw and uz are in S, it follows that $uw = g$ and $uz = h$. The common divisor u must divide $\gcd(g, h)$, which was assumed to be 1. Hence $u = 1$, which shows that f is irreducible. For instance, $X_1 + Y_1^2 X_2$ is irreducible and in D.

Let R be the localization of T at D. The elements of the ring R are the rational functions $\frac{f}{g}$ where $f \in T$ and $g \in D$. If f, g are both in D, obviously then $\frac{f}{g}$ is a unit of R with inverse $\frac{g}{f}$. Conversely, if $f \in T$ and $g \in D$ and $\frac{f}{g}$ is a unit of R, then f must be in D. To see this suppose $\frac{h}{k}$ with $h \in T$ and $k \in D$ is the inverse of $\frac{f}{g}$ in R. Thus $\frac{fh}{gk} = 1$, and then $fh = gk$. Since f divides gk and $gk \in D$, the saturation of D implies that $f \in D$.

We will show that all ideals of R are principal, while R admits no Euclidean function.

Suppose that J is an ideal of R, which can be assumed to be non-zero. The set $J \cap T$ is an ideal of T. Among the non-zero polynomials of $J \cap T$ pick an element g that has as few irreducible factors as possible. We first verify that such g generates $J \cap T$ as an ideal of T.

Let f be any non-zero element of $J \cap T$. Put $h = \gcd(g,f)$, and among the infinitely many variables X_i, select one, say X_n, that does not appear in any of the monomials of g nor of f. Now the polynomials $(\frac{g}{h})^2$ and $(\frac{f}{h})^2$ are even in the Y, coprime, and X_n does not appear in any of their monomials. As discussed after the definition of D, the polynomial

$$k = \left(\frac{g}{h}\right)^2 + \left(\frac{f}{h}\right)^2 X_n$$

is even in the Y and irreducible. Thereby $k \in D$, which puts $\frac{1}{k}$ inside R. The definition of k leads to the identity

$$h = g\left(\frac{1}{k}\frac{g}{h}\right) + f\left(\frac{1}{k}\frac{f}{h}X_n\right).$$

All of $\frac{1}{k}, \frac{g}{h}, \frac{f}{h}, X_n$ are in R, and since f, g are in J, it follows that $h \in J$. But $h \in T$ as well. Thus $h \in J \cap T$. As h divides g in T and g has the fewest irreducible factors among the non-zero elements of $J \cap T$, all irreducible factors that appear in g must remain in h without reduction in their multiplicities. Hence $g = h$, whereby g divides f in T, which shows that $J \cap T = \langle g \rangle$ as an ideal of T.

It follows readily that $J = \langle g \rangle$ as an ideal of R. This is because every element of J takes the form $\frac{f}{d}$ where $d \in D$ and $f \in T$. Thus $f = \frac{f}{d}d \in J \cap T$. As just shown $f = g\ell$ for some ℓ in T. Then $\frac{f}{d} = g\frac{\ell}{d}$. Since $\frac{\ell}{d} \in R$, the element $\frac{f}{d}$ lies in the principal ideal $\langle g \rangle$ of R.

To see that R admits no Euclidean function, suppose that $\phi : R \setminus \{0\} \to \mathbb{N}$ were such a function. Among the non-zero, non-unit elements of R select one, say $\frac{f}{g}$ where $f \in T$ and $g \in D$, such that its value $\phi\left(\frac{f}{g}\right)$ in $\mathbb{N}$ is minimal. Among the infinitely many variables Y_j pick one, say Y_n, that does not appear in any of the monomials of f. Since ϕ is a Euclidean function there is an h in T and a k in D such that

$$\phi\left(Y_n - \frac{f}{g}\frac{h}{k}\right) < \phi\left(\frac{f}{g}\right).$$

By the minimal choice of $\frac{f}{g}$, the element $Y_n - \frac{f}{g}\frac{h}{k}$, which of course equals $\frac{Y_n gk - fh}{gk}$, is 0 or a unit of R. Since $gk \in D$ and D is saturated

$$Y_n gk - fh \in D \cup \{0\}.$$

The multiplicity of Y_n in each monomial of gk is even, because $gk \in D$. So the multiplicity of Y_n in each monomial of $Y_n gk$ is odd. Let ℓ be the sum of those monomials of h in which the multiplicity of Y_n is odd. In the monomials of $h - \ell$ the multiplicity of Y_n is even. Since Y_n was chosen not to appear in any monomial of f, the multiplicity of Y_n in each monomial of $f\ell$ is odd, and the multiplicity of Y_n in each monomial of $f(h - \ell)$ is even. And we have

$$Y_n gk - fh = (Y_n gk - f\ell) - f(h - \ell) \in D \cup \{0\}.$$

If $Y_n gk - f\ell \neq 0$, this polynomial would have a monomial in which Y_n has odd multiplicity. That monomial would persist in $(Y_n gk - f\ell) - f(h-\ell)$ because Y_n has even multiplicity in the monomials of $f(h-\ell)$. This would contradict the fact that $(Y_n gk - f\ell) - f(h-\ell) \in D \cup \{0\}$ since the polynomials of D are even in the Y. Thus

$$Y_n gk - f\ell = 0.$$

The variable Y_n appears in every monomial of ℓ, and so $\frac{\ell}{Y_n} \in T$. The preceding equation gives

$$\frac{f}{g}\frac{\ell}{Y_n} = k.$$

Being in D the polynomial k is a unit of R. Thus its divisor $\frac{f}{g}$ in R is a unit of R. This contradicts the choice of $\frac{f}{g}$ as a non-unit of R. A Euclidean function ϕ is not possible for the domain R.

EXERCISES

1. If A is an integral domain and $A[X]$ is a unique factorization domain, show that A is a unique factorization domain. This is the converse of Proposition 5.34.
2. If A is a unique factorization domain, and P is a prime ideal of A, does the quotient ring A/P remain a unique factorization domain?
3. Show that $X^n - 75$ is irreducible in $\mathbb{Q}[X]$.
4. Show that the quotient ring $\mathbb{Q}[X]/\langle 3X^4 - 12X^3 + 8X^2 + 10\rangle$ is a field.
5. Prove that $X_n^2 + X_{n-1}^2 + \cdots + X_2^2 + X_1^2 + 1$ is irreducible in the ring $\mathbb{C}[X_1, X_2, \ldots, X_n]$ of polynomials in n variables.

   Hint. Use induction on the number of variables, starting with $n = 2$.
6. (a) Prove the following generalization of Eisenstein's criterion. Let $A[X]$ be a polynomial ring over a unique factorization domain A. Suppose $f(X) = a_n X^n + a_{n-1}X^{n-1} + \cdots + a_1 X + a_0 \in A[X]$, and p is a prime in A such that p divides all coefficients a_j except for a_n while p^2 does not divide some a_ℓ. If $f(X) = g(X)h(X)$ for some $g(X), h(X)$ in $A[X]$, show that one of h or g has degree at most ℓ.

   (b) Prove that $X^5 - 3X + 45$ is irreducible in $\mathbb{Q}[X]$.

   (c) Prove that $3X^{18} - 2X^2 + 4X - 20$ is irreducible in $\mathbb{Q}[X]$.
7. Prove that $XY^4 + 2X^2Y^3 + X^3Y^2 + Y^4 + XY^3 + X^3Y + 4X^4 + 3XY + X$ is irreducible in the polynomial ring $\mathbb{C}[X, Y]$.
8. Prove that 11 remains a prime in the ring $\mathbb{Z}[\sqrt{-5}]$.
9. If α is a complex root of $X^3 - X + 1$, show that 3 is a prime in $\mathbb{Z}[\alpha]$.
10. For each odd prime integer p show that the polynomial $X^2 + 1$ is irreducible modulo p if and only if $p \equiv 3 \bmod 4$.
11. Prove that 13 is a prime in the ring $\mathbb{Z}[\sqrt[3]{7}]$.

12. Show that $X^4 - 2$ is irreducible modulo 13, and deduce that 13 is a prime in $\mathbb{Z}[\sqrt[4]{2}]$.
13. If K is a field, the polynomial ring $R = K[X, Y]$ in two variables is Noetherian, as can be seen from Proposition 5.42. Show that the subring $K[X] + XK[X, Y]$ consisting of those polynomials of the form $f(X) + Xg(X, Y)$ is not Noetherian. This is yet another illustration that subrings of Noetherian rings need not be Noetherian.
14. Suppose that A is a Noetherian domain and that g is a monic polynomial in $A[X]$ that satisfies the assumptions of Eisenstein's criterion. Prove that g is a prime in $A[X]$.

    Hint. Let F be the fraction field of A, show that there is an injective homomorphism $\varphi : R[X]/gR[X] \to F[X]/gF[X]$, and apply Eisenstein for Noetherian domains.
15. The objective here is to show that $X^n - 7$ is an irreducible polynomial over the field $\mathbb{Q}[\sqrt{2}]$, regardless of n.
    (a) If a is an integer let $\overline{a}$ denote its residue in the finite field $\mathbb{Z}_7$. Show that the mapping $\varphi : \mathbb{Z}[\sqrt{2}] \to \mathbb{Z}_7$, defined by
    $$\varphi : a + b\sqrt{2} \mapsto \overline{a + 3b},$$
    is a surjective ring homomorphism.
    (b) Show that $\ker \varphi$ is the principal ideal generated by the element $3 - \sqrt{2}$. Deduce that $3 - \sqrt{2}$ is a prime in $\mathbb{Z}[\sqrt{2}]$.
    (c) Use Eisenstein to prove that $X^n - 7$ is irreducible over the field $\mathbb{Q}[\sqrt{2}]$.

6 Algebraic Field Extensions

Through field theory, insight into otherwise baffling problems can come to light. For example, it explains which regular polygons can be constructed with a compass and a straightedge, and why solutions of some polynomial equations cannot always be had by common formulas, such as the one used to solve quadratics. In addition, the finite fields have led to meaningful applications in cryptography and error correction.

The way to look at a field is through its relationship with its subfields, in other words, through the notion of a field extension. In order to approach field extensions, an understanding of vector spaces, bases, and dimension is essential.

6.1 Algebraic Elements and Degrees of Extensions

The following terminology will be pervasive.

Definition 6.1. If a field F is a subring of another field K, naturally we say that F is a ***subfield*** of K, and that K is a ***field extension*** of F. It is also common to say that K is a field ***over*** F. At times F is called the ***ground field*** of the extension K.

For instance, $\mathbb{C}$ is a field extension of $\mathbb{R}$, which in turn is a field extension of $\mathbb{Q}$. The set $\mathbb{Q}[\sqrt{2}]$ of numbers of the form $a + b\sqrt{2}$, where $a, b \in \mathbb{Q}$, is another field extension over $\mathbb{Q}$. For any field F, the field of rational functions $F(X)$, i.e. the fraction field of the polynomial ring $F[X]$, is a field extension of F.

The Characteristic of a Field

Every ring and thereby every field K has its characteristic, as discussed in Proposition 4.45. If K has characteristic 0, then K contains the integers $\mathbb{Z}$, and being a field, K contains the field of fractions of $\mathbb{Z}$, which is $\mathbb{Q}$. The other possibility is that K has a prime characteristic p, meaning that K contains the ring of residues $\mathbb{Z}_p$, which is itself a field. Thus, every field is an extension of either the field $\mathbb{Q}$ or one of the fields $\mathbb{Z}_p$ for some prime p.

6.1.1 The Degree of a Field Extension

An important thing to notice is that

every field extension K of F is a vector space over F.

Add the "vectors" in K in their normal way as elements of K, and scale them by multiplying them using the elements from F. For instance, the field $\mathbb{C}$ is a vector space over $\mathbb{R}$, whose dimension over $\mathbb{R}$ is two, while the rational function field $F(X)$ is a vector space over F whose dimension is infinite.

Definition 6.2. If the dimension of a field extension K over F is finite, i.e. K has a finite basis as a vector space over F, we say that K is a ***finite extension***[1] of F, and that dimension is called the ***degree of the extension.***[2] If K is a finite extension of F, the degree of the extension, i.e. the dimension of K as a vector space over F, is generally denoted by $[K : F]$.

For example, a basis for $\mathbb{Q}[\sqrt{2}]$ over $\mathbb{Q}$ consists of the elements $1, \sqrt{2}$. Thus $[\mathbb{Q}[\sqrt{2}] : \mathbb{Q}] = 2$. Also $[\mathbb{C} : \mathbb{R}] = 2$, since a basis of $\mathbb{C}$ as a vector space over $\mathbb{R}$ is the pair $1, i$. On the other hand the extension $\mathbb{R}$ is not finite over $\mathbb{Q}$. A quick way to see this (for those familiar with countable sets) is to note that, for every positive integer n, the vector space $\mathbb{Q}^n$ is countable while $\mathbb{R}$ is not countable. In other words there can be no bijection $\mathbb{R} \to \mathbb{Q}^n$. Thus $\mathbb{R}$ cannot have finite dimension over $\mathbb{Q}$. Alternatively, it is not too hard to produce arbitrarily long lists of real numbers that are linearly independent over $\mathbb{Q}$.

The Size of a Finite Field

If K is a finite field, then its smallest subfield must be one of the fields $\mathbb{Z}_p$ for some prime p. Since K is finite, so is the degree $[K : \mathbb{Z}_p]$. Say $\alpha_1, \alpha_2, \ldots, \alpha_n$ is a basis for K over $\mathbb{Z}_p$. Every element α in K has the representation

$$\alpha = a_1\alpha_1 + a_2\alpha_2 + \cdots + a_n\alpha_n,$$

where the coefficients a_j come from the smallest subfield $\mathbb{Z}_p$ and are uniquely determined by α. Since each a_j has p possibilities, α has exactly p^n possibilities. We have just bumped into something interesting about finite fields.

Proposition 6.3. *If K is a finite field of characteristic p and $[K : \mathbb{Z}_p] = n$, then K has exactly p^n elements.*

Naturally we should ask if there exist finite fields of size p^n for all primes p and all positive exponents n, and how many there are for each p^n. Up to isomorphism, there is exactly one field for each eligible size, as we will discover.

6.1.2 Algebraic Elements

Suppose that α is an element in a field extension K over F. The substitution map $\varphi_\alpha : F[X] \to K$, given by $f(X) \mapsto f(\alpha)$, is a homomorphism from the ring of polynomials

[1] More properly K ought to be called a finite-dimensional extension, but tradition has favored this more compact terminology. Note that the fields in question need not be themselves finite.

[2] Why the perfectly good term "dimension" gets replaced by "degree," when field extensions are concerned, is one of those accidents of history.

$F[X]$ into K. The image of φ_α is the smallest subring of K that includes F and contains α. This is the set of all polynomial expressions

$$a_0 + a_1\alpha + a_2\alpha^2 + \cdots + a_n\alpha^n,$$

where n can be any non-negative integer and the a_j run through F. This subring is denoted by $F[\alpha]$, and is called the ***ring generated*** by α over F. It is also common to say that the ring $F[\alpha]$ has been formed by ***adjoining*** the element α to F.

Definition 6.4. An element α in a field extension K of F is said to be ***algebraic***[3] over F provided $f(\alpha) = 0$ for some non-zero polynomial f in $F[X]$. If there is no such non-zero polynomial in $F[X]$, then α is called ***transcendental*** over F. If *every* α in K is algebraic over F, the extension K itself is called an ***algebraic field extension*** of F.

Algebraic elements over a field F are the roots of non-zero polynomials in $F[X]$. They are also the elements for which the kernel of the substitution map φ_α is non-zero. For example, every α in F is algebraic over F, being the root of the polynomial $X - \alpha$ in $F[X]$. The element $e^{\pi i/2} = 1/\sqrt{2} + i1/\sqrt{2}$ in the field $\mathbb{C}$ is algebraic over $\mathbb{Q}$, since it is a root of the polynomial $X^4 + 1$ in $\mathbb{Q}[X]$. On the other hand, the element X in the field of rational functions $F(X)$ is transcendental over the ground field F, simply because polynomials are designed with that in mind. A famous result due to Ferdinand von Lindemann is that the real number π is transcendental over $\mathbb{Q}$.

Finite Extensions Are Algebraic

Proposition 6.5. *If K is a field extension of F and the degree $[K : F]$ is finite, then every element of K is the root of a non-zero polynomial in $F[X]$ whose degree is at most $[K : F]$. Thus K is an algebraic extension of F.*

Proof. If $\alpha \in K$ and $n = [K : F]$, the list of $n + 1$ powers $1, \alpha, \alpha^2, \alpha^2, \ldots, \alpha^n$ is linearly dependent over F. The resulting dependency relation produces a non-zero polynomial in $F[X]$ of degree at most $[K : F]$ and having α as a root. □

For example, since $[\mathbb{C} : \mathbb{R}] = 2$, every complex number is the root of a polynomial in $\mathbb{R}[X]$ of degree at most 2 and is thereby algebraic over $\mathbb{R}$. All elements of a finite field are algebraic over its ground field $\mathbb{Z}_p$. More sophisticated examples will appear in the ensuing discussion. We shall see in due course that there exist algebraic extensions which are not of finite degree. Proposition 6.5 does not have a converse.

Building Fields by Adjunction of Algebraic Elements

Algebraic elements come with a significant feature.

[3] Algebra had its origins in attempts to solve equations of type $f(X) = 0$ where f is a polynomial, starting with linear and then quadratic polynomials. These came to be known as ***algebraic equations***. From that it does not take much to see why the roots of such polynomials are called algebraic.

Proposition 6.6. *An element α in a field extension K of F is algebraic over F if and only if the ring $F[\alpha]$ is already a field.*

Proof. The image of the substitution map $\varphi_\alpha : F[X] \to K$, given by $f \mapsto f(\alpha)$, is the ring $F[\alpha]$. The first isomorphism theorem, Proposition 4.43, gives

$$F[X]/\ker\varphi_\alpha \cong F[\alpha] \text{ where the residue } f + \ker\varphi_\alpha \text{ corresponds to } f(\alpha).$$

If α is algebraic over F, then the ideal $\ker\varphi_\alpha$ is non-zero. Furthermore, $\ker\varphi_\alpha$ is a prime ideal. This is so because the image $F[\alpha]$ of φ_α is an integral domain, which allows for the invocation of Proposition 4.56. All ideals in $F[X]$ are principal, and so by Proposition 5.11, $\ker\varphi_\alpha$ gets upgraded to a maximal ideal. Then Proposition 4.51 ensures that $F[X]/\ker\varphi_\alpha$ is a field, whence its isomorphic copy $F[\alpha]$ is a field too.

Conversely, if α is transcendental over F, the substitution map φ_α is injective because $\ker\varphi_\alpha$ is zero. Thus its image $F[\alpha]$ is isomorphic to the polynomial ring $F[X]$. Since $F[X]$ is not a field, neither is its isomorphic copy $F[\alpha]$. □

The proof of Proposition 6.6 relies on the fact that non-zero prime ideals in principal ideal domains are maximal. We shall come to see shortly how inverses in extensions formed by adjunction of an algebraic element can be found in practice.

6.1.3 The Minimal Polynomial of an Algebraic Element

Every algebraic element α over a field F comes with a distinguished polynomial, which is the workhorse for managing calculations in the field $F[\alpha]$.[4]

Proposition 6.7. *If K is a field extension of F and α in K is algebraic over F, then there is a polynomial g in $F[X]$, which is unique with the following properties:*

$$g \text{ is monic}, \quad g(\alpha) = 0, \text{ and } g \text{ is irreducible}.$$

Such g divides every polynomial in $F[X]$ that has α as root.

Proof. The substitution map

$$\varphi_\alpha : F[X] \to K \text{ given by } f \mapsto f(\alpha),$$

explains the matter. The ideal $\ker\varphi_\alpha$ is principal in $F[X]$, by Proposition 4.33. Its generator is unique up to an associate, i.e. up to a non-zero scalar multiple. Thus $\ker\varphi$ has just one generator that is monic. Call it g. By its very choice, g satisfies the first two of our three requirements.

In addition g is irreducible because the ideal $\ker\varphi_\alpha$, which g generates, is prime. For a direct argument that g is irreducible, suppose $g = fh$ for some polynomials f, h in $F[X]$. Apply the

[4] At this point an issue of notation arises. Do we use the ring extension notation $F[\alpha]$ (with square brackets) to denote the field extension generated by the algebraic α? Or do we use the more commonly adopted field extension notation $F(\alpha)$ to denote the smallest field containing F and α? In case α is algebraic over F, Proposition 6.6 says that these notations give the same creature. We have opted to use the square bracket notation, as a reminder that, when α is algebraic over F, the elements in the field $F[\alpha]$ are polynomial expressions in α.

substitution φ to see that $f(\alpha)h(\alpha) = 0$. Since K is a field, either $f(\alpha) = 0$ or $h(\alpha) = 0$. Say $f(\alpha) = 0$. Thus $f \in \ker\varphi$, and f divides the generator g of $\ker\varphi$. So g and f divide each other. Hence, the other factor h has to be a constant polynomial, i.e. a unit of $F[X]$. That shows g is irreducible.

Since g is a generator of the ideal $\ker\varphi$ and $\ker\varphi$ is precisely the ideal of polynomials in $F[X]$ that have α as a root, g divides all polynomials that vanish at α.

Regarding the uniqueness of g with the specified properties, note that any other monic, irreducible polynomial f in $F[X]$ having α as a root must lie in $\ker\varphi$. Thus $f = gh$ for some polynomial h in $F[X]$. Since f is irreducible, h has to be constant. Thus g and f are associates in $F[X]$, and since both are monic, they must be equal. □

Definition 6.8. If α is in a field extension K of F and α is algebraic over F, the ***minimal polynomial of α over F*** is the unique monic, irreducible polynomial g in $F[X]$ that has α as a root. The minimal polynomial divides every polynomial in $F[X]$ that vanishes at α.

A moment's consideration reveals that an algebraic element α already belongs to the ground field F if and only if its minimal polynomial is the linear polynomial $X - \alpha$. For another example, the minimal polynomial of the complex number i over $\mathbb{R}$ is $X^2 + 1$. The minimal polynomial of $e^{2\pi i/5}$ over $\mathbb{Q}$ is $X^4 + X^3 + X^2 + X + 1$. Indeed, as seen in Section 5.7, this polynomial is irreducible in $\mathbb{Q}[X]$. Also $e^{2\pi i/5}$ is a root of it, because $e^{2\pi i/5}$ is an obvious root of $X^5 - 1$, $X^5 - 1 = (X - 1)(X^4 + X^3 + X^2 + X + 1)$, and $e^{2\pi i/5}$ is not a root of $X - 1$.

The Minimal Polynomial of $\sqrt{2} + i$

For a more intricate example, the minimal polynomial of $\alpha = \sqrt{2} + i$ over $\mathbb{Q}$ can be obtained with a bit of squaring:

$$\begin{aligned}
\alpha &= \sqrt{2} + i \\
\alpha^2 &= 1 + 2\sqrt{2}i \\
\alpha^2 - 1 &= 2\sqrt{2}i \\
(\alpha^2 - 1)^2 &= -8 \\
\alpha^4 - 2\alpha^2 + 1 &= -8 \\
\alpha^4 - 2\alpha^2 + 9 &= 0.
\end{aligned}$$

Thus the minimal polynomial of α is an irreducible factor in $\mathbb{Q}[X]$ of $X^4 - 2X^2 + 9$. Now either $X^4 - 2X^2 + 9$ has a linear factor, or it has two irreducible quadratic factors, or is itself irreducible.

The first case does not happen as can be seen by using the rational roots test of Proposition 5.36. If $X^4 - 2X^2 + 9$ has two irreducible quadratic factors in $\mathbb{Q}[X]$, Gauss' lemma allows those factors to be in $\mathbb{Z}[X]$, and since $X^4 - 2X^2 + 9$ is monic, those quadratic factors can be taken to be monic. Thus we would have

$$X^4 - 2X^2 + 9 = (X^2 + aX + b)(X^2 + cX + d) \text{ where } a, b, c, d \in \mathbb{Z}.$$

By multiplying and comparing coefficients this leads to

$$a + c = 0,\ b + d + ac = -2,\ ad + bc = 0,\ bd = 9.$$

The last equation forces the integer pair (b, d) to be one of

$$(1,9), (-1,-9), (3,3), (-3,-3), (9,1), (-9,-1).$$

But every one of these six possibilities contradicts the remaining equations. For instance, if $b = d = 3$, then the second equation leads to $ac = -8$, and from $a + c = 0$ we get the impossible $-a^2 = -8$. The other five cases are similarly ruled out. Hence, the minimal polynomial of $\sqrt{2} + i$ is $X^4 - 2X^2 + 9$.

6.1.4 A Basis for a Singly Generated Algebraic Extension

The minimal polynomial of an algebraic element α, in a field extension K over F, establishes the degree of the algebraic extension $F[\alpha]$ over F.

Proposition 6.9. *If α in a field extension of F is algebraic over F with minimal polynomial g in $F[X]$, then $[F[\alpha] : F] = \deg g$. A basis for $F[\alpha]$ as a vector space over F consists of the powers*

$$1, \alpha, \alpha^2, \ldots, \alpha^{n-1} \text{ where } n = \deg g.$$

Thereby $F[\alpha]$ is an algebraic extension of F. Furthermore, the degree of the minimal polynomial of every element β in $F[\alpha]$ is at most $\deg g$.

Proof. Every element of $F[\alpha]$ takes the form $f(\alpha)$ for some polynomial f in $F[X]$. In the Euclidean domain $F[X]$ we have

$$f = gq + r, \text{ where } q, r \in F[X] \text{ and } \deg r < \deg g \text{ or } r = 0.$$

Apply the substitution $\varphi : F[X] \to F[\alpha]$, where $\varphi : X \mapsto \alpha$, on the preceding polynomial equation to get

$$f(\alpha) = g(\alpha)q(\alpha) + r(\alpha) = r(\alpha).$$

Since $\deg r < \deg g$, every element in $F[\alpha]$ is a linear combination of the powers $1, \alpha, \alpha^2, \ldots, \alpha^{n-1}$.

Furthermore, these powers of α are linearly independent over F. Otherwise, a dependency relation would yield a non-zero polynomial f in $F[X]$ such that $f(\alpha) = 0$ and $\deg f < \deg g$. But this would prevent g from dividing f, contrary to what minimal polynomials do.

The last two conclusions of this proposition flow from Proposition 6.5. □

Proposition 6.9 suggests the origin of the term "degree" to indicate the dimension of a field extension. In the case of an extension $F[\alpha]$, the dimension of $F[\alpha]$ over F is the degree of the minimal polynomial of α. Sometimes the word degree is applied to the algebraic element α itself.

Definition 6.10. The ***degree of an algebraic element*** over a field F is the same as the degree of its minimal polynomial in $F[X]$.

For instance, we saw prior to Proposition 6.9 that the minimal polynomial of $\alpha = \sqrt{2}+i$ over $\mathbb{Q}$ has degree 4. Thus the extension $\mathbb{Q}[\alpha]$ has degree 4 over $\mathbb{Q}$. A basis for the field extension $\mathbb{Q}[\sqrt{2}+i]$ over $\mathbb{Q}$ consists of $1,\alpha,\alpha^2,\alpha^3$. In addition $\sqrt{2}i = \frac{1}{2}\alpha^2 - 1 \in \mathbb{Q}[\alpha]$. Then $-\sqrt{2}+2i = \sqrt{2}i(\sqrt{2}+i) \in \mathbb{Q}[\alpha]$, and $i = \frac{1}{3}((-\sqrt{2}+2i)+(\sqrt{2}+i)) \in \mathbb{Q}[\alpha]$. Also $\sqrt{2} = (\sqrt{2}+i) - i \in \mathbb{Q}[\alpha]$. The minimal polynomial of i over $\mathbb{Q}$ is X^2+1, and the minimal polynomial of $\sqrt{2}$ over $\mathbb{Q}$ is X^2-2. Both have degree 2. In consonance with the last statement of Proposition 6.9, $2 \leq 4$.

Computing Inverses in the Field $F[\alpha]$

How are we to compute inverses in a field extension $F[\alpha]$ formed by adjoining an algebraic element α to F? If g is the minimal polynomial of α of degree n, we have a basis $1,\alpha,\ldots,\alpha^{n-1}$ of $F[\alpha]$, and we know that $g(\alpha) = 0$. Take a non-zero element β in $F[\alpha]$. Write β in terms of the basis as $\beta = a_0 + a_1\alpha + \cdots + a_{n-1}\alpha^{n-1}$, where $a_j \in F$. Since $\beta \neq 0$, the polynomial $f = a_0 + a_1X + \cdots + a_{n-1}X^{n-1}$ in $F[X]$ is non-zero and has degree less than n. Thus it is coprime with the irreducible polynomial g. Their greatest common divisor is 1. When the GCD-algorithm described in Section 5.5 is applied to the pair g,f, we obtain polynomials h,k in $F[X]$ such that

$$1 = gh + fk.$$

Substitute α for X to get

$$1 = g(\alpha)h(\alpha) + f(\alpha)k(\alpha) = f(\alpha)k(\alpha).$$

The inverse of $f(\alpha)$ is obviously $k(\alpha)$, another polynomial in α.

By way of illustration, take the monic polynomial $X^4 - 4X + 10$ in $\mathbb{Q}[X]$. By Eisenstein's criterion using the prime 2, this polynomial is irreducible. It has four roots in the field extension $\mathbb{C}$. Pick one of them, and call it α. By its very definition, α is algebraic over $\mathbb{Q}$, and its minimal polynomial is $X^4 - 4X + 10$. Then $\mathbb{Q}[\alpha]$ is a degree 4 field extension of $\mathbb{Q}$. Every element of $\mathbb{Q}[\alpha]$ is a unique linear combination of $1,\alpha,\alpha^2,\alpha^3$, with coefficients taken from $\mathbb{Q}$. This can be used to carry out field operations in $\mathbb{Q}[\alpha]$ based on the known operations in $\mathbb{Q}$ and the fact that $\alpha^4 = 4\alpha - 10$. For instance,

$$\begin{aligned}(7\alpha + 4\alpha^2)(2 - 5\alpha + \alpha^3) &= 14\alpha - 27\alpha^2 - 20\alpha^3 + 7\alpha^4 + 4\alpha^5 \\ &= 14\alpha - 27\alpha^2 - 20\alpha^3 + 7(4\alpha - 10) + 4\alpha(4\alpha - 10) \\ &= -70 + 2\alpha - 11\alpha^2 - 20\alpha^3.\end{aligned}$$

Now, let us find the inverse of $\alpha^2 + 1$. Since $X^4 - 4X + 10$ is irreducible, the polynomials $X^2 + 1$ and $X^4 - 4X + 10$ are coprime, meaning that their greatest common divisor is a constant in F. By the GCD-algorithm illustrated in Section 5.5, express their greatest common divisor as an $F[X]$-linear combination of X^2+1 and $X^4-4X+10$. To that end carry out suitable, admittedly messy, divisions and substitutions:

$$X^4-4X+10=(X^2+1)(X^2-1)-4X+11$$
$$X^2+1=(-4X+11)\left(-\frac{1}{4}X-\frac{11}{16}\right)+\frac{137}{16}.$$

The greatest common divisor of X^2+1 and $X^4-4X+10$ is $\frac{137}{16}$ (or any other non-zero constant in $\mathbb{Q}$ for that matter). From the last of the above equations observe:

$$\frac{137}{16}=(X^2+1)-(-4X+11)\left(-\frac{1}{4}X-\frac{11}{16}\right).$$

Substitute $-4X+11$ from the very first equation into the above to get:

$$\frac{137}{16}=(X^2+1)-\left((X^4-4X+10)-(X^2+1)(X^2-1)\right)\left(-\frac{1}{4}X-\frac{11}{16}\right).$$

Collect the respective factors of X^2+1 and $X^4-4X+10$ to get:

$$\frac{137}{16}=(X^2+1)\left(-\frac{1}{4}X^3-\frac{11}{16}X^2+\frac{1}{4}X+\frac{27}{16}\right)+(X^4-4X+10)\left(\frac{1}{4}X+\frac{11}{16}\right).$$

At this point apply the substitution map $F[X]\to F[\alpha]$ that sends X to α, note that $\alpha^4-4\alpha+10=0$, and obtain an equation in $F[\alpha]$:

$$\frac{137}{16}=(\alpha^2+1)\left(-\frac{1}{4}\alpha^3-\frac{11}{16}\alpha^2+\frac{1}{4}\alpha+\frac{27}{16}\right).$$

Finally, multiply through by $\frac{137}{16}$ to get:

$$1=(\alpha^2+1)\left(-\frac{4}{137}\alpha^3-\frac{11}{137}\alpha^2+\frac{4}{137}\alpha+\frac{27}{137}\right).$$

The inverse of α^2+1 is in plain sight.

6.1.5 Algebraic Elements with the Same Minimal Polynomial

The isomorphism type of an algebraic extension $F[\alpha]$ over F is determined by the minimal polynomial of α.

Proposition 6.11. *Let K, L be field extensions of a field F, and let $\alpha \in K, \beta \in L$. If α and β have the same minimal polynomial over F, then there is an isomorphism*

$$\sigma : F[\alpha] \to F[\beta] \text{ such that } \sigma : f(\alpha) \mapsto f(\beta) \text{ for every polynomial } f \text{ in } F[X].$$

In particular, $\sigma(a)=a$ for all a in F, and $\sigma(\alpha)=\beta$.

Proof. Once again the substitution map

$$\varphi : F[X] \to F[\alpha] \text{ where } \varphi : f \mapsto f(\alpha),$$

along with the first isomorphism theorem, Proposition 4.43, explain the matter. Accordingly, the mapping

$$\Phi : F[X]/\ker\varphi \to F[\alpha], \text{ where } f + \ker\varphi \mapsto f(\alpha),$$

is an isomorphism. Likewise, there is a substitution map

$$\psi : F[X] \to F[\beta] \text{ where } \psi : f \mapsto f(\beta),$$

and by the first isomorphism theorem, the mapping

$$\Psi : F[X]/\ker\psi \to F[\beta], \text{ where } f + \ker\psi \mapsto f(\beta),$$

is an isomorphism. The minimal polynomial of α is the unique monic generator of the ideal $\ker\varphi$ inside $F[X]$, and the minimal polynomial of β is the unique monic generator of $\ker\psi$ inside $F[X]$. Since these minimal polynomials are assumed equal, $\ker\varphi = \ker\psi$, and so $F[X]/\ker\varphi = F[X]/\ker\psi$. The composite mapping

$$\Psi \circ \Phi^{-1} : F[\alpha] \to F[X]/\ker\varphi = F[X]/\ker\psi \to F[\beta]$$

is the requisite isomorphism σ. □

For example, take the field extensions $\mathbb{Q}[\alpha]$ and $\mathbb{Q}[\beta]$ where $\alpha = \sqrt[3]{2}$ and $\beta = \sqrt[3]{2}e^{2\pi i/3}$. Both α and β have $X^3 - 2$ as their minimal polynomial over $\mathbb{Q}$. Hence the extensions are isomorphic. Furthermore, the isomorphism fixes $\mathbb{Q}$ and sends $\sqrt[3]{2}$ to $\sqrt[3]{2}e^{2\pi i/3}$.

6.1.6 The Tower Theorem

Do sums, products and quotients of algebraic elements remain algebraic? The next *very useful* result will help to explain why this is so, plus quite a bit more.

Definition 6.12. A sequence of nested subfields

$$F \subseteq F_1 \subseteq F_2 \subseteq \cdots \subseteq F_n$$

will be called a ***tower*** of fields.

Proposition 6.13 (Tower theorem). *Let $F \subseteq K \subseteq L$ be a tower of three fields. Suppose $x_1, x_1, \ldots, x_m$ is a basis for K as a vector space over F and $y_1, y_2, \ldots, y_n$ is a basis of L as a vector space over K. Then the set of all products x_iy_j forms a basis of L as a vector space over F. Consequently*

$$[L : F] = [L : K][K : F].$$

Proof. For the linear independence of the x_iy_j over the field F, suppose that

$$\sum_{i,j} a_{ij}x_iy_j = 0 \text{ for some coefficients } a_{ij} \text{ in } F.$$

Rearrange this summation to get $\sum_j(\sum_i a_{ij}x_i)y_j = 0$. For each j, the sum $\sum_i a_{ij}x_i$ lies in K, so the above is a linear combination of the y_j with coefficients in K. By the linear independence of the y_j in the vector space L over the field K, deduce that $\sum_i a_{ij}x_i = 0$ for all j. Since the $a_{ij} \in F$, the independence of the x_i over the field F implies that all $a_{ij} = 0$.

To show that the x_iy_j span L as a vector space over F, take any element z in L. Since the y_j span L as a vector space over K, there exist elements u_j in K such that $z = \sum_j u_jy_j$. Since the x_i span K as a vector space over F, each $u_j = \sum_i a_{ij}x_i$ for suitable a_{ij} in F. Substitute this into the expansion for z to obtain

$$z = \sum_j \left(\sum_i a_{ij}x_i \right) y_j = \sum_{i,j} a_{ij}x_iy_j,$$

which is a linear combination of the x_iy_j over the field F. □

For any tower $F \subseteq F_1 \subseteq F_2 \subseteq \cdots \subseteq F_n$ of finite extensions F_j over F_{j-1}, the degree $[F_n : F]$ equals the product of the degrees $[F_j : F_{j-1}]$. This can be seen from a routine induction argument based on Proposition 6.13.

A Severe Restriction on the Subfields of Finite Extensions

The tower theorem restricts the kinds of fields K that can sit between a ground field F and an extension L of finite degree over F. Those K whose degree $[K : F]$ is not a factor of $[L : F]$ are ineligible. For instance, if $[L : F] = 4$, then, except for L and F, every subfield between L and F has degree 2 over F. If $[L : F]$ is a prime, then the extension L over F has no proper intermediate fields, because the degree over F of such an intermediate field K would have to be either $[L : F]$ or 1. In the first case $K = L$, and in the second case $K = F$.

6.1.7 Fields Generated by Several Algebraic Elements

If a field K is an algebraic extension of F, and α, in a still bigger field L, is an algebraic element over K, does α remain algebraic over the bottom field F? If the degree $[K : F]$ is finite, this is easy to see. Proposition 6.9 forces $[K[\alpha] : K]$ to be finite, and the tower theorem, Proposition 6.13, shows that $[K[\alpha] : F]$ is also finite. Then α remains algebraic over F due to Proposition 6.5.

However, an algebraic extension K of F need not be of finite degree, as we shall learn. The preceding argument needs to be refined by means of a generalization to Propositions 6.6 and 6.9.

Definition 6.14. If $\alpha_1,\alpha_2,\ldots,\alpha_n$ is a list of elements in K, the smallest subring of K that contains F and includes all α_j is the set of all finite sums:

$$\sum_{i_1,\ldots,i_n} a_{i_1\ldots i_n}\alpha_1^{i_1}\cdots\alpha_n^{i_n},$$

where the i_j are non-negative integers and $a_{i_1 \dots i_n} \in F$. This is called the ***ring generated over F*** by $\alpha_1, \dots, \alpha_n$, and is denoted by

$$F[\alpha_1, \dots, \alpha_n].$$

Clearly, the smallest subring of K that contains F and all $\alpha_1, \dots, \alpha_n$ is the same subring as the smallest subring that contains the ring $F[\alpha_1, \alpha_2, \dots, \alpha_{n-1}]$ as well as α_n. Thus we see the equality

$$F[\alpha_1, \alpha_2, \dots, \alpha_n] = F[\alpha_1, a_2, \dots, \alpha_{n-1}][\alpha_n],$$

which lets us adjoin elements one at a time.

For example, $\mathbb{Q}[\sqrt{2}, \sqrt{3}] = \mathbb{Q}[\sqrt{2}][\sqrt{3}]$. This is the smallest subring of $\mathbb{C}$ which contains $\mathbb{Q}$ and includes $\sqrt{2}$ and $\sqrt{3}$. For all non-negative integers i, j, the products $\sqrt{2}^i \sqrt{3}^j$ collapse to integer multiples of $\sqrt{2}, \sqrt{3}$ and $\sqrt{6}$. Thus the ring $\mathbb{Q}[\sqrt{2}, \sqrt{3}]$ consists of all linear combinations

$$a + b\sqrt{2} + c\sqrt{3} + d\sqrt{6}, \text{ where } a, b, c, d \in \mathbb{Q}.$$

If the generators α_j are algebraic over F, the expected generalizations of Propositions 6.6 and 6.9 carry through.

Proposition 6.15. *If $\alpha_1, \dots, \alpha_n$ in a field K are algebraic over a subfield F, then*

- *the ring $F[\alpha_1, \dots, \alpha_n]$ is a field,*
- *the degree $[F[\alpha_1, \dots, \alpha_n] : F]$ is finite, and*
- *$F[\alpha_1, \dots, \alpha_n]$ is an algebraic extension of F.*

Every finite extension of F takes the form $F[\alpha_1, \dots, \alpha_n]$ for some elements $\alpha_1, \dots, \alpha_n$ which are algebraic over F.

Proof. First note that the third item in our claim is a consequence of Proposition 6.5 applied to the second item.

To prove the first two items, do an induction on the number n of generators. When $n = 0$ these items come down to saying the obvious, which is that F is a field and a finite extension of itself.

Assuming that the first two items hold for $n - 1$ generators, we have that $F[\alpha_1, \dots, \alpha_{n-1}]$ is a finite degree field extension of F. Obviously α_n remains algebraic over $F[\alpha_1, \dots, \alpha_{n-1}]$. By Propositions 6.6 and 6.9, $F[\alpha_1, \dots, \alpha_{n-1}][\alpha_n]$ is a finite degree field extension of $F[\alpha_1, \dots, \alpha_{n-1}]$. Then $F[\alpha_1, \dots, \alpha_{n-1}][\alpha_n]$ has finite degree over F because of the tower theorem, Proposition 6.13. Now simply observe that $F[\alpha_1, \dots, \alpha_{n-1}][\alpha_n] = F[\alpha_1, \dots, \alpha_{n-1}, \alpha_n]$.

The converse statement is trivial to see. If K is a finite extension of F, any basis $\alpha_1, \dots, \alpha_n$ of K over F will consist of algebraic elements that give $K = F[\alpha_1, \dots, \alpha_n]$. There could be abundant redundancy among these generators. □

An Example with Towers of Algebraic Extensions

If an integer a is not a perfect square, a basic exercise shows that $\sqrt{a} \notin \mathbb{Q}$. In that case the minimal polynomial of $\sqrt{a}$ cannot be linear, and therefore must be $X^2 - a$. Consequently the degree $[\mathbb{Q}[\sqrt{a}] : \mathbb{Q}] = 2$. Let us see how the degree goes up if we adjoin more such square roots.

Proposition 6.16. *If $a_1, a_2, \ldots, a_n$ are positive, pairwise coprime integers which are not perfect squares, then $[\mathbb{Q}[\sqrt{a_1}, \ldots, \sqrt{a_n}] : \mathbb{Q}] = 2^n$.*

Proof. As already noted, $[\mathbb{Q}[\sqrt{a_1}] : \mathbb{Q}] = 2$, whereby our result holds if $n = 1$. Looking towards an inductive proof, suppose that the result holds for all such lists involving fewer than n integers, and take our list $a_1, a_2, \ldots, a_n$ as specified. We can suppose $n \geq 2$.

Put

$$F = \mathbb{Q}[\sqrt{a_1}, \ldots, \sqrt{a_{n-2}}] \text{ and } K = \mathbb{Q}[\sqrt{a_1}, \ldots, \sqrt{a_n}] = F[\sqrt{a_{n-1}}][\sqrt{a_n}].$$

By the inductive assumption the degree $[F[\sqrt{a_{n-1}}] : \mathbb{Q}] = 2^{n-1}$. In order to get the desired $[K : \mathbb{Q}] = 2^n$, it is enough to have $\sqrt{a_n} \notin F[\sqrt{a_{n-1}}]$. This will ensure that the minimal polynomial of $\sqrt{a_n}$ over $F[\sqrt{a_{n-1}}]$ remains the quadratic polynomial $X^2 - b$. Then the tower theorem, Proposition 6.13, will yield

$$\begin{aligned}
[K : \mathbb{Q}] &= [F[\sqrt{a_{n-1}}][\sqrt{a_n}] : \mathbb{Q}] \\
&= [F[\sqrt{a_{n-1}}][\sqrt{a_n}] : F[\sqrt{a_{n-1}}]] \cdot [F[\sqrt{a_{n-1}}] : \mathbb{Q}] \\
&= 2^{n-1} \cdot 2 = 2^n.
\end{aligned}$$

Suppose to the contrary that $\sqrt{a_n} \in F[\sqrt{a_{n-1}}]$. This means that

$$\sqrt{a_n} = r + s\sqrt{a_{n-1}} \text{ for some } r, s \text{ in } F.$$

If $r = 0$, then $\sqrt{a_{n-1}a_n} = sa_{n-1} \in F$, and thereby $F[\sqrt{a_{n-1}a_n}] = F$, which shows that

$$[F[\sqrt{a_{n-1}a_n}] : \mathbb{Q}] = [F : \mathbb{Q}] = 2^{n-2},$$

by the inductive assumption. However, the list $a_1, a_2, \ldots, a_{n-2}, a_{n-1}a_n$ consists of $n - 1$ pairwise coprime, positive integers that are not perfect squares. By the inductive hypothesis again $[F[\sqrt{ab}] : \mathbb{Q}] = 2^{n-1}$. This contradiction gives $r \neq 0$.

If $s = 0$, then $\sqrt{a_n} \in F$. In this case, by reasoning exactly as above, it follows that $\left[F[\sqrt{a_n}] : \mathbb{Q}\right]$ would be both 2^{n-2} and 2^{n-1}. So both r and s are non-zero.

Now the equation $\sqrt{a_n} = r + s\sqrt{a_{n-1}}$ gives $a_n^2 = r^2 + s^2 a_{n-1} + 2rs\sqrt{a_{n-1}}$, from which $\sqrt{a_{n-1}} = (a_n^2 - r^2 - s^2 a_{n-1})/2rs \in F$. Once more as above, the degree $\left[F[\sqrt{a_{n-1}}] : \mathbb{Q}\right]$ would be both 2^{n-2} and 2^{n-1}.

The only way out is to have $\sqrt{a_n} \notin F[\sqrt{a_{n-1}}]$. □

At one point in the proof it was implicitly taken as fact that the positive integer $a_{n-1}a_n$ is coprime with the preceding $a_1, \ldots, a_{n-2}$ and that $a_{n-1}a_n$ is not a perfect square. If this is not immediately clear, the required mental exercise, based on unique factorization in $\mathbb{Z}$, needs to be carried out.

6.1.8 The Algebraic Closure inside a Field Extension

Two interesting consequences emerge from Proposition 6.15.

Proposition 6.17. *If K is a field extension of F, then the set of all elements in K that are algebraic over F is itself a subfield of K containing F.*

Proof. Suppose α and β are algebraic over F. According to Proposition 6.15, $F[\alpha,\beta]$ is a finite and thereby algebraic field extension of F. And so, the sum, difference, product and quotient of α and β, being in $F[\alpha,\beta]$, are algebraic over F. Thus the algebraic elements form a subfield of K. □

Definition 6.18. The elements in a field K that are algebraic over a subfield F form a field called the ***algebraic closure*** of F ***inside*** K.

Proposition 6.19. *If K is an algebraic extension of F and L is an algebraic extension of K, then L is an algebraic extension of F.*

Proof. Let $\alpha \in L$. Since α is algebraic over K, there is a non-zero polynomial f in $K[X]$ such that $f(\alpha) = 0$. Write $f = a_nX^n + \cdots + a_1X + a_0$, where the $a_j \in K$ and $a_n \neq 0$. These a_j are algebraic over F. By Proposition 6.15, $F[a_0, a_1, \ldots, a_n]$ is a finite extension of F. Clearly α remains algebraic over $F[a_0, a_1, \ldots, a_n]$ using the same polynomial f. By Proposition 6.9, the extension $F[a_0, a_1, \ldots, a_n][\alpha]$ is finite over $F[a_0, a_1, \ldots, a_n]$. Then the tower theorem, Proposition 6.13, causes $F[a_0, a_1, \ldots, a_n][\alpha]$ to be finite over F. So, α is algebraic over F by Proposition 6.5. □

Because of the two preceding results the algebraic closure A of a subfield F inside a bigger field K is indeed closed in the sense that K contains no elements that are algebraic over A, other than the elements of A itself. In other words, every element of the complement $K \setminus A$ is transcendental over A.

The Field of Algebraic Numbers

In number theory the algebraic closure of $\mathbb{Q}$ inside $\mathbb{C}$ is a highly studied field. This closure is called the ***field of algebraic numbers***, which we denote by $\mathbb{A}$. The field $\mathbb{A}$ illustrates that the converse of Proposition 6.5 fails. Indeed, for any positive integer n, Eisenstein's criterion shows that the polynomial $X^n - 2$ is irreducible in $\mathbb{Q}[X]$. A root for this polynomial is the real number $\alpha = \sqrt[n]{2}$. By Proposition 6.9 the field extension $\mathbb{Q}[\alpha]$ has $1, \alpha, \alpha^2, \ldots, \alpha^{n-1}$ as a basis over $\mathbb{Q}$. These numbers are in $\mathbb{A}$. Since there is no upper bound on n, the dimension of $\mathbb{A}$ over $\mathbb{Q}$ is infinite. Yet, every element of $\mathbb{A}$ is algebraic over $\mathbb{Q}$.

A natural question remains. Is every element of $\mathbb{C}$ algebraic over $\mathbb{Q}$?

6.1.9 The Tower Theorem and Composite Fields

Definition 6.20. Suppose K is a field containing two subfields E and L. The smallest subfield of K that contains both E and L is called the ***composite*** of E and L inside K, and is denoted by EL.

In the important case where E, L are finite extensions of another field F, Proposition 6.15 reveals that $E = F[\alpha_1, \dots, \alpha_m]$ and $L = F[\beta_1, \dots, \beta_n]$ for some α_i, β_j that are algebraic over F. Clearly

$$EL = F[\alpha_1, \dots, \alpha_m, \beta_1, \dots, \beta_n] = L[\alpha_1, \dots, \alpha_n] = E[\beta_1, \dots, \beta_n].$$

Thus EL is also a finite extension of F.

Let us enquire about the degree of EL over F.

Proposition 6.21. *Suppose K is a field containing subfields F, E, L, and that E, L are finite extensions of F of degree n, m, respectively. Then*

$$[EL : L] \leq n, \; [EL : E] \leq m \quad \text{and} \quad [EL : F] \leq mn.$$

Furthermore, if m, n are coprime, then

$$[EL : L] = n, \; [EL : E] = m \quad \text{and} \quad [EL : F] = mn.$$

Proof. A lattice diagram of these fields might help keep track of them.

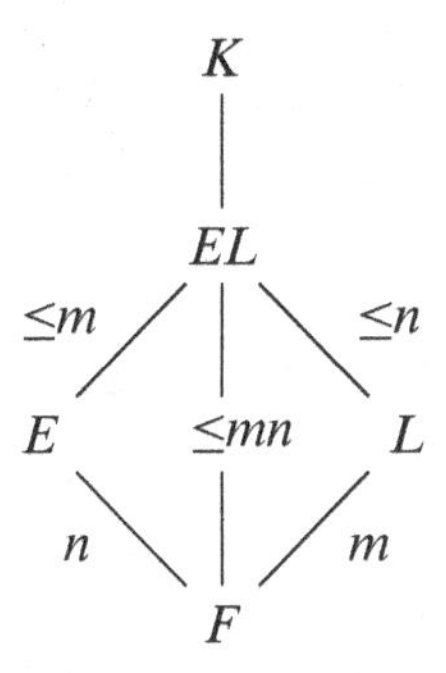

Let $\beta_1, \dots, \beta_m$ be a basis for L as a vector space over F. Then $EL = E[\beta_1, \dots, \beta_m]$. This means that every α in the field EL is an element of the form

$$\alpha = \sum_{i_1, \dots, i_m} e_{i_1, \dots, i_n} \beta_1^{i_1} \cdots \beta_m^{i_m},$$

where $i_1, \dots, i_m$ are non-negative integers and $e_{i_1, \dots, i_m} \in E$.

Because the β_j are chosen to generate L as a vector space over F, every monomial $\beta_1^{i_1} \cdots \beta_n^{i_m}$ in the β_j is an F-linear combination of the β_j. It follows that every α in EL is an E-linear combination of the β_j. Knowing that the β_j generate EL as a vector space over E, linear algebra tells us that

$$[EL : E] \leq m.$$

By a similar argument,

$$[EL : L] \leq n.$$

Then by the tower theorem, Proposition 6.13,

$$[EL : F] = [EL : E][E : F] \leq mn.$$

Furthermore, since E, L are subfields of EL, the tower theorem implies that m and n both divide the degree $[EL : F]$. If m, n are coprime, it follows that their product mn divides the inferior quantity $[EL : F]$. Consequently $[EL : F] = mn$. Finally, once more by the tower theorem, we get

$$mn = [EL : F] = [EL : E][E : F] = [EL : E]n,$$

whereby $[EL : E] = m$. Likewise $[EL : L] = n$. □

When m, n are not coprime, the inequalities of Proposition 6.21 could well be strict.

For example, with $\omega = e^{2\pi i/3}$, take the field extensions $E = \mathbb{Q}[\sqrt[3]{2}]$ and $L = \mathbb{Q}[\sqrt[3]{2}\omega]$ of $\mathbb{Q}$. The minimal polynomial of both $\sqrt[3]{2}$ and $\sqrt[3]{2}\omega$ is $X^3 - 2$. The irreducibility of this polynomial can be seen from Eisenstein's criterion using the prime 2. Thus our field extensions are finite of degree 3 over $\mathbb{Q}$. Their composite, taken inside $\mathbb{C}$, is the field

$$EL = \mathbb{Q}[\sqrt[3]{2}, \sqrt[3]{2}\omega] = E[\sqrt[3]{2}\omega].$$

Now the minimal polynomial of $\sqrt[3]{2}\omega$ over E does not remain $X^3 - 2$. Keeping in mind that $1 + \omega + \omega^2 = 0$, we see that the minimal polynomial of $\sqrt[3]{2}\omega$ over E collapses down to the quadratic:

$$(X - \sqrt[3]{2}\omega)(X - \sqrt[3]{2}\omega^2) = X^2 - \sqrt[3]{2}(\omega + \omega^2) + \sqrt[3]{2}^2\omega^3 = X^2 + \sqrt[3]{2}X + \sqrt[3]{2}^2.$$

Thus the degree of EL over E is 2, not 3. By the tower theorem, Proposition 6.13,

$$[EL : \mathbb{Q}] = [EL : E][E : \mathbb{Q}] = 2 \cdot 3 = 6, \text{ not } 9.$$

Finding Degrees of Extensions via Composite Fields

For an illustration of Proposition 6.21, pick a pair of primes p, q, and a positive integer a divisible by q but not by q^2.

Let $\alpha = \sqrt[p]{a}$, the real pth root of a. The minimal polynomial of α over the rational field $\mathbb{Q}$ is $X^p - a$. Its irreducibility can be seen by Eisenstein's criterion, Proposition 5.40, using the prime q. Thus the degree $[\mathbb{Q}[\alpha] : \mathbb{Q}] = p$.

Let $\zeta = e^{2\pi i/p}$. Clearly ζ is a root of $X^p - 1$, and

$$X^p - 1 = (X - 1)(X^{p-1} + X^{p-2} + \cdots + X + 1).$$

As ζ is not a root of $X - 1$, then ζ is a root of $X^{p-1} + X^{p-2} + \cdots + X + 1$. We saw just after Proposition 5.40 that this polynomial of degree $p - 1$ is irreducible. So the minimal polynomial of ζ is $X^{p-1} + X^{p-2} + \cdots + X + 1$. Hence the degree $[\mathbb{Q}[\zeta] : \mathbb{Q}] = p - 1$.

The composite field of $\mathbb{Q}[\alpha]$ and $\mathbb{Q}[\zeta]$, taken inside $\mathbb{C}$, is $\mathbb{Q}[\alpha, \zeta]$. Since $p, p-1$ are coprime, the degree of this composite field is $p(p-1)$, due to Proposition 6.21.

Now the full list of roots of $X^p - a$ is

$$\alpha, \alpha\zeta, \alpha\zeta^2, \ldots, \alpha\zeta^{p-1}.$$

Clearly all of these roots are in the composite field $\mathbb{Q}[\alpha, \zeta]$. Also, ζ lies in the field $\mathbb{Q}[\alpha, \alpha\zeta, \alpha\zeta^2, \ldots, \alpha\zeta^{p-1}]$ generated by the roots of $X^p - a$, because $\zeta = \alpha\zeta/\alpha$. Thus

$$\mathbb{Q}[\alpha, \zeta] = \mathbb{Q}[\alpha, \alpha\zeta, \alpha\zeta^2, \ldots, \alpha\zeta^{p-1}].$$

We have just shown that the field generated by the full set of roots of $X^p - a$ over the rational field $\mathbb{Q}$ has degree $p(p-1)$.

For instance, the degree over $\mathbb{Q}$ of the field generated by the roots of $X^5 - 2$ over $\mathbb{Q}$ is $5 \cdot 4 = 20$.

Field extensions which are generated by the complete set of roots of a polynomial are of paramount importance to an understanding of those roots. In the sections to follow, such extensions are pretty much all we will be discussing. They go by the name of ***splitting fields***.

6.1.10 The Tower Theorem Used in Conjunction with Eisenstein for Noetherian Domains

Before turning to splitting fields we indulge in an illustration of the tower theorem working together with Eisenstein's criterion, Proposition 5.45, for Noetherian domains and with Proposition 5.49. A review of these results is hereby recommended.

If $K = \mathbb{Q}[\sqrt[3]{7}, \sqrt[9]{13}]$, what is the degree $[K : \mathbb{Q}]$?

Letting $F = \mathbb{Q}[\sqrt[3]{7}]$, we can exploit the tower $\mathbb{Q} \subseteq F \subseteq K$. By Eisenstein's criterion, Proposition 5.40, the polynomial $X^3 - 7$ is irreducible over $\mathbb{Q}$, which makes it the minimal polynomial of $\sqrt[3]{7}$. Thus the degree $[F : \mathbb{Q}] = 3$. To use the tower theorem, the degree $[K : F]$ is needed.

Since $K = F[\sqrt[9]{13}]$, we need the minimal polynomial of $\sqrt[9]{13}$ over F. The first contender is $X^9 - 13$. This is irreducible over $\mathbb{Q}$, again by Eisenstein's criterion, Proposition 5.40. However, its irreducibility over F is not so clear. To that end consider the Noetherian domain $\mathbb{Z}[\sqrt[3]{7}]$, for which the field of fractions is F. The cubic polynomial $X^3 - 7$ remains irreducible modulo 13 because an inspection shows it has no root in the field $\mathbb{Z}_{13}$. By Proposition 5.49, the integer 13 remains a prime in $\mathbb{Z}[\sqrt[3]{7}]$. Then by Eisenstein, applied to the Noetherian domain $\mathbb{Z}[\sqrt[3]{7}]$, the polynomial $X^9 - 13$ is irreducible over F. Hence $[K : F] = 9$, and by the tower theorem, $[K : \mathbb{Q}] = 3 \cdot 9 = 27$. An inspection of the above argument, shows that 9 could be replaced by any positive integer n, to see that the degree $[\mathbb{Q}[\sqrt[3]{7}, \sqrt[n]{13}] : \mathbb{Q}] = 3n$.

If K is the field extension $\mathbb{Q}[\sqrt[4]{3}, \sqrt[6]{5}]$ over $\mathbb{Q}$, let us repeat our technique to verify that $[K : \mathbb{Q}] = 24$.

Let $F = \mathbb{Q}[\sqrt[4]{3}]$ and work with the tower $\mathbb{Q} \subseteq F \subseteq K$. By Eisenstein's criterion, Proposition 5.40, $X^4 - 3$ is irreducible in $\mathbb{Q}[X]$, and so $[F : \mathbb{Q}] = 4$. To get the degree $[K : \mathbb{Q}] = 24$, the tower theorem demands that $[K : F] = 6$. Since $K = F[\sqrt[6]{5}]$ it suffices to see that $X^6 - 5$ is the minimal polynomial of $\sqrt[6]{5}$ in $F[X]$. To that end we can use the Noetherian domain $\mathbb{Z}[\sqrt[4]{3}]$ for which the fraction field is F.

First we verify that the polynomial $X^4 - 3$ is irreducible modulo 5. Well, this polynomial has no root in $\mathbb{Z}_5[X]$ by inspection, and thus no linear or cubic factor in $\mathbb{Z}_5[X]$. If $X^4 - 5$ had a quadratic factor, then $X^4 - 3 = (X^2 + aX + b)(X^2 + cX + d)$ for some a, b, c, d in $\mathbb{Z}_5$. In that case multiply the quadratics to see that

$$a + c = 0, \quad b + d + ac = 0, \quad ad + bc = 0, \quad bd = -3 \text{ in } \mathbb{Z}_5.$$

If $a = 0$, then $c = 0$ and so $b + d = 0$, which leads to $b = -d$. Then $b^2 = 3$ which cannot be, since 3 is not a square in $\mathbb{Z}_5$. If $a \neq 0$, then $c \neq 0$. Use the fact $d = -3b^{-1} = 2b^{-1}$ to get $2ab^{-1} + bc = 0$, and then $2a + b^2c = 0$. Subtract this from the equation $2a + 2c = 0$ to get $b^2c = 2c$. Cancel the non-zero c and obtain $b^2 = 2$, which is impossible since 2 is not a square in $\mathbb{Z}_5$.

Since $X^4 - 3$ is irreducible in $\mathbb{Z}_5[X]$, the prime 5 remains prime in the Noetherian domain $\mathbb{Z}[\sqrt[4]{3}]$, due to Proposition 5.49. By Eisenstein's criterion for Noetherian domains, Proposition 5.45, $X^6 - 5$ is irreducible in $F[X]$. Then the tower theorem gives $[K : \mathbb{Q}] = 4 \cdot 6 = 24$.

The above argument can be replicated to see that $[\mathbb{Q}[\sqrt[4]{3}, \sqrt[n]{5}] : \mathbb{Q}] = 4n$ for any positive exponent n. In fact $\sqrt[n]{5}$ can be replaced by any root α of a polynomial $a_nX^n + a_{n-1}X^{n-1} + \cdots + a_1X + a_0$ in $\mathbb{Z}[X]$ such that 5 divides all a_j except for a_n and 5^2 does not divide a_0. What makes Eisenstein work is that 5 remains prime in the Noetherian domain $\mathbb{Z}[\sqrt[4]{3}]$.

EXERCISES

1. Show that there are no homomorphisms between two fields of different characteristic.
2. If F is a finite field, show that the product of all of its non-zero elements equals -1.
3. If $\alpha \in \mathbb{Q}[\sqrt[5]{3}\,] \setminus \mathbb{Q}$, show that $\sqrt[5]{3} \in \mathbb{Q}[\alpha]$.
4. Show that the extension $\mathbb{R}$ does not have finite degree over $\mathbb{Q}$.
5. If A is an integral domain containing a field F, and A has finite dimension as a vector space over F, prove that A is a field.
6. Suppose K is a finite field extension of a field F and that R is a subring of K which contains F. Show that R is a field.
7. Find the minimal polynomial in $\mathbb{Q}[X]$ of $\sqrt{2} + \sqrt{5}$.
8. Find the minimal polynomial in $\mathbb{Q}[X]$ of $\sqrt{3 + 2\sqrt{2}}$.
9. Verify that the polynomial $f = X^4 + X^3 + 1$ is irreducible in $\mathbb{Z}_2[X]$. Let K be a field extension of $\mathbb{Z}_2$ containing a root α of f. (We will see in the next section that such extensions exist.) Find the inverse of $\alpha^2 + 1$ inside K as a polynomial expression in α.
10. If α is in a field extension of a field F and its minimal polynomial has odd degree, show that $F[\alpha] = F[\alpha^2]$.
11. (a) Let $F \subset E \subset K$ be a tower of fields such that $[E : F]$ is finite. If $\alpha \in K$, show that $[E[\alpha] : F[\alpha]] \leq [E : F]$.

    (b) If $K = \mathbb{Q}[\sqrt[3]{2}, e^{2\pi i/3}]$, find a subfield E and α in K such that the degree $[E : \mathbb{Q}] = 3$ while $[E[\alpha] : \mathbb{Q}[\alpha]] = 2$.

12. Let $F(X)$ be the field of rational functions in X over a field F.

(a) Show that the only rational functions which are algebraic over F are the constant functions.

(b) If $\alpha \in F[X] \setminus F$, prove that X is algebraic over the field $F(\alpha)$ of rational expressions in α, i.e. the fraction field of $F[\alpha]$.

(c) Let $\alpha = \frac{X^2-1}{X^3+X+1}$ in K. Find the minimal polynomial of X over the field $F(\alpha)$.

13. Let $\mathbb{Q}(t)$ be the field of rational functions in the variable t and with coefficients in the rational field $\mathbb{Q}$. Let $\mathbb{Q}(t^3)$ be the subfield of $\mathbb{Q}(t)$ consisting of rational functions in t^3.

(a) Show that the degree $[\mathbb{Q}(t) : \mathbb{Q}(t^3)] = 3$.

Hint. Show that the minimal polynomial of t over the field $\mathbb{Q}(t^3)$ has degree 3.

(b) Part (a), along with Proposition 6.9, tells us that every rational function in $\mathbb{Q}(t)$ can be written in the form $a+bt+ct^2$ where a, b, c are in $\mathbb{Q}(t^3)$. Write the rational function $\frac{1}{1+t}$ in that form.

14. If $\alpha_1, \alpha_2, \ldots, \alpha_n$ are complex numbers such that $\alpha_j^2 \in \mathbb{Q}$, show that the field $\mathbb{Q}[\alpha_1, \ldots, \alpha_n]$ cannot contain $\sqrt[6]{2}$.

15. Let K be a field extension of a field F. Show that K is an algebraic extension of F if and only if every ring extension R of F such that $R \subseteq K$ is already a field.

16. If F is the finite field $\mathbb{Z}_5[X]/\langle X^2 + X + 1\rangle$, then a basis for F over $\mathbb{Z}_5$ consists of $1, \alpha$ where $\alpha^2 = -\alpha - 1$. Find the minimal polynomial of $2\alpha + 3$ over $\mathbb{Z}_5$.

17. Using the fact that $[\mathbb{C} : \mathbb{R}] = 2$ and that all polynomials of positive degree in $\mathbb{C}[X]$ have a root in $\mathbb{C}$, show that every irreducible polynomial in $R[X]$ has degree 1 or 2.

18. Prove that $\mathbb{Q}\left[\sqrt[5]{2}, e^{2\pi i/5}\right]$ coincides with the field generated over $\mathbb{Q}$ by the roots of $X^5 - 2$. Find the degree of $\left[\mathbb{Q}[\sqrt[5]{2}, e^{2\pi i/5}] : \mathbb{Q}\right]$.

19. Prove that $\mathbb{Q}\left[\sqrt{2} + \sqrt{3}\right]$ coincides with the field generated over $\mathbb{Q}$ by the roots of the polynomial $X^4 - 5X^2 + 6$. Find the degree of this field extension of $\mathbb{Q}$. Find the minimal polynomial of $\sqrt{2} + \sqrt{3}$ in $\mathbb{Q}[X]$.

20. Show that $\mathbb{Q}[\sqrt[3]{7} + 2i] = \mathbb{Q}[\sqrt[3]{7}, i]$. Use this fact to find the degree of $\sqrt[3]{7} + 2i$ over $\mathbb{Q}$. Then find the minimal polynomial of $\sqrt[3]{7} + 2i$ over $\mathbb{Q}$.

21. Show that $\cos(\pi/9)$ is algebraic over $\mathbb{Q}$, and then find its minimal polynomial in $\mathbb{Q}[X]$.

Hint. Take the DeMoivre identity $(\cos(\theta) + i\sin(\theta))^3 = \cos(3\theta) + i\sin(3\theta)$, expand the left side, equate real parts, and put $\theta = \pi/9$.

22. Let K be a field extension of F and let α, β in K be algebraic over F, with respective minimal polynomials f, g in $F[X]$. Show that f remains the minimal polynomial of α in $F[\beta][X]$ if and only if g remains the minimal polynomial of β in $F[\alpha][X]$.

23. If R is a commutative ring of prime characteristic p, the mapping $\varphi : R \to R$ defined by $\varphi(x) = x^p$ is called the Frobenius map on R.

(a) Show that the Frobenius map on R is a ring homomorphism.

(b) Show that the Frobenius map on every finite field of characteristic p is a bijection.

(c) Let $F = \mathbb{Z}_p(t)$ be the field of rational functions in the variable t with coefficients in the finite field $\mathbb{Z}_p$. Show that the Frobenius map on F using the prime p is injective but not surjective.

24. An automorphism of a field F is a bijective homomorphism $\varphi : F \to F$.

(a) Show that the real field $\mathbb{R}$ has just one automorphism, namely the identity map.

Hint. If φ is an automorphism of $\mathbb{R}$, first show that φ is the identity map on $\mathbb{Q}$. If $0 < x$ in $\mathbb{R}$, show that $0 < \varphi(x)$. Deduce that if $x < y$, then $\varphi(x) < \varphi(y)$. If x in $\mathbb{R}$ is such that $x < \varphi(x)$ or $\varphi(x) < x$, derive a contradiction.

(b) Prove there are only two automorphisms $\varphi : \mathbb{C} \to \mathbb{C}$ that fix $\mathbb{R}$, i.e. such that $\varphi(x) = x$ for all x in $\mathbb{R}$. What are those automorphisms?

(c) Find all automorphisms $\varphi : \mathbb{Q}[\sqrt{2}] \to \mathbb{Q}[\sqrt{2}]$.

25. Take it as given that the polynomial $X^5 + X^4 + X^2 + X + 1$ (note X^3 is not there) is irreducible in $\mathbb{Z}_2[X]$. Let α in $\mathbb{C}$ be a root of this polynomial. Show that the degree $[\mathbb{Q}[\alpha, \sqrt[5]{2}] : \mathbb{Q}] = 25$.

26. Suppose the field E is a degree 2 extension of F. Let f in $F[X]$ be irreducible of degree 6. Show that every irreducible factor of f in $E[X]$ has degree 3 or 6. Thus f is either irreducible or the product of two irreducibles of degree 3 in $E[X]$.

27. Find the degree $[\mathbb{Q}[\sqrt[4]{2}, \sqrt{3}] : \mathbb{Q}]$.

Hint. Decide if $\sqrt{3} \in \mathbb{Q}[\sqrt[4]{2}]$. An assortment of calculations will be needed.

28. Here we compute the degree of $\mathbb{Q}[\sqrt{-5}, \sqrt[8]{-29}]$ over $\mathbb{Q}$. The argument is based on Eisenstein's criterion for Noetherian domains along with the identification of a suitable prime in $\mathbb{Z}[\sqrt{-5}]$. If a is an integer, let $\bar{a}$ be its residue in the finite field $\mathbb{Z}_{29}$. It will be useful to notice that $\overline{13}^2 = \overline{-5}$ in $\mathbb{Z}_{29}$, also that $(3 + 2\sqrt{-5})(3 - 2\sqrt{-5}) = 29$.

(a) Verify that $\varphi : \mathbb{Z}[\sqrt{-5}] \to \mathbb{Z}_{29}$ defined by $a + b\sqrt{-5} \mapsto \overline{a + 13b}$ is a ring homomorphism.

(b) Show that $\ker \varphi$ is the principal ideal in $\mathbb{Z}[\sqrt{-5}]$ generated by $3+2\sqrt{-5}$, and deduce that $3 + 2\sqrt{-5}$ is prime in $\mathbb{Z}[\sqrt{-5}]$.

(c) Show that the polynomial $X^8 - 29$ is irreducible in $\mathbb{Q}[\sqrt{-5}][X]$.

Hint. Use Eisenstein for Noetherian domains.

(d) Find the degree of $\mathbb{Q}[\sqrt{-5}, \sqrt[8]{-29}]$ over $\mathbb{Q}$.

29. Let $K = F(t)$ be the field of rational functions in the variable t with coefficients in a field F. Select a (non-constant) function $h(t)$ from $K \setminus F$. Write $h(t) = f(t)/g(t)$ where $f(t), g(t)$ are coprime polynomials in t. Let E be the subfield of K generated by $h(t)$ over F, that is $E = F(h(t))$.

(a) Show that $h(t)$ is not algebraic over F.

(b) Show that t is a root of the polynomial $p(X) = h(t)g(X) - f(X)$ taken in $E[X]$. Deduce that t is algebraic over E.

(c) Explain why $K = E[t]$.

(c) Prove that $\deg p(X) = \max\{\deg f(t), \deg g(t)\}$.

(d) Prove that $p(X)$ is irreducible in $E[X]$.

Hint. Since $h(t)$ is not algebraic over F, put $Y = h(t)$ and observe that $F[Y]$ is a polynomial ring and that its field of fractions is $E = F(Y)$. Use the fact that $F[X, Y] = F[X][Y] = F[Y][X]$ to show $p(X)$ is irreducible in $F[Y][X]$.

(e) Deduce that $p(X)$ is the minimal polynomial of $h(t)$ over E, and find the degree $[K : E]$.

6.2 Splitting Fields

The problem with any polynomial is to extract its roots, or at least say something informative about them. For this we need to deal with the fields formed by such roots. At this point things start to get somewhat more intricate.

6.2.1 The Synthesis of Algebraic Elements

Algebraic elements α over a field F come with their irreducible, minimal polynomial. This all presupposes that there actually is a field extension containing α. Just as likely, all we might have is a field F and a monic, irreducible polynomial g in $F[X]$. In that case, it is fair to ask if there *exists* a field extension K and an element in K that is algebraic over F with the prescribed minimal polynomial g. The important result, known as ***Kronecker's theorem***, creates such a K.

Proposition 6.22 (Kronecker's theorem). *If F is a field and g is a monic irreducible polynomial in $F[X]$, then there exists a field extension K of F with an element α in K such that α is algebraic over F and g is its minimal polynomial.*

Proof. Since g is irreducible, Proposition 4.60 says that the quotient ring $F[X]/\langle g\rangle$ is a field, which we call K.

There is the canonical projection:

$$\varphi : F[X] \to K, \text{ where } f \mapsto f + \langle g\rangle \text{ in } K.$$

The restriction of φ to the field F of constant polynomials inside $F[X]$ is injective, because the kernel of φ restricted to F, being an ideal in the field F, is the zero ideal in F. Thus $\varphi(F)$ is a subfield of K isomorphic to F.

Next, put $\alpha = \varphi(X)$.

If $g = X^n + a_{n-1}X^{n-1} + \cdots + a_1X + a_0$ with the a_j in F, then

$$\begin{aligned}\alpha^n+\varphi(a_{n-1})\alpha^{n-1}+\cdots+\varphi(a_1)\alpha+\varphi(a_0) &= \varphi(X)^n+\varphi(a_{n-1})\varphi(X)^{n-1}+\cdots+\varphi(a_1)\varphi(X)+\varphi(a_0)\\ &= \varphi(X^n + a_{n-1}X^{n-1} + \cdots + a_1X + a_0)\\ &= \varphi(g) = 0.\end{aligned}$$

Thus α is a root of $X^n + \varphi(a_{n-1})X^{n-1} + \cdots + \varphi(a_1)X + \varphi(a_0)$, a polynomial which lies in $\varphi(F)[X]$.

At this point a leap of imagination goes a long way. Identify F with its image field $\varphi(F)$. In other words, declare that $a = \varphi(a)$ for every a in F. After this leap, K becomes a field containing F and α becomes a root of g. □

A valid criticism of the proof of Proposition 6.22 concerns the identification of F with its image $\varphi(F)$. In fact, K extends $\varphi(F)$, not F. This technical flaw can be fixed. Roughly, replace K by the set $L = (K \setminus \varphi(F)) \cup F$. Now L contains F, and there is an obvious bijection between K and L. Make L into a field extension of F by transporting the field operations in K onto L using the obvious bijection. The mundane but irksome details are best left omitted.

It is worth noting that the field extension K constructed in Kronecker's theorem is $F[\alpha]$, the field generated by α over F.

An easy but useful enhancement of Proposition 6.22 follows.

Proposition 6.23. *If F is a field and f in $F[X]$ is a non-constant polynomial (possibly reducible), then there is a field extension of F containing a root of f.*

Proof. Being non-constant, the polynomial f has at least one irreducible factor g, which can be taken as monic. By Kronecker's theorem F has a field extension K containing a root of g. That root is also a root of f. □

How to Make a Field of Size p^n

To illustrate Kronecker's theorem, take the field $\mathbb{Z}_p$ for any prime p, and any monic irreducible polynomial g of degree n in $\mathbb{Z}_p[X]$. According to Kronecker's theorem, Proposition 6.22, there is a field extension K over $\mathbb{Z}_p$ that contains a root α of g. By Proposition 6.9, the degree of $\mathbb{Z}_5[\alpha]$ over $\mathbb{Z}_p$ is $\deg g = n$, which causes $\mathbb{Z}_5[\alpha]$ to have precisely p^n elements.

For instance, X^3+X+1 is irreducible in $\mathbb{Z}_5[X]$, since it is cubic and none of the five residues $0, 1, 2, 3, 4$ is a root of it. Thus there exists an algebraic extension $\mathbb{Z}_5[\alpha]$ of $\mathbb{Z}_5$ where α is a root of our given polynomial. The field $\mathbb{Z}_5[\alpha]$ has precisely 125 elements.

As long as irreducible polynomials of arbitrary degree n over every $\mathbb{Z}_p$ are available, we can build finite fields of all possible sizes. The plenitude of such irreducibles will be shown in due course.

A Function Field

Take $F = \mathbb{C}(t)$, the field of rational functions in the indeterminate t with coefficients in the complex numbers. By Eisenstein's criterion, Proposition 5.40, using the prime t in the unique factorization domain $\mathbb{C}[t]$, the polynomial $X^4 - t$ is irreducible in $F[X]$. By Kronecker's theorem there is a field K containing a root α of $X^4 - t$. This root satisfies the equation $\alpha^4 = t$ inside K.

The indeterminate t can be interpreted as the identity function $z \mapsto z$ on $\mathbb{C}$. Thus we may view α as a function defined on a suitable region of $\mathbb{C}$ whose fourth power gives the identity function. In other words, we can think of α as the function $z \mapsto z^{1/4}$ defined on a suitable region of $\mathbb{C}$. Algebraic elements over $\mathbb{C}(t)$, such as our α, are known as ***algebraic functions***.

Using Kronecker's Theorem to Test for Irreducibility

It is a challenge to decide if a polynomial in several indeterminates is irreducible. For instance, with any field F and positive exponents m, n, consider the polynomial $f(t, X) = X^n - t^m$

in $F[t,X]$ in two indeterminates t and X. If m,n have a common factor $k \geq 2$, a proper factorization of f is easy. Write $n = ki, m = kj$ where $i,j \geq 1$, and then

$$f(t,X) = (X^i)^k - (t^j)^k$$
$$= \left(X^i - t^j\right)\left((X^i)^{k-1} + (X^i)^{k-2}t^j + \cdots + X^i(t^j)^{k-2} + (t^j)^{k-1}\right).$$

If m,n are coprime, it is not so clear that f is irreducible in $F[t,X]$.

To see that this is so, we can verify that f is irreducible as a polynomial in X with coefficients in the field $F(t)$, i.e. as a polynomial in the larger ring $F(t)[X]$. This can be done by showing that f is the minimal polynomial of some element α in a field extension of $F(t)$.

Using Proposition 6.23 let K be an extension of $F(t)$ containing a root α of f. We show that f is the minimal polynomial of this α. The minimal polynomial of α divides f in $F(t)[X]$. Since the degree of the minimal polynomial of α is the degree of the extension $F(t)[\alpha]$ over $F(t)$, it suffices to see that this extension has degree n.

Put $u = t^m$, which remains transcendental over F. Let $F(u)$ denote the smallest subfield of $F(t)$ containing both F and u. This coincides with the fraction field of the (bona fide) polynomial ring $F[u]$.

The polynomial $X^m - u$ in $F(u)[X]$ is irreducible, as can be seen by Eisenstein's criterion using the prime u in the unique factorization domain $F[u]$. Since $t^m - u = 0$, this polynomial is the minimal polynomial of t over $F(u)$, which reveals that $[F(u)[t] : F(u)] = m$. But $F(u)[t]$ is just the field $F(t)$, by inspection. Thus $F(t)$ is an extension of $F(u)$ of degree m.

Similarly by Eisenstein, the polynomial $X^n - u$ is irreducible over $F(u)$. Now α in K is a root of this polynomial because $\alpha^n - t^m = \alpha^n - u = 0$. And so $F(u)[\alpha]$ is an extension of $F(u)$ of degree n.

The composite field of $F(t)$ and $F(u)[\alpha]$ inside K is $F(t)[\alpha]$, whose degree over $F(t)$ we are seeking. The following lattice diagram keeps track of the fields involved.

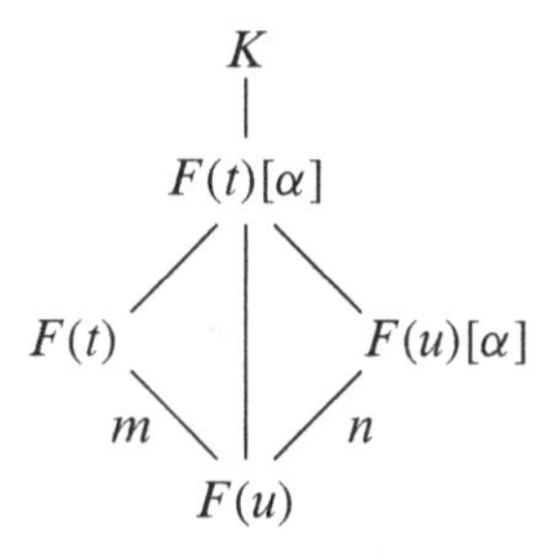

Since m,n are coprime, Proposition 6.21 gives $[F(t)[\alpha] : F(t)] = n$, as desired.

6.2.2 What Is a Splitting Field?

Definition 6.24. If K is a field, a non-constant polynomial f in $K[X]$ is said to ***split in*** $K[X]$ provided

$$f = a(X - \alpha_1)(X - \alpha_2)\cdots(X - \alpha_n)$$

for some a in K and some α_j in K. We also say that f ***splits over*** K.

Obviously these α_j are roots of f, possibly repeated, and a is the leading coefficient of f. There can be no other roots of f inside the field K, for if b were another root, the factor theorem, Proposition 4.35, would force $X - b$ to appear in the unique factorization of f.

For instance, X^2+1 does not split over $\mathbb{R}$, but splits over $\mathbb{C}$ as $(X-i)(X+i)$. The polynomial $X^3 - 1$ factors as $(X - 1)(X^2 + X + 1)$ in $\mathbb{Q}[X]$, but it does not split in $\mathbb{Q}[X]$. However $X^3 - 1 = (X - 1)(X - \omega)(X - \omega^2)$, where $\omega = e^{2\pi i/3}$. This shows that $X^3 - 1$ splits in $\mathbb{C}[X]$. The fundamental theorem of algebra says that every non-constant polynomial in $\mathbb{C}[X]$ splits over $\mathbb{C}$. A proof of this will have to wait.

If an arbitrary field F is before us and $f \in F[X]$, of course f might not split over F. We need a larger home for its roots, more precisely, a field extension K of F over which f can split.

Definition 6.25. If F is a field and f is a polynomial in $F[X]$ of positive degree, a ***splitting field of f over F*** is any field extension K of F such that

- f splits in $K[X]$, i.e. $f = a(X - \alpha_1)(X - \alpha_2) \cdots (X - \alpha_n)$, where the α_j provide a full set of roots of f in K, a is the leading coefficient of f, and
- $K = F[\alpha_1, \alpha_2, \ldots, \alpha_n]$.

A splitting field K of f is a *minimal* extension of F over which f splits. More precisely, if E is an extension of F but a proper subfield of K, that is $F \subseteq E \subsetneq K$, then some α_j is not in E whereby f cannot split in $E[X]$.

If K is a splitting field of a polynomial f over a field F, this means that K is generated over F by a finite number of elements that are algebraic over F (the finitely many roots of f). So K is a finite extension of F.

From here on out, all references to a splitting field for a polynomial will automatically assume that the polynomial has positive degree. Constant polynomials have nothing interesting to offer in this regard.

In essence, Galois theory is the study of splitting fields.

A Small Observation about Splitting Fields

The following recurs enough that it may be worthwhile to clear up at the outset.

Proposition 6.26. *If K is a splitting field of a polynomial f over a field F and E is a subfield of K containing F, i.e. $F \subseteq E \subseteq K$, then K remains a splitting field for f over E.*

Proof. By default $f \in E[X]$, and f splits over K, say with roots $\alpha_1, \ldots, \alpha_n$. What is needed is that $K = E[\alpha_1, \ldots, \alpha_n]$. Well, this is so since

$$K = F[\alpha_1, \ldots, \alpha_n] \subseteq E[\alpha_1, \ldots, \alpha_n] \subseteq K.$$

The first inclusion holds because F is inside E and the second inclusion holds because E and all α_j are inside K. □

A Few Examples of Splitting Fields

Clearly, if f splits over F, then F is already a splitting field of f.

For another easy example, take the quadratic polynomial $f = X^2 - a$ in $F[X]$. If L is an extension of F containing a root α of f, then a splitting field of f is the extension $F[\alpha]$. The other root $-\alpha \in F[\alpha]$ automatically. So f does split over $F[\alpha]$, and this is a minimal extension of F that includes the roots of f.

For $f = X^n - 2$ in $\mathbb{Q}[X]$, a suitable splitting field is $K = \mathbb{Q}[\alpha, \zeta]$, where $\zeta = e^{2\pi i/n}$ and $\alpha = \sqrt[n]{2}$, the real nth root of 2. Indeed, the roots of f in $\mathbb{C}$ are $\alpha, \alpha\zeta, \alpha\zeta^2, \ldots, \alpha\zeta^{n-1}$. So the field L generated over $\mathbb{Q}$ by these roots sits inside K. And since $\zeta = \alpha\zeta/\alpha \in L$, it becomes clear that $K = L$, a splitting field of $X^n - 2$.

A worthy question is whether every finite extension of F is a splitting field for some polynomial. We will come to see shortly that this is not the case.

Finite Fields Are Splitting Fields

When it comes to finite fields, Lagrange's theorem unlocks a wealth of information.

Proposition 6.27. *If p is a prime and K is a finite field extension of $\mathbb{Z}_p$ of degree n, then K consists of the roots of $X^{p^n} - X$, and is thereby a splitting field of this polynomial over $\mathbb{Z}_p$. Furthermore, $K = \mathbb{Z}_p[\alpha]$ for some element α in K.*

Proof. The set $K^\star$ of non-zero elements of K is a finite group of order $p^n - 1$. By Lagrange's theorem $\alpha^{p^n-1} = 1$ for all α in $K^\star$, and so $\alpha^{p^n} = \alpha$ for all α in K. Thus K consists precisely of the set of roots of $X^{p^n} - X$, viewed as a polynomial over $\mathbb{Z}_p$. Clearly K is a splitting field of this polynomial.

In addition, Proposition 4.37 says that the group $K^\star$ is cyclic. There is an element α in $K^\star$ whose powers exhaust all of $K^\star$. Clearly $K = \mathbb{Z}_p[\alpha]$. □

The Existence of Splitting Fields

If F is a subfield of $\mathbb{C}$, the fundamental theorem of algebra (which we take for granted for the time being) ensures that every polynomial in $F[X]$ splits over $\mathbb{C}$. But if F is any field and $f \in F[X]$, a field extension of F over which f can split needs to be built.

Proposition 6.28. *For every field F and every polynomial f in $F[X]$ of positive degree there exists a field extension K of F over which f splits.*

Proof. An induction on the degree of f can be used. If $\deg f = 1$, then f splits already in $F[X]$, so we can take $K = F$ as a field over which f splits. Suppose that the result holds for all polynomials of degree less than n over any field, and consider f in $F[X]$ of degree n.

By Proposition 6.23 there is an extension L of F containing a root α of f. By the factor theorem, Proposition 4.35, $f = (X - \alpha)g$ for some polynomial g of degree $n - 1$ in $L[X]$. The inductive assumption applied to g yields an extension K of L over which g splits. That is

$g = a(X-\alpha_2)\cdots(X-\alpha_n)$ for some α_j in K and a in L. Then $f = a(X-\alpha)(X-\alpha_2)\cdots(X-\alpha_n)$, which is a splitting of f over K. □

After a field extension K of F in which a polynomial f in $F[X]$ can split is secured, a splitting field for f is readily obtained as $F[\alpha_1, \alpha_2, \ldots, \alpha_n]$, where these α_j are the roots that f has in K. So a valuable corollary emerges.

Proposition 6.29. *Every polynomial of positive degree has a splitting field.*

6.2.3 Lifting Isomorphisms to Splitting Fields

If f in $F[X]$ splits in some field extension K of F, then there is precisely one splitting field of f sitting inside K, namely, the subfield of K generated over F by the full set of roots of f. But if L is another extension of F over which f splits, the unique splitting field of f sitting inside L is not the same set as the splitting field inside K. The task before us is to verify that such distinct splitting fields at least remain isomorphic. So we need access to isomorphisms between splitting fields. It is also important to discover that splitting fields are endowed with enough automorphisms.

Transferring Coefficients between Polynomial Rings

A little tool seems indispensable.

Suppose that $\varphi : F \to E$ is an isomorphism between fields. Then φ extends to the mapping between their polynomial rings:

$$\varphi : F[X] \to E[X]$$

given by

$$\varphi : a_nX^n + a_{n-1}X^{n-1} + \cdots + a_0 \mapsto \varphi(a_n)X^n + \varphi(a_{n-1})X^{n-1} + \cdots + \varphi(a_0).$$

The coefficients of each polynomial f in $F[X]$ are replaced by the coefficients in $E[X]$ that correspond to them under φ. The rings $F[X]$ and $E[X]$ contain F and E, respectively, as the constant polynomials. Since the above mapping on the polynomial rings extends the original isomorphism φ, it does no harm to also call this extended mapping by the same name φ. This is a special case of what was introduced as the natural extension of φ in Definition 5.38.

Definition 6.30. If $f \in F[X]$, the new polynomial $\varphi(f)$ will be called the ***φ-transfer*** of f into $E[X]$.

The mapping φ that transfers coefficients comes with properties that can readily be checked.

- $\varphi : F[X] \to E[X]$ is an isomorphism of rings, and $\deg f = \deg \varphi(f)$ for all f in $F[X]$.
- f divides g in $F[X]$ if and only if $\varphi(f)$ divides $\varphi(g)$ in $E[X]$.
- f is irreducible in $F[X]$ if and only if $\varphi(f)$ is irreducible in $E[X]$.
- f splits in $F[X]$ if and only if $\varphi(f)$ splits in $E[X]$.

An Isomorphism Builder

The next result, although seemingly artificial, is a truly useful generalization of Proposition 6.11. The implications of the result make it worthy of study.

Proposition 6.31. *Suppose*

- $\varphi : F \to E$ *is an isomorphism of fields,*
- α *is in a field extension of* F *with minimal polynomial* g *in* $F[X]$*, and*
- β *is in a field extension of* E *with minimal polynomial* $\varphi(g)$ *in* $E[X]$*.*

Then there is an isomorphism $\psi : F[\alpha] \to E[\beta]$ *such that*

$$\psi \text{ extends } \varphi, \text{ i.e. } \psi(a) = \varphi(a) \text{ for all } a \text{ in } F, \text{ and } \psi(\alpha) = \beta.$$

Proof. The following diagram should help keep track of the situation.

$$\begin{array}{ccc} F[\alpha] & \xrightarrow{\psi} & E[\beta] \\ | & & | \\ F & \xrightarrow{\varphi} & E \end{array}$$

The proof somewhat follows the lines of Proposition 6.11.

There are two surjective homomorphisms defined on the polynomial ring $F[X]$:

$$\sigma : F[X] \to F[\alpha] \text{ given by } f \mapsto f(\alpha)$$

and

$$\tau : F[X] \to E[\beta] \text{ given by } f \mapsto \varphi(f)(\beta).$$

The latter map τ is the composite of the transfer map φ from $F[X]$ to $E[X]$ and the substitution $h \mapsto h(\beta)$ from $E[X]$ to $E[\beta]$. Since g is the minimal polynomial of α, we get $\ker \sigma = \langle g \rangle$. Also $f \in \ker \tau$ if and only if $\varphi(g) \mid \varphi(f)$ in $E[X]$, because $\varphi(g)$ is the minimal polynomial of β. This is the same as having $g \mid f$ in $F[X]$. So $\ker \tau = \langle g \rangle = \ker \sigma$.

By the first isomorphism theorem there are two field isomorphisms:

$$\overline{\sigma} : F[X]/\langle g \rangle \cong F[\alpha] \text{ where } f + \langle g \rangle \mapsto f(\alpha),$$

and

$$\overline{\tau} : F[X]/\langle g \rangle \cong E[\beta] \text{ where } f + \langle g \rangle \mapsto \varphi(f)(\beta).$$

From this, our required isomorphism is

$$\psi = \overline{\tau} \circ \overline{\sigma}^{-1} : F[\alpha] \cong E[\beta].$$

Note that if $a \in F$, then

$$\psi(a) = \overline{\tau} \circ \overline{\sigma}^{-1}(a) = \overline{\tau}(a + \langle g \rangle) = \varphi(a)(\beta) = \varphi(a),$$

since $\varphi(a)$ is a constant polynomial in $E[X]$. Thus ψ extends φ. Also

$$\psi(\alpha) = \overline{\tau} \circ \overline{\sigma}^{-1}(\alpha) = \overline{\tau}(X + \langle g \rangle) = X(\beta) = \beta,$$

as is required of ψ. □

To illustrate Proposition 6.31, consider the field $K = \mathbb{Q}[\sqrt{2}, \omega]$ where $\omega = e^{2\pi i/3}$. Over $\mathbb{Q}$ the minimal polynomials of $\sqrt{2}$ and ω are $X^3 - 2$ and $X^2 + X + 1$, respectively. The polynomial $X^2 + X + 1$ remains the minimal polynomial ω over $\mathbb{Q}[\sqrt{2}]$, i.e. in $\mathbb{Q}[\sqrt{2}][X]$. This is so because its roots ω, ω^2 do not lie in $\mathbb{R}$, and thereby do not lie in $\mathbb{Q}[\sqrt{2}]$.

According to Proposition 6.11 there is an isomorphism

$$\varphi : \mathbb{Q}[\sqrt{2}] \to \mathbb{Q}[\sqrt{2}] \text{ such that } \varphi(\sqrt{2}) = -\sqrt{2}.$$

This is because $\pm\sqrt{2}$ have the same minimal polynomial over $\mathbb{Q}$. Looking at Proposition 6.31, we have $E = F = \mathbb{Q}[\sqrt{2}]$ and the above isomorphism φ. As noted the minimal polynomial of ω in $F[X] = \mathbb{Q}[\sqrt{2}][X]$ is $X^2 + X + 1$. Clearly the φ-transfer $\varphi(X^2 + X + 1) = X^2 + X + 1$. Now ω^2 has $X^2 + X + 1$ as its minimal polynomial $E[X] = \mathbb{Q}[\sqrt{2}][X]$.

By Proposition 6.31 there is an isomorphism

$$\psi : \mathbb{Q}[\sqrt{2}][\omega] \to \mathbb{Q}[\sqrt{2}][\omega] \text{ such that } \psi = \varphi \text{ on } \mathbb{Q}[\sqrt{2}] \text{ and } \psi(\omega) = \omega^2.$$

We have just built an automorphism ψ of the field $\mathbb{Q}[\sqrt{2}, \omega]$ such that

$$\psi(\sqrt{2}) = -\sqrt{(2)} \text{ and } \psi(\omega) = \omega^2.$$

The presence of such a mapping was not obvious.

The Isomorphism Lifting Theorem

We can use Proposition 6.31 as a piece of scaffolding with which to build isomorphisms between splitting fields. The next result is crucial, not only for showing that splitting fields of the same polynomial are isomorphic, but also for showing that every splitting field is generously endowed with automorphisms. We call it the ***isomorphism lifting theorem***.

Proposition 6.32 (Isomorphism lifting theorem). *If*

- $\varphi : E_1 \to E_2$ *is an isomorphism of fields,*
- f *is a polynomial in* $E_1[X]$ *of positive degree,*
- $\varphi(f)$ *is the* φ*-transfer of* f *into* $E_2[X]$,
- K_1 *is a splitting field for* f, *and* K_2 *is a splitting field for* $\varphi(f)$,

then φ *can be lifted (i.e. extended) to an isomorphism* $\psi : K_1 \to K_2$.

Proof. Let $\mathcal{F}$ be the family of all field homomorphisms $\sigma : L \to K_2$ where

- L is an intermediate field between E_1 and K_1, and
- the restriction of σ to E_1 is φ.

The goal is to see that $\mathcal{F}$ contains a mapping ψ whose domain is all of K_1.

To keep track of the fields with their related maps and inclusions, a diagram might help:

$$\begin{array}{ccc} K_1 & \xrightarrow{\psi} & K_2 \\ | & & \| \\ L & \xrightarrow{\sigma} & K_2 \\ | & & | \\ E_1 & \xrightarrow{\varphi} & E_2 \end{array}$$

Suppose $\sigma : L \to K_2$ is a mapping in $\mathcal{F}$ but that $L \subsetneq K_1$. By the nature of splitting fields this means f has a root α in $K_1 \setminus L$. If g in $L[X]$ is the minimal polynomial of α, then $g \mid f$ in $L[X]$ and thereby in $K_1[X]$. Hence $\sigma(g) \mid \sigma(f)$ in $K_2[X]$. Since σ extends φ and $f \in E_1[X]$, we see that $\sigma(f) = \varphi(f)$, which splits in $K_2[X]$. Therefore $\sigma(g)$ has a root, say β, in K_2. The polynomial $\sigma(g)$ is irreducible in $\sigma(L)[X]$ because g is irreducible in $L[X]$, and thus $\sigma(g)$ is the minimal polynomial of β over $\varphi(L)[X]$.

Proposition 6.31 says that the isomorphism $\sigma : L \to \sigma(L)$ has an extension isomorphism $\tau : L[\alpha] \to \sigma(L)[\beta]$. So τ is a proper extension of σ that continues to lie in $\mathcal{F}$. Since K_1 has finite degree over E_1, the above extension process, starting with φ, can be applied a finite number of times until a mapping $\psi : K_1 \to K_2$ inside $\mathcal{F}$ is obtained.

It only remains to show that such ψ is an isomorphism between K_1 and K_2. Well, ψ is certainly injective, since its domain is a field. To see that ψ is surjective, observe that $\psi(f)$ splits over the subfield $\psi(K_1)$ of K_2, since f splits over K_1. But $\psi(f) = \varphi(f)$, because $f \in E_1[X]$ and ψ restricted to E_1 is φ. The minimality of K_2 as a field over which $\varphi(f)$ splits forces $\psi(K_1)$ to equal K_2. □

6.2.4 The Uniqueness of Splitting Fields

The isomorphism lifting theorem has important consequences. To begin with, it shows that, up to isomorphism, splitting fields are unique.

Proposition 6.33. *If K_1, K_2 are splitting fields for the same polynomial f in $F[X]$, then there is an isomorphism $\psi : K_1 \cong K_2$ whose restriction to their common subfield F is the identity map.*

Proof. In the isomorphism lifting theorem, Proposition 6.32, take $E_1 = E_2 = F$, and take φ to be the identity mapping on F. □

Henceforth, we take the liberty of referring to a chosen splitting field of a polynomial f as ***the splitting field*** of f. The actual choice of splitting field will not matter.

6.2.5 Finite Fields of Equal Size Are Isomorphic

The uniqueness of the splitting field of a polynomial immediately adds to our insights on finite fields.

Proposition 6.34. *If two finite fields have equal size, then they are isomorphic.*

Proof. Say that the common size of the finite fields is p^n for some prime p and some exponent n. According to Proposition 6.27 each field is the splitting field over $\mathbb{Z}_p$ of the polynomial $X^{p^n} - X$. Being splitting fields of the same polynomial over $\mathbb{Z}_p$, they must be isomorphic. □

It turns out that for every prime power p^n, there does exist a finite field of that size, but that proof can wait a bit.

6.2.6 Splitting Fields Are Rich with Automorphisms

The isomorphism lifting theorem not only ensures that splitting fields are unique up to isomorphism, but also unveils a bounty of automorphisms on a splitting field.

Proposition 6.35. *If K is the splitting field of a polynomial f over a field F, with a subfield E that contains F, and $\varphi : E \to K$ is a homomorphism that restricts to the identity map on F (i.e. $\varphi(a) = a$ for all a in F), then φ has an extension to an automorphism $\psi : K \to K$.*

Proof. Here is a diagram to keep track of the fields and mappings involved:

$$\begin{array}{ccc} K & \xrightarrow{\psi} & K \\ | & & | \\ E & \xrightarrow{\varphi} & \varphi(E) \\ | & & | \\ F & = & F \end{array}$$

In the isomorphism lifting theorem, Proposition 6.32, take

$$K_1 = K_2 = K,\ E_1 = E,\ E_2 = \varphi(E).$$

Observe that $\varphi(f) = f$ since φ is the identity map on F. Also, from Proposition 6.26, K remains the splitting field of f over E as well as over $\varphi(E)$. □

One specialization of Proposition 6.35 is particularly noteworthy.

Proposition 6.36. *Let K be the splitting field of a polynomial over a field F. If $\alpha, \beta \in K$ and both have the same minimal polynomial in $F[X]$, then there is an automorphism $\psi : K \to K$ that restricts to the identity on F and maps α to β.*

Proof. Here is a diagram to describe the situation:

$$\begin{array}{ccc} K & \xrightarrow{\psi} & K \\ | & & | \\ F[\alpha] & \xrightarrow{\varphi} & F[\beta] \\ | & & | \\ F & = & F \end{array}$$

Proposition 6.11 (and also Proposition 6.31) gives an isomorphism $\varphi : F[\alpha] \to F[\beta]$ which acts as the identity map on F and maps α to β. Then Proposition 6.35 kicks in with $E = F[\alpha]$. □

6.2.7 Fixed Fields, F-Maps and F-Conjugates

Here is some more commonly used terminology.

Definition 6.37. If K, L are field extensions of F, a homomorphism $\varphi : K \to L$ which restricts to the identity mapping on F, i.e. such that $\varphi(a) = a$ for every a in F, is said to ***fix*** the field F. Such a homomorphism (which is automatically injective) will be called an ***F-homomorphism***, as well as an ***F-map***. If φ is bijective, we will call it an ***F-isomorphism***, and if $K = L$ and φ is bijective it will be called an ***F-automorphism*** of K.

A trivial exercise reveals that every F-map $\varphi : K \to L$ is also an F-linear transformation between the vector spaces K and L over the field F.

Definition 6.38. Two elements α, β in a field extension K of F are said to be ***conjugates over F*** when they have the same minimal polynomial over F. They are also known as ***F-conjugates***.

Proposition 6.36 says something subtle and interesting about splitting fields. If K is a splitting field over F and α, β in K are F-conjugates, then there exists an F-automorphism of the extension K, which sends α to β. For example, $\mathbb{C}$ is the splitting field of $X^2 + 1$ over $\mathbb{R}$. The roots $\pm i$ of X^2+1 are $\mathbb{R}$-conjugates. Proposition 6.36 ensures there is an $\mathbb{R}$-automorphism of $\mathbb{C}$ that maps i to $-i$. Of course, that automorphism is the familiar conjugation map $z \mapsto \bar{z}$.

For a more lively example, take the field $K = \mathbb{Q}[\alpha, \omega]$ where $\alpha = \sqrt[3]{2}$ and $\omega = e^{2\pi i/3}$. This is the splitting field over $\mathbb{Q}$ of the irreducible polynomial $X^3 - 2$ in $\mathbb{Q}[X]$. The roots of this polynomial are $\alpha, \alpha\omega, \alpha\omega^2$. According to Proposition 6.36, for any pair of these conjugates, there exists a $\mathbb{Q}$-automorphism of K that matches one root to the other in the pair. Since ω and ω^2 in K are conjugates over $\mathbb{Q}$ with minimal polynomial $X^2 + X + 1$, there is also a $\mathbb{Q}$-automorphism of K that maps ω to ω^2.

Let us turn to a basic property of F-maps.

F-Maps Permute Roots of Polynomials

If K, L are field extensions of F, the F-maps from K to L come with a fundamental constraint, which emerges from the following more general but easy result.

Proposition 6.39. *If $\varphi : K \to L$ is a homomorphism between fields and α in K is a root of a polynomial f in $K[X]$, then $\varphi(\alpha)$ in L is a root of the φ-transfer $\varphi(f)$ in $L[X]$.*

Proof. Let $f = a_nX^n + \cdots + a_1X + a_0$ where $a_j \in K$. The φ-transfer of f into $L[X]$ is $\varphi(f) = \varphi(a_n)X^n + \cdots + \varphi(a_1)X + \varphi(a_0)$, and so

$$\begin{aligned}\varphi(f)(\varphi(\alpha)) &= \varphi(a_n)\varphi(\alpha)^n + \cdots + \varphi(a_1)\varphi(\alpha) + \varphi(a_0)\\ &= \varphi(a_n\alpha^n + \cdots + a_1\alpha + a_0) \text{ since } \varphi \text{ is a homomorphism,}\\ &= \varphi(f(\alpha)) = \varphi(0) = 0 \quad \text{ since } \alpha \text{ is a root of } f. \end{aligned}$$ □

When K, L are extensions of a common field F and $\varphi : K \to L$ is an F-map, Proposition 6.39 specializes to something worth remembering.

Proposition 6.40. *Let K, L be field extensions of F and let $\varphi : K \to L$ be an F-map, i.e. φ fixes F. If $f \in F[X]$ with positive degree and α is a root of f inside K, then $\varphi(\alpha)$ is a root of f inside L.*

Proof. Since φ fixes F and $f \in F[X]$, we get that $\varphi(f) = f$. Then Proposition 6.39 kicks in. □

When $K = L$, we get an even more memorable specialization.

Proposition 6.41. *If L is a field extension of F, $f \in F[X]$ with positive degree and Δ is the set of roots of f which lie in L, then the restriction of every F-map $\varphi : L \to L$ to Δ is a permutation of Δ.*

Proof. Proposition 6.40 ensures that if $\alpha \in \Delta$, then $\varphi(\alpha) \in \Delta$. Since the set Δ is finite and φ is an injection, the restriction of φ to Δ is a bijection, i.e. a permutation of Δ. □

6.2.8 The Normality Theorem

We are in a position to answer an important question. If K is a finite extension of a field F, must K be the splitting field of some polynomial over F? The answer, which is *no*, comes from a stunning property shared only among splitting fields. The next major (and rather hard) result, which exploits the abundance of automorphisms on splitting fields, will be called the ***normality theorem***.

Proposition 6.42 (Normality theorem). *For a finite extension K of a field F, the following properties are equivalent.*

1. *The field K is the splitting field over F of some polynomial in $F[X]$.*
2. *For every extension L of K and every F-map $L \to L$, its restriction to K becomes an F-automorphism of K.*
3. *For every α in K, its minimal polynomial over F splits over K.*

Proof. We will prove that $1 \implies 2 \implies 3 \implies 1$.

$1 \implies 2$. Suppose $f \in F[X]$ with splitting field K and that $\varphi : L \to L$ is an F-map on a field extension L of K. Let $\Delta = \{\alpha_1, \dots, \alpha_n\}$ be the finite set of roots of f. By the definition of splitting fields these roots generate K, i.e. $K = F[\Delta]$. Also, according to Proposition 6.41, φ restricts to a permutation of Δ.

Every α in K is a polynomial in the α_j of the form $\alpha = \sum a_{i_1,\dots,i_n}\alpha_1^{i_1}\cdots\alpha_n^{i_n}$, where the i_j run through non-negative integers and the $a_{i_1,\dots,i_n}$ are taken from F. Because φ is an F-map,

$$\varphi(\alpha) = \sum \varphi(a_{i_1,\ldots,i_n})\varphi(\alpha_1)^{i_1} \cdots \varphi(\alpha_n)^{i_n} = \sum a_{i_1,\ldots,i_n}\varphi(\alpha_1)^{i_1} \cdots \varphi(\alpha_n)^{i_n}.$$

Since the $\varphi(\alpha_1),\ldots,\varphi(\alpha_n)$ constitute a permutation of the original $\alpha_1,\ldots,\alpha_n$, we see that $\varphi(\alpha) \in K$ and also that $\varphi(\alpha)$ picks up every element of K as α runs over K. So, the restriction of φ to K becomes an F-automorphism of K.

$2 \implies 3$. Let $\alpha \in K$ and let g in $F[X]$ be its minimal polynomial, which is expected to split over K.

Being of finite degree over F, the field $K = F[\alpha_1,\ldots,\alpha_n]$ for some α_j that are algebraic over F. For each α_j let h_j be its minimal polynomial over F, and let $f = gh_1h_2\cdots h_n$, the product in $F[X]$ of the minimal polynomials of α and all α_j.

By default, f is also in $K[X]$. Take L to be the splitting field of f over K. This ensures the inclusions $F \subseteq K \subseteq L$. It turns out that L is also the splitting field of f over F. Indeed, let Δ inside L be the complete set of roots of f. Since L is the splitting field of f over K and since $\alpha_1,\ldots,\alpha_n$ generate K over F, we obtain

$$L = K[\Delta] = F[\alpha_1,\ldots,\alpha_n][\Delta] = F[\Delta].$$

The last equality holds because the α_j are roots of f and thereby already lie in Δ.

Our g is a factor of f, and f splits over L. Thus the irreducible g in $F[X]$ splits over L. Recall that α in K is a root of g. If β in L is any root of g, Proposition 6.36 yields an F-map $\varphi : L \to L$ such that $\varphi(\alpha) = \beta$. By assumption, the restriction of φ to K becomes an automorphism of K. Since $\alpha \in K$, we must have $\varphi(\alpha) = \beta \in K$. The roots of g, which led to its splitting over L, are already in K. Therefore g splits over K as desired.

$3 \implies 1$. We need a polynomial f in $F[X]$ whose splitting field is K. Being a finite extension of F, the field $K = F[\alpha_1,\ldots,\alpha_n]$ for some α_j that are algebraic over F. For each α_j let h_j in $F[X]$ be its minimal polynomial, and put $f = h_1h_2\cdots h_n$, the product of these minimal polynomials. By assumption each h_j splits over K, and thus f splits over K. Since K is a field generated over F by some of the roots of f, the field K is certainly generated over F by all of the roots of f. Thus K is the splitting field of f. □

Definition 6.43. A field extension K of F that satisfies the third property of the normality theorem is known as a ***normal extension***.[5]

As we will come to discover, normal extensions have an intimate connection to normal subgroups. The normality theorem tells us that a finite extension K of F is normal if and only if K is the splitting field of a polynomial in $F[X]$. Our plan is to continue to refer to these extensions as splitting fields.

To illustrate the normality theorem, observe that the field $K = \mathbb{Q}\left[\sqrt{2}, \sqrt{3}\right]$ is the splitting field of $(X^2-2)(X^2-3)$ over $\mathbb{Q}$. Take $\alpha = \sqrt{2}+\sqrt{3}$ inside K. Now $\alpha^2 = 5+2\sqrt{6}$ and then

[5] The origins of the term "normal" in this context are not so clear.

$(\alpha^2-5)^2 = 24$, which comes down to $\alpha^4 - 10\alpha^2 + 1 = 0$. So, the minimal polynomial of α is a divisor in $\mathbb{Q}[X]$ of $X^4 - 10X^2 + 1$. A bit of checking reveals that this polynomial is irreducible in $\mathbb{Q}[X]$, and thereby it is the minimal polynomial of α. According to the normality theorem, all of its roots have to be in K. Well, its four roots $\pm\sqrt{2}, \pm\sqrt{3}$ are in K.

A Non-splitting Field

Let $E = \mathbb{Q}[\sqrt[3]{2}\,]$ where the generator is the real cube root of 2. The minimal polynomial of $\sqrt[3]{2}$ over $\mathbb{Q}$ is $X^3 - 2$. Another root of $X^3 - 2$ is $\sqrt[3]{2}e^{2\pi i/3}$. If this polynomial were to split over E, then this other root would have to be in E. However, E is a subfield of $\mathbb{R}$, while this other root is not in $\mathbb{R}$. Thus E fails to be normal over $\mathbb{Q}$. By the normality theorem, there is no polynomial over $\mathbb{Q}$ for which E is the splitting field.

To summarize, any finite extension that fails condition 3 of the normality theorem is not a splitting field.

The Normal Closure

Though not every finite extension E of F is a splitting field, something can be salvaged.

Proposition 6.44. *If E is a finite extension of a field F, then there is a finite extension K of E such that*

- *K is a splitting field of some polynomial in $F[X]$, and*
- *K is the only splitting field over F which extends E and sits inside K.*

Proof. Being a finite extension, $E = F[\alpha_1, \ldots, \alpha_n]$ for some α_j which are algebraic over F. Let g_j in $F[X]$ be the minimal polynomial of α_j, and let f be their product $g_1 g_2 \cdots g_n$. Since f lies in $F[X]$ it is also in $E[X]$. Let K be the splitting field of f over E. This ensures the inclusions $F \subseteq E \subseteq K$. If Δ inside K is the full set of roots involved in the splitting of f, we have

$$K = E[\Delta] = F[\alpha_1, \ldots, \alpha_n][\Delta] = F[\Delta].$$

The last equality holds because the α_j are roots of f, and thereby lie in Δ, making them redundant as generators of K over F. Thus K is a splitting field of f in $F[X]$.

For the second item suppose that L is a splitting field of F and such that

$$F \subseteq E \subseteq L \subseteq K,$$

where the inclusions indicate subfields. Clearly, each generator α_j of E over F lies in L. Since L is assumed to be a splitting field over F, the normality theorem implies that each minimal polynomial g_j splits over L. This implies that f splits over L. So, the full set of roots Δ of f is already inside L. Thus $K = F[\Delta] \subseteq L \subseteq K$, whereby $L = K$. □

If E is a finite extension of F, a splitting field K over F that satisfies the conditions of Proposition 6.44 is called a ***normal closure*** of the extension E.

6.2.9 A Control on the Degree of a Splitting Field

Here are a couple of constraints on the degree of a splitting field.

Proposition 6.45. *If f is a polynomial of positive degree n over a field F and K is the splitting field of f over F, then $[K : F]$ divides $n!$.*

Furthermore, if f is irreducible, then n divides $[K : F]$.

Proof. There is no harm in taking f to be monic.

The second claim for irreducible f follows readily from the tower theorem. Take α in K to be a root of f. We have the tower $F \subseteq F[\alpha] \subseteq K$. Since f is irreducible, $[F[\alpha] : F] = n$, the degree of the minimal polynomial of α. By Proposition 6.13, n divides $[K : F]$.

To see that $[K : F]$ divides $n!$ we use induction on the degree n of f. Certainly when $n = 1$, we have $[K : F] = 1$, which divides $1!$. Suppose that the result holds for all splitting fields of all polynomials of degree less than n, and consider f of degree n in $F[X]$ with splitting field K. The cases where f is irreducible or not need a different treatment.

Suppose f is irreducible, and write

$$f = (X - \alpha_1)(X - \alpha_2) \cdots (X - \alpha_n),$$

where the $\alpha_j \in K$ and the α_j (possibly repeated) generate K over F. Since f is the minimal polynomial of α_1 we have that $[F[\alpha_1] : F] = n$. Also $f = (X - \alpha_1)g$ where $g = (X - \alpha_2) \cdots (X - \alpha_n)$. Now $g \in F[\alpha_1][X]$, by the factor theorem, Proposition 4.35. Clearly, K is the splitting field of g over $F[\alpha_1]$. The inductive assumption applied to g of degree $n - 1$ yields that $[K : F[\alpha_1]]$ divides $(n - 1)!$. The tower theorem, Proposition 6.13, gives

$$[K : F] = [K : F[\alpha_1]]\,[F[\alpha_1] : F].$$

The first factor above divides $(n-1)!$, and the second factor equals n. Thus $[K : F]$ divides $n!$.

Now suppose that f reduces. Say $f = gh$ where g, h are monic polynomials in $F[X]$, and their respective degrees k, ℓ are less than n. Let E be the splitting field of g over F, and write $g = (X - \alpha_1) \cdots (X - \alpha_k)$, where the α_j are the roots of g in E. Then let K be the splitting field of h over E (note E, not F), and write $h = (X - \beta_1) \cdots (X - \beta_\ell)$, where the β_j are the roots of h in K. We also have

$$K = E[\beta_1, \dots, \beta_\ell] = F[\alpha_1, \dots, \alpha_k][\beta_1, \dots, \beta_\ell] = F[\alpha_1, \dots, \alpha_k, \beta_1, \dots, \beta_\ell].$$

Since the roots of g and h together make up the roots of f, we see that K is the splitting field of f over F.

The inductive hypothesis applied to g gives that $[E : F]$ divides $k!$, and the inductive hypothesis applied to h gives that $[K : E]$ divides $\ell!$. The tower theorem applied to the field tower $F \subseteq E \subseteq K$ yields $[K : F] = [K : E]\,[E : F]$. Taken together these facts show that $[K : F]$ divides $k!\,\ell!$. Since $k + \ell = n$ we also have that $k!\,\ell!$ divides n. Indeed, $\frac{n!}{k!\ell!}$ is a binomial coefficient. Thus $[K : F]$ divides $n!$. □

The degree of its splitting field is an important aspect of a polynomial. We cannot help but notice that the statement "$[K : F]$ divides $n!$" still leaves a lot of room for such possible degrees. One interesting but hard question is whether for each n there do exist polynomials of degree n whose splitting field over $\mathbb{Q}$ has degree precisely $n!$.

For an illustration of Proposition 6.45 consider an irreducible polynomial of degree 3 over F. Its splitting field K has degree at most $3! = 6$ over F, and this degree is divisible by 3. Therefore, K must have degree 3 or degree 6 over F. For an example of degree 6 over $\mathbb{Q}$, take the irreducible polynomial $X^3 - 2$. One of its roots is $\alpha = \sqrt[3]{2}$, while another root is $\alpha\omega$ where $\omega = e^{2\pi i/3}$. The latter root is not in the degree 3 extension $\mathbb{Q}[\alpha]$. So the splitting field of $X^3 - 2$ must have degree 6 over $\mathbb{Q}$. We leave examples of degree 3 splitting fields to the exercises.

6.2.10 Quadratic Extensions Are Splitting Fields

We conclude this section with some small observations on quadratic extensions.

Proposition 6.46. *A finite extension K of a field F has degree 2 if and only if $K = F[\alpha]$ where α is the root of an irreducible polynomial of degree 2.*

All degree 2 extensions are splitting fields.

If the characteristic of F is not 2, then every degree 2 extension K is of the form $F[\sqrt{d}\,]$ for some d in F.

Proof. If $[K : F] = 2$, take any α from $K \setminus F$. The degree $[F[\alpha] : F]$, being greater than 1 and at most 2, must be 2, and so $K = F[\alpha]$. Clearly, the minimal polynomial of α has degree 2. Just as clearly, if $K = F[\alpha]$ where α has minimal polynomial of degree 2, then the degree $[K : F] = 2$ by the frequently used Proposition 6.9.

Now suppose $K = F[\alpha]$, and such that the minimal polynomial of α in $F[X]$ is g of degree 2. Then $X - \alpha$ divides g in $K[X]$. Since g is monic of degree 2, the other factor of g must be $X - \beta$ for some β in K. Hence g splits in $K[X]$ as $(X - \alpha)(X - \beta)$. This makes K the splitting field of g.

Finally, suppose the characteristic of F is not 2 and that $K = F[\alpha]$ for some α in K with minimal polynomial $g = X^2 + bX + c$ in $F[X]$. Since division by 2 is allowed, we calculate

$$\left(\alpha + \frac{b}{2}\right)^2 = \alpha^2 + b\alpha + \frac{b^2}{4} = \frac{b^2}{4} - c.$$

With $d = \frac{b^2}{4} - c$ we have an element of F such that $\sqrt{d} = \alpha + \frac{b}{2} \in K$. Thus $K = F[\alpha] = F[\sqrt{d}\,]$. □

The assumption that F not have characteristic 2 in the third part of Proposition 6.46 is unavoidable. For example, take $F = \mathbb{Z}_2$, the finite field with two elements. The polynomial $X^2 + X + 1$ is irreducible over F. According to Proposition 6.46 its splitting field K is the degree 2 extension $F[\alpha]$ where $\alpha^2 + \alpha + 1 = 0$. The other root of $X^2 + X + 1$ is $\alpha + 1$. Yet the only possible d in F are 0 and 1, and in either case $F[\sqrt{d}\,] = F$, not $F[\alpha]$.

EXERCISES

1. Let K, L be field extensions of a field F, and $\varphi : K \to L$ a homomorphism. Show that φ is an F-map if and only if φ is an F-linear transformation of the vector spaces K, L.
2. If f is a cubic polynomial over a field F, show that the degree of its splitting field must be 1, 2, 3 or 6.
3. Which of the following fields are splitting fields over $\mathbb{Q}$? Give reasons.
$$\mathbb{Q}[\sqrt{2}+i],\ \mathbb{Q}[e^{2\pi i/7}],\ \mathbb{Q}[\sqrt[5]{5}],\ \mathbb{Q}[\sqrt{2},\sqrt{6}],\ \mathbb{Q}[\sqrt[3]{2},i].$$
4. Let $\mathbb{Z}_3(t)$ be the field of rational functions in the variable t and having coefficients in the finite field $\mathbb{Z}_3$. If α is a root of $X^3 - t$, is the field extension $\mathbb{Z}_3(t)[\alpha]$ a splitting field over $\mathbb{Z}_3(t)$?
5. Find the degree of the splitting field over $\mathbb{Q}$ of these polynomials,
$$X^3+3,\ X^4+1,\ X^4+X^2+1,\ X^6-1,\ X^6-4.$$
6. (a) If p is a prime, show that the splitting field K of $X^p - 2$ over $\mathbb{Q}$ is $\mathbb{Q}[\sqrt[p]{2}, \zeta)$ where $\zeta = e^{2\pi i/p}$.
   (b) Find the degree $[K : \mathbb{Q}]$.
   (c) Show that $X^p - 2$ remains irreducible over the field $\mathbb{Q}[\zeta]$.
7. Suppose that $F \subset E \subset K$ is a field tower. Which of these statements is true?
   (a) If K is normal over F, then E is normal over F.
   (b) If K is normal over F, then K is normal over E.
   (c) If E is normal over F and K is normal over E, then K is normal over F.
8. For each positive integer n find a splitting field over $\mathbb{Q}$ of degree 2^n.
9. Find the degree of $\mathbb{Q}[\sqrt[3]{2}, \sqrt{3}, i]$ over $\mathbb{Q}$. Is this extension a splitting field over $\mathbb{Q}$?
10. Let $\alpha = \sqrt[3]{7} + 2i$.
    (a) Show that $\mathbb{Q}[\alpha] = \mathbb{Q}[\sqrt[3]{7}, i]$.
    (b) Find the degree $\mathbb{Q}[\alpha : \mathbb{Q}]$.
    (c) Find the minimal polynomial of α.
    (d) Find the $\mathbb{Q}$-conjugates of α.
    (e) Find the splitting field of the minimal polynomial of α.
11. Let $f = X^3 - 3X + 1$ in $\mathbb{Q}[X]$. Put $\omega = e^{2\pi i/9}$.
    Observe that $\overline{\omega} = \omega^{-1} = \omega^8 = e^{-2\pi i/9}$, for future use.
    Let $\alpha = \omega + \overline{\omega}$, $\beta = \omega^2 + \omega^{-2}$, $\gamma = \omega^4 + \omega^{-4}$. Notice that $\alpha, \beta, \gamma \in \mathbb{R}$.
    (a) In a sketch, mark the points ω^j in the complex plane as j runs from 1 to 9. Also mark the locations of α, β, γ.
    (b) Show that f is irreducible.
    (c) Verify in turn that α, β, γ are the roots of f.
    (d) Verify that $\beta = \alpha^2 - 2$ and $\gamma = \beta^2 - 2 = \alpha^4 - 4\alpha^2 + 2$.
    (e) Show that $\mathbb{Q}[\alpha]$ is the splitting field of f over $\mathbb{Q}$, and deduce that $[\mathbb{Q}[\alpha] : \mathbb{Q}] = 3$.

(f) Find three distinct automorphisms of the splitting field of f, and explain why they are automorphisms.

Note. Every automorphism of $\mathbb{Q}[\alpha]$ is completely determined by its value at α.

12. Let $f = X^4 - 2X^2 + 3$ in $\mathbb{Q}[X]$. Let us find the degree of its splitting field as well as some automorphisms of the splitting field. The problem is broken into small steps.

(a) Show that f is irreducible.

Hint. Replace $f(X)$ by $f(X+1)$ and use Eisenstein on the latter polynomial.

(b) Let α be the complex number in the first quadrant for which $\alpha^2 = 1 + i\sqrt{2}$. In other words α is the square root $\sqrt{1 + i\sqrt{2}}$ taken in the first quadrant.

Verify that α is a root of f and that the other roots of f are $-\alpha, \overline{\alpha}, -\overline{\alpha}$, where $\overline{\alpha}$ denotes the complex conjugate of α.

Show that these roots lie on the circle of radius $\sqrt[4]{3}$ centred at 0, and sketch their position on that circle.

(c) Explain briefly why $\mathbb{Q}[\alpha, \overline{\alpha}]$ is the splitting field of f over $\mathbb{Q}$.

(d) Find a quadratic polynomial in $\mathbb{Q}[\alpha][X]$ for which $\overline{\alpha}$ is a root.

Deduce that the degree $[\mathbb{Q}[\alpha, \overline{\alpha}] : \mathbb{Q}]$ is either 4 or 8.

(e) Show that $\mathbb{Q}[\sqrt{3}, \sqrt{2}i]$ is a subfield of $\mathbb{Q}[\alpha, \overline{\alpha}]$, and that the degree of $\mathbb{Q}[\sqrt{3}, \sqrt{2}i]$ over $\mathbb{Q}$ is 4.

(f) Show that $\alpha \notin \mathbb{Q}[\sqrt{3}, \sqrt{2}i]$.

Deduce that the degree of the splitting field $\mathbb{Q}[\alpha, \overline{\alpha}]$ over $\mathbb{Q}$ is 8.

Hint. A basis for $\mathbb{Q}[\sqrt{3}, \sqrt{2}i]$ as a vector space over $\mathbb{Q}$ is given by $1, \sqrt{3}, \sqrt{2}i, \sqrt{6}i$. Suppose α is a rational combination of that basis and calculate $|\alpha|^2$ to get a contradiction.

(g) Construct an automorphism of the splitting field $\mathbb{Q}[\alpha, \overline{\alpha}]$ in accordance with the following steps.

Briefly explain why $\mathbb{Q}[\alpha, \overline{\alpha}] = \mathbb{Q}[\alpha, \sqrt{3}] = \mathbb{Q}[\overline{\alpha}, \sqrt{3}]$.

Explain why there is an isomorphism $\varphi : \mathbb{Q}[\alpha] \to \mathbb{Q}[\overline{\alpha}]$ such that $\varphi(\alpha) = \overline{\alpha}$.

Use Proposition 6.31 to show that there is an isomorphism

$$\sigma : \mathbb{Q}[\alpha][\sqrt{3}] \to \mathbb{Q}[\overline{\alpha}][-\sqrt{3}] \text{ such that } \sigma(\alpha) = \overline{\alpha} \text{ and } \sigma(\sqrt{3}) = -\sqrt{3}.$$

Deduce that σ is an automorphism of $\mathbb{Q}[\alpha, \overline{\alpha}]$ which permutes the roots of f as follows:

$$\sigma : \alpha \mapsto \overline{\alpha} \mapsto -\alpha \mapsto -\overline{\alpha} \mapsto \alpha.$$

(h) Show that there exists an isomorphism

$$\tau : \mathbb{Q}[\alpha][\sqrt{3}] \to \mathbb{Q}[\overline{\alpha}][\sqrt{3}] \text{ such that } \tau(\alpha) = \overline{\alpha} \text{ and } \tau(\sqrt{3}) = \sqrt{3}.$$

Deduce that τ is an automorphism of $\mathbb{Q}[\alpha, \overline{\alpha}]$ which permutes the roots of f as follows:

$$\tau : \alpha \mapsto \overline{\alpha} \mapsto \alpha, \quad -\alpha \mapsto -\overline{\alpha} \mapsto -\alpha.$$

(i) * Prove that the automorphisms σ, τ, when viewed as permutations of the roots of f, generate a group which is isomorphic to the dihedral group $\mathcal{D}_4$.

13. (a) Show that the polynomial $f = X^4 + X^3 + X^2 + X + 1$ in $\mathbb{Z}_2[X]$ is irreducible.

By Kronecker's theorem there is a field extension of $\mathbb{Z}_2$ that contains a root α of f. Then $F = \mathbb{Z}_2[\alpha]$ is an extension of degree 4 over $\mathbb{Z}_2$, with basis $1, \alpha, \alpha^2, \alpha^3$.

(b) The set $F^\star$ of non-zero elements of F is a group under multiplication. Find the order of α in this group.

(c) The group $F^\star$ is cyclic, according to Proposition 4.37. Find a generator for this group expressed as the sum of the elements $1, \alpha, \alpha^2, \alpha^3$.

14. Prove that every degree 2 field extension of $\mathbb{Q}$ equals $\mathbb{Q}[\sqrt{n}\,]$ for some *integer n*.

15. Suppose f is a monic, *irreducible* polynomial over a field F. Let α, β be roots of f in its splitting field K. If r is a positive integer, show that $(X - \alpha)^r$ divides f in $K[X]$ if and only if $(X - \beta)^r$ divides f in $K[X]$.

Deduce from this that the unique splitting of f into linear factors in $K[X]$ takes the form

$$f = ((X - \alpha_1)(X - \alpha_2) \cdots (X - \alpha_n))^m,$$

where the α_j are the *distinct* roots of f and m is a positive integer. In other words, for an irreducible f in $F[X]$ its linear factors in $K[X]$ all repeat with the same multiplicity m.

We shall see in the next section that if F has zero characteristic, then that multiplicity m is always 1. We shall also see examples in prime characteristic where $m > 1$.

Hint. Use Proposition 6.36.

16. Suppose that F is a subfield of a field K. Let f, g be polynomials in $F[X]$.

(a) Clearly, if $f \mid g$ in $F[X]$, then also $f \mid g$ in $K[X]$. The converse is less clear. Suppose $f, g \in F[X]$ and that $f = gh$ for some h in $K[X]$. Prove that $h \in F[X]$.

Hint. The ring $F[X]$ is a Euclidean domain, wherein quotients and remainders are unique.

(b) The greatest common divisor of f and g in $F[X]$ is the unique monic polynomial h such that $h \mid f$, $h \mid g$ and $h = sf + tg$ for some polynomials s, t in $F[X]$. Explain why h remains the greatest common divisor of f and g when they are viewed as polynomials in $K[X]$.

(c) Two polynomials f, g in $F[X]$ are coprime when $\gcd(f, g) = 1$. Alternatively, they are not coprime when $\deg(\gcd(f, g)) \geq 1$. Show that f, g are coprime in $F[X]$ if and only if they never have a root in common in any field extension K of F.

17. The polynomials $f = X^3 + X + 1$ and $g = X^3 + X^2 + 1$ are irreducible in $\mathbb{Z}_2[X]$. If α, β are respective roots of these polynomials in some field extension, the finite fields $\mathbb{Z}_2[\alpha]$ and $\mathbb{Z}_2[\beta]$ each have degree 3 over $\mathbb{Z}_2$, and therefore each consists of 8 elements. By Proposition 6.34, these fields are isomorphic.

Find an explicit isomorphism between them, and show that your mapping is an isomorphism.

Hint. The element α had better go to something in $\mathbb{Z}_2[\beta]$ with the same minimal polynomial as α.

18. Suppose K is a field extension of F and that E, L are subfields of K which are splitting fields over F for suitable polynomials.

(a) Show that the composite EL inside K is also a splitting field over F.

(b) Use the normality theorem to show that the intersection $E \cap L$ is a splitting field over F.

19.* Let F be a subfield of $\mathbb{C}$. Let $a \in F$ and p a prime. If $X^p - a$ has no root in F, prove that $X^p - a$ is irreducible in $F[X]$.

6.3 Separability

If K is a field extension of F and $\varphi : K \to K$ is an F-map, then, according to Proposition 6.41, φ maps every algebraic α in K to one of its F-conjugates in K, i.e. to one of the roots of the minimal polynomial of α. If K is generated over F by some elements $\alpha_1, \ldots, \alpha_n$, then every F-map is determined by its action on these elements. So, if no α_j has an F-conjugate in K other than itself, then the only possible F-map on K will be the identity map.

A dearth of F-conjugates for an element α in K can arise in two ways. It is possible that K is not a splitting field. By Proposition 6.42 such K will not be a normal extension of F, and then the minimal polynomial of α might not split over K. But it may also happen that K is normal over F, so that every minimal polynomial in $F[X]$ of every α in K splits over K. However, the roots that split such a minimal polynomial could be repeated. This also limits the possible actions that an F-map could attain.

For example, let $F = \mathbb{Z}_2(t)$, the field of rational functions in the indeterminate t with coefficients in the two element field $\mathbb{Z}_2$. The polynomial $X^2 - t$ is irreducible in $\mathbb{Z}_2(t)[X]$. For otherwise t would have a square root in $\mathbb{Z}_2(t)$. That is, $t = (p/q)^2$ for some polynomials p, q in $\mathbb{Z}_2[t]$. This would lead to $tq^2 = p^2$, which is impossible because the degree in t of the left side is odd, while the degree on the right is even. The splitting field of $X^2 - t$ is $F[\alpha]$ where α is a root of $X^2 - t$. The other root is $-\alpha$, which coincides with α since we are in characteristic 2. Thus $X^2 - t$ is an irreducible polynomial whose factorization in $F[\alpha][X]$ is given by $X^2 - t = (X - \alpha)^2$.

The hazard that the number of distinct roots of a minimal polynomial might be less than its degree requires attention.

6.3.1 Derivatives of Polynomials and Repeated Roots

A useful tool for detecting polynomials with repeated roots is the derivative.

Definition 6.47. If $f = a_0 + a_1X + a_2X^2 + a_3X^3 + \cdots + a_{n-1}X^{n-1} + a_nX^n$ is a polynomial of degree n over a field F, its ***derivative*** is the polynomial

$$f' = a_1 + 2a_2X + 3a_3X^2 + \cdots + (n-1)a_{n-1}X^{n-2} + na_nX^{n-1}.$$

For example, the derivative of $iX^4 - (2-i)X^3 + 5X^2 - \sqrt{3}X + 5$ taken in $\mathbb{C}[X]$ is $iX^3 + 3(X-i)X^2+10X-\sqrt{3}$. If F has prime characteristic p, the integer exponents $1, 2, 3, \ldots, n-1, n$, brought down in the familiar way as multiples of the coefficients a_j, are to be interpreted as being in the base subfield $\mathbb{Z}_p$ in the only sensible way. Thus the derivative of $6X^3+5X^2+4X+1$ taken in $\mathbb{Z}_7[X]$ is $12X^2+10X+4 = 5X^2+3X+4$. The derivative of $3X^5+4X^3+2$ in $\mathbb{Z}_5[X]$ is $15X^4 + 12X^2 = 2X^2$, and the derivative of $X^{10} + X^5 + 1$ in $\mathbb{Z}_5[X]$ is $10X^9 + 5X^4 = 0$. Evidently, when F has characteristic zero the derivative drops the degree by one, but in case of prime characteristic p the degree can drop by more than one, and indeed the derivative could well be the zero polynomial.

Derivatives come with their expected linearity and Leibniz properties, whose mundane verifications deserve to be left as an exercise.

Proposition 6.48. *If K is any field and $f, g \in K[X]$ then*

$$(f+g)' = f' + g' \text{ and } (fg)' = fg' + f'g.$$

We also observe the obvious fact that if K is a field extension of F, the derivative operator on $K[X]$ restricts to the derivative operator on the subring $F[X]$.

Definition 6.49. Let K be a field extension of a field F and let f be a non-constant polynomial in $F[X]$. An element θ in K is called a ***repeated root*** of f when $(X-\theta)^2$ divides f in $K[X]$.

Here is how derivatives detect repeated roots.

Proposition 6.50. *Let K be a field extension of F and let f in $F[X]$ be a non-constant polynomial. A root θ in K of f is a repeated root of f if and only if θ is also a root of its derivative f'.*

Proof. The factor theorem yields

$$f = (X-\theta)g \text{ for some } g \text{ in } K[X].$$

The Leibniz rule for the derivative of a product, applied to this factorization, gives

$$f' = (X-\theta)g' + g.$$

If θ is also a root of f', the second equation above shows that θ is a root of g. Thus $X-\theta$ divides g in $K[X]$. The first equation then shows that $(X-\theta)^2$ divides f, making θ a repeated root of f. Conversely, if θ is a repeated root of f, i.e. $(X-\theta)^2 \mid f$, the first equation above forces $X-\theta$ to divide g in $K[X]$. The second equation reveals that $X-\theta$ divides f' in $K[X]$, making θ a root of f'. □

6.3.2 Separable Polynomials and the Derivative

Definition 6.51. A non-constant polynomial f over a field F is ***separable*** [6] provided there is no field extension E of F containing a repeated root of f.

We immediately offer another way to think of separable polynomials.

Proposition 6.52. *A non-constant polynomial f over a field F is separable if and only if the number of distinct roots of f in its splitting field equals the degree of f.*

Proof. We can take f to be monic, without harm.

If f is separable as defined and K is its splitting field, then in the unique factorization

$$f = (X - \theta_1)(X - \theta_2)\cdots(X - \theta_n),$$

which splits f in $K[X]$, the roots θ_j must be distinct. It is thereby evident that the number of roots of f inside K equals $\deg f$.

Conversely suppose that f is not separable. Thus there is a field extension of F containing a repeated root θ of f. This means that $f = (X - \theta)^2 g$ for some monic g in $F[\theta][X]$. Let K be the splitting field of g over $F[\theta]$. So, there exist $\alpha_1, \ldots, \alpha_k$ in K such that

$$g = (X - \alpha_1)\cdots(X - \alpha_k).$$

Then f splits over K as $(X - \theta)^2(X - \alpha_1)\cdots(X - \alpha_k)$. Furthermore, since K is the splitting field of g over $F[\theta]$, we have

$$K = F[\theta][\alpha_1, \ldots, \alpha_k] = F[\theta, \alpha_1, \ldots, \alpha_k].$$

This shows that K is also the splitting field of f, with roots $\theta, \alpha_1, \ldots, \alpha_k$. By looking at the factorization of f in $K[X]$ it is clear that the number of roots of f in K is strictly less than $\deg f$. □

An Alternative Definition of Separability

There is a commonly used *alternative definition of separability*. Suppose the unique factorization of f in $F[X]$ is given by

$$f = g_1^{\ell_1} g_2^{\ell_2} \cdots g_n^{\ell_n},$$

where the g_j are distinct irreducible factors of f in $F[X]$. Some authors say that f is separable when each of the g_j has no repeated root in its splitting field. Under this alternative definition, a polynomial such as X^2 in $\mathbb{Q}[X]$ would remain separable, whereas it is not separable under the definition we are adopting. If the irreducible factors g_j in $F[X]$ never repeat, i.e. if all $\ell_j = 1$, the two definitions of separability coincide. That is because distinct irreducibles do not have a root in common. In particular, the two definitions coincide when f is itself

[6] This terminology suggests that the roots of f never sit on top of each other. They are separated.

irreducible. Furthermore, it is clear that the splitting field of f coincides with the splitting field of $g_1 g_1 \cdots g_n$ in which any duplication of the g_j is removed. So all splitting fields are splitting fields of polynomials whose irreducible factors do not repeat. For this reason, the decision on how to define a separable polynomial is immaterial when it comes to working with its splitting field.

It might be a matter of opinion as to which definition makes the story of splitting fields flow smoother. One advantage of the definition which we have adopted is that a polynomial is separable if and only if the number of its distinct roots in its splitting field equals its degree.

Detecting Separability with Derivatives

The next test for separable polynomials has the advantage that it refers only to the given ground field.

Proposition 6.53. *A non-constant polynomial f over a field F is separable if and only if f is coprime with its derivative.*

Proof. If f and f' are not coprime, there is a polynomial h in $F[X]$ of positive degree that divides both f and f'. The splitting field of h contains a root θ of h. Such θ will be a common root of f and f'. By Proposition 6.50 f is not separable.

Conversely, if f and f' are coprime, i.e. $\gcd(f,f') = 1$, then there exist polynomials k, ℓ in $F[X]$ such that

$$fk + f'\ell = 1.$$

This identity shows that there can be no field extension of F containing a common root of f and f'. By Proposition 6.50 there is no field extension containing a repeated root of f. □

To illustrate Proposition 6.53, take $f = X^4 + 4X^3 + 8X^2 + 8X + 4$ in $\mathbb{Q}[X]$. Its derivative is $f' = 4X^3 + 12X^2 + 16X + 8$. To test for the coprimeness of f and f' calculate their greatest common divisor in $\mathbb{Q}[X]$ using the Euclidean algorithm. Thus,

$$X^4 + 4X^3 + 8X^2 + 8X + 4 = (4X^3 + 12X^2 + 16X + 8)\left(\frac{1}{4}X + \frac{1}{4}\right) + X^2 + 2X + 2,$$

and then

$$4X^3 + 12X^2 + 16X + 8 = (X^2 + 2X + 2)(X + 1) + 0.$$

The greatest common divisor of f and f' is $X^2 + 2X + 2$. Since f and f' are not coprime, the polynomial f has a repeated root. Indeed, both roots of $X^2 + 2X + 2$ in $\mathbb{C}$ are repeated roots of f.

Proposition 6.53 becomes especially interesting when applied to irreducible polynomials.

Proposition 6.54. *An irreducible polynomial f over a field F is separable if and only if its derivative is not the zero polynomial.*

Proof. Being irreducible, f is coprime with every non-zero polynomial of lesser degree. In particular, f is coprime with f' if and only if $f' \neq 0$. According to Proposition 6.53 f is separable if and only if $f' \neq 0$. □

If F has characteristic 0, non-constant polynomials have non-zero derivative. Thus Proposition 6.54 leads to a worthy corollary.

Proposition 6.55. *Over a field of characteristic zero, every irreducible polynomial is separable.*

Over fields of prime characteristic, we do encounter polynomials that are simultaneously irreducible and not separable. The polynomial $X^2 - t$, over the field $\mathbb{Z}_2(t)$ of rational functions in t with coefficients in $\mathbb{Z}_2$, is a simple example that we saw at the outset of this section. In agreement with Proposition 6.54 its derivative is $2X = 0$.

6.3.3 Finite Fields of All Possible Sizes

We are in a position to close a circle of facts about finite fields. Every finite field K is an extension of $\mathbb{Z}_p$ for some prime p. The size of K is p^n where $n = [K : \mathbb{Z}_p]$. Such K is the splitting field of the polynomial $X^{p^n} - X$ over $\mathbb{Z}_p$. Since splitting fields are unique up to isomorphism, finite fields of equal size are isomorphic. For every prime power p^n, there is at most one finite field of size p^n (up to isomorphism).

There remains the question: does a finite field of every possible size p^n exist? To that end a small observation will be needed.

If R is any commutative ring of prime characteristic p and $a, b \in R$, then

$$(a + b)^p = a^p + b^p.$$[7]

Indeed, the binomial theorem gives

$$(a + b)^p = a^p + pa^{p-1}b + \cdots + \binom{p}{j} a^{p-j}b^j + \cdots + pab^{p-1} + b^p.$$

As j runs from 1 to $p - 1$, the coefficients $\binom{p}{j}$, being divisible by p, cause all terms $\binom{p}{j} a^{p-j}b^j$ to be 0 inside the ring R.

Even more obvious is the identity given by $(ab)^p = a^p b^p$.

Thus the mapping $R \to R$ given by $x \mapsto x^p$ is a ring homomorphism.

Definition 6.56. If R is a ring of prime characteristic p, the homomorphism

$$\sigma : R \to R, \text{ defined by } x \mapsto x^p,$$

is called the ***Frobenius map*** on R.

This is named in honor of F. G. Frobenius (1849–1917).

Now we establish the existence of all possible finite fields.

[7] This identity is sarcastically known among some jaded professors as the "freshman's dream."

Proposition 6.57. *For every prime p and every positive exponent n, there exists a finite field of size p^n.*

Proof. Consider the polynomial $f = X^{p^n} - X$ in $\mathbb{Z}_p[X]$. Its derivative is

$$f' = p^n X^{p^n-1} - 1 = -1,$$

which is clearly coprime with f. By Proposition 6.53, f is separable. Let K be a field over which f splits. Proposition 6.52 shows that f has precisely p^n roots in K.

The set Δ of roots of f turns out to be a subfield of K. Indeed, if $\sigma : K \to K$ is the Frobenius map, observe that α is a root of f if and only if $\sigma^n(\alpha) = \alpha$. That is, if and only if α is fixed by the field homomorphism σ^n. A simple exercise verifies that the fixed points of a field homomorphism on K form a subfield of K. In our case that subfield is Δ.

Since Δ has precisely p^n elements, we have our field of size p^n. The entire field is made up of roots of f. □

6.3.4 Factoring $X^{p^n} - X$ in $\mathbb{Z}_p[X]$

This is a good time to digress into a few remarks about the irreducible polynomials over $\mathbb{Z}_p$. To start with, Proposition 6.57 yields something which is not so obvious.

Proposition 6.58. *For any prime p and any positive integer n, there exists an irreducible polynomial f in $\mathbb{Z}_p[X]$ of degree n.*

Every irreducible polynomial in $\mathbb{Z}_p[X]$ of degree n divides $X^{p^n} - X$.

Proof. Using Proposition 6.57 take a finite field K of size p^n. The degree of K over $\mathbb{Z}_p$ is n. By Proposition 6.27 (which came from the fact the group of units in a finite field is cyclic) $K = \mathbb{Z}_p[\alpha]$ for some α in K. The degree of the minimal polynomial of this α in $\mathbb{Z}_p[X]$ equals the degree n of the extension K over $\mathbb{Z}_p$.

For the second claim, suppose g in $\mathbb{Z}_p$ is irreducible of degree n. We can take g to be monic. Let α be a root of g in some field extension of $\mathbb{Z}_p$. The degree of the field $\mathbb{Z}_p[\alpha]$ over $\mathbb{Z}_p$ is n. By Proposition 6.27 α is a root of $X^{p^n} - X$. Then the minimal polynomial g of α divides $X^{p^n} - X$. □

The preceding argument does not offer a method for finding irreducible polynomials over $\mathbb{Z}_p$ of degree n, but merely a reassurance that they are there to be found.

Prior to the next result a small observation is needed. Suppose j, n are positive integers and that $j \mid n$. Write $n = jk$ for some k. Then

$$p^n - 1 = (p^j)^k - 1 = (p^j - 1)\ell$$

for some positive integer ℓ. And then

$$X^{p^n} - X = (X^{p^n-1} - 1)X = ((X^{p^j-1})^\ell - 1)X = (X^{p^j-1} - 1)gX$$

for some polynomial g in $\mathbb{Z}_p[X]$. So, if $j \mid n$, then $X^{p^j} - X$ divides $X^{p^n} - X$ in $\mathbb{Z}_p[X]$.

Proposition 6.59. *Let p be a prime and n a positive integer. An irreducible polynomial f of degree j in $\mathbb{Z}_p[X]$ divides $X^{p^n} - X$ if and only if j divides n.*

Consequently $X^{p^n} - X$ is the product of all monic irreducible polynomials in $\mathbb{Z}_p[X]$ whose degrees divide n.

Each such irreducible appears only once in the factorization.

Proof. There is no harm in taking f to be monic.

If $j \mid n$, then $X^{p^j} - X$ divides $X^{p^n} - X$ as noted already. Proposition 6.58 states that f divides $X^{p^j} - X$. So, f divides $X^{p^n} - X$.

Conversely, suppose f divides $X^{p^n} - X$. Thus f has a root α in the splitting field K of $X^{p^n} - X$. The degree of the subfield $\mathbb{Z}_p[\alpha]$ over $\mathbb{Z}_p$ is j since f, being irreducible, is the minimal polynomial of α. Hence $j \mid n$, due to the tower theorem, Proposition 6.13.

The irreducible factors of $X^{p^n} - X$ appear only once in its unique factorization in $\mathbb{Z}_p[X]$, because this polynomial is separable. Its derivative is -1, which is coprime with $X^{p^n} - X$. □

6.3.5 Perfect Fields

Proposition 6.55 said that over any field F of characteristic zero all irreducible polynomials are separable. Over some fields of prime characteristic there are irreducible polynomials that are not separable, for instance over the field $\mathbb{Z}_2(t)$, as we saw. It might be interesting to explore which fields of prime characteristic are such that all irreducible polynomials over them are separable. A straightforward observation will be helpful.

Proposition 6.60. *Let F be a field of prime characteristic p. A polynomial f has zero derivative if and only if $f(X) = g(X^p)$ for some polynomial g in $F[X]$.*

Proof. The condition $f(X) = g(X^p)$ is saying that f is of the form $\sum_{j=0}^{k} a_j X^{pj}$. In other words, the only exponents of X that can appear in the expansion of f are those divisible by p.

If this condition is met, then we routinely see that

$$f' = \sum_{j=0}^{k} pja_j X^{pj-1} = 0,$$

because the coefficients pja_j are 0 in characteristic p.

For the converse, write $f = \sum_{j=0}^{n} a_i X^i$, and then $f' = \sum_{j=0}^{n} ia_i X^{i-1}$. If $f' = 0$, this means that all coefficients $ia_i = 0$ in F. If $p \nmid i$, then i becomes a unit in F and can thereby be cancelled to deduce $a_i = 0$. This means that only exponents i divisible by p can appear in the expansion of f. □

This brings us to an interesting link with the Frobenius map.

Proposition 6.61. *In a field F of characteristic p, every irreducible polynomial is separable if and only if the Frobenius map $\sigma : F \to F$ given by $a \mapsto a^p$ is surjective.*

Proof. If σ is not surjective, we need an irreducible polynomial in $F[X]$ that is not separable. The assumption means that there is some b in $F \setminus \sigma(F)$. Hence the polynomial $f = X^p - b$ has no root in F. Let θ be a root of f inside its splitting field K. Being in characteristic p, we get

$$f = X^p - b = X^p - \theta^p = (X - \theta)^p.$$

Thus f is not separable.

It remains to check that f is irreducible. For that it suffices for the minimal polynomial g of θ over F to have degree p. For then g, being a factor of f, must be f itself. Since $g \mid f$, we have $g = (X - \theta)^k$ where $k \leq p$. Also $2 \leq k$, because $\theta \notin F$. The expansion of g looks like

$$g = X^k - k\theta X^{k-1} + \cdots .$$

Now $g \in F[X]$, and thus $k\theta \in F$. If $k < p$, cancel k as an element of K to falsely deduce that $\theta \in F$. Hence $k = p$, and $g = f$.

For the converse, suppose that the Frobenius map is surjective. Looking for a contradiction suppose f in $F[X]$ is an irreducible polynomial with a repeated root in some extension of F. According to Proposition 6.54, $f' = 0$. And then by Proposition 6.60, $f(X) = g(X^p)$ for some g in $F[X]$. Thus f takes the form

$$f = \sum_{j=0}^{k} a_j X^{pj} \text{ where the } a_j \in F.$$

The surjectivity of σ reveals that each $a_j = b_j^p$ for some b_j in F. Consequently,

$$f = \sum_{j=0}^{k} b_j^p X^{pj} = \left(\sum_{j=0}^{k} b_j X^j \right)^p .$$

The second equation holds because the Frobenius map applied to the ring $F[X]$ is a homomorphism. We see the contradiction that f is reducible. □

For any finite field, the appropriate Frobenius map is surjective. So a useful corollary follows, which puts finite fields in the same league as fields of characteristic 0.

Proposition 6.62. *Over a finite field, every irreducible polynomial is separable.*

Definition 6.63. Fields in which all irreducible polynomials are separable are called ***perfect fields***.[8]

As we saw, all finite fields as well as all fields of characteristic 0 are perfect fields. There do exist infinite, perfect fields of prime characteristic.

Over perfect fields, the separable polynomials are identifiable from their unique factorization.

[8] Whether this property represents perfection may well be a matter of opinion.

Proposition 6.64. *A non-constant polynomial over a perfect field is separable if and only if its irreducible factors do not repeat.*

Proof. Say the perfect field is F, and that f in $F[X]$ is such that $f = g_1 g_2 \cdots g_n$ where the g_j are distinct irreducible polynomials in $F[X]$. If f is not separable, there is a field extension K of F containing a root θ of f such that $(X - \theta)^2$ divides f in $K[X]$. Thus either $X - \theta$ must appear twice in the unique factorization of one g_j in $K[X]$, or $X - \theta$ appears in the unique factorization of two different g_j. The second possibility does not occur because the distinct irreducible g_j are coprime and so they cannot share a root in any field extension. The first possibility does not occur either, because each irreducible g_j is separable over the given perfect field. Thus f is separable.

Conversely suppose $f \in F[X]$ and that some irreducible g in $F[X]$ appears more than once in the unique factorization of f. Every root θ of g is in the splitting field K of f. Since g^2 divides f and $X - \theta$ divides g in $K[X]$, it must be that $(X - \theta)^2$ divides f in $K[X]$. This means that f is not separable. □

The preceding result applies in particular to all finite fields and to all fields of characteristic zero.

EXERCISES

1. If f, g are coprime polynomials over a field F, show that f, g never share a common root in any field extension of F.
2. Verify the basic properties of the derivative operator on polynomials over a field F. Namely, show that

$$(f + g)' = f' + g', \quad (fg)' = fg' + f'g,$$

   for all f, g in $F[X]$.
3. Let $\mathbb{Z}_2(t)$ be the field of rational functions in the variable t with coefficients in $\mathbb{Z}_2$. If α is a root of the polynomial $X^2 - t$, find all $\mathbb{Z}_2(t)$-maps from $\mathbb{Z}_2(t)[\alpha]$ to $\mathbb{Z}_2(t)[\alpha]$.
4. (a) If p is prime and $f(X) \in \mathbb{Z}_p[X]$, show that $(f(X))^p = f(X^p)$.

   (b) If n, r are integers with $0 \leq r \leq n$, prove that the binomial coefficients $\binom{pn}{pr}, \binom{n}{r}$ are congruent modulo p.

   Hint. Compare binomial coefficients of $(1 + X^p)^n$ and $(1 + X)^{pn}$ as polynomials in $\mathbb{Z}_p[X]$.
5. If F is a finite field of characteristic p and $a \in F$, show that the polynomial $X^p - a$ is reducible.
6. Suppose F is a field of characteristic 0 and that f is the product of distinct irreducible polynomials in $F[X]$. Show that f is coprime with f'.
7. If F is a field of characteristic p and p does not divide n, show that the polynomial $X^n - 1$ has n distinct roots in its splitting field.

8. Show that an algebraically closed field of prime characteristic is perfect.

Note. An ***algebraically closed field*** F is such that every non-constant polynomial in $F[X]$ splits. We accept for now that algebraically closed fields of prime characteristic do exist.

9. Show that there is no finite, algebraically closed field.

10. For every commutative ring R of prime characteristic p, recall that the mapping

$$\sigma : R \to R, \text{ given by } a \mapsto a^p,$$

is known as the ***Frobenius map*** on R.

(a) Verify that the Frobenius map is a ring homomorphism.

(b) Since the characteristic of R is p, the field $\mathbb{Z}_p$ is a subring of R. Verify that the Frobenius map fixes $\mathbb{Z}_p$.

(c) If R is a finite field of size p^n for some $n > 1$, show that the Frobenius map is an automorphism which fixes only the subfield $\mathbb{Z}_p$.

(d) Find an infinite field of prime characteristic p for which the Frobenius map is not an automorphism.

11. Let $\sigma : F \to F$ given by $\varphi : a \mapsto a^p$ be the Frobenius map on a finite field F of size p^n for some prime p and some positive exponent n.

Prove that for $k = 1, 2, \ldots, n$ the automorphisms σ^k are distinct, and that this list picks up all automorphisms of F.

Hint. There is a single element α in F such that $F = \mathbb{Z}_p[\alpha]$. The elements $\sigma^k(\alpha)$ have the same minimal polynomial.

12. If p is a prime and f is an irreducible polynomial of degree n over the field $\mathbb{Z}_p$, and α is a root of f in some extension of $\mathbb{Z}_p$, show that f splits over $\mathbb{Z}_p[\alpha]$.

13. Show that for every positive integer n there exists an irreducible polynomial in $\mathbb{Q}[X]$ of degree n and such that its coefficients consist of only 0 or 1.

14. The polynomial $X^4 + 1$ is irreducible in $\mathbb{Q}[X]$ but reducible in $\mathbb{Z}_p[X]$ for all primes p. Reducibility over $\mathbb{Z}_2$ is obvious because $X^4 + 1 = (X + 1)^4$ in $\mathbb{Z}_2[X]$. Complete these steps to show that $X^4 + 1$ is reducible in $\mathbb{Z}_p[X]$ when p is an odd prime.

- Explain why 8 divides $p^2 - 1$, and then deduce that $X^8 - 1$ divides $X^{p^2-1} - 1$ in $\mathbb{Z}_p[X]$ (in fact in $\mathbb{Z}[X]$).
- Explain why there exists a degree 2 extension K of $\mathbb{Z}_p$.
- Note that $X^4 + 1$ divides $X^8 - 1$ and thereby divides $X^{p^2-1} - 1$ to conclude that $X^4 + 1$ splits over K.
- Deduce that $X^4 + 1$ is reducible over $\mathbb{Z}_p$.

15. If K is a finite extension of a field F, an element α in K is called a ***primitive generator*** of K over F provided $K = F[\alpha]$. A nice result, which we will cover eventually, is that finite extensions of characteristic 0 have a primitive generator. For now, we offer an exercise to show this is not the case when the characteristic is not 0.

Let s, t be indeterminates, and form the field $F = \mathbb{Z}_2(s,t)$ of all rational functions in the two variables s, t.

(a) Explain why the polynomial $X^2 - s$ is irreducible in $F[X]$.

(b) Let α be a root of $X^2 - s$ in some field extension and let $K = F[\alpha]$. Explain why $X^2 - t$ is irreducible in $K[X]$.

(c) Let β be a root of $X^2 - t$ in some extension of K, and form the field $L = K[\beta]$. What is $[L : F]$?

(d) Prove there is no γ in L such that $L = F[\gamma]$.

Hint. Write such a γ in terms of α and β and do not forget we are in characteristic 2.

16. An ***intermediate field*** between a field F and an extension K is a subfield E of K such that $F \subseteq E \subseteq K$.

(a) Suppose α is an algebraic element over a field F. Let E be an intermediate field between F and $F[\alpha]$. Clearly α is algebraic over E too, and the minimal polynomial of α over E divides the minimal polynomial of α over F inside $E[X]$.

Suppose $g = a_0 + a_1X + \cdots + a_nX^n$ is the minimal polynomial of α over E, so $g \in E[X]$. Prove that $E = F[a_0, a_1, \ldots, a_n]$. Thus the minimal polynomial of α over E determines E.

Hint. You have the tower of fields

$$F \subseteq F[a_0, a_1, \ldots, a_n] \subseteq E \subseteq F[\alpha].$$

What is the minimal polynomial of α over $F[a_0, a_1, \ldots, a_n]$? Then use the tower theorem.

(b) Show that there can only be finitely many intermediate fields between F and $F[\alpha]$.

17. Let p be a prime, n a positive integer, c a non-zero element of the finite field $\mathbb{Z}_p$, and K the splitting field of $X^n - c$ over $\mathbb{Z}_p$.

(a) If $p \nmid n$, prove that $X^n - c$ has n distinct roots in K.

(b) If $p \mid n$, find a formula, in terms of p and n, for the number of distinct roots of $X^n - c$.

18. If p is a prime and a is a non-zero element of $\mathbb{Z}_p$, show that the polynomial $f = X^p - X + a$ is irreducible in $\mathbb{Z}_p[X]$.

Hint. If α is a root of f in some extension of $\mathbb{Z}_p$, show that the other roots are $\alpha + j$ where j runs through the non-zero elements of $\mathbb{Z}_p$. Deduce that the irreducible factors of f all have the same degree.

19. Let F be a field, and let $f = x^4 + x + 1$ in $F[X]$. Prove that f is separable if and only if the characteristic of F is not 229.

20.* In this exercise we exploit some field theory to do some group theory. Recall Proposition 3.32 stating that if p^k is a prime power dividing the order of a finite group G, then G contains a subgroup of order p^k. This can be viewed as a partial converse of Lagrange's theorem. Let us now discover that this partial converse of Lagrange cannot be pushed any further. Namely, if a positive integer m is not a prime power, then there exists a finite group G such that m divides the order of G while G contains no subgroup of order m.

Two facts about finite fields are needed. The first is Proposition 6.57 stating that for every prime p and every positive exponent r there exists a unique field F containing exactly p^r elements. The second is Proposition 4.37 which reveals that the finite group $F^\star$ of non-zero elements under multiplication is cyclic.

First consider the case where m is divisible by precisely two primes.

(a) Suppose that p, q are distinct primes and that $m = p^k q^\ell$ where k, ℓ are positive exponents. There is no harm in supposing that $p^k < q^\ell$.

Explain why there exists an integer $n > 1$, such that $p^{kn} \equiv 1 \bmod q^\ell$.

(b) Let F be a finite field of order p^{kn} and let $F^\star$ be the group of non-zero elements of F under multiplication. The order of $F^\star$ is $p^{kn} - 1$. Also, under addition, F is an abelian group of order p^{kn}.

Explain why $F^\star$ contains a unique subgroup of order q^ℓ and why that subgroup is cyclic.

(c) Let A be the unique cyclic subgroup of order q^ℓ inside $F^\star$ and consider F as an abelian group of order p^{kn} using addition.

Let $\varphi : A \to \mathrm{Aut}(F)$ be the automorphism action given by $\varphi : s \mapsto \varphi_s = (x \mapsto sx)$. In other words, for every s in A the mapping φ_s multiplies every element of F by s. Verify that φ is an automorphism action as discussed in Section 3.6.1.

(d) Let G be the semi-direct product $F \rtimes_\varphi A$, where φ is the chosen automorphism action. Thus if $(x, s), (y, t) \in F \rtimes_\varphi A$, their product is given by

$$(x, s) \star (y, t) = (x + sy, st).$$

Explain why m divides the order of G, and then prove that G has no subgroup of order m. (The last part is the essence of this exercise.)

(e) In part (d) we showed that if $m = p^k q^\ell$, then there is a group G whose order is divisible by m and yet contains no subgroup of order m. To prove the same result for any m that has more than one prime factor, here is one way to approach it.

Suppose m is divisible by at least two primes p, q. Write $m = p^k q^\ell d$ where $k, \ell \geq 1$ and $p, q \nmid d$. Let G be a group with order divisible by $p^k q^\ell$ but having no subgroup of order $p^k q^\ell$. For any group H of order d, the order of the product group $G \times H$ is certainly m.

Show that $G \times H$ has no subgroup of order m.

Hint. See Proposition 2.68.

6.4 The Galois Group

If K, L are field extensions of a common field F, recall that a homomorphism $\varphi : K \to L$ which fixes F is known as an F-map. Every F-map is also an F-linear transformation between the vector spaces K and L. This reveals that if K is a finite extension of F, then every F-map $\varphi : K \to K$, being already injective, must be an F-automorphism of K. A routine verification confirms that the set of all F-automorphisms of K forms a group under the composition of mappings.

Definition 6.65. If K is a field extension of a field F, the group of all F-maps from K to K is known as the ***Galois group of the extension*** K over F, and will be denoted by

$$\mathrm{Gal}(K/F).$$

The Galois group is also known as the ***F-automorphism group*** of K.

The connections which Galois made between the splitting field of a polynomial and the group named in his honor resolved the classical problem of finding a formula for the roots of a polynomial, and set the stage for a new way of doing algebra.

6.4.1 Roots, Generators and Galois Groups

We make the small observation that F-maps, and thereby Galois groups, are determined by their effect on the generators of a field.

Proposition 6.66. *Let K, L be field extensions of F and let $\varphi, \psi : K \to L$ be F-maps. If $K = F[\alpha_1, \ldots, \alpha_n]$ for some α_j in K that are algebraic over F and φ, ψ agree on the finitely many α_j, then $\varphi = \psi$ as mappings on K.*

Proof. Every α in K is a polynomial in the α_j of the form

$$\alpha = \sum a_{i_1,\ldots,i_n} \alpha_1^{i_1} \cdots \alpha_n^{i_n},$$

where the i_j run through non-negative integers and the $a_{i_1,\ldots,i_n}$ are arbitrarily taken from F. Then

$$\begin{aligned}
\varphi(\alpha) &= \sum \varphi(a_{i_1,\ldots,i_n})\varphi(\alpha_1)^{i_1} \cdots \varphi(\alpha_n)^{i_n} && \text{since } \varphi \text{ is a homomorphism} \\
&= \sum a_{i_1,\ldots,i_n}\varphi(\alpha_1)^{i_1} \cdots \varphi(\alpha_n)^{i_n} && \text{since } \varphi \text{ fixes } F \\
&= \sum a_{i_1,\ldots,i_n}\psi(\alpha_1)^{i_1} \cdots \psi(\alpha_n)^{i_n} && \text{since } \varphi = \psi \text{ on the generators } \alpha_j \\
&= \sum \psi(a_{i_1,\ldots,i_n})\psi(\alpha_1)^{i_1} \cdots \psi(\alpha_n)^{i_n} && \text{since } \psi \text{ fixes } F \\
&= \psi(\alpha) && \text{since } \psi \text{ is a homomorphism.}
\end{aligned}$$

Thus $\varphi = \psi$. □

In the important case where $K = L$ and φ, ψ in $\mathrm{Gal}(K/F)$ agree on a set of generators of K, then $\varphi = \psi$ in $\mathrm{Gal}(K/F)$.

Galois Groups and Their Action on Roots

In preceding sections we have established a number of features of $\mathrm{Gal}(K/F)$, which we now assemble here. Recall that a group G ***acts on a set*** Δ when there is a group homomorphism

$\Phi : G \to \mathcal{S}(\Delta)$, where $\mathcal{S}(\Delta)$ is the group of permutations of Δ. If Φ is injective, the action is called ***faithful***, and we can view G as a subgroup of $\mathcal{S}(\Delta)$. The action Φ is called ***transitive*** provided for every pair of elements a, b in Δ there is a σ in G such that $\Phi(\sigma)(a) = b$. These notions have a significant bearing on Galois groups.

Proposition 6.67. *Let K be a field extension of F and let Δ be the set of roots in K for a non-constant polynomial f in $F[X]$.*

1. *If $\varphi \in \mathrm{Gal}(K/F)$, then its restriction $\varphi_{|\Delta}$ is a permutation of Δ. Hence the restriction mapping*

$$\Phi : \mathrm{Gal}(K/F) \to \mathcal{S}(\Delta) \textit{ given by } \sigma \mapsto \varphi_{|\Delta}$$

   is an action of $\mathrm{Gal}(K/F)$ *on* Δ.
2. *If K is the splitting field of the polynomial f, then the action Φ is faithful. That is, Φ embeds* $\mathrm{Gal}(K/F)$ *as a subgroup of* $\mathcal{S}(\Delta)$.
3. *If K is a splitting field (but not necessarily that of f) and f is irreducible over F, then the action of* $\mathrm{Gal}(K/F)$ *on Δ is transitive.*

Proof. Item 1, that $\varphi_{|\Delta}$ permutes the roots of Δ, is in essence what Proposition 6.41 says, but it can be re-checked quickly. Let $f = a_nX^n + \cdots + a_1X + a_0$ where the $a_j \in F$. Assuming $\alpha \in \Delta$, i.e. $f(\alpha) = 0$, and keeping in mind that φ is an F-map we get

$$\begin{aligned} f(\varphi(\alpha)) &= a_n\varphi(\alpha)^n + \cdots + a_1\varphi(\alpha) + a_0 \\ &= \varphi(a_n\alpha^n + \cdots + a_1\alpha + a_0) = \varphi(f(\alpha)) = \varphi(0) = 0. \end{aligned}$$

Thus $\varphi(\alpha) \in \Delta$, meaning that the restriction $\varphi_{|\Delta}$ is a permutation of Δ. Then it is routine to check that Φ is a group homomorphism, and thereby an action.

In item 2 the action is faithful because the splitting field K is generated by the roots of f. Thus Proposition 6.66, with $L = K$, applies.

Item 3, wherein f is irreducible and K is a splitting field, is nothing but Proposition 6.36, which itself is a significant consequence of the isomorphism lifting theorem. □

Just to repeat the main points of the preceding result:

the Galois group of the splitting field of a polynomial f is also a subgroup of the group of permutations of the roots of f. Furthermore, if f is irreducible, that subgroup acts transitively on the roots of f.

It is needlessly cumbersome to write $\Phi(\varphi)$ or even $\varphi_{|\Delta}$ as in Proposition 6.67. There is no loss of clarity in denoting the restriction of φ to a set of roots Δ as simply φ, which is what we will do.

Because the Galois group $\mathrm{Gal}(K/F)$ is fully determined by what its F-maps do to the generators of K, it is also good practice to describe the elements of the Galois group by their effect on a chosen set of generators. A good choice of generators can make this task more congenial.

We should also mention a related piece of terminology in common use.

Definition 6.68. For any polynomial f over a field F, the ***Galois group of f over F*** is taken to be the Galois group $\mathrm{Gal}(K/F)$ of the splitting field K of f.

6.4.2 Some Examples of Galois Groups

It is no small challenge to compute the Galois group of a polynomial, but we can at least begin with a few manageable examples.

The Galois Group of a Quadratic Extension

If the characteristic of F is not 2, Proposition 6.46 says that every quadratic extension K of F takes the form $K = F[\alpha]$, where the minimal polynomial of α takes the form $X^2 - d$ for some d in F. The other F-conjugate of α is $-\alpha$. By Proposition 6.67, $\mathrm{Gal}(K/F)$ must act as a transitive subgroup of the group of permutations of the two-element set $\{\pm\alpha\}$. Such a group is nothing but the two-element group determined by the pair of maps $\alpha \mapsto \alpha$ and $\alpha \mapsto -\alpha$.

If the characteristic of F is 2, the story can change. For example, take $F = \mathbb{Z}_2(t)$, the field of rational functions in the variable t with coefficients in $\mathbb{Z}_2$. The quadratic polynomial $X^2 - t$ is irreducible, as noted at the outset of Section 6.3. Let α be a root of f in some extension. Then the extension $F[\alpha]$ has degree 2 over F, and is the splitting field of f. The latter is true because in characteristic 2 we have the splitting:

$$f = X^2 - t = X^2 - \alpha^2 = (X - \alpha)^2.$$

Here the irreducible f is not separable, with repeated root α. Every F-map $\varphi : F[\alpha] \to F[\alpha]$ sends α to a root of f, but α is the only option. Since φ fixes α it must fix all of $F[\alpha]$. Here the Galois group of f is trivial despite the fact $F[\alpha]$ is the splitting field of an irreducible quadratic.

The Galois Group of $\mathbb{Q}[\sqrt[3]{2}\,]$ over $\mathbb{Q}$

Every $\mathbb{Q}$-map $\sigma : \mathbb{Q}[\sqrt[3]{2}] \to \mathbb{Q}[\sqrt[3]{2}]$ is determined by its value at $\sqrt[3]{2}$, and $\sigma\left(\sqrt[3]{2}\right)$ must be a conjugate of $\sqrt[3]{2}$, i.e. a root of the minimal polynomial of $\sqrt[3]{2}$. The other conjugates of $\sqrt[3]{2}$ are $\sqrt[3]{2}\omega$ and $\sqrt[3]{2}\omega^2$ where $\omega = e^{2\pi i/3}$. These other conjugates are not in $\mathbb{Q}[\sqrt[3]{2}]$, because they are complex numbers outside of $\mathbb{R}$. Thus $\sigma(\sqrt[3]{2})$ must equal $\sqrt[3]{2}$. Consequently σ is the identity on $\mathbb{Q}[\sqrt[3]{2}]$. This Galois group is trivial, because $\sqrt[3]{2}$ has no other conjugates in this field that it could go to.

The Galois Group of $(X^2 - 2)(X^2 - 3)$ over $\mathbb{Q}$

If $f = (X^2 - 2)(X^2 - 3)$, its splitting field K over $\mathbb{Q}$ is $\mathbb{Q}[\sqrt{2}, \sqrt{3}]$. The fact that $\sqrt{3} \notin \mathbb{Q}[\sqrt{2}]$ (an easy exercise), makes it clear that the minimal polynomial of $\sqrt{3}$ over $\mathbb{Q}[\sqrt{2}]$ remains

$X^2 - 3$. Apply the tower theorem to the tower $\mathbb{Q} \subset \mathbb{Q}[\sqrt{2}] \subset \mathbb{Q}[\sqrt{2}][\sqrt{3}] = K$, along with Proposition 6.9, to see that $[K : \mathbb{Q}] = 4$.

Each automorphism of K fixes $\mathbb{Q}$ and maps each of the generators $\sqrt{2}, \sqrt{3}$ to one of its conjugates. Thus every automorphism in $\mathrm{Gal}(K/\mathbb{Q})$ restricts to one of the following maps on the generators $\sqrt{2}, \sqrt{3}$ of K:

$$\begin{array}{cccc} \sqrt{3} \mapsto \sqrt{3} & \sqrt{3} \mapsto -\sqrt{3} & \sqrt{3} \mapsto \sqrt{3} & \sqrt{3} \mapsto -\sqrt{3} \\ \sqrt{2} \mapsto \sqrt{2} & \sqrt{2} \mapsto \sqrt{2} & \sqrt{2} \mapsto -\sqrt{2} & \sqrt{2} \mapsto -\sqrt{2}. \end{array}$$

By Proposition 6.67 each of the actions above is the restriction of at most one automorphism of K. The first action is the restriction of the identity automorphism to $\sqrt{3}, \sqrt{2}$. But it is not immediately clear that there exist automorphisms of K whose restrictions to $\sqrt{3}, \sqrt{2}$ give the other three maps.

Let us verify that there is an automorphism of K which executes the fourth map $\sqrt{2} \mapsto -\sqrt{2}$, $\sqrt{3} \mapsto -\sqrt{3}$. By Proposition 6.11 there is an automorphism φ of $\mathbb{Q}[\sqrt{2}]$ such that $\varphi : \sqrt{2} \mapsto -\sqrt{2}$. To get the desired automorphism of K use Proposition 6.31, with the variables therein interpreted as follows:

$$F = E = \mathbb{Q}[\sqrt{2}], \quad \varphi \text{ as above}, \quad \alpha = \sqrt{3}, \quad g = X^2 - 3, \quad \varphi(g) = X^2 - 3, \quad \beta = -\sqrt{3}.$$

Proposition 6.31 yields an automorphism ψ of $\mathbb{Q}[\sqrt{2}][\sqrt{3}]$ that maps $\sqrt{3}$ to $-\sqrt{3}$ and extends φ as above. This automorphism carries out the fourth action.

In a similar way we can show that the second and third actions on the generators of K are restrictions of automorphisms. So $\mathrm{Gal}(K/\mathbb{Q})$ is a group of order 4. As a permutation group on the set of roots $\pm\sqrt{2}$, $\pm\sqrt{3}$ it consists of these maps:

$$\begin{array}{l} \text{the identity} \\ \sqrt{3} \mapsto -\sqrt{3} \mapsto \sqrt{3}, \sqrt{2} \mapsto \sqrt{2}, -\sqrt{2} \mapsto -\sqrt{2} \\ \sqrt{3} \mapsto \sqrt{3}, -\sqrt{3} \mapsto -\sqrt{3}, \sqrt{2} \mapsto -\sqrt{2} \mapsto \sqrt{2} \\ \sqrt{3} \mapsto -\sqrt{3} \mapsto \sqrt{3}, \sqrt{2} \mapsto -\sqrt{2} \mapsto \sqrt{2}. \end{array}$$

By thinking of $\sqrt{3}, -\sqrt{3}$ as the letters 1, 2, and $-\sqrt{2}, -\sqrt{2}$ as the letters 3, 4 this group of permutations can be seen as the group $\{1, (1,2), (3,4), (1,2)(3,4)\}$, which is isomorphic to the Klein 4-group.

It might be informative to see how $\mathrm{Gal}(K/F)$ can be computed when *different* generators of K are chosen. Let $\alpha = \sqrt{2}+\sqrt{3}$. Clearly $\alpha \in K = \mathbb{Q}[\sqrt{2}, \sqrt{3}]$. On the other hand, the equation $(\alpha - \sqrt{2})^2 = 3$ leads to the observation that $\sqrt{2} = \frac{\alpha^2-1}{2\alpha} \in \mathbb{Q}[\alpha]$. Then $\sqrt{3} = \alpha - \sqrt{2} \in \mathbb{Q}[\alpha]$. From this it follows that $K = \mathbb{Q}[\alpha]$. We now have a single generator α for K.

Since the degree of K over $\mathbb{Q}$ is 4, the minimal polynomial of α has degree 4. We can find that polynomial with ease. From $(\alpha - \sqrt{2})^2 = 3$ we come to $\alpha^2 - 1 = 2\sqrt{2}\alpha$, and then $(\alpha^2 - 1)^2 = 8\alpha^2$, which gives $\alpha^4 - 6\alpha^2 + 1 = 0$. The minimal polynomial of α is $g = X^4 - 6X^2 + 1$. The other three roots of g are

$$\beta = \sqrt{2} - \sqrt{3}, \ -\beta = -\sqrt{2} + \sqrt{3}, \text{ and } -\alpha = -\sqrt{2} - \sqrt{3}.$$

It can be checked directly that each of these four roots lies in K (and in fact generates K), but this also follows from the normality theorem, Proposition 6.42, since we know K is a splitting field.

According to Proposition 6.11, for each of these four roots θ there exists an automorphism $\varphi : K = \mathbb{Q}[\alpha] \to \mathbb{Q}[\theta] = K$ such that $\varphi(\alpha) = \theta$. Thus we pick up four automorphisms of K. That is all that there are, because every automorphism is determined by its action on the generator α and the generator must go to one of its conjugates.

These four automorphisms restrict to permutations of the roots $\pm\alpha, \pm\beta$ of g. To conclude this example, let us examine how the Galois group permutes these roots.

Of course, the identity automorphism becomes the identity permutation.

What about the automorphism φ determined by $\varphi(\alpha) = \beta$? We have seen that $\sqrt{2} = \frac{\alpha^2-1}{2\alpha}$. Since φ is an automorphism

$$\varphi(\sqrt{2}) = \frac{\beta^2 - 1}{2\beta} = \frac{(\sqrt{2} - \sqrt{3})^2 - 1}{2(\sqrt{2} - \sqrt{3})} = \sqrt{2}.$$

The last equation is based on a routine simplification of its left side. (In fact, since $\varphi(\sqrt{2})$ must be a conjugate of $\sqrt{2}$, we could have predicted that the final calculation of $\varphi(\sqrt{2})$ would yield one of $\pm\sqrt{2}$.) Then

$$\varphi(\sqrt{3}) = \varphi(\alpha - \sqrt{2}) = \varphi(\alpha) - \varphi(\sqrt{2}) = \beta - \sqrt{2} = -\sqrt{3}.$$

So,

$$\varphi(\beta) = \varphi(\sqrt{2} - \sqrt{3}) = \sqrt{2} + \sqrt{3} = \alpha.$$

Our φ transposes α with β. Next,

$$\varphi(-\alpha) = -\varphi(\alpha) = -\beta, \text{ and } \varphi(-\beta) = -\varphi(\beta) = -\alpha.$$

The automorphism of K which sends α to β permutes the roots as follows:

$$\alpha \mapsto \beta \mapsto \alpha, \quad -\alpha \mapsto -\beta \mapsto -\alpha.$$

Working in a similar spirit we can check that the automorphism which sends α to $-\beta$ permutes the roots of g as follows:

$$\alpha \mapsto -\beta \mapsto \alpha, \quad -\alpha \mapsto \beta \mapsto -\alpha.$$

Finally, the automorphism which sends α to $-\alpha$ permutes the roots of g as follows:

$$\alpha \mapsto -\alpha \mapsto \alpha, \quad \beta \mapsto -\beta \mapsto \beta.$$

If we identify the roots $\alpha, \beta, -\beta, -\alpha$ with the letters $1, 2, 3, 4$, respectively, the four automorphisms of K correspond to the permutations

$$1 \text{ (the identity)}, (1,2)(3,4), (1,3)(2,4), (1,4)(2,3),$$

which is just another way to represent the Klein 4-group.

The Galois Group of $X^3 - 2$ over $\mathbb{Q}$

As noted after Proposition 6.36, the splitting field of $f = X^3 - 2$ over $\mathbb{Q}$ is $K = \mathbb{Q}[\alpha, \omega]$, where $\alpha = \sqrt[3]{2}$ and $\omega = e^{2\pi i/3}$. The set of roots of f is $\Delta = \{\alpha, \alpha\omega, \alpha\omega^2\}$ inside K. According to Proposition 6.67, $\mathrm{Gal}(K/\mathbb{Q})$ can be viewed as a subgroup of the six element group $\mathcal{S}(\Delta)$ of permutations of the three roots. Since f is irreducible, the Galois group must act transitively on Δ. In conjunction with Lagrange's theorem this already confirms that the order of $\mathrm{Gal}(K/\mathbb{Q})$ is either 3 or 6. In fact, $\mathrm{Gal}(K/\mathbb{Q})$ is all $\mathcal{S}(\Delta)$ of order 6, as we can show now.

The precursor to the isomorphism lifting theorem, Proposition 6.31, will be used twice. Since α and $\alpha\omega$ are conjugates over $\mathbb{Q}$ with minimal polynomial $X^3 - 2$, Proposition 6.31 gives a $\mathbb{Q}$-isomorphism

$$\varphi : \mathbb{Q}[\alpha] \to \mathbb{Q}[\alpha\omega] \text{ such that } \varphi(\alpha) = \alpha\omega.$$

The minimal polynomial of ω in $\mathbb{Q}[\alpha][X]$ remains $X^2 + X + 1$, while the minimal polynomial of ω^2 in $\mathbb{Q}[\alpha\omega][X]$ also remains $X^2 + X + 1$. The φ-transfer of $X^2 + X + 1$ happens to be itself. By Proposition 6.31 there is an isomorphism $\psi : \mathbb{Q}[\alpha][\omega] \to \mathbb{Q}[\alpha\omega][\omega^2]$ such that

$$\psi \text{ extends } \varphi \text{ and } \psi(\omega) = \omega^2.$$

But $\mathbb{Q}[\alpha][\omega] = K = \mathbb{Q}[\alpha\omega][\omega^2]$. Thus we pick up the automorphism ψ of K determined by

$$\psi(\alpha) = \alpha\omega \text{ and } \psi(\omega) = \omega^2.$$

By replacing ω^2 by ω in the preceding discussion we also get the automorphism τ of K such that

$$\tau(\alpha) = \alpha\omega \text{ and } \tau(\omega) = \omega.$$

The action of ψ on the roots Δ is given by

$$\begin{aligned} \psi &: \alpha \mapsto \alpha\omega \mapsto (\alpha\omega)(\omega^2) = \alpha\omega^3 = \alpha \\ \psi &: \alpha\omega^2 \mapsto (\alpha\omega)(\omega^2)^2 = \alpha\omega^5 = \alpha\omega^2, \end{aligned}$$

which is a 2-cycle. The action of τ on Δ becomes

$$\tau : \alpha \mapsto \alpha\omega \mapsto (\alpha\omega)\omega = \alpha\omega^2 \mapsto (\alpha\omega)(\omega)^2 = \alpha\omega^3 = \alpha,$$

which is a 3-cycle. Since $\mathrm{Gal}(K/\mathbb{Q})$ as a subgroup of $\mathcal{S}(\Delta)$ contains both a 3-cycle and a 2-cycle, it follows (by Lagrange) that $\mathrm{Gal}(K/\mathbb{Q})$ is all of $\mathcal{S}(\Delta)$. Every permutation of Δ arises from an automorphism of K.

The Galois Group of $X^p - 1$ over $\mathbb{Q}$

If n is a positive integer, the roots in $\mathbb{C}$ of $X^n - 1$ are $1, \zeta, \zeta^2, \ldots, \zeta^{n-1}$ where $\zeta = e^{2\pi i/n}$. Hence the splitting field of $X^n - 1$ over $\mathbb{Q}$ is $K = \mathbb{Q}[\zeta]$. When n is not a prime, the minimal polynomial of ζ will be discussed in the next chapter.

Here we consider $X^p - 1$ where p is prime. The minimal polynomial of ζ is the cyclotomic polynomial $f = X^{p-1} + X^{p-2} + \cdots + X + 1$ as demonstrated after Definition 5.41. The extension $\mathbb{Q}[\zeta]$ has degree $p-1$ over $\mathbb{Q}$, and is the splitting field of f.

The Galois group $\mathrm{Gal}(K/\mathbb{Q})$ acts faithfully and transitively on the set $\Delta = \{\zeta, \zeta^2, \ldots, \zeta^{p-1}\}$ of roots of the irreducible f. In particular, for every power ζ^j, where $j = 1, 2, \ldots, p-1$, there is an automorphism φ in $\mathrm{Gal}(K/\mathbb{Q})$ such that $\varphi(\zeta) = \zeta^j$. This already accounts for $p-1$ automorphisms. Since every automorphism is determined by its effect on ζ and must map ζ to one of its conjugates ζ^j, the elements of $\mathrm{Gal}(K/\mathbb{Q})$ are all accounted for.

Having seen that $\mathrm{Gal}(K/\mathbb{Q})$ has order $p-1$, it might be interesting to learn that this group is isomorphic to the group of units $\mathbb{Z}_p^\star$ of the finite field $\mathbb{Z}_p$. For any pair of integers i,j, Proposition 2.5 lets us see that $i \equiv j \bmod p$ if and only if $\zeta^i = \zeta^j$. Thus for every pair of (non-zero) residues $[i], [j]$ in $\mathbb{Z}_p^\star$, we see that $[i] = [j]$ if and only if $\zeta^i = \zeta^j$. If j is an integer such that $p \nmid j$, i.e. $[j] \in \mathbb{Z}_p^\star$, let σ_j be the automorphism of K that maps ζ to ζ^j. Since automorphisms are completely determined by their effect on ζ it follows that $\sigma_i = \sigma_j$ if and only if $[i] = [j]$ in $\mathbb{Z}_p^\star$. This establishes the (well-defined) bijection

$$\Phi : \mathbb{Z}_p^\star \to \mathrm{Gal}(K/\mathbb{Q}) \text{ where } [j] \mapsto \sigma_j.$$

Furthermore Φ is an isomorphism between these groups. Indeed, this comes down to showing that $\sigma_{ij} = \sigma_i\sigma_j$ for all integers i,j not divisible by p. In turn this equality only needs to be tested on the generator ζ of K. Well,

$$\sigma_{ij}(\zeta) = \zeta^{ij} = (\zeta^i)^j = (\sigma_i(\zeta))^j = \sigma_i(\zeta^j) = \sigma_i(\sigma_j(\zeta)).$$

Note that the fourth equality holds because σ_i is an automorphism of K. So, the Galois group $\mathrm{Gal}(K/\mathbb{Q})$ is isomorphic to the group $\mathbb{Z}_p^\star$. We also see from Proposition 4.37 that this group of units in a finite field is cyclic.

EXERCISES

1. If f in $F[X]$ is a polynomial with at least two distinct irreducible factors and Δ is its set of roots in its splitting field K, show that the action of $\mathrm{Gal}(K/F)$ on Δ is not transitive.
2. Let g, h be distinct irreducible polynomials over a field F and let K be the splitting field of their product gh. Let A, B be the sets of roots of g, h in K.

   Explain why A, B are disjoint.

   If $\mathcal{S}(A), \mathcal{S}(B)$ are the permutation groups on the sets A, B, respectively, find an injective group homomorphism $\mathrm{Gal}(K/F) \to \mathcal{S}(A) \times \mathcal{S}(B)$.

   Give an example with polynomials g, h of degree 2 where $\mathrm{Gal}(K/F) \cong \mathcal{S}(A) \times \mathcal{S}(B)$, and an example where $\mathrm{Gal}(K/F) \ncong \mathcal{S}(A) \times \mathcal{S}(B)$.
3. Find the Galois group of the quadratic polynomial $X^2 + 1$ over the field $\mathbb{Z}_3$.
4. Find the Galois group of the polynomial $X^2 + X + 1$ over $\mathbb{Z}_5$.
5. Let p, q be distinct primes, let $f = (X^2 - p)(X^2 - q)$ over $\mathbb{Q}$, and let K be the splitting field of f.

(a) Find the degree $[K : \mathbb{Q}]$.
(b) Compute the elements of $\text{Gal}(K/\mathbb{Q})$ as permutations of the roots $\pm\sqrt{p}, \pm\sqrt{q}$ of f.
(c) Show that $K = \mathbb{Q}[\sqrt{p} + \sqrt{q}]$.
(d) Find the minimal polynomial g of $\sqrt{p} + \sqrt{q}$.
(e) What are the roots of g and how does $\text{Gal}(K/\mathbb{Q})$ permute them?

6. Let $f = X^4 - 4X^2 + 2$ in $\mathbb{Q}[X]$.
(a) Show that f is irreducible in $\mathbb{Q}[X]$.
(b) Find the four roots of f.
(c) Show that the splitting field K of f can be generated by just one of the four roots of f.
(d) What is the degree $[K : \mathbb{Q}]$?
(e) Find all automorphisms in $\text{Gal}(K/\mathbb{Q})$.
(f) Show that $\text{Gal}(K/\mathbb{Q})$ is cyclic.

7. Let $f = X^4 - 8X^2 + 15$ in $\mathbb{Q}[X]$.
(a) Find the four roots of f.
(b) Show that the splitting field K of f can be generated by a single element.
Hint. Consider taking the sum of a pair of roots of f as the generator.
(d) What is the degree $[K : \mathbb{Q}]$?
(e) Find all automorphisms in $\text{Gal}(K/\mathbb{Q})$.
(f) Show that $\text{Gal}(K/\mathbb{Q})$ is isomorphic to the Klein 4-group.

8. Let $f = X^4 + 1$ over $\mathbb{Q}$. Specify what each automorphism of the Galois group does to a suitable set of generators of the splitting field of f. Then specify the permutations which the Galois group implements on the set of roots of f.

9. Let $f = X^4 - X^2 + 4$ in $\mathbb{Q}[X]$. Show that $K = \mathbb{Q}[\sqrt{5}, i\sqrt{3}]$ is the splitting field of f over $\mathbb{Q}$. Then find the Galois group of K over $\mathbb{Q}$.
Hint. The identity $X^4 - X^2 + 4 = (X^2 + 2)^2 - 5X^2$ helps get the roots.

10. Find the Galois group of $X^2 + 32$ over $\mathbb{Q}$.

6.5 The Core of Galois Theory

The renowned Galois correspondence can be approached along varied lines. Here is our take on this intricate but fundamental theory.

6.5.1 The Independence of Characters

Start with a powerful general result, which in essence comes down to linear algebra.

Definition 6.69. If G is any group and E any field, a ***character on the group G with values in the field E*** is any group homomorphism $\varphi : G \to E^\star$, where $E^\star$ is the group of non-zero elements in E.

For example, the restriction of any field homomorphism $\varphi : K \to E$ to the group of units $K^\star$ is a character on $K^\star$. For another example, take $G = \mathbb{Z}$ under addition and pick a non-zero α in E. Then

$$\varphi : \mathbb{Z} \to E^\star \text{ where } n \mapsto \alpha^n$$

is a character on $\mathbb{Z}$.

Characters belong to the vector space of all functions from G to E. The next result, attributed to Richard Dedekind (1831–1916), establishes that distinct characters, no matter how many, are linearly independent in this space of functions.

Proposition 6.70 (Independence of characters). *Distinct characters on a group G with values in a field E, are linearly independent as functions from G to E.*

Proof. In search of a contradiction, suppose the result fails. There must then be a list of distinct characters $\varphi_1, \ldots, \varphi_n$ that is linearly dependent over E and as short as possible. Since a list of just one character, being a non-zero function on G, is linearly independent, it must be that $n \geq 2$.

Then there exist $e_1, \ldots, e_n$ in E which are not all zero and such that

$$e_1\varphi_1 + e_2\varphi_2 + \cdots + e_n\varphi_n = 0, \text{ the zero function on } G.$$

The minimality of the list of characters ensures that *all* $e_j \neq 0$.

Select a v in G such that $\varphi_1(v) \neq \varphi_2(v)$. For all u in G, the assumed linear dependency of the characters gives

$$e_1\varphi_1(vu) + e_2\varphi_2(vu) + \cdots + e_n\varphi_n(vu) = 0.$$

Since the φ_j are characters, this identity becomes

$$e_1\varphi_1(v)\varphi_1(u) + e_2\varphi_2(v)\varphi_2(u) + \cdots + e_n\varphi_n(v)\varphi_n(u) = 0,$$

for all u in G. That is

$$e_1\varphi_1(v)\varphi_1 + e_2\varphi_2(v)\varphi_2 + \cdots + e_n\varphi_n(v)\varphi_n = 0, \text{ the zero function on } G.$$

Multiply the identity at the start by $\varphi_1(v)$ to see that

$$e_1\varphi_1(v)\varphi_1 + e_2\varphi_1(v)\varphi_2 + \cdots + e_n\varphi_1(v)\varphi_n = 0, \text{ again the zero function on } G.$$

Subtract these latter two identities to get

$$e_2(\varphi_2(v) - \varphi_1(v))\varphi_2 + \cdots + e_n(\varphi_n(v) - \varphi_1(v))\varphi_n = 0.$$

In this dependency relation the first coefficient $e_2(\varphi_2(v) - \varphi_1(v)) \neq 0$.

This dependency among $\varphi_2, \ldots, \varphi_n$ contradicts the minimal length of the dependent list $\varphi_1, \varphi_2, \ldots, \varphi_n$. □

6.5.2 A Bound on the Order of a Galois Group

If K is a finite extension of F, the Galois group $\mathrm{Gal}(K/F)$ must be finite. Indeed, write $K = F[\alpha_1, \dots, \alpha_n]$ for some α_j in K, and then recall that every F-automorphism is determined by its effect on the α_j and that each α_j must go to one of its finitely many F-conjugates in K. But the next result goes quite a bit further.

Proposition 6.71. *If K, L are extension fields of F and $[K : F]$ is finite, then the number of distinct F-maps from K to L is at most $[K : F]$. Consequently, the order of* $\mathrm{Gal}(K/F)$ *is at most the degree $[K : F]$.*

Proof. If $\alpha_1, \dots, \alpha_m$ is a basis of K as a vector space over F, and $\varphi_1, \dots, \varphi_n : K \to L$ are distinct F-maps, we require that $n \le m$.

Consider the homogeneous system of m equations in n unknowns over the field L given by

$$\sum_{j=1}^{n} \varphi_j(\alpha_i)x_j = 0 \text{ where } i = 1, \dots, m.$$

Let $(x_1, \dots, x_n)$ in L^n be a solution to this system. For any α in K we can write $\alpha = \sum_{i=1}^{m} a_i\alpha_i$ where the $a_i \in F$. Since each $\varphi_j : K \to L$ is an F-map we get that

$$\begin{aligned}\sum_{j=1}^{n} \varphi_j(\alpha)\, x_j &= \sum_{j=1}^{n} \varphi_j\left(\sum_{i=1}^{m} a_i\alpha_i\right) x_j = \sum_{j=1}^{n}\left(\sum_{i=1}^{m} \varphi_j(a_i)\varphi_j(\alpha_i)\right) x_j \\ &= \sum_{j=1}^{n}\left(\sum_{i=1}^{m} a_i\varphi_j(\alpha_i)\right) x_j = \sum_{i=1}^{m} a_i\left(\sum_{j=1}^{n} \varphi_j(\alpha_i)x_j\right) = \sum_{i=1}^{m} a_i \cdot 0 = 0.\end{aligned}$$

Hence the combination $\sum_{j=1}^{n} x_j\varphi_j$ is the zero function on $K^\star$. The φ_j are also distinct characters, and so by Proposition 6.70 all $x_j = 0$. Thereby the homogeneous system at the start, with m equations and n unknowns, only has the trivial solution. From basic linear algebra this implies that the number of unknowns is at most the number of equations, i.e. $n \le m$.

The conclusion on the order of $\mathrm{Gal}(K/F)$ comes from putting $L = K$. □

It might be informative to examine a different proof of Proposition 6.71.

Alternative proof of Proposition 6.71. Let us do an induction based on the number of generators used to get K as an extension of F. If $K = F$, with no additional generators needed, the number of F-maps from F to L is 1, which of course equals $[F : F]$. Suppose the result applies to all extensions of F that are generated with fewer than n generators, and let $K = F[\alpha_1, \dots, \alpha_n]$ where the $\alpha_j \in K$. Put $E = F[\alpha_1, \dots, \alpha_{n-1}]$. Then $K = E[\alpha_n]$, and by assumption the number of F-maps $\varphi : E \to L$ is at most $[E : F]$.

Since $K = E[\alpha_n]$, every lifting of an F-map $\varphi : E \to L$ to an F-map $\psi : K \to L$ is determined by the value $\psi(\alpha_n)$. Hence the number of liftings of φ to K is at most the number of possible choices of $\psi(\alpha_n)$. If g is the minimal polynomial of α_n in $E[X]$, then $\psi(\alpha_n)$ must be a root of the φ-transfer $\varphi(g)$ in $\varphi(E)[X]$. But $\deg \varphi(g) = \deg g = [E[\alpha_n] : E] = [K : E]$. Therefore the number of possible choices of $\psi(\alpha_n)$ is at most $[K : E]$.

We have just seen that the number of liftings of an F-map $\varphi : E \to L$ to an F-map $\psi : K \to L$ is at most $[K : E]$. Since every F-map $\psi : K \to L$ is a lifting of an F-map $\varphi : E \to L$ and since the number of such φ is at most $[E : F]$, it follows that the number of F-maps from K to L is at most $[K : E][E : F] = [K : F]$. □

6.5.3 The Fixed Field of an Automorphism Group

Definition 6.72. If K is a field and H is a group of automorphisms of K (under composition of mappings), the set of elements in K that are fixed by every element in H, i.e.

$$\{\alpha \in K : \varphi(\alpha) = \alpha \text{ for all } \varphi \text{ in } H\},$$

is known as the ***fixed field*** of H and will be denoted by $\mathrm{Fix}(H)$.

It is a triviality to check that $\mathrm{Fix}(H)$ is a subfield of K.

If H is the trivial group, consisting of just the identity automorphism, then $\mathrm{Fix}(H) = K$. For instance, when $K = \mathbb{Q}[\sqrt[3]{2}]$, we saw that $\mathrm{Gal}(K/\mathbb{Q})$ is trivial, and so $\mathrm{Fix}(\mathrm{Gal}(K/\mathbb{Q})) = K$.

The Fixed Field of the Galois Group of $X^3 - 2$

For a more interesting fixed field, take K to be the splitting field of $f = X^3 - 2$ over $\mathbb{Q}$. We have seen that its Galois group $\mathrm{Gal}(K/\mathbb{Q})$ consists of the full group of permutations of the roots $\alpha, \alpha\omega, \alpha\omega^2$ of f. Here $\alpha = \sqrt[3]{2}$ in $\mathbb{R}$ and $\omega = e^{2\pi i/3}$. Trivially, $\mathbb{Q} \subseteq \mathrm{Fix}(\mathrm{Gal}(K/\mathbb{Q}))$. Let us see that equality holds in fact.

Since $K = \mathbb{Q}[\alpha, \omega]$, a basis for K as a vector space over $\mathbb{Q}$ consists of

$$1, \alpha, \alpha^2, \omega, \alpha\omega, \alpha^2\omega.$$

If $\beta \in K$, then

$$\beta = a_1 + a_2\alpha + a_3\alpha^2 + a_4\omega + a_5\alpha\omega + a_6\alpha^2\omega \text{ where the } a_j \in \mathbb{Q}.$$

Supposing $\beta \in \mathrm{Fix}(\mathrm{Gal}(K/\mathbb{Q}))$, deduce what this implies about the a_j.

In $\mathrm{Gal}(K/\mathbb{Q})$ there is an automorphism φ such that

$$\varphi : \alpha \mapsto \alpha \text{ and } \alpha\omega \mapsto \alpha\omega^2 \mapsto \alpha\omega.$$

Since $1 + \omega + \omega^2 = 0$ we see that

$$\varphi(\omega) = \varphi(\alpha\omega/\alpha) = \varphi(\alpha\omega)/\varphi(\alpha) = \alpha\omega^2/\alpha = \omega^2 = -1 - \omega.$$

And then

$$\begin{aligned}\varphi(\beta) &= a_1 + a_2\alpha + a - 3\alpha^2 + a_4(-1-\omega) + a_5\alpha(-1-\omega) + a_6\alpha^2(-1-\omega) \\ &= (a_1 - a_4) + (a_2 - a_5)\alpha + (a_3 - a_6)\alpha^2 - a_4\omega - a_5\alpha\omega - a_6\alpha^2\omega.\end{aligned}$$

By comparing coefficients of β with those of $\varphi(\beta)$ the requirement $\varphi(\beta) = \beta$ forces $a_4 = a_5 = a_6 = 0$, and then $\beta = a_1 + a_2\alpha + a_3\alpha^2$.

In $\mathrm{Gal}(K/\mathbb{Q})$ there is also a σ such that $\sigma : \alpha \mapsto \alpha\omega$. And so

$$\begin{aligned}\sigma(\beta) &= a_1 + a_2\alpha\omega + a_3(\alpha\omega)^2 \\ &= a_1 + a_2\alpha\omega + a_3\alpha^2(-1-\omega) \\ &= a_1 - a_3\alpha^2 + a_2\alpha\omega - a_3\alpha^2\omega.\end{aligned}$$

Now the condition $\sigma(\beta) = \beta$ forces $a_2 = a_3 = 0$, and then $\beta = a_1 \in \mathbb{Q}$.

Thus $\mathrm{Fix}(\mathrm{Gal}(K/\mathbb{Q})) = \mathbb{Q}$.

The Fixed Field of the Galois Group of $X^p - 1$

Let p be an odd prime. As seen at the end of Section 6.4 the minimal polynomial over $\mathbb{Q}$ of the complex number $\zeta = e^{2\pi i/p}$ is $f = 1 + X + \cdots + X^{p-1}$ of degree $p-1$. The splitting field of f is $\mathbb{Q}[\zeta]$. The set Δ of roots of f is given by $\Delta = \{\zeta, \zeta^2, \ldots, \zeta^{p-1}\}$.

By Proposition 6.9 a basis for $\mathbb{Q}[\zeta]$ over $\mathbb{Q}$ consists of the $p-1$ elements $1, \zeta, \ldots, \zeta^{p-2}$. By multiplying these by ζ we see that Δ forms an alternative basis for $\mathbb{Q}[\zeta]$ over $\mathbb{Q}$.

Now we enquire about the fixed field of $\mathrm{Gal}(\mathbb{Q}[\zeta]/\mathbb{Q})$. Each element β of $\mathbb{Q}[\zeta]$ takes the form $\beta = a_1\zeta + a_2\zeta^2 + \cdots + a_{p-1}\zeta^{p-1}$ where the a_j are unique coefficients in $\mathbb{Q}$. Suppose β is fixed by $\mathrm{Gal}(\mathbb{Q}[\zeta]/\mathbb{Q})$. According to Proposition 6.67, the Galois group permutes the set Δ of roots of f and the permutation action is transitive. So, for every j from 1 to $p-1$, there is an automorphism σ in the Galois group such that $\sigma(\zeta) = \zeta^j$. Then

$$\sigma(\beta) = a_1\sigma(\zeta) + a_2\sigma(\zeta^2) + \cdots + a_{p-1}\sigma(\zeta^{p-1}).$$

Keeping in mind that the $\sigma(\zeta^i)$ give a permutation of Δ and $\sigma(\zeta) = \zeta^j$, the assumed equation $\beta = \sigma(\beta)$ implies that each $a_j = a_1$. Thus

$$\beta = a_1(\zeta + \zeta^2 + \cdots + \zeta^{p-1}) = -a_1 \in \mathbb{Q}.$$

The second equality above holds because the minimal polynomial of ζ is $1 + X + \cdots + X^{p-1}$. We have verified that the fixed field of $\mathrm{Gal}(\mathbb{Q}[\zeta]/\mathbb{Q})$ is the base field $\mathbb{Q}$.

6.5.4 Galois Extensions

We come to an amazing result which harmonizes the notions of splitting fields, separable polynomials, normal extensions, fixed fields and the order of the Galois group. We will refer to this pivotal result as the ***characterization of Galois extensions***.

Proposition 6.73 (Characterization of Galois extensions). *The following statements, pertaining to a finite extension K of a field F, are equivalent.*

1. *The extension K is the splitting field of a separable polynomial in $F[X]$.*
2. *The order of $\mathrm{Gal}(K/F)$ equals the degree $[K : F]$.*
3. $\mathrm{Fix}(\mathrm{Gal}(K/F)) = F$.
4. *For every α in K its minimal polynomial in $F[X]$ is separable and splits over K.*

Proof. We will prove that $1 \implies 2 \implies 3 \implies 4 \implies 1$.

$1 \implies 2$. Suppose to the contrary that there is an extension K of F which satisfies 1 but not 2. We know from Proposition 6.71 that $|\text{Gal}(K/F)| \leq [K:F]$. We can then take K to be an extension of F such that

- K is the splitting field of a separable polynomial f in $F[X]$,
- $|\text{Gal}(K/F)| < [K:F]$, and
- $[K:F]$ is as low as possible among extensions with the preceding two properties.

In that case, K is a proper extension of F, meaning that f has an irreducible factor g in $F[X]$ of degree $m > 1$. Being a factor of f, this g splits over K too. Let α in K be a root of g. Put $E = F[\alpha]$, an intermediate field between F and K such that $[E:F] = m$. Putting $n = [K:F]$, the tower theorem, Proposition 6.13, implies that $[K:E] = n/m$, which is less than n.

Now K remains the splitting field of f over E. Furthermore f remains separable as a polynomial in $E[X]$. After all, the number of roots of f in K still equals the degree of f. By the minimality of $[K:F]$, it must be that $|\text{Gal}(K/E)| = [K:E] = n/m$. The new Galois group $\text{Gal}(K/E)$ is a subgroup of $\text{Gal}(K/F)$, because E-maps are automatically F-maps.

Since f is separable, so is its irreducible factor g separable, and thus g has exactly m distinct roots. Call them $\alpha = \alpha_1, \alpha_2, \ldots, \alpha_m$. Because K is a splitting field over F, Proposition 6.36 ensures that for every $j = 1, 2, \ldots, m$ there is an F-automorphism $\varphi_j : K \to K$ such that $\varphi_j(\alpha) = \alpha_j$. The separability of g ensured that these α_j are distinct for different j. Thus $\varphi_i(\alpha) \neq \varphi_j(\alpha)$ when $i \neq j$, and then $\varphi_j^{-1}\varphi_i(\alpha) \neq \alpha$. So $\varphi_j^{-1}\varphi_i \notin \text{Gal}(K/E)$. This means that the cosets $\varphi_i \text{Gal}(K/E)$ and $\varphi_j \text{Gal}(K/E)$ are distinct inside $\text{Gal}(K/F)$. Therefore $\text{Gal}(K/E)$ has at least m left cosets inside the bigger group $\text{Gal}(K/F)$.

Since these m cosets inside $\text{Gal}(K/F)$ are disjoint, we come to the contradiction:

$$n = [K:F] > |\text{Gal}(K/F)| \geq m|\text{Gal}(K/E)| = m \cdot \frac{n}{m} = n.$$

$2 \implies 3$. Suppose $|\text{Gal}(K/F)| = [K:F]$ and let $E = \text{Fix}(\text{Gal}(K/F))$. Of course $F \subseteq E$, but we want equality. The definition of E makes it clear that an automorphism $\varphi : K \to K$ fixes F if and only if φ fixes E, i.e. E is fixed by the maps that fix F. Thus $\text{Gal}(K/F) = \text{Gal}(K/E)$.

From the bound on the size of a Galois group given in Proposition 6.71:

$$[K:F] = |\text{Gal}(K/F)| = |\text{Gal}(K/E)| \leq [K:E].$$

By the tower theorem, Proposition 6.13:

$$1 \leq [E:F] = \frac{[K:F]}{[K:E]} \leq 1.$$

So $[E:F] = 1$, and then $E = F$ as desired.

$3 \implies 4$. If $\alpha \in K$ and g in $F[X]$ is its minimal polynomial, we require that g split over K with no repeated roots. The set $\Delta = \{\sigma(\alpha) : \sigma \in \text{Gal}(K/F)\}$ is the orbit of α under $\text{Gal}(K/F)$. Since F-maps preserve F-conjugates, Δ consists of roots of g, and clearly $\alpha \in \Delta$. Also $\tau(\Delta) = \Delta$ for every τ in $\text{Gal}(K/F)$, simply by the definition of Δ. So if

$\alpha = \alpha_1, \alpha_2, \ldots, \alpha_m$ is an enumeration of the orbit Δ without repetitions and $\tau \in \mathrm{Gal}(K/F)$, then $\tau(\alpha_1), \tau(\alpha_2), \ldots, \tau(\alpha_m)$ is a permutation of the preceding list.

The polynomial

$$h = (X - \alpha_1)(X - \alpha_2) \cdots (X - \alpha_m)$$

lies in $K[X]$, has no repeated root, and has $\alpha = \alpha_1$ as a root. For each τ in $\mathrm{Gal}(K/F)$ the τ-transfer of h turns out to be

$$\tau(h) = (X - \tau(\alpha_1))\,(X - \tau(\alpha_2)) \cdots (X - \tau(\alpha_m)) = h.$$

The first equality holds because τ acting on the ring $K[X]$ is an automorphism, and the second equality holds because τ permutes the listing of the orbit Δ. This implies that the coefficients of h lie in $\mathrm{Fix}(\mathrm{Gal}(K/F))$, which by assumption is F. So $h \in F[X]$.

Since α is a root of h and g is the minimal polynomial of α, it follows that g divides h in $F[X]$. On the other hand $\deg h \le \deg g$, because the roots of h are all roots of g. Therefore $g = h$, which splits over K with the roots α_j not repeated.

$4 \implies 1$. The finite extension K can be written as $K = F[\alpha_1, \ldots, \alpha_n]$ where the α_j are algebraic over F. By the assumption, the minimal polynomial g_j in $F[X]$ of each α_j splits over K without repeated roots. Let $h = g_1 g_2 \cdots g_n$, the product of these minimal polynomials. Each g_j splits over K, and thereby h splits over K. Also K is generated over F by the α_j which comprise some of the roots of h. Thus K is also generated over F by all of the roots of h. This shows that K is the splitting field of h.

The polynomial h will not be separable if some g_j equals another g_i. In that case discard all but one of the duplicated g_j, and let f be the product of the now distinct g_j that are retained. The roots of f coincide with the roots of h. Thus K is also the splitting field of f. And since distinct irreducible polynomials in $F[X]$ can never share a root in K, the roots of f are never repeated. So f is a separable polynomial whose splitting field is K. □

Definition 6.74. A field extension K of a field F which satisfies any and thereby all of the conditions of Proposition 6.73 is known as a ***Galois extension*** of F.[9]

A Galois extension is nothing but the splitting field of a separable polynomial over a field F. We might note that when F is a perfect field, for example when F is finite or when F has characteristic zero, *every* splitting field is the splitting field of a separable polynomial. Indeed, let f be the polynomial in $F[X]$ that made the splitting field K. Factor $f = g_1^{d_1} g_2^{d_2} \cdots g_k^{d_k}$ into irreducible factors g_j each with d_j possible repetitions. Clearly, K is also the splitting field of $h = g_1 g_2 \cdots g_k$ where each irreducible factor g_j only appears once. This h is separable, as seen from Proposition 6.64.

[9] For each of the four equivalent properties of a Galois extension, there exist authors who *define* a "Galois extension" to be a field extension with that property.

A Way to Obtain Minimal Polynomials

It might be worthwhile to squeeze a bit more out of the proof of Proposition 6.73, in particular, the proof that property 3 implies property 4. Suppose we have established that K is the splitting field of a separable polynomial over a field F, and that we have an understanding of what the Galois group $\mathrm{Gal}(K/F)$ does to K by its action on some chosen generators of the extension K. If $\alpha \in K$, we can enumerate the orbit of α under the action of $\mathrm{Gal}(K/F)$ as $\alpha = \alpha_1, \alpha_2, \ldots, \alpha_m$, *without repetitions*. The proof that property 3 implies property 4 includes the fact that the product $(X - \alpha_1)(X - \alpha_2) \cdots (X - \alpha_m)$ lies in $F[X]$, and that this product is the minimal polynomial of α over the field F. Let us make note of this fact.

Proposition 6.75. *If K is a Galois extension over a field F and $\alpha \in K$, then the minimal polynomial of α in $F[X]$ is the product*

$$(X - \alpha_1)(X - \alpha_2) \cdots (X - \alpha_m)$$

where the distinct elements $\alpha_1, \alpha_2, \ldots, \alpha_m$ constitute the orbit of α under the action of the Galois group $\mathrm{Gal}(K/F)$.

Before exploiting the characterization of Galois extensions, we need an alternative way to obtain them.

6.5.5 Artin's Theorem

Here is a stunning result observed by Emil Artin (1898–1962), whose influence in the organization of Galois theory, as currently understood, is paramount.

Proposition 6.76. *If H is a finite group of automorphisms of a field K and $F = \mathrm{Fix}(H)$, then*

- *the degree $[K : F]$ is finite and equal to the order of H,*
- *K is a Galois extension of F, and*
- $\mathrm{Gal}(K/F) = H$.

Proof. For the first item let n be the order of H. The subtle thing is to see that $[K : F] \leq n$. To that end we suppose $\beta_1, \beta_2, \ldots, \beta_m \in K$ where $m > n$, and we prove that the β_j are linearly dependent over F.

Over the field K, the homogeneous system of linear equations

$$\varphi(\beta_1)x_1 + \varphi(\beta_2)x_2 + \cdots + \varphi(\beta_m)x_m = 0,$$

taken over all φ in H, has more unknowns in K than equations. Thus the system has a non-zero solution in K^m.

If $(x_1, x_2, \ldots, x_m)$ is such a solution and $\psi \in H$, then $(\psi(x_1), \psi(x_2), \ldots, \psi(x_m))$ remains a solution of the system. Indeed,

$$\begin{aligned} &\psi\varphi(\beta_1)\psi(x_1) + \psi\varphi(\beta_2)\psi(x_2) + \cdots + \psi\varphi(\beta_m)\psi(x_m) \\ &\quad = \psi(\varphi(\beta_1)x_1 + \varphi(\beta_2)x_2 + \cdots + \varphi(\beta_m)x_m) = \psi(0) = 0. \end{aligned}$$

But as φ runs over H, so does $\psi\varphi$ run over H, which reveals that the solution space of the system is ψ-invariant as well as non-trivial.

Select a non-zero solution of the system to contain a minimal number of non-zero entries, and by rearranging the β_j the non-zero entries of the minimal non-zero solution can be made to appear before the zero entries (if any). (This last step is just to keep notations uncluttered.) Thus the minimal non-zero solution takes the form

$$(x_1, x_2, \dots, x_r, 0, \dots, 0), \text{ where } r \geq 1 \text{ and all } x_j \neq 0.$$

Actually $r \geq 2$. For otherwise we would have $\varphi(\beta_1)x_1 = 0$ for all φ, which certainly would force x_1 to be 0 too.

After rescaling by $1/x_1$, the minimal non-zero solution takes the form

$$(1, x_2, \dots, x_r, 0, \dots, 0) \text{ where } r \geq 2, \text{ and the } x_j \neq 0.$$

It turns out that all $x_2, \dots, x_r \in F$. Indeed, suppose on the contrary that some $x_j \notin F$. By rearranging the $\beta_2, \dots, \beta_r$ we can say that $x_2 \notin F$. (This keeps notations uncluttered.) Since $F = \text{Fix}(H)$, there is some ψ in H such that $\psi(x_2) \neq x_2$. As noted

$$(\psi(1), \psi(x_2), \dots, \psi(x_r), \psi(0), \dots, \psi(0)) = (1, \psi(x_2), \dots, \psi(x_r), 0, \dots, 0)$$

remains a solution of the system. Subtract this solution from $(1, x_2, \dots, x_r, 0, \dots, 0)$ to get the solution $(0, x_2 - \psi(x_2), \dots, x_r - \psi(x_r), 0, \dots, 0)$. Since $x_2 - \psi(x_2) \neq 0$, this solution remains non-zero but has fewer non-zero entries than the solution with the minimal number r of non-zero entries. Consequently, all $x_j \in F$, which reveals that the β_j are linearly dependent over F. Thus $[K : F] \leq n = |H|$.

Clearly H is a subgroup of $\text{Gal}(K/F)$ which, along with Proposition 6.71 and the preceding inequality, gives

$$|H| \leq |\text{Gal}(K/F)| \leq [K : F] \leq |H|.$$

So the above three quantities are equal. In particular $H = \text{Gal}(K/F)$. Also condition 2 of Proposition 6.73 reveals that K is a Galois extension of F. □

6.5.6 The Galois Correspondence

The result coming up forms the major part of what is generally known as the ***fundamental theorem of Galois theory***. The heavy lifting needed for this theorem has already been done.

If K is a field extension of F, let $\mathcal{E}$ denote the family of all intermediate subfields E that sit between F and K, and let $\mathcal{H}$ denote the family of all subgroups H of $\text{Gal}(K/F)$.

For each E in $\mathcal{E}$, the group $\text{Gal}(K/E)$ is a subgroup of $\text{Gal}(K/F)$, because automorphisms that fix E also fix F. And for each H in $\mathcal{H}$, the field $\text{Fix}(H)$ sits between F and K, because H is part of the group $\text{Gal}(K/F)$ that fixed F.

Thus we have two correspondences

$$\text{Gal}(K/-) : \mathcal{E} \to \mathcal{H} \text{ given by } E \mapsto \text{Gal}(K/E),$$

and

$$\mathrm{Fix} : \mathcal{H} \to \mathcal{E} \text{ given by } H \mapsto \mathrm{Fix}(H).$$

These mappings constitute the ***Galois correspondence***.

An observation to keep in mind is that the Galois correspondence *reverses* inclusions. If $E, L \in \mathcal{E}$ and $E \subseteq L$, then $\mathrm{Gal}(K/E) \supseteq \mathrm{Gal}(K/L)$, because automorphisms that fix L also fix the smaller field E. And if $H, N \in \mathcal{H}$ and $H \subseteq N$, then $\mathrm{Fix}(H) \supseteq \mathrm{Fix}(N)$, because field elements fixed by N are also fixed by H.

6.5.7 The Fundamental Theorem of Galois Theory

More profoundly, when K is a Galois extension of F, the fundamental theorem of Galois theory says that the mappings of the Galois correspondence are mutual inverses, and thereby they are bijections. The upcoming proof pretty much rests on the characterization of Galois extensions, Proposition 6.73, and on Artin's theorem, Proposition 6.76.

An additional observation helps. If K is a Galois extension of F and E is an intermediate field between F and K, i.e. $F \subseteq E \subseteq K$, this lesser field need not be a Galois extension of F. However, K remains a Galois extension of E. Indeed, K is the splitting field of a separable polynomial f in $F[X]$. By Proposition 6.26 our K is the splitting field of f over E. And clearly f remains separable as a polynomial in $E[X]$.

Proposition 6.77 (Fundamental theorem of Galois theory). *Let the field K be a Galois extension of F. If E is an intermediate field between F and K, then*

$$\mathrm{Fix}(\mathrm{Gal}(K/E)) = E \text{ and } |\mathrm{Gal}(K/E)| = [K : E].$$

If H is a subgroup of $\mathrm{Gal}(K/F)$, *then*

$$\mathrm{Gal}(K/\mathrm{Fix}(H)) = H \text{ and } [K : \mathrm{Fix}(H)] = |H|.$$

Thereby, the Galois correspondence implements an inclusion reversing bijection between the family $\mathcal{E}$ of all fields between F and K and the family $\mathcal{H}$ of all subgroups of $\mathrm{Gal}(K/F)$.

Proof. Since K is also a Galois extension of E, item 3 in the characterization of Galois extensions, Proposition 6.73, confirms that $\mathrm{Fix}(\mathrm{Gal}(K/E)) = E$, while item 2 confirms that $|\mathrm{Gal}(K/E)| = [K : E]$.

Because H is a finite group of automorphisms of K, Artin's theorem confirms that $\mathrm{Gal}(K/\mathrm{Fix}(H)) = H$ as well as $[K : \mathrm{Fix}(H)] = |H|$. □

For an illustration of this major insight take K to be the splitting field of $X^3 - 2$ over $\mathbb{Q}$. We have seen that $\mathrm{Gal}(K/\mathbb{Q})$ is the group $\mathcal{S}_3$ of permutations of the three roots of this polynomial. This group contains precisely 6 subgroups: the trivial group, the three cyclic subgroups generated by its transpositions, one cyclic subgroup generated by either of its 3-cycles, and the full group $\mathcal{S}_3$. The fundamental theorem reveals that K contains exactly 6 subfields. The six subfields of K can be readily calculated as the fixed fields of each of the six subgroups of S_3. More generally, the fundamental theorem implies that every Galois extension K of F admits only finitely many fields between F and K. The number of such subfields coincides with the number of subgroups of $\mathrm{Gal}(K/F)$. Without the fundamental theorem this lovely fact is far from easy to obtain.

Of course, the fundamental theorem is meant to solve problems regarding polynomials, which is what we will do in subsequent sections, but first we need to look into another aspect of the Galois correspondence.

6.5.8 The Normality Connection

We continue with K being a Galois extension of F. The field K remains a Galois extension of every intermediate field E that sits between F and K. Here we explore what it takes for the intermediate field E to be a Galois extension of F, and thereby reveal another aspect of the Galois correspondence.

For every F-map φ in $\mathrm{Gal}(K/F)$, the image $\varphi(E)$ is once again a field between F and K. A clean connection between the subgroups $\mathrm{Gal}(K/E)$ and $\mathrm{Gal}(K/\varphi(E))$ of $\mathrm{Gal}(K/F)$ prevails.

Proposition 6.78. *Let K be a Galois extension of F and E a field between F and K. If $\varphi \in \mathrm{Gal}(K/F)$, then*

$$\mathrm{Gal}(K/\varphi(E)) = \varphi\,\mathrm{Gal}(K/E)\varphi^{-1}.$$

Proof. For each σ in $\mathrm{Gal}(K/F)$ we see that

$$\begin{aligned}
\sigma \in \mathrm{Gal}(K/E) &\iff \sigma(\alpha) = \alpha \text{ for all } \alpha \text{ in } E \\
&\iff \sigma(\varphi^{-1}\varphi(\alpha)) = \varphi^{-1}\varphi(\alpha) \text{ for all } \alpha \text{ in } E \\
&\iff \sigma(\varphi^{-1}(\beta)) = \varphi^{-1}(\beta) \text{ for all } \beta \text{ in } \varphi(E) \\
&\iff \varphi\sigma\varphi^{-1}(\beta) = \beta \text{ for all } \beta \text{ in } \varphi(E) \\
&\iff \varphi\sigma\varphi^{-1} \in \mathrm{Gal}(K/\varphi(E)).
\end{aligned}$$

This says that $\mathrm{Gal}(K/\varphi(E)) = \varphi\,\mathrm{Gal}(K/E)\varphi^{-1}$. □

If H is a group of automorphisms of a field K and E is a subfield of K, we say that E is ***invariant under the group*** H when $\varphi(E) = E$ for all φ in H. (This is weaker than saying E is fixed by H.)

Proposition 6.79. *Let K be a Galois extension of F. A field E between F and K is invariant under* $\mathrm{Gal}(K/F)$ *if and only if the subgroup* $\mathrm{Gal}(K/E)$ *is normal in* $\mathrm{Gal}(K/F)$.

Proof. Recall that the subgroup $\mathrm{Gal}(K/E)$ is normal in $\mathrm{Gal}(K/F)$ if and only if $\varphi\,\mathrm{Gal}(K/E)\varphi^{-1} = \mathrm{Gal}(K/E)$, for all φ in $\mathrm{Gal}(K/F)$. By Proposition 6.78 this is the same as saying that $\mathrm{Gal}(K/\varphi(E)) = \mathrm{Gal}(K/E)$. Since the Galois correspondence is a bijection, the latter is the same as saying that $\varphi(E) = E$ for all φ in $\mathrm{Gal}(K/F)$. □

Let us explore further what it means for E to be invariant under $\mathrm{Gal}(K/F)$.

Proposition 6.80. *Let K be a Galois extension of F. A field E between F and K is itself a Galois extension of F if and only if E is invariant under* $\mathrm{Gal}(K/F)$.

Proof. If E is Galois over F, then E is a normal extension of F in the sense that the extension E satisfies Proposition 6.42. According to the second item in that proposition (with E replacing K and K replacing L) every φ in $\mathrm{Gal}(K/F)$ restricts to an automorphism of E, which just says that $\varphi(E) = E$. A proof based on Proposition 6.73 is also possible, but the above is quite efficient.

Conversely, suppose E is invariant under $\mathrm{Gal}(K/F)$. One way to see that E is a Galois extension of F is to verify that the extension E satisfies condition 4 in Proposition 6.73. So take α in E with minimal polynomial g in $F[X]$. This irreducible g is separable and splits over K, by condition 4 of Proposition 6.73 applied to the Galois extension K in which α also lies. We need to see that g actually splits over E. For that let β in K be another root of g. Proposition 6.36 (a by-product of the isomorphism lifting theorem) provides an F-map φ in $\mathrm{Gal}(K/F)$ such that $\varphi(\alpha) = \beta$. Since E is invariant under $\mathrm{Gal}(K/F)$ and $\alpha \in E$, it must be that $\beta \in E$ too. All roots of g are in E, meaning that g splits over E. By item 4 in Proposition 6.73, E is a Galois extension of F. □

In summary, given a Galois extension K of F and an intermediate field E, the following three properties are equivalent.

- E is a Galois extension of F.
- E is invariant under $\mathrm{Gal}(K/F)$.
- The subgroup $\mathrm{Gal}(K/E)$ is normal in $\mathrm{Gal}(K/F)$.[10]

The Restriction of $\mathrm{Gal}(K/F)$ to an Invariant Subfield

If E is an intermediate field in a Galois extension K over F and E is itself Galois over F (with the three equivalent properties just above), we obtain the restriction mapping

$$\Gamma : \mathrm{Gal}(K/F) \to \mathrm{Gal}(E/F) \text{ given by } \sigma \mapsto \sigma_{|E}.$$

It is trivial to see that Γ is a group homomorphism. Furthermore, Γ is surjective. That more subtle fact comes from Proposition 6.35, telling us in particular that every F-map $\varphi : E \to E$ lifts to an F-map $\psi : K \to K$. In other words, every φ in $\mathrm{Gal}(E/F)$ is the restriction of a ψ in $\mathrm{Gal}(K/F)$.

What is $\ker\Gamma$? Well, $\sigma \in \ker\Gamma$ if and only if σ fixes E, which is the same as saying that $\sigma \in \mathrm{Gal}(K/E)$. So $\ker\Gamma = \mathrm{Gal}(K/E)$. This reveals once more that if E is Galois over F, then $\mathrm{Gal}(K/E)$ is a normal subgroup of $\mathrm{Gal}(K/F)$.

The foregoing discussion establishes something else about the Galois correspondence.

[10] Being a Galois extension of F implies that E is a normal extension of F in the sense of Proposition 6.42. So the term "normal" as used in the setting of field extensions and as used in the setting of groups are intimately connected. Surely the adoption of this term in one setting gave birth to its adoption in the other. However, Galois did not use the term "normal." The term came into use later on.

Proposition 6.81. *Let K be a Galois extension of F. A field E between F and K is itself a Galois extension of F if and only if* $\mathrm{Gal}(K/E)$ *is a normal subgroup of* $\mathrm{Gal}(K/F)$. *In that case,*

$$\mathrm{Gal}(E/F) \cong \mathrm{Gal}(K/F)/\,\mathrm{Gal}(K/E).$$

Proof. The restriction map

$$\Gamma : \mathrm{Gal}(K/F) \to \mathrm{Gal}(E/F), \text{ given by } \sigma \mapsto \sigma_{|E},$$

is a surjective group homomorphism whose kernel is $\mathrm{Gal}(K/E)$. By the first isomorphism theorem for groups, Proposition 2.62, the result follows. □

A Summary of the Galois Correspondence

For every Galois extension K of F, the mappings Fix and $\mathrm{Gal}(K/-)$ are mutually inversive bijections between the set of subgroups of $\mathrm{Gal}(K/F)$ and the family of fields between F and K. The bijections are inclusion reversing. The order of a subgroup H equals the degree of K over its corresponding field $\mathrm{Fix}(H)$. The degree of K over an intermediate field E equals the order of $\mathrm{Gal}(K/E)$. An intermediate field E is itself Galois over F if and only if its corresponding subgroup $\mathrm{Gal}(K/E)$ is normal in $\mathrm{Gal}(K/F)$. In that case $\mathrm{Gal}(E/F) \cong \mathrm{Gal}(K/F)/\,\mathrm{Gal}(K/E)$.

EXERCISES

1. If G is a group of automorphisms of a field K and S is a set of generators of the group G, and α is an element of K, show that $\alpha \in \mathrm{Fix}(G)$ if and only if α is fixed by the generators in S.
2. If K is a Galois extension of $\mathbb{Q}$, show that the conjugation mapping $z \mapsto \bar{z}$ on $\mathbb{C}$ restricts to an element of $\mathrm{Gal}(K/\mathbb{Q})$.
3. If K is a Galois extension of F of prime degree p, show that $\mathrm{Gal}(K/F)$ is cyclic.
4. Show that every Galois extension K of F of degree 4 must have either three or five subfields (counting K and F).
5. Find a tower of fields $F \subset E \subset K$ such that E is a Galois extension of F, K is a Galois extension of E, but K is not a Galois extension of F.
6. Let K be a finite extension of F. Let $\mathcal{E}$ be the set of subfields of K which contain F, and let $\mathcal{H}$ be the set of subgroups of $\mathrm{Gal}(K/F)$. If K is a Galois extension of F, then the Galois correspondence $\mathrm{Gal}(K/-) : \mathcal{E} \to \mathcal{H}$ given by $E \mapsto \mathrm{Gal}(K/E)$ is a bijection.
   (a) Show that $\mathrm{Gal}(K/-)$ remains onto $\mathcal{H}$ even if K is not a Galois extension of F.
   (b) If K is not a Galois extension of F, show that $\mathrm{Gal}(K/-)$ is not one-to-one.
   Hint. If L is the fixed field of $\mathrm{Gal}(K/F)$, Artin's theorem says that K is a Galois extension of L with all the benefits of the Galois correspondence.
   This exercise reveals that Galois extensions are precisely those finite extensions for which the Galois correspondence is bijective between $\mathcal{E}$ and $\mathcal{H}$.
7. Let $f = X^8 - 2$ in $\mathbb{Q}[X]$, and let K be its splitting field.
   (a) Briefly explain why K is a Galois extension of $\mathbb{Q}$.

(b) Show that $K = \mathbb{Q}[\alpha, i]$, where $\alpha = \sqrt[8]{2}$.
(c) Show that $[K : \mathbb{Q}] = 16$. Then find the order of $\mathrm{Gal}(K/\mathbb{Q})$.
(d) List all elements of $\mathrm{Gal}(K/\mathbb{Q})$ by specifying their action of α and i.

8. If K is a finite field of characteristic p, show that K is a Galois extension of $\mathbb{Z}_p$.

9. Let f be a polynomial in $\mathbb{Q}[X]$ with distinct roots $\alpha_1, \alpha_2, \ldots, \alpha_n$ in $\mathbb{C}$.
(a) Show that $\frac{1}{\alpha_1^5} + \frac{1}{\alpha_2^5} + \cdots + \frac{1}{\alpha_n^5} \in \mathbb{Q}$.
(b) Show that the squared product $\left(\prod_{i<j}(\alpha_j - \alpha_i)\right)^2 \in \mathbb{Q}$.

10. (a) Explain why $\mathbb{Q}(\sqrt{2}, \sqrt{3})$ is a Galois extension of $\mathbb{Q}$, find its Galois group, and find all of its subfields.
(b) Repeat the preceding question for $\mathbb{Q}(\sqrt{2}, \sqrt{3}, \sqrt{5})$.

11. If $\alpha_1, \ldots, \alpha_n$ are non-zero elements in a field E and $e_1, \ldots, e_n$ in E are such that

$$e_1\alpha_1^k + \cdots + e_n\alpha_n^k = 0, \text{ for all integers } k,$$

show that all $e_j = 0$.

12. In Section 6.4.2 the Galois group of $X^3 - 2$ over $\mathbb{Q}$ was shown to be the full permutation group of the three roots of this polynomial. Let K be the splitting field of $X^3 - 2$ over $\mathbb{Q}$. Use the Galois correspondence to find all subfields of K.
Which of these subfields is itself a Galois extension of $\mathbb{Q}$?

13. Let $K = \mathbb{Q}[\sqrt{2 + \sqrt{2}}]$.
(a) Find the degree $[K : \mathbb{Q}]$.
(b) Show that K is a Galois extension of $\mathbb{Q}$.
(c) Show that $\mathrm{Gal}(K/\mathbb{Q})$ is cyclic, and give its generator by its action on $\sqrt{2 + \sqrt{2}}$.

14. Let $f = X^3 - 3X + 1$ in $\mathbb{Q}[X]$. Find the Galois group of the splitting field of f. How many subfields does the splitting field have?
Hint. First complete Exercise 10 in Section 6.2.

15. Let $f = X^4 - 2X^2 + 3$ in $\mathbb{Q}[X]$. Find the Galois group of f. How many subfields does the splitting field of f have?
Hint. First complete Exercise 11 in Section 6.2.

16. If K is the splitting field of a polynomial over $\mathbb{Q}$ and if the degree $[K : \mathbb{Q}]$ is odd, show that K is a subfield of $\mathbb{R}$.
Hint. The field K sits inside $\mathbb{C}$, and the latter field has the conjugation automorphism $z \mapsto \bar{z}$.

17. Let K be the splitting field of $X^7 - 1$ over $\mathbb{Q}$.
(a) Explain why $\mathrm{Gal}(K/\mathbb{Q})$ is a cyclic group, and find a generator for that group.
Hint. See Section 6.4.2.
(b) Show that K has precisely 4 subfields, and find these subfields using suitable generators over $\mathbb{Q}$.

18. Let K be the splitting field of $X^{29} - 1$ over $\mathbb{Q}$.

(a) Explain why $\mathrm{Gal}(K/\mathbb{Q})$ is a cyclic group, and find a generator for that group.
(b) How many subfields does K have?
Find generators for each subfield of K.

19. Let K be the splitting field of $f = X^4 - 2$ over $\mathbb{Q}$. The roots of f are $\alpha = \sqrt[4]{2}, -\alpha, \alpha i$ and $-\alpha i$. They sit at the vertices of a square in the complex plane.

(a) Show that $K = \mathbb{Q}[\sqrt[4]{2}, i]$ and determine $[K : \mathbb{Q}]$.
(b) List the elements of $\mathrm{Gal}(K/\mathbb{Q})$, by describing their action on the field generators α and i.
Be sure to explain how the characterization of Galois extensions ensures that your actions actually arise from the automorphisms in $\mathrm{Gal}(K/\mathbb{Q})$.
(c) Show how each of the automorphisms in part (b) behaves as permutations of the four roots $\alpha, -\alpha, \alpha i, -\alpha i$. This is your presentation of $\mathrm{Gal}(K/\mathbb{Q})$ as a group of permutations of the roots.
(d) Plot the four roots as the vertices of a square in the complex plane. In terms of the geometry of this square, what familiar group does the table in part (c) embody?
(e) One of the subgroups H in part (c) is cyclic of order 4. Find $\mathrm{Fix}(H)$ by specifying a generator for this field as an extension of $\mathbb{Q}$. Is $\mathrm{Fix}\, H$ a Galois extension of $\mathbb{Q}$?
(f) Find a subfield of K which is not a Galois extension of $\mathbb{Q}$.
(g) How many subfields does K have?

20. Let $f = X^6 - 4$ in $\mathbb{Q}[X]$.

(a) Find generators for the splitting field K of f.
(b) What is the degree $[K : \mathbb{Q}]$ and the order of the Galois group $\mathrm{Gal}(K/\mathbb{Q})$?
(c) Show that $\mathrm{Gal}(K/\mathbb{Q}) \cong S_3$.
(d) Find generators for each subfield of K.

21. Suppose K is a Galois extension of F such that $\mathrm{Gal}(K/F) \cong S_3$. Show that K is the splitting field of an irreducible cubic over F.
Hint. There is an intermediate field E between F and K such that $[E : F] = 3$. Pick α in $E \setminus F$, and let f in $F[X]$ be its minimal polynomial. Show that K is the splitting field of f. Keep the normality connection in mind.

22. Let $K = L(t)$ be the field of rational functions in t over a field K.

(a) Verify that the mapping $\sigma : K \to K$ given by $f(t) \mapsto f(1/t)$ is an automorphism of K.
(b) Let H be the cyclic group of automorphisms generated by σ, and let $F = \mathrm{Fix}(H)$. What is the degree $[K : F]$?
(c) Give an explicit description of the field F.
(d) Find the minimal polynomial of t in $F[X]$.
(e) Answer the preceding questions for the mapping $\tau : f(t) \mapsto f(1 - t)$.

23. Let $K = L(t)$ be the field of rational functions in the variable t over a field L. Let σ, τ be the automorphisms of K given by

$$\sigma : f(t) \mapsto f(1/t), \quad \tau : f(t) \mapsto f(1-t).$$

Put $\rho = \sigma\tau$. Let G be the group of automorphisms of K generated by σ and τ, and put $F = \mathrm{Fix}(G)$.

(a) Show that $G = \{1, \rho, \rho^2, \sigma, \sigma\rho, \sigma\rho^2\}$, and that G is isomorphic to S_3, the symmetric group on three letters.

(b) Show that $\mathrm{Fix}(G) = L(h)$ where $h = \dfrac{(t^2 - t + 1)^3}{t^2(t-1)^2}$.

Hints. First check that h and thereby $L(h)$ is fixed by G. Just check that the generators of G fix h. This will show that $L(h) \subseteq F = \mathrm{Fix}(G)$. The element t is algebraic over F, because Artin's theorem says that $[K : F]$ equals the order of G. To get $L(h) = F$ it is enough to show that the minimal polynomial of t over $L(h)$ is the same as the minimal polynomial of t over $L(h)$. Figure out the minimal polynomial of t over F and see that its coefficients are already in $L(h)$. Recall that the minimal polynomial of t is the product of the linear polynomials $X - r$ where r runs through the conjugates of t.

According to Artin's theorem, you have obtained that K is a Galois extension of $F = L(h)$ whose Galois group is isomorphic to S_3.

24. Let $\mathbb{F}_4$ be the finite field with 4 elements, and let $K = \mathbb{F}_4(t)$ be the field of rational functions in the variable t with coefficients in $\mathbb{F}_4$.

(a) For each a in $\mathbb{F}_4$ let $\varphi_a : K \to K$ be the mapping defined by $f(t) \mapsto f(t+a)$. Show that the four φ_a constitute a group G of automorphisms of K. Explain why $G \cong \mathbb{Z}_2 \times \mathbb{Z}_2$, i.e. the Klein 4-group.

(b) Let $E = \mathrm{Fix}(G)$, the subfield of K that is fixed by G. Explain why K is a Galois extension of E, and why $[K : E] = 4$.

(c) The element t is algebraic over E. Find the minimal polynomial g of t in $E[X]$, and justify your answer.

Hint. Check that $t^4 + t$ is fixed by G, i.e. $t^4 + t \in E$.

(d) Explain why $K = E[t]$.

(e) If α in K is a root of g, what are the other roots of g? Factor g into linear factors in $K[X]$, and deduce that K is the splitting field of g over E.

(f) For each intermediate F between E and K find an element β in K such that $F = E[\beta]$.

25. Let $F = \mathbb{Z}_{37}(t)$ be the field of rational functions in t with coefficients in the field $\mathbb{Z}_{37}$. Let $f = X^9 - t$ in $F[X]$, and take a root α of this polynomial to form the extension $K = F[\alpha]$.

(a) Show that f is irreducible and separable.

(b) Show that K is the splitting field of f by finding the other eight roots of f inside K.

Hint. You need some λ in K such that $\lambda^9 = 1$.

(c) Show that $\mathrm{Gal}(K/F)$ has order 9, and that this group is cyclic.

(d) If H is the unique subgroup of $\mathrm{Gal}(K/F)$ other than the whole group and the trivial group, find Fix H.

Hint. To get the fixed field recall that a basis of K over F consists of the list $1, \alpha, \alpha^2, \ldots, \alpha^8$.

26.* This intricate exercise provides an example of a Galois extension of $\mathbb{Q}$ whose Galois group is the quaternion group $\mathcal{Q}$ of order 8 discussed in Section 2.4.3.

If c is a positive real number, $\sqrt{c}$ will stand for its positive square root. Select one of the two complex square roots of $-3+\sqrt{3}$, and for this exercise denote it by $\sqrt{-3+\sqrt{3}}$. Observe that $\left(\sqrt{6}/\sqrt{-3+\sqrt{3}}\right)^2 = -3-\sqrt{3}$. Since $\sqrt{6}/\sqrt{-3+\sqrt{3}}$ is a complex square root of $-3-\sqrt{3}$, we take the liberty of denoting $\sqrt{6}/\sqrt{-3+\sqrt{3}}$ by $\sqrt{-3-\sqrt{3}}$. With these notations $\left(\sqrt{-3+\sqrt{3}}\right)\left(\sqrt{-3-\sqrt{3}}\right) = \sqrt{6}$.

Put

$$\alpha = \left(\sqrt{2+\sqrt{2}}\right)\left(\sqrt{-3+\sqrt{3}}\right) \text{ and } K = \mathbb{Q}[\alpha].$$

Follow the steps to check that the extension K is Galois of degree 8 over $\mathbb{Q}$, and that $\mathrm{Gal}(K/\mathbb{Q})$ is isomorphic to the quaternion group.

One fact to keep in mind is that the field extension $\mathbb{Q}[\sqrt{2},\sqrt{3}]$ is Galois of degree 4 over $\mathbb{Q}$, and that its Galois group is given by the four automorphisms which on the generators act according to:

$$\sqrt{2} \mapsto \pm\sqrt{2}, \sqrt{3} \mapsto \pm\sqrt{3}.$$

(a) Show that $\mathbb{Q}[\alpha^2] = \mathbb{Q}[\sqrt{2},\sqrt{3}]$.

Hint. If not, the minimal polynomial of α^2 over $\mathbb{Q}$ would be quadratic. Do some calculations to verify this is not the case. A useful fact could be that $1, \sqrt{2}, \sqrt{3}, \sqrt{6}$ form a basis of $\mathbb{Q}[\sqrt{2},\sqrt{3}]$ over $\mathbb{Q}$.

(b) Explain why $\mathbb{Q}[\alpha^2] \subsetneq \mathbb{Q}[\alpha]$, and deduce that $[\mathbb{Q}[\alpha] : \mathbb{Q}] = 8$.

(c) Let $\alpha_1, \alpha_2, \ldots, \alpha_8$ be the eight complex numbers

$$\pm\left(\sqrt{2\pm\sqrt{2}}\right)\left(\sqrt{-3\pm\sqrt{3}}\right).$$

Note that α is one of these eight. Put

$$f = (X-\alpha_1)(X-\alpha_2)\cdots(X-\alpha_8).$$

Prove that $f \in \mathbb{Q}[X]$.

Hint. Confirm that

$$f = \left(X^2-(2+\sqrt{2})(-3+\sqrt{3})\right)\left(X^2-(2-\sqrt{2})(-3+\sqrt{3})\right)$$
$$\left(X^2-(2+\sqrt{2})(-3-\sqrt{3})\right)\left(X^2-(2-\sqrt{2})(-3-\sqrt{3})\right),$$

revealing that $f \in \mathbb{Q}[\sqrt{2},\sqrt{3}][X]$. Then show that each automorphism σ of $\mathbb{Q}[\sqrt{2},\sqrt{3}]$, when extended naturally to an automorphism of $\mathbb{Q}[\sqrt{2},\sqrt{3}][X]$, is such that $\sigma(f) = f$. Deduce from this that $f \in \mathbb{Q}[X]$.

(d) Explain why f is the minimal polynomial of α.

(e) Let $\beta = \left(\sqrt{2-\sqrt{2}}\right)\left(\sqrt{-3+\sqrt{3}}\right)$, one of the other roots of f. Verify that $\alpha\beta = \sqrt{2}(-3+\sqrt{3}) \in \mathbb{Q}[\alpha]$, and deduce that $\beta \in \mathbb{Q}[\alpha]$. Repeat a similar verification for all roots of f.

(f) Explain why $\mathbb{Q}[\alpha]$ is a Galois extension of $\mathbb{Q}$.

(g) Explain why $\mathrm{Gal}(\mathbb{Q}[\alpha]/\mathbb{Q})$ has order 8.

Why is it that for each root γ of f there is an automorphism σ of $\mathbb{Q}[\alpha]$ such that $\sigma : \alpha \mapsto \gamma$?

(h) With β as in part (e), let φ be the automorphism of $\mathbb{Q}[\alpha]$ defined by $\alpha \mapsto \beta$. Prove that

$$\varphi(\sqrt{2}) = -\sqrt{2},\ \varphi(\sqrt{3}) = \sqrt{3},\ \varphi(\alpha\beta) = -\alpha\beta,\ \varphi(\beta) = -\alpha.$$

Hint. Use the fact $\varphi(\alpha^2) = \beta^2$, along with the fact that $\varphi(\sqrt{2}) = \pm\sqrt{2}, \varphi(\sqrt{3}) = \pm\sqrt{3}$.

(i) Evaluate $\varphi^2(\alpha)$ and $\varphi^3(\alpha)$, and deduce that φ has order 4.

This also shows that φ^3 is the inverse of φ, also of order 4.

(j) Let $\gamma = \left(\sqrt{2+\sqrt{2}}\right)\left(\sqrt{-3-\sqrt{3}}\right)$, another root of f. Let ψ be the automorphism of $\mathbb{Q}[\alpha]$ such that $\alpha \mapsto \gamma$. As in the preceding parts verify that ψ has order 4.

Since the inverse of ψ is ψ^3, at this point we have four automorphisms of order 4.

(k) Let $\delta = \left(\sqrt{2-\sqrt{2}}\right)\left(\sqrt{-3-\sqrt{3}}\right)$, yet another root of f. Let η be the automorphism of $\mathbb{Q}[\alpha]$ such that $\alpha \mapsto \gamma$. As in the preceding parts verify that η has order 4.

Show that $\mathrm{Gal}(\mathbb{Q}[\alpha]/\mathbb{Q})$ has six automorphisms of order 4, one automorphism of order 2, plus the identity of order one.

(l) Verify that $\eta = \psi\varphi$.

(m) Explain why $\mathrm{Gal}(\mathbb{Q}[\alpha]/\mathbb{Q})$ is isomorphic to the quaternion group $\mathcal{Q}$.

(n) How many subfields does $\mathbb{Q}[\alpha]$ have?

7 Applications of Galois Theory

The development of Galois theory would be for naught without seeing its application in the advancement of algebra. We examine a variety of situations that fit into the framework of the Galois correspondence.

7.1 Three Classical Theorems

The primitive generator theorem, the fundamental theorem of algebra and the symmetric function theorem are major results which can be proven in varied ways. Their proofs by means of the Galois correspondence are worthy of attention.

7.1.1 The Primitive Generator Theorem

A field K is a finite extension of F if and only if $K = F[\alpha_1, \ldots, \alpha_n]$ for some α_j that are algebraic over F. The choice of generators α_j can vary a great deal. Naturally we might ask for a minimal number of generators for the field K. For instance, the field $\mathbb{Q}[\sqrt{2}, \sqrt{3}]$, presented with two generators, coincides with $\mathbb{Q}[\sqrt{2} + \sqrt{3}]$ where only one generator is needed. It turns out that in many important cases, such as in characteristic 0, all finite extensions can be built from just one generator. If K can be presented as $F[\gamma]$, where γ is a single algebraic element, then γ is often known as a ***primitive generator*** for K, and the extension K over F is called a ***simple extension.***[1]

It turns out that the existence of a primitive element for a finite extension relates to the number of subfields that sit between the base field and the extension.

Proposition 7.1. *A finite extension K of a field F has a primitive generator if and only if there are only finitely many fields between F and K.*

Proof. Suppose that K has only finitely many subfields that contain F. The case where F is finite gets a different treatment from the case where F is infinite.

If the field F is finite, then the field K is finite also. By Proposition 4.37, the group $K^\star$ of non-zero elements in K is cyclic. If β is a generator for this cyclic group, then clearly $K = F[\beta]$, as desired.

If F is infinite, then K is generated over F by finitely many algebraic elements. Suppose to start with that $K = F[\alpha, \beta]$ using only two generators α and β. As θ runs over the infinite F,

[1] It could also be argued that algebraists are quite primitive in their use of the term "simple."

the assumption forces some of the subfields $F[\alpha + \theta\beta]$ to duplicate themselves. Thus there are distinct θ_1, θ_2 in F such that $F[\alpha + \theta_1\beta] = F[\alpha + \theta_2\beta]$. As both $\alpha + \theta_1\beta$ and $\alpha + \theta_2\beta$ are in $F[\alpha + \theta_1\beta]$, their difference $(\theta_2 - \theta_1)\beta \in F[\alpha + \theta_1\beta]$. Since $0 \neq \theta_2 - \theta_1 \in F$ it follows that $\beta \in F[\alpha + \theta_1\beta]$. Also then $\alpha = (\alpha + \theta_1\beta) - \theta_1\beta \in F[\alpha + \theta_1\beta]$. This shows that $K = F[\alpha, \beta] = F[\alpha + \theta_1\beta]$, making $\alpha + \theta_1\beta$ a suitable primitive generator for K.

In the general case where $K = F[\alpha_1, \alpha_2, \ldots, \alpha_n]$ with any number of generators, note that $F[\alpha_1, \alpha_2]$ still has only finitely many subfields, and by the case just done $F[\alpha_1, \alpha_2] = F[\beta_1]$ using a primitive β_1 for this field. Then $K = F[\beta_1, \alpha_3, \ldots, \alpha_n]$, where the number of needed generators has been knocked down by one. Repeat the preceding argument until a single generator for K is obtained.

The converse, although less important for our purposes, has a slightly more subtle proof. Suppose that $K = F[\gamma]$ for a single γ algebraic over F. For any intermediate field E between F and K, we still have $K = E[\gamma]$, and the minimal polynomial f of γ over E has degree $[K : E]$. Write $f = X^n + a_{n-1}X^{n-1} + \cdots + a_1X + a_0$ where the $a_j \in E$, and let $L = F[a_0, a_1, \ldots, a_{n-1}]$. Clearly $F \subseteq L \subseteq E \subseteq K$, and $f \in L[X]$. Since f is irreducible in $E[X]$, it remains irreducible in the smaller ring $L[X]$. Now K equals $L[\gamma]$, forcing $[K : L]$ to equal $\deg f$ too. The tower theorem says that $[K : L] = [K : E][E : L]$, whereby $[E : L] = 1$ and then $E = L$.

We have just seen that every intermediate E between K and F is completely determined by the minimal polynomial of γ over E. These minimal polynomials vary over different intermediate subfields, but they all divide the minimal polynomial of γ over the base field F. The division occurs in $K[X]$. So, the number of subfields of K that contain F is at most the number of monic divisors of the minimal polynomial of γ over F, and that upper bound is finite. □

It follows from the preceding result that every Galois extension K of F can be generated over F by a single algebraic element. That is because its Galois group $\mathrm{Gal}(K/F)$, being finite, has only finitely many subgroups. By the Galois correspondence in Proposition 6.77, there are only finitely many subfields between F and K. So, Proposition 7.1 kicks in. It also follows from this that if E extends F and E sits inside some Galois extension K of F, then E too has a primitive generator. This provides a line of attack for the primitive generator theorem.

Proposition 7.2 (Primitive generator). *Every finite extension E of a field F of characteristic* 0 *has a primitive generator.*

Proof. According to Proposition 6.44, E is a subfield of a splitting field K over F. Being in characteristic 0 the polynomial for which K splits can be made separable. By Proposition 6.73 K is a Galois extension. Thus, K has only finitely many subfields containing F. Hence, E has only finitely many subfields containing F. By Proposition 7.1 E has a primitive generator. □

One can also prove the primitive generator theorem directly, without using the Galois correspondence, and then use this theorem to prove that the Galois correspondence is bijective.

7.1.2 The Fundamental Theorem of Algebra

Every non-constant polynomial in $\mathbb{C}[X]$ splits over $\mathbb{C}$. This result, first proven by C. F. Gauss (1777–1855) and commonly known as the ***fundamental theorem of algebra***, comes with a wide variety of proofs. Together with the fact that every polynomial of odd degree in $\mathbb{R}[X]$ has a real root, plus a bit of Sylow's theorem, Proposition 3.30, the Galois correspondence can be used to provide one of those proofs.

Proposition 7.3 (Fundamental theorem of algebra). *The only algebraic extension of $\mathbb{C}$ is $\mathbb{C}$ itself. Thus every non-constant polynomial in $\mathbb{C}[X]$ splits over $\mathbb{C}$.*

Proof. Suppose to the contrary that L is an algebraic extension of $\mathbb{C}$, but that $\mathbb{C} \subsetneq L$.

Since $\mathbb{C}$ is algebraic over $\mathbb{R}$, it follows that L is also algebraic over $\mathbb{R}$. Choose θ in $L \setminus \mathbb{C}$ and let g be the minimal polynomial of θ over $\mathbb{R}$ (not over $\mathbb{C}$). Then let K be the splitting field of g over $\mathbb{C}$ (not over $\mathbb{R}$). Thus

$$\mathbb{R} \subsetneq \mathbb{C} \subsetneq K.$$

It should be clear that K is also the splitting field of $(X^2+1)g$ over $\mathbb{R}$. So, in addition to being a Galois extension of $\mathbb{C}$, our K is a Galois extension of $\mathbb{R}$.

Write the order of $\mathrm{Gal}(K/\mathbb{R})$ as

$$|\mathrm{Gal}(K/\mathbb{R})| = 2^j m \text{ for some } j \geq 0 \text{ and some odd } m.$$

Sylow's theorem, Proposition 3.30, provides a subgroup H of $\mathrm{Gal}(K/\mathbb{R})$ such that $|H| = 2^j$.

For $E = \mathrm{Fix}(H)$, the Galois correspondence, Proposition 6.77, gives $[K : E] = 2^j$, and then the tower theorem gives $[E : \mathbb{R}] = m$, an odd integer.

It turns out that $m = 1$. Indeed, let $\alpha \in E$ with minimal polynomial h over $\mathbb{R}$. The degree of h divides m because $\deg g = [\mathbb{R}[\alpha] : \mathbb{R}]$, which divides $[E : \mathbb{R}] = m$. So $\deg h$ is odd. But then h has a root in $\mathbb{R}$, using ordinary calculus. Because h is also irreducible over $\mathbb{R}$, h must be linear. So $\alpha \in \mathbb{R}$, whereby $E = \mathbb{R}$ and $m = 1$.

Thus the order of $\mathrm{Gal}(K/\mathbb{R})$ is a power of 2, and the order of its subgroup $\mathrm{Gal}(K/\mathbb{C})$ is also 2^ℓ for some ℓ. This $\ell \geq 1$ due to the fact $\mathbb{C} \subsetneq K$. By Proposition 3.32 the 2-group $\mathrm{Gal}(K/\mathbb{C})$ contains a subgroup, call it N, of order $2^{\ell-1}$. From the Galois correspondence, Proposition 6.77, $[K : \mathrm{Fix}(N)] = 2^{\ell-1}$, and thereby $[\mathrm{Fix}(N) : \mathbb{C}] = 2$.

So, if $\mathbb{C}$ has a proper algebraic extension, then $\mathbb{C}$ has an extension of degree 2. Consequently there must be an irreducible quadratic polynomial in $\mathbb{C}[X]$. Namely, the minimal polynomial of any element in the quadratic extension that is not in $\mathbb{C}$. Since the familiar quadratic formula provides roots in $\mathbb{C}$ to every quadratic polynomial in $\mathbb{C}[X]$, a contradiction ensues.

Every non-constant polynomial in $\mathbb{C}[X]$ now splits over $\mathbb{C}$ because the splitting field of such a polynomial must be $\mathbb{C}$. □

7.1.3 The Symmetric Function Theorem

Here let $K = L(t_1, t_2, \ldots, t_n)$ be the field of rational functions in n indeterminates t_j and having coefficients in any base field L. The symmetric group S_n of permutations on n

letters is realized as a group of L-automorphisms of K as follows. Suppose $\sigma \in S_n$ and $f = f(t_1, t_2, \ldots, t_n) \in K$, and let

$$\sigma f = f(t_{\sigma(1)}, t_{\sigma(2)}, \ldots, t_{\sigma(n)}).$$

Thus, σ permutes the variables t_j in accordance with how σ permutes their subscripts. For instance, if σ is the cycle $(1, 2, 4)$ in S_4 then $\sigma : f(t_1, t_2, t_3, t_4) \mapsto f(t_2, t_4, t_3, t_1)$. Even more concretely $\sigma : t_1^2 t_2 + t_3 t_4^3 - t_1 t_2 t_3 t \mapsto t_2^2 t_4 + t_3 t_1^3 - t_2 t_4 t_3$.

Since S_n is a finite group of automorphisms of K, Artin's theorem, Proposition 6.76, establishes that K is a Galois extension of $\mathrm{Fix}(S_n)$. Let $F = \mathrm{Fix}(S_n)$ for brevity. A rational function $f(t_1, t_2, \ldots, t_n)$ lies in the fixed field F if and only if

$$f(t_{\sigma(1)}, t_{\sigma(2)}, \ldots, t_{\sigma(n)}) = f(t_1, t_2, \ldots, t_n) \text{ for every } \sigma \text{ in } S_n.$$

Definition 7.4. A function of this sort is called a ***symmetric function***.

Here is an important list of symmetric functions:

- the constant function $s_0 = 1$,
- the sum of all the variables $s_1 = t_1 + t_2 + \cdots + t_n$,
- the sum of the paired products $s_2 = \sum_{j_1 < j_2} t_{j_1} t_{j_2}$,
- the sum of the triple products $s_3 = \sum_{j_1 < j_2 < j_3} t_{j_1} t_{j_2} t_{j_3}$,
- in general, the sum of the r-fold products $s_r = \sum_{j_1 < j_2 < \cdots < j_r} t_{j_1} t_{j_2} \cdots t_{j_r}$,
- and finally, the product of all the variables $s_n = t_1 t_2 \cdots t_n$.

For instance with $n = 4$,

$$s_2 = t_1 t_2 + t_1 t_3 + t_1 t_4 + t_2 t_3 + t_2 t_4 + t_3 t_4$$

and

$$s_3 = t_1 t_2 t_3 + t_1 t_2 t_4 + t_1 t_3 t_4 + t_2 t_3 t_4.$$

Definition 7.5. The above functions are called the ***elementary symmetric functions***.

It should be evident that these s_r are fixed under all permutations of the list of variables t_j, i.e. that $s_r \in F$.

The number of products to be added in forming the general s_r equals the binomial coefficient $\binom{n}{r}$ which counts the number of ways to choose r variables from the set of n variables.

Here is a symmetric function that is not elementary

$$\frac{t_1^3 + t_2^3 + t_3^3 + t_4^3}{t_1 t_2 t_3 t_4},$$

and there are infinitely many more.

Together with the base field L the elementary symmetric functions generate a subfield of F, namely

$$E = L(s_0, s_1, s_2, \ldots, s_n).$$

(Here s_0 is redundant since it already lies in the base field L.) So we get a tower of fields

$$L \subset E = L(s_0, s_1, \ldots, s_n) \subseteq F = \mathrm{Fix}(\mathcal{S}_n) \subset K = L(t_1, t_1, \ldots, t_n).$$

As already noted, K is a Galois extension of F due to Artin's theorem. But in addition, K is a Galois extension of E. This holds because K is the splitting field of the separable polynomial

$$f = (X - t_1)(X - t_2)(X - t_3) \cdots (X - t_n).$$

Indeed, at first glance f lies in $K[X]$, but the expansion of f comes out to

$$f = X^n - s_1 X^{n-1} + s_2 X^{n-2} + \cdots + (-1)^r s_r X^{n-r} + \cdots + (-1)^{n-1} s_{n-1} X + (-1)^n s_n,$$

which reveals that $f \in E[X]$. The t_j are now algebraic over E because they are the roots of f, which is in $E[X]$. Clearly $K = E[t_1, t_2, \ldots, t_n]$. Furthermore f is separable since its roots are the distinct t_j.

We have enough information for a lovely result.

Proposition 7.6 (Symmetric function theorem). *If $K = L(t_1, t_2, \ldots, t_n)$ is the field of rational functions in n variables with coefficients in a field L, F is the subfield consisting of the symmetric functions, and E is the subfield of F generated over L by the elementary symmetric functions, then $E = F$.*

More concisely, every symmetric function is a rational expression in the elementary symmetric functions.

Proof. There is the tower $E = L(s_0, s_1, s_2, \ldots, s_n) \subseteq F \subset K$. The field K is a Galois extension of both $F = \mathrm{Fix}(\mathcal{S}_n)$ and the smaller field E. The first claim comes from Artin's theorem, Proposition 6.76, which also tells us that

$$[K : F] = |\mathrm{Gal}(K/F)| = |\mathcal{S}_n| = n!\,.$$

The second claim comes from the fact, as noted above, that K is the splitting field of a separable polynomial of degree n over E. According to Proposition 6.45, which bounds the degree of a splitting field, $[K : E] \leq n!$. From the inclusions $E \subseteq F \subset K$ it follows that $n! = [K : F] \leq [K : E] \leq n!$. Thus $[K : F] = [K : E]$, whereby $E = F$. □

For example, take the symmetric function $r = \frac{1}{t_1} + \frac{1}{t_2} + \frac{1}{t_3} + \frac{1}{t_4}$ in four indeterminates. Rewrite this to get $r = \dfrac{t_2 t_3 t_4 + t_1 t_3 t_4 + t_1 t_2 t_4 + t_1 t_2 t_3}{t_1 t_2 t_3 t_4} = \dfrac{s_3}{s_4}$, which is a rational expression in the elementary symmetric functions.

For another example, take the symmetric function $r = t_1^2 + t_2^2 + t_3^2$ in three variables. Notice that

$$s_1^2 = (t_1 + t_2 + t_3)^2 = t_1^2 + t_2^2 + t_3^2 + 2(t_1 t_2 + t_1 t_3 + t_2 t_3) = r + 2s_2.$$

Thus $r = s_1^2 - 2s_2$, a polynomial in the elementary symmetric functions.

For one more example, take the symmetric function $r = (t_1 - t_2)^2$ in two variables. Then

$$r = t_1^2 + t_2^2 - 2t_1t_2 = (t_1 + t_2)^2 - 4t_1t_2 = s_1^2 - 4s_2.$$

Incidentally note that

$$(X - t_1)(X - t_2) = X^2 - (t_1 + t_2)X + t_1t_2 = X^2 - s_1X + s_2.$$

Together with the preceding identity this shows that the familiar discriminant $s_1^2 - 4s_2$ of the quadratic polynomial $X^2 - s_1X + s_2$ equals the square of the difference of its roots.

EXERCISES

1. Suppose K is a Galois extension of F and let $\alpha \in K$. Show that $K = F[\alpha]$ if and only if the only automorphism in $\mathrm{Gal}(K/F)$ which fixes α is the identity map.
2. Let $K = L(t_1, t_2, t_3)$, the field of rational functions in the variables t_1, t_2, t_3 over a field L. Let s_0, s_1, s_2, s_3 be the elementary symmetric functions in these variables. Then $F = L(s_0, s_1, s_2, s_3)$ is the fixed field of the group $\mathcal{S}_3$ of automorphisms which permutes the variables. Show that the function $\alpha = t_1 + 2t_2 + 3t_3$ is such that $K = F[\alpha]$.
3. Suppose $K = \mathbb{Q}(t_1, t_2, \ldots, t_n)$, where the t_j are indeterminates and let $\mathcal{S}_n$, the group of permutations on n letters, act on K by permuting the t_j in the obvious way. The fixed field F consists of the symmetric functions. According to Artin's theorem K becomes a Galois extension of F of degree $n!$. The primitive element theorem ensures that this extension K over F has a primitive generator. Find a primitive generator of K over F.

   Give a recipe for finding all primitive generators.
4. Let F be a field of characteristic zero and let K be an extension of F, which is algebraic but not finite. Show that there exist elements in K whose minimal polynomials over F have arbitrarily high degree.
5. If K is a Galois extension of a field F and H is a subgroup of $\mathrm{Gal}(K/F)$, show that there is an element α in K such that $H = \{\sigma \in \mathrm{Gal}(K/F) : \sigma(\alpha) = \alpha\}$, i.e. such that H is the stabilizer of α.
6. Let K be the field $\mathbb{Z}_2(s, t)$ of rational functions in the indeterminates s, t and having coefficients in the two element field $\mathbb{Z}_2$. Let $F = \mathbb{Z}_2(s^2, t^2)$, the subfield of K generated by s^2, t^2.
   (a) Show that s, t are algebraic over F, that $K = F[s, t]$ and $[K : F] = 4$.
   (b) Prove that there is no element α in K such that $K = F[\alpha]$.
   (c) Find infinitely many intermediate fields between F and K.

   This example illustrates that the primitive generator theorem does not apply to all finite algebraic extensions.
7. Let t_1, t_2, t_3, t_4 be indeterminates and let

$$f(t_1, t_2, t_3, t_4) = (t_1 + t_2 - t_3 - t_4)(t_1 + t_3 - t_2 - t_4)(t_1 + t_4 - t_2 - t_3)$$

   a polynomial in these four variables over $\mathbb{Q}$.

(a) Show that f is a symmetric function.

Hint. The group $\mathcal{S}_4$ is generated by the permutation $(1,2)$ and the subgroup of permutations that fix 1.

(b) If g is a degree 4 polynomial in $\mathbb{Q}[X]$ with distinct roots r_1, r_2, r_3, r_4, show that $f(r_1, r_2, r_3, r_4) \in \mathbb{Q}$.

7.2 Special Extensions and Their Galois Groups

Here we examine the Galois groups of finite field extensions, of cubics, and of the so-called cyclotomic polynomials.

7.2.1 The Galois Correspondence in Finite Fields

Every finite field K contains $\mathbb{Z}_p$ for some prime p, and has size p^n, where $n = [K : \mathbb{Z}_p]$. For each prime power p^n, there is just one field of that size, up to isomorphism. That field is the splitting field of the separable polynomial $f = X^{p^n} - X$ in $\mathbb{Z}_p[X]$. Let us denote the unique field of order p^n by $\mathbb{F}_{p^n}$. In particular, $\mathbb{Z}_p = \mathbb{F}_p$.

The polynomial f is separable because its derivative is -1, which is certainly coprime with f. The separability is also clear because the splitting field of f is precisely the set of roots of f and the number of such roots is p^n, which coincides with $\deg f$. Being the splitting field of a separable polynomial over $\mathbb{F}_p$, the finite field $\mathbb{F}_{p^n}$ is a Galois extension of $\mathbb{F}_p$ as laid out in Proposition 6.73. Since the degree $[\mathbb{F}_{p^n} : \mathbb{F}_p] = n$, the order of $\mathrm{Gal}(\mathbb{F}_{p^n}/\mathbb{F}_p)$ is also n. So, what is this group of order n like?

Well, $\mathrm{Gal}(\mathbb{F}_{p^n}/\mathbb{F}_p)$ contains the Frobenius map

$$\sigma : \mathbb{F}_{p^n} \to \mathbb{F}_{p^n}, \text{ where } \alpha \mapsto \alpha^p.$$

Suppose j is the order of σ. By Lagrange j divides n. For every α in $\mathbb{F}_{p^n}$ we have

$$\alpha^{p^j} = \sigma^j(\alpha) = \alpha.$$

This means that every α in $\mathbb{F}_{p^n}$ is a root of $X^{p^j} - X$. Since K has p^n elements, it must be that $j = n$ to allow for this polynomial to have p^n roots. The group $\mathrm{Gal}(\mathbb{F}_{p^n}/\mathbb{F}_p)$ of order n contains an automorphism of order n, which proves the following.

Proposition 7.7. *Every finite field $\mathbb{F}_{p^n}$ is a Galois extension of $\mathbb{F}_p$ for some prime p, and $\mathrm{Gal}(\mathbb{F}_{p^n}/\mathbb{F}_p)$ is a cyclic group generated by the Frobenius map on $\mathbb{F}_{p^n}$.*

If G is a cyclic group of order n with generator σ and j is a divisor of n, then the subgroup generated by σ^j has order n/j. According to Proposition 2.41, all subgroups of G are accounted for in this way.

In particular, the unique subgroup H of order n/j inside $\mathrm{Gal}(\mathbb{F}_{p^n}/\mathbb{F}_p)$ is the subgroup generated by σ^j where σ is the Frobenius map. For this cyclic subgroup of $\mathrm{Gal}(\mathbb{F}_{p^n}/\mathbb{F}_p)$ an element α in $\mathbb{F}_{p^n}$ is fixed by H if and only if α is fixed by the generator σ^j of H. Thus

$$\mathrm{Fix}(H) = \{\alpha \in \mathbb{F}_{p^n} : \sigma^j(\alpha) = \alpha\} = \{\alpha \in \mathbb{F}_{p^n} : \alpha^{p^j} = \alpha\}.$$

When j divides n, we learn that the fixed field $\mathrm{Fix}(H)$ is the set of roots of the polynomial $X^{p^j} - X$. This fixed field is a finite field of size p^j. Inside $\mathbb{F}_{p^n}$ there can only be one subfield of size p^j. That subfield is the set of roots of $X^{p^j} - X$. This is in concert with the fundamental theorem of Galois theory, Proposition 6.77, which tells us that the subfields of K are precisely the fixed fields $\mathrm{Fix}(H)$ as H runs over the subgroups of $\mathrm{Gal}(K/\mathbb{Z}_p)$.

This discussion can be summarized in the following result.

Proposition 7.8. *Let $\mathbb{F}_{p^n}$ be the finite field of size p^n for some prime p and some exponent n, and let σ be the Frobenius automorphism on $\mathbb{F}_{p^n}$. For each positive divisor j of n, the set of roots of $X^{p^j} - X$ is the fixed subfield of the subgroup of* $\mathrm{Gal}(K/\mathbb{Z}_p)$ *generated by σ^j. That fixed field is $\mathbb{F}_{p^j}$. All subfields of $\mathbb{F}_{p^n}$ are accounted for in this way.*

For an illustration of Proposition 7.8, take $\mathbb{F}_{p^{24}}$. The Galois group $\mathrm{Gal}(\mathbb{F}_{p^{24}}/\mathbb{F}_p)$ is cyclic of order 24. If $\sigma : \alpha \mapsto \alpha^p$ is the Frobenius map on $\mathbb{F}_{p^{24}}$, the subgroups of $\mathrm{Gal}(\mathbb{F}_{p^{24}}/\mathbb{F}_p)$ are the eight cyclic groups generated respectively by

$$\sigma = \sigma^1, \sigma^2, \sigma^3, \sigma^4, \sigma^6, \sigma^8, \sigma^{12} \text{ and } \sigma^{24} = 1 \text{ (the identity map).}$$

The subfields of $\mathbb{F}_{p^{24}}$ are the eight subfields fixed by each of the above generators. Here is the lattice of subgroups of $\mathrm{Gal}(\mathbb{F}_{p^{24}}/\mathbb{F}_p)$ beside its corresponding (inverted) lattice of fixed fields.

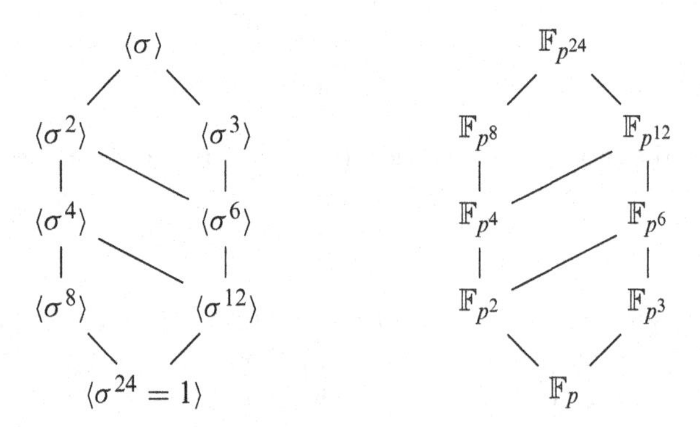

We can pause to observe a few things. The fixed field of $\mathrm{Gal}(\mathbb{F}_{p^{24}}/\mathbb{F}_p)$ (which is generated by the Frobenius map σ) is the bottom field $\mathbb{F}_p$. The fixed field of the trivial group $\langle\sigma^{24}\rangle$ is all of $\mathbb{F}_{p^{24}}$. The order of $\langle\sigma^2\rangle$ is 12. Its fixed field is $\mathbb{F}_{p^2}$. The degree of this field over $\mathbb{F}_p$ is 2. In accordance with Artin's theorem, the degree of $\mathbb{F}_{p^{24}}$ over the fixed field $\mathbb{F}_{p^2}$ is 12, which is also consistent with the tower theorem. Similar observations apply to all of the subfields of $\mathbb{F}_{p^{24}}$.

7.2.2 The Galois Groups of Cubics

If f is a monic irreducible quadratic over a field F, Proposition 6.46 says that its splitting field K has degree 2 over F, and if that quadratic is separable, K is a Galois extension of F. When the characteristic of F is not 2, the separability of f is assured. Indeed, let $f = X^2 + bX + c$. Then $f' = 2X + b$, which is not zero because 2 is not zero in F. In this quadratic case the characterization of Galois extensions, Proposition 6.73, shows that $\mathrm{Gal}(K/F)$ has order 2. And, up to isomorphism, there is only one group of order 2.

With that out of the way, let us explore the possibilities for the Galois group of a separable cubic polynomial f in $F[X]$ of degree 3. Of course, if f splits over F, then the Galois group is trivial. If f has a linear and an irreducible quadratic factor, then its Galois group is the same as the Galois group of the quadratic factor, which is a group of order 2.

We are left with the case of an irreducible and separable cubic $f = X^3 + aX^2 + bX + c$ over a field F. We might note that if the characteristic of F is not 3, the irreducibility of f forces its separability because $f' = 3X^2 + 2aX + b \neq 0$. Let K be the splitting field of this f. Proposition 6.45 shows that $[K : F]$ divides 6, and is divisible by 3. Hence $[K : F]$ is either 3 or 6. By the characterization of Galois extensions, Proposition 6.73, $\mathrm{Gal}(K/F)$ has order 3 or order 6.

To say a bit more let $\Delta = \{\alpha_1, \alpha_2, \alpha_3\}$ be the set of distinct roots of f. These roots generate K over F. We know from Proposition 6.67 that $\mathrm{Gal}(K/F)$ acts as a subgroup of the group $\mathcal{S}(\Delta)$ of permutations of the three roots. Being a group of permutations on three letters (i.e. the roots), $\mathcal{S}(\Delta)$ is the same as $\mathcal{S}_3$ of order 6. The only subgroup of $\mathcal{S}_3$ having order 3 is the alternating group $\mathcal{A}_3$ of even permutations and the only subgroup of order 6 is $\mathcal{S}_3$ itself. This discussion gets us the following result.

Proposition 7.9. *The Galois group of a separable irreducible cubic polynomial is (isomorphic to) either the alternating group $\mathcal{A}_3$ or the full symmetric group $\mathcal{S}_3$.*

It would be satisfying to discover by a simple test on the polynomial f whether the Galois group of f is $\mathcal{A}_3$ or $\mathcal{S}_3$. The answer lies in the nature of what is known as the discriminant of a cubic.

By way of motivation recall that for a quadratic $X^2 + bX + c$, its discriminant is $b^2 - 4c$. If this quadratic splits as $(X - \alpha_1)(X - \alpha_2)$, then

$$b = -(\alpha_1 + \alpha_2), \quad c = \alpha_1\alpha_2,$$
$$b^2 - 4c = (-(\alpha_1 + \alpha_2))^2 - 4\alpha_1\alpha_2 = (\alpha_1 - \alpha_2)^2.$$

The discriminant of a quadratic is the square of the difference of its roots.

Definition 7.10. For a cubic $f = X^3 + aX^2 + bX + c$ in $F[X]$ with splitting field K and factorization $(X - \alpha_1)(X - \alpha_2)(X - \alpha_3)$ in $K[X]$, the ***discriminant*** of f is the quantity

$$D = (\alpha_1 - \alpha_2)^2(\alpha_1 - \alpha_3)^2(\alpha_2 - \alpha_3)^2.$$

By inspection it becomes clear that the definition of D does not depend on how the roots have been ordered. For polynomials of higher degree there is an analogous definition, but here we focus on cubics. Evidently $D \in K$. But in fact $D \in F$. Indeed, if f is not separable, then $D = 0$. If f is separable, then K is a Galois extension of F. By the characterization of Galois extensions, Proposition 6.73, $\text{Fix}(\text{Gal}(K/F)) = F$. Since $\text{Gal}(K/F)$ restricts to a group of permutations of the three roots, we see that D is fixed by $\text{Gal}(K/F)$, and so $D \in F$.

By its very definition, D is the square of an element of K, and as seen $D \in F$. However, its square roots need not be in F.

Proposition 7.11. *Let f be a separable irreducible cubic over a field F of characteristic not 2. The square roots of the discriminant of f are in F if and only if* $\text{Gal}(K/F)$ *is isomorphic to* $\mathcal{A}_3$.

Proof. Let $\alpha_1, \alpha_2, \alpha_3$ be the three roots of f in its splitting field K. One of the square roots of D that lie in K is $d = (\alpha_1 - \alpha_2)(\alpha_1 - \alpha_3)(\alpha_2 - \alpha_3)$. The other root $-d$ is not the same as d since we are not in characteristic 2. By Proposition 7.9 the only options for $\text{Gal}(K/F)$ are $\mathcal{S}_3$ and $\mathcal{A}_3$.

If $\text{Gal}(K/F) = \mathcal{S}_3$, there is an automorphism σ of K that transposes α_1 with α_2 and fixes α_3. Then $\sigma(d) = -d \neq d$. Since $\text{Gal}(K/F)$ fixes F, it follows that $d \notin F$.

If $\text{Gal}(K/F) = \mathcal{A}_3$, then the automorphism of K determined by the 3-cycle $\sigma : \alpha_1 \mapsto \alpha_2 \mapsto \alpha_3 \mapsto \alpha_1$ generates $\mathcal{A}_3$. By inspection $\sigma(d) = d$. Thus $d \in \text{Fix}(\text{Gal}(K/F)) = F$. The last equality comes from the characterization of Galois extensions, Proposition 6.73. □

In order for Proposition 7.11 to be applicable, a formula for the discriminant based solely on the coefficients of f, such as we have for quadratics, would be desirable. The formula is complicated. To improve matters, a clever substitution can be used. In addition to avoiding characteristic 2 we also assume that the characteristic of our field is not 3. Given the monic cubic $f(X) = X^3 + aX^2 + bX + c$, let $g(X) = f(X - a/3)$. This substitution is known as a ***Tschirnhausen transform***, after Ehrenfried von Tschirnhaus (1651–1708). If $\alpha_1, \alpha_2, \alpha_3$ are the roots of f in its splitting field K, the roots of g are

$$r_1 = \alpha_1 + a/3,\ r_2 = \alpha_2 + a/3,\ r_3 = \alpha_3 + a/3.$$

Clearly g has the same splitting field as f, and just as clearly the discriminant of g equals the discriminant of f. But g comes with an advantage. An explicit calculation of g gives

$$\begin{aligned} g(X) &= (X - a/3)^3 + a(X - a/3)^2 + b(X - a/3) + c \\ &= X^3 + (b - a^2/3)X + a^3/9 - ab/3 + c. \end{aligned}$$

So g is a polynomial with the same splitting field and same discriminant as f, but the coefficient of X^2 in g is 0.

Thus, in the calculation of the discriminant, it is harmless to consider only cubics of the form $g = X^3 + pX + q$ where $p, q \in F$. Still, some calculations are required.

Let r_1, r_2, r_3 be the roots of such g in its splitting field K. Since

$$g = X^3 + pX + q = (X - r_1)(X - r_2)(X - r_3),$$

we obtain

$$g' = 3X^2 + p = (X - r_1)(X - r_2) + (X - r_1)(X - r_3) + (X - r_2)(X - r_3).$$

In turn, substitute r_1, r_2, r_3 for X into the above to obtain

$$\begin{aligned} 3r_1^2 + p &= (r_1 - r_2)(r_1 - r_3) \\ 3r_2^2 + p &= (r_2 - r_1)(r_2 - r_3) \\ 3r_3^2 + p &= (r_3 - r_1)(r_3 - r_2). \end{aligned}$$

Equate the product of the left side with that of the right to see that

$$(3r_1^2 + p)(3r_2^2 + p)(3r_3^2 + p) = -D.$$

Expand the left side to obtain

$$D = -(27r_1^2r_2^2r_3^2 + 9p(r_1^2r_2^2 + r_1^2r_3^2 + r_2^2r_3^2) + 3p^2(r_1^2 + r_2^2 + r_3^2) + p^3).$$

Expand the factorization of g to get

$$g = X^3 - (r_1 + r_2 + r_3)X^2 + (r_1r_2 + r_1r_3 + r_2r_3)X - r_1r_2r_3.$$

Thus

$$\begin{aligned} r_1 + r_2 + r_3 &= 0 \\ r_1r_2 + r_1r_3 + r_2r_3 &= p \\ r_1r_2r_3 &= -q. \end{aligned}$$

And then

$$r_1^2 + r_2^2 + r_3^2 = (r_1 + r_2 + r_3)^2 - 2(r_1r_2 + r_1r_3 + r_2r_3) = -2p,$$

as well as

$$r_1^2r_2^2 + r_1^2r_3^2 + r_2^2r_3^2 = (r_1r_2 + r_1r_3 + r_2r_3)^2 - 2(r_1 + r_2 + r_3)r_1r_2r_2 = p^2.$$

Put these values into the expression for D up above to get

$$D = -4p^3 - 27q^2.$$

Let us gather what we have found.

Proposition 7.12. *If F is a field whose characteristic is neither 2 nor 3, then every irreducible cubic f in $F[X]$ is separable, and there is an irreducible cubic of the form $g = X^3 + pX + q$ whose splitting field and discriminant coincide with those of f. The discriminant D of such g equals $-4p^3 - 27q^2$. The Galois group of g is the cyclic group $\mathcal{A}_3$ if and only if D is a square in F. Otherwise, the Galois group is $\mathcal{S}_3$.*

To illustrate Proposition 7.12, take the polynomial X^3-3X+1 over the rationals $\mathbb{Q}$. Having no roots in $\mathbb{Q}$, this cubic is irreducible over $\mathbb{Q}$. Its discriminant is $-4(-3)^3 - 27(1)^2 = 81$, which is a square in $\mathbb{Q}$. The Galois group of this polynomial is $\mathcal{A}_3$.

On the other hand, $X^3 - 4X + 2$ is also irreducible over $\mathbb{Q}$. Calculating the discriminant we get $-4(-4)^3 - 27(2)^2 = 148$, which has no square root in $\mathbb{Q}$. The Galois group of this polynomial is the symmetric group $\mathcal{S}_3$.

For an irreducible polynomial of type $X^3 + q$, the discriminant is $-27q^2$, which has no square root in $\mathbb{Q}$. The Galois group of such a polynomial is $\mathcal{S}_3$. We have confirmed this in great detail for $X^3 - 2$.

Suppose F is the finite field $\mathbb{Z}_r$ where the prime r is neither 2 nor 3, and $X^3+pX+q \in \mathbb{Z}_r[X]$ is an irreducible cubic. The splitting field of this polynomial is a degree 3 extension of $\mathbb{Z}_r$. From our knowledge of finite fields, the Galois group of a degree 3 extension of $\mathbb{Z}_r$ is cyclic of order 3, generated by the Frobenius map. In other words, the Galois group is $\mathcal{A}_3$. By Proposition 7.12 the discriminant $-4p^3 - 27q^2$ is a square in $\mathbb{Z}_r$. Putting this in reverse, we come to the curious statement that if $p, q \in \mathbb{Z}_r$ and $-4p^3 - 27q^2$ is not a square in $\mathbb{Z}_r$, then the cubic polynomial $X^3 + pX + q$ has a root in $\mathbb{Z}_r$.

7.2.3 Cyclotomic Extensions

If p is a prime, the splitting field of $X^p - 1$ over $\mathbb{Q}$ is $\mathbb{Q}[\zeta]$, where ζ is the root $e^{2\pi i/p}$. The minimal polynomial of ζ is $X^{p-1} + X^{p-2} + \cdots + X + 1$. So the degree of $\mathbb{Q}[\zeta]$ over $\mathbb{Q}$ is $p-1$ and the Galois group of this splitting field is isomorphic to the group of units $\mathbb{Z}_p^\star$. The latter group is cyclic of order $p-1$. This is all explained in Section 6.4.2.

It is time to enquire about the Galois group of $X^n - 1$ over $\mathbb{Q}$, when n is a positive integer that need not be prime.

The roots of $X^n - 1$ are

$$1, \zeta_n, \zeta_n^2, \ldots, \zeta_n^{n-1} \text{ where } \zeta_n = e^{2\pi i/n}.$$

The set Δ_n of these roots is a cyclic group of order n generated by ζ_n. According to Proposition 2.11 the other generators of Δ_n are the powers ζ_n^k where the k are coprime with n. The number of these generators of Δ_n equals $\phi(n)$, where ϕ is the Euler function. The elements of Δ_n are typically called ***nth roots of unity***, and the generators of Δ_n are known as ***primitive nth roots of unity***. The set of primitive nth roots of unity will be denoted by Λ_n. Since every root of $X^n - 1$ is a power of the root ζ_n, the field $\mathbb{Q}[\zeta_n]$, taken inside $\mathbb{C}$, is the splitting field of $X^n - 1$. Any other primitive nth root of unity will also generate K.

Definition 7.13. The extension $\mathbb{Q}[\zeta_n]$ is commonly known as a ***cyclotomic extension***.[2]

[2] This is the smallest field that contains the roots of $X^n - 1$. The roots of $X^n - 1$ cut the circle into n equal parts. The term "cyclotomic" originates from the Greek "kuklos" for circle and "tomia" for cutting. So for instance, the term "atomic" meant "uncuttable." After the term settled in, the presumably indivisible atom did get split.

In order to grasp the Galois group of $X^n - 1$, the degree of its splitting field $\mathbb{Q}[\zeta_n]$ over $\mathbb{Q}$ is required. For that we seek the minimal polynomial of ζ_n. If $d \mid n$, then $X^d - 1$ divides $X^n - 1$, which shows that the group Δ_d of dth roots of unity is a subgroup of Δ_n. Inside Δ_d there is the set Λ_d of primitive dth roots of unity. This Λ_d coincides with the set of elements in Δ_n whose order is d. Since the order of every element in the cyclic group Δ_n divides n we see that

$$\Delta_n = \bigcup_{d|n} \Lambda_d, \text{ and this union is disjoint.}$$

Thus in $K[X]$ we can arrange the splitting of $X^n - 1$ like so:

$$X^n - 1 = \prod_{\zeta \in \Delta_n} (X - \zeta) = \prod_{d|n} \left(\prod_{\zeta \in \Lambda_d} (X - \zeta) \right).$$

Definition 7.14. For every positive integer n the polynomial

$$\Phi_n = \prod_{\zeta \in \Lambda_n} (X - \zeta)$$

is known as the nth ***cyclotomic polynomial***. This is the product of the linear factors corresponding to the primitive nth roots of unity.

With these notations

$$X^n - 1 = \prod_{d|n} \Phi_d = \left(\prod_{d|n, d<n} \Phi_d \right) \Phi_n,$$

and so

$$\Phi_n = \frac{X^n - 1}{\prod_{d|n, d<n} \Phi_d}.$$

A recursive method for calculating cyclotomic polynomials reveals itself. Here are the first few cyclotomic polynomials:

$$\begin{aligned}
\Phi_1 &= X - 1 \\
\Phi_2 &= \frac{X^2 - 1}{\Phi_1} = \frac{X^2 - 1}{X - 1} = X + 1 \\
\Phi_3 &= \frac{X^3 - 1}{\Phi_1} = \frac{X^3 - 1}{X - 1} = X^2 + X + 1 \\
\Phi_4 &= \frac{X^4 - 1}{\Phi_1 \Phi_2} = \frac{X^4 - 1}{(X - 1)(X + 1)} = X^2 + 1 \\
\Phi_5 &= \frac{X^5 - 1}{\Phi_1} = \frac{X^5 - 1}{X - 1} = X^4 + X^3 + X^2 + X + 1.
\end{aligned}$$

These early examples can also be found directly from their definition. For instance, with $n = 4$ the primitive 4th roots of unity are $\pm i$. And so $\phi_4 = (X - i)(X + i) = X^2 + 1$. If p is a prime, then

$$\Phi_p = \frac{X^p - 1}{\Phi_1} = \frac{X^p - 1}{X - 1} = X^{p-1} + X^{p-2} + \cdots + X + 1,$$

which is the minimal polynomial of $\zeta_p = e^{2\pi i/p}$. Here are a few more examples:

$$\Phi_6 = \frac{X^6 - 1}{\Phi_1\Phi_2\Phi_3} = \frac{X^6 - 1}{(X-1)(X+1)(X^3 + X + 1)} = X^2 - X + 1$$
$$\Phi_7 = \frac{X^7 - 1}{\Phi_1} = \frac{X^7 - 1}{(X-1)} = X^6 + X^5 + \cdots + X + 1$$
$$\Phi_8 = \frac{X^8 - 1}{\Phi_1\Phi_2\Phi_4} = \frac{X^8 - 1}{(X-1)(X+1)(X^2+1)} = X^4 + 1$$
$$\Phi_9 = \frac{X^4 - 1}{\Phi_1\Phi_3} = \frac{X^9 - 1}{(X-1)(X^2 + X + 1)} = X^6 + X^3 + 1.$$

The degree of Φ_n is the size of Λ_n which is $\phi(n)$, the Euler function evaluated at n. By inspection we might infer that the Φ_n actually lie in $\mathbb{Z}[X]$. This is indeed the case, as we shall see momentarily. We might also infer that the coefficients of Φ_n can only come from $0, 1, -1$. This is not the case, however. Cyclotomic polynomials are out there with arbitrarily large coefficients. Also, the early examples might indicate that cyclotomic polynomials are irreducible over $\mathbb{Q}$. This last observation is true, and by far the most significant matter before us.

For the next couple of results, recall the first item in Gauss' lemma, Proposition 5.32. It implies that if f, g, h are monic polynomials with $f, g \in \mathbb{Z}[X]$ and $f = gh$, then $h \in \mathbb{Z}[X]$ too.

Proposition 7.15. *The cyclotomic polynomials Φ_n lie in $\mathbb{Z}[X]$.*

Proof. Do an induction on n. For $n = 1$ we have $\Phi_1 = X - 1 \in \mathbb{Z}[X]$. Suppose $\Phi_d \in \mathbb{Z}[X]$ for all $d < n$. Let $g = \prod_{d|n, d<n} \Phi_d$, which is also in $\mathbb{Z}[X]$. We have seen that

$$X^n - 1 = g\Phi_n.$$

Since $X^n - 1, g \in \mathbb{Z}[X]$, their quotient $\Phi_n \in \mathbb{Q}[X]$. By item 1 of Gauss' lemma, Proposition 5.32, $\Phi_n \in \mathbb{Z}[X]$. □

There are a number of proofs that cyclotomic polynomials are irreducible over $\mathbb{Q}$. The now standard proof is an adaptation of the ingenious approach of Richard Dedekind (1831–1916). The proof is quite demanding.

Proposition 7.16. *For each positive integer n the minimal polynomial over $\mathbb{Q}$ of the primitive nth root of unity $e^{2\pi i/n}$ is the cyclotomic polynomial Φ_n. Thereby the cyclotomic polynomials are irreducible over $\mathbb{Q}$.*

Proof. Let $\zeta_n = e^{2\pi i/n}$ and let g in $\mathbb{Q}[X]$ be its minimal polynomial. Clearly g divides Φ_n in $\mathbb{Q}[X]$. In order for Φ_n to equal g, it suffices to see that every root of Φ_n is a root of g. The roots of Φ_n, being the primitive nth roots of unity, are powers ζ_n^k where k is coprime with n. Each such $k = p_1p_2\cdots p_j$, where the p_i are primes that do not divide n. To see that ζ_n^k is a root of g, it suffices to verify that the iterated powers

$$\zeta_n^{p_1}, (\zeta_n^{p_1})^{p_2}, ((\zeta_n^{p_1})^{p_2})^{p_3}, \ldots$$

are all roots of g.

For the above, it is enough to show that if ζ is any root of g and p is any prime that does not divide n, then ζ^p remains a root of g. In search of a contradiction, suppose that ζ is a root of g, but that ζ^p is not a root of g. Since g divides Φ_n this ζ is also a root of Φ_n, meaning that ζ is a primitive nth root of unity. Given that $p \nmid n$, the power ζ^p is a primitive nth root of unity as well. This means that ζ^p is a root of Φ_n, but not a root of g.

Write

$$\Phi_n = gh \text{ for some polynomial } h \text{ in } \mathbb{Q}[X].$$

This h is monic because g and Φ_n are monic. Because g, h are monic and in $\mathbb{Q}[X]$, Gauss' lemma, Proposition 5.32, reveals that $g, h \in \mathbb{Z}[X]$.

Since ζ^p is a root of Φ_n but not of g, it must be that $h(\zeta^p) = 0$. Define f in $\mathbb{Z}[X]$ by $f(X) = h(X^p)$. Then $f(\zeta) = h(\zeta^p) = 0$. The fact that g is irreducible and ζ is a root of g imply that g divides f in $\mathbb{Q}[X]$. So we have

$$f = gk \text{ for some } k \text{ in } \mathbb{Q}[X].$$

By item 1 of Gauss' lemma, Proposition 5.32, $k \in \mathbb{Z}[X]$.

We shall need the reduction homomorphism $\mathbb{Z}[X] \to \mathbb{Z}_p[X]$, given in Definition 5.47, which reduces the coefficients of a polynomial modulo p. If $m \in \mathbb{Z}[X]$, let $\overline{m}$ denote its reduction. If $h = \sum b_j X^j$ in $\mathbb{Z}[X]$, the above $f = \sum b_j(X^{pj})$. Using Fermat's theorem for $\mathbb{Z}_p$ and the Frobenius map on $\mathbb{Z}_p[X]$ we see that

$$\overline{f} = \sum \overline{b_j} X^{pj} = \sum \overline{b_j}^p (X^j)^p = \left(\sum \overline{b_j} X^j\right)^p = \overline{h}^p.$$

The reduction modulo p applied to the equation $f = gk$ along with the preceding calculation gives

$$\overline{h}^p = \overline{f} = \overline{g}\overline{k} \text{ in } \mathbb{Z}_p[X].$$

Since $\overline{g}$ is not a constant polynomial (because g is monic and not constant), this reduced polynomial has an irreducible factor, call it $\overline{\ell}$, in $\mathbb{Z}_p[X]$. Now $\overline{\ell}$ divides $\overline{h}^p$ and thus also divides $\overline{h}$ in $\mathbb{Z}_p[X]$.

Reduce coefficients modulo p in the equation $\Phi_n = gh$ to get

$$\overline{\Phi_n} = \overline{g}\overline{h} \text{ inside } \mathbb{Z}_p[X].$$

This reveals that $\overline{\ell}^2$ divides $\overline{\Phi_n}$ in $\mathbb{Z}_p[X]$.

But Φ_n divides $X^n - 1$ in $\mathbb{Z}[X]$. (The other factor is $\prod_{d|n, d<n} \Phi_d$.) This implies that $\overline{\Phi_n}$ divides $X^n - \overline{1}$ in $\mathbb{Z}_p[X]$. And so

$$X^n - \overline{1} = \overline{\ell}^2 \overline{q} \text{ for some polynomial } \overline{q} \text{ in } \mathbb{Z}_p[X].$$

Taking derivatives we get

$$\overline{n}X^{n-1} = \overline{\ell}\left(\overline{\ell}\overline{q}' + \overline{2}\overline{\ell}'\overline{q}\right).$$

We now come to our desired contradiction. Note that $\overline{\ell}$ has non-zero roots because it divides $X^n - \overline{1}$. Thus $\overline{n}X^{n-1}$ has non-zero roots. This can only happen if $\overline{n} = 0$, but $p \nmid n$ and thus $\overline{n} \neq 0$. □

Proposition 7.16 opens up a bounty of insights.

Since the minimal polynomial of ζ_n is Φ_n, the set Λ_n of primitive nth roots of unity comprises the conjugates of ζ_n. Thus Φ_n is the minimal polynomial of any primitive nth root of unity. The degree of $\mathbb{Q}[\zeta_n]$ over $\mathbb{Q}$ is the degree of Φ_n, which is the size of Λ_n, namely the value $\phi(n)$ of the Euler function at n. The field $\mathbb{Q}[\zeta_n]$ is the splitting field of $X^n - 1$, and since we are in characteristic zero, this is a Galois extension of $\mathbb{Q}$. Thus $\mathrm{Gal}(\mathbb{Q}[\zeta_n]/\mathbb{Q})$ is a group of order $\phi(n)$. Every automorphism in $\mathrm{Gal}(\mathbb{Q}[\zeta_n]/\mathbb{Q})$ must map ζ_n to one of its conjugates in Λ_n. The order of $\mathrm{Gal}(\mathbb{Q}[\zeta_n]/\mathbb{Q})$ equals the number of conjugates of ζ_n. For this reason the mappings

$$\sigma_k : \mathbb{Q}[\zeta_n] \to \mathbb{Q}[\zeta_n] \text{ determined by } \sigma_k : \zeta_n \mapsto \zeta_n^k,$$

where k is coprime with n and $1 \leq k < n$, pick up all of $\mathrm{Gal}(\mathbb{Q}[\zeta_n]/\mathbb{Q})$.

Another group of order $\phi(n)$ is $\mathbb{Z}_n^\star$, the group of units modulo n. It might come as no surprise that $\mathrm{Gal}(\mathbb{Q}[\zeta_n]/\mathbb{Q})$ is isomorphic to $\mathbb{Z}_n^\star$. The argument below essentially duplicates the one found in Section 6.4.2, where the cyclotomic extensions for a prime were discussed.

Proposition 7.17. *The Galois group of $X^n - 1$ over $\mathbb{Q}$ is isomorphic to the group of units $\mathbb{Z}_n^\star$ inside the ring $\mathbb{Z}_n$. Consequently, the Galois group of every cyclotomic extension is abelian.*

Proof. The splitting field of $X^n - 1$ is the cyclotomic extension $\mathbb{Q}[\zeta_n]$. For each integer k, coprime with n, let σ_k be the automorphism of $\mathbb{Q}[\zeta_n]$ determined by

$$\sigma_k : \zeta_n \mapsto \zeta_n^k.$$

All automorphisms of $\mathbb{Q}[\zeta_n]$ are picked up in this way. Let $\overline{k}$ be the residue of k modulo n. This residue sits in $\mathbb{Z}_n^\star$, because k is coprime with n. For k, ℓ each coprime with n, we have that

$$\overline{k} = \overline{\ell} \iff k \equiv \ell \bmod n \iff \zeta_n^k = \zeta_n^\ell \iff \sigma_k = \sigma_\ell.$$

The second implication holds because the order of ζ_n in the group Δ_n is n, and the third implication holds because automorphisms are determined by what they do to ζ_n. This establishes the bijection

$$\Gamma : \mathbb{Z}_n^\star \to \mathrm{Gal}(\mathbb{Q}[\zeta_n]/\mathbb{Q}), \text{ well-defined by } \overline{k} \mapsto \sigma_k.$$

All that remains is to see that Γ is a homomorphism between these groups. Well, for k, ℓ coprime with n, we easily see that

$$\sigma_k\sigma_\ell(\zeta_n) = \sigma_k(\zeta_n^\ell) = (\zeta_n^k)^\ell = \zeta_n^{k\ell} = \sigma_{k\ell}(\zeta_n).$$

The second equation holds because σ_k is an automorphism. Since automorphisms of $\mathbb{Q}[\zeta_n]$ are determined by their action on the field generator ζ_n, it follows that $\sigma_k\sigma_\ell = \sigma_{k\ell}$. Hence,

$$\Gamma(\overline{k}\,\overline{\ell}) = \Gamma(\overline{k\ell}) = \sigma_{k\ell} = \sigma_k\sigma_\ell = \Gamma(k)\Gamma(\ell),$$

and that does it. □

For example, the Galois group of $X^{15}-1$ is isomorphic to the group $\mathbb{Z}_{15}^\star$ of units modulo 15. The elements of $\mathbb{Z}_{15}$ are $1, 2, 4, 7, 8, 11, 13, 14$ with multiplication modulo 15. So, the Galois group of $X^{15}-1$ is an abelian group of order 8. All finite abelian groups are direct products of cyclic groups. Up to isomorphism, the possible abelian groups of order 8 are

$$C_8, C_4 \times C_2, C_2 \times C_2 \times C_2,$$

where C_j denotes the cyclic group of order j. Our $\mathbb{Z}_{15}^\star$ has no element of order 8, four elements of order 4, three elements of order 2 and one element of order 1. Thus our Galois group is isomorphic to $C_4 \times C_2$. In the Galois group proper the four elements of order 4 are given by

$$\zeta_{15} \mapsto \zeta_{15}^2, \quad \zeta_{15} \mapsto \zeta_{15}^7, \quad \zeta_{15} \mapsto \zeta_{15}^8, \quad \zeta_{15} \mapsto \zeta_{15}^{13}.$$

The exponents come from the elements of order 4 in $\mathbb{Z}_{15}^\star$.

EXERCISES

1. Find the first 15 cyclotomic polynomials.
2. How many subfields does $\mathbb{Q}[e^{2\pi i/15}]$ possess?
3. Find $\Phi_{20}(X)$. Then factor $X^{20}-1$ using monic irreducible polynomials in $\mathbb{Q}[X]$.
4. This exercise is intended to demonstrate that every quadratic extension of $\mathbb{Q}$ is a subfield of the cyclotomic extension $\mathbb{Q}[\zeta_n]$ for some n. Here $\zeta_n = e^{2\pi i/n}$.
   (a) Show that $\sqrt{2}, \sqrt{-2} \in \mathbb{Q}[\zeta_8]$.
   (b) If p is an odd prime show that $\sqrt{p} \in \mathbb{Q}[\zeta_{4p}]$.
5. If p is prime, how many subfields does a field of size p^{40} contain?
6. If K is a field extension of F with a primitive generator and E is a subfield of K that also extends F, show that E has a primitive generator.
7. Express the symmetric function $(t_1-t_2)^2(t_1-t_3)^2(t_2-t_3)^2$ as a polynomial in the elementary symmetric functions $s_1 = t_1+t_2+t_3, s_2 = t_1t_2+t_1t_3+t_2t_3, s_3 = t_1t_2t_3$.
8. Show that every finite group is the Galois group of some polynomial over some field.
   Note. If the base field you start with is $\mathbb{Q}$, you will have solved a profound research problem.
   Hint. For each positive integer n find an extension whose Galois group is S_n. Then use Cayley's representation theorem for finite groups.
9. Suppose p is a prime other than 2 or 3 and $f = X^3+aX+b$ is an irreducible polynomial in $\mathbb{Z}_p[X]$. Show that its discriminant is a perfect square in $\mathbb{Z}_p$.
   Prove that this holds as well for $p = 2, 3$.

10. Show that for every finite field F and every positive integer n, there exist irreducible polynomials of degree n in $F[X]$.
11. If p, q are rational numbers, $p > 0$, and α is a root of $X^2 + pX + q$, find the degree of $\mathbb{Q}[\alpha]$ over $\mathbb{Q}$.

7.3 Solvability of Equations by Radicals

The problem of finding a formula for the roots of a polynomial has a venerable history, with contributions from the likes of del Ferro (1465–1526), Tartaglia (1500–1557), Cardano (1522–1565), Lagrange (1746–1813), Ruffini (1765–1822) and Abel (1802–1829). This culminated with the revolutionary work of Galois (1811–1831) who settled the matter by showing that the presence of a formula hinges on what has come to be known as the Galois group of the polynomial.

Throughout this section we assume that our fields have characteristic zero. Thus, splitting fields coincide with Galois extensions, because every splitting field is also the splitting field of a separable polynomial.

7.3.1 Cardano's Formula

In 1545 Cardano published his formula for the roots of a cubic, along the lines of what had been discovered earlier by del Ferro and Tartaglia. The formula in terms of square and cube roots is significant for its impact on the development of algebra more than for its current utility. Yet it remains worth seeing. Finding it is a display of insightful substitutions.

Suppose F is a subfield of $\mathbb{C}$, for instance $F = \mathbb{Q}$, and $f = X^3 + aX^2 + bX + c$ is an irreducible polynomial over F. As noted in Section 7.2.2 the Tschirnhausen transform given by $g(X) = f(X - a/3)$ causes g to take the form

$$g = X^3 + pX + q \text{ where } p, q \text{ are polynomials in } a, b, c.$$

The splitting field K of g coincides with that of f. If $\theta_1, \theta_2, \theta_3$ are the roots of g, then the roots of f are $\theta_1 - a/3, \theta_2 - a/3, \theta_3 - a/3$. Thus a formula for the roots of g will instantly give the formula for those of the general irreducible f.

If $p = 0$, the roots of g are the three complex cube roots of $-q$. If $\sqrt[3]{-q}$ is one of them, the other two are $\omega\sqrt[3]{-q}$ and $\omega^2\sqrt[3]{-q}$, where $\omega = e^{2\pi i/3}$. In this case the formula for the roots of g is in plain sight.

From here on, suppose $p \neq 0$.

An ingenious identity in $\mathbb{C}(X)$ plays a key role:

$$\begin{aligned} X^3\left(\left(X - \frac{p}{3X}\right)^3 + p\left(X - \frac{p}{3X}\right) + q\right) &= X^3\left(X^3 - pX + \frac{p^2}{3X} - \frac{p^3}{27X^3} + pX - \frac{p^2}{3X} + q\right) \\ &= X^6 + qX^3 - \frac{p^3}{27}. \end{aligned}$$

Letting

$$h = X^6 + qX^3 - \frac{p^3}{27},$$

the preceding identity says that

$$h(X) = X^3 g\left(X - \frac{p}{3X}\right).$$

Since $p \neq 0$, the roots in $\mathbb{C}$ of h are non-zero, and by the above identity every such root α will cause $\alpha - p/3\alpha$ to be a root of g.

But now this h has a formula for its roots. Indeed, if α is a root of h, then α^3 is a root of the quadratic

$$k = X^2 + qX - \frac{p^3}{27},$$

which by the quadratic formula means that

$$\alpha^3 = \frac{1}{2}\left(-q \pm \sqrt{q^2 + \frac{4p^3}{27}}\right) = -\frac{q}{2} \pm \sqrt{\frac{q^2}{4} + \frac{p^3}{27}}.$$

Here $\sqrt{\frac{q^2}{4} + \frac{p^3}{27}}$ stands for a chosen square root of $\frac{q^2}{4} + \frac{p^3}{27}$.

The two roots of k are not repeated. Indeed, the discriminant of the quadratic k becomes

$$\frac{q^2}{4} + \frac{p^3}{27} = \frac{1}{4 \cdot 27}(4p^3 + 27q^2) = -\frac{D}{4 \cdot 27},$$

where $D = -4p^3 - 27q^2$, which is the discriminant of the irreducible cubic g defined in Section 7.2.2. Being irreducible, g is separable with three distinct roots, and since D is the square of the product of differences of these roots, it must be that $D \neq 0$. From that, the discriminant of k is non-zero, meaning that k's roots are distinct.

Thus, h has six roots in $\mathbb{C}$. Three are the cube roots of $-\frac{q}{2}+\sqrt{\frac{q^2}{4} + \frac{p^3}{27}}$, while the other three are the cube roots of $-\frac{q}{2} - \sqrt{\frac{q^2}{4} + \frac{p^3}{27}}$. As α runs over these six roots, the values $\alpha - p/3\alpha$ will give roots of g with some duplication, since g only has three roots.

Let α be a chosen cube root of $-\frac{q}{2} + \sqrt{\frac{q^2}{4} + \frac{p^3}{27}}$, which can be presented by the familiar notation

$$\alpha = \sqrt[3]{-\frac{q}{2} + \sqrt{\frac{q^2}{4} + \frac{p^3}{27}}}.$$

The other cube roots of $-\frac{q}{2} + \sqrt{\frac{q^2}{4} + \frac{p^3}{27}}$ are $\omega\alpha$ and $\omega^2\alpha$, where $\omega = e^{2\pi i/3}$. We know that $\alpha - p/3\alpha$ is a root of the original cubic g. The preceding radical formula for α, in terms of the coefficients of g, will then provide a radical formula for the root $\alpha - p/3\alpha$ of g. It is just a matter of finding it.

Well, $-p/3\alpha$ happens to be another root of h. To see this use the identity for h given at the outset to get

$$h\left(\frac{-p}{3\alpha}\right) = \left(\frac{-p}{3\alpha}\right)^3 g\left(\frac{-p}{3\alpha} - \frac{p}{3\left(\frac{-p}{3\alpha}\right)}\right) = \left(\frac{-p}{3\alpha}\right)^3 g\left(\alpha - \frac{p}{3\alpha}\right) = 0.$$

And so $(-p/3\alpha)^3$ is a root of the quadratic k, as was α^3. The product of the distinct roots of k is $-p^3/27$, which also equals the product of α^3 and $(-p/3\alpha)^3$. Since $\alpha^3 = -\frac{q}{2} + \sqrt{\frac{q^2}{4} + \frac{p^3}{27}}$, it must be that $\left(\frac{-p}{3\alpha}\right)^3 = -\frac{q}{2} - \sqrt{\frac{q^2}{4} + \frac{p^3}{27}}$, the other root of k. This means that $-p/3\alpha$ is one of the cube roots of $-\frac{q}{2} - \sqrt{\frac{q^2}{4} + \frac{p^3}{27}}$. For brevity let $\beta = -p/3\alpha$, which can be expressed by the familiar notation

$$\beta = \sqrt[3]{-\frac{q}{2} - \sqrt{\frac{q^2}{4} + \frac{p^3}{27}}}.$$

Hence one of the roots of g is

$$\alpha + \beta = \sqrt[3]{-\frac{q}{2} + \sqrt{\frac{q^2}{4} + \frac{p^3}{27}}} + \sqrt[3]{-\frac{q}{2} - \sqrt{\frac{q^2}{4} + \frac{p^3}{27}}}.$$

The other two roots of g are given by

$$\omega\alpha - \frac{p}{3\omega\alpha} = \omega\alpha + \omega^2\frac{-p}{3\alpha} = \omega\alpha + \omega^2\beta,$$

and

$$\omega^2\alpha - \frac{p}{3\omega^2\alpha} = \omega^2\alpha + \omega\frac{-p}{3\alpha} = \omega^2\alpha + \omega\beta.$$

The roots of g can also be written in terms of the discriminant D of g. Indeed, $\frac{q^2}{4} + \frac{p^3}{27} = -\frac{D}{4\cdot 27}$, and since $\omega = \frac{i\sqrt{3}-1}{2}$, we have $i\sqrt{3} = 2\omega + 1$. So

$$-\frac{q}{2} \pm \sqrt{\frac{q^2}{4} + \frac{p^3}{27}} = -\frac{q}{2} \pm \sqrt{-3\frac{D}{4\cdot 81}} = -\frac{q}{2} \pm \frac{i\sqrt{3}}{18}\sqrt{D} = -\frac{q}{2} \pm \frac{2\omega+1}{18}\sqrt{D}.$$

Here $\sqrt{D}$ can be either square root of D, and $\sqrt{3}$ is the positive square root of 3. Then our α, β from above take the form

$$\alpha = \sqrt[3]{-\frac{q}{2} + \frac{2\omega+1}{18}\sqrt{D}} \quad \text{and} \quad \beta = \sqrt[3]{-\frac{q}{2} - \frac{2\omega+1}{18}\sqrt{D}}.$$

As before, the roots of g are

$$\theta_1 = \alpha + \beta, \ \theta_2 = \omega\alpha + \omega^2\beta, \ \theta_3 = \omega^2\alpha + \omega\beta.$$

Since $1+\omega+\omega^2=0$, another insightful formula comes into play:

$$\theta_1+\omega^2\theta_2+\omega\theta_3=(\alpha+\beta)+\omega^2(\omega\alpha+\omega^2\beta)+\omega(\omega^2\alpha+\omega\beta)=3\alpha.$$

If $K=F[\theta_1,\theta_2,\theta_3]$ is the splitting field of g, we now see that α is in the extension $K[\omega]$, and thus $F[\omega,\alpha]\subseteq K[\omega]$. On the other hand, since $\beta=-p/3\alpha\in F[\omega,\alpha]$, an inspection of the roots $\theta_1,\theta_2,\theta_3$ of g reveals that $K[\omega]\subseteq F[\omega,\alpha]$. So $K[\omega]=F[\omega,\alpha]$. Since $\sqrt{D}=\left(\alpha^3+\frac{q}{2}\right)\frac{18}{2\omega+1}$, we also have that $F[\omega,\sqrt{D},\alpha]=F[\omega,\alpha]$.

This reveals that the extension $K[\omega]$ of the splitting field of g sits atop of the following tower:

$$F\subseteq F[\omega]\subseteq F[\omega,\sqrt{D}]\subseteq F[\omega,\sqrt{D},\alpha]=K[\omega].$$

Furthermore,

$$\begin{aligned}\omega^3&=1\in F\\ \sqrt{D}^2&=-4p^3-27q^2\in F[\omega]\text{ (actually }\sqrt{D}^2\in F)\\ \alpha^3&=-\frac{q}{2}-\frac{2\omega+1}{18}\sqrt{D}\in F[\omega,\sqrt{D}].\end{aligned}$$

Each successive field is obtained from its preceding field by the adjunction of a single square or cube root of something in the preceding field. This observation points the way to what it means for a polynomial to be solvable by radicals. Informally, it says that the roots of the cubic are obtained by a repeated application of the operations of $+$, $-$, $\times$ and $\div$, along with the radical operations $\sqrt{\cdot}$ and $\sqrt[3]{\cdot}$, starting with the elements of F.

Here is a small illustration of Cardano's formula. Let $g=X^3+3X-1$, an irreducible cubic over $\mathbb{Q}$. With $p=3, q=-1$ we get

$$\frac{-q}{2}+\sqrt{\frac{q^2}{4}+\frac{p^3}{27}}=\frac{1+\sqrt{5}}{2}.$$

Taking real cube roots we then get

$$\alpha=\sqrt[3]{\frac{1+\sqrt{5}}{2}}\text{ and }\beta=\frac{-p}{3\alpha}=\sqrt[3]{\frac{1-\sqrt{5}}{2}}\text{ (after a bit of simplification).}$$

The roots of g become

$$\alpha+\beta,\ \ \omega\alpha+\omega^2\beta,\ \ \omega^2\alpha+\omega\beta.$$

The first of these three roots is the real root where the cubic cuts the x-axis.

To illustrate how the Tschirnhausen transform works, here are the roots of the cubic $f=X^3+X^2+\frac{10}{3}X+\frac{1}{27}$. The transform says to look for the roots of $f\left(X-\frac{1}{3}\right)$ first. Well,

$$f\left(X-\frac{1}{3}\right)=X^3+3X-1,$$

which is the g whose roots we just computed. With α, β as above, the roots of f are

$$\alpha + \beta - \frac{1}{3}, \ \omega\alpha + \omega^2\beta - \frac{1}{3}, \ \omega^2\alpha + \omega\beta - \frac{1}{3}.$$

Having trudged through to a formula for the roots of a cubic, the next step might be to seek a formula for quartics. We shall be content to declare that a formula based on taking radical expressions from the coefficients exists and move on instead to a general discussion of radical extensions.

7.3.2 Extensions by a Single Radical and Cyclic Galois Groups

Definition 7.18. If a is in a field F and α in a field extension of F is such that $\alpha^n = a$ for some positive integer n, we say that α is a ***radical*** [3] of a. We refer to an extension $F[\alpha]$, where α is the radical of some a in F, as an ***extension of F by a single radical***. Even though α is a root of $X^n - a$, we are not assuming this is the minimal polynomial of α.

A common way to express that α is a radical of a is by the notation $\alpha = \sqrt[n]{a}$. Still we must be cautious in our use of these root symbols because a will have multiple radicals.

Definition 7.19. If K is a Galois extension of F and the Galois group $\mathrm{Gal}(K/F)$ is cyclic, K is called a ***cyclic extension*** of F.

In the next two results we shall see that as long as the ground field is endowed with suitable roots of unity, extensions by a single radical coincide with cyclic extensions.

Proposition 7.20. *Let n be a positive integer and F a field over which the polynomial $X^n - 1$ splits. If K is an extension of F by a single radical α such that $\alpha^n \in F$, then K is a cyclic extension of F, and the degree of $[K : F]$ divides n.*

Proof. We have that $K = F[\alpha]$ and $\alpha^n \in F$. The field F contains the splitting field of $X^n - 1$ over $\mathbb{Q}$, and thus F contains a primitive nth root of unity ζ, as discussed in Section 7.2.3. The full list of roots of unity is $1, \zeta, \zeta^2, \ldots, \zeta^{n-1}$, all inside F. By assumption, the polynomial $X^n - \alpha^n \in F[X]$, and α is a root of this polynomial. Thus its complete list of n roots is $\alpha, \zeta\alpha, \zeta^2\alpha, \ldots, \zeta^{n-1}\alpha$, all of them in K. Being the splitting field of $X^n - \alpha^n$, the field K is a Galois extension of F.

We will need the minimal polynomial g of α. Say $d = \deg g$, which of course equals $[K : F]$ because $K = F[\alpha]$. Since g divides $X^n - \alpha^n$, the roots of g take the form $\zeta^j\alpha$ for various exponents j. Up to a sign, the constant term of g is the product of its roots. Therefore the constant term of g equals $\pm\zeta^\ell\alpha^d$, where ℓ is some exponent. It follows that $\alpha^d \in F$, because the constant term of g and ζ^ℓ are in F. Now $X^d - \alpha^d \in F[X]$ and has α as root, which shows that g divides $X^d - \alpha^d$. Since $d = \deg g$, it must be that $g = X^d - \alpha^d$.

To see that $d \mid n$, write $n = dq + r$ where $0 \leq r < d$. Then $\alpha^n = (\alpha^d)^q\alpha^r$, and since α^n and α^d are in F, so is α^r in F. Thus $X^r - \alpha^r \in F[X]$ and has α as a root. Because g is the

[3] From Latin for "root."

polynomial of least degree in $F[X]$ having α as a root, it has to be that $r = 0$. Thus d, which equals $[K : F]$, divides n.

It remains to show that $\mathrm{Gal}(K/F)$ is cyclic. Well, since $d \mid n$, F contains a primitive dth root of unity, namely $\zeta^{n/d}$. Call it η. As $g = X^d - \alpha^d$, another root of g is $\eta\alpha$. According to Proposition 6.67, $\mathrm{Gal}(K/F)$ acts transitively on the roots of g, and so there is an F-automorphism σ of K such that $\sigma : \alpha \mapsto \eta\alpha$. Then we see that

$$\sigma(\alpha) = \eta\alpha,\ \sigma^2(\alpha) = \eta^2\alpha, \ldots, \sigma^{d-1}(\alpha) = \eta^{d-1}\alpha,\ \sigma^d(\alpha) = \eta^d\alpha = \alpha.$$

Since $\eta^j\alpha \neq \alpha$ until $j = d$, we see that the order of σ is d, which is the order of $\mathrm{Gal}(K/F)$. Thus $\mathrm{Gal}(K/F)$ is cyclic, with σ as its generator. □

For an illustration of Proposition 7.20 let

$$\zeta = e^{2\pi i/6},\ F = \mathbb{Q}[\zeta],\ \alpha = \sqrt[6]{4},\ K = F[\alpha].$$

Since $\alpha = \sqrt[6]{2^2} = \sqrt[3]{2}$, the minimal polynomial of α is $X^3 - 2$. Then K is a cyclic extension of degree 3 over F. The Galois group is generated by the map $\alpha \mapsto \eta\alpha$ where $\eta = e^{2\pi i/3}$.

Here is something close to the converse of the above result.

Proposition 7.21 (Lagrange resolvent). *If K is a cyclic extension of F of degree d and $X^d - 1$ splits over F, then $K = F[\alpha]$ for some α such that $\alpha^d \in F$.*

Proof. The procurement of the needed α may seem out of the blue.

The splitting field of $X^d - 1$ over $\mathbb{Q}$ sits inside F, and so F contains a primitive dth root of unity. Call it ζ. Let σ be a generator for the cyclic group $\mathrm{Gal}(K/F)$ whose order is also d. With 1 denoting the identity map on K, Dedekind's theorem, Proposition 6.70, on the independence of characters shows that the F-linear map on K given by

$$1 + \zeta^{d-1}\sigma + \zeta^{d-2}\sigma^2 + \cdots + \zeta^2\sigma^{d-2} + \zeta\sigma^{d-1}$$

is not the zero map. So there is a β in K such that

$$\beta + \zeta^{d-1}\sigma(\beta) + \zeta^{d-2}\sigma^2(\beta) + \cdots + \zeta^2\sigma^{d-2}(\beta) + \zeta\sigma^{d-1}(\beta) \neq 0.$$

The above non-zero sum in K is our candidate α. Calculate $\sigma(\alpha)$ to get

$$\begin{aligned}\sigma(\alpha) &= \sigma(\beta) + \zeta^{d-1}\sigma^2(\beta) + \zeta^{d-2}\sigma^3(\beta) + \cdots + \zeta^2\sigma^{d-1}(\beta) + \zeta\sigma^d(\beta)\\ &= \zeta^d\sigma(\beta) + \zeta^{d-1}\sigma^2(\beta) + \zeta^{d-2}\sigma^3(\beta) + \cdots + \zeta^2\sigma^{d-1}(\beta) + \zeta\beta\\ &= \zeta\left(\zeta^{d-1}\sigma(\beta) + \zeta^{d-2}\sigma^2(\beta) + \zeta^{d-3}\sigma^3(\beta) + \cdots + \zeta\sigma^{d-1}(\beta) + \beta\right)\\ &= \zeta\alpha.\end{aligned}$$

This α, known as a ***Lagrange resolvent*** for ζ and σ, turns out to be an eigenvector for σ with eigenvalue ζ.

To see that $K = F[\alpha]$, it suffices to check that the minimal polynomial g of α has degree d. This suffices since $[K : F] = d$. Well, $\deg g \leq d$ because $F[\alpha]$ is a subfield of K. By Proposition 6.40 the values

$$\alpha,\ \sigma(\alpha) = \zeta\alpha,\ \sigma^2(\alpha) = \zeta^2\alpha, \ldots, \sigma^{d-1}(\alpha) = \zeta^{d-1}\alpha$$

are roots of g. They are distinct since ζ is a primitive dth root of unity and $\alpha \neq 0$, and the number of them is d. Since $\deg g \leq d$, we must have $\deg g = d$.

Furthermore,

$$\sigma(\alpha^d) = (\sigma(\alpha))^d = (\zeta\alpha)^d = \zeta^d\alpha^d = \alpha^d.$$

Being fixed by the generator σ of $\mathrm{Gal}(K/F)$, the power α^d is fixed by all of $\mathrm{Gal}(K/F)$, and because K is a Galois extension of F, we get $\alpha^d \in F$. □

7.3.3 Radical Towers

It is time to capture what it means for a polynomial to be solvable by radicals.

Definition 7.22. A tower of fields

$$F = F_0 \subseteq F_1 \subseteq F_2 \subseteq \cdots \subseteq F_k = K$$

is called a ***radical tower*** provided that for every index j there exists an α_j in F_{j+1} and a positive exponent n_j such that

$$F_{j+1} = F_j[\alpha_j] \text{ and } \alpha_j^{n_j} \in F_j.$$

In other words, each field F_{j+1} is an extension of its preceding field F_j by a single radical. The top field K in a radical tower is called a ***radical extension***[4] of the bottom field F.

Obviously, every field in a radical tower is a radical extension of the fields below it.

Definition 7.23. If a polynomial f in $F[X]$ splits over a radical extension of a field F, we say that f is ***solvable by radicals***.

This says, in a precise way, that the roots of f are obtainable starting from the elements of F by a successive application of the operations $+$, $-$, $\times$ and $\div$, along with the radical operations $\sqrt[n]{\cdot}$ for various n.

Quadratics $X^2 + bX + c$ in $F[X]$ are solvable by radicals, because their splitting field $F\left[\sqrt{b^2 - 4c}\right]$ is a radical extension of F formed by adjoining a single radical. A reducible cubic in $F[X]$ is solvable by radicals because its splitting field is either F or the splitting field of one of its irreducible quadratic factors in $F[X]$. Cardano's formula showed that every irreducible cubic is solvable by radicals. With a bit of work it can be shown that every quartic polynomial is solvable by radicals.

The puzzle remained for polynomials of degree five and up. It was Abel and Ruffini who showed that a polynomial of degree five is not in general solvable by radicals. We propose

[4] Some authors refer to any subfield of what we call a radical extension as being a radical extension.

to show what Galois did, which is to identify precisely what it takes for a polynomial of any degree to be solvable by radicals.

Of course, the roots of a polynomial f in $F[X]$ are inside a radical extension of F if and only if the splitting field of f is a subfield of that radical extension of F. However, it need not be that the splitting field itself is a radical extension of F. For instance, if the splitting field K of a cubic has degree three over $\mathbb{Q}$, then K is not a radical extension of $\mathbb{Q}$, even though the bigger field $K[e^{2\pi i/3}]$ is a radical extension of $\mathbb{Q}$. This remark is left as an interesting exercise.

Improving a Radical Tower to a Galois Extension

In order to bring the fundamental theorem of Galois theory into the picture, we need to enlarge radical towers so that the top field becomes a Galois extension.

Proposition 7.24. *Every radical extension K of F is a subfield of another extension L of F that is both radical and Galois over F.*

Proof. Since we are in characteristic zero, the primitive generator theorem, Proposition 7.2, applies to yield an element γ_1 in K such that $K = F[\gamma_1]$. Let g be the minimal polynomial of γ_1 over F. Take L to be the splitting field of g viewed as a polynomial over K. This ensures the inclusions $F \subseteq K \subseteq L$. It is clear that L is also the splitting field of g over F. Thus L is a Galois extension of F. Our L is known as the *normal closure* of the extension K over F. The task is to build a radical tower running from F to L.

Let $\gamma_2, \dots, \gamma_n$ be the other roots of g inside L, so that $L = F[\gamma_1, \gamma_2, \dots, \gamma_n]$. These roots provide the tower

$$F \subseteq K = F[\gamma_1] \subseteq F[\gamma_1, \gamma_2] \subseteq F[\gamma_1, \gamma_2, \gamma_3] \subseteq \cdots \subseteq F[\gamma_1, \gamma_2, \gamma_3, \dots, \gamma_n] = L.$$

This tower from F to L need not be radical, but we will refine it to a tower that is radical.

By assumption there is a radical tower running from F to K. It suffices to show that for every j from 2 to n, there is a radical tower running from $F[\gamma_1, \gamma_2, \dots, \gamma_{j-1}]$ to $F[\gamma_1, \gamma_2, \dots, \gamma_{j-1}, \gamma_j]$. The obvious concatenation of these radical towers will provide the radical tower from F to L that we want.

Here is how to do this for $j = 3$, just to keep notations simpler. The method is the same for the other j.

As stated in Proposition 6.67, the Galois group of L over F acts transitively on the roots. Thus there is an F-automorphism σ of L such that $\sigma(\gamma_1) = \gamma_3$. Also there exist elements $\alpha_1, \alpha_2, \dots, \alpha_k$ in K such that the tower

$$F \subseteq F[\alpha_1] \subseteq F[\alpha_1, \alpha_2] \subseteq F[\alpha_1, \alpha_2, \alpha_3] \subseteq \cdots \subseteq F[\alpha_1, \alpha_2, \alpha_3, \dots, \alpha_k] = K$$

is radical. This means there are positive exponents $n_1, n_2, \dots, n_k$ such that

$$\alpha_i^{n_i} \in F[\alpha_1, \alpha_2, \dots, \alpha_{i-1}].$$

Here is a tower starting at $F[\gamma_1,\gamma_2]$:

$$\begin{aligned} F[\gamma_1,\gamma_2] &\subseteq F[\gamma_1,\gamma_2,\sigma(\alpha_1)] \\ &\subseteq F[\gamma_1,\gamma_2,\sigma(\alpha_1),\sigma(\alpha_2)] \\ &\subseteq \dots \\ &\subseteq F[\gamma_1,\gamma_2,\sigma(\alpha_1),\sigma(\alpha_2),\dots,\sigma(\alpha_{i-1})] \\ &\subseteq F[\gamma_1,\gamma_2,\sigma(\alpha_1),\sigma(\alpha_2),\dots,\sigma(\alpha_i)] \\ &\subseteq \dots \\ &\subseteq F[\gamma_1,\gamma_2,\sigma(\alpha_1),\sigma(\alpha_2),\dots,\sigma(\alpha_{k-1}),\sigma(\alpha_k)]. \end{aligned}$$

This tower is radical. Indeed, $\alpha_i^{n_i} \in F[\alpha_1,\dots,\alpha_{i-1}]$ for every i from 1 to k, and since σ is an F-automorphism, we obtain

$$\sigma(\alpha_i)^{n_i} = \sigma(\alpha_i^{n_i}) \in F[\sigma(\alpha_1),\dots,\sigma(\alpha_{i-1})] \subseteq F[\gamma_1,\gamma_2,\sigma(\alpha_1),\sigma(\alpha_2),\dots,\sigma(\alpha_{i-1})].$$

(When $i = 1$, the last field should be interpreted as $F[\gamma_1,\gamma_2]$.) Furthermore, the above tower ends at $F[\gamma_1,\gamma_2,\gamma_3]$. To see this, notice that $F[\alpha_1,\alpha_2,\dots,\alpha_k] = K = F[\gamma_1]$. Since σ is an F-automorphism, it follows that $F[\sigma(\alpha_1),\dots,\sigma(\alpha_k)] = F[\sigma(\gamma_1)] = F[\gamma_3]$, and thus

$$F[\gamma_1,\gamma_2,\sigma(\alpha_1),\sigma(\alpha_2),\dots,\sigma(\alpha_{k-1}),\sigma(\alpha_k)] = F[\gamma_1,\gamma_2,\gamma_3].$$

So we are done. □

Allowing Roots of Unity to Play a Role

Propositions 7.20 and 7.21 reveal that, in the presence of suitable roots of unity, an extension K over F is Galois and cyclic if and only if K extends F by a single radical. This is a crucial link. But the presence of such roots of unity is not typically granted. To remedy this, they have to be adjoined to the fields in question.

Every field E of characteristic zero is contained in the splitting field L of $X^n - 1$ over E. Since L contains the splitting field of $X^n - 1$ over $\mathbb{Q}$, it follows that L contains a primitive nth root of unity, say ζ. Clearly $L = E[\zeta]$ since all the other roots of $X^n - 1$ are powers of ζ. In this sense we can adjoin a primitive nth root of unity to any field E of characteristic zero.

Improving Radical Extensions with Roots of Unity

We have seen in Proposition 7.24 that every radical tower can be expanded to a radical tower in which the top field is a Galois extension of the bottom field. But we can do even better by adjoining roots of unity.

Proposition 7.25. *For every radical extension K of F there exists a radical tower*

$$F = E_0 \subseteq E_1 \subseteq E_2 \subseteq \dots \subseteq E_n = L$$

starting with F and such that

- *K is a subfield of L,*
- *L is Galois over F,*
- *each field E_j is Galois over its preceding E_{j-1}, and*
- *each Galois group $\mathrm{Gal}(E_j/E_{j-1})$ is abelian.*

Proof. By Proposition 7.24 we can suppose that K is also a Galois extension of F. Since K is a radical extension of F, this means there is a tower

$$F = F_0 \subseteq F_1 \subseteq F_2 \subseteq \cdots \subseteq F_k = K$$

where each F_j is an extension of its preceding F_{j-1} by a single radical. So, for each j from 1 to k, there is an element α_j in F_j and a positive integer n_j such that

$$F_j = F_{j-1}[\alpha_j] \text{ and } \alpha_j^{n_j} \in F_{j-1}.$$

Let n be the product $n_1 n_2 \cdots n_k$, and let ζ be a primitive nth root of unity. Then build the alternative tower

$$F = F_0 \subseteq F_0[\zeta] \subseteq F_1[\zeta] \subseteq F_2[\zeta] \subseteq \cdots \subseteq F_k[\zeta] = K[\zeta].$$

The fields in this tower are the E_j that we need, as we now check.

Evidently K is a subfield of $K[\zeta]$ which we take as our desired L. Being Galois over F, the field K is the splitting field of some polynomial f in $F[X]$. Then $K[\zeta]$ becomes the splitting field of $(X^n - 1)f$ over F. Thus $K[\zeta]$ is Galois over F.

Next examine the extension $F_0[\zeta] = F[\zeta]$ over $F_0 = F$ at the top of the first inclusion. Since $\zeta^n = 1 \in F$, this extension is by a single radical, namely $\sqrt[n]{1}$. The next thing is to check that $F[\zeta]$ is Galois over F and that the Galois group $\mathrm{Gal}\,(F[\zeta]/F)$ is abelian. The extension is the splitting field of $X^n - 1$, so it is Galois. Every F-automorphism σ of $F[\zeta]$ maps ζ to one of its conjugates ζ^ℓ, and σ is determined by what it does to ζ. So if σ, τ are F-automorphisms of $F[\zeta]$ given by

$$\sigma : \zeta \mapsto \zeta^\ell, \quad \tau : \zeta \mapsto \zeta^m,$$

the composite $\sigma\tau$ is the F-automorphism that is determined by

$$\sigma\tau(\zeta) = \sigma(\zeta^m) = \sigma(\zeta)^m = (\zeta^\ell)^m = \zeta^{\ell+m}.$$

Now it is clear that $\sigma\tau = \tau\sigma$ because $\ell+m = m+\ell$. The Galois group $\mathrm{Gal}(F[\zeta]/F)$ is abelian.

Finally, check, for j from 1 to k, that each $F_j[\zeta]$ is an extension of $F_{j-1}[\zeta]$ by a single radical, that the extension is Galois, and that $\mathrm{Gal}(F_j[\zeta]/F_{j-1}[\zeta])$ is abelian. This falls into place by Proposition 7.20. Clearly

$$F_j[\zeta] = F_{j-1}[\alpha_j][\zeta] = F_{j-1}[\zeta][\alpha_j] \text{ and } \alpha_j^{n_j} \in F_{j-1} \subseteq F_{j-1}[\zeta].$$

Thus $F_j[\zeta]$ extends $F_{j-1}[\zeta]$ by a single radical. Proposition 7.20 applies to this extension. This is because n_j divides n. Consequently ζ^{n/n_j} is a primitive root inside $F_{j-1}[\zeta]$ which implies that $X^{n_j} - 1$ splits over $F_{j-1}[\zeta]$. By Proposition 7.20 the extension $F_j[\zeta]$ is not only Galois over $F_{j-1}[\zeta]$ but its Galois group is cyclic, which is of course abelian. □

7.3.4 Solvable Polynomials Have Solvable Galois Groups

We are in a position to demonstrate the first half of Galois' great theorem, which says that if a polynomial is solvable by radicals then its Galois group is a solvable group. Needless to say, to appreciate the theorem it is essential to have at hand an understanding of solvable groups as discussed in Section 3.7. A group G is solvable provided there is a chain of subgroups

$$G = G_0 \supset G_1 \supset G_2 \supset \cdots \supset G_n = \langle 1 \rangle,$$

where $\langle 1 \rangle$ is the trivial subgroup, each G_j is a normal subgroup of its preceding group G_{j-1}, and the quotient G_{j-1}/G_j is abelian. Obviously abelian groups are solvable. Proposition 3.42 says that subgroups and homomorphic images of solvable groups remain solvable. By Proposition 3.21, all p-groups are solvable when p is prime. For $n \geq 5$ the symmetric groups $\mathcal{A}_n$ of even permutations on n letters are not solvable. So, if $n \geq 5$ the full symmetric groups $\mathcal{S}_n$ are not solvable either because their subgroups $\mathcal{A}_n$ are not solvable.

Here is the reason why solvable groups matter.

Proposition 7.26 (Galois' great theorem, part 1). *If a polynomial is solvable by radicals, then its Galois group is a solvable group.*

Proof. If K is the splitting field of the solvable polynomial over some field F, the assumption says that K is a subfield of a radical extension L of F. By Proposition 7.25 we can take L to be a Galois extension of F and such that L sits at the top of a tower

$$F = E_0 \subset E_1 \subset E_2 \subset \cdots \subset E_n = L,$$

where each E_j is Galois over E_{j-1} and $\mathrm{Gal}(E_j/E_{j-1})$ is abelian.

The Galois correspondence reverses the inclusions of this tower to give a chain of subgroups

$$\mathrm{Gal}(L/F) \supset \mathrm{Gal}(L/E_1) \supset \mathrm{Gal}(L/E_2) \supset \cdots \supset \mathrm{Gal}(L/E_n) = \mathrm{Gal}(L/L) = \langle 1 \rangle.$$

The group $\mathrm{Gal}(L/F)$ is solvable provided that each $\mathrm{Gal}(L/E_j)$ is a normal subgroup of $\mathrm{Gal}(L/E_{j-1})$ and that the quotient $\mathrm{Gal}(L/E_{j-1})/\mathrm{Gal}(L/E_j)$ is abelian. Being Galois over F, the field L is also Galois over E_{j-1} and over E_j. Furthermore E_j is Galois over E_{j-1}. By Proposition 6.81 applied to the short tower $E_{j-1} \subset E_j \subset L$, the group $\mathrm{Gal}(L/E_j)$ is normal inside $\mathrm{Gal}(L/E_{j-1})$ and

$$\mathrm{Gal}(L/E_{j-1})/\mathrm{Gal}(L/E_j) \cong \mathrm{Gal}(E_j/E_{j-1}).$$

Since the Galois group on the right is abelian, so is the group on the left. This proves that $\mathrm{Gal}(L/F)$ is a solvable group.

But we need $\mathrm{Gal}(K/F)$ to be solvable. For that use the tower $F \subseteq K \subseteq L$. Here L is Galois over F and over K, and K is Galois over F. By Proposition 6.81, $\mathrm{Gal}(L/K)$ is a normal subgroup of $\mathrm{Gal}(L/F)$ and

$$\mathrm{Gal}(L/F)/\mathrm{Gal}(L/K) \cong \mathrm{Gal}(K/F).$$

Being a homomorphic image of the solvable group $\mathrm{Gal}(L/F)$, the group $\mathrm{Gal}(K/F)$ is solvable too. □

The origins of the term *solvable group* are now self-evident.

Finding Unsolvable Polynomials

Proposition 7.26 teaches us that a polynomial cannot be solved by radicals if its Galois group is not solvable. For example, any polynomial whose Galois group is isomorphic to $\mathcal{S}_5$ cannot be solved by radicals.

The next result points a way to obtain such polynomials over the field $\mathbb{Q}$ of rational numbers.

Proposition 7.27. *Suppose f is an irreducible polynomial over $\mathbb{Q}$ of prime degree p where $p \geq 5$. If all but two of the roots of f are real numbers, then the Galois group of f is isomorphic to $\mathcal{S}_p$, and thereby f is not solvable by radicals.*

Proof. Since f is irreducible, its degree p divides the order of its splitting field, which we call K. Indeed, for any root α of f in K, the subfield $F[\alpha]$ of K has degree p over F, and so we invoke the tower theorem. Because the extension K is Galois, the order of $\mathrm{Gal}(K/F)$ equals $[K : F]$. Thus p divides the order of $\mathrm{Gal}(K/F)$. Then Cauchy's theorem, Proposition 3.23, provides an element σ of order p inside $\mathrm{Gal}(K/F)$.

Let Δ be the set of roots of f. There are p of them. The group $\mathcal{S}(\Delta)$ of permutations of Δ is nothing but the symmetric group $\mathcal{S}_p$. The restriction mapping

$$\Gamma : \mathrm{Gal}(K/F) \to \mathcal{S}(\Delta) \text{ given by } \Gamma : \varphi \mapsto \varphi_{|\Delta}$$

is injective. We need to show it is surjective.

The usual conjugation map $\theta : \mathbb{C} \to \mathbb{C}$ given by $z \mapsto \bar{z}$ restricts to an automorphism of the subfield K. That is because K satisfies the normality theorem, Proposition 6.42. Thus $\Gamma(\theta) \in \mathcal{S}(\Delta)$. Since all but two of these roots in Δ are real, $\Gamma(\theta)$ fixes all but two of them and transposes the two non-real roots. In other words $\Gamma(\theta)$ is a transposition in $\mathcal{S}(\Delta)$.

Since σ from above has order p, its restriction $\Gamma(\sigma)$ has order p in $\mathcal{S}(\Delta)$. The only elements of order p are p-cycles. This is because the order of any permutation is the least common multiple of the lengths of the cycles in its cycle decomposition. Since p is prime, the only cycles that can appear in the cycle decomposition of σ are those of length p, and clearly there can only be one of them.

So the image of Γ contains both a transposition and a p-cycle. A routine exercise reveals that if a subgroup of $\mathcal{S}_p$ contains a transposition and a p-cycle, then that subgroup is the entire symmetric group. Thus Γ is surjective. □

An Unsolvable Polynomial

Consider $f = X^5 - 10X^3 - 135X + 5$ in $\mathbb{Q}[X]$. Eisenstein's criterion shows that f is irreducible. The derivative of f is

$$f' = 5X^4 - 30X^2 - 135 = 5(X^2 - 9)(X^2 + 3).$$

From basic calculus, f increases from $-\infty$ to -3, decreases from -3 to 3 and increases from 3 to ∞. Also $f(-3) > 0$ and $f(3) < 0$. Thus f has precisely three real roots, one to the left of -3, one between -3 and 3 and one to the right of 3. The other two roots are non-real complex numbers. By Proposition 7.27 the Galois group of f is the unsolvable group $\mathcal{S}_5$. According to the first part of Galois' theorem, f is not solvable by radicals.

That such a polynomial exists would have amazed Cardano and the mathematicians of his time. Actually it has continued to amaze.

7.3.5 Polynomials with Solvable Galois Groups Are Solvable

It was already known to Ruffini and Abel that not all quintics could be solved by radicals. Galois made the connection to what we now call the Galois group of the polynomial. His great theorem is that the solvability of the polynomial by radicals is *equivalent* to the solvability of its Galois group.

The proof of this half of Galois' theorem requires Proposition 3.44. Namely, if a finite group G is solvable, then there is a chain of subgroups

$$G = H_0 \supset H_1 \supset H_2 \supset \cdots \supset H_n = \langle 1 \rangle,$$

such that each H_j is normal in its preceding group H_{j-1} and the quotients H_{j-1}/H_j are not only abelian but actually of prime order and thereby cyclic.

Proposition 7.21 is also used. In order to use it, we need to introduce roots of unity into Galois extensions. If K is a field (still in characteristic zero), and ζ is a primitive nth root of unity, keep in mind that the extension $K[\zeta]$ is nothing but the splitting field of $X^n - 1$ over K.

Proposition 7.28. *If K is a Galois extension of F of degree n and ζ is a primitive nth root of unity, then the extension $K[\zeta]$ is a Galois extension of $F[\zeta]$ and* $\mathrm{Gal}(K[\zeta]/F[\zeta])$ *is isomorphic to a subgroup of* $\mathrm{Gal}(K/F)$.

Proof. If f in $F[X]$ is the polynomial whose splitting field is K, then $K[\zeta]$ is also the splitting field of f over the larger field $F[\zeta]$. Hence, $K[\zeta]$ is Galois over $F[\zeta]$. Every $F[\zeta]$-automorphism φ of $K[\zeta]$ is also an F-automorphism. The normality of K coming from Proposition 6.42 ensures that $\varphi(K) = K$. This gives rise to the group homomorphism given by the restriction mapping:

$$\Gamma : \mathrm{Gal}(K[\zeta]/F[\zeta]) \to \mathrm{Gal}(K/F) \text{ where } \Gamma : \varphi \mapsto \varphi_{|K}.$$

Now, Γ is injective, because every $F[\zeta]$-map φ on $K[\zeta]$ fixes ζ. So, if φ restricts to the identity on K it must be the identity on $K[\zeta]$. This means that $\ker \Gamma$ is trivial. So $\mathrm{Gal}(K[\zeta]/F[\zeta])$ is isomorphic to a subgroup of $\mathrm{Gal}(K/F)$. □

In the preceding result there is nothing special about adjoining ζ. Any number of algebraic elements could be adjoined to F and to K with no essential change in the result nor the proof.

Proposition 7.29 (Galois' great theorem, part 2). *If the Galois group of a polynomial is solvable, then the polynomial is solvable by radicals.*

Proof. Let f over some field F be the polynomial in question, and let K be its splitting field. The group $\mathrm{Gal}(K/F)$ is assumed solvable. We need to prove that K is a subfield of a radical extension of F. Let n be the degree of K over F, and create the extension $K[\zeta]$ of $F[\zeta]$, where ζ is a primitive nth root of unity. This $K[\zeta]$ is the desired radical extension of F, as we now verify.

By Proposition 7.28, $\mathrm{Gal}(K[\zeta]/F[\zeta])$ is isomorphic to a subgroup of $\mathrm{Gal}(K/F)$, and since subgroups of solvable groups are solvable, $\mathrm{Gal}(K[\zeta]/F[\zeta])$ is a solvable group. By Proposition 3.44, there is a chain of subgroups

$$\mathrm{Gal}(K[\zeta]/F[\zeta]) = H_0 \supset H_1 \supset H_2 \supset \cdots \supset H_k = \langle 1 \rangle,$$

where each H_i is normal in H_{i-1} and the quotients H_{i-1}/H_i are cyclic groups of prime order. The Galois correspondence applied to the Galois extension $K[\zeta]$ over $F[\zeta]$ provides the tower

$$\begin{aligned} F[\zeta] &= \mathrm{Fix}\,(\mathrm{Gal}(K[\zeta]/F[\zeta])) \\ &= \mathrm{Fix}(H_0) \subset \mathrm{Fix}(H_1) \subset \mathrm{Fix}(H_2) \subset \cdots \subset \mathrm{Fix}(H_{i-1}) \subset \mathrm{Fix}(H_i) \subset \cdots \subset \mathrm{Fix}(H_k) = K[\zeta]. \end{aligned}$$

This tower is radical. Indeed, for each i examine the inclusions

$$\mathrm{Fix}(H_{i-1}) \subset \mathrm{Fix}(H_i) \subset K[\zeta].$$

Being Galois over $F[\zeta]$, the extension $K[\zeta]$ is Galois over $\mathrm{Fix}(H_{i-1})$ and over $\mathrm{Fix}(H_i)$. The intermediate field $\mathrm{Fix}(H_i)$ is also Galois over $\mathrm{Fix}(H_{i-1})$. This is because the fundamental theorem of Galois theory, Proposition 6.77, gives

$$\mathrm{Gal}\,(K[\zeta]/\,\mathrm{Fix}(H_i)) = H_i \text{ and } \mathrm{Gal}\,(K[\zeta]/\,\mathrm{Fix}(H_{i-1})) = H_{i-1}.$$

Since H_i is a normal subgroup of H_{i-1}, Proposition 6.81 causes $\mathrm{Fix}(H_i)$ to be a Galois extension of $\mathrm{Fix}(H_{i-1})$. Furthermore, Proposition 6.81 reveals that

$$\mathrm{Gal}\,(K[\zeta]/\,\mathrm{Fix}(H_{i-1}))\,/\,\mathrm{Gal}\,(K[\zeta]/\,\mathrm{Fix}(H_i)) \cong H_{i-1}/H_i,$$

a group which is cyclic by assumption.

Let d be the degree of the extension $\mathrm{Fix}(H_i)$ over $\mathrm{Fix}(H_{i-1})$. At this point the Lagrange resolvent theorem, Proposition 7.21, would apply to this extension as long as $X^d - 1$ splits over $\mathrm{Fix}(H_{i-1})$. Well, Proposition 7.28 together with Lagrange's theorem reveals that the order of $\mathrm{Gal}(K[\zeta]/F[\zeta])$ divides the order of $\mathrm{Gal}(K/F)$. By the Galois correspondence, Proposition 6.77, the degree $[K[\zeta] : F[\zeta]]$ divides $[K : F]$, which is n. Since $\mathrm{Fix}(H_{i-1})$ and $\mathrm{Fix}(H_i)$ are intermediate fields between $K[\zeta]$ and $F[\zeta]$, the degree d of $\mathrm{Fix}(H_i)$ over $\mathrm{Fix}(H_{i-1})$ divides $[K[\zeta] : F[\zeta]]$. Thus $d \mid n$. Consequently $X^d - 1$ divides $X^n - 1$. But $F[\zeta]$ is built to have $X^n - 1$ split over it. So its factor $X^d - 1$ splits over $F[\zeta]$. Since $\mathrm{Fix}(H_{i-1})$ is an extension of $F[\zeta]$, it follows that $X^d - 1$ splits over $\mathrm{Fix}(H_{i-1})$.

From the above, invoke Proposition 7.21 to see that $\mathrm{Fix}(H_i)$ is an extension of $\mathrm{Fix}(H_{i-1})$ obtained by adjoining a single radical. Thus we obtain a radical tower from $F[\zeta]$ to $K[\zeta]$.

To get a radical tower that runs from F to $K[\zeta]$, note that the extension $F[\zeta]$ of F is itself obtained by adjoining a single radical to F, namely ζ, where $\zeta^n = 1 \in F$. By tacking the inclusion $F \subseteq F[\zeta]$ at the bottom of the radical tower running from $F[\zeta]$ to $K[\zeta]$, the field $K[\zeta]$ is seen to be a radical extension of F. □

On looking back, the connection that Galois made between the solvability of a polynomial by radicals and the nature of the automorphism group of its splitting field can only be called a tour-de-force and a testament to human ingenuity.

The explicit formulas in terms of radicals for polynomials of degree up to four are known of course. Galois' discovery Proposition 7.29 corroborates the fact polynomials of degree up to four are solvable by radicals, because their Galois groups must be groups of permutations of at most four roots and because the groups $\mathcal{S}_2, \mathcal{S}_3, \mathcal{S}_4$ and all of their subgroups are solvable.

We shall have occasion to further exploit Galois' great theorem, Proposition 7.29, in a subsequent discussion of geometric constructions.

EXERCISES

1. If f is an irreducible polynomial over a field F and one of its roots lies in a radical extension of F, show that all roots of f lie in a radical extension of F.
2. Suppose that f is an irreducible cubic in $\mathbb{Q}[X]$ and that its splitting field K has degree 3 over $\mathbb{Q}$. Show that f is solvable by radicals, but that K is not the top field of a radical tower.
   Give an example of such a polynomial in $\mathbb{Q}[X]$.
3. Prove that every polynomial of degree 4 over a field is solvable.
4. Which of these polynomials is solvable by radicals:
   (a) $2X^5 - 5X^4 + 5$
   (b) $X^6 + 2X^3 + 1$
   (c) $3X^5 - 15X + 5$.
5. Find a polynomial in $\mathbb{Q}[X]$ of degree 7 and whose Galois group is $\mathcal{S}_7$.
6. Let $f = X^3 - 3X + 1$ in $\mathbb{Q}[X]$, and let β be one of its roots.
   (a) Show that $\mathbb{Q}[\beta]$ is a degree 3 Galois extension of $\mathbb{Q}$.
   (b) If $\omega = e^{2\pi i/3}$, explain why $\mathbb{Q}[\omega, \beta]$ remains a degree 3 Galois extension of $\mathbb{Q}[\omega]$.
   (c)* Since $X^3 - 1$ splits over $\mathbb{Q}[\omega]$, Proposition 7.21 implies that $\mathbb{Q}[\omega, \beta] = \mathbb{Q}[\omega][\alpha]$ for some α such that $\alpha^3 \in \mathbb{Q}[\omega]$.
   Find such an α.

7.4 Ruler and Compass Constructions

Euclid taught us how to perform various geometric constructions with an unmarked ruler and a compass, such as build an equilateral triangle on a given base, erect the right bisector of a line segment, bisect an angle, or build a pentagon. We may subsequently have wondered if it is possible to trisect an angle, reproduce the area of a circle as a square, or build regular polygons of all sorts, by using only these instruments. A rigorous discussion of these matters comes down to translating these problems into the language of field theory. Then we can use its tools to provide answers.

7.4.1 Constructible Points, Lines and Circles

Suppose S is a subset of the Euclidean plane $\mathbb{R}^2$. For brevity we shall refer to any line passing through a pair of the points in S as an ***S-line***. Any circle centred at a point of S and passing through another point of S will be called an ***S-circle***. Ruler and compass constructions, based on a set of points S, must begin with some S-lines and S-circles. Indeed, if S is what we have to work with, rulers can only draw lines between two points of S, and compasses can only draw circles with each tip of the compass on a point of S. These are the construction rules handed down from Euclid.

If a point p in $\mathbb{R}^2$ lies in S, or is the intersection of two S-lines or two S-circles or an S-line with an S-circle, we shall say that p is ***constructible from S in at most one step***.

Now *start with a two-point set A*. We might as well take those points to be the origin $(0,0)$ and the point $(1,0)$ on the x-axis. As the picture below shows, such A spawns one A-line (the x-axis), two A-circles, and six points constructible from A in at most one step. The six points are

$$(0,0), (1,0), (2,0), (-1,0), (1/2, \sqrt{3}/2), (1/2, -\sqrt{3}/2).$$

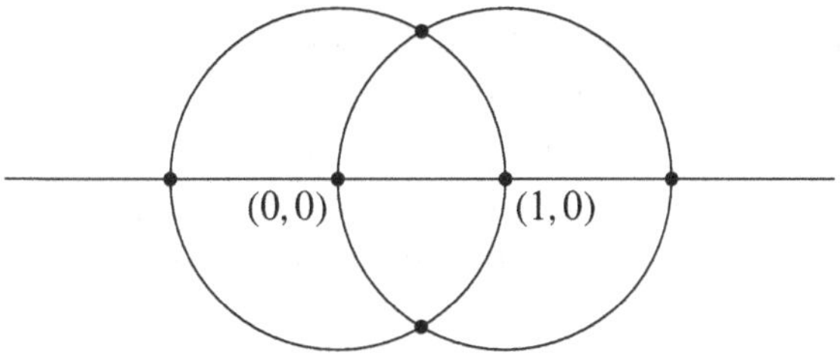

(0,0)
(1,0)

If A' denotes the set of six points constructible from A in at most one step, we can readily construct the A'-lines and A'-circles, and then obtain the set A'' of points constructible from A' in at most one step. Naturally, the points of A'' are said to be ***constructible from A in at most two steps***. While A'' is finite, the number of points therein is already quite large. It might be a good test of one's patience to count how many there are.

By repeating this procedure, create the larger, but still finite set A''' of points constructible from A in at most three steps, and so on.

Definition 7.30. The points in $\mathbb{R}^2$ which are constructible from A in at most a finite number of steps are known as the ***constructible points***.

The lines and circles, whose intersections give the constructible points, are themselves called ***constructible lines*** and ***constructible circles***.

Clearly, every constructible line passes through two constructible points, and every constructible circle passes through a constructible point and has a constructible centre.

The following high school constructions could well be familiar.

Constructing the Right Bisector of a Line Segment

If p,q are constructible points, the line that bisects the segment from p to q orthogonally (the so-called right bisector) can be constructed. Indeed, the circle with centre p and passing through q meets the circle with centre q and passing through p at the points r,s, which are thereby constructible. Then the line joining r to s is constructible, cuts the line joining p to q orthogonally and bisects the segment from p to q, as desired.

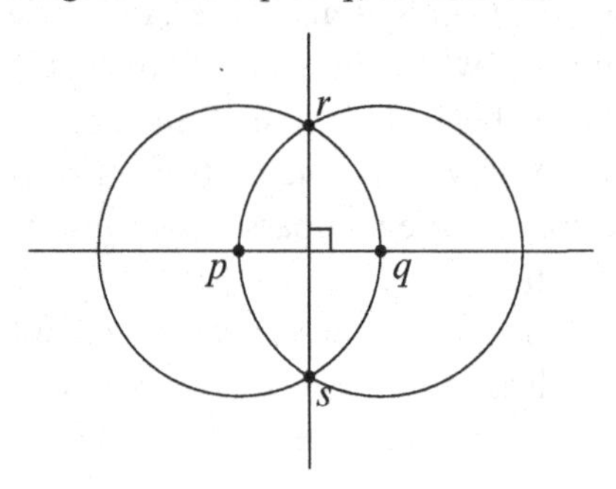

Constructing the Orthogonal Projection from a Point to a Line

If ℓ is a constructible line and p is a constructible point not on ℓ, the line k passing through p and perpendicular to ℓ is constructible. Indeed, ℓ has a constructible point q on it (actually at least two). The circle with centre p and passing through q cuts ℓ at another point r, which is thereby constructible. The right bisector k of the segment from q to r is constructible, passes through p, and is perpendicular to the line ℓ. The point s where this right bisector meets ℓ is called the ***orthogonal projection*** of p onto ℓ, and clearly s is constructible. Some refer to this process as dropping a perpendicular from p to ℓ.

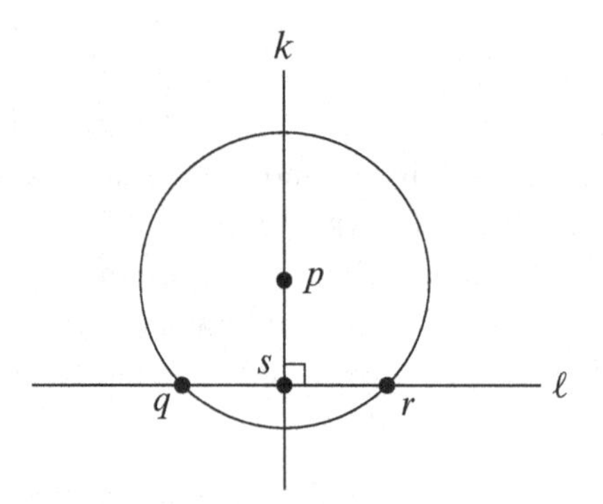

Raising a Perpendicular from a Point on a Line

If p is a constructible point on a constructible line ℓ, the line k passing through p and perpendicular to ℓ is constructible. Indeed, ℓ contains at least one other constructible point q. The circle with centre p and passing through q meets ℓ at another point r. The right bisector k of the segment from q to r is constructible and passes through p.

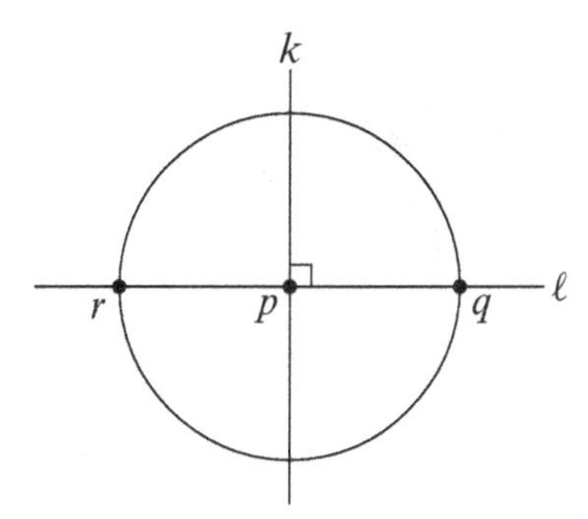

Constructing a Line through a Point and Parallel to Another Line

If ℓ is a constructible line and p is a constructible point not on ℓ, the line j passing through p and parallel to ℓ is constructible. Indeed, the line k through p that is perpendicular to ℓ is constructible, as already demonstrated. The line j perpendicular to k and passing through p is also constructible. This line j is parallel to ℓ.

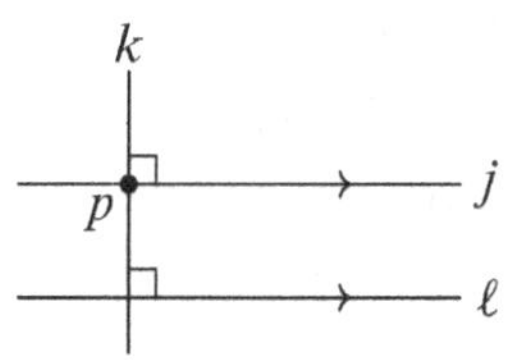

Lifting the Compass

If p, q, r are three constructible points, the circle with centre r and radius equal to the distance from p to q is constructible.

Before going to a proper construction, here is an attempt that presumes too much. Place the compass with the centre tip on p and writing tip on q. Then *lift* the compass without changing the determined distance between p and q. Place the centre tip of the compass on r and draw your circle. Unfortunately, it is not clear that such a direct lift is permitted by the construction rules. Here is a way to do this properly.

- Construct the line ℓ through p and r.
- Construct the line k passing through q and parallel to ℓ.
- Construct the circle b with centre p and passing through q.
- The circle b meets line k at another point s, which is thereby constructible.
- Construct the line m passing through p and s.
- Through r construct the line n which is parallel to the constructed line m.
- The point t where line n meets line k is constructible, and the distance of t to r equals the distance of p to q.
- Construct the circle c with centre r and passing through t.

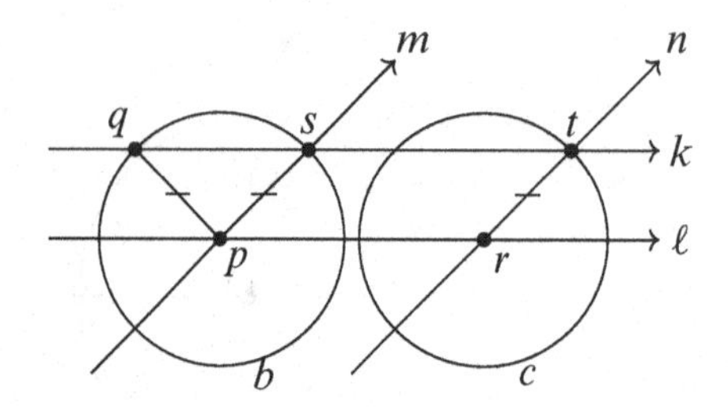

Constructing Square Roots

If p, q are constructible points at a distance d, then there exists a constructible point t whose distance to p is $\sqrt{d}$. Here is how to construct it.

- Construct the line ℓ through p and q.
- The initial points $(0,0), (1,0)$ at a distance 1 to each other are constructible. By lifting the compass construct a circle of radius 1 with centre p (not shown in the diagram), and let r be the point where this circle meets ℓ. Choose r so that p is between r and q. Such r is a constructible point on line ℓ and at a distance 1 from p.
- Construct the mid-point s of the segment from r to q.
- Construct the line m through p and orthogonal to line ℓ.
- Construct the circle k with centre s and passing through q. Circle k also passes through r.
- Let t be the constructible point where circle k meets line m.
- The distance between t and p is $\sqrt{d}$.

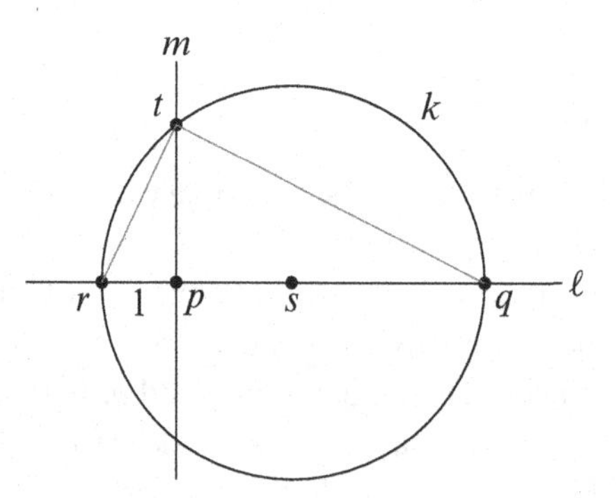

To see the claim, note that the triangle formed by r, q, t has a right angle at t. Recall that d is the given distance from p to q. If a is the distance from t to r, and b is the distance from t to q, and c is the distance from t to p, the Pythagorean theorem gives

$$(1+d)^2 = a^2 + b^2 = (1^2 + c^2) + (c^2 + d^2).$$

This equation simplifies to $d = c^2$, whence $c = \sqrt{d}$.

7.4.2 Constructible Real Numbers

The points $(0,0)$ and $(1,0)$ are constructible by decree. Thus the x-axis is constructible, and, being perpendicular to the x-axis and running through $(0,0)$, the y-axis is constructible too.

Definition 7.31. A real number a will be called a ***constructible real number*** when $(a,0)$ on the x-axis is a constructible point.

If $p = (a,b)$ is a constructible point, its orthogonal projections $(a,0)$ and $(0,b)$ onto the x-axis and the y-axis, respectively, are constructible points. The circle with centre $(0,0)$ and passing through $(0,b)$ cuts the x-axis at $(\pm b,0)$, which are thereby constructible points. Thus both coordinates of every constructible point are constructible numbers.

Conversely, suppose a,b are constructible numbers. The circle centred at $(0,0)$ and passing through $(b,0)$ meets the constructible y-axis at the point $(0,b)$. The perpendiculars to the x-axis taken at $(a,0)$ and the perpendicular to the y-axis taken at $(0,b)$ are constructible and meet at the point (a,b). Thus every pair of constructible numbers gives the coordinates of a constructible point.

Furthermore, if p,q are constructible points whose distance between them is d, then the circle with centre at the origin $(0,0)$ and radius d is constructible and cuts the x-axis at the point $(d,0)$. Hence, the distance between any two constructible points is a constructible number.

Let $\mathbb{E}$ denote the set of constructible real numbers.

Proposition 7.32. *The set $\mathbb{E}$ of constructible numbers is a subfield of $\mathbb{R}$. Furthermore, if $a \in \mathbb{E}$ and $a \geq 0$, then $\sqrt{a} \in \mathbb{E}$.*

Proof. We need to see that $\mathbb{E}$ is closed under the ring operations, and that every non-zero number in $\mathbb{E}$ has its inverse in $\mathbb{E}$.

Obviously $0,1 \in \mathbb{E}$. Let $a,b \in \mathbb{E}$. We need that $a \pm b, ab \in \mathbb{E}$, and if $a \neq 0$ that $1/a \in \mathbb{E}$. Since $\mathbb{E}$ is clearly closed under negations (just look at the circle with centre at the origin and through a constructible point $(c,0)$), there is no harm in supposing that $0 < a < b$ and proving that $a+b, ab, 1/a \in \mathbb{E}$.

To see that $a+b \in \mathbb{E}$, notice that the circle with centre $(b,0)$ and radius equal to a cuts the x-axis at $(a+b,0)$. The point $(a+b,0)$ is constructible because the compass can be lifted, as proven. Hence $a+b \in \mathbb{E}$.

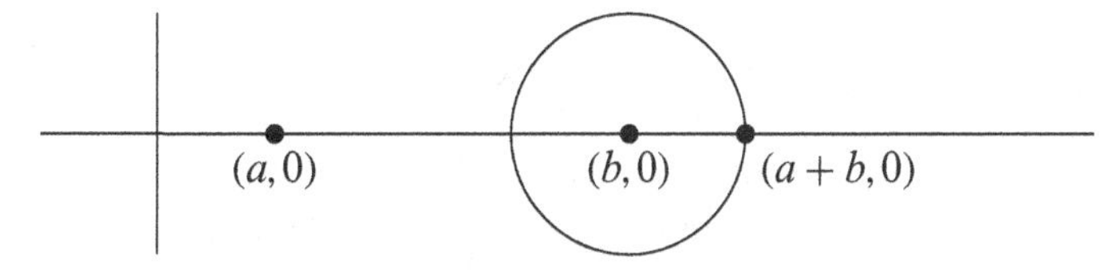

To see that $ab \in \mathbb{E}$, construct the line ℓ through $(a,0)$ and $(0,1)$. Through the constructible point $(0,b)$ construct the line m parallel to ℓ. The line m meets the x-axis at a constructible point $(c,0)$. By observing similar triangles, $c/b = a/1$, and thus $ab = c \in \mathbb{E}$.

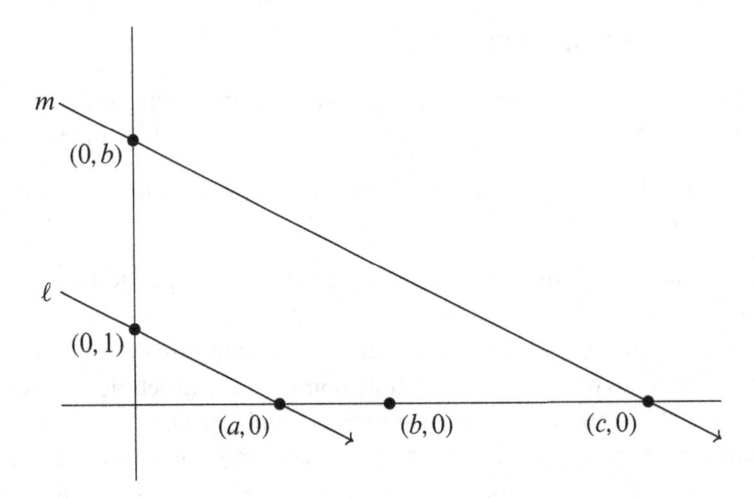

To see that $1/a \in \mathbb{E}$, draw the line ℓ through $(a,0)$ and $(0,1)$. Through $(1,0)$ construct the line m parallel to ℓ. Line m meets the y-axis at the constructible point $(0,c)$. Using similar triangles, $1/a = c/1 = c \in \mathbb{E}$.

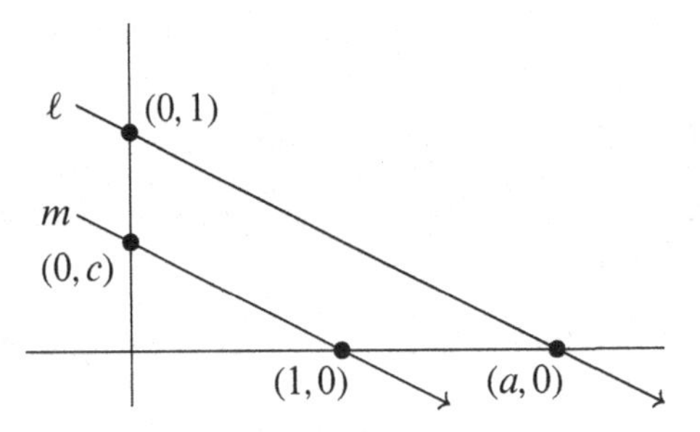

To see that $\sqrt{a} \in \mathbb{E}$ whenever $a \in \mathbb{E}$ and $a \geq 0$, note that a is the distance between the constructible point $(a,0)$ and $(0,0)$. As demonstrated in the preceding section, there is a constructible point whose distance to $(0,0)$ is $\sqrt{a}$, and all distances between constructible points are constructible numbers. Thus $\sqrt{a} \in \mathbb{E}$. □

Since $\mathbb{E}$ is a field, $\mathbb{Q} \subseteq \mathbb{E}$ meaning that every rational number is constructible. By Proposition 7.32 the numbers

$$\sqrt{2}, \sqrt{\sqrt{2}}, \ldots, \sqrt[2^n]{2}, \ldots$$

are constructible. The minimal polynomial of $\sqrt[2^n]{2}$ over $\mathbb{Q}$ is $X^{2^n} - 2$. Thus $\mathbb{E}$ contains extensions of $\mathbb{Q}$ that are of arbitrarily high degree. So, the degree of $\mathbb{E}$ over $\mathbb{Q}$ is infinite.

Constructing a Regular Pentagon

For an application of Proposition 7.32, let us verify that a regular pentagon can be constructed by ruler and compass. We can take the pentagon to have as vertices the complex numbers $1, \zeta, \zeta^2, \zeta^3, \zeta^4$ where $\zeta = e^{2\pi i/5} = \cos(2\pi/5) + i\sin(2\pi/5)$. Here we identify $\mathbb{R}^2$ with $\mathbb{C}$.

It will be enough to see that the real and imaginary parts $\cos(2\pi/5)$ and $\sin(2\pi/5)$ of ζ are constructible numbers.

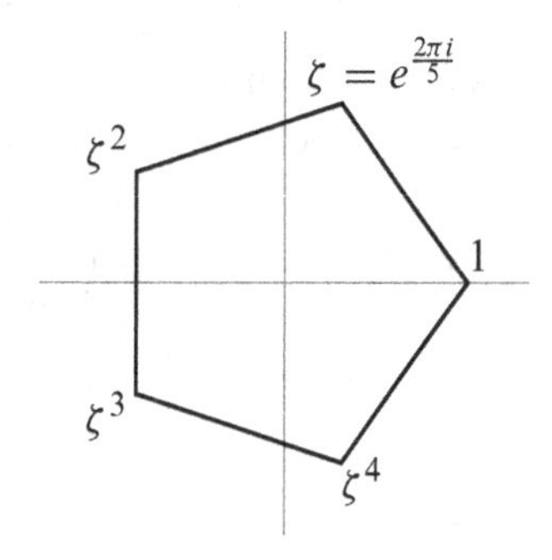

The minimal polynomial of ζ is $X^4 + X^3 + X^2 + X + 1$. Keeping in mind that $\zeta + \zeta^{-1} = 2\cos(2\pi/5)$, we get

$$\begin{aligned}
0 &= 1 + \zeta + \zeta^2 + \zeta^3 + \zeta^4 \\
&= 1 + (\zeta + \zeta^4) + (\zeta^2 + \zeta^3) \\
&= 1 + (\zeta + \zeta^{-1}) + (\zeta^2 + \zeta^{-2}) \\
&= 1 + (\zeta + \zeta^{-1}) + (\zeta + \zeta^{-1})^2 - 2 \\
&= 4\cos^2(2\pi/5) + 2\cos(2\pi/5) - 1.
\end{aligned}$$

Thus $\cos(2\pi/5)$ is the positive root of the quadratic $4X^2 + 2X - 1$, which turns out to be $\frac{\sqrt{5}-1}{4}$. Since the constructible numbers $\mathbb{E}$ form a field that contains the positive square roots of all of its positive elements, $\cos(2\pi/5)$ must be a constructible number.

Also $\sin(2\pi/5)$ is constructible because $\sin(2\pi/5) = \sqrt{1-\cos^2(2\pi/5)}$, which allows for Proposition 7.32 to be invoked once more. And so the vertex ζ of the pentagon is a constructible point. The other vertices of the pentagon are constructible by similar algebraic arguments. But we can also construct them instantly with a compass starting with the constructible points 1 (on the x-axis) and ζ. Indeed, the circle with centre at the origin and passing through 1 (i.e. the unit circle) meets the circle centred at ζ and passing through 1 at ζ^2. Thus ζ^2 is constructible, Then the circle centred at ζ^2 and passing through ζ meets the unit circle at ζ^3, and likewise to get to ζ^4.

7.4.3 Constructible Real Numbers and Field Towers

Definition 7.33. If

$$\mathbb{Q} \subseteq F_1 \subseteq F_2 \subseteq \cdots \subseteq F_n$$

is a tower of fields where each F_j is a subfield of $\mathbb{C}$ and the degrees $[F_j : F_{j-1}] = 2$, we shall call it a ***quadratic tower***. Furthermore, if the F_j are subfields of $\mathbb{R}$, we shall refer to the tower as a ***quadratic real tower***.

Quadratic towers are special types of radical towers. Proposition 7.32 implies that every number in the top field of a quadratic real tower is a constructible number. Indeed, if F is a subfield of $\mathbb{E}$ and F has a quadratic extension K that lies inside $\mathbb{R}$, then such K takes the form $F[\sqrt{a}]$ where a is a positive real number inside F and thus inside $\mathbb{E}$. By Proposition 7.32 $\sqrt{a} \in \mathbb{E}$, which forces $K \subseteq \mathbb{E}$. Apply this argument inductively to each inclusion of the quadratic real tower to see that all numbers in the top field of the tower are constructible.

Conversely, as we shall now prove, every constructible number belongs to a quadratic real tower. Recall that for a subset A of the plane $\mathbb{R}^2$, the lines joining any two of its points are called A-lines and the circles centred at the points of A and passing through other points of A are called A-circles.

Proposition 7.34. *If F is a subfield of $\mathbb{R}$, and p is a point in $\mathbb{R}^2$ which is the intersection of two F^2-lines, or the intersection of an F^2-line with an F^2-circle or the intersection of two F^2-circles, then the coordinates of p lie in F or in a subfield of $\mathbb{R}$ which is a quadratic extension of F.*

Proof. The equation of the line passing through two points $(a_1, b_1), (a_2, b_2)$ in F^2 is $(Y - b_1)(a_2 - a_1) = (X - a_1)(b_2 - b_1)$, which simplifies to the form

$$aX + bY + c = 0 \text{ where } a, b, c \in F.$$

The equation of the F^2-circle with centre at (a_1, b_1) and passing through (a_2, b_2) is given by $(X - a_1)^2 + (Y - b_1)^2 = (a_2 - a_1)^2 + (b_2 - b_1)^2$, which reduces to the form

$$X^2 + Y^2 + dX + eY + f = 0, \text{ where } d, e, f \in F.$$

The coordinates of the intersection of two such F^2-lines come from solving two linear equations in two unknowns over the field F. Such solutions belong to the field F.

When solving for the intersection of an F^2-line with an F^2-circle, write X or Y in the equation of the F^2-line as a linear polynomial in the other variable. Say Y is written in terms of X. Substitute that Y into the equation of the F^2-circle. The resulting equation is at most quadratic in X with coefficients in F. The solution, say α, for this equation is the first coordinate of p. Then the extension $F[\alpha]$ of F must be either F itself or a quadratic extension of F. Also since $\alpha \in \mathbb{R}$, the extension $F[\alpha]$ is a subfield of $\mathbb{R}$. The other coordinate of p also lies in $F[\alpha]$ because it is a linear polynomial in α.

When solving for the intersection of two F^2-circles, subtract their equations to get the equation of a line of the form $aX + bY + c = 0$ where $a, b, c \in F$. The intersection of this line with either F^2-circle will give the points of intersection of the two circles. As seen from the argument just above, the coordinates of these intersection points lie in F or in a subfield of $\mathbb{R}$ which is a quadratic extension of F. □

Now comes the key result about the field $\mathbb{E}$ of constructible numbers.

Proposition 7.35. *There is an infinite tower of field extensions*

$$\mathbb{Q} = F_0 \subseteq F_1 \subseteq F_2 \subseteq \cdots \subseteq F_n \subseteq \cdots$$

such that every degree $[F_j : F_{j-1}] = 2$ and $\mathbb{E} = \bigcup_{j=0}^{\infty} F_n$.

Proof. The constructible numbers are the real coordinates of the constructible points. The definition of constructible points in $\mathbb{R}^2$ at the outset of this section reveals that they can be enumerated as

$$p_0, p_1, p_2, \ldots, p_n, \ldots$$

where $p_0 = (0,0), p_1 = (1,0)$ and, after that, each p_n is an intersection of two $\{p_0, p_1, \ldots, p_{n-1}\}$-lines/circles.

First put $F_0 = \mathbb{Q}$. Next, for each positive integer n, build a field F_n inductively such that

- $F_n \subseteq \mathbb{R}$,
- the coordinates of p_n lie in F_n, and
- F_n is a degree 1 or degree 2 extension of F_{n-1}.

Since the coordinates of p_1 are in $\mathbb{Q}$, take $F_1 = \mathbb{Q}$ as well. Suppose that the fields $F_0, F_1, \ldots, F_{n-1}$ have been built to the above specifications. Thus we have the tower

$$\mathbb{Q} = F_0 \subseteq F_1 \subseteq F_2 \subseteq \cdots \subseteq F_{n-1},$$

where all $p_0, p_1, \ldots, p_{n-1} \in F_{n-1}$ and F_{n-1} is a subfield of $\mathbb{R}$.

Clearly p_n is an intersection of two F_{n-1}-lines, or the intersection of an F_{n-1}-line with an F_{n-1}-circle, or the intersection of two F_{n-1}-circles. By Proposition 7.34, the coordinates of p_n are in F_{n-1} or else in a subfield of $\mathbb{R}$ which is a quadratic extension of F_{n-1}. This extension, call it F_n, of degree 1 or degree 2 over F_{n-1} fulfills the specifications, and the infinite tower

$$\mathbb{Q} = F_0 \subseteq F_1 \subseteq F_2 \subseteq \cdots \subseteq F_n \subseteq \cdots,$$

of extensions of degree one or two, covers $\mathbb{E}$.

Every F_n lies inside $\mathbb{E}$ by Proposition 7.32. Thus the union of the F_n picks up precisely the field of constructible numbers $\mathbb{E}$. We can drop any F_n in the tower if it equals one or more of its predecessors, and the remaining tower will consist of extensions that are strictly quadratic. Since $\mathbb{E}$ is not a finite extension of $\mathbb{Q}$, the remaining tower will still have an infinite number of inclusions. □

From Propositions 7.32 and 7.35 there readily emerges a way to detect constructible real numbers purely in terms of field extensions.

Proposition 7.36. *A real number is constructible if and only if it lies in the top field of a finite quadratic real tower.*

From this there emerges a consequence pointing to an abundance of non-constructible numbers.

Proposition 7.37. *Every constructible real number is algebraic over $\mathbb{Q}$, and the degree of its minimal polynomial is a power of 2.*

Proof. Say a is that constructible number. According to Proposition 7.35, a lies in the top field F of a quadratic real tower starting with $\mathbb{Q}$. By the tower theorem, Proposition 6.13, the degree of F over $\mathbb{Q}$ is a power of 2. The degree $[\mathbb{Q}[a] : \mathbb{Q}]$ divides this power of 2, and is thereby itself a power of 2, which is the degree of the minimal polynomial of a over $\mathbb{Q}$. □

Squaring Circles, Doubling Cubes and Trisecting Angles

If the real number a is not algebraic over $\mathbb{Q}$, or if its minimal polynomial is not a power of 2, then the point $(a, 0)$ cannot be constructed, as we just proved. This leads to the solution of a trio of ancient problems on constructibility.

The number π is not algebraic over $\mathbb{Q}$. This is a famous result proven by Ferdinand von Lindemann in 1882. Hence, neither is $\sqrt{\pi}$ algebraic. By Proposition 7.37 the point $(\sqrt{\pi}, 0)$ is not constructible. A square of area π has side length $\sqrt{\pi}$. Such a square cannot be made using ruler and compass. Otherwise, we could lift the compass with a radius of $\sqrt{\pi}$ taken from the side of the square, put the centre at the origin and mark off the point $(\sqrt{\pi}, 0)$ on the x-axis. So, the area of a circle of radius 1 cannot be copied by ruler and compass in the form of a square.[5]

Next, the minimal polynomial of $\sqrt[3]{2}$ over $\mathbb{Q}$ is $X^3 - 2$. Since 3 is not a power of 2, Proposition 7.37 shows that $\sqrt[3]{2}$ is not constructible. Now, a cube of volume 1 has edge lengths equal to 1. In order to double its volume by ruler and compass, it is necessary to construct an edge length equal to $\sqrt[3]{2}$. Because this is not possible, the volume of a cube cannot be doubled by ruler and compass construction.

Also consider the trisection of angles by ruler and compass. We can easily trisect a $90°$ angle. Simply construct an equilateral triangle and then bisect one of its $60°$ angles. However, there is no *general* ruler and compass method for trisecting all angles.

For instance, a $60°$ angle cannot be trisected by ruler and compass. Suppose, on the contrary that a $60°$ angle permitted a trisection. Then the line through the origin making a $20°$ angle with the x-axis would be constructible. This line meets the circle of centre $(0, 0)$ and radius 1 at the point $(\cos(\pi/9), \sin(\pi/9))$. (Here the $20°$ angle is expressed in radians.) Thus $\cos(\pi/9)$ would be a constructible number.

But let us find the minimal polynomial of $\cos(\pi/9)$. Start with $2\cos(\pi/9) = e^{i\pi/9} + e^{-i\pi/9}$. Cube both sides and remember that $\cos(\pi/3) = 1/2$ to get

$$\begin{aligned} 8\cos^3(\pi/9) &= e^{i\pi/3} + e^{-i\pi/3} + 3e^{i\pi/9} + 3e^{-i\pi/9} \\ &= 2\cos(\pi/3) + 6\cos(\pi/9) = 1 + 6\cos(\pi/9). \end{aligned}$$

Thus $\cos(\pi/9)$ is a root of the cubic polynomial $8X^3 - 6X - 1$. This cubic polynomial is irreducible over $\mathbb{Q}$, because it has no rational root. So it is the minimal polynomial of $\cos(\pi/9)$. By Proposition 7.37, the number $\cos(\pi/9)$ is not constructible, which prevents the trisection of a $60°$ angle using ruler and compass.

[5] This is likely the origin of the idiom that one "cannot square the circle."

A Non-constructible Number of Degree Four

Proposition 7.37 is neat because it answers historic questions. Here is an example to show that its converse fails.

The polynomial $f = X^4 - 4X + 2$ is irreducible over $\mathbb{Q}$, by Eisenstein's criterion. Its derivative is $4(X^3 - 1)$, which reveals that f decreases to the left of 1 and increases to the right of 1. By inspection, $f(1) < 0$ while $f(0) > 0, f(2) > 0$. Thus $f(X)$ has exactly two real roots, one between 0 and 1 and one between 1 and 2. Let us show that at least one of the roots is not constructible.

Let r, s be the roots of f. Then

$$X^4 - 4X + 2 = (X - r)(X - s)(X^2 + aX + b) \text{ for some } a, b \text{ in } \mathbb{R}.$$

Expand this factorization and equate coefficients to get

$$\begin{aligned} a - (r + s) &= 0 \\ b - a(r + s) + rs &= 0 \\ ars - b(r + s) &= -4 \\ brs &= 2. \end{aligned}$$

The first and fourth equations are the same as

$$r + s = a \quad \text{and} \quad rs = \frac{2}{b}.$$

Substitute these values of $r + s$ and rs into the second and third equations to get

$$a^2 = b + \frac{2}{b} \quad \text{and} \quad a\left(b - \frac{2}{b}\right) = 4, \text{ respectively.}$$

Squaring the second of these last equations leads to

$$a^2 = b + \frac{2}{b} \quad \text{and} \quad a^2\left(\left(b + \frac{2}{b}\right)^2 - 8\right) = 16.$$

With $c = a^2$ we get

$$c(c^2 - 8) = 16, \text{ which becomes } c^3 - 8c - 16 = 0.$$

The cubic polynomial $X^3 - 8X - 16$ is irreducible because it has no rational roots. Hence it is the minimal polynomial of c. Being the root of an irreducible cubic, c is not constructible, by Proposition 7.37. However, $c = a^2 = (r + s)^2$. If both r and s were constructible, then c would be constructible, since constructible numbers form a field. Thus one of r or s is not constructible.

The preceding example leaves something to be desired. Namely, which root of $X^4 - 4X + 2$ is non-constructible: r, s or both? The upcoming Proposition 7.42 will reveal that both are non-constructible.

A Degree Four Extension with No Proper Intermediate Fields

The preceding example provides as well a degree 4 field extension K of $\mathbb{Q}$, having no proper intermediate fields. Namely, take $K = \mathbb{Q}[r]$ where r is a non-constructible real number whose minimal polynomial has degree 4 over $\mathbb{Q}$. Any field F properly between $\mathbb{Q}$ and K would have degree 2 over $\mathbb{Q}$, by the tower theorem. That would put r inside a quadratic tower of field extensions. Then Proposition 7.36 would yield a contradiction.

This example also provides a degree 4 extension of $\mathbb{Q}$ which is not a Galois extension. Indeed, Galois extensions of degree 4 have Galois groups of order 4. Every group of order 4 has a proper non-trivial subgroup. By the Galois correspondence, the extension of degree 4 would have to have proper intermediate fields.

7.4.4 Constructible Complex Numbers

Identify the Euclidean plane $\mathbb{R}^2$ with the field of complex numbers $\mathbb{C}$, in the familiar way.

Definition 7.38. The complex numbers that are constructible as points of $\mathbb{R}^2$ are called ***constructible complex numbers***.

From the observations at the start of Section 7.4.2 a complex number $a+bi$ is constructible if and only if its real and imaginary parts a, b are constructible real numbers. Consequently, the set of constructible complex numbers is nothing but the field extension $\mathbb{E}[i]$ of the field $\mathbb{E}$ of constructible real numbers.

Proposition 7.39. *If $\alpha \in \mathbb{E}[i]$, then its square roots are in $\mathbb{E}[i]$.*

Proof. This can be done with a bit of trigonometry.

Ignoring the trivial $\alpha = 0$, the polar form of α is $re^{i\theta} = r\cos(\theta) + ir\sin(\theta)$, where r, θ are real numbers and $r \geq 0$. By the definition of constructible complex numbers $r\cos(\theta), r\sin(\theta) \in \mathbb{E}$. Furthermore, $r \in \mathbb{E}$ because r is the distance from the constructible point α to the origin. Thus $\cos(\theta), \sin(\theta) \in \mathbb{E}$.

The square roots of α are

$$\pm\left(\sqrt{r}e^{i\theta/2}\right) = \pm\left(\sqrt{r}\cos(\theta/2) + i\sqrt{r}\sin(\theta/2)\right).$$

As shown in Proposition 7.32, $\sqrt{r} \in \mathbb{E}$. Recalling the (somewhat obscure) half-angle formulas we see that

$$\cos(\theta/2) = \pm\sqrt{\frac{1+\cos(\theta)}{2}} \text{ and } \sin(\theta/2) = \pm\sqrt{\frac{1-\cos(\theta)}{2}}.$$

Since $\frac{1\pm\cos(\theta)}{2}$ are non-negative real numbers in $\mathbb{E}$, their square roots are in $\mathbb{E}$ as well. It follows that the real and imaginary parts of the square roots of α are in $\mathbb{E}$. This means that the square roots of α are in $\mathbb{E}[i]$. □

Here is the analogue of Proposition 7.36 for constructible complex numbers.

Proposition 7.40. *A complex number is constructible if and only if it lies in the top field of a quadratic tower.*

Proof. Suppose that

$$\mathbb{Q} \subset F_1 \subset F_2 \subset \cdots \subset F_k$$

is a quadratic tower. So, there is an α_1 in F_1 such that $F_1 = \mathbb{Q}[\alpha_1]$ and $\alpha_1^2 \in \mathbb{Q}$. By Proposition 7.39 $\alpha_1 \in \mathbb{E}[i]$, and thus $F_1 \subseteq \mathbb{E}[i]$. Likewise, there is an α_2 in F_2 such that $F_2 = F_1[\alpha_2]$ and $\alpha_2^2 \in F_1$. By Proposition 7.39 $\alpha_2 \in \mathbb{E}[i]$, and thus $F_2 \subseteq \mathbb{E}[i]$. Repeat this argument going up the tower to see that $F_k \subseteq \mathbb{E}[i]$.

Conversely suppose $\alpha \in \mathbb{E}[i]$. The real and imaginary parts, say a, b, of α are in $\mathbb{E}$. By Proposition 7.35 there is a real quadratic tower

$$\mathbb{Q} \subset F_1 \subset F_2 \subset \cdots \subset F_k$$

such that both $a, b \in F_k$. Then $\alpha = a + bi \in F_k[i]$, which puts α at the top of the slightly taller quadratic tower

$$\mathbb{Q} \subset F_1 \subset F_2 \subset \cdots \subset F_k \subset F_k[i]. \qquad \square$$

The analogue of Proposition 7.37 carries over to constructible complex numbers, with an identical proof in no need of repetition.

Proposition 7.41. *Every constructible complex number is algebraic over $\mathbb{Q}$ and the degree of its minimal polynomial is a power of 2.*

The Splitting Field of a Constructible Complex Number

If a complex number α is algebraic over $\mathbb{Q}$, with minimal polynomial g, the splitting field K of g is a subfield of $\mathbb{C}$ that contains α. The degree $[K : \mathbb{Q}]$ is all that it takes to detect if α is constructible. Galois theory plays a role.

Proposition 7.42. *If α is a constructible complex number with minimal polynomial g in $\mathbb{Q}[X]$ and β is another root of g, then β is also constructible.*

Proof. By Proposition 7.40 there is a quadratic tower

$$\mathbb{Q} \subset F_1 \subset F_2 \subset \cdots \subset F_k \text{ with } \alpha \in F_k.$$

We need a quadratic tower for β.

By Proposition 6.44 there is a field L which is both an extension L of F_k and a splitting field over $\mathbb{Q}$. (To brush up on this matter, let γ be a primitive generator of F_k over $\mathbb{Q}$. Thus $F_k = \mathbb{Q}[\gamma]$. Let h be the minimal polynomial of γ over $\mathbb{Q}$. The splitting field of h inside $\mathbb{C}$ is a suitable L.)

Both α and β are in L, because of the normality theorem, Proposition 6.42. By Proposition 6.36 there is an automorphism σ in $\text{Gal}(L/\mathbb{Q})$ such that $\sigma(\alpha) = \beta$. The tower

$$\mathbb{Q} = \sigma(\mathbb{Q}) \subset \sigma(F_1) \subset \sigma(F_2) \subset \cdots \subset \sigma(F_k)$$

remains quadratic by a routine exercise, and now $\beta \in \sigma(F_k)$. □

Another way to put the preceding result is that the roots of an irreducible polynomial in $\mathbb{Q}[X]$ are either all constructible or all non-constructible. We are in a position to prove a nice theorem.

Proposition 7.43. *A complex number is constructible if and only if it is algebraic over $\mathbb{Q}$ and the degree over $\mathbb{Q}$ of the splitting field of its minimal polynomial is a power of 2.*

Proof. Let α be the number in question.

Say α is algebraic over $\mathbb{Q}$ with minimal polynomial g having splitting field K such that $[K : \mathbb{Q}] = 2^n$ for some n. For α to be constructible it suffices to see that α lies in the top field of a quadratic tower.

Well, 2^n is also the order of $\mathrm{Gal}(K/\mathbb{Q})$ since K is a Galois extension of $\mathbb{Q}$. Proposition 3.21, applied to the 2-group $\mathrm{Gal}(K/\mathbb{Q})$, provides a strictly descending chain of normal subgroups

$$\mathrm{Gal}(K/\mathbb{Q}) \supset H_1 \supset H_2 \supset \cdots \supset H_n = \langle 1 \rangle.$$

The fundamental theorem of Galois theory turns this chain into the field tower

$$\mathbb{Q} = \mathrm{Fix}(\mathrm{Gal}(K/\mathbb{Q})) \subset \mathrm{Fix}(H_1) \subset \mathrm{Fix}(H_2) \subset \cdots \subset \mathrm{Fix}(\langle 1 \rangle) = K.$$

By the tower theorem the degree 2^n of K over $\mathbb{Q}$, equals the product of the field extension degrees $[\mathrm{Fix}(H_j) : \mathrm{Fix}(H_{j-1})]$. (Here $H_0 = \mathrm{Gal}(K/\mathbb{Q})$.) These degrees are powers of 2 and the number of such degrees is n. It follows that each $[\mathrm{Fix}(H_j) : \mathrm{Fix}(H_{j-1})] = 2$. Thus the above field tower is quadratic, and $\alpha \in K$ at the top of such a tower.

Conversely, suppose α is constructible. By Proposition 7.43 all conjugates of α are constructible, and so the entire splitting field K consists of constructible numbers, i.e. $K \subseteq \mathbb{E}[i]$. By the primitive generator theorem, Proposition 7.2, there is a γ in K such that $K = \mathbb{Q}[\gamma]$. Since γ is constructible, the degree of its minimal polynomial is a power of 2, due to Proposition 7.41. This power of 2 equals $[K : \mathbb{Q}]$. □

The proof of the preceding result is an apt illustration of the influence of Galois theory. As a consequence we are rewarded with a stunning revelation regarding the constructibility of regular polygons.

Constructible Polygons

Which regular polygons are constructible? We can easily construct equilateral triangles and squares, and we have seen that regular pentagons are constructible because $\cos(2\pi/5)$ is quadratic over $\mathbb{Q}$. A hexagon can readily be constructed, but not a heptagon, as will become apparent. In 1796 Gauss showed how to construct a regular 17-sided polygon, and then he identified the regular polygons which can be constructed. Here is that remarkable discovery.

Proposition 7.44. *A regular polygon with n sides is constructible by ruler and compass if and only if the value $\phi(n)$ of the Euler function taken at n is a power of 2.*

Proof. We might as well take that polygon to be the one whose vertices are $1, \zeta, \zeta^2, \ldots, \zeta^{n-1}$, where $\zeta = e^{2\pi i/n}$. These vertices are constructible complex numbers if and only if the primitive root of unity ζ is constructible. By remarks following Proposition 7.16, the splitting field of the minimal polynomial of ζ is $\mathbb{Q}[\zeta]$, and $[\mathbb{Q}[\zeta] : \mathbb{Q}] = \phi(n)$. By Proposition 7.42, ζ is constructible if and only if $[\mathbb{Q}[\zeta] : \mathbb{Q}]$, and thereby $\phi(n)$, is a power of 2. □

Here are the values of $\phi(n)$ for n from 3 to 17:

n	3	4	5	6	7	8	9	10	11	12	13	14	15	16	17
$\phi(n)$	2	2	4	2	6	4	6	4	10	4	12	6	8	8	16

Thus we see for instance that the regular polygons with 7, 9, 11, 13 or 14 sides are non-constructible. The constructible cases up to 16 were known to the ancient Greek mathematicians. After two millennia Gauss showed that a regular 17-gon is constructible.[6]

In case n is a prime, $\phi(n) = n - 1$, and so $\phi(n)$ is a power of 2 if and only if $n = 2^k + 1$ for some k. Primes of this sort are known as ***Fermat primes***. The known Fermat primes are

$$2 = 2^0 + 1,\ 3 = 2^1 + 1,\ 5 = 2^2 + 1,\ 17 = 2^4 + 1,\ 257 = 2^8 + 1 \text{ and } 65537 = 2^{16} + 1.$$

Whether there are any more remains an open problem in number theory.

From the unique factorization of a positive integer n, it is routine to detect when $\phi(n)$ is a power of 2. Namely, $\phi(n)$ is a power of 2 (meaning that the n-gon is constructible) if and only if the only odd primes that divide n are Fermat primes each appearing in its factorization without repetition. The verification of this can be left as an exercise.

EXERCISES

1. Which of the following numbers are constructible:

$$\sqrt[6]{2},\ \sqrt[5]{27},\ \sqrt[6]{3+\sqrt{5}},\ \sqrt[4]{2-\sqrt[4]{5}},\ \sqrt[3]{7-5\sqrt{2}}?$$

2. Show that a number $\cos\theta$ is constructible if and only if $\sin\theta$ is constructible.
3. If a is a positive real number of degree 4 over $\mathbb{Q}$, show that a is constructible if and only if $\mathbb{Q}[a]$ contains a subfield properly between itself and $\mathbb{Q}$.
4. Show that the polynomial $X^4 - 6X + 3$ is irreducible over $\mathbb{Q}$ and has a real, non-constructible root.
5. As demonstrated after Proposition 7.37 the polynomial $X^4 - 4X + 2$ in $\mathbb{Q}[X]$ is irreducible, has two real roots r, s and two non-real complex roots. Show that the degree over $\mathbb{Q}$ of the

[6] Legend has it that Gauss asked for the regular 17-gon to be inscribed on his tombstone. Even though this did not occur, a monument of Gauss in his hometown of Braunschweig displays the configuration.

splitting field of this polynomial is 24. What is the Galois group of this polynomial as a group of permutations of the four roots?

6. Suppose that the factorization of an integer n is given by $n = 2^e p_1^{e_1} p_2^{e_2} \cdots p_k^{e_k}$ where the p_j are odd primes, $e \geq 1$ and all $e_j \geq 1$. Prove that a regular n-gon is constructible if and only if every p_j is a Fermat prime and all $e_j = 1$.

7.5 The Inverse Galois Problem over $\mathbb{Q}$

Is every finite group isomorphic to the Galois group of a Galois extension of $\mathbb{Q}$? This esoteric and still open question is known as the *inverse Galois problem*. Our modest goal here is to show that every finite *abelian* group can be realized as the Galois group of a polynomial over $\mathbb{Q}$.

We know from Proposition 7.17 that if ζ is a primitive nth root of unity, then the extension $\mathbb{Q}[\zeta]$ is Galois over $\mathbb{Q}$ and that $\text{Gal}(\mathbb{Q}[\zeta]/\mathbb{Q})$ is isomorphic to $\mathbb{Z}_n^\star$, the group of units modulo n. Our approach will be to show that every finite abelian group H is the homomorphic image of some $\mathbb{Z}_n^\star$. That will put us in position to use Proposition 6.81 and show that there is an intermediate field E between $\mathbb{Q}$ and $\mathbb{Q}[\zeta]$ such that $\text{Gal}(E/\mathbb{Q}) \cong H$. A few results from number theory will be needed.

The Evaluation of Cyclotomic Polynomials at Integers

Recall from Proposition 7.16 that the minimal polynomial of a primitive nth root of unity is the cyclotomic polynomial Φ_n. As seen in Section 7.2.3, Φ_n is a monic polynomial with coefficients in $\mathbb{Z}$, and the factorization into irreducibles of $X^n - 1$ in $\mathbb{Q}[X]$ is given by

$$X^n - 1 = \prod_{d|n} \Phi_d.$$

This lets us see that for every integer a the prime factors of the integer $\Phi_n(a)$ are severely constrained.

Proposition 7.45. *If n, a are integers with n positive, and p is a prime such that*

$$p \mid \Phi_n(a) \text{ but } p \nmid n,$$

then

$$p \equiv 1 \bmod n.$$

Proof. For each integer b let $\overline{b}$ denote its residue in $\mathbb{Z}_p$, and for each polynomial $f = \sum a_j X^j$ in $\mathbb{Z}[X]$, let $\overline{f} = \sum \overline{a_j} X^j$, the polynomial in $\mathbb{Z}_p[X]$ whose coefficients are reduced modulo p. These reduction maps are ring homomorphisms, and it should be clear that $\overline{f(b)} = \overline{f}\left(\overline{b}\right)$ for every b in $\mathbb{Z}$ and every f in $\mathbb{Z}[X]$.

It suffices to show that the residue $\overline{a}$ is in the group $\mathbb{Z}_p^\star$ and that its order is n. Then Lagrange will give that n divides the order $p - 1$ of $\mathbb{Z}_p^\star$.

Substitute a into the identity $X^n - 1 = \prod_{d|n} \Phi_d$ and reduce modulo p to get

$$\overline{a}^n - \overline{1} = \prod_{d|n} \overline{\Phi_d(a)} = \overline{0} \text{ in } \mathbb{Z}_p.$$

The second equation holds because $\overline{\Phi_n(a)}$ is one of the factors in the above product and because $p \,|\, \Phi_n(a)$, meaning that $\overline{\Phi_n(a)} = \overline{0}$. Since $\overline{a}^n = \overline{1}$, the residue $\overline{a}$ is in the group $\mathbb{Z}_p^\star$, and its order, say k, divides n. We want $k = n$.

There is also the identity $X^k - 1 = \prod_{d|k} \Phi_d$. Substitute a into this identity and reduce modulo p to get

$$\overline{0} = \overline{a}^k = \prod_{d|k} \overline{\Phi_d(a)} \text{ in } \mathbb{Z}_p.$$

Hence $\overline{\Phi_e(a)} = \overline{0}$ for some divisor e of k.

The factorization $X^n - 1 = \prod_{d|n} \Phi_d$ in $\mathbb{Z}[X]$ leads to the factorization $X^n - \overline{1} = \prod_{d|n} \overline{\Phi_d}$ in $\mathbb{Z}_p[X]$. The polynomial $\overline{\Phi_n}$ appears in this factorization, and so does $\overline{\Phi_e}$, since $e \,|\, k$ and $k \,|\, n$. By assumption $\overline{a}$ is a root of $\overline{\Phi_n}$, and we saw that $\overline{a}$ is a root of $\overline{\Phi_e}$. If the order k of $\overline{a}$ is less than n, these two factors of $X^n - \overline{1}$ must be distinct, whereby $\overline{a}$ is a repeated root of $X^n - \overline{1}$. However, the derivative of $X^n - \overline{1}$ is $\overline{n}X^{n-1}$. Since $p \nmid n$, this derivative has degree $n - 1$ and shares no root with $X^n - \overline{1}$. By Proposition 6.53, $X^n - \overline{1}$ has no repeated root. So, the order of $\overline{a}$ must be n. □

Let us test the above result using Φ_6, which equals $X^2 - X + 1$. Use $a = 5$ to get $\Phi_6(5) = 21 = 3 \cdot 7$. The prime 7 divides $\Phi_6(5)$ but does not divide 6, and sure enough $7 \equiv 1 \bmod 6$. Use $a = 8$ to get $\Phi_6(8) = 57 = 3 \cdot 19$. Now 19 divides $\Phi_6(8)$ but does not divide 6, and once more $19 \equiv 1 \bmod 6$.

Let us test the result for Φ_8, which equals $X^4 + 1$. Plug in 5 to get $\Phi_8(5) = 626 = 2 \cdot 313$. The prime 313 divides $\Phi_8(5)$ and does not divide 8. As expected $313 \equiv 1 \bmod 8$. Plug in 10 to get $\Phi_8(10) = 10001 = 73 \cdot 137$. Here the primes 73 and 137 divide $\Phi_8(10)$ but do not divide 8. Again $73 \equiv 1 \bmod 8$ and $137 \equiv 1 \bmod 8$.

Primes in Arithmetic Progressions – a Special Case

For a positive integer n we might expect that the number of primes appearing as divisors in the sequence of integers

$$\Phi_n(1), \Phi_n(2), \Phi_n(3), \ldots$$

is infinite. If this were the case, Proposition 7.45, together with the fact n has only a finite number of prime divisors, would show that there are infinitely many primes p such that $p \equiv 1 \bmod n$. This would be a decent improvement on the fact there are infinitely many primes. A far reaching and deep result in number theory is *Dedekind's theorem*. It says that for every pair of coprime positive integers n, k there are infinitely many primes p such that $p \equiv k \bmod n$.

The special case where $k = 1$ will be what is needed to realize every abelian group as the Galois group of a Galois extension of $\mathbb{Q}$.

Here is a general result about polynomials with integer coefficients, with a rather clever proof.

Proposition 7.46. *If f is a non-constant polynomial in $\mathbb{Z}[X]$, then the number of primes appearing in the factorizations of the integers*

$$f(1), f(2), f(3), \ldots$$

is infinite.

Proof. It comes down to showing that if $p_1, p_2, \ldots, p_k$ is a finite list of primes, then there exists a prime p not in the list and a positive integer b such that $p \mid f(b)$.

Since f has but finitely many zeroes, select an integer a such that $f(a) \neq 0$, put $m = f(a)$, and let

$$g(X) = \frac{1}{m} f(a + mp_1p_2 \cdots p_kX).$$

An inspection of the polynomial g reveals that

- $\deg g = \deg f \geq 1$,
- the constant coefficient of g is $g(0) = \frac{1}{m} f(a) = 1$,
- every non-constant coefficient of $f(a+mp_1p_2 \cdots p_kX)$ is divisible by m, and thus $g \in \mathbb{Z}[X]$,
- all coefficients of g, except for $g(0)$ which is 1, are divisible by $p_1p_2 \cdots p_k$.

Thus for every integer n, the product $p_1p_2 \cdots p_k$, divides $g(n) - 1$. Since the degree of g is positive, there is a positive integer n such that $|g(n)| > 1$. Let p be a prime factor of this $g(n)$. This p cannot be one of the p_j, for otherwise such p_j would divide 1. Then p also divides $mg(n) = f(a + mp_1p_2 \cdots p_kn)$. The positive b we seek is $a + mp_1p_2 \cdots p_kn$. □

A weak, but still interesting, version of Dedekind's theorem readily follows from the above two results.

Proposition 7.47. *For every positive integer n there exist infinitely many primes p such that $p \equiv 1 \bmod n$.*

Proof. Let S be any finite set of primes each congruent to 1 modulo n. The set of primes that divide n is also finite. According to Proposition 7.46, there must be a prime p and an integer a such that

$$p \notin S, \; p \nmid n \text{ and } p \mid \Phi_n(a).$$

By Proposition 7.45, $p \equiv 1 \bmod n$. In this way S can be augmented by an additional prime congruent to 1 modulo n. □

Abelian Groups as Galois Groups over $\mathbb{Q}$

Before going to the Galois theory, here is a curiosity about finite abelian groups.

Proposition 7.48. *If A is a finite abelian group, then there is a positive integer n such that A is a homomorphic image of the group $\mathbb{Z}_n^\star$.*

Proof. According to Proposition 2.92 on the structure of finite abelian groups,

$$A \cong C_{n_1} \times C_{n_2} \times \cdots \times C_{n_k},$$

where each C_{n_j} is a cyclic group of order n_j.

By the weak version of Dedekind's theorem, Proposition 7.47, there are infinitely many primes congruent to 1 modulo each n_j. Thus there are *distinct* primes $p_1, p_2, \ldots, p_k$ such that each $p_j \equiv 1 \bmod n_j$.

The order of the group $\mathbb{Z}_{p_j}^\star$ is $p_j - 1$, and $p_j - 1 = n_j \ell_j$ for some ℓ_j. Furthermore this group, being a finite group of units in a field, is cyclic, as shown in Proposition 4.37. By Proposition 2.41, $\mathbb{Z}_{p_j}^\star$ has a subgroup L_j of order ℓ_j. The quotient group $\mathbb{Z}_{p_j}^\star / L_j$ is a cyclic group, and Lagrange says its order is n_j. Since all cyclic groups of equal order are isomorphic, there results the obvious surjective homomorphism $\varphi_j : \mathbb{Z}_{p_j}^\star \to \mathbb{Z}_{p_j}^\star / L_j \cong C_{n_j}$. Put these maps together to get the surjective homomorphism

$$\varphi : \mathbb{Z}_{p_1}^\star \times \mathbb{Z}_{p_2}^\star \times \cdots \times \mathbb{Z}_{p_k}^\star \to C_{n_1} \times C_{n_2} \times \cdots \times C_{n_k},$$

where

$$(x_1, x_2, \ldots, x_k) \mapsto (\varphi_1(x_1), \varphi_2(x_2), \ldots, \varphi_k(x_k)).$$

Let n be the product $p_1 p_2 \cdots p_k$. Because the primes p_j are distinct, there is the ring isomorphism

$$\mathbb{Z}_n \cong \mathbb{Z}_{p_1} \times \mathbb{Z}_{p_2} \times \cdots \times \mathbb{Z}_{p_k}.$$

This comes from the obvious generalization of Proposition 4.48. Under this isomorphism the respective groups of units become isomorphic, and thus

$$\mathbb{Z}_n^\star \cong \mathbb{Z}_{p_1}^\star \times \mathbb{Z}_{p_2}^\star \times \cdots \times \mathbb{Z}_{p_k}^\star.$$

The desired homomorphism is the above group isomorphism, followed by φ, followed by the isomorphism with A at the outset. □

We come to our desired result.

Proposition 7.49. *If A is a finite abelian group, then there is a Galois extension E of $\mathbb{Q}$ such that* $\mathrm{Gal}(E/\mathbb{Q}) \cong A$. *Furthermore, E is a subfield of a cyclotomic extension of $\mathbb{Q}$.*

Proof. Take a positive integer n such that A is a homomorphic image of $\mathbb{Z}_n^\star$. If ζ is a primitive nth root of unity, the extension $\mathbb{Q}[\zeta]$ is Galois over $\mathbb{Q}$, and by Proposition 7.17,

$\operatorname{Gal}(\mathbb{Q}[\zeta]/\mathbb{Q}) \cong \mathbb{Z}_n^\star$. Thus there is a surjective homomorphism $\varphi : \operatorname{Gal}(\mathbb{Q}[\zeta]/\mathbb{Q}) \to A$. If $H = \ker\varphi$ and $E = \operatorname{Fix}(H)$, the Galois correspondence gives $\operatorname{Gal}(\mathbb{Q}[\zeta]/E) = H$. Since H is normal in $\operatorname{Gal}(\mathbb{Q}[\zeta]/\mathbb{Q})$, the extension E is Galois over $\mathbb{Q}$ and Proposition 6.81 gives

$$\operatorname{Gal}(E/\mathbb{Q}) \cong \operatorname{Gal}(\mathbb{Q}[\zeta]/\mathbb{Q})/H \cong A.$$

The latter isomorphism comes from the first isomorphism theorem for groups. □

The Cyclic Group of Order 5 as a Galois Group

By way of illustration, with A the cyclic group of order 5, let us follow the foregoing train of thought to find a Galois extension of $\mathbb{Q}$ whose Galois group is isomorphic to A. By inspection, the prime 11 satisfies $11 \equiv 1 \bmod 5$. So let ζ be a primitive 11th root of unity. The Galois extension $\mathbb{Q}[\zeta]$ has degree 10 over $\mathbb{Q}$. Its Galois group is isomorphic to the cyclic group $\mathbb{Z}_{11}^\star$ of order 10. And $\mathbb{Q}[\zeta]$ contains a subfield E such that $\operatorname{Gal}(E/\mathbb{Q}) \cong A$. Let us find that E.

The subgroup H of $\operatorname{Gal}(\mathbb{Q}[\zeta]/\mathbb{Q})$ generated by the automorphism σ of $\mathbb{Q}[\zeta]$, determined by $\sigma : \zeta \mapsto \zeta^{10} = \zeta^{-1}$, has order 2. The desired Galois extension of $\mathbb{Q}$ is $\operatorname{Fix}(H)$, as the preceding discussion assures. Alternatively, the degree $[\mathbb{Q}[\zeta] : \operatorname{Fix}(H)] = 2$ and so $[\operatorname{Fix}(H) : \mathbb{Q}] = 5$. The order of $\operatorname{Gal}(\operatorname{Fix}(H)/\mathbb{Q})$ is also 5, and, since 5 happens to be prime, this Galois group has to be isomorphic to A.

It might be good practice to find a primitive generator for $\operatorname{Fix}(H)$. Let us take a basis for $\mathbb{Q}[\zeta]$ over $\mathbb{Q}$ to be the list of powers $\zeta, \zeta^2, \ldots, \zeta^{10}$. If $\alpha = a_1\zeta + a_2\zeta^2 + \cdots + a_9\zeta^9 + a_{10}\zeta^{10}$, with the a_j in $\mathbb{Q}$, is an element of $\mathbb{Q}[\zeta]$, such $\alpha \in \operatorname{Fix}(H)$ if and only if $\sigma(\alpha) = \alpha$. Since $\sigma(\alpha) = a_1\zeta^{10} + a_2\zeta^9 + \cdots + a_9\zeta^2 + a_{10}\zeta$, we see that $\sigma(\alpha) = \alpha$ if and only if

$$a_1 = a_{10},\ a_2 = a_9,\ a_3 = a_8,\ a_4 = a_7,\ a_5 = a_6.$$

Keeping in mind that $\zeta^j = \zeta^{11-j}$, the elements of $\operatorname{Fix}(H)$ are those of the form

$$b_1(\zeta + \zeta^{-1}) + b_2(\zeta^2 + \zeta^{-2}) + b_3(\zeta^3 + \zeta^{-3}) + b_4(\zeta^4 + \zeta^{-4}) + b_5(\zeta^5 + \zeta^{-5}).$$

Now with $\gamma = \zeta + \zeta^{-1}$ a routine calculation based on expanding the binomials $\zeta^2, \zeta^3, \zeta^4$ and ζ^5 reveals the following formulas:

$$\begin{aligned}
\zeta^2 + \zeta^{-2} &= \gamma^2 - 2 \\
\zeta^3 + \zeta^{-3} &= \gamma^3 - 3\gamma \\
\zeta^4 + \zeta^{-4} &= \gamma^4 - 4\gamma^2 + 2 \\
\zeta^5 + \zeta^{-5} &= \gamma^5 - 5\gamma^3 + 5\gamma.
\end{aligned}$$

It follows that $\operatorname{Fix}(H) = \mathbb{Q}[\gamma]$. These formulas also show how to calculate the minimal polynomial of γ. Indeed, note that the minimal polynomial of ζ is the cyclotomic polynomial $1 + X + X^2 + \cdots + X^9 + X^{10}$. This means that

$$1 + (\zeta + \zeta^{-1}) + (\zeta^2 + \zeta^{-2}) + (\zeta^3 + \zeta^{-3}) + (\zeta^4 + \zeta^{-4}) + (\zeta^5 + \zeta^{-5}) = 0.$$

By substituting the preceding identities into the above we come up with

$$1 + \gamma + (\gamma^2 - 2) + (\gamma^3 - 3\gamma) + (\gamma^4 - 4\gamma^2 + 2) + (\gamma^5 - 5\gamma^3 + 5\gamma) = 0,$$

which reduces to

$$\gamma^5 + \gamma^4 - 4\gamma^3 - 3\gamma^2 + 3\gamma + 1 = 0.$$

The minimal polynomial of γ is

$$X^5 + X^4 - 4X^3 - 3X^2 + 3X + 1.$$

Incidentally, if ζ is the primitive root $e^{2\pi i/11}$, then $\gamma = \zeta + \zeta^{-1} = 2\cos(2\pi/11)$, whose minimal polynomial has just been found. The Galois group of this polynomial is cyclic of order 5.

Another Small Example

Let us find an extension of $\mathbb{Q}$ whose Galois group is isomorphic to the product $C_2 \times C_4$ of cyclic groups. Note that 3 is prime with $3 \equiv 1 \bmod 2$ and 5 is prime with $5 \equiv 1 \bmod 4$. According to the proofs in Propositions 7.48 and 7.49, the extension $\mathbb{Q}[\zeta]$, where ζ is a primitive 15th root of unity, contains a subfield E such that $\mathrm{Gal}(E/\mathbb{Q}) \cong C_2 \times C_4$. But the degree $[\mathbb{Q}[\zeta] : \mathbb{Q}] = \varphi(15) = 8$. The degree $[E : \mathbb{Q}]$ equals the order of $C_2 \times C_4$, which is also 8. Thus, in this case, $E = \mathbb{Q}[\zeta]$. Since $\mathrm{Gal}(\mathbb{Q}[\zeta]/\mathbb{Q}) \cong \mathbb{Z}_{15}^\star$, we also confirm in passing the simple fact that $\mathbb{Z}_{15}^\star \cong C_2 \times C_4$.

EXERCISES

1. Show that for every finite group G there are fields F, K such that K is a Galois extension of F and $\mathrm{Gal}(K/F) \cong G$.

   Hint. By Cayley's theorem G is isomorphic to a subgroup of the symmetric group $\mathcal{S}_n$ for some n. Find a Galois extension of some field for which the Galois group is $\mathcal{S}_n$, and then use the fundamental theorem of Galois theory. Unfortunately, the field F cannot be forced to be $\mathbb{Q}$.
2. Find a Galois extension of $\mathbb{Q}$ whose Galois group is isomorphic to the product $C_5 \times C_5$, where C_5 is the cyclic group of order 5.
3. Find a Galois extension E of $\mathbb{Q}$ whose Galois group is isomorphic to the cyclic group C_7. Specify a primitive generator for E.

8 Modules over Principal Ideal Domains

At times, seemingly disparate enterprises in mathematics are best understood when they are seen as special cases of one unifying concept. Groups, rings and fields are classic illustrations of this trend towards unification. And once a good concept has been articulated, it becomes an object of study in its own right. The idea of a ***module*** illustrates once more that a general approach, if properly conceived, can lead to multiple rewards.

While some of the early concepts to follow apply readily to non-commutative rings, in this chapter we shall consider only rings that are commutative, and we ultimately specialize to principal ideal domains.

8.1 The Language and Tools of Modules

Definition 8.1. Let R be a commutative ring and M an abelian group *using the operation of addition*. We say that M is an ***R-module***, alternatively a ***module over*** R, when there is a mapping

$$R \times M \to M \text{ denoted by } (r, v) \mapsto rv,$$

which obeys the following conditions for all r, s in R and all u, v in M:

- $r(u + v) = ru + rv$
- $(r + s)v = rv + sv$
- $(rs)v = r(sv)$
- $1v = v$.

The mapping $R \times M \to M$ is at times called the ***scalar multiplication***.

The first two laws are the ***distributive laws***. The third law is the ***associative law***, which lets us unambiguously write rsv for any r, s in R and v in M, without the use of brackets. Since R is assumed to be commutative we also have $rsv = srv$. The order of doing successive scalar multiplications does not matter.

A brief exercise with the module axioms reveals that

$$a0 = 0 \text{ and } 0v = 0 \text{ for every } a \text{ in } R \text{ and } v \text{ in } M,$$

and also for all v that

$$(-1)v = -v, \text{ the additive inverse of } v.$$

We can see plainly that if K is a field, a K-module is nothing but a vector space over K. An R-module is like a vector space, except that the scalar multiplication $R \times M \to M$ uses a ring. This relaxation of the potential scalars greatly increases the complexity that modules can have. For instance, if F is a field and V is a vector space over F, we know that if $a \in F$ and $v \in V$ and $av = 0$, then either $a = 0$ or $v = 0$. Such a simple fact can fail even when R is an integral domain, as we shall observe shortly. We shall also learn that, unlike a vector space, a module need not have a basis.

8.1.1 Examples of Modules

For each example below the verification that it is a module is routine.

- Every abelian group M, using addition, is a $\mathbb{Z}$-module. The scalar multiplication of each x in M by an integer a is given by the multiple notation ax introduced in Section 2.1 and used throughout. For example, the finite abelian group $\mathbb{Z}_5$ under addition is a $\mathbb{Z}$-module. Notice that even though $5 \neq 0$ in $\mathbb{Z}$ and $1 \neq 0$ in $\mathbb{Z}_5$, we do have that the scalar multiplication $5 \cdot 1 = 0$ in $\mathbb{Z}_5$.
- Every ideal J of a commutative ring R is an R-module. The scalar multiplication $R \times J \to J$ is given by $(r, a) \mapsto ra$, where the latter product is the one given in R. In particular R is an R-module. Note that if J is a subring of R, the above scalar multiplication does not make J into an R-module, because ra need not lie in J as it does with ideals.
- If $\varphi : R \to S$ is a homomorphism of commutative rings, then S is an R-module with the scalar multiplication

$$R \times S \to S \text{ given by } (r, s) \mapsto \varphi(r)s.$$

  In particular, if S is a commutative ring extension of a commutative ring R, then S is an R-module by means of the inclusion map of R into S. Also in particular, if J is an ideal of R, then R/J is an R-module by means of the canonical projection $\varphi : R \to R/J$. Here the scalar multiplication $R \times R/J \to R/J$ given by $(a, x + J) \mapsto (a + J)(x + J) = ax + J$.
- Suppose that V is a vector space over a field K. The set $\mathcal{L}(V)$ of K-linear operators $V \to V$ is a ring under addition and composition. We can view each a in K as the K-linear operator $a : V \to V$ where $v \mapsto av$. In this way $\mathcal{L}(V)$ contains K as a subring. For a variable X there is the substitution map on the polynomial ring $K[X]$ given by

$$K[X] \to \mathcal{L}(V) \text{ where } f(X) \mapsto f(V).$$

  Then for each fixed T in $\mathcal{L}(V)$ the space V becomes a $K[X]$-module according to the multiplication

$$K[X] \times V \to V \text{ where } v \mapsto f(T)(v) \text{ for every } v \text{ in } V,$$

  using the much enlarged ring of "scalars" $K[X]$. In other words if $f(X) \in K[X]$ and $v \in V$, the definition of our $K[X]$-module says that

$$f(X)v = f(T)(v).$$

For instance, $Xv = T(v)$, $(X^2 + 1)v = (T^2 + 1)(v)$, and for each a in K the notation av means the same whether we view V as a vector space over K or as a $K[X]$-module.

Let us call this module the ***$K[X]$-module inherited from the linear operator*** T. This important example, suggests that $K[X]$-modules may be a fruitful way to think about linear operators.

- If R is a commutative ring and $M_1, \dots, M_n$ are R-modules, the n-fold Cartesian product $M_1 \times \cdots \times M_n$ is an abelian group under point-wise addition of coordinates. The point-wise scalar multiplication

$$R \times (M_1 \times \cdots \times M_n) \to (M_1 \times \cdots \times M_n) \text{ defined by } ((a, (v_1, \dots, v_n)) \mapsto (av_1, \dots, av_n),$$

makes $M_1 \times \cdots \times M_n$ into another R-module. This module is called the ***direct product***, and often the ***direct sum***[1] of the M_j.

In particular, the n-fold Cartesian product $R^n = R \times \cdots \times R$, of R taken as a module over itself, is once more an R-module.

8.1.2 Module Homomorphisms

Definition 8.2. If R is a commutative ring and M, N are R-modules, a mapping $\varphi : M \to N$ is called an ***R-module homomorphism***, or more briefly an ***R-map***, provided φ is a homomorphism between the underlying abelian groups M, N and also satisfies

$$\varphi(av) = a\varphi(v) \text{ for every } a \text{ in } R \text{ and every } v \text{ in } M.$$

When K is a field, a K-module homomorphism is nothing but a K-linear transformation.

Every abelian group using addition is also a $\mathbb{Z}$-module, as already noted. If A, B are such abelian groups, the $\mathbb{Z}$-maps from A to B coincide with the group homomorphisms from A to B. Indeed, if $\varphi : A \to B$ is a group homomorphism, a moment's reflection reveals that $\varphi(nx) = n\varphi(x)$ for every n in $\mathbb{Z}$ and x in A.

If R is a commutative ring and M is an A-module, the R-maps $R \to M$ are easy to come by. For any a in M, the mapping $\varphi_a : R \to M$ defined by $r \mapsto ra$ is an R-map by routine verification, and this accounts for all of them. Indeed, if $\varphi : R \to M$ is an R-map, put $a = \varphi(1)$. Then for every r in R we get $\varphi(r) = \varphi(r1) = r\varphi(1) = ra$.

We should note from the above that ring homomorphisms $R \to R$ are not the same as R-maps when the ring R is viewed as a module over itself. The requirement $\varphi(rs) = \varphi(r)\varphi(s)$ for a ring homomorphism differs from the requirement $\varphi(rs) = r\varphi(s)$ for an R-map. If $\varphi : R \to R$ is a ring homomorphism, then $\varphi(1) = 1$, whereas if φ is an R-map, then $\varphi(1)$ could be any element of R.

[1] The term "sum" can be justified because every n-tuple $(x_1, \dots, x_n)$ from $M_1 \times M_2 \times \cdots \times M_n$ is the sum $(x_1, 0, 0, \dots, 0) + (0, x_2, 0, \dots, 0) + \cdots + (0, 0, 0, \dots, x_n)$.

For more examples, let M, N be R-modules and let $M \times N$ be their direct sum module. The projection mappings $M \times N \to M$ and $M \times N \to N$ given by $(x,y) \mapsto x$ and $(x,y) \mapsto y$, respectively, are R-maps.

An interesting example of module homomorphisms comes from linear algebra. Let $S, T : V \to V$ be commuting linear operators on a vector space V over a field K. Take V to be the $K[X]$-module inherited from T. The mapping S is an abelian group homomorphism for sure. But S is also a $K[X]$-map. Indeed, for every polynomial $f(X)$ in $K[X]$ and every v in V:

$$S(f(X)v) = S(f(T)(v)) = f(T)(S(v)) = f(X)S(v).$$

We used the fact that since S and T commute, S also commutes with every polynomial in T. Conversely, every $K[X]$-map $S : V \to V$ must be a K-linear transformation that commutes with T. Indeed, if $v \in V$, then $S(T(v)) = S(Xv) = XS(v) = T(S(v)$, whereby S and T commute.

Definition 8.3. When an R-module homomorphism $\varphi : M \to N$ is a bijection, we say that φ is an ***isomorphism of modules*** and that M, N are ***isomorphic modules***. We write $M \cong N$ to denote that M and N are isomorphic.

If $R = \mathbb{Z}$, the isomorphism problem for $\mathbb{Z}$-modules comes down to that of abelian groups. This was answered for finite abelian groups in Propositions 2.92 and 2.93.

For another illustration of isomorphism, suppose that M, N are $K[X]$-modules inherited from transformations $S, T : V \to V$, respectively. This means that $M = N = V$ as vector spaces over the field K. However, for a polynomial $f(X)$ in $K[X]$ and a vector v in V, the scalar multiplication for M is given by $f(X)v = f(S)(v)$, while the scalar multiplication for N is given by $f(X)v = f(T)(v)$. An isomorphism $\varphi : M \to N$ is an isomorphism from V to V as an abelian group, which also satisfies the property

$$\varphi(f(S)(v)) = f(T)(\varphi(v)) \text{ for all } f(X) \text{ in } K[X] \text{ and all } v \text{ in } V.$$

In particular,

$$\varphi(av) = a\varphi(v) \text{ for each } a \text{ in } K \text{ and } \varphi(S(v)) = T(\varphi(v)).$$

This reveals that φ is a K-linear bijection on V such that under composition of operators:

$$\varphi \circ S = T \circ \varphi.$$

In other words $S = \varphi^{-1} \circ T \circ \varphi$, which is the statement that S and T are ***similar operators*** on V. As we may know, in the case of finite-dimensional V, a linear operator on V corresponds to a matrix. Thus the fundamental problem of deciding on the similarity of two matrices is the same as the isomorphism problem for two $K[X]$-modules.

8.1.3 Submodules

Definition 8.4. If M is an R-module, a subset N of M is called an ***R-submodule*** provided N is closed under the module operations. Namely,

- $0 \in N$,
- $v + w \in N$ for every v, w in N, and
- $av \in N$ for every a in R and every v in N.

A bit of reflection shows that submodules are modules in their own right. For instance, to check that N is closed under negation, and is thereby a subgroup of M, simply observe that $-v = (-1)v$ for all v in N and then use the third closure property above.

In case M is a vector space over a field, the submodules of M are the same as the subspaces of M. When M is a $\mathbb{Z}$-module (i.e. an abelian group), its submodules are its subgroups. In the case of R as module over itself, the submodules coincide with its ideals.

Finitely Generated and Cyclic Modules

Definition 8.5. Suppose S is a non-empty subset of M. The finite sums of the form

$$\sum a_j v_j \text{ where } a_j \in R \text{ and } v_j \in S$$

are called ***R-linear combinations*** of the elements of S. By inspection the set N of such combinations is a submodule of M, which we call the ***submodule generated by*** S.

Definition 8.6. If a module M is generated by a *finite* set, we say that M is a ***finitely generated module***.

In the special case where S is the singleton $\{v\}$, the submodule generated by v is the set

$$Rv = \{av : a \in R\}$$

consisting of all R-multiples of v.

Definition 8.7. Modules that are generated by a single element v are known as ***cyclic R-modules***.[2]

In the case of a field K, the finitely generated K-modules are the familiar finite-dimensional spaces, which come with their all important bases. We might also recall that a Noetherian ring R is one whose submodules (i.e. ideals) are finitely generated. The cyclic submodules of a commutative ring R, viewed as a module over itself, are nothing but the principal ideals of R.

If a finite-dimensional vector space V, over a field K, is the $K[X]$-module inherited from a linear operator $T : V \to V$, then V is a finitely generated $K[X]$-module. Any vector space basis of V will do as a finite set of $K[X]$-module generators of V. The classical problem in this case is to obtain generators that can detect when $K[X]$-modules are isomorphic.

8.1.4 Free Modules

When R is viewed as a module over itself, the product module R^n consists of all n-tuples $(a_1, \ldots, a_n)$ where the $a_j \in R$ with the addition and scalar multiplication taken point-wise. The n-tuple $e_j = (0, 0, \ldots, 1, \ldots, 0)$, with zero entries except for a 1 in the jth spot, is known

[2] This terminology is borrowed from the case of finite cyclic groups, where (using addition) the integer multiples of the generator do cycle repeatedly though the group.

as the jth ***standard n-tuple***. Clearly every element v of R^n has a representation as the R-linear combination $v = a_1e_1 + a_2e_2 + \cdots + a_ne_n$, where the $a_j \in R$. Furthermore, the a_j are uniquely determined by v. That is, a change in any of the a_j represents a different element of R^n.

The above example inspires the following definition.

Definition 8.8. A list of elements $v_1, \ldots, v_n$ in an R-module M is called a ***module basis*** of M, provided that for every v in M there is a *unique* n-tuple $(a_1, \ldots, a_n)$ in R^n such that

$$v = a_1v_1 + a_2v_2 + \cdots + a_nv_n.$$

In other words, every v in M is a unique R-linear combination of $v_1, \ldots, v_n$. A module having a basis will be called ***free***.

As noted above, R^n is free. More precisely such modules should be called finitely generated and free. We could also speak of modules with infinite bases, but they will not come up. So, for us a free module is one having a finite basis.

Not All Modules Are Free

The first thing we learn about finitely generated vector spaces over a field is that they have bases, and also that the number of vectors in a basis equals the number in any other basis. But not all finitely generated modules M come with a basis. The following property of free modules over integral domains can be exploited to come up with an abundance of non-free modules.

Proposition 8.9. *If R is an integral domain and M is a free R-module, then for every non-zero v in M and non-zero a in R, the scalar product $av \neq 0$.*

Proof. Let $v_1, v_2, \ldots, v_n$ form a basis of M and say $v = a_1v_1 + a_2v_2 + \cdots + a_nv_n$, where the a_j are the unique coefficients in R used to represent v in terms of this basis. Since $v \neq 0$, some coefficient is non-zero, say $a_1 \neq 0$. Then $av = aa_1v_1 + aa_2v_2 + \cdots + aa_nv_n$. In the domain R the product $aa_1 \neq 0$. Thus $av \neq 0$, because the v_j form a basis of M. □

Modules with the property of Proposition 8.9 are called ***torsion-free***.[3] We just proved that free modules over integral domains are torsion-free.

Now finite abelian groups (i.e. $\mathbb{Z}$-modules) are not torsion-free because each element has finite order. Thus, finite $\mathbb{Z}$-modules are not free. In fact any abelian group with a non-zero element of finite order is not torsion-free and thereby not free.

A Torsion-Free but Still Not Free Module

Here is an example of a module that is torsion-free but not free. For the ring take $R = \mathbb{Q}[X, Y]$, the polynomials over $\mathbb{Q}$ in two variables X, Y, and for the R-module take

[3] The word "torsion" seems to have crept into the lexicon of algebra from the realm of topology where certain abelian groups arising from the twisted structure of a space have elements of finite order.

the ideal $M = \mathbb{Q}[X, Y]X + \mathbb{Q}[X, Y]Y$ generated by X, Y inside $\mathbb{Q}[X, Y]$. If f is a non-zero polynomial in $\mathbb{Q}[X, Y]$ and g is a non-zero polynomial in M, then the product $fg \neq 0$, which means that M is torsion-free. In fact, by this argument, every ideal inside an integral domain is torsion-free. We might recall (or can check) that M is not a principal ideal, i.e. not a cyclic R-module. Thus, if M were free, its R-module basis $g_1, g_2, \ldots, g_n$ would contain at least two polynomials. With g_1, g_2 alternating in their roles as scalars in R and as module elements in M, we see that

$$g_2g_1 - g_1g_2 = 0 = 0g_1 + 0g_2 + \cdots + 0g_n.$$

We thereby witness two distinct representations of the zero element in M, in contradiction to the g_j forming a basis.

Submodules of Free Modules Need Not Be Free

The preceding example reveals as well that a submodule of a free module need not be free. Indeed, R is free as a module over itself with basis consisting of the single element 1, and M is a submodule of R.

An Alternative Way to View a Free Module

Definition 8.10. A list of elements $v_1, \ldots, v_n$ in an R-module M is ***independent*** when the only coefficients $a_1, \ldots, a_n$ in R that give $a_1v_1 + \cdots + a_nv_n = 0$ are $a_1 = a_2 = \cdots = a_n = 0$.

If $v_1, \ldots, v_n$ form a basis for a free module M, the v_j are independent because the element 0 can only have the representation $0v_1 + \cdots + 0v_n$.

Conversely if $v_1, \ldots, v_n$ are independent and generate M, then M is free with the v_j as its basis. To see the uniqueness of representations in terms of these v_j, suppose

$$a_1v_1 + \cdots + a_nv_n = b_1v_1 + \cdots + b_nv_n,$$

with all a_j, b_j in R. The rules for modules lead to $(a_1 - b_1)v_1 + \cdots + (a_n - b_n)v_n = 0$. By the independence of the v_j, it follows that $a_1 - b_1 = \cdots = a_n - b_n = 0$, and then $a_j = b_j$ for all j.

To summarize, $v_1, \ldots, v_n$ form a basis of M if and only if they generate M and are independent, just like it is with vector spaces.

Homomorphisms on Free Modules

The following result virtually copies what is well known about bases in vector spaces, and affords some justification for the term "free."

Proposition 8.11. *Let M, N be R-modules and let M be free with basis $v_1, \ldots, v_n$. If $w_1, \ldots, w_n$ is an arbitrary list of elements of N (allowing possible repetitions), then there exists a unique R-map*

$$\varphi : M \to N \text{ such that } \varphi(v_j) = w_j.$$

Proof. Any R-map $\varphi : M \to N$ that sends v_j to w_j must be such that

$$\varphi(a_1v_1 + \cdots + a_nv_n) = a_1\varphi(v_1) + \cdots + a_n\varphi(v_n) = a_1w_1 + \cdots + a_nw_n,$$

for every list of scalars $a_1, \ldots, a_n$ in R. So, if such an R-map exists, it must be given by the above recipe.

For the existence of such an R-map $\varphi : M \to N$, all that is needed is to confirm that the recipe $\varphi : a_1v_1 + \cdots + a_nv_n \mapsto a_1w_1 + \cdots + a_nw_n$ is well defined, R-linear and sends each v_j to w_j. That φ is well defined comes from the fact the v_j form a basis of M, which ensures that each element of M is a unique R-linear combination of the basis. The verifications that such φ is an R-map and sends v_j to w_j are trivial and probably best omitted. □

With a basis for an R-module M at hand, Proposition 8.11 allows for the construction of R-maps from M at will. From this remark it follows that every free R-module M with basis $v_1, \ldots, v_n$ is isomorphic to the direct sum module R^n. The isomorphism is the one that sends v_j to the standard n-tuple $(0, 0, \ldots, 1, \ldots, 0)$ with 1 in the jth position.

8.1.5 Quotient Modules and the First Isomorphism Theorem

Suppose that N is a submodule of an R-module M. In particular, N is an abelian subgroup of M, which lets us form the quotient group M/N consisting of cosets $x + N$ where $x \in M$. The quotient group M/N becomes an R-module under the scalar multiplication

$$a(x + N) = ax + N \text{ for every } a \text{ in } R \text{ and every } x \text{ in } M.$$

The scalar multiplication is well defined. Indeed, suppose that $x + N = y + N$ for some x, y in M. Then $x - y \in N$, and since N is a submodule so is $a(x - y)$ in N. Hence $ax - ay \in N$, which means that $ax + N = ay + N$. The R-module requirements for M/N are readily verifiable.

Definition 8.12. The module M/N is called the ***quotient module*** of M by N. The mapping

$$\varphi : M \to M/N, \text{ given by } x \mapsto x + M,$$

is called the ***canonical projection*** of M onto M/N.

The scalar multiplication for the quotient module is defined to ensure that the ***canonical projection*** is a module homomorphism.

In the case of R as a module over itself, a submodule J of R is an ideal. In this case the quotient R/J is both a ring and an R-module as constructed above. This quotient is alternatively an R-module by virtue of the quotient map $\varphi : R \to R/J$ as a ring homomorphism. This alternative scalar multiplication is defined by

$$a(x + J) = \varphi(a)(x + J), \text{ for every } a \text{ in } R \text{ and every } x + J \text{ in } R/J.$$

Since $\varphi(a)(x+J) = (a+J)(x+J) = ax+J$, we see that the scalar multiplication coming from the ring homomorphism $\varphi : R \to R/J$ coincides with the scalar multiplication defined for R/J as a quotient module. The two seemingly different ways to make R/J into an R-module in fact coincide.

Definition 8.13. For any R-map $\varphi : M \to P$ its ***kernel*** is the familiar set

$$\ker \varphi = \{x \in M : \varphi(x) = 0\}.$$

The ***image*** of M under φ is the equally familiar set

$$\varphi(M) = \{\varphi(x) : x \in M\}.$$

Not only is $\ker \varphi$ a subgroup of M, but it is also an R-submodule of M, by an easy verification. The image is a submodule of P, by inspection.

Since an R-map $\varphi : M \to P$ is by default a homomorphism of the underlying abelian groups, we also have the well worn fact that φ is injective if and only if $\ker \varphi$ is the zero submodule $\{0\}$.

With the above items in place the ***first isomorphism theorem for modules*** prevails just as it did for groups and for rings. We present a small generalization of this basic theorem.

Proposition 8.14. *If $\varphi : M \to P$ is a homomorphism of R-modules, and N is a submodule of M contained in $\ker \varphi$, then*

$$\overline{\varphi} : M/N \to P \text{ given by } x + N \mapsto \varphi(x)$$

is a well defined R-map and $\ker \overline{\varphi} = (\ker \varphi)/N$. If $N = \ker \varphi$, then $\overline{\varphi}$ is an injection and thereby

$$M/\ker \varphi \cong \varphi(M).$$

Proof. To see that $\overline{\varphi}$ is well defined, suppose $x + N = y + N$. Thus $x - y \in N \subseteq \ker \varphi$, which leads to $\varphi(x - y) = 0$ and then $\varphi(x) = \varphi(y)$. The proof that $\overline{\varphi}$ is an R-map can be omitted because it is a matter of following the pertinent definitions. To see that $\ker \overline{\varphi} = (\ker \varphi)/N$, suppose $x + N \in \ker \overline{\varphi}$. Then

$$0 = \overline{\varphi}(x + N) = \varphi(x),$$

whence $x \in \ker \varphi$ and then $x + N \in (\ker \varphi)/N$. Conversely, suppose $x + N \in (\ker \varphi)/N$. Thus $x + N = y + N$ where $y \in \ker \varphi$. But $x - y \in N \subseteq \ker \varphi$. So, $x \in \ker \varphi$ already, and by the definition of $\overline{\varphi}$, we get $x + N \in \ker \overline{\varphi}$.

The last statement of the proposition follows immediately from the one preceding it, because $(\ker \varphi)/(\ker \varphi)$ is the zero module which implies that $\overline{\varphi}$ is injective. □

The Correspondence Theorem for Surjections

Here is yet another general fact that can be useful at times.

Proposition 8.15. *Suppose $\varphi : M \to N$ is a surjective homomorphism between modules over a commutative ring R. For each submodule P of N, its inverse image $\varphi^{-1}(P)$ is a submodule of M containing $\ker \varphi$. The inverse image establishes a bijective correspondence between the family of submodules of N and the family of submodules of M that contain $\ker \varphi$.*

Proof. If P is a submodule of N, it is trivial to see that $\varphi^{-1}(P)$ is a submodule of M containing $\ker \varphi$. The correspondence $P \mapsto \varphi^{-1}(P)$ from submodules of N to submodules of

M containing $\ker\varphi$ is injective because φ is surjective. This ensures that $\varphi(\varphi^{-1}(P)) = P$ for every submodule of P of N.

To see that this correspondence is surjective, suppose Q is a submodule of M containing $\ker\varphi$. Clearly $\varphi(Q)$ is a submodule of N. It is enough to see that $\varphi^{-1}(\varphi(Q)) = Q$. Certainly $\varphi^{-1}(\varphi(Q)) \supseteq Q$. So, take v in $\varphi^{-1}(\varphi(Q))$. Then $\varphi(v) \in \varphi(Q)$. That is $\varphi(v) = \varphi(w)$ for some w in Q. From this we get that $v - w \in \ker\varphi \subseteq Q$. Since $w \in Q$ it must be that $v \in Q$ also. This verifies that $\varphi^{-1}(\varphi(Q)) = Q$. □

8.1.6 Cyclic Modules and Annihilator Ideals

If the commutative ring R is taken as a module over itself and J is an ideal of R, then the quotient module R/J is cyclic with generator $1 + J$, because

$$a + J = a \cdot 1 + J = a(1 + J),$$

for every a in R. The first isomorphism theorem for modules will show us that for cyclic modules this is the way by which they all arise. First another definition comes into play.

Definition 8.16. If M is a module over R, the ***annihilator*** of M is the set

$$\operatorname{ann}(M) = \{a \in R : av = 0 \text{ for all } v \text{ in } M\}.$$

By routine verification the annihilator is an ideal of R.

As a special case let M be a finite abelian group. As a $\mathbb{Z}$-module M has an annihilator, which is an ideal of $\mathbb{Z}$. Recalling that in $\mathbb{Z}$ all ideals are principal, let $\operatorname{ann}(M) = \mathbb{Z}r$ for some positive integer r. Every integer n such that $nx = 0$ for all x in M lies in $\mathbb{Z}r$, i.e. n is divisible by r. In particular r divides the order of M. The generator of $\operatorname{ann}(M)$ is the smallest positive integer that annihilates all of M. For instance, the annihilator of $\mathbb{Z}_6 \times \mathbb{Z}_{15}$ is the ideal $\mathbb{Z}30$.

The annihilator is an entity that isomorphic modules have in common, a so-called isomorphism invariant.

Proposition 8.17. *Isomorphic modules over R have identical annihilators.*

Proof. Let $\varphi : M \to N$ be an isomorphism of R-modules.

If $r \in \operatorname{ann}(N)$ and $u \in M$, then

$$0 = r\varphi(u) = \varphi(ru).$$

Since φ is injective $ru = 0$. Thus $r \in \operatorname{ann}(M)$.

If $r \in \operatorname{ann}(M)$ and $v \in N$, then $v = \varphi(u)$ for some u in M, since φ is surjective. Then

$$rv = r\varphi(u) = \varphi(ru) = \varphi(0) = 0.$$

Thus $r \in \operatorname{ann}(N)$. □

In case the modules are cyclic, the above result has a converse. In that regard we might observe the triviality that if M is the cyclic module Rv, an element r is in $\operatorname{ann}(M)$ if and only if $rv = 0$.

Proposition 8.18. *If M is a cyclic R-module, then M is isomorphic to $R/\operatorname{ann}(M)$. Cyclic modules with the same annihilator are isomorphic.*

Proof. Let $M = Rv$ for some generator v in V. The mapping

$$\varphi : R \to M \text{ given by } a \mapsto av$$

is a surjective R-map. By the first isomorphism theorem $M \cong R/\ker\varphi$. An inspection shows that $\ker\varphi = \operatorname{ann}(M)$, and so $M \cong R/\operatorname{ann}(M)$.

For the second statement of the proposition suppose M, N are cyclic and such that $\operatorname{ann}(M) = \operatorname{ann}(N)$. From what was just shown

$$M \cong R/\operatorname{ann}(M) = R/\operatorname{ann}(N) \cong N.$$

The modules M, N are isomorphic. □

For instance, suppose R is a module over itself and that v in R generates the cyclic submodule Rv (i.e. a principal ideal). Then Rv is isomorphic to R/J where J is the kernel of the mapping $R \to Rv$ given by $a \mapsto av$. So, J consists of all a in R for which $av = 0$. In case R is an integral domain and $v \neq 0$, we see that J is the zero ideal, and that the mapping $a \mapsto av$ is an isomorphism of R-modules. If R is a domain and $v \neq 0$, then $Rv \cong R$ as R-modules.

8.1.7 Direct Sums of Submodules

The definitions to follow provide a language for understanding the structure of modules.

Definition 8.19. If $N_1, \ldots, N_n$ are submodules of an R-module M, their ***sum*** $N_1+N_2+\cdots+N_n$ is the set of all sums of elements $v_1 + v_2 + \cdots + v_n$ where each v_j runs over N_j. We further say that M is the ***direct sum*** of the N_j provided for every v in M there is a *unique* n-tuple $(v_1, v_2, \ldots, v_n)$ in the direct product $N_1 \times N_2 \times \cdots \times N_n$ such that

$$v = v_1 + v_2 + \cdots + v_n.$$

After a moment's reflection it becomes apparent that the sum of submodules is another submodule of M. At times this sum is expressed in the more compact sigma notation $\sum_{j=1}^{n} N_j$. If M is the direct sum of submodules $N_1, N_2, \ldots, N_n$, then the mapping

$$\varphi : N_1 \times N_2 \times \cdots \times N_n \to M \text{ given by } (v_1, v_2, \ldots, v_n) \mapsto v_1 + v_2 + \cdots + v_n$$

is an isomorphism of R-modules. Indeed, the statement that φ is an isomorphism could be taken as the definition of direct sum. This also suggests why the Cartesian product $N_1 \times N_2 \times \cdots \times N_n$ is at times itself called a direct sum. To denote that M is the direct sum of the N_j we write:

$$M = N_1 \oplus N_2 \oplus \cdots \oplus N_n,$$

and in more compact notation $M = \bigoplus_{j=1}^{n} N_j$.

Definition 8.20. If $M = N_1 \oplus N_2 \oplus \cdots \oplus N_n$ we say that M has a ***direct sum decomposition*** into submodules N_j. If $v = v_1 + v_2 + \cdots + v_n$ with unique v_j in N_j, then v_j is called the ***component*** of v inside N_j.

Let S be a subset of the set of indices $\{1, 2, \ldots, n\}$. An element v of M lies in the partial sum $\sum_{i \in S} N_i$ if and only if for each j not in S the component of v inside N_j is 0.

If M is a free module with basis $v_1, \ldots, v_n$, then $M = Rv_1 \oplus \cdots \oplus Rv_n$. However, if we simply have a module M which is the direct sum of cyclic submodules, we cannot conclude that M is free. For instance, the abelian group $\mathbb{Z}_6$ is the direct sum of the two element subgroup $N_1 = \{0, 3\}$ and the three element subgroup $N_2 = \{0, 2, 4\}$, but there is no free $\mathbb{Z}$-module in sight. A small exercise reveals that if the direct cyclic summands were torsion-free, then M would be a free module.

Here is a way to recognize that a module is a direct sum of submodules.

Proposition 8.21. *An R-module M is the direct sum of submodules $N_1, \ldots, N_n$ if and only if $M = \sum_{j=1}^{n} N_j$ and for each j from 1 to n, the intersection*

$$N_j \cap \left(\sum_{i \neq j} N_i \right) = (0).$$

Proof. Say $M = N_1 \oplus N_2 \oplus \cdots \oplus N_n$. For each j from 1 to n suppose v lies in the intersection $N_j \cap \left(\sum_{i \neq j} N_i \right)$. The fact that $v \in N_j$ says that the components of v in the other N_i are 0. And the fact that $v \in \sum_{i \neq j} N_i$ says that the jth component of v is 0. Thus all components of v are 0, meaning that $v = 0$.

Conversely suppose that M is the sum of the N_j and that the above submodule intersections are zero. We need to show that for every v in M there is a unique n-tuple $(v_1, \ldots, v_n)$ in $N_1 \times N_2 \times \cdots \times N_n$ such that $v = v_1 + \cdots + v_n$. Well, the existence of such an n-tuple is given by the fact M is the sum of its submodules. Suppose next that $v_j, w_j \in N_j$ and that

$$v = v_1 + v_2 + \cdots + v_n = w_1 + w_2 + \cdots + w_n.$$

For each j from 1 to n we obtain

$$v_j - w_j = \sum_{i \neq j} w_i - v_i.$$

Clearly $v_j - w_j \in N_j$, while this same element, when examined on the right of this equality, lies in $\sum_{i \neq j} N_i$. Since $N_j \cap \left(\sum_{i \neq j} N_i \right) = (0)$, it follows that $v_j - w_j = 0$, and thus $v_j = w_j$. Hence, there can only be one n-tuple in $N_1 \times \cdots \times N_n$ whose components add up to v. Thus $M = N_1 \oplus N_2 \oplus \cdots \oplus N_n$. □

In the special case of only two submodules N, P of M, Proposition 8.21 says that $M = N \oplus P$ if and only if $M = N + P$ and $N \cap P$ is the zero module.

Definition 8.22. In case N, P are submodules of M such that $M = N \oplus P$, we say that N, P are ***direct summands*** of M.

We will make heavy use of the preceding concepts.

8.1.8 Free Modules Are Projective

The next definition seems obscure at first glance, but it has long proven its value.

Definition 8.23. A module M over a commutative ring R is called ***projective***[4] provided that for every surjective R-map $\varphi : N \to M$ there is an R-map $\psi : M \to N$ such that $\varphi \circ \psi$ is the identity map on M.

We will not delve much into this important concept except to note something we will use.

Proposition 8.24. *Every free module is projective. In more detail, if $\varphi : N \to M$ is a surjective homomorphism of R-modules and M is a free module, then there is an injective homomorphism $\psi : M \to N$ such that $\varphi \circ \psi$ is the identity map on M. Consequently*

$$N = \ker \varphi \oplus \psi(M).$$

Proof. Let $v_1, \ldots, v_n$ be a basis of M, and let w_j in N be such that $\varphi(w_j) = v_j$. By Proposition 8.11 there is an R-map $\psi : M \to N$ such that $\psi(v_j) = w_j$. Clearly $\varphi \circ \psi(v_j) = \varphi(w_j) = v_j$. Since the R-map $\varphi \circ \psi$ is the identity on the basis of M, it is also the identity on M. The injectivity of ψ follows instantly from the fact that $\varphi \circ \psi$ is the identity map on M.

For the second claim, write every x in N as

$$x = \psi(\varphi(x)) + (x - \psi(\varphi(x))).$$

Then

$$\psi(\varphi(x)) \in \psi(M) \text{ and } \varphi(x - \psi(\varphi(x)) = \varphi(x) - \varphi(\psi(\varphi(x))) = \varphi(x) - \varphi(x) = 0.$$

This shows that $N = \ker \varphi + \psi(M)$.

To see that this sum is direct suppose $x \in \psi(M) \cap \ker \varphi$. Thus $x = \psi(y)$ for some y in M, and $\varphi(\psi(y)) = 0$. Since $\varphi \circ \psi$ is the identity on M, it follows that $y = 0$, and then $x = 0$. Thus $N = \ker \varphi \oplus \psi(M)$, by Proposition 8.21. □

The proof of Proposition 8.24 assumed a finite basis for M only because we have not raised the concept of free modules with infinite bases, but the proof could readily be adapted to all free modules regardless of the size of their basis.

A Non-free Projective Module

For the sake of curiosity, it might be satisfying to see a projective module which is not free. To that end the following improvement on Proposition 8.24 is helpful.

Proposition 8.25. *If R is a commutative ring and M is a direct summand of a free R-module, then M is projective.*

Proof. Let F be a free module containing M as a direct summand with complementary submodule P giving $F = M \oplus P$. For every surjective homomorphism of R-modules

[4] This concept was introduced in the influential book *Homological Algebra* by Samuel Eilenberg and Henri Cartan in 1956.

$\varphi : N \to M$, we need an R-map $\psi : M \to N$, such that $\varphi \circ \psi$ is the identity map on M. This is obtained by taking a detour of R-maps through the free module F.

We have the surjective R-maps

$$\alpha : N \times P \to M \oplus P = F \text{ given by } (n,p) \mapsto \varphi(n) + p$$

and

$$\pi : N \times P \to N \text{ given by } (n,p) \mapsto n.$$

There is also the inclusion R-map

$$i : M \to M \oplus P = F \text{ given by } i : m \mapsto m.$$

The free R-module F is projective according to Proposition 8.24. So, there is an R-map $\beta : F \to N \times P$ such that $\alpha \circ \beta$ is the identity map on F. The requisite $\psi : M \to N$ is the composite R-map $\pi \circ \beta \circ i$. The following diagram keeps track of the various R-maps:

$$\begin{array}{ccc} N \times P & \underset{\beta}{\overset{\alpha}{\rightleftarrows}} & M \oplus P = F \\ \downarrow{\scriptstyle \pi} & & \uparrow{\scriptstyle i} \\ N & \underset{\psi = \pi \circ \beta \circ i}{\longleftarrow} & M \end{array} \qquad \alpha \circ \beta = \text{identity on } F.$$

For each m in M we need that $\varphi \circ \psi(m) = m$. Now $\beta(m) = (n,p)$ for some n in N and some p in P. The fact that $\alpha \circ \beta$ is the identity map on F gives

$$m = \alpha(\beta(m)) = \alpha(n,p) = \varphi(n) + p.$$

Since F is the direct sum of M and P, it follows that $p = 0$ and $\varphi(n) = m$. Thus $\beta(m) = (n,0)$ inside $N \times P$ and $\varphi(n) = m$. Consequently

$$\varphi \circ \psi(m) = \varphi \circ \pi \circ \beta \circ i(m) = \varphi \circ \pi \circ \beta(m) = \varphi \circ \pi(n,0) = \varphi(n) = m,$$

as desired. □

The converse of Proposition 8.25 holds. Consequently, projective modules coincide with modules which are isomorphic to direct summands of free modules. We leave this converse as an exercise based on Proposition 8.24.

Proposition 8.25 makes it easy to come up with a projective module that is not free. Let A, B be any pair of commutative rings and let R be their product ring $A \times B$. As with all rings, R is a free R-module. Clearly R is the direct sum of its ideals $A \times \{0\}$ and $\{0\} \times B$. These ideals are projective R-modules, in accordance with Proposition 8.25. Yet $A \times \{0\}$ is not free. This can be seen by noting that every list $(a_1, 0), (a_2, 0), \ldots, (a_k, 0)$ in $A \times \{0\}$ is burdened with the R-dependency relation

$$(0,1)(a_1,0) + (0,1)(a_2,0) + \cdots + (0,1)(a_k,0) = (0,0),$$

where the non-zero $(0,1) \in A \times B = R$. Thus no basis for $A \times \{0\}$ is possible.

EXERCISES

1. If an R-module M is generated by a set of elements S, show that a ring element r lies in $\text{ann}(M)$ if and only if $rv = 0$ for all v in S.
2. If M is a module over a ring R and $v \in m$, show that $(-1)v$ is the inverse (under addition) of v.
3. Show that a cyclic module is free if and only if it is torsion-free.
4. Show that a commutative ring R is a principal ideal domain if and only if all of its ideals are free R-modules.
5. Show that every finitely generated module over a commutative ring R is free if and only if R is a field.

   Note. For one direction of the proof some linear algebra will be needed.
6. (a) Show that the only independent sets in the $\mathbb{Z}$-module $\mathbb{Q}$ are those consisting of one non-zero element.

   (b) If A is a torsion-free $\mathbb{Z}$-module that properly contains $\mathbb{Q}$ as a submodule, show that A has an independent subset consisting of two elements.
7. In our discussion of free modules we only considered those which have a finite basis. The definition can be extended to free modules with infinite bases. Let M be an R-module and B a subset (possibly infinite) of M. We say that B is a basis for M when every element v of M has unique coefficients a_b where $b \in B$, all but finitely many of them are 0, and $v = \sum_{b \in B} a_b b$. A module with a basis is called free. Show that the abelian group $\mathbb{Q}$ (under addition) is not a free $\mathbb{Z}$-module.
8. Show that every finitely generated projective module is a direct summand of a free module having a finite basis.
9. Prove that a direct summand of a projective module is a projective module.
10. If M, N are projective R-modules, is their product module $M \times N$ projective?
11. Prove that a submodule P of an R-module M is a direct summand of M if and only if there exists an R-map $\psi : M/P \to M$ such that $\psi(x + P) + P = x + P$.

    If $\varphi : M \to M/P$ is the canonical projection, the latter condition is saying that $\varphi \circ \psi$ is the identity map on M/P.
12. Let R be the polynomial ring $\mathbb{Q}[X, Y]$, and let M be the submodule of $R \times R$ generated by the elements $(X, 0), (Y, Y), (0, X)$. Find a direct sum decomposition $M = P \oplus Q$ into non-zero submodules P, Q.
13. Find the annihilator of the $\mathbb{Z}$-module $\mathbb{Z}_{24} \times \mathbb{Z}_{15} \times \mathbb{Z}_{50}$.

8.2 Modules over Integral Domains

In order to put to use the generalities of the preceding section, we specialize by having the ring be an integral domain.

8.2.1 Torsion Elements and Modules

Definition 8.26. If R is an integral domain and v in an R-module M is such that $av = 0$ for some *non-zero* a in R we say that v has ***torsion***, and also that v is a ***torsion element*** of M.

The zero element of M has torsion, because $1 \cdot 0 = 0$. In case M is a vector space over a field K, the only torsion element is the zero vector. In case M is a finite $\mathbb{Z}$-module (i.e. a finite abelian group using addition) every element has torsion. That is because every v in M has finite order, by Proposition 2.5, meaning that $nv = 0$ for some positive integer n. In any abelian group the elements of finite order are the torsion elements.

For a richer example, take $T : V \to V$ to be a linear operator on a finite-dimensional vector space V over a field K. Then V is also the $K[X]$-module inherited from the linear operator T. In this case every v in V has torsion. Here is the proof. Say V has dimension n over K. The list $v, T(v), \ldots, T^n(v)$ is linearly dependent over K. Hence there exist $a_0, a_1, \ldots, a_n$ in K not all zero and such that

$$(a_0 + a_1 T + \cdots + a_n T^n)(v) = a_0 v + a_1 T(v) + \cdots + a_n T^n(v) = 0.$$

The polynomial $g(X) = a_0 + a_1 X + \cdots + a_n X^n$ is non-zero in $K[X]$, and as just observed $g(X)v = g(T)(v) = 0$.

If an R-module M has a non-zero annihilator, then every element of M has torsion. In this case there exists a non-zero a in R such that $av = 0$ for all v in M.

Torsion and Torsion-Free Modules

We continue with an integral domain R.

Definition 8.27. An R-module in which all elements have torsion is called a ***torsion module***. At the opposite extreme, an R-module in which the only torsion element is 0 is called ***torsion-free***.

A torsion-free module is such that if $a \in R$ and $v \in M$ and $av = 0$, then either $a = 0$ or $v = 0$. For instance, all vector spaces over a field are torsion-free. We already showed in Proposition 8.9 that free R-modules are torsion-free. In particular, the integral domain R as a module over itself is torsion-free.

It is clear from the definition that every submodule of a torsion-free module is torsion-free.

The Torsion Part of a Module

Definition 8.28. The ***torsion part*** of a module M over an integral domain R is the set of all torsion elements of M. Denote this set by $\mathrm{t}(M)$. More explicitly

$$\mathrm{t}(M) = \{v \in M : av = 0 \text{ for some non-zero } a \text{ in } R\}.$$

The set $\mathrm{t}(M)$ is a submodule of M, as we check in the next proposition. The submodule $\mathrm{t}(M)$ allows us to separate the study of modules into two types: those which are torsion and those which are torsion-free.

Proposition 8.29. *If M is a module over an integral domain R, then*

- $\mathrm{t}(M)$ *is an R-submodule of M,*
- *the quotient $M/\mathrm{t}(M)$ is a torsion-free module, and*
- *every R-map $\varphi : M \to N$ sends $\mathrm{t}(M)$ into $\mathrm{t}(N)$.*

Proof. To check that $\mathrm{t}(M)$ is a submodule of M, note first that $0 \in \mathrm{t}(M)$. Next, supposing that $v, w \in \mathrm{t}(M)$ and that $a \in R$, we need both $v+w$ and av to be in $\mathrm{t}(M)$. There exist non-zero b, c in R such that $bv = 0$ and $cw = 0$. Then $bc \neq 0$ since R is an integral domain, and

$$bc(v + w) = c(bv) + b(cw) = c0 + b0 = 0.$$

Also

$$b(av) = a(bv) = a0 = 0.$$

Hence $v + w, av \in \mathrm{t}(M)$.

To see that $M/\mathrm{t}(M)$ is torsion-free let $a \in R, v + \mathrm{t}(M) \in M/\mathrm{t}(M)$ and suppose $a(v + \mathrm{t}(M)) = 0 + \mathrm{t}(M)$. This means that $av \in \mathrm{t}(M)$. Thus there is a non-zero b in R such that $bav = 0$. If $a \neq 0$, then $ba \neq 0$ because R is an integral domain, and thereby $v \in \mathrm{t}(M)$. So $v + \mathrm{t}(M) = 0 + \mathrm{t}(M)$, meaning that $M/\mathrm{t}(M)$ is torsion-free.

For the third item suppose $v \in \mathrm{t}(M)$. Thus $av = 0$ for some non-zero a in R. Consequently $a\varphi(v) = \varphi(av) = \varphi(0) = 0$, which puts $\varphi(v)$ inside $\mathrm{t}(N)$. □

8.2.2 Rank

In a finitely generated vector space V over a field, the number of vectors in a basis of V does not depend on the choice of basis. That is what defines the concept of dimension. The same principle holds for a free module over an integral domain.

Recall that a list of elements $w_1, \ldots, w_m$ in an R-module M is ***independent*** when the only m-tuple $(a_1, \ldots, a_m)$ in R^m that gives $a_1 w_1 + \cdots + a_m w_m = 0$ is the zero m-tuple $(0, \ldots, 0)$. A basis of M (should it exist) is an independent list of w_j that generate M.

Proposition 8.30. *If R is an integral domain and M is a free R-module with basis $v_1, \ldots, v_n$ and if $w_1, \ldots, w_m$ is an independent list of elements of M, then $m \leq n$. Furthermore, if the w_j form a basis of M, then $m = n$.*

Proof. Since the free module M is isomorphic to R^n, it does no harm to say that M is R^n.

Let K be the field of fractions of R. Then R is a subring of K, and R^n sits inside the vector space K^n in the obvious way. The elements $w_1, \ldots, w_m$ remain K-linearly independent as vectors in K^n. To see this claim, suppose that $a_1/b_1, \ldots, a_m/b_m$ are fractions in K with the a_j, b_j in R and $b_j \neq 0$, and that

$$\sum_{j=1}^{m} \frac{a_j}{b_j} w_j = 0.$$

Clear the denominators to get

$$\sum_{j=1}^{m} \left(\prod_{i \neq j} b_i \right) a_j w_j = 0.$$

The coefficients of the w_j now lie in R and the w_j are assumed independent as elements of R^n. Consequently all $(\prod_{i \neq j} b_i)a_j = 0$, and since R is an integral domain and the b_i are not 0 it must be that all $a_j = 0$. Thus all $a_j/b_j = 0$, which shows the K-linear independence of the w_j as vectors in K^n. Since K^n has dimension n as vector space over K, the basics of linear algebra reveal that $m \leq n$.

The second claim of the proposition can be seen by interchanging the roles of the v_j and w_j. □

What is called the dimension of a vector space gets a new name in the case of free modules.

Definition 8.31. The ***rank of a free module*** is the unique number of elements in any of its bases.

Proposition 8.30 ensures that the concept of rank is properly defined. It follows from Proposition 8.30 that the free modules R^n and R^m are isomorphic if and only if $n = m$. Indeed, if an isomorphism $R^m \cong R^n$ prevailed with $n < m$, then a basis for R^m would get mapped by the isomorphism into an independent list of elements inside R^n, contradicting Proposition 8.30.

Submodules of Free Modules over Integral Domains

Every free module is torsion-free. The submodules of a free module need not be free, but they do remain torsion-free. Conversely, if R is an integral domain and M is a finitely generated torsion-free module, then M is isomorphic to a submodule of a free module. To get this result we need an extension of Proposition 8.30 to all finitely generated modules.

Proposition 8.32. *If R is an integral domain and M is an R-module generated by n elements, then every independent list in M can contain at most n elements.*

Proof. Let $v_1, \ldots, v_n$ be generators for M, and let $e_1, \ldots, e_n$ be the standard n-tuples of the free module R^n. According to Proposition 8.11 there is an R-map $\varphi : R^n \to M$ such that $\varphi : e_j \mapsto v_j$. Since the v_j generate M, this φ is surjective.

Now suppose $w_1, \ldots, w_m$ are independent in M. Using the fact φ is surjective, let $f_1, \ldots, f_m$ be pre-images inside R^n of $w_1, \ldots, w_m$, respectively. These f_j are independent in the free module R^n. For otherwise a dependency relation among the f_j would map down by φ to a dependency relation among the w_j. By Proposition 8.30 applied to R^n, it follows that $m \leq n$. □

In the next result we learn that finitely generated torsion-free modules over an integral domain coincide with the submodules of free modules.

Proposition 8.33. *If R is an integral domain and M is a torsion-free R-module generated by n elements, then M is isomorphic to a submodule of a free module of rank at most n.*

Proof. Let $v_1, \ldots, v_n$ be generators of M. Let $w_1, \ldots, w_m$ be a maximal independent list of elements of M. By Proposition 8.32 such a list exists and $m \leq n$. Let N be the submodule generated by the w_j. Since these generators are independent, N is a free module of rank m.

For every x in M the list $w_1, \ldots, w_m, x$ has a non-trivial dependency relation due to the maximality of the w_j. From this relation it follows that there exists a non-zero a in R such that $ax \in N$. In particular, for each j from 1 to n there is a non-zero a_j in R such that $a_j v_j \in N$. Let $b = a_1 a_2 \cdots a_n$. This product is non-zero because R is an integral domain. Clearly $bv_j \in N$ for all j, and thus $bx \in N$ for all x in M. The mapping

$$\varphi : M \to N \text{ given by } x \mapsto bx$$

is easily seen to be an R-map. This mapping is also injective because M is torsion-free and $b \neq 0$. Thus we have

$$M \cong \varphi(M) \subseteq N,$$

as desired. □

EXERCISES

1. Find a free module with a finite basis and a submodule that is not finitely generated.
2. Let R be an integral domain and let K be its field of fractions. Since R is a subring of K, the abelian group K is an R-module using the obvious scalar multiplication. If K is free as an R-module, show that R must be a field.

   Show that R is a field if and only if the only torsion-free R-modules are the free modules.
3. If R is an integral domain, prove that every projective module is torsion-free.
4. Find an integral domain R and an R-module M in which every element has torsion but whose annihilator is the zero ideal.

8.3 Modules over Principal Ideal Domains

When the ring is further restricted to be a principal ideal domain, we come to an explicit understanding of the structure of its modules as direct sums of standard examples. The result we are after is called the *structure theorem for finitely generated modules over principal ideal domains*. This will yield at least two applications. One is the extension of Propositions 2.92 and 2.93 to all finitely generated abelian groups. The other is the development of the canonical forms for matrices. As the results in this section are more profound, the proofs are commensurably more intricate.

8.3.1 Splitting Torsion from Torsion-Free

For a module M over an integral domain R, the torsion part $\mathrm{t}(M)$ is a torsion module and the quotient $M/\,\mathrm{t}(M)$ is a torsion-free module. We can put this information to good use when all ideals of R are principal.

Submodules of Free Modules over a Principal Ideal Domain Are Free

Proposition 8.34. *If R is a principal ideal domain, then every submodule of a free R-module of finite rank n, is also a free module of rank at most n.*

Proof. Since all free modules of rank n are isomorphic to R^n, it is enough to show that the submodules of R^n are free of rank at most n. Use induction on n.

If $n = 1$, a submodule of R is a principal ideal, say Ra for some a in R. Because R is an integral domain the mapping $R \to Ra$ given by $x \mapsto xa$ is either the zero map when $a = 0$, or else it is an R-module isomorphism. In either case Ra is free of rank at most 1. (We take the zero module to be free with an empty basis, and thereby of rank 0.)

Suppose $n > 1$ and the submodules of R^{n-1} are free of rank at most $n - 1$. Take M to be a submodule of R^n. Let $\varphi : R^n \to R^{n-1}$ be the R-map given by $\varphi : (a_1, a_2, \ldots, a_n) \mapsto (a_2, \ldots, a_n)$, and let $N = \varphi(M)$. The R-module N sits inside the free module R^{n-1}. By the inductive hypothesis, N is a free module of rank at most $n - 1$. By Proposition 8.24, N is projective. In particular, the restriction $\varphi_{|M} : M \to N$, which is obviously surjective, comes with an R-map $\psi : N \to M$ such that $\varphi_{|M} \circ \psi$ is the identity on N and

$$M = \ker \varphi_{|M} \oplus \psi(N).$$

For every x in N such that $\psi(x) = 0$, we have $x = \varphi_{|M}(\psi(x)) = \varphi_{|M}(0) = 0$, which shows that ψ is injective. Thus $\psi(N)$ is free of rank at most $n - 1$, since it is isomorphic to the free module N.

By the definition of φ we see that the module $\ker \varphi_{|M}$ is the intersection of M with the submodule $\{(x_1, 0, \ldots, 0) : x_1 \in R\}$ inside R^n. Since $\ker \varphi_{|M}$ is a submodule of a module that is clearly isomorphic to R, it follows from the first step of our induction that $\ker \varphi_{|M}$ is a free module of rank 0 or 1.

The direct sum of two free modules is another free module of rank equal to the sum of the ranks of the direct summands. So, our M is a free module with rank equal to the sum of the rank of $\ker \varphi_{|M}$ and the rank of $\psi(N)$. This sum is at most $1 + (n - 1) = n$. □

In Proposition 8.34 it is possible for a proper submodule of a free module to be proper and still have the same rank as the parent module. For instance, take $\mathbb{Z}$ as a free module over itself of rank 1. The submodule $\mathbb{Z}2$ is free, proper but still of rank 1. This anomaly does not happen with vector spaces.

When the integral domain R has ideals which are not principal, Proposition 8.34 always fails. Indeed, R is free as a module over itself. Now suppose M is a non-principal ideal of R.

For any two distinct v, w in M, the dependency relation $wv - vw = 0$ shows that any basis of M can only consist of one element. So M would have to be a principal ideal.

Submodules of Finitely Generated Modules over a Principal Ideal Domain

The next result holds for all Noetherian rings, but all we are going to need is for the ring to be a principal ideal domain.

Proposition 8.35. *Every submodule of a finitely generated module over a principal ideal domain remains finitely generated.*

Proof. Say R is our principal ideal domain and that M is an R-module generated by $v_1, \ldots, v_n$. Let N be a submodule of M. The module R^n is free with its standard basis $e_1, \ldots, e_n$. Proposition 8.11 yields an R-map $\varphi : R^n \to M$ such that $\varphi : e_j \mapsto v_j$. This φ is surjective. The inverse image $\varphi^{-1}(N)$ is a submodule of R^n. By Proposition 8.34, $\varphi^{-1}(N)$ is free of rank at most n, and is thereby finitely generated. Since φ is surjective, $\varphi(\varphi^{-1}(N)) = N$. The finitely many generators of $\varphi^{-1}(N)$ get mapped by φ to generators of N. Thus N is finitely generated. □

To see an integral domain in which the above result fails, take R to be any non-Noetherian domain, for instance, the ring of polynomials in infinitely many variables with coefficients in some field. This ring is generated by 1 as a module over itself. But, being non-Noetherian, it contains non-finitely generated ideals, i.e. non-finitely generated submodules.

Finitely Generated Torsion-Free Modules over a Principal Ideal Domain Are Free

Proposition 8.36. *If R is a principal ideal domain and M is a torsion-free R-module generated by n elements, then M is a free R-module of rank at most n.*

Proof. By Proposition 8.33 M is isomorphic to a submodule of a free module of rank at most n. Then by Proposition 8.34 that submodule, and thereby M itself, is a free module of rank at most n. □

Breaking the Module into Two Parts

We are in a position to split up every finitely generated module over a principal ideal domain into a torsion and a torsion-free part.

Proposition 8.37. *If R is a principal ideal domain and M is a finitely generated R-module, then M is the direct sum of its torsion part $\mathrm{t}(M)$ and a free module isomorphic to $M/\mathrm{t}(M)$.*

Proof. Since $M/\mathrm{t}(M)$ is torsion-free and also finitely generated, Proposition 8.36 ensures that $M/\mathrm{t}(M)$ is free. By Proposition 8.24 $M/\mathrm{t}(M)$ is projective. If $\varphi : M \to M/\mathrm{t}(M)$ is the canonical projection, there is an R-map $\psi : M/\mathrm{t}(M) \to M$ such that $\varphi \circ \psi$ is the identity map on $M/\mathrm{t}(M)$. Furthermore, since $\ker \varphi = \mathrm{t}(M)$, Proposition 8.24 also gives

$$M \cong \mathrm{t}(M) \oplus \psi(M/\mathrm{t}(M)).$$

Finally, the submodule $\psi(M/\mathrm{t}(M))$ inside M is isomorphic to the free module $M/\mathrm{t}(M)$ because ψ is injective. □

Proposition 8.38. *Two finitely generated modules M, N over a principal ideal domain R are isomorphic if and only if*

$$\mathrm{t}(M) \cong \mathrm{t}(N) \text{ and } M/\mathrm{t}(M) \cong N/\mathrm{t}(N).$$

Proof. Suppose $\varphi : M \to N$ is an isomorphism. By Proposition 8.29, applied to φ and to its inverse, $\varphi(\mathrm{t}(M)) = \mathrm{t}(N)$. Thus $\mathrm{t}(M) \cong \mathrm{t}(N)$.

Also if $\overline{\varphi} : M \to N/\mathrm{t}(N)$ is the surjective homomorphism given by $v \mapsto \varphi(v) + \mathrm{t}(N)$, its kernel is made up of all v in M such that $\varphi(v) \in \mathrm{t}(N)$. That is, all v such that $\varphi(v) \in \varphi(\mathrm{t}(M))$. So, $v \in \ker\overline{\varphi}$ if and only if $v \in \mathrm{t}(M)$. By the first isomorphism theorem $M/\mathrm{t}(M) \cong N/\mathrm{t}(N)$.

The above general argument made no use of the fact R is a principal ideal domain nor the fact that M, N are finitely generated. But the reverse implication does.

Suppose that $\mathrm{t}(M) \cong \mathrm{t}(N)$ and $M/\mathrm{t}(M) \cong N/\mathrm{t}(N)$. By Proposition 8.37

$$M = \mathrm{t}(M) \oplus P \text{ and } N = \mathrm{t}(N) \oplus Q,$$

where P is a free module isomorphic to $M/\mathrm{t}(M)$ and Q is a free module isomorphic to $N/\mathrm{t}(N)$. By assumption $M/\mathrm{t}(M) \cong N/\mathrm{t}(N)$. Hence $P \cong Q$. An isomorphism from $\mathrm{t}(M) \oplus P$ to $\mathrm{t}(N) \oplus Q$ is obtained by stitching together, in a self evident way, the given isomorphism from $\mathrm{t}(M)$ to $\mathrm{t}(N)$ and the just shown isomorphism from P to Q. Thus $M \cong N$. □

Proposition 8.37 matters because, for principal ideal domains at least, the classification of finitely generated modules breaks down into the separate classifications of torsion modules and torsion-free modules. As we have just seen, the latter modules are free of finite rank, and it follows from Proposition 8.30 that two such free modules are isomorphic if and only if they have equal rank.

We are left with the not so simple task of understanding the torsion modules.

8.3.2 Torsion Modules over a Principal Ideal Domain

A module M over a principal ideal domain R is a torsion module provided that for every v in M there is a *non-zero* a in R such that $av = 0$. In case M is also finitely generated we can say something stronger. For a in R the notation aM stands for the submodule $\{av : v \in M\}$. When this is the zero submodule we shall write $aM = 0$. Recall that the annihilator $\mathrm{ann}(M)$ is the ideal consisting of all a in R such that $aM = 0$.

Proposition 8.39. *A finitely generated module M over a principal ideal domain R is torsion if and only if its annihilator is a non-zero ideal of R.*

Proof. As every ideal of R is principal, let $\mathrm{ann}(M) = Rb$ for some b in R.

If $b \neq 0$, then M is a torsion module because $bM = 0$.

Conversely suppose M is torsion. Say $v_1, \ldots, v_n$ generate M. Since M is torsion, for each v_j there is a non-zero a_j in R such that $a_j v_j = 0$. The non-zero product $a = a_1 a_2 \cdots a_n$ is such that $av_j = 0$ for all v_j, and thereby for all v in M. Thus $aM = 0$. This means that $0 \neq a \in \operatorname{ann}(M)$, which causes $\operatorname{ann}(M)$ to be non-zero. □

Following Definition 8.26 we saw that if V is a finite-dimensional vector space over a field K and T is a linear operator on V, then V is a torsion $K[X]$-module. Since $K[X]$ is a principal ideal domain, Proposition 8.39 ensures that $\operatorname{ann}(V)$ is a non-zero ideal, say $K[X]g$ for some monic polynomial g. This g is known as the ***minimal polynomial*** of T. It is the monic polynomial g of least degree such that $gv = g(T)(v) = 0$ for all v in V. Being the generator of $\operatorname{ann}(V)$ the minimal polynomial divides all f in $K[X]$ such that $f(T)(V) = 0$.

The Primary Components of a Torsion Module

It may help to assemble some known facts about principal ideal domains. For every principal ideal domain R, primes coincide with irreducible elements. Two primes p, q are associated when one is a unit multiple of the other. This is equivalent to saying that the principal ideals Rp and Rq are equal. If two primes are not associated, then their only common divisors are units of R. Every non-zero, non-unit a in R has a factorization as

$$a = up_1^{e_1} \cdots p_k^{e_k},$$

where u is a unit, the p_j are pairwise non-associated primes, and the exponents $e_j \geq 1$. The primes p_j and their multiplicities e_j in the factorization are uniquely determined except for possible replacement of the primes by associates. For any $a_1, \ldots, a_n$ in R, the ideal $Ra_1 + \cdots + Ra_n$ is principal, say Rb for some b. This b is a common divisor of all a_j. If the a_j have only units as common divisors, then b is a unit, which we might as well take to be 1. In that case there exist x_j in R such that $a_1 x_1 + \cdots + a_n x_n = 1$. When the a_j have only units as common divisors of them all, we say that the a_j are ***coprime***.

For any module M over R and any prime p in R, let

$$M_p = \{v \in M : p^k v = 0 \text{ for some exponent } k\}.$$

By routine verification M_p is a submodule of M. In this definition, the exponent k depends on the element v. However, when M is finitely generated, so is M_p according to Proposition 8.35, and then there is a single power of p, say p^n, such that $p^n M_p = 0$. Indeed, let $v_1, \ldots, v_\ell$ be generators for M_p and for each j from 1 to ℓ let k_j be an exponent by which $p^{k_j} v_j = 0$. If n is the maximum of these k_j, then p^n will annihilate all v_j, and thereby annihilate all of M_p.

Definition 8.40. If p is a prime in the principal ideal domain R, a module M is said to be ***p-primary*** when for every v in M there is a pth power p^k such that $p^k v = 0$. When the module is called ***primary***, it is to be understood that this is with respect to an unspecified prime in R.

The submodules M_p of a module M are p-primary by their very definition.

For example, take a nilpotent linear operator $T : V \to V$ on a finite-dimensional space V over a field K. This means that $T^n = 0$ for some exponent n. With V as the module over $K[X]$ inherited from T, we see that $X^nV = T^n(V) = 0$. Since the polynomial X is a prime in the principal ideal domain $K[X]$, the module V is X-primary.

In case M is a finitely generated, torsion R-module let us find the annihilator of each p-primary submodule M_p in terms of the annihilator of M.

Proposition 8.41. *Let M be a finitely generated, torsion module over a principal ideal domain R, and let $\operatorname{ann}(M) = Rb$ for some non-zero b in R. If p is a prime in R and p^e is the highest power of p that divides b, then*

$$\operatorname{ann}(M_p) = Rp^e \text{ and } M_p = \{v \in M : p^e v = 0\}.$$

Proof. We have $b = p^e c$, where $e \geq 0, c \in R$ and $p \nmid c$. It turns out that

$$cM \subseteq M_p \text{ and } cM_p = M_p.$$

The first claim holds because $0 = bM = p^e cM$, whereby p^e annihilates cM and puts cM inside M_p. For the second claim let n be an exponent sufficiently high that $p^n M_p = 0$. As noted just prior to this proposition, such p^n exists because M, and thus M_p, is finitely generated. Since p^n and c are coprime, there exist x, y in R such that $xp^n + cy = 1$. For every v in M_p we get

$$v = 1v = xp^n v + cyv = cyv,$$

which shows that $cM_p = M_p$.

Now $bM = 0$, and so $bM_p = 0$. Hence

$$p^e M_p = p^e cM_p = bM_p = 0.$$

Thus $p^e \in \operatorname{ann}(M_p)$. Consequently any generator of $\operatorname{ann}(M_p)$ divides p^e, which means there is a non-negative exponent $k \leq e$ such that $\operatorname{ann}(M_p) = Rp^k$. The fact that $cM \subseteq M_p$ yields

$$p^k cM \subseteq p^k M_p = 0.$$

Therefore $p^k c \in \operatorname{ann}(M) = Rb = Rp^e c$. This means that $p^e c$ divides $p^k c$. Since $p \nmid c$, it follows that p^e divides p^k. In other words $e \leq k$, whereby $e = k$ and $\operatorname{ann}(M_p) = Rp^e$.

The second conclusion of the proposition follows immediately, because for any v in M, the condition $p^e v = 0$ implies $v \in M_p$ by the definition of M_p. Conversely if $v \in M_p$, then $p^e v = 0$ because p^e annihilates M_p, as just shown. □

An interesting corollary emerges.

Proposition 8.42. *Let M be a finitely generated, torsion module over a principal ideal domain R and let $\operatorname{ann}(M) = Rb$ for some non-zero b in R. For every prime p in R, the p-primary submodule M_p is non-zero if and only if $p \mid b$.*

Proof. Working with the contrapositive we show that $M_p = 0$ if and only if $p \nmid b$. Evidently $M_p = 0$ if and only if $\operatorname{ann}(M_p) = R$. By Proposition 8.41 $\operatorname{ann}(M_p) = Rp^e$, where p^e is the highest power of p that divides b. Now $Rp^e = R$ if and only if $p^e = 1$. In other words, if and only if $e = 0$, which is the same as $p \nmid b$. □

Definition 8.43. If p is a prime in a principal ideal domain R and M is a finitely generated torsion R-module, the submodule M_p will be called a ***p-primary component*** of M whenever M_p is *non-zero*.

By Proposition 8.42 the primary components correspond to the prime divisors of the generator b of $\operatorname{ann}(M)$. If p, q are associated primes, i.e. $q = cp$ for some unit c, then $M_p = M_q$, simply because $p^k v = 0$ if and only if $q^k v = 0$ for any v in M and any exponent k. Thus the choice of non-associated primes p_j used to factor b has no effect on the primary components. If $b = up_1^{e_1} \cdots p_k^{e_k}$ where the p_j are suitably chosen non-associated primes, then M has precisely k primary components.

The next result tells us why the primary components matter.

Primary Decomposition of a Finitely Generated, Torsion Module

Proposition 8.44 (Primary decomposition). *Over a principal ideal domain R, every finitely generated, torsion module M is the direct sum of its primary components.*

Proof. Suppose b generates $\operatorname{ann}(M)$ and let $p_1, \ldots, p_k$ be non-associated primes such that $b = up_1^{e_1} \cdots p_k^{e_k}$ with unique positive exponents e_j and some unit u. For each j from 1 to k let

$$r_j = \frac{b}{p_j^{e_j}} = u \prod_{i \neq j} p_i^{e_i}.$$

Since $p_j^{e_j}(r_j M) = bM = 0$, the submodule $r_j M$ sits inside the primary component M_{p_j}.

No p_i divides all r_j. Therefore, the r_j are coprime, and there exist $x_1, \ldots, x_k$ in R such that $r_1 x_1 + \cdots + r_k x_k = 1$. Then every v in M decomposes as follows:

$$v = 1v = (r_1 x_1 + \cdots + r_k x_k)v = r_1 x_1 v + \cdots + r_k x_k v.$$

Each $r_j x_j v$ lies in $r_j M$ and thereby, as observed, in M_{p_j}. This shows that

$$M = M_{p_1} + \cdots + M_{p_k}.$$

To see that the above sum is direct, we can verify that for every j:

$$M_{p_j} \cap \left(\sum_{i \neq j} M_{p_i} \right) = 0,$$

in accordance with Proposition 8.21. Well, if v lies in the above intersection, Proposition 8.41 reveals that

$$p_j^n v = 0 \quad \text{and} \quad \left(\prod_{i \neq j} p_i^n\right) v = 0,$$

for a sufficiently high exponent n. For each j the elements p_j^n and $\prod_{i \neq j} p_i^n$ are coprime, whereby $xp_j^n + y\prod_{i \neq j} p_i^n = 1$ for some x, y in R. So,

$$v = 1v = \left(xp_j^n + y\prod_{i \neq j} p_i^n\right) v = xp_j^n v + y\left(\prod_{i \neq j} p_i^n\right) v = x0 + y0 = 0,$$

as desired. □

Proposition 8.44 was proven already for finite abelian groups as Proposition 2.90.

Module Isomorphism Boils down to that of Primary Submodules

Proposition 8.45. *Two finitely generated modules M, N over a principal ideal domain R are isomorphic if and only if for every prime p in R the p-primary submodules M_p and N_p are isomorphic.*

Proof. Let $\varphi : M \to N$ be a module isomorphism, and let p be a prime in R. If $v \in M$ and $p^e v = 0$ for some exponent e, then $p^e \varphi(v) = \varphi(p^e v) = \varphi(0) = 0$. Given that φ is a bijection we see that $v \in M_p$ if and only if $\varphi(v) \in N_p$. That is $\varphi(M_p) = N_p$, which means that $M_p \cong N_p$.

Conversely suppose $M_p \cong N_p$ for all primes p. Consequently $M_p \neq 0$ if and only if $N_p \neq 0$. If a generates $\operatorname{ann}(M)$ and b generates $\operatorname{ann}(N)$, Proposition 8.42 shows that a and b share the same prime divisors, say $p_1, \ldots, p_k$. Proposition 8.44 yields that

$$M = M_{p_1} \oplus \cdots \oplus M_{p_k} \text{ and } N = N_{p_1} \oplus \cdots \oplus N_{P_k}.$$

If $\varphi_j : M_{p_j} \to N_{p_j}$ are the given isomorphisms, an isomorphism $\varphi : M \to N$ comes from pasting together the φ_j according to the rule

$$\varphi : v_1 + \cdots + v_k \to \varphi_1(v_1) + \cdots + \varphi_k(v_k),$$

where the v_j are the unique components in M_{p_j} of the elements of M. The verification that φ is an isomorphism of R-modules is routine. □

8.3.3 Primary Modules over a Principal Ideal Domain

Proposition 8.37 led us to consider only finitely generated, torsion modules over principal ideal domains. And now, because of Proposition 8.44, we come to the consideration of primary modules.

To recap, for p a prime in a principal ideal domain R, an R-module M is p-primary provided every v in M is such that $p^k v = 0$ for suitable exponent k. If M is also finitely generated, there will be a single power p^n such that $p^n M = 0$. The annihilator $\operatorname{ann}(M)$ is a principal ideal Ra for some a in R. Clearly $p^n \in \operatorname{ann}(M)$ and thus $a \mid p^n$. By unique factorization we can take a to be p^e for some exponent e. Thus $\operatorname{ann}(M) = Rp^e$ where e is the least exponent such that $p^e M = 0$.

Every submodule N of M remains p-primary, and if M is finitely generated, so is N, due to Proposition 8.35. Let $\operatorname{ann}(N) = Rp^d$ for some exponent d. Clearly $\operatorname{ann}(M) \subseteq \operatorname{ann}(N)$, which means that p^d divides p^e, i.e. $d \leq e$. We can see that the annihilators of submodules of M are to be found in the increasing chain

$$Rp^e \subset Rp^{e-1} \subset Rp^{e-2} \subset \cdots \subset Rp \subset R.$$

Different submodules of M can have the same annihilator. However, if the module M is also cyclic, then the submodules of M correspond precisely to the above chain of ideals in R. Indeed, suppose $M = Rv$ for some v in M. The R-map $\varphi : R \to M$ given by $x \mapsto xv$ is surjective with $\ker \varphi = \operatorname{ann}(M) = Rp^e$. By Proposition 8.15, the submodules of M correspond bijectively with the submodules (i.e. ideals) of R containing Rp^e. But the above chain lists all the ideals of R that contain Rp^e. For each exponent $d \leq e$, we have $\varphi(Rp^d) = Rp^d v$, which now gives an explicit description of all submodules of M.

Cyclic Summands of Primary Modules over a Principal Ideal Domain

Our goal is to prove that every finitely generated primary module over a principal ideal domain is the direct sum of submodules that are cyclic. In the literature one can find a variety of approaches to this theorem, all of which are intricate. We begin by showing that such a module has at least one non-zero, cyclic direct summand.

But first a basic observation about modules over any commutative ring. The proof, a blatant imitation of the proof of Proposition 5.13, can be omitted.

Proposition 8.46. *If M is a module over a commutative ring R such that every submodule of M is finitely generated, then every chain of submodules*

$$N_1 \subseteq N_2 \subseteq N_3 \cdots \subseteq N_k \subseteq \cdots$$

terminates. That is, $N_m = N_{m+1} = N_{m+2} = \ldots$, starting with some index m. Furthermore every non-empty family of submodules contains a module that is maximal (with respect to inclusion) within the family.

Definition 8.47. A module for which all submodules are finitely generated, and thereby satisfies the ascending chain condition of Proposition 8.46, is called a ***Noetherian module***.

A ring R is Noetherian as a ring when its ideals are finitely generated. That coincides precisely with R being Noetherian as a module over itself.

Getting back to a finitely generated p-primary module M over a principal ideal domain R, let $\operatorname{ann}(M) = Rp^e$ for some positive exponent e. There must be an element v in M such that

$p^e v = 0$ while $p^{e-1} v \neq 0$, for otherwise p^e would not be the lowest power of p to annihilate M. In other words, there is an element v in M such that

$$\operatorname{ann}(Rv) = \operatorname{ann}(M).$$

Here is a rather deep result.

Proposition 8.48. *Let R be a principal ideal domain, p a prime in R, and M a finitely generated, non-zero, p-primary module, and let v in M be such that* $\operatorname{ann}(Rv) = \operatorname{ann}(M)$. *Then Rv is a direct summand of M. Consequently all such M have a non-zero cyclic direct summand.*

Proof. Say $\operatorname{ann}(M) = \operatorname{ann}(Rv) = Rp^e$ for some exponent e. To see that Rv is a direct summand of M, let N be maximal among the submodules of M that satisfy the condition $N \cap Rv = 0$. Such N exists by Proposition 8.46. If we can show that $M = N + Rv$, it will follow from Proposition 8.21 that $M = N \oplus Rv$, which is what we want.

Seeking a contradiction suppose that $N + Rv \subsetneq M$.

We shall first prove that there is an element

$$w \text{ in } M \setminus (N + Rv) \text{ such that } pw \in N.$$

Take u in $M \setminus (N + Rv)$. Since $p^e u = 0$ there must be some exponent k such that $p^k u \in M \setminus (N + Rv)$ while $p^{k+1} u \in N + Rv$. By replacing u by $p^k u$ we can choose u from $M \setminus (N + Rv)$ such that $pu \in N + Rv$. Let n in N and r in R be such that $pu = n + rv$. The equalities

$$p^{e-1} rv = p^{e-1}(pu - n) = p^e u - p^{e-1} n = -p^{e-1} n$$

show that $p^{e-1} rv \in N \cap Rv$, meaning that $p^{e-1} rv = 0$. Thus $p^{e-1} r \in \operatorname{ann}(Rv)$, which is Rp^e, the annihilator of M. Consequently $p^e \mid p^{e-1} r$, and it follows that $p \mid r$.

Let s in R be such that $r = ps$. Then the equalities $pu = n + rv = n + psv$ show that

$$p(u - sv) = n \in N.$$

Note that $u - sv \notin N + Rv$ because $u \notin N + Rv$. The w we sought after is $u - sv$.

Now the submodule $N + Rw$ properly contains N. Hence $(N + Rw) \cap Rv \neq 0$, by the maximal choice of N. Thus there exist m in N and t in R such that

$$0 \neq m + tw \in Rv.$$

Note that $p \nmid t$. For if $p \mid t$, the fact that $pw \in N$ would force $m + tw$ to be in N, and consequently $N \cap Rv$ would contain the non-zero $m + tw$.

As p and t are coprime, there exist a, b in R such that $ap + bt = 1$. Hence

$$w = 1w = (ap + bt)w = apw + btw.$$

Observe that $tw \in N + Rv$ because $m \in N$ and $m + tw \in Rv$. Also $pw \in N$. Therefore $w \in N + Rv$, contradicting the fact that $w \notin N + Rv$.

So $M = N \oplus Rv$ as desired. □

Primary Modules Are Direct Sums of Cyclic Submodules

For our next result recall that finitely generated modules over a principal ideal domain are Noetherian, by Proposition 8.35.

Proposition 8.49. *Every non-zero, finitely generated, primary module over a principal ideal domain is the direct sum of cyclic submodules.*

Proof. Let M be such a module over a principal ideal domain R. According to Proposition 8.48, there is a non-zero element v_1 in M and a submodule N_1 such that

$$M = Rv_1 \oplus N_1.$$

If N_1 is the zero module, we are done. If N_1 is not zero, then N_1 is itself primary and finitely generated, by Proposition 8.35. Again by Proposition 8.48, N_1 has a non-zero element v_2 and a submodule N_2 such that $N_1 = Rv_2 \oplus N_2$. Hence

$$M = Rv_1 \oplus Rv_2 \oplus N_2.$$

If N_2 is the zero module, the desired direct sum is obtained. If N_2 is not the zero module, apply the preceding argument to N_2 to get a non-zero element v_3 and a submodule N_3 such that $N_2 = Rv_3 \oplus N_3$. And then

$$M = Rv_1 \oplus Rv_2 \oplus Rv_3 \oplus N_3.$$

Continue the above argument using N_3, and so on to pick up non-zero elements v_j and submodules N_j such that

$$M = Rv_1 \oplus Rv_2 \oplus Rv_3 \oplus \cdots \oplus Rv_k \oplus N_k.$$

At some point N_k must be the zero module. For otherwise an infinite, strictly ascending chain

$$Rv_1 \subset Rv_1 \oplus Rv_2 \subset Rv_1 \oplus Rv_2 \oplus Rv_3 \subset \cdots \subset Rv_1 \oplus Rv_2 \oplus \cdots \oplus Rv_k \subset \cdots$$

results, contrary to the fact M is a Noetherian module. As soon as $N_k = 0$ we obtain the direct sum decomposition:

$$M = Rv_1 \oplus Rv_2 \oplus Rv_3 \oplus \cdots \oplus Rv_k,$$

where the non-zero summands are obviously cyclic. □

The Elementary Divisors for a Cyclic Decomposition

Suppose that M is a p-primary R-module for some prime p in a principal ideal domain R, and that $M = Rv_1 \oplus \cdots \oplus Rv_k$ is a direct sum decomposition of M into non-zero, cyclic submodules. Let $\text{ann}(M) = Rp^e$ for some positive exponent e. Clearly $\text{ann}(M) \subseteq \text{ann}(Rv_j)$ for each summand Rv_j. Thus $\text{ann}(Rv_j) = Rp^{e_j}$ for some exponent $e_j \leq e$. It is customary to arrange the cyclic direct summands so that

$$p^e \geq p^{e_1} \geq p^{e_2} \geq \cdots \geq p^{e_k}.$$

Definition 8.50. If M is a p-primary module over a principal ideal domain R and M has a direct sum decomposition into non-zero, cyclic submodules Rv_j, the list of pth powers p^{e_j} which generate the annihilators of Rv_j is called the ***list of elementary divisors*** for the cyclic decomposition of M.

Here is a small observation about the largest elementary divisor.

Proposition 8.51. *If M is a p-primary module for some prime p in a principal domain R and $M = Rv_1 \oplus \cdots \oplus Rv_k$ is a direct sum decomposition of M into cyclic submodules, then the largest elementary divisor for this decomposition is the generator of* $\operatorname{ann}(M)$.

Proof. Let $\operatorname{ann}(M) = Rp^e$, and for each cyclic summand Rv_j let Rp^{e_j} be its annihilator. For convenience arrange the elementary divisors in decreasing order. Thus p^{e_1} is the largest elementary divisor. Since every p^{e_j} divides p^{e_1} it is clear that p^{e_1} annihilates every cyclic summand and thereby annihilates M. The lowest power of p that annihilates M is p^e, and $p^{e_1} \leq p^e$. Hence $p^e = p^{e_1}$ as desired. □

8.3.4 Structure of Finitely Generated Modules over a Principal Ideal Domain

Suppose M is any finitely generated module over a principal ideal domain R. Here M is assumed to be neither torsion nor primary. We have come to a rewarding piece of information.

Proposition 8.52. *Every finitely generated module M over a principal ideal domain R is the direct sum of cyclic submodules. Each summand is either a primary torsion module or else it is isomorphic to R.*

Proof. Proposition 8.37 ensures that $M = F \oplus \operatorname{t}(M)$, where F is a free module. Being free, $F \cong R^n$ for some exponent n, and so F is a direct sum of n cyclic modules that are isomorphic to R. By Proposition 8.44 the torsion part $\operatorname{t}(M)$ is the direct sum of its primary components $\operatorname{t}(M)_{p_j}$ where the p_j run over suitable non-associated primes. By Proposition 8.49 each p_j-primary component is the direct sum of cyclic submodules, which are themselves p_j-primary. □

Proposition 8.52 gives the recipe for constructing all possible finitely generated modules over a principal ideal domain R, up to isomorphism. Keep in mind that a cyclic module Rv, say, is isomorphic to $R/\operatorname{ann}(Rv)$ according to Proposition 8.18. Furthermore if Rv is p-primary for some prime p, then $\operatorname{ann}(Rv) = Rp^e$ for some exponent e. Thus the modules R/Rp^e, as p varies over the primes of R and e varies over positive exponents, will pick up all possible cyclic, primary R-modules.

Here is the recipe to build an arbitrary M.

- Pick the free module R^n for some non-negative integer n, interpreting R^0 to be the zero module. This chooses the rank of the torsion-free part of the module to be built.
- Pick some non-associated primes $p_1, \ldots, p_k$ from R.
- For each p_j pick a list of positive exponents (which can be listed in descending order):

$$e_{j1} \geq e_{j2} \geq \cdots \geq e_{j\ell_j}.$$

- Using the direct sum notation, form the Cartesian product

$$M = R^n \oplus \bigoplus_{j=1}^{k} \bigoplus_{i=1}^{\ell_j} R/Rp_j^{e_{ji}}.$$

For instance, take the ring of integers $\mathbb{Z}$. Pick $n = 9$ for the rank, and choose the primes 3 and 7. For the prime 3 take the decreasing exponents $4, 2, 1$. For the prime 7 take exponents $8, 8, 5$. Then (keeping in mind the notation $\mathbb{Z}_r = \mathbb{Z}/\mathbb{Z}r$) write

$$M = \mathbb{Z}^9 \oplus \mathbb{Z}_{34} \oplus \mathbb{Z}_{32} \oplus /\mathbb{Z}_3 \oplus \mathbb{Z}_{78} \oplus /\mathbb{Z}_{78} \oplus /\mathbb{Z}_{75}.$$

With the added possibility of having a few copies of $\mathbb{Z}$ thrown in, the above illustrates the general structure of finite abelian groups already given in Section 2.8. One can readily adapt this example to build all possible finitely generated abelian groups.

A Cancellation Theorem for Direct Summands

After knowing how to build all finitely generated modules over a principal ideal domain, one hurdle remains. Namely, to see that the cyclic summands of a module given by Proposition 8.52 are uniquely determined up to isomorphism. In other words, to see that any alteration in the parameters of the preceding recipe for constructing a finitely generated module will produce a module that is not isomorphic to the original.

These cyclic summands cannot be absolutely unique. For example, take the $\mathbb{Z}$-module $M = \mathbb{Z}_3 \times \mathbb{Z}_9$. Clearly $M = \mathbb{Z}(1,0) \oplus \mathbb{Z}(0,1)$, but there is an alternative decomposition given by $M = \mathbb{Z}(1,3) \oplus \mathbb{Z}(1,1)$, which can be seen by a simple inspection. While these latter two summands differ from the former two, the isomorphisms $\mathbb{Z}(1,0) \cong \mathbb{Z}(1,3)$ and $\mathbb{Z}(0,1) \cong \mathbb{Z}(1,1)$ prevail. This is true because $\text{ann}(1,0) = \text{ann}(1,3) = \mathbb{Z}3$, and $\text{ann}(1,1) = \text{ann}(0,1) = \mathbb{Z}9$, which permit us to apply Proposition 8.18.

The main challenge comes down to the uniqueness of the cyclic decomposition of primary modules. To get to the uniqueness of cyclic summands up to isomorphism, we need yet one more subtle result, which picks up from Proposition 8.48. That proposition showed that if v in a primary module M is such that $\text{ann}(Rv) = \text{ann}(M)$, then Rv is a direct summand of M. There could be other w in M such that $\text{ann}(Rw) = \text{ann}(M)$. In that case $Rv \cong Rw$ because these cyclic modules have the same annihilator. As with Rv, the cyclic module Rw is a direct summand of M. The remarkable fact is that the complementary summands of Rv and Rw are also isomorphic.

Proposition 8.53. *Suppose that p is a prime in a principal ideal domain R and that M is a finitely generated p-primary R-module. If v, w in M are such that*

$$\text{ann}(Rv) = \text{ann}(Rw) = \text{ann}(M)$$

and N, L are submodules such that

$$M = N \oplus Rv = L \oplus Rw,$$

then $N \cong L$.

Proof. First we will use the given condition that $\operatorname{ann}(Rw) = \operatorname{ann}(M)$ to prove there exists a submodule Q of M such that

$$Q \cong N \text{ and } M = Q \oplus Rw.$$

Let e be the minimal exponent by which $p^e M = 0$. That is, $\operatorname{ann}(M) = Rp^e$. Given that $M = N \oplus Rv$, there exist u in N and a in R such that

$$w = u + av.$$

The cases where $p \mid a$ and $p \nmid a$ will be treated differently.

Suppose $p \nmid a$. In this case there exist s, t in R such that $sa + tp^e = 1$. Then

$$v = (sa + tp^e)v = sav + tp^e v = sav,$$

which reveals that $Rv = Rav$. So $M = N \oplus Rav$.

In this case it turns out that $M = N \oplus Rw$. Indeed, for each m in M there exist x in N and b in R such that

$$m = x + bav = x + b(w - u) = (x - bu) + bw,$$

which puts m in $N + Rw$. Hence $M = N + Rw$. For the sum to be direct we need to check that $N \cap Rw = 0$. To that end suppose r in R is such that $rw \in N$. Therefore $r(u+av) = ru+rav \in N$, and since $ru \in N$, it follows that $rav \in N$. Then $rav = 0$ because $N \cap Rav = 0$. Consequently

$$r \in \operatorname{ann}(Rav) = \operatorname{ann}(Rv) = \operatorname{ann}(M).$$

Thus $rw = 0$, because r annihilates all of M.

We come to the case where $p \mid a$. Now $p^{e-1}av = 0$ because $p^e v = 0$. This implies that $\operatorname{ann}(Ru) = \operatorname{ann}(M)$. For otherwise it must be that $p^k u = 0$ for some $k \leq e - 1$, whereby $p^{e-1}u = 0$ as well. Then

$$p^{e-1}w = p^{e-1}(u + av) = p^{e-1}u + p^{e-1}av = 0 + 0 = 0,$$

which contradicts the assumption that $\operatorname{ann}(Rw) = \operatorname{ann}(M)$. Since

$$\operatorname{ann}(N) \subseteq \operatorname{ann}(Ru) = \operatorname{ann}(M) \subseteq \operatorname{ann}(N),$$

we see that $\operatorname{ann}(Ru) = \operatorname{ann}(N)$. Then Proposition 8.48 yields a submodule P of N such that $N = P \oplus Ru$, and thus

$$M = P \oplus Ru \oplus Rv.$$

In this case it turns out that

$$M = (P + Rv) \oplus Rw.$$

Indeed, for each m in M there exist x in P and b, c in R such that

$$m = x + bu + cv = x + b(w - av) + cv = x + (c - ba)v + bw.$$

This places m inside $(P + Rv) + Rw$, and shows $M = (P + Rv) + Rw$.

We also need that $(P + Rv) \cap Rw = 0$. Well, suppose $r \in R$ is such that $rw \in P + Rv$. Thus

$$ru + rav = r(u + av) = rw \in P + Rv,$$

and since $rav \in Rv \subseteq P+Rv$, it follows that $ru \in P+Rv$. But the intersection $(P+Rv)\cap Ru = 0$, according to Proposition 8.21 applied to the direct sum $P \oplus Ru \oplus Rv$. Thus $ru = 0$, meaning that $r \in \operatorname{ann}(Ru) = \operatorname{ann}(M)$. Hence $rw = 0$, which shows that $(P + Rv) \cap Rw = 0$.

We have just seen that $M = (P + Rv) \oplus Rw$. Now the sum $P + Rv$ is direct because P is a submodule of N. Also Rv, Ru have the same annihilator, namely $\operatorname{ann}(M)$. Thus $Ru \cong Rv$. Using the obvious mapping it follows that $P \oplus Rv \cong P \oplus Ru = N$.

Both in the case where $p \nmid a$ and in the case where $p \mid a$ we have found a submodule Q such that $N \cong Q$ and $M = Q \oplus Rw$. To finish the proof recall that $M = L \oplus Rw$ as given. The restriction of the canonical projection $M \to M/Rw$, first to Q and then to L leads to isomorphisms

$$N \cong Q \cong R/Rw \cong L,$$

as desired. □

Uniqueness of Cyclic Decomposition of Primary Modules up to Isomorphism

Proposition 8.53 leads to a significant result.

Proposition 8.54 (Uniqueness of cyclic decomposition). *Let R be a principal ideal domain, p a prime in R, and M, N finitely generated p-primary R-modules. Suppose that*

$$M = Rv_1 \oplus Rv_2 \oplus \cdots \oplus Rv_k \text{ and } N = Rw_1 \oplus Rw_2 \oplus \cdots \oplus Rw_\ell,$$

for some non-zero v_j in M and non-zero w_j in N. Let

$$\operatorname{ann}(Rv_j) = Rp^{e_j}, \ \operatorname{ann}(Rw_j) = Rp^{d_j}$$

using suitable exponents e_j, d_j, and arrange the v_j and w_j so that

$$e_1 \geq e_2 \geq \cdots \geq e_k \text{ and } d_1 \geq d_2 \geq \cdots \geq d_\ell.$$

The following statements are equivalent:

1. *$M \cong N$,*
2. *$k = \ell$ and $e_j = d_j$ for all j,*
3. *$k = \ell$ and $Rv_j \cong Rw_j$ for all j.*

Proof. We show that $1 \implies 2 \implies 3 \implies 1$.

$1 \implies 2$. Suppose $\varphi : N \to M$ is our module isomorphism. Clearly then $\operatorname{ann}(M) = \operatorname{ann}(N)$. By Proposition 8.51,

$$\operatorname{ann}(M) = \operatorname{ann}(Rv_1) = Rp^{e_1} \text{ and } \operatorname{ann}(N) = \operatorname{ann}(Rw_1) = Rp^{d_1}.$$

Thus $Rp^{e_1} = Rp^{d_1}$, which forces $e_1 = d_1$. The isomorphism φ also gives

$$M = \varphi(Rw_1 \oplus \cdots \oplus Rw_\ell) = \varphi(Rw_1) \oplus \cdots \oplus \varphi(Rw_\ell) = R\varphi(w_1) \oplus \cdots \oplus R\varphi(w_\ell).$$

Since $Rw_1 \cong \varphi(Rw_1) = R\varphi(w_1)$ we see that

$$\operatorname{ann}(R\varphi(w_1)) = \operatorname{ann}(Rw_1) = \operatorname{ann}(N) = \operatorname{ann}(M).$$

Proposition 8.53 applied to the alternative direct sum decompositions

$$M = Rv_1 \oplus Rv_2 \oplus \cdots \oplus Rv_k = R\varphi(w_1) \oplus R\varphi(w_2) \oplus \cdots \oplus R\varphi(w_\ell)),$$

results in the isomorphism

$$Rv_2 \oplus \cdots \oplus Rv_k \cong R\varphi(w_2) \oplus \cdots \oplus R\varphi(w_\ell)).$$

Because $R\varphi(w_2) \oplus \cdots \oplus R\varphi(w_\ell)$ and $Rw_2 \oplus \cdots \oplus Rw_\ell$, are isomorphic we get

$$Rv_2 \oplus \cdots \oplus Rv_k \cong Rw_2 \oplus \cdots \oplus Rw_\ell.$$

Apply the preceding argument to the above isomorphism to obtain

$$e_2 = d_2 \text{ and } Rv_3 \oplus \cdots \oplus Rv_k \cong Rw_3 \oplus \cdots \oplus Rw_\ell.$$

Continue in this fashion to conclude that $e_j = d_j$ until j reaches the lesser of k and ℓ. If it were the case that $k < \ell$, say, the resulting contradiction would be that $Rw_{k+1} \oplus \cdots \oplus R_{w_\ell} \cong 0$. Thus $k = \ell$.

2 $\Longrightarrow$ 3. It follows obviously that $\operatorname{ann}(Rv_j) = \operatorname{ann}(Rw_j)$. Since the cyclic modules Rv_1, Rw_j have the same annihilator, they are isomorphic.

3 $\Longrightarrow$ 1. Let $\varphi_j : Rv_j \to Rw_j$ be the assumed isomorphisms. The desired isomorphism $\varphi : M \to N$ comes naturally. Each element of M equals the sum $x_1 + x_2 + \cdots + x_k$ where the $x_j \in Rv_j$, and these components are unique. Then define φ by

$$\varphi : x_1 + x_2 + \cdots + x_k \mapsto \varphi_1(x_1) + \varphi_2(x_2) + \cdots + \varphi_k(x_k).$$

Such φ is an isomorphism by routine verification. □

Proposition 8.54 is saying in particular that if a primary module has two direct sum decompositions, then the list of elementary divisors for one decomposition is identical to the list of elementary divisors for the other decomposition. Thus we can speak of the ***elementary divisors of a primary module***, rather than a particular cyclic decomposition of the module.

Uniqueness of Cyclic Decompositions of General Torsion Modules

According to Proposition 8.52 all finitely generated, torsion modules are direct sums of cyclic, primary modules. Now Proposition 8.54 combines with Proposition 8.45 to show that such a cyclic, primary decomposition is unique, up to isomorphism.

Proposition 8.55. *Suppose that M, N are finitely generated, torsion modules over a principal ideal domain R and that*

$$M = \bigoplus_{k=1}^{n} Rv_k \text{ and } N = \bigoplus_{k=1}^{m} Rw_k$$

give respective direct sum decompositions into cyclic, primary submodules Rv_k and Rw_k. The modules M, N are isomorphic if and only if $n = m$ and, after a suitable rearrangement of the indices, each $Rv_k \cong Rw_k$.

Proof. If $m = n$ and all $Rv_k \cong Rw_k$ after suitably rearranging indices, the isomorphism of M and N comes by matching up the isomorphisms of the cyclic summands in a natural way.

The proof of the converse is a matter of using the established results along with a bit of (possibly irksome) bookkeeping. Suppose then that $M \cong N$. Thus $\operatorname{ann}(M) = \operatorname{ann}(N)$, and let us say this common annihilator is Ra for some a in R. By Proposition 8.45 the primary modules M_p and N_p are isomorphic for all primes p. In particular $M_p \neq 0$ if and only if $N_p \neq 0$, which, by Proposition 8.42, occurs if and only if p divides a. Say $p_1, \ldots, p_r$ are the non-associated prime divisors of a.

Every Rv_k, in the direct sum for M, is p_j-primary for some prime p_j. For each prime p_j, let $v_{j1}, \ldots, v_{jn_j}$ be those generators v_k whose modules Rv_k are p_j-primary. Thus, as each j runs from 1 to r and i runs from 1 to n_j, the v_{ji} constitute a rearrangement of the v_k, whence

$$M = \bigoplus_{j=1}^{r} \left(\bigoplus_{i=1}^{n_j} Rv_{ji} \right).$$

If M_{p_j} is the p_j-primary component of M the primary decomposition of M gives

$$M = \bigoplus_{j=1}^{r} M_{p_j}.$$

Furthermore $\bigoplus_{i=1}^{n_j} Rv_{ji} \subseteq M_{p_j}$ because each $Rv_{ji} \subseteq M_{p_j}$. It follows from these alternative direct sum decompositions of M and the preceding inclusion that

$$M_{p_j} = \bigoplus_{i=1}^{n_j} Rv_{ji}.$$

Similarly, for each j from 1 to m let $w_{j1}, \ldots, w_{jm_j}$ be those generators w_k in the decomposition of N whose modules Rw_k are p_j-primary. As was the case with M the p_j primary component N_{p_j} is given by

$$N_{p_j} = \bigoplus_{i=1}^{m_j} Rv_{ji}.$$

But $M_{p_j} \cong N_{p_j}$, and these modules are p_j-primary. It follows from Proposition 8.54 that $n_j = m_j$, and after a suitable rearrangement of the indices i from 1 to n_j, we also get

$Rv_{ji} \cong Rw_{ji}$. Because n equals the sum of the n_j and m equals the sum of the m_j, we see that $n = m$ and, after all the necessary rearrangements of indices, $Rv_k \cong Rw_k$. □

Pulling the Threads Together

We can summarize our findings as what can fairly be called the ***structure of finitely generated modules over a principal ideal domain***. Its proof can be found in the propositions which have appeared in this section.

Proposition 8.56. *For each finitely generated module M over a principal ideal domain R, the following facts apply.*

1. *The module M is the direct sum of its torsion part* $\mathrm{t}(M)$ *and a free module which is isomorphic to* $M/\mathrm{t}(M)$ *(Proposition 8.37).*
2. *All bases of the free module* $M/\mathrm{t}(M)$ *have equal size known as* $\mathrm{rank}(M)$ *(Proposition 8.30).*
3. *A finitely generated module N is isomorphic to M if and only if* $\mathrm{t}(N) \cong \mathrm{t}(M)$ *and* $\mathrm{rank}(N) = \mathrm{rank}(M)$ *(Propositions 8.38, 8.37 and 8.30).*
4. *If the annihilator* $\mathrm{ann}(\mathrm{t}(M)) = Ra$ *for some a in R, then* $\mathrm{t}(M)$ *is the direct sum of its p-primary components where p runs over the non-associated primes that divide a (Proposition 8.44).*
5. *A finitely generated module N is such that* $\mathrm{t}(N) \cong \mathrm{t}(M)$ *if and only if for each prime p in R the primary submodules* $\mathrm{t}(M)_p$ *and* $\mathrm{t}(N)_p$ *are isomorphic (Proposition 8.45).*
6. *If a in R generates* $\mathrm{ann}(\mathrm{t}(M))$ *and p is a prime in R, the p-submodule* $\mathrm{t}(M)_p$ *is non-zero if and only if p divides a. In that case* $\mathrm{t}(M)_p$ *is the direct sum of cyclic submodules (Proposition 8.49).*
7. *The cyclic summands Rv of each primary component* $\mathrm{t}(M)_p$ *of* $\mathrm{t}(M)$ *are uniquely determined up to isomorphism by* $\mathrm{t}(M)_p$ *(Proposition 8.54).*
8. *If N is a finitely generated module and p is a prime in R, then the p-primary components* $\mathrm{t}(M)_p$ *and* $\mathrm{t}(N)_p$ *of* $\mathrm{t}(M)$ *and* $\mathrm{t}(N)$*, respectively, are isomorphic if and only if they have the same list of elementary divisors (Proposition 8.54).*

To know such a module we are obliged to find its rank and the list of elementary divisors for each primary component of its torsion part. Of course, the actual computation of these invariants is yet another matter.

EXERCISES

1. Show that a finitely generated, torsion $\mathbb{Z}$-module is actually finite.
2. If R is a principal ideal domain and M is a cyclic R-module, show that every submodule of M is cyclic.
3. If R is an integral domain which is not a field, find a torsion-free module which is not a free module.

   Explain why such an example does not contradict Proposition 8.36.

4. Find a module M over an integral domain R, and submodules L, N, P, Q such that

$$M = N \oplus Q = L \oplus P, \quad N \cong L \text{ but } P \ncong Q.$$

5. If R is a principal ideal domain and p is a prime in R and e is a positive exponent, show that the submodules of the cyclic module R/Rp^e are Rp^k/Rp^e where k runs from 0 to e.
6. If K is a field and M is a module over the ring $K[X]$, then M is also a vector space over the subring K. Show that if M is finitely generated and torsion, then M is finite dimensional as a vector space over K.
7. We saw in Proposition 8.37 that the torsion part of a finitely generated module M over a principal ideal domain R is a direct summand of M. This exercise develops a counter-example to this result when the assumption that all ideals of R are principal is dropped.

   For the ring take $R = \mathbb{Z}_2[X, Y, Z]$. This polynomial ring in three variables and coefficients in $\mathbb{Z}_2$ is a unique factorization domain, but some ideals are not principal. The free module $R \times R$ of rank 2 has the cyclic submodule $R \cdot (XY, XZ)$ generated by the element (XY, XZ). (Here the "dot" is put in for the scalar multiplication merely to separate the visual clutter.) Let M be the quotient R-module

$$(R \times R)/R \cdot (XY, XZ).$$

   There is the canonical projection $R \times R \to (R \times R)/R \cdot (XY, XZ)$ given by $(f, g) \mapsto (f, g) + R \cdot (XY, XZ)$, and we denote $(f, g) + R \cdot (XY, XZ)$ by the more compact $\overline{(f, g)}$. Fill in the ensuing steps to show that $\mathrm{t}(M)$ is not a direct summand of M.
   (a) Verify that the cyclic submodule $R \cdot \overline{(Y, Z)} \subseteq \mathrm{t}(M)$.
   (b) If (p, q) in $R \times R$ is such that $\overline{(p, q)} \in \mathrm{t}(M)$, prove that $\overline{(p, q)} \in R \cdot \overline{(Y, Z)}$, and deduce that $\mathrm{t}(M)$ is the cyclic submodule $R \cdot \overline{(Y, Z)}$ generated by $\overline{(Y, Z)}$.
   Hint. The assumption leads to $r(p, q) = s(XY, XZ)$ for some non-zero r, s in R. From this deduce that $(p, q) = t(Y, Z)$ for some t in R.
   (c) This is a general principle to be used in the next part. If P is a direct summand of a module L over a commutative ring A, show there is an A-map $\varphi : L \to P$ such that the restriction $\varphi_{|P}$ is the identity map on P. That is, such that $\varphi(p) = p$ for all p in P.
   (d) Suppose that $\mathrm{t}(M)$ is a direct summand of M, and work towards a contradiction. Let $\varphi : M \to \mathrm{t}(M)$ be an R-map which restricts to the identity on $\mathrm{t}(M)$.
   Why is $\varphi\left(\overline{(1, 0)}\right) = p\overline{(Y, Z)}$ and $\varphi\left(\overline{(0, 1)}\right) = q\overline{(Y, Z)}$ for some p, q in R?
   Use the R-map φ to show that $\overline{(Y, Z)} = Yp\overline{(Y, Z)} + Zq\overline{(Y, Z)}$.
   Use the definition of M to derive your contradiction.
   (e) Find a finitely generated R-module N such that $\mathrm{t}(M) \cong \mathrm{t}(N)$ and $M/\mathrm{t}(M) \cong N/\mathrm{t}(N)$ but $M \ncong N$.
8. This item develops a counter-example to Proposition 8.37 when the assumption that the module be finitely generated is dropped. This can be done with $\mathbb{Z}$-modules, i.e. abelian groups using addition.

   Let $S = \{2, 3, 5, 7, 11, \ldots\}$, the set of positive primes. Let M be the full Cartesian product $\prod_{p \in S} \mathbb{Z}_p = \mathbb{Z}_2 \times \mathbb{Z}_3 \times \mathbb{Z}_5 \times \cdots \times \mathbb{Z}_p \times \cdots$. This is the group of all functions from $f : S \to \cup_{p \in S} \mathbb{Z}_p$ such that $f(p) \in \mathbb{Z}_p$ for every prime p. Write such functions as

S-tuples $(x_p)_{p\in S}$ where $x_p \in \mathbb{Z}_p$. The addition is taken point-wise. Let N be the subgroup consisting of all $(x_p)_{p\in S}$ such that all but finitely many $x_p = 0$. The group M is usually called the ***direct product*** of the $\mathbb{Z}_p$ while the subgroup N is called the ***direct sum*** of the $\mathbb{Z}_p$.

(a) Prove that $N = \mathrm{t}(M)$.

Hint. An element x_p in $\mathbb{Z}_p$ is 0 if and only if its order is 1. Otherwise its order is p.

(b) If N is a direct summand of M, show that M contains a non-zero subgroup isomorphic to M/N.

(c) Show that for every element $\overline{(x_p)_{p\in S}}$ in M/N and every positive integer n, there exists a $\overline{(y_p)_{p\in S}}$ in M/N such that $n\overline{(y_p)_{p\in S}} = \overline{(x_p)_{p\in S}}$. This says that M/N is a so-called ***divisible group***.

Hint. The integer n is a unit mod p except for a finite number of p.

(d) Prove that for every non-zero $(x_p)_{p\in S}$ in M there is a positive integer n such that the equation $n(y_p)_{p\in S} = (x_p)_{p\in S}$ has no solution $(y_p)_{p\in S}$.

(e) Deduce that $\mathrm{t}(M)$ is not a direct summand of M.

9. Let M be a torsion-free module over an integral domain R. If $v \in M$ define the subset $[v]$ of M as follows:

$$[v] = \{u \in M : av = bu \text{ for some non-zero } a, b \text{ in } R\}.$$

For instance when $v = 0$, the subset $[v]$ is the torsion part $\mathrm{t}(M)$.

(a) Show that $[v]$ is a submodule of M containing the cyclic module Rv.

(b) In case M is the $\mathbb{Z}$-module $\mathbb{Z} \times \mathbb{Q}$, find the submodules $[(0,2)]$ and $[(2,3)]$.

(c) If P is a direct summand of M and $v \in P$, show that $[v] \subseteq P$.

(d) If P is a direct summand of M and v is a non-zero element of P such that $[v] \subsetneq P$, show that P contains at least two elements which are independent.

In other words, a direct summand P of M such that P does not contain a two element independent set is such that $[v] = P$ for every non-zero v in P.

10.* A finitely generated, torsion-free module M over a principal ideal domain is free. If n is the rank of the module, then M is the direct sum of n non-zero, cyclic submodules, each one generated by one of the basis elements of M. This exercise constructs an example of a torsion-free module over a unique factorization domain such that the module contains two independent elements and yet no proper non-zero direct summands. Such a module is certainly not free.

Let R be the unique factorization domain $F[X, Y, Z]$ of polynomials in three variables over a field F. Let M be the submodule of the free, rank-2 module $R \times R$ that is generated by

$$(X,0), (Y,Z), (0,X).$$

Prove that M is an indecomposable R module. That is, show that M has no proper, non-zero direct summand P.

Hint. Suppose that $M = P \oplus Q$ where P, Q are non-zero submodules of M. For every non-zero (f,g) in P we have $[(f,g)] = P$ in accordance with the previous exercise. Likewise for $[(f,g)]$ in Q. Focus on the parts of f and g that are F-linear combinations of X, Y and Z.

8.4 Linear Algebra and Modules

When the structure of finitely generated modules over principal ideal domains is applied to the polynomial ring $K[X]$ over a field K, some interesting information about linear operators and matrices emerges.

Let V be a vector space over a field K, and let $T : V \to V$ be a K-linear operator on V. With such T the space V becomes a module over the principal ideal domain $K[X]$ of polynomials in one variable X. The action of any polynomial f in $K[X]$ on V is given by evaluating the linear operator $f(T)$ and putting

$$fv = f(T)(v) \text{ for every } v \text{ in } V.$$

In particular $Xv = T(v)$. If f is the constant polynomial with value a in K, then $fv = av$, the usual scalar multiplication coming from K. For another example, if $f = X^2 - X + 1$, then $fv = (T^2 - T + I)(v)$, where I is the identity operator.

Definition 8.57. With this enhanced scalar multiplication coming from $K[X]$, the space V will be called the $K[X]$-***module inherited from*** T.

Every $K[X]$-module V arises from a K-linear operator on V. Indeed, since such V is a $K[X]$-module and K is a subring of $K[X]$, the abelian group V is also a vector space over K. If $T : V \to V$ is the mapping defined by $v \mapsto Xv$, then T is K-linear, and the original $K[X]$-module V is easily seen to be the module inherited from T.

Invariant Subspaces Are the Submodules

A subspace W of V is called T-***invariant*** when $T(W) \subseteq W$. The T-invariant subspaces of V coincide with the $K[X]$-submodules of V. To check this suppose W is a $K[X]$-submodule of V. Then W is a subspace of V because it is closed under the scalar multiplication by constant polynomials. Also for every v in W we have $T(v) = Xv \in W$, which makes W a T-invariant subspace. Conversely if W is a subspace of V and W is T-invariant, then W is also invariant under every operator $f(T)$ where $f \in K[X]$. Thus for any v and any f in $K[X]$, we get $fv = f(T)(v) \in W$, which shows that W is a $K[X]$-submodule of V.

$K[X]$-Module Isomorphisms and Similarity of Operators

Suppose V, W are $K[X]$-modules inherited from linear operators T, S, respectively. If $\varphi : V \to W$ is a $K[X]$-module homomorphism, then for every constant polynomial a and every v in V we have:

$$\varphi(av) = a\varphi(v) \text{ and } \varphi(T(v)) = \varphi(Xv) = X\varphi(v) = S(\varphi(v)).$$

The first equality confirms that φ is also a K-linear map from V to W. The second equality says that $\varphi \circ T = S \circ \varphi$. Conversely, it is easy to see that if $\varphi : V \to W$ is a K-linear map such that $\varphi \circ T = S \circ \varphi$, then φ is a $K[X]$-module homomorphism.

When φ is a module isomorphism, we have $T = \varphi^{-1} \circ S \circ \varphi$. Two operators that are so related are called ***similar operators***.

Two operators are similar if and only if the respective $K[X]$-modules arising from them are isomorphic.

The Minimal Polynomial when V Is Finite Dimensional

Suppose henceforth that V is finite dimensional, and let $v_1, \ldots, v_n$ be a basis of V as a vector space over K. This finite set also serves as a set of generators for V as a $K[X]$-module. Indeed, every v in V is a linear combination $\sum a_j v_j$ for some a_j in K. Taking the a_j as constant polynomials in $K[X]$, this says that v is a $K[X]$-linear combination of the finitely many v_j.

When V is finite dimensional, V is no longer a free $K[X]$-module. In fact V is a torsion module. Indeed, for every v in V the vectors $v, T(v), T^2(v), \ldots, T^n(v)$ are linearly dependent. So there is a K-linear combination $\sum a_j T^j(v) = 0$ where some $a_j \neq 0$. Then the non-zero polynomial $f = \sum a_j X^j$ is such that $fv = 0$, which means that V is a torsion $K[X]$-module. Since V is finitely generated and torsion, Proposition 8.39 says that the annihilator of V is a non-zero principal ideal.

Definition 8.58. The unique monic polynomial that generates $\operatorname{ann}(V)$ is commonly known as the ***minimal polynomial*** of T. This is the unique monic polynomial g of least degree such that $g(T)$ is the zero operator.

Primary Decomposition into Invariant Subspaces

Let

$$g = p_1^{e_1} p_2^{e_2} \cdots p_k^{e_k}$$

give the unique factorization of the minimal polynomial of the linear operator T into monic irreducible polynomials p_j in $K[X]$ with multiplicities e_j. According to Propositions 8.41 and 8.42 the primary components of V are the non-zero submodules

$$V_{p_j} = \{v \in V : p_j^{e_j} v = 0\}.$$

Since $p_j^{e_j} v = p_j^{e_j}(T)(v)$, the primary components of V are the null spaces of the linear operators $p_j^{e_j}(T)$. That is $V_{p_j} = \ker\left(p_j^{e_j}(T)\right)$. As long as g can be found and factored, standard row operation techniques with matrices will compute these kernels. Proposition 8.44 gives the direct sum decomposition

$$V = V_{p_1} \oplus V_{p_2} \oplus \cdots \oplus V_{p_k}.$$

This direct sum is known as the ***primary decomposition of the operator*** T.

Eigenvalues and the Minimal Polynomial

Here is how to interpret the roots of the minimal polynomial.

Proposition 8.59. *Let g be the minimal polynomial of a linear operator on a finite-dimensional space V over a field K. An element λ in K is a root of g if and only if $T(v) = \lambda v$ for some non-zero v in V.*

Proof. If λ is a root of g, then $X - \lambda$ is one of the irreducible factors of g, and the $(X - \lambda)$-primary component of the $K[X]$-module V is the non-zero T-invariant subspace $V_{X-\lambda} = \ker\big((T - \lambda_j I)^e\big)$, where e is the multiplicity of the factor $X - \lambda$ in g. By Proposition 8.41, $\text{ann}(V_{X-\lambda}) = K[X](X - \lambda)^e$. Thus $(X - \lambda)^{e-1}$ does not annihilate $V_{X-\lambda}$. So there is a w in $V_{X-\lambda}$ such that $(X-\lambda)^{e-1}w \neq 0$. With $v = (X-\lambda)^{e-1}w$ inside $V_{X-\lambda}$, we see that $(X-\lambda)v = 0$. This means $(T - \lambda)(v) = 0$, and then $T(v) = \lambda v$.

Conversely, suppose that $\lambda \in K$ is such that $T(v) = \lambda v$ for some non-zero v in V. This means $(X-\lambda)v = 0$. Consequently the primary module $V_{X-\lambda}$ corresponding to the irreducible $X-\lambda$ in $K[X]$ is non-zero. Proposition 8.42 ensures that $X - \lambda$ divides the generator of $\text{ann}(V)$, which is the minimal polynomial g of T. Thus λ is one of the roots of g. □

Definition 8.60. The elements λ of Proposition 8.59 are called the ***eigenvalues*** of T. The primary components $V_{X-\lambda} = \ker((T - \lambda)^e)$ are known as ***generalized eigenspaces*** of T. The subspaces $\ker(T - \lambda)$ of the primary components are called the eigenspaces of T.

The Matrix of a Linear Operator

Suppose that $v_1, v_2, \ldots, v_n$ is a basis for the vector space V over a field K, and that $T : V \to V$ is a K-linear operator on V. For each v_j there exist unique coefficients a_{ij} in K such that

$$T(v_j) = \sum_{i=1}^{n} a_{ij} v_i.$$

The $n \times n$ matrix $A = [a_{ij}]$ whose jth column lists the coefficients of $T(v_j)$ in terms of the basis $v_1, \ldots, v_n$ is called the ***matrix of T using the basis*** $v_1, \ldots, v_n$. We also say that A ***represents*** T using the v_j as basis. Matrices are the tools by which linear operators are described.

Suppose that $S, T : V \to V$ are linear operators represented by matrices A, B, respectively, using a basis $v_1, \ldots, v_n$ of V. It can be routinely verified that the sum and composite operators $S + T$ and $S \circ T$ are represented by the matrix sum $A + B$ and the familiar matrix product AB, respectively. Let $\mathcal{L}(V)$ denote the (non-commutative) ring of linear operators on V and let $M_n(K)$ be the ring of $n \times n$ matrices over K. The preceding remarks say that, after a basis $v_1, \ldots, v_n$ of V has been selected, the mapping $\psi : \mathcal{L}(V) \to M_n(K)$, which sends an operator T to its representing matrix A, is a ring homomorphism between these rings. Under ψ the scalar operators λI, where $\lambda \in K$, go to the scalar matrices λI. Furthermore, $\ker \psi$ is trivial because the equations $\sum_{i=1}^{n} a_{ij} v_i = 0$, for all j, force all $a_{ij} = 0$. In addition, every $A = [a_{ij}]$

represents the operator T defined on the basis of v_j by the formula $T(v_j) = \sum_{i=1}^{n} a_{ij}v_i$. In summary, ψ is an isomorphism of rings.

It is this isomorphism which permits the liberal interchange between linear transformations and matrices.

Similar Matrices and Linear Operators

Here is a routine piece of bookkeeping.

Proposition 8.61. *Two $n \times n$ matrices A, B over a field K represent the same linear operator T on a vector space V, if and only if $B = P^{-1}AP$ for some invertible matrix P.*

Proof. Suppose that $A = [a_{ij}]$ represents T using a basis $v_1, \dots, v_n$ and $B = [b_{ij}]$ represents T using a basis $w_1, \dots, w_n$.

For each w_j there exist unique coefficients p_{ij} in K such that $w_j = \sum_{i=1}^{n} p_{ij}v_i$. The matrix $P = [p_{ij}]$ whose jth column lists the coefficients that yield w_j in terms of the basis $v_1, \dots, v_n$ can be called ***the matrix that writes the basis*** $w_1, \dots, w_n$ ***in terms of the basis*** $v_1, \dots, v_n$.

For each index k expand $T(w_k)$ in terms of the v_i in two different ways. All indices below run from 1 to n. First,

$$T(w_k) = \sum_j b_{jk}w_j = \sum_j b_{jk}\left(\sum_i p_{ij}v_i\right) = \sum_i \left(\sum_j p_{ij}b_{jk}\right) v_i.$$

But also,

$$T(w_k) = T\left(\sum_j p_{jk}v_j\right) = \sum_j p_{jk}T(v_j)$$

$$= \sum_j p_{jk}\left(\sum_i a_{ij}v_i\right) = \sum_i \left(\sum_j a_{ij}p_{jk}\right) v_i.$$

These calculations reveal that

$$\sum_j p_{ij}b_{jk} = \sum_j a_{ij}p_{jk},$$

for all i, k from 1 to n. This is nothing but the statement that $PB = AP$, where these are the usual matrix products. Since the matrix P is easily seen to be invertible, and the inverse writes the v_i in terms of the w_j, we deduce that

$$B = P^{-1}AP.$$

Conversely suppose A is the matrix of T using the basis of v_i and $B = P^{-1}AP$ for some invertible matrix P. The vectors w_j defined by $w_j = \sum_{i=1}^{n} p_{ij}v_i$ form a basis of V, and B is the matrix of T using the basis of w_j. The routine checking can be omitted. □

Definition 8.62. Matrices A, B that satisfy the conditions of Proposition 8.61 are called ***similar matrices***.

Similar Operators and Matrices

Here is another piece of bookkeeping from linear algebra. Recall that two linear operators S, T on respective spaces V, W are similar when the $K[X]$-modules inherited from them are isomorphic. This is equivalent to saying that $T = \varphi^{-1} \circ S \circ \varphi$ for some linear bijection $\varphi : V \to W$.

Proposition 8.63. *Suppose that $T : V \to V$ and $S : W \to W$ are linear operators on finite-dimensional spaces over a field K, and that A is the matrix of T using a basis $v_1, \ldots, v_n$ of V. The same matrix A is the matrix of S using some basis of W if and only if $V \cong W$ as $K[X]$-modules, i.e. if and only if S and T are similar operators.*

Proof. As already noted a K-linear map $\varphi : V \to W$ is a $K[X]$-map if and only if $\varphi \circ T = S \circ \varphi$.

Suppose that $\varphi : V \to W$ is an isomorphism of $K[X]$-modules. Let the entries of A be a_{ij} and put $w_j = \varphi(v_j)$. The w_j now form a basis of W. Then

$$S(w_j) = S(\varphi(v_j)) = \varphi(T(v_j)) = \varphi\left(\sum_{i=1}^{n} a_{ij} v_i\right) = \sum_{i=1}^{n} a_{ij}\varphi(v_i) = \sum_{i=1}^{n} a_{ij} w_i,$$

which verifies that A is the matrix of S using the w_j.

Conversely suppose that $A = [a_{ij}]$ is also the matrix of S using a basis $w_1, \ldots, w_n$ of W. Let $\varphi : V \to W$ be the vector space isomorphism given by $\varphi(v_j) = w_j$ for all j from 1 to n. To see that φ is an isomorphism of $K[X]$-modules it is enough to check that $\varphi \circ T = S \circ \varphi$, and for that we check that these linear maps agree on the basis of V. From the fact A is the matrix both of T and of S we get:

$$\varphi(T(v_j)) = \varphi\left(\sum_{i=1}^{n} a_{ij} v_i\right) = \sum_{i=1}^{n} a_{ij}\varphi(v_i) = \sum_{i=1}^{n} a_{ij} w_i = S(w_j) = S(\varphi(v_j)),$$

as desired. □

Proposition 8.63 allows us to define the ***minimal polynomial of a matrix*** A to be the minimal polynomial of any linear operator T that A represents. The choice of T does not matter because the isomorphic $K[X]$-modules arising from similar operators all have the same annihilator, whose monic generator is the minimal polynomial.

It also allows us to declare that λ in K is an ***eigenvalue of a matrix*** A, whenever λ is an eigenvalue of any operator T that A represents. Again the choice of T does not matter. Indeed, if A also represents an operator S on a space W, then $T = \varphi^{-1} \circ S \circ \varphi$ for some linear bijection $\varphi : V \to W$. If λ is an eigenvalue of T with non-zero eigenvector v, we get

$$\varphi^{-1} \circ S \circ \varphi(v) = T(v) = \lambda v \text{ and then } S(\varphi(v)) = \lambda\varphi(v),$$

which shows that λ remains an eigenvalue of S.

Jordan Blocks

When the $K[X]$-module V inherited from an operator T is primary and cyclic with annihilator generated by the power of a linear polynomial, there is an elegant matrix representation of T.

Proposition 8.64. *Suppose that T is a K-linear operator on V with minimal polynomial of the form $(X-\lambda)^e$ for some λ in K and some positive exponent e. If V is a cyclic module with generator v, i.e. $V = K[X]v$, then the vectors*

$$v, (X-\lambda)v, (X-\lambda)^2 v, \ldots, (X-\lambda)^{e-1}v$$

form a basis of V, and the $e \times e$ matrix of T using the above basis is

$$\begin{bmatrix} \lambda & 0 & 0 & \ldots & 0 & 0 \\ 1 & \lambda & 0 & \ldots & 0 & 0 \\ 0 & 1 & \lambda & \ldots & 0 & 0 \\ \vdots & \vdots & \vdots & \ddots & \vdots & \vdots \\ 0 & 0 & 0 & \ldots & \lambda & 0 \\ 0 & 0 & 0 & \ldots & 1 & \lambda \end{bmatrix}.$$

Proof. First an observation. The mapping $K[X] \to K[X]$ defined by $h(X) \mapsto h(X-\lambda)$ is an automorphism of the ring $K[X]$. Furthermore $\deg h(X) = \deg h(X-\lambda)$ for every h in $K[X]$. Thus every polynomial in $K[X]$ can be written in the form

$$f = b_0 + b_1(X-\lambda) + b_2(X-\lambda)^2 + \cdots + b_m(X-\lambda)^m,$$

with the coefficients b_j in K uniquely determined by f.

To test for linear independence of the specified vectors suppose that

$$a_0 + a_1(X-\lambda)v + a_2(X-\lambda)^2 v + \cdots + a_{e-1}(X-\lambda)^{e-1}v = 0$$

for some a_j in K. Hence the polynomial $a_0 + a_1(X-\lambda) + \cdots + a_{e-1}(X-\lambda)^{e-1}$ annihilates v, and thereby the full cyclic module V. Consequently, the annihilator $(X-\lambda)^e$ of V divides the preceding polynomial. This forces the preceding polynomial to be zero, for otherwise its degree would be less than the degree of its divisor. It follows that all $a_j = 0$.

To see that the specified vectors span V as a vector space over K, let $w \in V$. As V is a cyclic $K[X]$-module with generator v, there is an f in $K[X]$ such that $w = fv$. By Euclidean division there are polynomials q, r such that

$$f = (X-\lambda)^e q + r \text{ where } r = 0 \text{ or } \deg r < e.$$

Since $(X-\lambda)^e$ annihilates V we get that

$$w = ((X-\lambda)^e q + r)v = (X-\lambda)^e qv + rv = rv.$$

If $r = 0$, there is nothing to check. If $r \neq 0$, write

$$r = b_0 + b_1(X-\lambda) + b_2(X-\lambda)^2 + \cdots + b_\ell(X-\lambda)^\ell,$$

where $\ell < e$ and the $b_i \in K$. Clearly then rv, which is w, lies in the span of the specified vectors.

To get the matrix of T using the just determined basis, evaluate T at the basis and expand the values in terms of the basis. It helps to observe that T is itself a $K[X]$-map. For each j from 0 to $e-1$ we get:

$$\begin{aligned} T((X-\lambda)^j v) &= (X-\lambda)^j T(v) \\ &= (X-\lambda)^j((T-\lambda)v + \lambda v) \\ &= \lambda(X-\lambda)^j v + (X-\lambda)^{j+1} v. \end{aligned}$$

If $j = e-1$, then $(X-\lambda)^{j+1}v = 0$, and we can read from the above calculation that the desired matrix is the matrix of T using the basis obtained. □

Definition 8.65. A matrix of the type given in Proposition 8.64 is called a ***Jordan block*** of size e with eigenvalue λ, and is denoted by

$$J(\lambda, e).$$

This is named in honor of Camille Jordan (1838–1922).

A small verification reveals the converse of Proposition 8.64.

Proposition 8.66. *If a Jordan block $J(\lambda, e)$ is the matrix of a linear operator T on V using a basis $v_0, v_1, \ldots, v_{e-1}$, then the $K[X]$-module V inherited from T is cyclic with v_0 as its generator and $K[X](X-\lambda)^e$ as its annihilator.*

Proof. The Jordan block informs us that

$$T(v_j) = \lambda v_j + v_{j+1}$$

for every j from 0 to $e-2$, as well as

$$T(v_{e-1}) = \lambda v_{e-1}.$$

In terms of the $K[X]$-action on V this gives

$$\begin{aligned} v_1 &= (X-\lambda)v_0 \\ v_2 &= (X-\lambda)v_1 = (X-\lambda)^2 v_0 \\ v_3 &= (X-\lambda)v_2 = (X-\lambda)^3 v_0 \\ &\vdots \\ v_{e-1} &= (X-\lambda)v_{e-2} = (X-\lambda)^{e-1} v_0 \\ 0 &= (X-\lambda)v_{e-1} = (X-\lambda)^e v_0. \end{aligned}$$

For any $a_0, a_1, \ldots, a_{e-1}$ in K, the ensuing identity

$$\begin{aligned} &a_0 v_0 + a_1 v_1 + a_2 v_2 + \cdots + a_{e-1} v_{e-1} \\ &= \left(a_0 + a_1(X-\lambda) + a_2(X-\lambda)^2 + \cdots + a_{e-1}(X-\lambda)^{e-1}\right) v_0 \end{aligned}$$

reveals that $V = K[X]v_0$, the cyclic module generated by v_0.

The condition $(X-\lambda)^e v_0 = 0$ shows that $(X-\lambda)^e$ annihilates V. Hence $\operatorname{ann}(V)$ is generated by $(X-\lambda)^k$ for some $k \le e$. But if $k < e$, then $(X-\lambda)^k v_0$ would be 0. In other words, $v_k = 0$, which is not so. Thus the annihilator of V is $K[X](X-\lambda)^e$. □

Block-Diagonal Matrices and Direct Sum Decompositions

If W is a submodule of the $K[X]$-module V inherited from a linear operator T, then the operator on W that gives rise to the module W is clearly the restriction of T to the T-invariant subspace W.

Suppose that U, W are submodules of the $K[X]$-module V arising from a K-linear operator T on a finite-dimensional space V, i.e. suppose U, W are T-invariant. Let $u_1, \dots, u_k$ be a basis for U and $w_1, \dots, w_\ell$ be a basis for W. Take A to be the $k \times k$ matrix of the restriction $T|_U$ using the basis $u_1, \dots, u_k$ of U, and take B to be the $\ell \times \ell$ matrix of $T|_W$ using the basis $w_1, \dots, w_\ell$ of W. If $V = U \oplus W$, then a basis for V is the full list $u_1, \dots, u_k, w_1, \dots, w_\ell$. The dimension n of V is $k + \ell$ and the $n \times n$ matrix of T using this basis of V is the matrix

$$\begin{bmatrix} A & 0 \\ 0 & B \end{bmatrix},$$

typically known as a ***block-diagonal*** matrix with blocks A and B.

A bit of reflection on the definition of a representing matrix reveals the converse observation. Namely, if a block-diagonal matrix, made up of a $k \times k$ block A and an $\ell \times \ell$ block B such as above, represents an operator T on a space V, then the $K[X]$-module V arising from T is the direct sum $V = U \oplus W$ where $\dim U = k$ and $\dim W = \ell$. Furthermore U is the span of the first k elements of this basis while W is the span of the remaining ℓ elements of this basis.

The preceding remarks generalize in the obvious way to direct sum decompositions of V into any finite number of $K[X]$-submodules.

The Jordan Normal Form

The Jordan normal form is a matrix based interpretation of the elementary divisors arising from the structure theorem, Proposition 8.56, applied to the principal ideal domain $K[X]$.

Proposition 8.67. *Suppose T is a linear operator on a finite-dimensional vector space V over a field K and that the minimal polynomial of T splits as*

$$(X-\lambda_1)^{e_1}(X-\lambda_2)^{e_2}\cdots(X-\lambda_k)^{e_k},$$

where the λ_j are the eigenvalues of T. For each irreducible $X - \lambda_j$ let

$$(X-\lambda_j)^{e_{j1}}, (X-\lambda_j)^{e_{j2}}, \dots, (X-\lambda_j)^{e_{j\ell_j}},$$

be the unique list of elementary divisors of the primary component $V_{X-\lambda_j}$.

Then there is a basis of V such that the matrix of T using this basis is block-diagonal with Jordan blocks $J(\lambda_j, e_{ji})$.

Proof. For each j from 1 to k Proposition 8.44 gives the primary decomposition:

$$V = V_{X-\lambda_1} \oplus \cdots \oplus V_{X-\lambda_k}.$$

Since each $V_{X-\lambda_j}$ is an $(X - \lambda_j)$-primary module, Proposition 8.49 gives a further direct sum decomposition

$$V_{X-\lambda_j} = V_{j1} \oplus V_{j2} \oplus \cdots \oplus V_{j\ell_j},$$

into cyclic submodules V_{ji}. Assemble the preceding direct sum decompositions together to obtain the refined decomposition

$$V = \bigoplus_{j=1}^{k} \bigoplus_{i=1}^{\ell_j} V_{ji},$$

where each V_{ji} is a cyclic, $(X - \lambda_j)$-primary submodule of V. For each j the annihilators of V_{ji} are the ideals $K[X](X - \lambda_j)^{e_{ji}}$, whose generators are the unique elementary divisors of $V_{X-\lambda_j}$.

By Proposition 8.64 the restriction of T to the invariant space V_{ji} has a Jordan block $J(\lambda_j, e_{ji})$ as its matrix using a suitable basis of V_{ji}. From the preceding direct sum decomposition the totality of all these suitable bases provides a basis for V such that the representing matrix of T has block-diagonal form with Jordan blocks $J(\lambda_j, e_{ji})$. □

The splitting of the minimal polynomial of T in Proposition 8.67 is of course assured when the field K is algebraically closed, for example when $K = \mathbb{C}$.

Definition 8.68. The matrix representing T using Jordan blocks as in Proposition 8.67 is called a ***Jordan normal form*** of T.

Uniqueness of the Jordan Normal Form

Proposition 8.69. *Suppose $T : V \to V$ and $S : W \to W$ are linear operators on vector spaces V, W over a field K, such that their minimal polynomials split over K, and let matrices J_T, J_S be Jordan normal forms for T, S, respectively. If $V \cong W$ as $K[X]$-modules, then $J_T = J_S$, after a suitable rearrangement of the Jordan blocks in these matrices.*

Proof. The proof amounts to an interpretation of Proposition 8.55. Let

$$J(\lambda_1, e_1), \ldots, J(\lambda_n, e_n)$$

be the list of Jordan blocks appearing down the diagonal of J_T, and let

$$J(\mu_1, d_1), \ldots, J(\mu_m, d_m)$$

be the list of Jordan blocks appearing down the diagonal of J_S.

Each $J(\lambda_k, e_k)$ is the matrix of T restricted to a submodule V_k of V. Since J_T is block-diagonal with blocks $J(\lambda_k, e_k)$ this informs us that $V = \bigoplus_{k=1}^{n} V_k$. By Proposition 8.66 each submodule V_k is cyclic and primary. Likewise each $J(\mu_k, d_k)$ is the matrix of S restricted to a submodule W_k of W. Also $W = \bigoplus_{j=1}^{m} W_k$, and each W_k is cyclic and primary.

Since $V \cong W$, Proposition 8.55 reveals that $n = m$ and $V_k \cong W_k$ as $K[X]$-modules, after a suitable rearrangement of indices. By Proposition 8.66

$$\operatorname{ann}(V_k) = K[X](X - \lambda_k)^{e_k} \text{ and } \operatorname{ann}(W_k) = K[X](X - \mu_k)^{d_k}.$$

These annihilators are equal because $V_k \cong W_k$. Hence $(X - \lambda_k)^{e_k} = (X - \mu_k)^{d_k}$. This means that $\lambda_k = \mu_k$ and $e_k = d_k$. In other words $J(\lambda_k, e_k) = J(\mu_k, d_k)$, after a suitable reindexing of these blocks. □

An immediate consequence of Proposition 8.69 is that every linear operator T with a minimal polynomial that splits into linear factors has only one Jordan normal form, except for a possible rearrangement of the Jordan blocks. Indeed, just take $S = T$ in the above proposition.

The Jordan Form and Matrix Similarity

Every $n \times n$ matrix A represents a linear operator T on an n-dimensional space V. All T represented by the same A are similar operators according to Proposition 8.63. In other words, the $K[X]$-modules arising from such T are isomorphic. If the minimal polynomial of such T splits into linear factors, Proposition 8.69 says that such T have the same Jordan normal form, up to a permutation of the Jordan blocks. The ***Jordan normal form of a matrix*** A whose minimal polynomial splits is defined to be the unique Jordan normal form J of any operator T which A represents. Since A and J represent the same operator, this means that the matrix A and its Jordan normal form J are similar.

A classical fact about similar matrices emerges.

Proposition 8.70. *Two square matrices whose minimal polynomials split into linear factors are similar if and only if they have the same Jordan normal form.*

Proof. If A, B are similar matrices, then, by Proposition 8.61, they represent the same linear operator T. The Jordan normal form of T is the Jordan normal form of A and B. Conversely, if A, B have the same Jordan normal form J, then both A and B are similar to J, and thus similar to each other. □

The Characteristic Polynomial

Since the Jordan normal form determines the similarity of matrices, the question of computing the Jordan form needs some attention. One tool that is somewhat helpful is the characteristic polynomial of a matrix. To that end, a knowledge of determinants is needed, and will be assumed.

Let $\det(A)$ denote the determinant of an $n \times n$ matrix with entries in a field K. Among its properties we note that if A is an upper triangular matrix, i.e. with all zeroes above the main diagonal, then $\det(A)$ equals the product of the diagonal entries of A. The same applies to lower triangular matrices, such as the Jordan normal forms. Another property is that $\det(AB) = \det(A)\det(B)$ for all $n \times n$ matrices A, B. Thus the restriction of the determinant to the group of invertible matrices is a group homomorphism into the group of non-zero elements of K. Furthermore, similar matrices have the same determinant. This allows the determinant of any linear operator on a finite-dimensional space to be defined, without ambiguity, as the determinant of any matrix that represents the operator.

A matrix A with entries in a field K can be viewed as a matrix with entries in the rational function field $K(X)$. With I the identity matrix, $XI - A$ is another matrix with entries in $K(X)$.

Definition 8.71. The ***characteristic polynomial*** of the matrix A is defined to be $\det(XI - A)$.

From any definition of determinants it can be seen that $\det(XI - A)$ is a monic polynomial of degree n. If $B = P^{-1}AP$ for some invertible matrix P, then $XI - B = P^{-1}(XI - A)P$, and so

$$\det(P^{-1}(XI - B)P) = \det(XI - A).$$

Thus the characteristic polynomial is the same for all similar matrices. This allows us to define the ***characteristic polynomial of a linear operator*** T to be the characteristic polynomial of any matrix A used to represent T.

For example, suppose A is an $e \times e$ matrix that is similar to a Jordan block $J(\lambda, e)$ as displayed in Proposition 8.64. Then

$$XI - J(\lambda, e) = \begin{bmatrix} X-\lambda & 0 & 0 & \dots & 0 & 0 \\ -1 & X-\lambda & 0 & \dots & 0 & 0 \\ 0 & -1 & X-\lambda & \dots & 0 & 0 \\ \vdots & \vdots & \vdots & \ddots & \vdots & \vdots \\ 0 & 0 & 0 & \dots & X-\lambda & 0 \\ 0 & 0 & 0 & \dots & -1 & X-\lambda \end{bmatrix}.$$

This is an $e \times e$ lower triangular matrix with the repeated entry $X - \lambda$ on its diagonal. Hence

$$\det(XI - A) = \det(XI - J(\lambda, e)) = (X - \lambda)^e.$$

Recall that the minimal polynomial of a matrix is the minimal polynomial of any operator that the matrix represents. By Proposition 8.66 the minimal polynomial of $J(\lambda, e)$ also equals $(X - \lambda)^e$. From these observations it follows that if the Jordan normal form of a matrix A consists of one Jordan block $J(\lambda, e)$, then the characteristic polynomial of A equals the minimum polynomial of A.

The Minimum Polynomial Divides the Characteristic Polynomial

The next result affords an entry point for calculating the minimum polynomial of a matrix.

Proposition 8.72. *If A is an $n \times n$ matrix over a field K and the minimal polynomial of A splits into linear factors, then the characteristic polynomial of A also splits and the minimum polynomial of A divides the characteristic polynomial of A. Furthermore the roots of the characteristic polynomial of A coincide with those of its minimal polynomial.*

Proof. Since the Jordan normal form J of A has the same minimum and the same characteristic polynomial as A, replace A by J.

Let us review how J came to be. Take T to be any linear operator on an n-dimensional space V represented by the matrix A, and let $(X - \lambda_1)^{e_1} \cdots (X - \lambda_k)^{e_k}$ be the minimal polynomial of T. The distinct λ_j are the eigenvalues of T. There is the primary decomposition of V given by

$$V = V_{X-\lambda_1} \oplus \cdots \oplus V_{X-\lambda_k},$$

where the submodules $V_{X-\lambda_j}$ have $K[X](X - \lambda_j)^{e_j}$ as their annihilator. For each j there is a further decomposition

$$V_{X-\lambda_j} = V_{j1} \oplus V_{j2} \oplus \cdots \oplus V_{j\ell_j},$$

where the submodules V_{ji} are cyclic and $(X - \lambda_j)$-primary. By Proposition 8.64 the minimal polynomial of T restricted to V_{ji} is $(X - \lambda_j)^{e_{ji}}$ where e_{ji} is the dimension of V_{ji}. Also the restriction of T to the invariant subspace V_{ji} is represented by the Jordan block $J(\lambda_j, e_{ji})$. Thus the Jordan blocks of J can be arranged into a table as shown:

$$\begin{array}{llll}
J(\lambda_1, e_{11}) & J(\lambda_1, e_{12}) & \ldots & J(\lambda_1, e_{1\ell_1}) \\
\vdots & & & \\
J(\lambda_j, e_{j1}) & J(\lambda_j, e_{j2}) & \ldots & J(\lambda_j, e_{j\ell_j}) \\
\vdots & & & \\
J(\lambda_k, e_{k1}) & J(\lambda_k, e_{k2}) & \ldots & J(\lambda_k, e_{k\ell_k}).
\end{array}$$

In addition, for each j we can safely arrange the V_{ji} so that the sizes of the blocks are decreasing. That is

$$e_{j1} \geq e_{j2} \geq e_{j3} \geq \cdots \geq e_{j\ell_j}.$$

For each j Proposition 8.51 tells us that $e_j = e_{j1}$. Thus the minimal polynomial of T is

$$(X - \lambda_1)^{e_{11}} (X - \lambda_2)^{e_{21}} \cdots (X - \lambda_k)^{e_{k1}}.$$

We see that the minimal polynomial of T appears as the product of the minimal polynomials of the Jordan blocks appearing in the first column of our table. These are the largest blocks that go with each eigenvalue λ_j.

By inspecting the lower triangular matrix $XI - J$ we can see that the characteristic polynomial of J is the full product

$$\prod_{j=1}^{k}\prod_{i=1}^{\ell_j}(X-\lambda_j)^{e_{ji}}.$$

This is the product of the minimal polynomials of all the Jordan blocks of J.

Having examined how the minimal and characteristic polynomials of J are recovered from the Jordan blocks of J, our desired claims are plain to see. □

Actually the minimal polynomial of a matrix divides the characteristic polynomial even when the minimal polynomial does not split over the field K. For an explanation of why this is so, suppose A is a matrix with entries in K. Let F be a field extension of K over which the minimal polynomial of A splits. It is clear that the characteristic polynomial of A viewed as a matrix with entries in F coincides with the characteristic polynomial of A as a matrix with entries in K. Let us now explain why the minimal polynomial of A as a matrix over F coincides with its minimal polynomial as a matrix over K.

Suppose $f = a_0 + a_1X + \cdots + a_{k-1}X^{k-1} + X^k$ in $F[X]$ is the minimal polynomial of A viewed as a matrix with entries in F. The fact that $f(A) = 0$ means the matrix equation

$$x_0 + x_1A + \cdots + x_{k-1}A^{k-1} = A^k$$

has a unique solution $-a_0, -a_1, \ldots, -a_{k-1}$ in F. But this matrix equation amounts to a system of n^2 linear equations in k unknowns. The coefficients of this system arise from the powers A^j, which puts the coefficients inside K. After that the well established techniques for solving linear systems using row operations reveal that since the coefficients of the system lie in the field K, so must its unique solution $-a_0, -a_1, \ldots, -a_{k-1}$ lie in K. Thus $f \in K[X]$ after all. Now if g is the minimal polynomial of A inside $K[X]$, we have $g \,|\, f$ in $K[X]$. Since $g(A) = 0$ it also holds that $f \,|\, g$ in $F[X]$, and since these polynomials are monic it follows that $g = f$.

By Proposition 8.72 the minimal polynomial g of A (regardless of which field is used) divides the characteristic polynomial h of A inside $F[X]$. Since g and h are in $K[X]$ it must be that $f \,|\, h$ inside $K[X]$ already.

The Cayley–Hamilton Theorem

Let f be the characteristic polynomial of a matrix A and let g be its minimal polynomial in $K[X]$. Since $g \,|\, f$ and $g(A) = 0$, it follows that $f(A) = 0$. The characteristic polynomial of a matrix evaluated at the matrix is the zero matrix. This is a famous result known as the ***Cayley–Hamilton theorem***, in honor of Arthur Cayley (1821–1895) and William Rowan Hamilton (1805–1865).

Finding the Jordan Normal Form (Sometimes)

The relationship between the characteristic and the minimal polynomial can be helpful in finding the Jordan normal form J of a matrix A. We suppose that the characteristic polynomial f of A is found using determinants, and (being greedy) that it is factored into linear factors as

$$f = (X - \lambda_1)^{d_1}(X - \lambda_2)^{d_2} \cdots (X - \lambda_k)^{d_k},$$

where the λ_j are the distinct eigenvalues of A. By Proposition 8.72 the minimal polynomial of A is

$$g = (X - \lambda_1)^{e_1}(X - \lambda_2)^{e_2} \cdots (X - \lambda_k)^{e_k}, \text{ where } 1 \leq e_j \leq d_j.$$

According to Proposition 8.72 the minimal polynomial can be found by substituting A into the finitely many eligible monic divisors of f. The divisor of least degree which vanishes at A is the minimal polynomial of A.

When the Jordan blocks appearing in J are put into a table as displayed in the proof of Proposition 8.72, a few constraints on the parameters d_j, e_j, e_{ji} and ℓ_j come to light.

- Since $\deg f = d_1 + \cdots + d_k$, this sum equals the size of A, which is the dimension of any space on which an operator represented by A acts.
- The characteristic polynomial f is the full product $\prod_{j=1}^{k} \prod_{i=1}^{\ell_j} (X - \lambda_j)^{e_{ji}}$. Thus

$$(X - \lambda_j)^{d_j} = \prod_{i=1}^{\ell_j} (X - \lambda_j)^{e_{ji}} = (X - \lambda_j)^{e_{j1} + e_{j2} + \cdots + e_{j\ell_j}}.$$

  So for each j we have

$$d_j = e_{j1} + e_{j2} + \cdots + e_{j\ell_j}.$$

  The sizes of the Jordan blocks that go with a single eigenvalue λ_j form a partition of d_j, in the sense discussed in Section 2.2.2. This partition reveals that the number of times that an eigenvalue λ_j appears on the diagonal of J is d_j.
- For each eigenvalue λ_j the table of Jordan blocks is arranged so that

$$e_{j1} \geq e_{j2} \geq \cdots \geq e_{j\ell_j}.$$

  This ensures that $e_{j1} = e_j$, as noted in the proof of Proposition 8.72. In fact this an application of Proposition 8.51. Thus for each eigenvalue λ_j the Jordan block $J(\lambda_j, e_j)$ always appears in J, and it is the largest of the blocks in J having eigenvalue λ_j.
- The number ℓ_r of Jordan blocks that go with each eigenvalue λ_r turns out to be the nullity of the matrix $A - \lambda_r I$ i.e. the dimension of the space of eigenvectors of A that go with the eigenvalue λ_r. We might recall that the nullity of a matrix is the dimension of the solution space of the homogeneous system of linear equations determined by the matrix, a quantity which is readily computable by well worn row operation techniques. To see why ℓ_r equals this nullity, observe that similar matrices have equal nullities. If J is the Jordan normal

form of A, the matrix $J - \lambda_r I$ is similar to $A - \lambda_r I$, and so it suffices to see that the nullity $J - \lambda_r I$ is ℓ_r.

Well, the matrix $J - \lambda_r I$ is another block-diagonal matrix whose blocks are given by

$$J(\lambda_j - \lambda_r, e_{ji}), \text{ where } j \text{ runs from 1 to } k \text{ and } i \text{ runs from 1 to } \ell_j.$$

A simple exercise reveals that the nullity of a block-diagonal matrix is the sum of the nullities of its blocks. Thus the nullity of $J - \lambda_r I$ is the sum of the nullities of the above blocks. If $j \neq r$, the block $J(\lambda_j - \lambda_r, e_{ji})$ is lower triangular with the repeated non-zero entry $\lambda_j - \lambda_r$ on the diagonal. Such a matrix is non-singular and its nullity is 0. If $j = r$, we get the blocks $J(0, e_{ri})$. These blocks each have nullity 1, by a simple inspection. Since the number of such blocks is ℓ_r, it follows that ℓ_r is the nullity of $J - \lambda_r I$.

For a given matrix A the characteristic and minimal polynomials along with the nullities of the matrices $A - \lambda I$ for each eigenvalue λ can yield the Jordan normal form of A in a number of special cases.

For instance, we could have the possibility that $e_j = d_j$ for some eigenvalue λ_j. Since the e_{ji} form a partition of d_j and $e_{j1} = e_j$ the Jordan form J could only have the one Jordan block $J(\lambda_j, d_j)$ attached to the eigenvalue λ_j. Should it happen that $g = f$, meaning that $e_j = d_j$ for all eigenvalues λ_j, the Jordan normal form of A would consist precisely of the Jordan blocks $J(\lambda_1, d_1), \ldots, J(\lambda_k, d_k)$.

At the other extreme we could have the possibility that $e_j = 1$ for some eigenvalue λ_j. Since $1 = e_j = e_{j1} \geq e_{j2} \geq \cdots \geq e_{j\ell_j} \geq 1$, we see in this case that all Jordan blocks for the eigenvalue λ_j are 1×1 matrices containing the single entry λ_j. Since the e_{ji}, all of them equal to 1, form a partition of d_j, we see that $\ell_j = d_j$. In case the minimal polynomial g has no repeated roots, this says that all $e_j = 1$. Now the Jordan form J becomes a diagonal matrix with each eigenvalue λ_j repeated precisely d_j times on the diagonal. Thus we see that a matrix is diagonalizable if and only if its minimal polynomial has no repeated roots.

A Few Examples of Jordan Normal Forms

To do some worked examples it would expedite the various matrix operations by using a suitable computer algebra system.

Let us use the constraints of the preceding discussion to find the Jordan normal form J of the matrix:

$$A = \begin{bmatrix} 2 & -1 & 0 & 1 \\ 0 & 3 & -1 & 0 \\ 0 & 1 & 1 & 0 \\ 0 & -1 & 0 & 3 \end{bmatrix}.$$

The characteristic polynomial of A is computed to be

$$\det(XI - A) = (X - 2)^3(X - 3).$$

It follows from Proposition 8.72 that the minimal polynomial of A is one of

$$(X-2)(X-3),\ (X-2)^2(X-3),\ (X-2)^3(X-3).$$

The substitution of A for X gives

$$(A-2I)(A-3I)=\begin{bmatrix}0&-1&1&0\\0&-1&1&0\\0&-1&1&0\\0&-1&1&0\end{bmatrix}\text{ and }(A-2I)^2(A-3I)=\begin{bmatrix}0&0&0&0\\0&0&0&0\\0&0&0&0\\0&0&0&0\end{bmatrix}.$$

Hence the minimal polynomial of A is $(X-2)^2(X-3)$. From this we see that the eigenvalue 3 appears just once on the diagonal of J as the Jordan block $J(3,1)$. The 2×2 block $J(2,2)$ appears as the largest block for the eigenvalue 2. Since the sizes of the blocks for the eigenvalue 2 form a partition of 3 (which is the multiplicity of the eigenvalue 2), there is only room for another 1×1 block $J(2,1)$. Thus

$$J=\begin{bmatrix}2&0&0&0\\1&2&0&0\\0&0&2&0\\0&0&0&3\end{bmatrix}.$$

Of course this J is unique only up to a permutation of the three blocks. For this example, the characteristic and minimal polynomials were sufficient to get J.

Let us find the Jordan normal form J of the matrix:

$$A=\begin{bmatrix}3&2&-4&4\\-6&-3&8&-12\\-3&-2&5&-6\\-1&-1&2&-1\end{bmatrix}.$$

The characteristic polynomial of A is computed to be

$$\det(XI-A)=X^4-4X^3+6X^2-4X+1=(X-1)^4.$$

Since this polynomial splits over the field $\mathbb{Q}$, the Jordan normal form J is also a matrix over $\mathbb{Q}$. For each Jordan block of J the only eigenvalue is 1.

The minimal polynomial of A is one of $X-1,(X-1)^2,(X-1)^3$ or $(X-1)^4$. Now

$$A-I=\begin{bmatrix}2&2&-4&4\\-6&-4&8&-12\\-3&-2&4&-6\\-1&-1&2&-2\end{bmatrix}.$$

Clearly $A-I$ is not the zero matrix, but $(A-I)^2$ is the zero matrix, by a routine calculation. The minimal polynomial of A is $(X-1)^2$. Hence the 2×2 Jordan block $J(1,2)$ appears in J.

The remaining blocks of J could be another $J(1,2)$, or they could be two $J(1,1)$ blocks. We know that the total number of blocks is the nullity of $A-I$. With a simple row reduction the nullity of $A - I$ is seen to be 2. This means that J has just two Jordan blocks, and thereby J has another $J(1,2)$. With these calculations we see that

$$J = \begin{bmatrix} 1 & 0 & 0 & 0 \\ 1 & 1 & 0 & 0 \\ 0 & 0 & 1 & 0 \\ 0 & 0 & 1 & 1 \end{bmatrix}.$$

Let us find the Jordan normal form of the matrix:

$$A = \begin{bmatrix} 11 & 6 & -10 & 9 & -1 & 3 \\ -16 & -10 & 20 & -18 & 2 & -5 \\ 7 & 6 & -8 & 9 & -1 & 2 \\ 14 & 12 & -20 & 20 & -2 & 4 \\ 14 & 12 & -14 & 18 & -3 & 4 \\ -18 & -12 & 20 & -18 & 2 & -4 \end{bmatrix}.$$

The characteristic polynomial of A (obtained with the help of computer algebra software) is:

$$\det(XI - A) = X^6 - 6X^5 + 9X^4 + 8X^3 - 24X^2 + 16 = (X+1)^2(X-2)^4.$$

Consequently the minimal polynomial of A is one of

$$(X+1)^i(X-2)^j \text{ where } 1 \le i \le 2 \text{ and } 1 \le j \le 4.$$

Substitute A into each of these possibilities to discover that the polynomial of least degree which annihilates A is $(X+1)^2(X-2)^2$. We deduce from our constraints discussed above that the Jordan normal form J of A contains $J(-1,2)$ as the only block with eigenvalue -1. The block $J(2,2)$ appears in J as the largest block with eigenvalue 2. The possibility remains of having an additional 2×2 block $J(2,2)$ appearing in J or maybe the 1×1 block $J(2,1)$ appearing twice in J. The nullity of $A - 2I$ counts the number of blocks with eigenvalue 2. That nullity is readily computed to be 3, which forces two $J(2,1)$ blocks to appear. Therefore

$$J = \begin{bmatrix} 2 & 0 & 0 & 0 & 0 & 0 \\ 1 & 2 & 0 & 0 & 0 & 0 \\ 0 & 0 & 2 & 0 & 0 & 0 \\ 0 & 0 & 0 & 2 & 0 & 0 \\ 0 & 0 & 0 & 0 & -1 & 0 \\ -0 & 0 & 0 & 0 & 1 & -1 \end{bmatrix}.$$

Let us find the Jordan normal form of the matrix:

$$A = \begin{bmatrix} 7 & 2 & 6 & -3 & 0 & 0 \\ -6 & -1 & -12 & 6 & 0 & 1 \\ 4 & 3 & 9 & -3 & 0 & -1 \\ 8 & 6 & 12 & -3 & 0 & -2 \\ 8 & 6 & 10 & -5 & 3 & -2 \\ -8 & -4 & -12 & 6 & 0 & 3 \end{bmatrix}.$$

The characteristic polynomial of A (obtained with the help of computer algebra software) is:

$$\det(XI - A) = X^6 - 18X^5 + 135X^4 - 540X^3 + 1215X^2 - 1458X + 729 = (X - 3)^6.$$

Consequently the minimal polynomial of A is one of

$$(X - 3)^i \text{ where } 1 \leq i \leq 6.$$

Substitute A into each of these possibilities to discover that the polynomial of least degree which annihilates A is $(X - 3)^3$. We deduce from our constraints discussed just prior to these examples that the Jordan normal form J of A contains only blocks with eigenvalue 3 and that $J(3, 3)$ appears in J as its largest block. The possibilities for the remaining blocks are to have an additional $J(3, 3)$ block, or maybe one $J(3, 2)$ and one $J(3, 1)$, or else three $J(3, 1)$ blocks. The nullity of $A - 3I$ counts the number of blocks with eigenvalue 3. That nullity is readily computed to be 3. This shows that $J(3, 3), J(3, 2)$ and $J(3, 1)$ are the blocks appearing in J. In other words

$$J = \begin{bmatrix} 3 & 0 & 0 & 0 & 0 & 0 \\ 1 & 3 & 0 & 0 & 0 & 0 \\ 0 & 1 & 3 & 0 & 0 & 0 \\ 0 & 0 & 0 & 3 & 0 & 0 \\ 0 & 0 & 0 & 1 & 3 & 0 \\ 0 & 0 & 0 & 0 & 0 & 3 \end{bmatrix}.$$

Admittedly, the preceding examples were such that their characteristic polynomials factored readily. With larger matrices, computations with the aid of computer algebra systems will be hard to avoid. On the other hand, with a computer algebra system a direct command to get the Jordan normal form will also work. The techniques on which the preceding examples are based are effective as long as each of the roots of the characteristic polynomial of the matrix A has multiplicity 6 or less. To handle all possible matrices there is an algorithm, based on doing row operations to the matrix $XI - A$, which computes the Jordan normal form over a field in which the characteristic polynomial splits. But perhaps we can pause at this point.

EXERCISES

1. If the Jordan blocks of a matrix A are

$$J(0, 4), J(0, 4), J(0, 2), J(1, 3), J(1, 1), J(2, 3), J(2, 3), J(2, 3),$$

write the minimal and characteristic polynomials of A.

2. If A is an 8×8 matrix with X^8 as its characteristic polynomial, X^5 as its minimal polynomial and 4 as its nullity, find the Jordan canonical form of A.
3. Show that every Jordan block $J(\lambda, e)$ is similar to its transpose matrix. Use this to show that every $n \times n$ matrix over an algebraically closed field is similar to its transpose matrix.

4. Find the Jordan normal form of the matrix $\begin{bmatrix} 5 & 0 & 0 & 1 \\ -6 & 3 & 0 & -3 \\ 7 & 0 & 2 & 2 \\ -5 & 0 & 1 & 2 \end{bmatrix}$.
5. Suppose that A, B are square matrices over a field K, and that A, B have the same characteristic and the same minimal polynomial and that these polynomials split over K. Also suppose that each root λ of the characteristic polynomial has multiplicity no more than 6, and that the matrices $A - \lambda I, B - \lambda I$ have the same nullity. Show that A, B are similar matrices.
6. Find two 7×7 matrices A, B over some field, such that A, B have the same characteristic polynomial and the same minimal polynomial, and such that for every scalar λ the matrices $A - \lambda I, B - \lambda I$ have the same nullities, and yet A, B are not similar.

9 Division Algorithms

This chapter is about doing algorithms in some rings. These can get quite lengthy. Yet they afford a reassurance that some hard questions can be answered.

As discussed in Section 5.3, a Euclidean domain R comes with a function ϕ from $R \setminus \{0\}$ to the set of non-negative integers $\mathbb{N}$, such that for every b in R and every non-zero a in R, there exist q, r in R such that

$$b = aq + r \text{ and } \phi(r) < \phi(a) \text{ or } r = 0.$$

Such ϕ are called ***Euclidean functions***.

All ideals in a Euclidean domain are principal. If the ideal J is non-zero, its generator is obtained by taking a non-zero element in J of minimal ϕ value. This typically involves the Euclidean algorithm using ϕ to get the greatest common divisor of a given set of generators. If a generates J, the problem of deciding whether an element $b \in J$ is answered by computing a remainder r using the Euclidean function to divide a into b. Indeed, if $r = 0$, then certainly $b \in J$. While if $b \in J$, then $b = ak$ for some k in R. If we divide using the Euclidean function ϕ, then $b = aq + r$ where $q, r \in R$ and $\phi(r) < \phi(a)$. Consequently $r = a(k - q) \in J$, and since a is a non-zero element of J having minimal ϕ value, it follows that $r = 0$. The remainder r computed by means of the Euclidean function ϕ becomes the test for membership in J.

To have a mechanism for detection of ideal membership for more general rings, such as $\mathbb{Q}[X, Y]$, a much more general kind of division is needed. This problem was investigated by Pierre Samuel (1921–2009) in an interesting paper entitled “About Euclidean rings” (Journal of Algebra, 19 (1971) 282). One generalization might be to replace the target set $\mathbb{N}$ of our Euclidean function by any well-ordered set (to be defined shortly). Masayoshi Nagata (1927–2008) discovered domains which have Euclidean functions into a well-ordered set but not into $\mathbb{N}$. The ideals of such domains remain principal, and so important rings such as $\mathbb{Q}[X, Y]$ are excluded.

It became apparent that the target of our Euclidean function should be what we shall call a ***well-partially ordered set*** (also known as a partially well-ordered set). In that case Samuel showed that a Euclidean function can be replaced by one whose values are in a well-ordered set, which puts us no further ahead. As a consequence the division using such a Euclidean function had to be restricted to suitable pairs a, b in R.

With the proper definitions and conditions in place, the class of Euclidean domains can be enlarged to what we shall call the family of ***Gröbner domains***. They encompass the important rings of polynomials in several variables, and more. Wolfgang Gröbner (1899–1980) was an Austrian mathematician whose name recurs in the computational theory of polynomials in several variables and algebraic geometry. His student Bruno Buchberger did much to develop this theory and to hallow his supervisor's name into the mathematical lexicon.

Our primary example of a Gröbner domain will be the ring of polynomials in several variables over a field. A discussion of the many other Gröbner domains would take us too far afield. Yet we might hope that an abstract beginning will cut through the maze of details that seem to bedevil this subject, as well as lay the groundwork for future examples of such rings.

Gröbner theory entails severe computations, whose complexity is an important problem. However, the subject at hand is algebra, which is where our focus will be directed.

9.1 Well-Partial Orders

If M is a non-empty set, a subset P of the Cartesian product $M \times M$ is sometimes called a ***binary relation*** on M. Such P is known as a ***partial order*** on M when the following properties hold:

- *reflexivity* $(a,a) \in P$ for all a in M,
- *transitivity* if both $(a,b) \in P$ and $(b,c) \in P$, then $(a,c) \in P$,
- *anti-symmetry* if both $(a,b) \in P$ and $(b,a) \in P$, then $a = b$.

The set M is then called a ***partially ordered set***. It is common to forsake these formal set notations and write $a \leq b$ to mean that $(a,b) \in P$, using the inequality symbol $\leq$ or some variant of it. The above properties translate into the familiar properties of reflexivity, transitivity and anti-symmetry, which all inequalities are expected to have. In a partially ordered set write

$$a < b \text{ to mean that } a \leq b \text{ but } a \neq b.$$

We also use the reverse notation $b \geq a$ to mean $a \leq b$. To describe this situation we adopt common terminology such as *a is less than or equal to b*, *b is greater than or equal to a*, *b is above a*, *a is below b*, *a is lower than b*, and so on.

Any subset M of $\mathbb{R}$, with the usual ordering, affords an example. For another example, take the set $\mathbb{P}$ of positive integers. Declare that $a \preceq b$ whenever a divides b. A brief inspection shows that this ***order by divisibility*** is a partial order on $\mathbb{P}$.

If a,b are in a partially ordered set M, we say that a,b are ***comparable*** provided $a \leq b$ or $b \leq a$. If M is such that *all* pairs a,b are comparable, we say that A is ***totally ordered*** and that the relation $\leq$ is a ***total order***. Any subset of $\mathbb{R}$ with the usual order is totally ordered. But with the order by divisibility on $\mathbb{P}$, the elements $2, 3$ are not comparable. This indicates why the term "partial" is used. Despite the awkwardness of the terminology, totally ordered sets are considered to be special types of partially ordered sets.

9.1.1 Well-Ordered Sets: Total and Partial

If S is a non-empty subset of a totally ordered set M, there need not be a ***least element*** in S. Namely, an element a such that

$$a \in S \text{ and } a \leq b \text{ for all } b \text{ in } S.$$

For instance, $\mathbb{Z}$ with its usual ordering has no least element. But the set of non-negative integers $\mathbb{N}$, with its usual ordering, is such that every non-empty subset S has a least element, also known as the ***first element*** in S. If a totally ordered set M is such that every non-empty subset has a least element, the set A is said to be ***well-ordered***. The order relation itself is known as a ***well-order***. Subsets of well-ordered sets are well-ordered, with the original order restricted to the elements of the subset.

For another example of a well-ordered set, take the product $\mathbb{N} \times \{0, 1\}$, which can be viewed as the disjoint union of two copies of $\mathbb{N}$. With the usual order $\leq$ on $\mathbb{N}$, declare that

$$(a, 0) < (b, 1) \text{ for all } a, b \text{ in } \mathbb{N},$$

and

$$(a, 0) \leq (b, 0) \text{ if and only if } a \leq b, \text{ and } (a, 1) \leq (b, 1) \text{ if and only if } a \leq b.$$

This is one copy of $\mathbb{N}$ stacked above another copy of $\mathbb{N}$. Here is the total order on full display:

$$(0, 0) < (1, 0) < (2, 0) < (3, 0) < \cdots < (0, 1) < (1, 1) < (2, 1) < (3, 1) < \cdots .$$

Let S be a non-empty subset of $\mathbb{N} \times \{0, 1\}$. If some $(a, 0) \in S$, the least element of S is $(b, 0)$ where b is minimal among the a for which $(a, 0) \in S$. If S only has elements of the form $(a, 1)$, the least element of S is $(b, 1)$, where b is minimal among the a for which $(a, 1) \in S$.

Here is a well-order on $\mathbb{N} \times \mathbb{N}$. With the usual order on $\mathbb{N}$, declare that

$$(a, b) < (c, d) \text{ provided } a < c, \text{ or } a = c \text{ and } b < d.$$

This is known as the ***lexicographic order*** on $\mathbb{N} \times \mathbb{N}$. This total order looks like so:

$$\begin{aligned} &(0,0) < (0,1) < (0,2) < (0,3) < (0,4) < \cdots \\ &< (1,0) < (1,1) < (1,2) < (1,3) < (1,4) < \cdots \\ &< (2,0) < (2,1) < (2,2) < (2,3) < (2,4) < \cdots \\ &< (3,0) < (3,1) < (3,2) < (3,3) < (3,4) < \cdots \\ &< (4,0) < (4,1) < \cdots \\ &\vdots \end{aligned}$$

If S is a non-empty subset of $\mathbb{N} \times \mathbb{N}$, let a be the least element of $\mathbb{N}$ such that $(a, c) \in S$ for some c in $\mathbb{N}$. Then let b be least among all c such that $(a, c) \in S$. The element (a, b) is the least element of S.

An alternative way to see that a totally ordered set is well-ordered will lead to an important generalization.

Proposition 9.1. *A totally ordered set M is well-ordered if and only if every infinite subset S contains an infinite strictly ascending chain*

$$a_1 < a_2 < a_3 \cdots < a_n < \cdots .$$

Proof. Suppose M is well-ordered and that S is an infinite subset of M. Let a_1 be its first element. Then let a_2 be the first element of $S \setminus \{a_1\}$. Clearly $a_1 < a_2$. Let a_3 be the first element of $S \setminus \{a_1, a_2\}$. Clearly $a_1 < a_2 < a_3$. Proceed in this way to get ever longer ascending chains inside S. The chain goes on indefinitely since S is assumed to be infinite.

Conversely, suppose that M is not well-ordered. In this case M has a non-empty subset S with no first element. Pick an a_1 from S. Since S is totally but not well-ordered, there must be an a_2 in S such that $a_1 > a_2$. Likewise there is an a_3 in S such that $a_1 > a_2 > a_3$. By this line of reasoning there is an infinite descending chain

$$a_1 > a_2 > a_3 > \cdots > a_n > \cdots$$

all inside S. Now the infinite set $\{a_1, a_2, \ldots, a_n, \ldots\}$ cannot possibly contain an infinite ascending chain. Indeed, each element of this set only has finitely many elements above it. □

Proposition 9.1 inspires a useful concept that applies to any *partially* ordered set.

Definition 9.2. A partially ordered set M is ***well-partially ordered*** provided every infinite subset S of M contains an infinite strictly increasing chain

$$b_1 < b_2 < \cdots < b_n < \cdots .$$

The order relation itself is a ***well-partial order***.

By Proposition 9.1 every well-order, such as the usual on $\mathbb{N}$, is a well-partial order. On the other hand, the set $\mathbb{P}$ of positive integers with the partial order by divisibility, is not a well-partial order. Indeed, the infinite set of primes is a subset of $\mathbb{P}$ in which no two elements are comparable, and there is no infinite ascending chain of primes.

Well-Partially Ordered Sets Are Artinian

If M is well-partially ordered, then there is no infinite descending chain $a_1 > a_2 > a_3 > \cdots$ inside M. For if there were such a chain, the infinite set of a_j would contain an infinite ascending chain, which cannot be since every a_j has only finitely many a_i above it. A partially ordered set with no infinite descending chains is called ***Artinian*** in honor of Emil Artin (1898–1962). As we just saw, the set of positive integers $\mathbb{P}$ with the partial order by divisibility illustrates a set which is not well-partially ordered. But $\mathbb{P}$ is Artinian because every positive integer has only finitely many divisors.

A Well-Partial Order that Is Not Total

Take the set $\mathbb{N}$ again with $\leq$ denoting the usual order. Declare that

$$a \preceq b \text{ when either } a = b \text{ or } 2a \leq b.$$

Under this relation, $2 \preceq 4$, but $2 \not\preceq 3$ and $3 \not\preceq 4$. Clearly $a \preceq b$ implies $a \leq b$, but not conversely. This reveals that the ordering $\preceq$ is anti-symmetric. It is obviously reflexive. The verification of transitivity is routine. Thus we have a partial order. Now if S is an infinite subset of $\mathbb{N}$, such S is unbounded with the usual ordering. Pick a non-zero a_1 in S, then pick an a_2 in S such that $2a_1 \leq a_2$. Thus $a_1 \preceq a_2$. In like fashion pick a_3 in S such that $a_1 \preceq a_2 \preceq a_3$, and continue like this to pick up an infinite increasing chain in S. The order $\preceq$ on $\mathbb{N}$ is a well-partial order, which is truly partial.

A Technical Variant on Well-Partial Orders

Here is a slightly more technical way to think about well-partial orders, which can be useful at times.

Proposition 9.3. *A partial order $\leq$ on a set M is a well-partial order if and only if every sequence in M, which assumes infinitely many values, has a strictly ascending subsequence.*

Proof. Suppose that every sequence assuming infinitely many values in M has a strictly increasing subsequence. Let S be an infinite subset of M. The fact S is infinite means that there is a sequence of distinct elements $a_1, a_2, a_3, \ldots, a_n, \ldots$ taken from S. By assumption, there are indices $j_1 < j_2 < j_3 \cdots < j_k < \cdots$ such that $a_{j_1} < a_{j_2} < a_{j_3} < \cdots < a_{j_k} < \cdots$. This ascending chain in M shows that M is well-partially ordered.

Conversely, suppose M is well-partially ordered. Let $a_1, a_2, a_3, \ldots$ be a sequence in M which takes on infinitely many values. The infinite set of a_j taken as a set S contains a strictly ascending chain $b_1 < b_2 < b_3 < \cdots$. This might not constitute a subsequence of the a_n, but a subsequence of the b_j does.

Indeed, $b_1 = a_{j_1}$ for some index j_1. To get a pattern started let $i_1 = 1$ so that $b_1 = b_{i_1}$. The set of successors of a_{j_1} is co-finite in S, and must thereby meet the infinite set of successors of b_{i_1}. So, there is an index $i_2 > i_1$ and an index $j_2 > j_1$ such that $b_{i_2} = a_{j_2}$. Then

$$a_{j_1} = b_{i_1} < b_{i_2} = a_{j_2}.$$

The set of successors of a_{j_2} remains co-finite in S, and thereby meets the infinite set of successors of b_{j_2}. So, there is an index $i_3 > i_2$ and an index $j_3 > j_2$ such that $b_{i_3} = a_{j_3}$. Then

$$a_{j_1} < a_{j_2} = b_{i_2} < b_{i_3} = a_{j_3}.$$

Repeat the above argument with the co-finite set of successors of a_{j_3} and the infinite set of successors of b_{i_3} to pick up $i_4 > i_3$ and $j_4 > j_3$ such that $b_{i_4} = a_{j_4}$. Thus

$$a_{j_1} < a_{j_2} < a_{j_3} = b_{i_3} < b_{i_4} = a_{j_4}.$$

By repeating this argument, a strictly ascending subsequence of a_n emerges. □

The Product Order

Starting with a pair of well-partially ordered sets, their Cartesian product is well-partially ordered in a natural way.

Proposition 9.4. *If M and N are well-partially ordered by $\leq_1$ and $\leq_2$, respectively, then the relation $\leq$ on the product set $M \times N$, given by*

$$(a,b) \leq (c,d) \text{ if and only if } a \leq_1 c \text{ and } b \leq_2 d,$$

is a well-partial order.

Proof. That $\leq$ is a partial order on $M \times N$ is routine to verify, so we focus on the wellness part.

Suppose S is an infinite subset of $M \times N$. Let

$$S_M = \{a \in M : (a,b) \in S \text{ for some } b \text{ in } N\},$$

and likewise

$$S_N = \{b \in N : (a,b) \in S \text{ for some } a \text{ in } M\}.$$

Since S is infinite, one of S_M or S_N is infinite. Say S_M is infinite. Because M is well-partially ordered, S_M contains in infinite, ascending chain

$$a_1 <_1 a_2 <_1 a_3 <_1 a_4 <_1 \cdots .$$

Let b_j in N be such that $(a_j, b_j) \in S$.

If the sequence of b_j takes infinitely many values, then by Proposition 9.3 there is an increasing list of indices $j_1 < j_2 < j_3 < \cdots$ such that

$$b_{j_1} <_2 b_{j_2} <_2 b_{j_3} <_2 \cdots .$$

If the sequence of b_j takes on only finitely many values, there must be an increasing list of indices $j_1 < j_2 < j_3 < \cdots$ such that

$$b_{j_1} = b_{j_2} = b_{j_3} = \cdots .$$

In either case there is an increasing list of indices $j_1 < j_2 < j_3 < \cdots$ such that

$$b_{j_1} \leq_2 b_{j_2} \leq_2 b_{j_3} \leq_2 \cdots .$$

Since the indices j_i increase we also get

$$a_{j_1} <_1 a_{j_2} <_1 a_{j_3} <_1 \cdots .$$

These chains in N and M, respectively, give the increasing chain

$$(a_{j_1}, b_{j_1}) < (a_{j_2}, b_{j_2}) < (a_{j_3}, b_{j_3}) < \cdots$$

inside S. □

A Well-Partial Order on Monomials in Several Variables

From Proposition 9.4 it follows by induction that if $\mathbb{N}$ is well-ordered in the usual way, and if the n-fold Cartesian product $\mathbb{N}^n = \mathbb{N} \times \mathbb{N} \times \cdots \times \mathbb{N}$ is ordered by the rule:

$$(a_1, a_2, \ldots, a_n) \leq (b_1, b_2, \ldots, b_n) \text{ if and only if } a_i \leq b_i \text{ for all } i,$$

then this Cartesian product is well-partially ordered. Let us call this the ***product order*** on $\mathbb{N}^n$.

For any field K, take the polynomial ring $R = K[X_1, X_2, \ldots, X_n]$ in n variables. The polynomials

$$X_1^{d_1} X_2^{d_2} \cdots X_n^{d_n} \text{ where } d_j \in \mathbb{N},$$

which are pure products of the variables, are called ***monomials***. The correspondence

$$(d_1, d_2, \ldots, d_n) \mapsto X_1^{d_1} X_2^{d_2} \cdots X_n^{d_n}$$

from $\mathbb{N}^n$ to the set M of monomials is a bijection. Clearly,

$$X_1^{d_1} X_2^{d_2} \cdots X_n^{d_n} \text{ divides } X_1^{e_1} X_2^{e_2} \cdots X_n^{e_n} \text{ in the ring } R,$$

if and only if,

$$(d_1, d_2, \ldots, d_n) \leq (e_1, e_2, \ldots, e_n) \text{ in the product ordering on } \mathbb{N}^n.$$

By the above correspondence and Proposition 9.4, the following result emerges.

Proposition 9.5. *The relation $\preceq$ on the set M of monomials in the variables $X_1, \ldots, X_n$ prescribed by*

$$X_1^{d_1} \cdots X_n^{d_n} \preceq X_1^{e_1} \cdots X_n^{e_n} \textit{ whenever } X_1^{d_1} X_2^{d_2} \cdots X_n^{d_n} \textit{ divides } X_1^{e_1} X_2^{e_2} \cdots X_n^{e_n},$$

is a well-partial order.

We shall say that the monomial set M is ***ordered by divisibility***.

9.1.2 The Dickson Basis

Every non-empty subset of a (totally) well-ordered set has a least element. An analogous result holds for well-partially ordered sets.

Let S be a non-empty subset of a well-partially ordered set M. An element a of S is called ***minimal*** in S provided there is no b in S such that $b < a$. If M is totally and well-ordered, the first element of S is its only minimal element.

Proposition 9.6. *If S is a non-empty subset of a well-partially ordered set and D is the set of minimal elements of S then D is finite, non-empty, and for every element a of S, there exists b in D such that $b \leq a$.*

Proof. No two elements of D are comparable. Thus D cannot contain an infinite ascending chain $b_1 < b_2 < b_3 < \cdots$. Since M is well-partially ordered, D must be finite.

Suppose $a \in S$, but a does not sit above any of the minimal elements of D. Consequently a is not minimal, which means there is some a_1 in S such that $a > a_1$. By the assumption on a there is a_2 in S such that $a > a_1 > a_2$. Repeat this argument to build an infinite descending chain $a > a_1 > a_2 > \cdots > a_n > \cdots$, which cannot occur because well-partially ordered sets are Artinian.

Also, D is non-empty, since each element of S sits above an element of D. □

Definition 9.7. If S is a non-empty subset of a well-partially ordered set M, the finite set of minimal elements of S is called the ***Dickson basis*** of S.

This is named after the American algebraist L. E. Dickson (1874–1954).

In case M is the set of monomials in the variables $X_1, \dots, X_n$, partially ordered by division, Proposition 9.6 says that if S is a non-empty subset of M, then its set D of minimal elements is finite, non-empty and such that for every monomial f in S there is a monomial g in D that divides f.

For example, the Dickson basis of the set S of non-constant monomials (i.e. those other than 1) is the above set of variables. If the variables are called X, Y, and S is the set of monomials $X^i Y^j$ where $i + j \geq 3$, the Dickson basis of S consists of X^3, X^2Y, XY^2, Y^3. On the set $\mathbb{N}$ impose the well-partial order that $a \preceq b$ if and only if $a = b$ or $2a \leq b$. If $S = \{3, 5, 7, 9, \dots\}$, its Dickson basis is the set $\{3, 5\}$.

9.1.3 Extensions of Well-Partial Orders

If $\preceq$ is a partial order on a set M, an ***extension*** of $\preceq$ is another partial order $\leq$ on M such that $a \preceq b$ implies $a \leq b$. An extension causes more pairs of elements from M to be comparable. It does not enlarge the set M.

For instance, take $\preceq$ to be the partial order by divisibility on the set of positive integers $\mathbb{P}$. Let $\leq$ be the usual partial order on $\mathbb{P}$. If $a \preceq b$, i.e. $a \mid b$, then $a \leq b$. So $\leq$ extends $\preceq$.

Every extension $\leq$ of a well-partial order $\preceq$ on M remains a well-partial order. That is because every infinite set S in M contains an ascending chain using $\preceq$, which remains ascending using $\leq$. If the extension $\leq$ happens to be a total order on M, Proposition 9.1 reveals that M becomes well-ordered.

Monomial Orders

Definition 9.8. A partial order $\leq$ on the set M of monomials in several variables is called a ***monomial order*** provided:

- $\leq$ is a total order,
- $1 \leq f$ for every f in M, and
- whenever $f, g, h \in M$ and $f \leq g$, then also $fh \leq gh$.

We shall present examples shortly. In the meantime, let us note that every monomial order extends the order on M given by divisibility.

Proposition 9.9. *Every monomial order $\leq$ on the set M of monomials in several variables, extends the order $\preceq$ given by divisibility. Thus every monomial order on M is a well-order.*

Proof. Suppose $f,g \in M$ and that $f \preceq g$ (i.e. $f \mid g$). Thus $g = fh$ for some monomial h. The monomial order satisfies $1 \leq h$, and then

$$f = f1 \leq fh = g.$$

Thus a monomial order extends the divisibility order. Since, by Proposition 9.5 the divisibility order is a well-partial order on M, its extension to a monomial order must be a well-order. □

The Lexicographic Order on Monomials

Our first example of a monomial order is known as the ***lexicographic order***.

Two monomials in n variables $f = X_1^{d_1}X_2^{d_2}\cdots X_n^{d_n}$ and $g = X_1^{e_1}X_2^{e_2}\cdots X_n^{e_n}$ differ if and only if there is a first variable X_j whose exponent e_j exceeds d_j, or vice versa.[1] With that in mind, declare that $f \leq g$ provided $f = g$ or there is a j from 1 to n such that

$$d_1 = e_1, \ldots, d_{j-1} = e_{j-1} \text{ but } d_j < e_j.$$

Of course, if j happens to be 1, the equalities above do not occur. The reflexivity and anti-symmetry of this relation are quite evident. Transitivity can be checked by routine bookkeeping with the definition. It is also clear that the order is total.

For instance, $X_1^2X_2^2X_3^4X_4^6 < X_1^2X_2^2X_3^5X_4^1$. Also,

$$X_1^0X_2^0X_3^1 < X_1^0X_2^1X_3^0 < X_1^1X_2^0X_3^0, \text{ i.e. } X_3 < X_2 < X_1.$$

Here is part of the lexicographic order using the variables X as X_1 and Y as X_2, i.e. with $Y < X$.

$$\begin{aligned}
&1 < Y < Y^2 < Y^3 < Y^4 < \cdots \\
&< X < XY < XY^2 < XY^3 < XY^4 < \cdots \\
&< X^2 < X^2Y < X^2Y^2 < X^2Y^3 < X^2Y^4 < \cdots \\
&< X^3 < X^3Y < X^3Y^2 < X^3Y^3 < X^3Y^4 < \cdots \\
&< X^4 < X^4Y < X^4Y^2 < X^4Y^3 < X^4Y^4 < \cdots.
\end{aligned}$$

Given that a monomial X^iY^j corresponds to the pair (i,j) in $\mathbb{N}^2$, one sees that the lexicographic order on the set of monomials in two variables is nothing but a copy of the lexicographic order on $\mathbb{N} \times \mathbb{N}$ discussed in Section 9.1.1.

The lexicographic order is a monomial order. Indeed, it is clear that $1 \leq f$ for every monomial f. Next suppose that

$$f = X_1^{d_1}\cdots X_n^{d_n}, \quad g = X_1^{e_1}\cdots X_n^{e_n}, \quad h = X_1^{\ell_1}\cdots X_n^{\ell_n} \text{ are in } M, \text{ and that } f < g.$$

[1] Two dictionary words differ whenever there is a first letter at which they differ. The word with the earlier first letter in the alphabet comes first in the dictionary. A lexicographer is one who compiles dictionaries.

This means that there is a first index j such that $d_j < e_j$, which is also the first index such that $d_j+\ell_j < e_j+\ell_j$. Since $fh = X_1^{d_1+\ell_1} \cdots X_n^{d_n+\ell_n}$ and $gh = X_1^{e_1+\ell_1} \cdots X_n^{e_n+\ell_n}$, we see that $fh < gh$.

Having seen that the lexicographic order $\leq$ is a monomial order, Proposition 9.9 reveals that $\leq$ extends the well-partial order of M given by divisibility. Thus the lexicographic order is a (total) well-order on M, which of course we can also see directly.

The Degree Lexicographic Order

Here is a variant of the lexicographic order. Continue with the set M of monomials in the variables $X_1, \cdots, X_n$. If $f = X_1^{d_1} X_2^{d_2} \cdots X_n^{d_n}$ the ***degree*** of f is the sum $d_1 + d_2 + \cdots + d_n$ of the exponents in each variable. Write this degree as $\deg f$.

For now let $\leq_\ell$ denote the lexicographic order on M. If $f, g \in M$, declare that $f \leq g$ when

$$\deg f < \deg g, \text{ or } \deg f = \deg g \text{ and } f \leq_\ell g.$$

This is known as the ***degree lexicographic order*** on M. First, order monomials according to their degree, and within equal degree, order them lexicographically. It is routine to verify that this is a monomial order on M.

For instance, X_1, X_2, X_3 have degree 1, and thus under the degree lexicographic order we have $X_1 > X_2 > X_3$ since that is how they are ordered under $\leq_\ell$. With $X = X_1, Y = X_2$ here is a part of the degree lexicographic order:

$$\begin{aligned} 1 &< Y < X < Y^2 < XY < X^2 \\ &< Y^3 < XY^2 < X^2Y < X^3 \\ &< Y^4 < XY^3 < X^2Y^2 < X^3Y < X^4 \\ &< Y^5 < XY^4 < X^2Y^3 < X^3Y^2 < X^4Y < Y^5 < \cdots . \end{aligned}$$

The degree lexicographic order mimics the usual order of the natural numbers, while the lexicographic order looks like an infinitude of copies of $\mathbb{N}$ stacked above one another.

Infinitely many examples of monomial orders abound. They are left to the exercises. We shall content ourselves in working only with the lexicographic and the degree lexicographic order.

EXERCISES

1. Let $\mathbb{N} \times \mathbb{N}$ be well-partially ordered according to the usual product order. Find an infinite subset with precisely 3 elements in its Dickson basis. Find an infinite subset with 4 elements in its Dickson basis.
2. Using the lexicographic ordering on the monomials in X, Y, Z with $X > Y > Z$, place the monomials

$$X^2Z^2, X^2YZ^2, X^2, X^3Z^2, X^3Y, XY^2Z, X^2Y^2Z, X^2Z$$

in increasing order. Place them in increasing order using the degree lexicographic ordering with $X > Y > Z$. Repeat using $Z > Y > X$.

3. Let M be the set of monomials in X, Y, Z well-partially ordered by division. Find the Dickson basis of the set of monomials $X^i Y^j Z^k$ for which $i + j + k \geq 2$.
4. Let M be the set of monomials in X_1, X_2. If $f, g \in M$, declare that $f \leq g$ if either $f = g$ or $\deg f \leq \deg g$. Show that $\leq$ extends the order on M by divisibility, but that $\leq$ is not a monomial order.
5. Let $\leq$ be a monomial order on the set M of monomials in n variables. If $f, g, p, q \in M$ and $f \leq g, p \leq q$, show that $fp \leq gq$. If $f, g \in M$ and $f \leq g$, show that $f^m \leq g^m$ for every positive integer m.
6. Verify that the lexicographic and the degree lexicographic orders are monomial orders on the set of monomials in the variables $X_1, X_2, \ldots, X_n$.
7. Suppose that $\leq$ is a monomial order on the set M of monomials in several variables. Define a new order $\preceq$ on M by the rule

$$f \preceq g \text{ provided } \deg f < \deg g \text{ or } \deg f = \deg g \text{ and } f \leq g.$$

Verify that $\preceq$ is a monomial order on M.

8. Show that in one variable X, there is only one possible monomial order. Namely, $1 < X < X^2 < X^3 < X^4 < \cdots$.
9. Let $\leq$ be a monomial order on the set M of monomials in $X_1, X_2, \ldots, X_n$. If Y is a new variable, every monomial in the variables $X_1, X_2, \ldots, X_n, Y$ can be uniquely written as $Y^e h$ where the exponent $e \geq 0$ and $h \in M$. If $Y^e h, Y^d k$ are two such monomials, declare that

$$Y^e h \preceq Y^d k \text{ provided } Y^e h = Y^d k, \text{ or } e < d, \text{ or } e = d \text{ together with } h \leq k.$$

(a) Show that $\preceq$ is a monomial order on this larger set of variables.
(b) Show that the restriction of $\preceq$ to M equals the original $\leq$ on M.
(c) Show that every monomial h in M is such that $h \prec Y$.

10. This exercise shows that a polynomial ring with more than one variable can take on infinitely many monomial orders.

(a) Let M be the set of monomials in the variables X, Y. Let α be a negative, irrational number. If $(a, b), (c, d) \in \mathbb{N} \times \mathbb{N}$, declare

$$X^a Y^b \preceq_\alpha X^c Y^d \text{ provided } (a, b) = (c, d) \text{ or } (b - d) < \alpha(a - c).$$

Observe that the condition $(b - d) < \alpha(a - c)$ is saying, in geometric terms, that the line with irrational slope α and passing through (a, b) is below the parallel line with irrational slope α and passing through (c, d). Note that since α is irrational, the line through (a, b) never coincides with the line through (c, d) unless $(a, b) = (c, d)$.

Show that $\preceq_\alpha$ is a monomial order on M, as in Definition 9.8.

(b) If α, β are negative, irrational numbers such that $\alpha < \beta$, show there exist monomials $X^a Y^b$ and $X^c Y^d$ such that

$$X^a Y^b \prec_\beta X^c Y^d \text{ while } X^c Y^d \prec_\alpha X^a Y^b.$$

Deduce that there are infinitely many monomial orders on M.

(c) If N is the set of monomials in the variables $X_1, X_2, X_3, \ldots, X_n$, where $n \geq 3$, show that there are infinitely many monomial orders on N.

Hint. There are infinitely many monomial orders on the monomials in X_1, X_2. Extend each of these orders to a monomial order on N, following the preceding exercise.

11. Suppose that M is a set with a partial order $\leq$. Let a, b in M be such that they are not comparable. That is, neither $a \leq b$ nor $b \leq a$ holds. Define a new relation $\preceq$ on M by the rule

$$u \preceq v \text{ provided } u = v, \text{ or } u < v, \text{ or } u \leq a \text{ and } b \leq v.$$

Show that the relation $\preceq$ is anti-symmetric and transitive, and thereby deduce that $\preceq$ is a partial order on M which extends the given order $\leq$.

12.* Suppose that $\leq$ is a partial order on a set M and that $a_1, a_2, a_3, \ldots, a_n, \ldots$ is an infinite list of pairwise incomparable elements of M. Prove that the order $\leq$ has an extension $\preceq$ satisfying

$$a_1 \succ a_2 \succ a_3 \succ \cdots \succ a_n \succ a_{n+1} \succ \cdots.$$

13.* Prove that a partial order $\leq$ on a set M is a well-partial order if and only if every extension of $\leq$ is Artinian.

Note. Here the relation $\leq$ is extended, not the set M.

Hint. Use the preceding exercise.

14. Define a total order $\prec$ on the set M of monomials in X, Y as follows:

$$X^a Y^b \prec X^c Y^d$$

if and only if

$$\max\{a, b\} < \max\{c, d\}$$

or $\max\{a, b\} = \max\{c, d\}$ and $X^a Y^d$ are ordered lexicographically taking $X < Y$. With such a total ordering we have

$$1 \prec X \prec Y \prec XY \prec X^2 \prec X^2 Y \prec Y^2 \prec XY^2 \prec X^2 Y^2 \prec X^3 \prec \cdots.$$

(a) Prove that the total order $\prec$ extends the well-partial order by divisibility on M.

(b) Show that $\prec$ is *not* a monomial order.

15. Let M be the set of monomials in $X_1, X_2, \ldots, X_n$. If f, g are monomials and f does not divide g, show that there exists a monomial ordering $\preceq$ on M such that $g \prec f$. If $n > 1$, show there are infinitely many such orderings.

16.* This exercise is about the notion of an Archimedean monomial order.

Suppose M is the set of monomials in the variables $X_1, X_2, \ldots, X_k$, and that there is a monomial order $\leq$ on M. Define the relation $\approx$ on the set of variables by:

$$X_i \approx X_j$$

if and only if there is a positive integer n such that

$$X_i < X_j^n \text{ and } X_j < X_i^n.$$

(a) Verify that $\approx$ gives an equivalence relation on the set of variables.

(b) How many equivalence classes does a lexicographic order have?

(c) How many equivalence classes does a degree lexicographic order have?

We say that the monomial order is Archimedean if there is only one equivalence class. That is, if $X_i \approx X_j$ for all variables X_i, X_j. From here on suppose the monomial order is Archimedean.

(d) If f, g are monomials in M, other than 1, prove that there is a positive integer n such that $f < g^n$ and $g < f^n$.

(e) Let Y be the product of the variables $X_1 X_2 \cdots X_k$. For any monomial g and any positive integer t, let $n(t)$ be the unique positive integer such that

$$Y^{n(t)-1} < g^t \leq Y^{n(t)}.$$

Prove that the sequence $n(t)/t$ is a Cauchy sequence, and thereby has a limit as $n \to \infty$.

Hint. For any positive integers s, t, derive the inequalities

$$(n(t) - 1)s \leq n(s)t \text{ and } (n(s) - 1)t \leq n(t)s.$$

If $t \leq s$, deduce that

$$\left| \frac{n(t)}{t} - \frac{n(s)}{s} \right| \leq \frac{1}{t}.$$

(f) For every monomial g let $n(t)$ be defined as in part (e), and put

$$D(g) = \lim_{t \to \infty} \frac{n(t)}{t}.$$

If f, g are monomials, prove that $D(fg) = D(f) + D(g)$.

If $f = X_1^{e_1} X_2^{e_2} \cdots X_k^{e_k}$ in M, deduce that $D(f) = \sum_{j=1}^{k} e_j D(X_j)$.

(g) If $g \neq 1$, prove that $D(g)$ is positive.

(h) If f, g are monomials such that $D(f) < D(g)$, prove that $f < g$.

The upshot is that Archimedean monomial orders come with a degree function D which behaves much like the normal degree function on monomials, and carries out a preliminary ordering on the monomials much like the degree lexicographic ordering does.

9.2 Gröbner Domains

Throughout this section, for any integral domain R, the notation $R^\star$ will stand for the set of non-zero elements of R.

Well-partial orders are a tool which enable the expansion of the family of integral domains on which Euclidean-like functions are possible. The notion of a Gröbner basis for ideals in the ring of polynomials in several variables is a well accepted computational tool in algebraic geometry. We offer an abstract point of view based on our preceding discussion of well-partial orders, which enlarges the concept of Gröbner basis to a wider class of rings.

9.2.1 Gröbner Functions

We need to become acclimatized to the following terminology.

Definition 9.10. Start with a well-partial order $\preceq$ on a set M and an extension of $\preceq$ to a total order $\leq$ on M. If R is an integral domain, a function $\phi : R^\star \to M$ will be called a ***Gröbner function with respect to the orders*** $\preceq$ and $\leq$, provided that for every non-zero a, b in R satisfying the restriction

$$\phi(a) \preceq \phi(b),$$

there exist q, r in R such that

$$b = aq + r, \text{ where } \phi(r) < \phi(b) \text{ or } r = 0.$$

An integral domain having a Gröbner function will be called a ***Gröbner domain***.

When $\phi(a) \preceq \phi(b)$, we say that b is ***reducible modulo*** a, or that a ***reduces*** b. The element r will be called a ***reduction*** (alternatively a ***remainder***) of b ***modulo*** a, with ***quotient*** q.

The definition of Gröbner functions is reminiscent of the definition of Euclidean functions given in Section 5.3. Yet we should observe the interplay between the well-partial order and its extension to a total order. Note that the condition $\phi(r) < \phi(b)$ is correct, and should not be $\phi(r) < \phi(a)$.

Gröbner Functions when the Well-Ordered Set $\mathbb{N}$ Is Used

If $\mathbb{N}$, with its usual order, is used as the well-partially ordered set in the above definition, then Gröbner functions coincide with Euclidean functions.

Proposition 9.11. *Suppose that both the well-partial order and its total order extension on the set $\mathbb{N}$ of natural numbers are taken to be the usual order $\leq$. In that case, a function $\phi : R^\star \to \mathbb{N}$ is a Gröbner function if and only if ϕ is Euclidean.*

Proof. Recall that ϕ is Euclidean provided that for every a, b in R with $a \neq 0$, there exist q, r in R such that $b = aq + r$, where $\phi(r) < \phi(a)$ or $r = 0$.

Supposing ϕ is Euclidean, take non-zero elements a,b in R such that $\phi(a) \leq \phi(b)$. Then there exist, q,r in R such that $b = aq + r$, where $\phi(r) < \phi(a)$ or $r = 0$. But since $\phi(a) \leq \phi(b)$, we also get $\phi(r) < \phi(b)$, as desired for ϕ to be a Gröbner function.

Conversely, suppose ϕ is Gröbner. To see that ϕ is Euclidean, take a,b in R with $a \neq 0$. If $b = 0$ or if $\phi(b) < \phi(a)$, we have $b = a0 + b$. So, with $q = 0$ and $r = b$ the desired inequality $\phi(r) < \phi(a)$ prevails.

There remains the case where $\phi(a) \leq \phi(b)$.

Since ϕ is Gröbner, there exist q_1, r_1 in R such that

$$b = aq_1 + r_1, \text{ where } \phi(r_1) < \phi(b) \text{ or } r_1 = 0.$$

But for ϕ to be Euclidean the inequality $\phi(r_1) < \phi(a)$ is needed. Well, if $\phi(r_1) < \phi(a)$, we have our desired quotient and remainder q_1, r_1.

On the other hand, if $\phi(a) \leq \phi(r_1)$, reapply the fact ϕ is Gröbner to get q_2, r_2 in R such that

$$r_1 = aq_2 + r_2, \text{ where } \phi(r_2) < \phi(r_1) \text{ or } r_2 = 0.$$

Then

$$b = aq_1 + r_1 = a(q_1 + q_2) + r_2, \text{ where } \phi(r_2) < \phi(r_1) \text{ or } r_2 = 0.$$

Now, if $\phi(r_2) < \phi(a)$, the elements $q_1 + q_2$ and r_2 are suitable quotients and remainders for ϕ to be Euclidean.

On the other hand if $\phi(a) \leq \phi(r_2)$, reapply the fact ϕ is Gröbner to get q_3, r_3 in R such that

$$r_2 = aq_3 + r_3, \text{ where } \phi(r_3) < \phi(r_2) \text{ or } r_3 = 0.$$

Then

$$b = a(q_1 + q_2) + r_2 = a(q_1 + q_2 + q_3) + r_3 \text{ where } \phi(r_3) < \phi(r_2) < \phi(r_1) \text{ or } r_3 = 0.$$

Once more, if $\phi(r_3) < \phi(a)$, we have our desired quotient $q_1 + q_2 + q_3$ and remainder r_3 to ensure ϕ is Euclidean.

But if $\phi(a) \leq \phi(r_3)$, repeat the above argument to pick up q_4, r_4 in R such that

$$b = a(q_1 + q_2 + q_3 + q_4) + r_4,$$

where $\phi(r_4) < \phi(r_3) < \phi(r_2) < \phi(r_1)$ or $r_4 = 0$.

Continue with this argument. At some point a suitable quotient $q_1 + q_2 + \cdots + q_n$ and suitable remainder r_n, where $\phi(r_n) < \phi(a)$ or $r_n = 0$, must be attained. For otherwise we would get an infinite descending chain

$$\phi(r_1) > \phi(r_2) > \phi(r_3) > \phi(r_4) > \cdots,$$

which cannot happen in $\mathbb{N}$. □

As just proven, every domain with a Gröbner function into $\mathbb{N}$ is a Euclidean domain. For instance, the usual Gröbner function for $\mathbb{Z}$ is the absolute value:

$$|\cdot| : \mathbb{Z}^\star \to \mathbb{N} \text{ where } n \mapsto |n|.$$

To check this let a, b be integers such that b is reducible modulo a. This means that $0 < |a| \leq |b|$. With $q = 1$ or $q = -1$ chosen appropriately, we get $|b - aq| < |b|$. Put $r = b - aq$ to obtain

$$b = aq + r, \text{ where } |r| < |b| \text{ or } r = 0.$$

Then by Proposition 9.11 the absolute value function is Euclidean as well, which of course we knew already.

For the ring $K[X]$ of polynomials in one variable X over a field K, an exercise reveals that the degree function $\deg : K[X]^\star \to \mathbb{N}$ is the common Gröbner function to take.

Non-uniqueness of Reductions

A reduction of b modulo a need not be uniquely determined by a, b. For example, take the Gaussian integers $\mathbb{Z}[i]$. As verified in Section 5.3 the function $\phi : \mathbb{Z}[i] \to \mathbb{N}$, where $\phi : a + ib \mapsto a^2 + b^2$, is Euclidean. By Proposition 9.11, ϕ is a Gröbner function.

Now take $a = 1 + i$ and $b = 2 + i$. Since $\phi(1 + i) = 2 < 5 = \phi(2 + i)$, the element $2 + i$ is reducible modulo $1 + i$. Explicitly, with $q = 1, r = 1$, we get

$$2 + i = (1 + i)q + r \text{ where } \phi(r) = 1 < 2 = \phi(1 + i).$$

Alternatively with $q_1 = 2$ and $r_1 = -i$, we also get

$$2 + i = (1 + i)q_1 + r_1 \text{ where } \phi(r_1) = 1 < 2 = \phi(1 + i).$$

Even more obviously take $\mathbb{Z}$ with the absolute value as the Gröbner function. Since $|2| \leq |7|$, the integer 7 is reducible modulo 2. Then we have $7 = 2 \cdot 1 + 5$ with $|5| < |7|$, as well as $7 = 2 \cdot 2 + 3$ with $|3| < |7|$, and also $7 = 2 \cdot 3 + 1$ with $|1| < |7|$.

To sidestep this ambiguity, we assume for each Gröbner domain that there is an established algorithm that determines for each pair a, b, a quotient q and the resulting reduction $r = b - aq$. Thus for all non-zero a, b we can safely refer to r as ***the reduction*** of b modulo a.

9.2.2 The Gröbner Basis of an Ideal

With a Gröbner function in hand the notion of Gröbner basis can be formulated.

Let $\preceq$ be a well-partial order on a set M. For any non-empty subset S of M, recall that its Dickson basis is the finite set of minimal elements of S. Every element of S sits above (not necessarily strictly) some element of its Dickson basis.

If E is a subset of S such that every element of S sits above some element of E, then E contains the Dickson basis D. For if this were not so, then some a in D would be outside of E

but (by assumption) sit above some element b of E. That would force $b < a$, contrary to the minimality of a. Thus, a subset E of S contains the Dickson basis of S if and only if every element of S sits above some element of E.

Definition 9.12. Let R be a Gröbner domain with Gröbner function $\phi : R^\star \to M$ with respect to a total order extension $\leq$ of a well-partial order $\preceq$ on M. If J is a non-zero ideal of R, the subset $\phi(J^\star)$ of M has a Dickson basis (i.e. a minimal set) D, using the well-partial order $\preceq$. A *finite* subset B of $J^\star$ is called a ***Gröbner basis*** of J, using the function ϕ, whenever $\phi(B)$ contains the Dickson basis D.

If, in addition, B is chosen so that ϕ restricted to B gives a bijection onto D, the basis B is called a ***minimal Gröbner basis*** of J.

As seen from our preceding remarks, a finite subset B of $J^\star$ is a Gröbner basis of J if and only if for every element b of $J^\star$ there is an a in B such that $\phi(a) \preceq \phi(b)$. In other words, every element of $J^\star$ is reducible modulo some element of the Gröbner basis B.

Clearly, every non-zero ideal of R has a Gröbner basis. In practice it takes serious effort to find it.

A minimal Gröbner basis of J is obtained by selecting, for each element m in the finite Dickson basis D of $\phi(J^\star)$, an element b in $J^\star$ such that $\phi(b) = m$.

Gröbner Bases Generate Their Ideal

The term "basis" would be pointless without the following result.

Proposition 9.13. *Let $\preceq$ be a well-partial order on a set M with a total order extension $\leq$, and let $\phi : R^\star \to M$ be a Gröbner function on a domain R with respect to these orders. If J is a non-zero ideal of R with Gröbner basis B using the function ϕ, then B is a finite set of generators of the ideal.*

Proof. Suppose to the contrary that the ideal $\langle B\rangle$ generated by B is not all of J. If $b \in J \setminus \langle B\rangle$, then there is an a in the Gröbner basis B such that $\phi(a) \preceq \phi(b)$. This comes from the definition of the Gröbner basis. Since ϕ is a Gröbner function, there exist q, r in R such that

$$b = aq + r, \text{ where } \phi(r) < \phi(b) \text{ or } r = 0.$$

Now $r \in J$ because $b, a \in J$, and $r \notin \langle B\rangle$ because $a \in \langle B\rangle$ while $b \notin \langle B\rangle$. Hence $r \in J \setminus \langle B\rangle$ and $\phi(r) < \phi(b)$. This proves that the non-empty set $\phi(J \setminus \langle B\rangle)$ has no minimal elements using $\leq$, contrary to the fact every non-empty subset of M has minimal elements (actually a unique least element) using $\leq$. □

Since every non-zero ideal in a Gröbner domain has a (finite) Gröbner basis and such a basis generates the ideal, the next result is immediate.

Proposition 9.14. *Every Gröbner domain is Noetherian.*

The computational utility of Gröbner bases will be seen in due course.

9.2.3 Polynomials in Several Variables Are Gröbner Domains

There exist numerous examples of Gröbner domains, some of which we leave to the exercises.

Yet the most significant example is $K[X_1, X_2, \ldots, X_m]$, the ring of polynomials in the variables X_j and having coefficients in a field K. This will be the focus of our attention. Here is how it works.

The set M of monomials in the X_j has the well-partial order given by divisibility. For f, g in M, the inequality $f \preceq g$ means $f \mid g$. More explicitly, for $f = X_1^{d_1} X_2^{d_2} \cdots X_m^{d_m}$ and $g = X_1^{e_1} X_2^{e_2} \cdots X_m^{e_m}$, we have $f \preceq g$ if and only if $d_j \leq e_j$ for all j. Any monomial order $\leq$ on M is a total order extension of $\preceq$. For instance, $\leq$ could be the lexicographic order or possibly the degree lexicographic order already discussed.

Leading Monomials, Terms and Coefficients

For brevity let R be our ring of polynomials in several variables X_j over a field K.

To get a Gröbner function $\phi : R^\star \to M$ using the well-partial order $\preceq$ by divisibility and its extension to a monomial order $\leq$, a bit of notation and nomenclature will be helpful.

The set M of monomials is a basis for R as vector space over the field K. Therefore, for every non-zero polynomial f there is a unique set of monomials $f_1, f_2, \ldots, f_k$ in M, and there are unique non-zero coefficients $\lambda_1, \ldots, \lambda_k$ in K such that

$$f = \lambda_1 f_1 + \lambda_2 f_2 + \cdots + \lambda_k f_k.$$

These unique monomials f_j needed to represent f (with non-zero coefficients λ_j) are said to ***appear*** in f. They are also called the monomials ***of*** f. If f_j appears in f, the non-zero scalar multiple $\lambda_j f_j$ is called a ***term*** of f. The distinct f_j which appear in f will be written from highest to lowest

$$f_1 > f_2 > \cdots > f_k,$$

using the chosen monomial order on M. The highest monomial f_1 appearing in the representation of f will be known as the ***leading monomial*** of f, and the term $\lambda_1 f_1$ will be called the ***leading term*** of f. The leading monomial of f will be denoted by $\mathrm{lm}(f)$. The non-zero coefficient λ_1 is called the ***leading coefficient*** of f. The leading term of f will be denoted by $\mathrm{lt}(f)$, and the leading coefficient by $\mathrm{lc}(f)$. Obviously

$$\mathrm{lt}(f) = \mathrm{lc}(f)\,\mathrm{lm}(f).$$

For example, take the lexicographic order on the monomials of $\mathbb{Q}[X, Y]$ with the convention that $X > Y$ (i.e. $X = X_1, Y = X_2$). Here is a polynomial whose monomials descend lexicographically from highest to lowest.

$$7X^3Y + 2X^2Y^2 + 5XY^4 + 4XY^3 + 6XY + X + Y^4 + 9.$$

The leading monomial is X^3Y, and the leading term is $7X^3Y$.

If the degree lexicographic order is used instead, the same polynomial with monomials descending from highest to lowest looks like so:

$$5XY^4 + 7X^3Y + 2X^2Y^2 + 5XY^3 + Y^4 + 6XY + X + 9.$$

Now the leading monomial is XY^4, and the leading term is $5XY^4$.

The definition of $\operatorname{lm}(f)$ depends on the monomial order, even though our notation does not reflect this fact.

Two Properties of the Leading Monomial Function

By definition, a monomial order $\leq$ on the set M of monomials in the ring R of polynomials in several variables is such that for f, g, h in M the condition $f \leq g$ implies $fh \leq gh$. As seen in Proposition 9.9 this property ensures that monomial orders extend the order $\preceq$ on M given by divisibility. It also ensures that the leading monomial function is multiplicative, as we now check.

Proposition 9.15. *If $\leq$ is a monomial order on the set of monomials in the ring R of polynomials in several variables and f, g are non-zero polynomials in R, then*

$$\operatorname{lm}(fg) = \operatorname{lm}(f)\operatorname{lm}(g).$$

In addition, if $f + g \neq 0$, then

$$\operatorname{lm}(f + g) \leq \max(\operatorname{lm}(f), \operatorname{lm}(g)),$$

with equality when $\operatorname{lm}(f) \neq \operatorname{lm}(g)$.

Proof. The product fg is a linear combination of monomials mn, where m is a monomial appearing in f and n is a monomial appearing in g. Some monomials mn might no longer appear in fg, due to cancellations. But the monomial $\operatorname{lm}(f)\operatorname{lm}(g)$ does appear in fg. Indeed, the other monomial products mn arise by having

$$m \leq \operatorname{lm}(f), \quad n < \operatorname{lm}(g) \quad \text{or} \quad m < \operatorname{lm}(f), \quad n \leq \operatorname{lm}(g).$$

The defining property of monomial orders applied twice to each of these two cases yields:

$$mn \leq \operatorname{lm}(f)n < \operatorname{lm}(f)\operatorname{lm}(g) \quad \text{and} \quad mn < m\operatorname{lm}(g) \leq \operatorname{lm}(f)\operatorname{lm}(g).$$

This reveals that $\operatorname{lm}(f)\operatorname{lm}(g)$ does not cancel out and thereby persists as a monomial of fg. It also reveals that $\operatorname{lm}(f)\operatorname{lm}(g)$ is the leading monomial of fg, as desired.

Next suppose $f + g \neq 0$, and say $\operatorname{lm}(f) \leq \operatorname{lm}(g)$. If m is a monomial appearing in f or in g, then $m \leq \operatorname{lm}(g)$. The monomials m appearing in $f + g$ are among the monomials that appear in f or in g. Thus such $m \leq \operatorname{lm}(g)$. In particular, $\operatorname{lm}(f + g) \leq \operatorname{lm}(g) = \max(\operatorname{lm}(f), \operatorname{lm}(g))$.

If $\operatorname{lm}(f) \neq \operatorname{lm}(g)$, say $\operatorname{lm}(f) < \operatorname{lm}(g)$. Then $\operatorname{lm}(g)$ is greater than all monomials appearing in f, and greater than all the other monomials appearing in g. Thus $\operatorname{lm}(g)$ appears as a monomial of $f + g$ and exceeds all monomials of $f + g$ other than itself. This proves that $\operatorname{lm}(f + g) = \max(\operatorname{lm}(f), \operatorname{lm}(g))$. □

The Leading Monomial Function Is a Gröbner Function

Leading monomial functions turn polynomial rings into Gröbner domains. We remind ourselves that $\mathrm{lt}(f)$ stands for the leading term (leading monomial times leading coefficient) of a non-zero polynomial.

Proposition 9.16. *Let R be the ring of polynomials in several variables over a field, and let M be its set of monomials with the well-partial order given by divisibility and extended to a monomial order $\leq$. If f, g in $R^\star$ are such that $\mathrm{lm}(g)$ divides $\mathrm{lm}(f)$, then the polynomials q, r defined by*

$$q = \frac{\mathrm{lt}(f)}{\mathrm{lt}(g)}, \quad r = f - gq = f - \frac{\mathrm{lt}(f)}{\mathrm{lt}(g)} g$$

will give $r = 0$ or else $\mathrm{lm}(r) < \mathrm{lm}(f)$. Consequently the leading monomial function

$$\mathrm{lm} : R^\star \to M, \text{ where } f \mapsto \mathrm{lm}(f)$$

is a Gröbner function, and R is a Gröbner domain.

Proof. By assumption $\dfrac{\mathrm{lm}(f)}{\mathrm{lm}(g)}$ is a monomial. By Proposition 9.15,

$$\mathrm{lm}\left(\frac{\mathrm{lm}(f)}{\mathrm{lm}(g)} g\right) = \mathrm{lm}\left(\frac{\mathrm{lm}(f)}{\mathrm{lm}(g)}\right) \mathrm{lm}(g) = \frac{\mathrm{lm}(f)}{\mathrm{lm}(g)} \mathrm{lm}(g) = \mathrm{lm}(f).$$

The same identity carries over to leading terms, namely $\mathrm{lt}\left(\dfrac{\mathrm{lt}(f)}{\mathrm{lt}(g)} g\right) = \mathrm{lt}(f)$. Due to cancellation of leading terms the monomials, if any, appearing in $f - \dfrac{\mathrm{lt}(f)}{\mathrm{lt}(g)} g$ are lower than $\mathrm{lm}(f)$. In other words

$$f = gq + r, \text{ where } \mathrm{lm}(r) < \mathrm{lm}(f) \text{ or } r = 0,$$

as required of a Gröbner function. □

For an illustration, take the ring of polynomials $\mathbb{Q}[X, Y]$ in the variables X, Y, and place the degree lexicographic order $\leq$ on the set of monomials with $X > Y$. Let $g = Y^2$ and $f = XY^3 + X^2Y + XY^2$, with its monomials written from highest to lowest using $\leq$. Here $\mathrm{lm}(g) = Y^2$ and $\mathrm{lm}(f) = XY^3$. Since $\mathrm{lm}(g)$ divides $\mathrm{lm}(f)$, the polynomial f is reducible modulo g. Indeed,

$$f = XY^3 + X^2Y + XY^2 = Y^2XY + (X^2Y + XY^2) = gXY + (X^2Y + XY^2).$$

With $q = XY$ and $r = X^2Y + XY^2$ we have $\mathrm{lm}(r) = X^2Y < XY^3 = \mathrm{lm}(f)$.

Suppose instead that the lexicographic order (without considering the degree) is imposed on the monomials of $\mathbb{Q}[X, Y]$. In this case, f as above with its monomials written in descending order takes the appearance $X^2Y + XY^3 + XY^2$. Again $\mathrm{lm}(g) = Y^2$, but $\mathrm{lm}(f) = X^2Y$. Since $\mathrm{lm}(g)$ does not divide $\mathrm{lm}(f)$, the polynomial f is not reducible modulo g. The choice of monomial order makes a difference.

How to Think of Gröbner Bases in Polynomial Rings

With the Gröbner function lm on our polynomial ring R as discussed above, here is what Gröbner bases boil down to. A finite set B of polynomials in R is a Gröbner basis of an ideal J in which they sit if and only if for every f in $J^\star$ there is a g in B such that $\text{lm}(g)$ divides $\text{lm}(f)$. This is the well established definition of Gröbner bases in the case of polynomial rings. From this starting point it is less obvious that Gröbner bases even exist. Another issue that needs to be resolved is how to actually test that a Gröbner basis is present, given that ideals have infinitely many elements.

The selection of monomial order determines the definition of the leading monomial function, and in turn this selection will affect whether or not some finite set B is a Gröbner basis. To be more precise we should be speaking of Gröbner bases with respect to a monomial order. Despite this oversight there is little risk of confusion.

Gröbner Bases for Principal Ideals in Polynomial Rings

Suppose J is a principal non-zero ideal of the polynomial ring R, and let g generate J. The polynomials of J are the multiples fg, where $f \in R$. According to Proposition 9.15, $\text{lm}(fg) = \text{lm}(f)\,\text{lm}(g)$, regardless of which monomial order is used. Thus the leading monomial of g divides the leading monomial of every non-zero polynomial of J. This means that the original generator g constitutes a Gröbner basis of J.

In a subsequent section we shall address the important matter of finding Gröbner bases.

The Hilbert Basis Theorem

Since the ring of polynomials in several variables is a Gröbner domain, Proposition 9.14 makes it a Noetherian domain. Thus we have an alternative proof of an important corollary to the Hilbert basis theorem, Proposition 5.42.

Proposition 9.17. *The ring of polynomials in several variables over a field is Noetherian.*

9.2.4 A Division Algorithm for Complete Reductions

Return to R, an integral domain with Gröbner function $\phi : R^\star \to M$, where the set M carries a well-partial order $\preceq$ and a total order extension $\leq$.

If $a, b \in R^\star$ and b is reducible modulo a, then in accordance with the definition of Gröbner functions, there is a quotient q and a reduction (also known as a remainder) r in R such that

$$b = aq + r \text{ where } \phi(r) < \phi(b) \text{ or else } r = 0.$$

We can call this process a ***division*** of b by the ***divisor*** a.

Now if $r \neq 0$ and r remains reducible modulo a, we can repeat the division process to obtain a new quotient q_1 and reduction r_1 such that

$$r = aq_1 + r_1 \text{ where } \phi(r_1) < \phi(r) \text{ or else } r_1 = 0.$$

By substitution we get

$$b = aq + r = aq + (aq_1 + r_1) = a(q + q_1) + r_1 \text{ where } \phi(r_1) < \phi(r) < \phi(b) \text{ or } r_1 = 0.$$

If a reduces r_1 the division can be reapplied to obtain a further reduction r_2 modulo a. This can continue until we come to a reduction that is either 0 or no longer reducible modulo a.

An Illustration with Polynomials

Take the ring $\mathbb{Q}[X, Y]$ of polynomials in two variables. On the set M of monomials take the lexicographic order using $X > Y$. This order extends the well-partial order on M given by divisibility. The Gröbner function is the leading monomial function lm. The polynomial

$$f = X^3Y + 3X^2Y^2 + XY^3 \text{ is reducible modulo } g = X^2Y + XY^4,$$

because $\mathrm{lm}(g) = X^2Y$, which divides $\mathrm{lm}(f) = X^3Y$.

As specified in Proposition 9.16, the reduction of f modulo g comes by taking the quotient $q = \dfrac{\mathrm{lt}(f)}{\mathrm{lt}(g)} = X$. Then

$$r = f - gq = f - Xg = -X^2Y^4 + 3X^2Y^2 + XY^3.$$

Using the lexicographic order on M we confirm that

$$\phi(r) = X^2Y^4 < X^3Y = \phi(f),$$

as expected of the reduction of f modulo g.

Observe that r remains reducible modulo g, because $\mathrm{lm}(g) = X^2Y$ divides $\mathrm{lm}(r) = X^2Y^4$. According to Proposition 9.16 with quotient $q_1 = -Y^3$, the reduction of r modulo g is

$$r_1 = r - gq_1 = r + gY^3 = 3X^2Y^2 + XY^7 + XY^3.$$

In the lexicographic order on M we have

$$\mathrm{lm}(r_1) = X^2Y^2 < X^2Y^4 = \mathrm{lm}(r).$$

Since $\mathrm{lm}(r_1) = X^2Y^2$ and this is divisible by $X^2Y = \mathrm{lm}(g)$, the polynomial r_1 remains reducible modulo g. As in Proposition 9.16, take $q_2 = 3Y$, to get the reduction

$$r_2 = r_1 - gq_2 = r_1 - 3Yg = XY^7 - 3XY^5 + XY^3$$

of r_1 modulo g. Once more

$$\mathrm{lm}(r_2) = XY^7 < X^2Y^2 = \mathrm{lm}(r_1).$$

Since $\mathrm{lm}(r_2) = XY^7$, which is not divisible by $\mathrm{lm}(g) = X^2Y$, this final remainder r_2 is no longer reducible modulo g.

Putting the above bits together, we get

$$f = gq + r = gq + (gq_1 + r_1) = gq + gq_1 + (gq_2 + r_2) = g(q + q_1 + q_2) + r_2,$$

where r_2 is not reducible modulo g and

$$\operatorname{lm}(r_2) < \operatorname{lm}(r_1) < \operatorname{lm}(r) < \operatorname{lm}(f).$$

By substituting the appropriate polynomials we get

$$X^3Y + 3X^2Y^2 + XY^3 = (X^2Y + XY^4)(X - Y^3 + 3Y) + XY^7 - 3XY^5 + XY^3,$$

where the remainder $XY^7 - 3XY^5 + XY^3$ is not reducible by $X^2Y + XY^4$.

Long Division by Several Divisors

The upcoming result gives an extension of the preceding process in a way that incorporates several divisors at once.

Proposition 9.18 (Division algorithm). *Suppose that R is an integral domain, that M is a set with a well-partial order $\preceq$ and a total order extension $\leq$, and that there is a Gröbner function $\phi : R^\star \to M$.*

1. *For any list of elements $a_1, a_2, \ldots, a_n, b$ in $R^\star$, there exist $r, q_1, q_2, \ldots, q_n$ in R such that*

$$b = a_1q_1 + a_2q_2 + \cdots + a_nq_n + r,$$

where either $r = 0$ or

$$r \text{ is not reducible modulo any } a_j \text{ and } \phi(r) \leq \phi(b).$$

2. *Furthermore if ϕ is such that*

$$\phi(u + v) \leq \max(\phi(u), \phi(v)) \text{ and } \phi(-v) = \phi(v) \text{ when all of } u, v, u + v \in R^\star,$$

then $\phi(a_jq_j) \leq \phi(b)$ for all non-zero a_jq_j.

Proof. The proof is an algorithm for obtaining the requisite q_j and r.

If b is not reducible modulo any a_j, the conclusions follow by putting all $q_j = 0$ and $r = b$.

Otherwise, let d_1 be the first a_j that reduces b. We could take any other a_j that reduces b, but by choosing the first one the process becomes well determined. Using the Gröbner function take c_1, r_1 in R such that

$$b = d_1c_1 + r_1 \text{ where } \phi(r_1) < \phi(b) \text{ or else } r_1 = 0.$$

If $r_1 = 0$ or if no a_j reduces r_1, stop the process. Otherwise, let d_2 be the first a_j that reduces r_1. Then take c_2, r_2 in R such that

$$r_1 = d_2c_2 + r_2 \text{ where } \phi(r_2) < \phi(r_1) \text{ or else } r_2 = 0.$$

Substitute this expression for r_1 into the preceding expression for b to get

$$b = d_1c_1 + d_2c_2 + r_2 \text{ where } \phi(r_2) < \phi(r_1) < \phi(b) \text{ or else } r_2 = 0.$$

If $r_2 = 0$ or if no a_j reduces r_2, stop the process. Otherwise, let d_3 be the first a_j that reduces r_2. Then take c_3, r_3 in R such that

$$r_2 = d_3c_3 + r_3 \text{ where } \phi(r_3) < \phi(r_2) \text{ or else } r_3 = 0.$$

Substitute this expression for r_2 into the preceding expression for b to get

$$b = d_1c_1 + d_2c_2 + d_3c_3 + r_3 \text{ where } \phi(r_3) < \phi(r_2) < \phi(r_1) < \phi(b) \text{ or else } r_3 = 0.$$

Once again the process stops if $r_3 = 0$ or if no a_j reduces r_3. Otherwise, pick up a d_4 equal to one of the a_j and c_4, r_4 in R such that

$$b = d_1c_1 + d_2c_2 + d_3c_3 + d_4c_4 + r_4$$
$$\text{where } \phi(r_4) < \phi(r_3) < \phi(r_2) < \phi(r_1) < \phi(b) \text{ or else } r_4 = 0.$$

Unless the process stops, continue in this way to pick up $d_1, d_2, \ldots, d_\ell$ from among the a_j, along with $c_1, c_2, \ldots, c_\ell$ and $r_1, r_2, \ldots, r_\ell$ in R such that

$$b = d_1c_1 + d_2c_2 + \cdots + d_\ell c_\ell + r_\ell$$
$$\text{where } \phi(r_\ell) < \phi(r_{\ell-1}) < \cdots < \phi(r_2) < \phi(r_1) < \phi(b) \text{ or else } r_\ell = 0.$$

The process has to stop, for if not, as we can see directly above, there would result an infinite descending chain in the well-ordered set M. Say it stops at r_ℓ. This means that either $r_\ell = 0$ or no a_j reduces r_ℓ, and the above expression for b holds.

Now, $d_1, d_2, \ldots, d_\ell$ is just a listing of a_j with possible repetitions. By collecting all terms d_ic_i where $d_i = a_1$, and all d_ic_i where $d_i = a_2$ and so on, the expression for b takes the form

$$b = a_1 \sum_{d_i=a_1} c_i + a_2 \sum_{d_i=a_2} c_i + \cdots + a_n \sum_{d_i=a_n} c_i + r_\ell.$$

Putting

$$q_1 = \sum_{d_i=a_1} c_i,\ q_2 = \sum_{d_i=a_2} c_i, \ldots, q_n = \sum_{d_i=a_n} c_i, \text{ as well as } r = r_\ell,$$

we see that

$$b = a_1q_1 + a_2q_2 + \cdots + q_na_n + r$$

where either $r = 0$ or

$$r \text{ is not reducible modulo any } a_j \text{ and } \phi(r) \leq \phi(b).$$

This completes the proof of item 1.

Regarding item 2, suppose ϕ is such that $\phi(u + v) \leq \max(\phi(u), \phi(v))$ and $\phi(-v) = \phi(v)$ whenever $u, v, u + v \in R^\star$.

By inspecting our algorithm, we see that each $d_i c_i$ is a difference of two r_k. We can put $r_0 = b$ in the case of $d_1 c_1$. And since each $a_j q_j$ is given by

$$a_j q_j = a_j \sum_{d_i = a_j} c_i = \sum_{d_i = a_j} d_i c_i,$$

it follows that

$$a_j q_j = \sum \pm r_k,$$

where the above represents a finite sum of suitable r_k or their negatives. If $a_j q_j$ is not 0, we can take all $\pm r_k$, needed to add up to $a_j q_j$, to be non-zero. Say that the maximum ϕ value of these $\pm r_k$ is $\phi(\pm r_m)$ for some index m. The assumption on ϕ reveals that

$$\phi(a_j q_j) \leq \phi(\pm r_m) = \phi(r_m) \leq \phi(b).$$

The last inequality holds because $\phi(b)$ dominates the ϕ value of all non-zero r_k.

This completes the proof of item 2. □

Complete Reduction

Definition 9.19. Suppose R is an integral domain with Gröbner function $\phi : R^\star \to M$. If $b, a_1, \ldots, a_n \in R^\star$ and there exist $q_1, \ldots, q_n, r$ in R such that

$$b = a_1 q_1 + \cdots + a_n q_n + r$$

and either $r = 0$, or else

$$r \text{ is not reducible modulo any } a_j \text{ and } \phi(r) \leq \phi(b),$$

we say that r is a ***complete reduction of** b **modulo** $a_1, \ldots, a_n$ **with quotients*** $q_1, \ldots, q_n$.

Proposition 9.18 outlines an algorithm, based on the operations of R and the Gröbner function ϕ, which obtains a complete reduction r of every non-zero b modulo any non-zero $a_1, \ldots, a_n$.

Note that

- b is its own complete reduction if and only if b is not reducible modulo any a_j, and
- 0 is a complete reduction of b modulo $a_1, \ldots, a_n$ if and only if b lies in the ideal generated by the a_j.

Complete Reduction in Polynomial Rings

Suppose that $R = K[X_1, \ldots, X_m]$ is a polynomial ring in m variables with coefficients in a field K. The set M of monomials is well-partially ordered by divisibility. Extend this to a monomial order $\leq$, such as for example, the lexicographic order or the degree lexicographic order. As seen in Proposition 9.16 the leading monomial function $\mathrm{lm} : R \to M$ is a Gröbner function. To avoid repetition we make the understanding that in polynomial rings R such a Gröbner function lm is given.

We saw in Proposition 9.15 that $\text{lm}(fg) = \text{lm}(f)\,\text{lm}(g)$ for all non-zero polynomials f, g, and that $\text{lm}(f + g) \le \max(\text{lm}(f), \text{lm}(g))$ whenever $f + g$ is also non-zero. Furthermore, it is obvious that $\text{lm}(-f) = \text{lm}(f)$. With these observations, Proposition 9.18 specializes to the following result.

Proposition 9.20. *If $f, g_1, g_2, \ldots, g_n$ are non-zero polynomials in $K[X_1, \ldots, X_m]$, then there exist polynomials $r, q_1, q_2, \ldots, q_n$ such that*

$$f = g_1q_1 + g_2q_2 + \cdots + g_nq_n + r$$

and such that either $r = 0$ or else

$$\text{lm}(r) \textit{ is not divisible by any } \text{lm}(g_j) \textit{ and } \text{lm}(r) \le \text{lm}(f).$$

In addition, $\text{lm}(g_jq_j) \le \text{lm}(f)$ for every non-zero g_jq_j.

The last statement in the above result is telling us that every monomial in every g_jq_j is dominated by the leading monomial of f by means of the monomial order at hand.

Proposition 9.20 can be strengthened as follows.

Proposition 9.21. *If $f, g_1, g_2, \ldots, g_n$ are non-zero polynomials in $K[X_1, \ldots, X_m]$, then there exist polynomials $r, q_1, q_2, \ldots, q_n$ such that*

$$f = g_1q_1 + g_2q_2 + \cdots + g_nq_n + r,$$

and either $r = 0$ or else

$$\textit{no } \text{lm}(g_i) \textit{ divides any monomial appearing in } r \textit{ and } \text{lm}(r) \le \text{lm}(f).$$

Proposition 9.21 is known as the division algorithm throughout much of the literature. We will get by nicely with Proposition 9.20, and omit the proof of the above, which is a variation of the proof of Proposition 9.18.

An Illustration of the Division Algorithm in Obtaining Complete Reductions

Take $\mathbb{Q}[X, Y]$ with lexicographic order of the set on monomials, with $X > Y$. With monomials written in descending order, let us carry out the division algorithm, as specified in the proof of Proposition 9.18, to obtain a complete reduction of

$$f = X^2Y^3 - X^2Y + XY^3 + XY^2 - X - Y^2$$

modulo the polynomials

$$g_1 = XY - 1,\ g_2 = Y^2 - 1.$$

- Since $\text{lm}(g_1) = XY$, $\text{lm}(f) = X^2Y^3$ and $XY \mid X^2Y^3$, the first g_j to reduce f is g_1. So put

$$c_1 = \text{lt}(f)/\,\text{lt}(g_1) = X^2Y^3/XY = XY^2$$

  to get the first remainder

$$r_1 = f - g_1c_1 = -X^2Y + XY^3 + 2XY^2 - X - Y^2.$$

- Since $\mathrm{lm}(g_1) = XY$, $\mathrm{lm}(r_1) = X^2Y$ and $XY \mid X^2Y$, the first g_j to reduce r_1 is g_1 again. So put

$$c_2 = \mathrm{lt}(r_1)/\,\mathrm{lt}(g_1) = -X^2Y/XY = -X$$

to get the second remainder

$$r_2 = r_1 - g_1c_2 = XY^3 + 2XY^2 - 2X - Y^2.$$

- Since $\mathrm{lm}(g_1) = XY$, $\mathrm{lm}(r_2) = XY^3$ and $XY \mid XY^3$, the first g_j to reduce r_2 remains g_1. So put

$$c_3 = \mathrm{lt}(r_2)/\,\mathrm{lt}(g_1) = XY^3/XY = Y^2$$

to get the third remainder

$$r_3 = r_2 - g_1c_3 = 2XY^2 - 2X.$$

- Since $\mathrm{lm}(g_1) = XY$, $\mathrm{lm}(r_3) = 2XY^2$ and $XY \mid 2XY^2$, the first g_j to reduce r_3 is still g_1. So put

$$c_4 = \mathrm{lt}(r_3)/\,\mathrm{lt}(g_1) = 2XY^2/XY = 2Y$$

to get the fourth remainder

$$r_4 = r_3 - g_1c_4 = -2X + 2Y.$$

At this point the algorithm stops because neither $\mathrm{lm}(g_1)$ nor $\mathrm{lm}(g_2)$ divides $\mathrm{lm}(r_4)$. In other words, neither g_1 nor g_2 reduces r_4.

It seems that in this example we were able to obtain a complete reduction r_4 modulo g_1, g_2 without recourse to division by g_2.

- From the equations

$$f = g_1c_1 + r_1,\ r_1 = g_1c_2 + r_2,\ r_2 = g_1c_2 + r_3,\ r_3 = g_1c_4 + r_4$$

we see that

$$f = g_1(c_1 + c_2 + c_3 + c_4) + g_20 + r_4,$$

which explicitly becomes

$$f = (XY - 1)(XY^2 - X + Y^2 + 2Y) + (Y^2 - 1)0 + (-2X - 2Y),$$

as required in item 1 of Proposition 9.18.

We might also observe that in the chosen lexicographic order

$$X < XY^2 < XY^3 < X^2Y < X^2Y^3.$$

In other words

$$\mathrm{lm}(r_4) < \mathrm{lm}(r_3) < \mathrm{lm}(r_2) < \mathrm{lm}(r_1) < \mathrm{lm}(f)$$

as the division algorithm demands.

Variations in the Division Algorithm

In each step of the example above we chose the *first* g_j that reduced the preceding remainder as the divisor of the preceding remainder. That turned out to always be g_1. But at each step we could have taken *any* g_j that reduced the preceding remainder. With the same

$$f = X^2Y^3 - X^2Y + XY^3 + XY^2 - X - Y^2,\ g_1 = XY - 1,\ g_2 = Y^2 - 1$$

as in the preceding example, let us run through the algorithm again, only this time taking as the divisor in each step the *last* g_j that reduces the preceding remainder.

- The last g_j that reduces f is g_2, because $\text{lm}(g_2) = Y^2$, $\text{lm}(f) = X^2Y^3$ and $Y^2 \mid X^2Y^3$. So put

$$c_1 = \text{lt}(f)/\,\text{lt}(g_2) = X^2Y^3/Y^2 = X^2Y$$

  to get the first remainder

$$r_1 = f - g_2c_1 = XY^3 + XY^2 - X - Y^2.$$

- The last g_j that reduces r_1 is g_2, because $\text{lm}(g_2) = Y^2$, $\text{lm}(r_1) = XY^3$ and $Y^2 \mid XY^3$. So put

$$c_2 = \text{lt}(r_1)/\,\text{lt}(g_2) = XY^3/Y^2 = XY$$

  to get the second remainder

$$r_2 = r_1 - g_2c_2 = XY^2 + XY - X - Y^2.$$

- Once more the last g_j that reduces r_2 is g_2, because $\text{lm}(g_2) = Y^2$, $\text{lm}(r_2) = XY^2$ and $Y^2 \mid XY^2$. So put

$$c_3 = \text{lt}(r_2)/\,\text{lt}(g_2) = XY^2/Y^2 = X$$

  to get the third remainder

$$r_3 = r_2 - g_2c_3 = XY - Y^2.$$

- Now $\text{lm}(g_2) \nmid \text{lm}(r_3)$. The last g_j that reduces r_1 is g_1, because $\text{lm}(g_1) = XY$, $\text{lm}(r_3) = XY$ and $XY \mid XY$. So put

$$c_4 = \text{lt}(r_3)/\,\text{lt}(g_1) = XY/XY = 1$$

  to get the fourth remainder

$$r_4 = r_3 - g_1c_4 = -Y^2 + 1.$$

- The last g_j that reduces r_4 is g_2, because $\text{lm}(g_2) = Y^2$, $\text{lm}(r_2) = Y^2$ and $Y^2 \mid Y^2$. So put

$$c_5 = \text{lt}(r_4)/\,\text{lt}(g_2) = -Y^2/Y^2 = -1$$

  to get the fifth remainder

$$r_5 = r_4 - g_2c_5 = 0.$$

  With the fifth remainder of 0 the algorithm stops.

- From the equations

$$f = g_2c_1 + r_1,\ r_1 = g_2c_2 + r_2,\ r_2 = g_2c_3 + r_3,\ r_3 = g_1c_4 + r_4,\ r_4 = g_2c_5 + r_5$$

we see that

$$f = g_1c_4 + g_2(c_1 + c_2 + c_3 + c_5) + r_5,$$

which explicitly becomes

$$f = (XY - 1)1 + (Y^2 - 1)(X^2Y + XY + X - 1) + 0.$$

The Division Algorithm Alone Cannot Determine Ideal Membership

With f, g_1, g_2 as in the last run of the division algorithm we see that 0 is a complete reduction of f modulo g_1, g_2. Thus f lies in the ideal generated by g_1, g_2. However, the earlier run of the division algorithm produced $-2X - 2Y$ as a complete reduction of f modulo g_1, g_2. This reveals that a polynomial can have different complete reductions modulo the same set of divisors. It also reveals that the division algorithm cannot determine ideal membership. The division algorithm may well yield a non-zero complete reduction of a polynomial modulo some divisors, while that polynomial still lies in the ideal generated by those divisors.

Complete Reduction Modulo Gröbner Bases

If 0 is a complete reduction of a non-zero element b modulo some non-zero elements $a_1, \ldots, a_n$ in a Gröbner domain R, then b lies in the ideal generated by the a_j. Since a complete reduction of b can be obtained by the division algorithm, one might hope that the division algorithm could provide an effective tool to decide on the so-called ***ideal membership problem***. Namely, given generators $a_1, \ldots, a_n$ for an ideal J and an element b inside a Gröbner domain R, how can one tell if $f \in J$? The limitation of the division algorithm is that complete reductions of b modulo $a_1, \ldots, a_n$ are not unique. As we saw in the preceding examples, the division algorithm could yield a non-zero complete reduction for b even when $b \in J$. This is where Gröbner bases come to the rescue.

Proposition 9.22. *Let R be a Gröbner domain, and let $a_1, \ldots, a_n$ be non-zero elements which lie in an ideal J of R. These elements form a Gröbner basis of J if and only if every complete reduction of every non-zero element of J is* 0.

Proof. Suppose the a_j form a Gröbner basis of J. Let $b \in J^*$ and let r be a complete reduction of b modulo the a_j. Thus there exist $q_1, \ldots, q_n$ in R such that

$$b = a_1q_1 + \cdots + a_nq_n + r, \text{ where either } r = 0 \text{ or } r \text{ is not reducible modulo any } a_j.$$

Since $a_1, \ldots, a_n$ form a Gröbner basis of J, every non-zero element of J is reducible modulo some a_j. It follows that r, which is clearly in J, must be 0.

Conversely suppose that $a_1, \ldots, a_n$ do not form a Gröbner basis of J. Thus there exists a non-zero b in J which is not reducible modulo any a_j. Taking $q_1 = \cdots = q_n = 0$, we see that this non-zero b is its own complete reduction modulo $a_1, \ldots, a_n$. □

The Significance of Gröbner Bases

Proposition 9.22 brings to the fore the importance of having a Gröbner basis for each ideal of a Gröbner domain R. If J is an ideal with Gröbner basis $a_1, \ldots, a_n$ and $b \in R^\star$, a method to decide if $b \in J$ is at hand. Carry out the division algorithm to compute a complete reduction r of b modulo $a_1, \ldots, a_n$. Now, $b \in J$ if and only if $r = 0$.

An Illustration of Ideal Membership Using Gröbner Bases

Let $R = \mathbb{Q}[X, Y, Z]$ be the polynomial ring in three variables X, Y, Z, with the lexicographic order placed on the set of monomials, taking $X > Y > Z$, and with the leading monomial function as its Gröbner function. The polynomials

$$g_1 = X + Z,\ g_2 = Y - Z$$

constitute a Gröbner basis for the ideal J that they generate. To see that, suppose $X+Z, Y-Z$ do not constitute a Gröbner basis of J. Hence, in J there is a non-zero polynomial f whose leading monomial is divisible by neither X nor Y. But since neither X nor Y appears in the leading monomial of f, then neither X nor Y can appear in any monomial of f. This is because, under the specified lexicographic order, monomials having an X or a Y in them would be higher than the leading monomial with no X or Y in it. Now take the substitution map

$$\sigma : \mathbb{Q}[X, Y, Z] \to \mathbb{Q}[Z] \text{ given by } \ell(X, Y, Z) \mapsto \ell(-Z, Z, Z).$$

Evidently $X + Z, Y - Z \in \ker \sigma$, and thereby $f \in \ker \sigma$. However, the monomials in f have no X or Y in them. For this reason $\sigma(f) = f \neq 0$ inside $\mathbb{Q}[Z]$. This contradiction shows that $X + Z, Y - Z$ form a Gröbner basis of J.

Now let

$$f = X^3Z + X^2Z^2 - XY + Y^3 - Y^2Z - Z^3,$$

and let us implement the division algorithm as described in the proof of Proposition 9.18, to decide if $f \in J$.

- The first g_j that reduces f is g_1, because $\text{lm}(g_1) = X$, $\text{lm}(f) = X^3Z$ and $X \mid X^3Z$. Put

  $$c_1 = \text{lt}(f)/\text{lt}(g_1) = X^3Z/X = X^2Z,$$

  to obtain the first remainder

  $$r_1 = f - g_1c_1 = -XY + Y^3 - Y^2Z - Z^3.$$

- The first g_j that reduces r_1 is g_1, because $\mathrm{lm}(g_1) \mid \mathrm{lm}(r_1)$. Put

$$c_2 = \mathrm{lt}(r_1)/\,\mathrm{lt}(g_1) = -XY/X = -Y,$$

  to obtain the second remainder

$$r_2 = r_1 - g_1c_2 = Y^3 - Y^2Z + YZ - Z^3.$$

- Now g_1 does not reduce r_2 because $\mathrm{lm}(g_1) \nmid \mathrm{lm}(r_2)$. But $\mathrm{lm}(g_2) \mid \mathrm{lm}(r_2)$, and thereby g_2 is the first g_j that reduces r_2. Put

$$c_3 = \mathrm{lt}(r_2)/\,\mathrm{lt}(g_2) = Y^3/Y = Y^2,$$

  to obtain the third remainder

$$r_3 = r_2 - g_2c_3 = YZ - Z^3.$$

- By inspection as above we see that g_2 is the first g_j that reduces r_3. So put

$$c_4 = \mathrm{lt}(r_3)/\,\mathrm{lt}(g_2) = YZ/Y = Z,$$

  to obtain the fourth remainder

$$r_4 = r_3 - g_2c_4 = -Z^3 + Z^2.$$

  The division algorithm stops here because neither g_1 nor g_2 reduces r_4.
- From the equations

$$f = g_1c_1 + r_1,\ r_1 = g_1c_2 + r_2,\ r_2 = g_2c_3 + r_3,\ r_3 = g_2c_4 + r_4$$

  we see that

$$f = g_1(c_1 + c_2) + g_2(c_3 + c_4) + r_4,$$

  which in explicit form becomes

$$f = (X + Z)(X^2Z - Y) + (Y - Z)(Y^2 + Z) + (-Z^3 + Z^2).$$

Since $-Z^3 + Z^2$ is a non-zero complete reduction of f modulo the Gröbner basis g_1, g_2, it follows from Proposition 9.22 that f is not in the ideal generated by g_1, g_2.

Had the algorithm concluded with a zero remainder, the opposite conclusion would have prevailed.

EXERCISES

1. For any polynomial ring $K[X]$ over a field K, verify that the usual degree function $\deg : K[X]^\star \to \mathbb{N}$ is a Gröbner function. Is the degree function on $\mathbb{Z}[X]^\star$ a Gröbner function?
2. Suppose that R is a ring of polynomials in several variables with coefficients in a field, and that the set of monomials carries a monomial order. If J is an ideal of R generated by a finite set of monomials, show that those monomials form a Gröbner basis of J.

3. In the polynomial ring $K[X, Y, Z]$ take the lexicographic order on the set of monomials using $X > Y > Z$, and the resulting leading monomial function as the Gröbner function. Let J be the ideal generated by the polynomials $X + Z$, $Y - Z$. We have verified that they constitute a Gröbner basis of J. If f is the polynomial

$$X^3Z + X^2YZ - 2XY^2Z - 2YZ^3,$$

decide if $f \in J$.

4. Let M be a set well-partially ordered by $\preceq$ and having a totally ordered extension $\leq$. Suppose that a domain R has a Gröbner function $\phi : R^\star \to M$ such that $\phi(a) \preceq \phi(ab)$ for all a, b in $R^\star$.
   (a) Show that every a in $R^\star$ gives a minimal Gröbner basis for the principal ideal J that it generates.
   (b) If J is a principal ideal and B is a minimal Gröbner basis of J, show that B is a singleton which generates J.
   (c) On the set $\mathbb{P}$ of positive integers define a partial order $\preceq$ according to

$$n \preceq k \text{ if and only if } 2 \leq n \leq k \text{ and } 1 \preceq n \text{ for all } n \neq 2.$$

   Verify that $\preceq$ is a well-partial order on $\mathbb{P}$, with the usual order $\leq$ on $\mathbb{P}$ as its total order extension.

   Show that $\phi : \mathbb{Z}^\star \to \mathbb{P}$ defined by $n \mapsto |n|$ is a Gröbner function on $\mathbb{Z}$.

   Show that the generator 1 of the full ideal $\mathbb{Z}$ does not form a Gröbner basis of $\mathbb{Z}$, but that a Gröbner basis consists of $1, 2$.

5. Let $\phi : R^\star \to M$ be a Gröbner function on a domain R into a set M that is well-partially ordered by $\preceq$ and has a total order extension $\leq$. Define another function

$$\psi : R^\star \to M \text{ by } \psi(a) = \min\{\phi(au) : u \text{ is a unit of } R\}.$$

The minimum here is calculated using the total order $\leq$ on M, and is taken over all units of R.

Verify that ψ is a Gröbner function such that $\psi(a) = \psi(au)$ for all a in $R^\star$ and all units u.

6. Take the product ordering $\preceq$ on the Cartesian product $\mathbb{N} \times \mathbb{N}$. As shown in Proposition 9.4, this is a well-partial order.
   (a) Verify that the order $\leq$ on $\mathbb{N} \times \mathbb{N}$, defined by

$$(a, b) \leq (c, d) \text{ provided } b < d, \text{ or } b = d \text{ and } a \leq c,$$

   is a total order extension of the product order.
   (b) If f is a non-zero polynomial in $\mathbb{Z}[X]$ let $\mathrm{lc}(f)$ be its leading coefficient and let $\deg(f)$ be its degree. Prove that the function

$$\phi : \mathbb{Z}[X]^\star \to \mathbb{N} \times \mathbb{N}, \text{ given by } f \mapsto (|\mathrm{lc}(f)|, \deg(f)),$$

   is a Gröbner function.

(c) Verify that the polynomials $X + 3, 7$ form a Gröbner basis for the ideal that they generate inside $\mathbb{Z}[X]$ with the given Gröbner function.

7.* Suppose that M is well-partially ordered by $\preceq$ and has a total order extension $\leq$. Let $\phi : R^\star \to M$ be a Gröbner function on R such that

$$\phi(a) \leq \phi(ab) \text{ for all } a, b \text{ in } R^\star,$$

and

$$\phi(a+b) \leq \max\{\phi(a), \phi(b)\} \text{ whenever } a, b, a+b \in R^\star.$$

(a) Prove that u is a unit of R if and only if $\phi(au) = \phi(a)$ for some a in $R^\star$.
(b) Prove that u is a unit of R if and only if $\phi(u) = \phi(1)$.
(c) Prove that R is a finitely generated algebra over a field.
Hint. The field is the set of units of R along with the zero element.

8. Let R be a domain with Gröbner function $\phi : R^\star \to M$, where M is well-partially ordered by $\preceq$ and has a total order extension $\leq$. Suppose that D is a denominator set of R as in Definition 4.61, and that R_D is the localization of R at D. Assume that every divisor of every element of D is a unit multiple of an element of D. Such a denominator set is called ***saturated***. Take the function $\psi : R_D^\star \to M$ defined at every a in $R_D^\star$ by

$$\psi(a) = \min\{\phi(x) : x \in R^\star \text{ and } as = xt \text{ for some } s, t \text{ in } D\}.$$

The minimum here is taken using the well-order extension $\leq$.

(a) Prove that ψ is a Gröbner function, in other words that localizations of Gröbner domains are Gröbner domains.
Hint. First observe that for every a, b in $R_D^\star$ there exist c, d in $R^\star$ such that $\psi(a) = \phi(c)$ and $\psi(b) = \phi(d)$.
(b) The absolute value function $\phi : \mathbb{Z}^\star \to \mathbb{N}$ is a Gröbner function. What is the function on $\mathbb{Q}^\star$ induced by ϕ in accordance with our definition?

9.3 Buchberger's Algorithm

An ideal of a Gröbner domain is typically given by presenting a finite set of generators. Bruno Buchberger (born 1942) gave an algorithm for extracting a Gröbner basis from that set of generators, in the case of a polynomial ring over a field. We now turn to this ingenious technique.

9.3.1 Detecting Gröbner Bases via S-Polynomials

Because of the importance of Gröbner bases, we need a method for obtaining them, which brings us to the essence of Buchberger's work. This deals with the important ring $R = K[X_1, \dots, X_n]$ of polynomials in several variables over a field K.

Throughout this section we have the set M of monomials, well-partially ordered by division and with a chosen monomial order extension as discussed in Section 9.1.3. The leading monomial function is the Gröbner function in effect.

S-Polynomials

The primary tool used by Buchberger is what he calls an ***S-polynomial***, which we now define. If we have monomials $s = X_1^{d_1} \cdots X_n^{d_n}$ and $t = X_1^{e_1} \cdots X_n^{e_n}$, their ***least common multiple***, denoted by $\mathrm{lcm}(s,t)$, is the monomial

$$X_1^{\max(d_1,e_1)} \cdots X_n^{\max(d_n,e_n)}.$$

Evidently $\mathrm{lcm}(s,t)$ is divisible by both s and t and $\mathrm{lcm}(s,t)$ divides every monomial which is divisible by s and t.

We need to distinguish between ***monomials***, which are products of powers of the X_j, and ***terms***, which are non-zero scalar multiples of monomials. Every monomial is a term, but not conversely. A non-zero polynomial f whose leading term is a monomial, i.e. such that $\mathrm{lc}(f) = 1$, will be called ***monic***.

If $f, g \in R^\star$, let

$$p = \mathrm{lcm}(\mathrm{lm}(f), \mathrm{lm}(g)).$$

By Proposition 9.15, which carries over readily to leading terms,

$$\mathrm{lt}\left(\frac{p}{\mathrm{lt}(f)} f\right) = \mathrm{lt}\left(\frac{p}{\mathrm{lt}(f)}\right) \mathrm{lt}(f) = \frac{p}{\mathrm{lt}(f)} \mathrm{lt}(f) = p.$$

The same identity holds for g. Thus $\frac{p}{\mathrm{lt}(f)} f$ and $\frac{p}{\mathrm{lt}(g)} g$ are monic with the same leading term, namely the monomial p.

Definition 9.23. The difference

$$S(f,g) = \frac{p}{\mathrm{lt}(f)} f - \frac{p}{\mathrm{lt}(g)} g,$$

is called the ***S-polynomial*** of f and g.

Due to cancellation of the leading term p, every monomial that appears in $S(f,g)$ is strictly below p.

For example, in $\mathbb{Q}[X,Y]$ with lexicographic order on monomials and $X > Y$ we get:

$$\begin{aligned} &S(-3X^2Y^3 - 4XY^4 + X, 7X^4Y + XY^2 - 2) \\ &= \frac{X^4Y^3}{-3X^2Y^3}(-3X^2Y^3 - 4XY^4 + X) - \frac{X^4Y^3}{7X^4Y}(7X^4Y + XY^2 - 2) \end{aligned}$$

$$= -\frac{1}{3}X^2(-3X^2Y^3 - 4XY^4 + X) - \frac{1}{7}Y^2(7X^4Y + XY^2 - 2)$$
$$= \left(X^4Y^3 + \frac{4}{3}X^3Y^4 - \frac{1}{3}X^3\right) - \left(X^4Y^3 + \frac{1}{7}XY^4 - \frac{2}{7}Y^2\right)$$
$$= \frac{4}{3}X^3Y^4 - \frac{1}{3}X^3 - \frac{1}{7}XY^4 + \frac{2}{7}Y^2.$$

The least common multiple of the leading monomials of the given polynomials is X^4Y^3, and this is higher than all of the monomials appearing in their S-polynomial.

The terms $u = \frac{p}{\text{lt}(f)}$ and $v = \frac{p}{\text{lt}(g)}$ were chosen so that uf and vg are monic with the same leading monomial. With that in mind, here is a useful property of the S-polynomial.

Proposition 9.24. *If $f, g \in R^\star$ and u, v are terms in R such that uf, vg are monic with the same leading monomial, then $uf - vg = qS(f, g)$ for some monomial q.*

Proof. Let m be the common leading term of uf and vg. Since uf, vg are monic, m is a monomial. By Proposition 9.15, which carries over readily to leading terms:

$$m = \text{lt}(uf) = \text{lt}(u)\,\text{lt}(f) = u\,\text{lt}(f) \text{ and } m = \text{lt}(vg) = \text{lt}(v)\,\text{lt}(g) = v\,\text{lt}(g).$$

These equations reveal that $\text{lm}(f)$ and $\text{lm}(g)$ divide the monomial m. Hence $\text{lcm}(\text{lm}(f), \text{lm}(g))$, which we call p, divides m. Write $m = pq$ for some monomial q. Thus

$$u\,\text{lt}(f) = pq \text{ and } v\,\text{lt}(g) = pq,$$

which leads to

$$u = \frac{p}{\text{lt}(f)}q \text{ and } v = \frac{p}{\text{lt}(g)}q.$$

And then

$$uf - vg = \frac{p}{\text{lt}(f)}qf - \frac{p}{\text{lt}(g)}qg = qS(f, g).$$

□

Standard Expressions

The upcoming result is the key to this whole subject. It leads to algorithms for detecting as well as constructing Gröbner bases.

But first yet more terminology. Let us recall Proposition 9.20. It tells us that if $f, g_1, \ldots, g_n \in R^\star$, then there exist $r, q_1, \ldots, q_n$ in R such that

$$f = \sum_i q_i q_i + r \text{ and either } r = 0 \text{ or else } \text{lm}(r) \text{ is not divisible by any } \text{lm}(g_i),$$

and every monomial s appearing in each of the $g_i q_i$ satisfies $s \leq \text{lm}(f)$.

The division algorithm will find suitable q_i.

Definition 9.25. We shall say that the above summation with the q_i, r, taken to satisfy *both* of the preceding specifications, is a ***standard expression for f modulo*** $g_1, \ldots, g_n$ ***with complete reduction*** r.

Buchberger's Theorem

If $g_1, \ldots, g_n$ are non-zero elements in R, each S-polynomial $S(g_i, g_j)$, for $i < j$, is defined as a difference of two multiples of g_i and g_j. However, this defining expression for $S(g_i, g_j)$ may not be standard modulo $g_1, \ldots, g_n$. Nevertheless, when $S(g_i, g_j)$ is non-zero, a standard expression for it modulo $g_1, \ldots, g_n$ does exist, and the division algorithm will find it.

Without the next result, the notion of Gröbner bases would be rather pointless.

Proposition 9.26 (Buchberger's theorem). *Suppose $g_1, \ldots, g_n$ are non-zero elements generating an ideal J of a polynomial ring R. These polynomials constitute a Gröbner basis of J if and only if for each pair i,j such that $1 \leq i < j \leq n$, the S-polynomial $S(g_i, g_j)$ either is 0 or has a standard expression modulo $g_1, \ldots, g_n$ with complete reduction 0.*

Proof. Assuming the g_j constitute a Gröbner basis of J, every non-zero polynomial f in in J has a standard expression with complete reduction 0 modulo $g_1, \ldots, g_n$. The standard expression for f comes from the division algorithm, and the zero complete reduction comes from Proposition 9.22. The S-polynomials $S(g_i, g_j)$ lie in J. Thus either they are 0 or they have a standard expression modulo $g_1, \ldots, g_n$ with complete reduction 0.

For the converse, take a non-zero f in J. We need to show that $\mathrm{lm}(f)$ is divisible by some $\mathrm{lm}(g_j)$. Since $f \in J$, there exist polynomials $h_1, \ldots, h_n$ in R such that

$$f = h_1 g_1 + h_2 g_2 + \cdots + h_n g_n.$$

By dropping some g_i if need be, we can suppose that all $h_j g_j \neq 0$. Let p be the highest monomial appearing inside all of the polynomials $h_j g_j$. In other words, $p = \max_{j=1}^{n}\{\mathrm{lm}(h_j g_j)\}$.

By Proposition 9.15, $\mathrm{lm}(f) \leq p$. If equality occurs, this means that $\mathrm{lm}(f) = \mathrm{lm}(h_j g_j)$ for some j. By Proposition 9.15 once more, $\mathrm{lm}(h_j g_j) = \mathrm{lm}(h_j)\,\mathrm{lm}(g_j)$, from which it follows that $\mathrm{lm}(g_j)$ divides $\mathrm{lm}(f)$.

We are left to deal with the case where $\mathrm{lm}(f) < p$.

By rearranging indices we can suppose that for some index k between 1 and n the polynomials $h_1 g_1, \ldots, h_k g_k$ all have p as their leading monomial, while the monomials in each of $h_{k+1} g_{k+1}, \ldots, h_n g_n$ are all strictly below p.

For j from 1 to k, let $u_j = \mathrm{lt}(h_j)$, the leading term of h_j. Then $f = f_1 + f_2$ where

$$f_1 = u_1 g_1 + \cdots + u_k g_k$$

and

$$f_2 = (h_1 - u_1) g_1 + \cdots + (h_k - u_k) g_k + h_{k+1} g_{k+1} + \cdots + h_n g_n.$$

All monomials appearing in every summand of f_2 are strictly below p. This is because any monomials appearing in $h_j - u_j$ are strictly below $\mathrm{lm}(h_j)$, and thereby monomials appearing in $(h_j - u_j) g_j$ are strictly below $\mathrm{lm}(h_j g_j)$, which is p.

Furthermore, for j from 1 to k, the leading monomial of each summand $u_j g_j$ of f_1 is p, while at the same time $\mathrm{lm}(f_1) < p$. Let

$$\lambda_j = \mathrm{lc}(u_j g_j) \text{ and } t_j = \frac{u_j}{\lambda_j}.$$

Thus t_j is another term, every $t_j g_j$ is monic with leading monomial p, and $u_j g_j = \lambda_j t_j g_j$. The fact that $\mathrm{lm}(f_1) < p$, means that $\sum_{j=1}^{k} \lambda_j = 0$. The following identity for f_1 ensues:

$$\begin{aligned}
f_1 &= u_1 g_1 + u_2 g_2 + \cdots + u_k g_k \\
&= \lambda_1 t_1 g_1 + \lambda_2 t_2 g_2 + \cdots + \lambda_k t_k g_k \\
&= \lambda_1 (t_1 g_1 - t_2 g_2) \\
&\quad + (\lambda_1 + \lambda_2)(t_2 g_2 - t_3 g_3) \\
&\quad + (\lambda_1 + \lambda_2 + \lambda_3)(t_3 g_3 - t_4 g_4) \\
&\quad + (\lambda_1 + \lambda_2 + \lambda_3 + \lambda_4)(t_4 g_4 - t_5 g_5) \\
&\quad + \cdots \\
&\quad + (\lambda_1 + \lambda_2 + \cdots + \lambda_{k-1})(t_{k-1} g_{k-1} - t_k g_k) \\
&\quad + (\lambda_1 + \lambda_2 + \cdots + \lambda_k) t_k g_k,
\end{aligned}$$

where the last summand $(\lambda_1 + \lambda_2 + \cdots + \lambda_k) t_k g_k = 0$.

For j from 1 to $k-1$, all monomials appearing in $t_j g_j - t_{j+1} g_{j+1}$ are strictly below p, because the leading monomial p in each of $t_j g_j, t_{j+1} g_{j+1}$ cancels out. Proposition 9.24 applied to these differences yields monomials q_j such that

$$t_j g_j - t_{j+1} g_{j+1} = q_j S(g_j, g_{j+1}).$$

And because of this equation, all monomials appearing in $q_j S(g_j, g_{j+1})$ are strictly below p.

By assumption, each S-polynomial

$$S(g_j, g_{j+1}) = \sum_{i=1}^{n} t_{ji} g_i \text{ for some } t_{ji} \text{ in } R,$$

where every non-zero $t_{ji} g_i$ is such that $\mathrm{lm}(t_{ji} g_i) \leq \mathrm{lm}(S(g_j, g_{j+1}))$. Such an expansion also occurs trivially in the cases where $S(g_i, g_j) = 0$. Then for each j from 1 to $k-1$:

$$t_j g_j - t_{j+1} g_{j+1} = q_j S(g_j, g_{j+1}) = \sum_{i=1}^{n} q_j t_{ji} g_i,$$

and every non-zero $q_j t_{ji} g_i$ is such that

$$\mathrm{lm}(q_j t_{ji} g_i) = \mathrm{lm}(q_j)\,\mathrm{lm}(t_{ji} g_i) \leq \mathrm{lm}(q_j)\,\mathrm{lm}(S(g_j, g_{j+1})) = \mathrm{lm}(q_j S(g_j, g_{j+1})) < p.$$

From the identity for f_1 as a linear combination of the differences $t_j g_j - t_{j+1} g_{j+1}$, it follows that f_1 can be rewritten as a summation $\sum_{i=1}^{n} s_i q_i$ for some s_i in R and such that any monomials appearing in $s_i q_i$ are strictly below p. Since f_2 is also such a summation, the same applies to their sum $f_1 + f_2$, which is the original f.

From a summation $f = h_1 g_1 + \cdots + h_n g_n$ with p the highest monomial appearing among all of the $h_j g_j$, we have shown that if $\mathrm{lm}(f) < p$, then there is an alternative summation $f = \ell_1 g_1 + \cdots + \ell_n g_n$ for some ℓ_j in R with all monomials appearing in each $\ell_j g_j$ strictly below p.

We could have chosen the h_j in the original representation of f so that p was minimal using the monomial order. And if we had that $\mathrm{lm}(f) < p$, the foregoing argument affords a new representation $f = \sum \ell_j g_j$ whose largest monomial appearing among all $\ell_j g_j$ is strictly below p, in contradiction to the minimality of p.

Thus there is a representation of $f = \sum h_j g_j$ such that $\mathrm{lm}(f) = \mathrm{lm}(h_j g_j) = \mathrm{lm}(h_j)\,\mathrm{lm}(g_j)$ for some j, which shows that $g_1, \ldots, g_n$ form a Gröbner basis of J. □

A Practical Test for Gröbner Bases

Take a generating set $g_1, \ldots, g_n$ for an ideal J of the polynomial ring R. For $i < j$ apply the division algorithm to each S-polynomial $S(g_i, g_j)$. This will provide a standard expression for $S(g_i, g_j)$ with complete reduction r modulo $g_1, \ldots, g_n$. If $r = 0$, then Buchberger's theorem, Proposition 9.26, assures that $g_1, \ldots, g_n$ form a Gröbner basis of J.

Conversely, suppose that $g_1, \ldots, g_n$ form a Gröbner basis of J. Since the $S(g_i, g_j)$ lie in J, Proposition 9.22 assures that a complete reduction r obtained for each $S(g_i, g_j)$ by the division algorithm is 0.

In summary we have the following.

Proposition 9.27. *A list of generators $g_1, \ldots, g_n$ for an ideal J in a polynomial ring R forms a Gröbner basis of J if and only if a complete reduction of every S-polynomial $S(g_i, g_j)$ modulo $g_1, \ldots, g_n$ obtained by the division algorithm is* 0.

This result is useful because the S-polynomials $S(g_i, g_j)$ are easy to determine, and there are only finitely many, in fact $\frac{n(n-1)}{2}$ of them. This is unlike the test of Proposition 9.22, wherein complete reductions of all non-zero elements of J need to be shown to be 0.

Bases with Pairwise Coprime Leading Monomials Are Gröbner

Two monomials f, g in a polynomial ring $K[X_1, \ldots, X_m]$ are coprime if and only if there is no variable X_j which appears in both f and g. A small but useful consequence of Buchberger's theorem, Proposition 9.26, is that a list of polynomials whose leading monomials are pairwise coprime will be a Gröbner basis for the ideal that they generate.

Proposition 9.28. *If polynomials f, g in a polynomial ring R, with a suitable monomial order, are such that their leading monomials are coprime, then $S(f, g)$ either is* 0 *or has a standard expression modulo f, g with complete reduction* 0.

Proof. The standard expression comes directly from the definition of $S(f, g)$. Write

$$f = \lambda\,\mathrm{lm}(f) + p \text{ and } g = \mu\,\mathrm{lm}(g) + q,$$

where λ, μ are the leading coefficients of f, g, respectively, and where the monomials of p are strictly below $\mathrm{lm}(f)$ and the monomials of q are strictly below $\mathrm{lm}(g)$. Since $\mathrm{lm}(f)$ is coprime with $\mathrm{lm}(g)$, their least common multiple is their product $\mathrm{lm}(f)\,\mathrm{lm}(g)$. Consequently,

$$\begin{aligned} S(f,g) &= \frac{\mathrm{lm}(f)\,\mathrm{lm}(g)}{\lambda\,\mathrm{lm}(f)} f - \frac{\mathrm{lm}(f)\,\mathrm{lm}(g)}{\mu\,\mathrm{lm}(g)} g \\ &= \frac{\mu\,\mathrm{lm}(g)}{\lambda\mu} f - \frac{\lambda\,\mathrm{lm}(f)}{\lambda\mu} g \\ &= \frac{g-q}{\lambda\mu} f - \frac{f-p}{\lambda\mu} g \\ &= \frac{p}{\lambda\mu} g - \frac{q}{\lambda\mu} f. \end{aligned}$$

This gives a standard expression for $S(f,g)$ modulo f, g with complete reduction 0. To verify this claim, in accordance with Definition 9.25 it suffices to see that the monomials of $\frac{p}{\lambda\mu}g$ and $\frac{q}{\lambda\mu}f$ do not exceed $\mathrm{lm}(S(f,g))$. This is the case because the leading terms of $\frac{p}{\lambda\mu}g$ and $\frac{q}{\lambda\mu}f$ are distinct, and thereby cannot cancel each other. Indeed, if the leading terms were the same, we would have that

$$\mathrm{lm}(p)\,\mathrm{lm}(g) = \mathrm{lm}(pg) = \mathrm{lm}(fq) = \mathrm{lm}(f)\,\mathrm{lm}(q).$$

Since $\mathrm{lm}(f)$ is coprime with $\mathrm{lm}(g)$, it would follow that $\mathrm{lm}(f)$ divides $\mathrm{lm}(p)$, a contradiction because the monomials of p are strictly below $\mathrm{lm}(f)$. □

Proposition 9.29. *If $g_1, \ldots, g_n$ generate an ideal J inside a polynomial ring R having some monomial order, and for every pair g_i, g_j their leading monomials are coprime, then $g_1, \ldots, g_n$ constitute a Gröbner basis of J.*

Proof. For each pair g_i, g_j, Proposition 9.28 yields a standard expression for $S(g_i, g_j)$ modulo g_i, g_j with complete reduction 0. This is automatically also a standard expression for $S(g_i, g_j)$, with complete reduction 0, modulo the full list $g_1, \ldots, g_n$. By Buchberger's theorem, Proposition 9.26, the list $g_1, \ldots, g_n$ gives a Gröbner basis of J. □

For example, in the polynomial ring $K[X, Y]$ with degree lexicographic order taking $X > Y$, the polynomials $X^4 + 2X^3Y - XY^2 + XY$, $Y^3 - X^2 + 3XY + X$ provide a Gröbner basis for the ideal that they generate.

For another example, take $R = \mathbb{Q}[X, Y, Z]$ with the lexicographic order on the set of monomials taking $X > Y > Z$. We have seen in an earlier example that $X + Z, Y - Z$ constitute a Gröbner basis for the ideal that they generate. We can now reconfirm this using Proposition 9.29, because their leading monomials X, Y are certainly coprime.

9.3.2 Buchberger's Algorithm

Not only does Buchberger's theorem detect Gröbner bases, but it can be used to build them, in a finite number of steps, from an initial set of generators of an ideal. Here is how the now famous ***Buchberger algorithm*** goes.

- Start with a list of non-zero generators $g_1, \ldots, g_m$ for an ideal J of a polynomial ring R. If $m = 1$, then J is principal with generator g_1. This g_1 forms a Gröbner basis of J, because every polynomial f in J is such that $f = g_1 h$ for some h in R. If $f \neq 0$, we have $\mathrm{lm}(f) = \mathrm{lm}(g_1 h) = \mathrm{lm}(g_1)\,\mathrm{lm}(h)$, which reveals that $\mathrm{lm}(g_1)$ divides $\mathrm{lm}(f)$.
- We can now suppose $m \geq 2$. For $1 \leq i < j \leq m$, compute the S-polynomials $S(g_i, g_j)$. Then, using the division algorithm compute their standard expressions and complete reductions modulo $g_1, \ldots, g_m$. If all complete reductions are 0, Buchberger's theorem assures that $g_1, \ldots, g_m$ form a Gröbner basis of J, and we need go no further.
- Otherwise, the complete reduction of some $S(g_i, g_j)$, where $1 \leq i < j \leq m$, is not zero. Denote this complete non-zero reduction modulo $g_1, \ldots, g_m$ by g_{m+1}. At this point the algorithm allows for some choice of which $S(g_i, g_j)$ with non-zero complete reduction is used to define g_{m+1}. A systematic way of choosing could be adopted, but we will not worry about that here.

  Due to the nature of standard expressions and complete reductions, $g_{m+1} \in J$. Also $\mathrm{lm}(g_j) \nmid \mathrm{lm}(g_{m+1})$ for every j from 1 to m.
- Repeat the above argument on the larger list of generators $g_1, \ldots, g_m, g_{m+1}$. Again, if now $1 \leq i < j \leq m + 1$ and the complete reductions of all $S(g_i, g_j)$ are zero, Buchberger's theorem assures us that $g_1, \ldots, g_m, g_{m+1}$ form a Gröbner basis of J, and we stop our search. Otherwise, the list of non-zero generators of J grows to $g_1, \ldots, g_m, g_{m+1}, g_{m+2}$. Furthermore, $\mathrm{lm}(g_j) \nmid \mathrm{lm}(g_{m+2})$ for every j from 1 to $m + 1$.
- Repeat the above steps. Either a Gröbner basis of J is obtained in a finite (but possibly lengthy) number of steps, or else there results an infinite sequence of non-zero elements in J:

  $$g_1, \ldots, g_m, g_{m+1}, \ldots, g_\ell, \ldots,$$

  such that if $m < \ell$ and $1 \leq j < \ell$, then $\mathrm{lm}(g_j) \nmid \mathrm{lm}(g_\ell)$.

  The set of leading monomials

  $$\mathrm{lm}(g_1), \ldots, \mathrm{lm}(g_m), \mathrm{lm}(g_{m+1}), \ldots, \mathrm{lm}(g_\ell), \ldots$$

  is infinite. On the other hand the order by divisibility on the set of monomials is a well-partial order. This implies that the infinite set of $\mathrm{lm}(g_\ell)$ contains an infinite strictly ascending chain using the order by divisibility. In particular, there are indices j, ℓ such that $m < \ell$ and $j < \ell$ such that $\mathrm{lm}(g_j)$ divides $\mathrm{lm}(g_\ell)$, in contradiction to the way the complete reductions were created.
- After a finite but possibly large number of steps, Buchberger's algorithm will produce a Gröbner basis of J.

An Illustration of Buchberger's Algorithm

Let us work out an example to illustrate Buchberger's algorithm.

Take the ring $\mathbb{Q}[X, Y]$, with lexicographic order on the set of monomials such that $X > Y$, and seek a Gröbner basis for the ideal J generated by the polynomials

$$g_1 = X^3 - 2XY, \; g_2 = X^2Y + X - 2Y^2.$$

Although we describe the steps in some detail, some verifications by the reader will be required.

1. The initial S-polynomial is

$$S(g_1,g_2) = \frac{X^3Y}{X^3}g_1 - \frac{X^3Y}{X^2Y}g_2 = -X^2.$$

Since neither $\mathrm{lm}(g_1)$ nor $\mathrm{lm}(g_2)$ divides $\mathrm{lm}(-X^2)$, the complete reduction of $-X^2$ modulo g_1, g_2 is the non-zero $-X^2$ itself. Thus g_1, g_2 do not constitute a Gröbner basis.

Expand the basis of J by putting $g_3 = -X^2$ and adjoining it to g_1, g_2.

2. The three S-polynomials $S(g_i, g_j)$ for $1 \le i < j \le 3$ are $S(g_1, g_2)$ given above as well as

$$S(g_1,g_3) = -2XY,\ S(g_2,g_3) = X - 2Y^2.$$

Since $-X^2 = g_1 0 + g_2 0 + g_3 1 + 0$, the complete reduction of $S(g_1,g_2)$ is 0. However, none of $\mathrm{lm}(g_1)$, $\mathrm{lm}(g_2)$, $\mathrm{lm}(g_3)$ divides $\mathrm{lm}(-2XY)$. So, $-2XY$ is the complete reduction of $S(g_1,g_3)$ modulo g_1, g_2, g_3.

Since g_1, g_2, g_3 is still not a Gröbner basis of J expand the basis by putting $g_4 = -2XY$ and adjoining g_4 to the basis of J.

3. The six S-polynomials $S(g_i, g_j)$ for $1 \le i < j \le 4$ are the three already listed as well as

$$S(g_1,g_4) = -2XY^2, \quad S(g_2,g_4) = X - 2Y^2, \quad S(g_3,g_4) = 0.$$

Taking $S(g_2,g_3) = X - 2Y^2$, which happens to coincide with $S(g_2,g_4)$, we see that none of $\mathrm{lm}(g_i)$ for i from 1 to 4 divides $\mathrm{lm}(X - 2Y^2)$. So $X - 2Y^2$ is the complete reduction of $S(g_2,g_3)$ modulo g_1, g_2, g_3, g_4.

Expand the basis by adjoining $g_5 = X - 2Y^2$ to it.

4. The ten S-polynomials $S(g_i, g_j)$ for $1 \le i < j \le 5$ are the six already listed as well as

$$S(g_1,g_5) = 2X^2Y^2 - 2XY \qquad S(g_2,g_5) = 2XY^3 + X - 2Y^2$$
$$S(g_3,g_5) = 2XY^2 \qquad S(g_4,g_5) = -2Y^3.$$

Because of its simplicity, examine $S(g_4,g_5) = -2Y^3$. None of the leading monomials of g_j for j from 1 to 5 divides Y^3, the leading monomial of $S(g_4,g_5)$. Hence $-2Y^3$ is the complete reduction of $S(g_4,g_5)$ modulo $g_1, \ldots, g_5$.

Expand the basis by adjoining $g_6 = -2Y^3$.

5. The fifteen S-polynomials $S(g_i, g_j)$ for $1 \le i < j \le 6$ are the ten already listed along with

$$S(g_1,g_6) = -2X^4Y \qquad S(g_2,g_6) = XY^2 - 2Y^4 \qquad S(g_3,g_6) = 0$$
$$S(g_4,g_6) = 0 \qquad S(g_5,g_6) = -2Y^5.$$

Standard expressions for these fifteen S-polynomials modulo $g_1, \ldots, g_6$ are exhibited below. Fortunately these can be seen by inspection, without recourse to the division algorithm.

$$
\begin{aligned}
S(g_1,g_2) &= -X^2 = 0g_1 + 0g_2 + 1g_3 + 0g_4 + 0g_5 + 0g_6 + 0 \\
S(g_1,g_3) &= -2XY = 0g_1 + 0g_2 + 0g_3 + 1g_4 + 0g_5 + 0g_6 + 0 \\
S(g_2,g_3) &= X - 2Y^2 = 0g_1 + 0g_2 + 0g_3 + 0g_4 + 1g_5 + 0g_6 + 0 \\
S(g_1,g_4) &= -2XY^2 = 0g_1 + 0g_2 + 0g_3 + Yg_4 + 0g_5 + 0g_6 + 0 \\
S(g_2,g_4) &= X - 2Y^2 = 0g_1 + 0g_2 + 0g_3 + 0g_4 + 1g_5 + 0g_6 + 0 \\
S(g_3,g_4) &= 0 \\
S(g_1,g_5) &= 2X^2Y^2 - 2XY = 0g_1 + 0g_2 + 0g_3 + (XY-1)g_4 + 0g_5 + 0g_6 + 0 \\
S(g_2,g_5) &= 2XY^3 + X - 2Y^2 = 0g_1 + 0g_2 + 0g_3 - Y^2g_4 + 1g_5 + 0g_6 + 0 \\
S(g_3,g_5) &= 2XY^2 = 0g_1 + 0g_2 + 0g_3 - Yg_4 + 0g_5 + 0g_6 + 0 \\
S(g_4,g_5) &= -2Y^3 = 0g_1 + 0g_2 + 0g_3 + 0g_4 + 0g_5 + 1g_6 + 0 \\
S(g_1,g_6) &= -2X^4Y = 0g_1 + 0g_2 + 0g_3 + X^3g_4 + 0g_5 + 0g_6 + 0 \\
S(g_2,g_6) &= XY^2 - 2Y^4 = 0g_1 + 0g_2 + 0g_3 + 0g_4 + Y^2g_5 + 0g_6 + 0 \\
S(g_3,g_6) &= 0 \\
S(g_4,g_6) &= 0 \\
S(g_5,g_6) &= -2Y^5 = 0g_1 + 0g_2 + 0g_3 + 0g_4 + 0g_5 + Y^2g_6 + 0.
\end{aligned}
$$

The preceding chart shows that either these S-polynomials are 0 or their complete reductions modulo $g_1, \ldots, g_6$ are 0. By Buchberger's theorem a Gröbner basis for the ideal generated by the initial polynomials g_1, g_2 is given by these six polynomials.

Since $\text{lm}(g_5) = X$ and X divides the lead monomials of g_1, g_2, g_3, g_4, these four generators are redundant. Thus a minimal Gröbner basis of the ideal generated by the original g_1, g_2 is given by g_5, g_6. We can without harm rescale these polynomials to be monic. Hence a Gröbner basis is given by the pair $X - 2Y^2, Y^3$.

For a bit more practice with ideal membership, let $f = X^3Y + XY^2 + 2Y$ and apply the division algorithm to find a complete reduction of f modulo $X - 2Y^2, Y^3$, which will tell us if f is in the ideal J or not. Here are the calculations of the algorithm without further explanations. We chose our divisors at each step as shown.

$$
\begin{aligned}
X^3Y + XY^2 + 2Y &= (X - 2Y^2)X^2Y + 2X^2Y^3 + XY^2 + 2Y \\
2X^2Y^3 + XY^2 + 2Y &= Y^3(2X^2) + XY^2 + 2Y \\
XY^2 + 2Y &= (X - 2Y^2)Y^2 + 2Y^4 + 2Y \\
2Y^4 + 2Y &= Y^3(2Y) + 2Y.
\end{aligned}
$$

The algorithm stops at $2Y$ because this polynomial is not strictly reducible modulo $X - 2Y^2$ nor modulo Y^3. The division algorithm reveals that the non-zero $2Y$ is the complete reduction of f modulo $X - 2Y^2, Y^3$. Thus f is not in the ideal J of this example.

Had we applied the division algorithm to get a non-zero complete reduction of f modulo the original generators g_1, g_2 of J, this would not guarantee that $f \notin J$, because the original g_1, g_2 do not form a Gröbner basis of J.

An Illustration of Buchberger's Algorithm Using a Finite Field

Taking the polynomial ring $\mathbb{Z}_5[X, Y]$ with lexicographic order on the set of monomials where $X > Y$, let us calculate a Gröbner basis for the ideal J generated by the polynomials

$$g_1 = X^2 + Y^2 + 1, \; g_2 = X^2Y + 2XY + X.$$

The upcoming calculations involving the division algorithm to obtain standard expressions are rather long and tedious, and are best left for the reader to replicate. This ought to elicit an appreciation for the computer algebra systems normally used to carry out such divisions. By and large we present only the standard expressions which result from the division algorithm.

Keep in mind that all polynomial coefficients are in the field $\mathbb{Z}_5$, and are thereby subject to arithmetic modulo 5.

1. The initial S-polynomial is

$$S(g_1, g_2) = Yg_1 - g_2 = 3XY + 4X + Y^3 + Y.$$

Since neither $\mathrm{lm}(g_1)$ nor $\mathrm{lm}(g_2)$ divides $\mathrm{lm}(S(g_1, g_2))$, which is XY, the complete reduction of $S(g_1, g_2)$ modulo g_1, g_2 is $S(g_1, g_2)$ itself. Thus g_1, g_2 do not constitute a Gröbner basis of J.

As discussed in the algorithm, let

$$g_3 = 3XY + 4X + Y^3 + Y$$

and adjoin g_3 to the basis g_1, g_2 of J.

2. The three S-polynomials $S(g_i, g_j)$ for $1 \le i < j \le 3$ are $S(g_1, g_2)$ given above as well as

$$S(g_1, g_3) = Yg_1 - 2Xg_3 = 2X^2 + 3XY^3 + 3XY + Y^3 + Y$$
$$S(g_2, g_3) = g_2 - 2Xg_3 = 2X^2 + 3XY^3 + X.$$

The complete reductions of these two S-polynomials modulo g_1, g_2, g_3 can be computed using the division algorithm 9.18. Start with $S(g_1, g_3)$. The steps shown follow the proof in the division algorithm with a suitable choice of divisors at each step. Keep arithmetic modulo 5 in mind.

$$2X^2 + 3XY^3 + 3XY + Y^3 + Y$$
$$= (X^2 + Y^2 + 1)2 + (3XY^3 + 3XY + Y^3 + 3Y^2 + Y + 3)$$
$$3XY^3 + 3XY + Y^3 + 3Y^2 + Y + 3$$
$$= (X^2Y + 2XY + X)0 + (3XY^3 + 3XY + Y^3 + 3Y^2 + Y + 3)$$

$$\begin{aligned}
&3XY^3 + 3XY + Y^3 + 3Y^2 + Y + 3 \\
&\quad = (3XY + 4X + Y^3 + Y)Y^2 + (XY^2 + 3XY + 4Y^5 + 3Y^2 + Y + 3) \\
&XY^2 + 3XY + 4Y^5 + 3Y^2 + Y + 3 \\
&\quad = (X^2 + Y^2 + 1)0 + (XY^2 + 3XY + 4Y^5 + 3Y^2 + Y + 3) \\
&XY^2 + 3XY + 4Y^5 + 3Y^2 + Y + 3 \\
&\quad = (X^2Y + 2XY + X)0 + (XY^2 + 3XY + 4Y^5 + 3Y^2 + Y + 3) \\
&XY^2 + 3XY + 4Y^5 + 3Y^2 + Y + 3 \\
&\quad = (3XY + 4X + Y^3 + Y)2Y + (4Y^5 + 3Y^4 + Y^2 + Y + 3).
\end{aligned}$$

By collecting quotients we come to

$$S(g_1, g_3) = 2g_1 + 0g_2 + (Y^2 + 2Y)g_3 + 4Y^5 + 3Y^4 + Y^2 + Y + 3.$$

Since none of $\text{lm}(g_1)$, $\text{lm}(g_2)$, $\text{lm}(g_3)$ divides Y^5, we have here a standard expression for $S(g_1, g_2)$ with complete reduction $4Y^5 + 3Y^4 + Y^2 + Y + 3$ modulo g_1, g_2, g_3.

Still g_1, g_2, g_3 do not constitute a Gröbner basis of J. As described in the algorithm, expand the basis of J by adjoining

$$g_4 = 4Y^5 + 3Y^4 + Y^2 + Y + 3$$

to the generators g_1, g_2, g_3.

3. The six S-polynomials $S(g_i, g_j)$ for $1 \le i < j \le 4$ are the three given above as well as

$$\begin{aligned}
S(g_1, g_4) &= Y^5 g_1 + X^2 g_4 = 3X^2Y^4 + X^2Y^2 + X^2Y + 3X^2 + Y^7 + Y^5 \\
S(g_2, g_4) &= Y^4 g_2 + X^2 g_4 = 3X^2Y^4 + X^2Y^2 + X^2Y + 3X^2 + 2XY^5 + XY^4 \\
S(g_3, g_4) &= 2Y^4 g_3 + X g_4 = XY^4 + XY^2 + XY + 3X + 2Y^7 + 2Y^5.
\end{aligned}$$

Standard expressions for these six S-polynomials modulo $g_1, \ldots, g_4$ are exhibited below. The first standard expression can be written by inspection. The others come from application of the division algorithm. All can be readily verified.

$$\begin{aligned}
S(g_1, g_2) &= 3XY + 4X + Y^3 + Y \\
&= 0g_1 + 0g_2 + 1g_3 + 0g_4 + 0 \\
S(g_1, g_3) &= 2X^2 + 3XY^3 + 3XY + Y^3 + Y \\
&= 2g_1 + 0g_2 + (Y^2 + 2Y)g_3 + 1g_4 + 0 \\
S(g_2, g_3) &= 2X^2 + 3XY^3 + X \\
&= 2g_1 + 0g_2 + (Y^2 + 2Y + 4)g_3 + 1g_4 + 0 \\
S(g_1, g_4) &= 3X^2Y^4 + X^2Y^2 + X^2Y + 3X^2 + Y^7 + Y^5 \\
&= 3Y^4 g_1 + Y g_2 + (2X + Y^2 + 3Y)g_3 + 4Y^2 g_4 + 0
\end{aligned}$$

$$\begin{aligned}S(g_2,g_4) &= 3X^2Y^4 + X^2Y^2 + X^2Y + 3X^2 + 2XY^5 + XY^4\\&= 3Y^4g_1 + Yg_2 + (2X + 4Y^4 + Y^2 + 3Y)g_3 + 4Y^2g_4 + 0\\S(g_3,g_4) &= XY^4 + XY^2 + XY + 3X + 2Y^7 + 2Y^5\\&= 0g_1 + 0g_2 + (2Y^3 + 4Y^2 + 2)g_3 + (3Y^2 + Y)g_4 + 0.\end{aligned}$$

Since the complete reductions of the $S(g_i, g_j)$ for $1 \leq 1 < j \leq 4$ are 0, these four g_i form a Gröbner basis of the ideal J generated by g_1, g_2.

The leading monomial XY of g_3 divides the leading monomial X^2Y of g_2. This makes g_2 redundant, and leaves us with g_1, g_3, g_4 as a minimal Gröbner basis of J. We can rescale g_3, g_4 to be monic and obtain the polynomials

$$f_1 = X^2 + Y^2 + 1, f_2 = XY + 3X + 2Y^3 + 2Y, f_3 = Y^5 + 2Y^4 + 4Y^2 + 4Y + 2$$

as a minimal Gröbner basis of J. It might be worth observing from this example that the number of elements in a minimal Gröbner basis can exceed the original number of generators of an ideal.

Let us put our newly found Gröbner basis to use. Let

$$f = X^3Y + X^2 + XY + 3X.$$

Is $f \in J$?

If we apply the division algorithm to f using the original (non-Gröbner) basis g_1, g_2 of J, starting the division with g_1, we come up with the standard expression

$$f = (XY + 1)g_1 + 0g_2 + 4XY^3 + 3X + 4Y^2 + 4.$$

Since the complete reduction $4XY^3 + 3X + 4Y^2 + 4$ modulo g_1, g_2 is non-zero, we might surmise that $f \notin J$. However, there is no such test when the basis is not Gröbner. Instead, let us get a complete reduction of f modulo the reduced Gröbner basis f_1, f_2, f_3. By applying the division algorithm we discover the following standard expression:

$$f = (XY + 1)f_1 + (4Y^2 + 3Y + 1)f_2 + 2f_3 + 0.$$

Since the complete reduction of f modulo the Gröbner basis f_1, f_2, f_3 is 0, we see that $f \in J$. In fact, a clever inspection reveals that $f \in J$ because $f = (X+3)g_2$. However, clever inspections have sporadic utility, while the tools of Gröbner bases and the division algorithm, despite their complexity, are always available.

EXERCISES

1. Let R be the Gröbner domain $\mathbb{R}[X, Y, Z]$ using lexicographic order on the monomials taking $X > Y > Z$. Show that $X - Z^2, Y - Z^3$ form a Gröbner basis for the ideal of R that they generate.

2. Take the degree lexicographic order on the monomials of $\mathbb{Q}[X, Y]$ with $X > Y$. Show that the polynomials $Y^3 - 2XY, 2X^3 - XY^2 - Y$ form a Gröbner basis for the ideal that they generate.
3. Take the lexicographic ordering with $X > Y$ on the polynomial ring $K[X, Y]$. Decide if the polynomials

$$g_1 = X^3Y - XY^2 + 1, \qquad g_2 = X^2Y^2 - Y^3 - 1$$

form a Gröbner basis for the ideal that they generate.

Do they form a Gröbner basis using the lexicographic ordering with $Y > X$?

What if we use the degree lexicographic ordering with $X > Y$? What about with $Y > X$?

4. In the ring $\mathbb{Z}_2[X, Y]$, with lexicographic order on the monomials taking $Y > X$, find a Gröbner basis for the ideal $\langle X^3 + Y^2, XY + 1\rangle$. Find a Gröbner basis taking $X > Y$.
5. Let R be a polynomial ring over a field endowed with a monomial order. Show that any finite set of monomials in R constitutes a Gröbner basis for the ideal that they generate.
6. For a polynomial ring R in several variables over a field with a suitable leading monomial function, show that $S(hf, hg) = hS(f, g)$ for every non-zero f, g, h in R.
7. Take the lexicographic monomial ordering with $X > Y$ on the monomials of $\mathbb{Q}[X, Y]$.
   (a) Use Buchberger's theorem to verify that

$$X^2 + XY^5 + Y^4, \quad XY^6 - XY^3 + Y^5 - Y^2, \quad XY^5 - XY^2, \quad Y^5 - Y^2$$

   form a Gröbner basis of the ideal that they generate.

   (b) Find a minimal Gröbner basis for this ideal.
   (c) With the degree lexicographic ordering and $X > Y$, show that

$$X^3 - Y^3, \quad Y^4 + XY^2 + X^2, \quad XY^3 + X^2Y + Y^2$$

   give a Gröbner basis for this ideal.

8. Let J be the ideal in $\mathbb{Z}_5[X, Y]$ generated by

$$X^2 + Y^2 + 1 \text{ and } X^2Y + 2XY + X.$$

Is $3X^2Y + XY^2$ in J?

9. Let $R = \mathbb{Q}[X, Y, Z, T]$ with the lexicographic ordering taking $X > Y > Z > T$. Suppose $\lambda \in \mathbb{Q}$, and let $\varphi_\lambda : R \to \mathbb{Q}[X, Y, Z]$ be the homomorphism given by

$$X \mapsto X, \quad Y \mapsto Y, \quad Z \mapsto Z \text{ and } T \mapsto \lambda.$$

Let J be the ideal of R generated by

$$(T - 1)X^2Y, \ (T - 2)XY^2 + Z.$$

Find a Gröbner basis G for this ideal.

If $\lambda \neq 0, 1, 2$, prove that $\varphi_\lambda(G)$ is a Gröbner basis of the ideal $\varphi_\lambda(J)$ in $\mathbb{Q}[X, Y, Z]$.

Show that $\varphi_2(G)$ is not a Gröbner basis of $\varphi_2(J)$.

10.* This is a modification of the construction of S-polynomials. This is a challenging theoretical problem. Let R be a polynomial ring over a field, in several variables and endowed with a monomial ordering. For each pair of monic f, g in R select monic polynomials a, b such that

$$\mathrm{lm}(a) = \frac{\mathrm{lcm}(\mathrm{lm}(f), \mathrm{lm}(g))}{\mathrm{lm}(f)} \text{ and } \mathrm{lm}(b) = \frac{\mathrm{lcm}(\mathrm{lm}(f), \mathrm{lm}(g))}{\mathrm{lm}(g)}.$$

Put $S(f, g) = af - bg$. The difference between this construction and the traditional one is that the a, b need not be monomials.

Suppose that $g_1, g_2, \ldots, g_n$ generate an ideal J of R. Also suppose that for each pair of indices i, j from 1 to n such that $i < j$, the polynomial $S(g_i, g_j)$ either is 0 or has a standard expression modulo $g_1, \ldots, g_n$ with complete reduction 0. Prove that $g_1, \ldots, g_n$ constitute a Gröbner basis of J.

11. Let $R = F[X_1, \ldots, X_k, Y]$ be a polynomial ring in $k + 1$ variables over a field F, and take a monomial ordering such that $Y > X_j^n$ for all j and all positive exponents n. If a, b, c, d are monic polynomials in $F[X_1, \ldots, X_k]$ such that

$$\gcd(\mathrm{lm}(ad), \mathrm{lm}(bc)) = 1,$$

prove that

$$aY + c, \quad bY + d, \quad ad - bc$$

constitute a Gröbner basis of the ideal generated by $aY + c, bY + d$.

Hint. Use the generalized S-polynomials of the preceding exercise.

9.4 Applications of Gröbner Bases

We have seen that Gröbner bases offer a method for deciding whether f in a polynomial ring belongs to an ideal generated by a given set of polynomials. First, use Buchberger's algorithm to construct a Gröbner basis of the ideal. Then use the division algorithm to find a complete reduction of f modulo that Gröbner basis. The polynomial f is in the ideal if and only if that complete reduction is 0.

Since ideal membership can be determined using Gröbner bases, ideal inclusion can also be determined. If I is an ideal generated by $f_1, \ldots, f_k$ and an ideal J is generated by $g_1, \ldots, g_n$, the inclusion $I \subseteq J$ is decided by computing a Gröbner basis of J and then using this to decide the membership of each f_i in J. Equality of I and J could then be determined by establishing the inclusion $J \subseteq I$. Of course calculations can get lengthy, and computer algebra systems are a godsend.

Other applications of Gröbner bases, to which we now turn, include finding bases for ideal intersections, and developing computational tools for finitely generated algebras over fields.

9.4.1 Gröbner Bases for Ideal Intersections

The ideals of a polynomial ring $K[X]$ in one variable X over a field K are principal. If $\langle f\rangle, \langle g\rangle$ are two such ideals, how does one compute a generator h for the intersection ideal $\langle f\rangle \cap \langle g\rangle$? Well, the polynomials k in $\langle f\rangle \cap \langle g\rangle$, are those such that $f \mid k$ and $g \mid k$. In turn, a generator h of the intersection divides all such k. Thus h is a common multiple of f, g which divides all other common multiples of f, g. There is only one such h if we insist that it be monic, and we call h the ***least common multiple*** of f and g, denoted by $\operatorname{lcm}(f,g)$.

As discussed in Section 5.5 the greatest common divisor $\gcd(f,g)$ of f, g is the unique monic polynomial that divides both f and g and is divisible by all other common divisors of f and g. It is also the unique monic generator of the ideal $\langle f, g\rangle$. The Euclidean algorithm as discussed in Section 5.5 provides a way to calculate $\gcd(f,g)$. By adapting the ideas of Section 5.5 based on unique factorization we can discover that

$$\gcd(f,g)\operatorname{lcm}(f,g) = fg.$$

Thus, a way to compute the generator $\operatorname{lcm}(f,g)$ of the intersection $\langle f\rangle \cap \langle g\rangle$ has revealed itself. Compute $\gcd(f,g)$ by the Euclidean algorithm and write $\operatorname{lcm}(f,g) = fg/\gcd(f,g)$.

When the polynomial ring has more than one variable, Gröbner bases become the tool for calculating generators for the intersection of ideals. Let us see how that can be done.

Gröbner Bases for Subrings of Gröbner Domains

Even though we shall use it only for rings of polynomials, the upcoming result might be more transparent when stated for general Gröbner domains.

A subring of a Gröbner domain need not be a Gröbner domain. For example, every field K is a Gröbner domain. In fact, every function $K^\star \to \mathbb{N}$ is Euclidean. Indeed, for all a, b in K where $a \neq 0$ we have $b = a\frac{b}{a} + 0$. In particular, the field $\mathbb{Q}(X_1, X_2, \ldots, X_n, \ldots)$ of rational functions in infinitely many variables is Gröbner. However, its subring, $\mathbb{Q}[X_1, X_2, \ldots, X_n, \ldots]$, of polynomials in those variables cannot be Gröbner because it is not Noetherian.

Yet there is a useful case where subrings of Gröbner domains remain Gröbner.

Proposition 9.30. *Suppose $\preceq$ is a well-partial order with a total order extension $\leq$ on a set M, and that $\phi : R^\star \to M$ is a Gröbner function on an integral domain R. If A is a subring of R such that*

$$\phi(a) < \phi(c) \text{ and } ac \in R \setminus A \text{ for all } a \text{ in } A^\star \text{ and all } c \text{ in } R \setminus A,$$

then A is a Gröbner domain whose Gröbner function is the restriction of ϕ to $A^\star$. Furthermore, if B is a Gröbner basis for an ideal J inside R, then $B \cap A$ is a Gröbner basis for the ideal $J \cap A$ inside A.

Proof. Suppose $a, b \in A^\star$ and that $\phi(a) \preceq \phi(b)$. Since ϕ is a Gröbner function on R, there exist q, r in R such that

$$b = aq + r \text{ where } \phi(r) < \phi(b) \text{ or } r = 0.$$

It follows that $r \in A$, for otherwise an element of $R \setminus A$ would sit below an element of A, contrary to assumption. Then $aq \in A$, and since $a \in A$ it follows from the second assumption about A that $q \in A$. Thus the restriction of ϕ to A is a Gröbner function.

Next, if $a \in (J \cap A)^\star$, then there is an element b in the Gröbner basis B of J such that $\phi(b) \preceq \phi(a)$. The same inequality holds with the total order extension of $\preceq$, namely $\phi(b) \leq \phi(a)$. It follows that $b \in A$, because the alternative that $b \in B \setminus A$ would, according to our assumption, imply $\phi(a) < \phi(b)$. Such b in $B \cap A$ constitute a Gröbner basis of $J \cap A$. □

Application to Polynomial Rings

The primary illustration of Proposition 9.30 comes from polynomials. Suppose that $R = K[X_1, X_2, \dots, X_k, Y]$ is a polynomial ring in several variables X_j and an extra variable Y. Let A be the subring $K[X_1, \dots, X_k]$ of polynomials involving only the X_j. Suppose as well that the monomial order on R is such that all monomials in which a positive power of Y appears exceed all monomials in which only X_j appear. That is, the monomials inside $R \setminus A$ are higher than the monomials inside A. We shall refer to such a monomial order as ***having dominant monomials in*** Y.

For example, the lexicographic order taking $Y > X_1 > X_2 > \cdots > X_k$ has dominant monomials in Y. More generally, we can take any monomial order $\leq$ on the monomials of the subring A and extend this to a monomial order $\leq$ (using the same inequality symbol) on the monomials of R as follows. Write the monomials of R in the form $Y^e h$, where $e \geq 0$ and h is a monomial of A. Then declare that

$$Y^{e_1} h_1 \leq Y^{e_2} h_2$$

if and only if

$$e_1 < e_2, \text{ or } e_1 = e_2 \text{ together with } h_1 \leq h_2.$$

We omit the routine verification that this is a monomial order on R. This order is custom built to have dominant monomials in Y.

As we now confirm, monomial orders with dominant monomials in Y satisfy Proposition 9.30.

Proposition 9.31. *If the polynomial ring $R = K[X_1, \dots, X_k, Y]$ has a monomial order with dominant monomials in Y and A is the subring $K[X_1, \dots, X_k]$, then, using the same monomial order, the restriction to $A^\star$ of the leading monomial function on $R^\star$ remains a Gröbner function. Furthermore if J is an ideal of R with Gröbner basis B, then a Gröbner basis of $J \cap A$ is $B \cap A$.*

Proof. The subring A satisfies the assumptions of Proposition 9.30, and thereby its conclusions. Indeed, suppose $f \in A^\star$ and $g \in R \setminus A$. Since monomials with a Y in them are higher than those without, and since g has monomials with a Y in them, it must be that $\mathrm{lm}(g)$ has a Y in it. And because $\mathrm{lm}(f)$ has no Y in it, we see that $\mathrm{lm}(f) < \mathrm{lm}(g)$. Furthermore the equality $\mathrm{lm}(fg) = \mathrm{lm}(f)\,\mathrm{lm}(g)$ reveals that $\mathrm{lm}(fg)$ has a Y in it, which then puts fg inside $R \setminus A$. □

The ideal $J \cap A$ eliminates from J all polynomials that have a Y in their monomials. Proposition 9.31 provides a Gröbner basis for such an ideal, which is typically called an ***elimination ideal***.

Realization of Ideal Intersections as Elimination Ideals

Here is a quirky result which interprets the intersection of two ideals in a polynomial ring as an elimination ideal. We remind ourselves that if I, J are ideals in a commutative ring R, and $a, b \in R$, then $aI + bJ$ is the ideal consisting of elements of the form $ax + by$ where $x \in I$ and $y \in J$.

Proposition 9.32. *If A is any commutative ring and $A[Y]$ is the ring of polynomials in the variable Y with coefficients in A and if I, J are ideals of A, then*

$$I \cap J = (YI + (1 - Y)J) \cap A.$$

Proof. Suppose $x \in I \cap J$. Since x is also in A and $x = Yx + (1 - Y)x$, evidently x is in $(YI + (1 - Y)J) \cap A$.

For the reverse inclusion suppose $x \in (YI + (1 - Y)J) \cap A$. Thus there exist u in I and v in J such that

$$x = Yu + (1 - Y)v = Y(u - v) + v.$$

Because $x \in A$ it follows that $u - v = 0$. Then $x = v$, which puts x inside J, and since $u = v$, also inside I. □

An Algorithm for Gröbner Bases of Ideal Intersections

Propositions 9.31 and 9.32, working together, provide an algorithm for computing a Gröbner basis for the intersection of two ideals in a polynomial ring.

Suppose $A = K[X_1, \dots, X_k]$ with a selected monomial order, and I, J are non-zero ideals of A with generating sets $B = \{f_1, \dots, f_n\}, C = \{g_1, \dots, g_m\}$, respectively. Here is how to obtain a Gröbner basis for the intersection $I \cap J$.

- Following Proposition 9.32, take a new variable Y, and let R be the larger polynomial ring $A[Y]$, which of course is the same as $K[X_1, \dots, X_k, Y]$.
- Extend the monomial order on A to the monomials of R so that it has dominant monomials in Y.
- By Proposition 9.32, $I \cap J = (YI + (1 - Y)J) \cap K[X_1, \dots, X_k]$. By Proposition 9.31 applied to the ideal $YI + (1 - Y)J$ inside R, a Gröbner basis of $(YI + (1 - Y)J) \cap K[X_1, \dots, X_k]$, and thereby a Gröbner basis of $I \cap J$, is the intersection of a Gröbner basis of the ideal $YI + (1 - Y)J$ with A.
- This brings it down to finding a Gröbner basis of $YI + (1 - Y)J$. To do that, observe that a set of generators for $YI + (1 - Y)J$ consists of the polynomials

  $$Yf_1, Yf_2, \dots, Yf_n, \ (1 - Y)g_1, (1 - Y)g_2, \dots, (1 - Y)g_m$$

  inside $K[X_1, \dots, X_k, Y]$.

- The monomial order in $K[X_1, \ldots, X_k, Y]$ is the one having dominant monomials in Y initially imposed on this polynomial ring. Using this monomial order carry out Buchberger's algorithm on the above set of generators and from the resulting Gröbner basis pick out those that are in the ring $K[X_1, \ldots, X_k]$.

An Illustration of the Ideal Intersection Algorithm

Here is a demonstration of how the ideal intersection algorithm works. The calculations, as is typical when Buchberger's algorithm is invoked, can go on for a long time. We take an example which can be done in reasonable time by hand calculations. For most situations it might be wise to resort to a computer algebra system.

On the monomials of $\mathbb{Q}[X, Y]$ impose the lexicographic order with $X > Y$. Let I, J be the principal ideals of $\mathbb{Q}[X, Y]$ generated by $f = X + Y, g = XY$ respectively. To get a Gröbner basis of $I \cap J$, introduce a new variable, say Z, and work in the larger ring $\mathbb{Q}[X, Y, Z]$.

We adopt the lexicographic order on monomials with $Z > X > Y$, since this order has dominant monomials in Z and extends the original lexicographic order on the monomials of $\mathbb{Q}[X, Y]$. The polynomials to follow are written with monomials in descending order.

- To get a Gröbner basis of $I \cap J$ by our algorithm, let

$$h_1 = Z(X + Y) = ZX + ZY, \; h_2 = (1 - Z)XY = -ZXY + XY,$$

  and compute a Gröbner basis for the ideal $N = \langle h_1, h_2 \rangle$ inside $\mathbb{Q}[X, Y, Z]$.
- Following Buchberger's algorithm

$$S(h_1, h_2) = ZY^2 + XY.$$

  By inspection, this polynomial is its own complete reduction modulo h_1, h_2. So, h_1, h_2 do not form a Gröbner basis of N. Buchberger's algorithm says to put

$$h_3 = ZY^2 + XY,$$

  and adjoin this to the basis h_1, h_2 of N.
- Next calculate

$$S(h_1, h_3) = ZY^3 - X^2Y \text{ and } S(h_2, h_3) = -X^2Y - XY^2.$$

  Since $S(h_2, h_3)$ is already its own complete reduction modulo h_1, h_2, h_3, we know that h_1, h_2, h_3 is not a Gröbner basis for the ideal N. Buchberger's algorithm says to put

$$h_4 = -X^2Y - XY^2,$$

  and adjoin this to the basis h_1, h_2, h_3 of N.

  At this point we could have calculated the complete reduction of $S(h_1, h_3)$ modulo h_1, h_2, h_3, using the division algorithm. After finding that it is non-zero, we could have adjoined that reduction instead to the basis h_1, h_2, h_3. But the chosen h_4 was slightly easier to come by.

- Then obtain

$$S(h_1,h_4) = 0,\ S(h_2,h_4) = -ZXY^2 - X^2Y,\ S(h_3,h_4) = -ZXY^3 + X^3Y.$$

- At this point it turns out that the complete reductions of the S-polynomials $S(h_i,h_j)$ with $1 \le i < j \le 4$ are all 0. Here are standard expressions for the $S(h_i,h_j)$ obtained either by inspection or by the division algorithm.

$$\begin{aligned}
S(h_1,h_2) &= ZY^2 + XY = 0h_1 + 0h_2 + 1h_3 + 0h_4 + 0\\
S(h_1,h_3) &= ZY^3 - X^2Y = 0h_1 + 0h_2 + Yh_3 + 1h_4 + 0\\
S(h_2,h_3) &= -X^2Y - XY^2 = 0h_1 + 0h_2 + 0h_3 + 1h_4 + 0\\
S(h_1,h_4) &= 0\\
S(h_2,h_4) &= -ZXY^2 - X^2Y = 0h_1 + Yh_2 + 0h_3 + 1h_4 + 0\\
S(h_3,h_4) &= -ZXY^3 + X^3Y = 0h_1 + Y^2h_2 + 0h_3 + (-X + Y)h_4 + 0.
\end{aligned}$$

  Buchberger's algorithm stops.

  A Gröbner basis of N consists of h_1,h_2,h_3,h_4. Since $\mathrm{lm}(h_1)$ divides $\mathrm{lm}(h_2)$, the polynomial h_2 is redundant, and a minimal Gröbner basis of N consists of h_1,h_3,h_4.
- According to the intersection algorithm, a Gröbner basis for $I \cap J$ is the intersection of the above Gröbner basis for N with $\mathbb{Q}[X, Y]$. By inspection that intersection is the polynomial $h_4 = -X^2Y - XY^2$, which for appearances we replace by $X^2Y + XY^2$.

Computing the lcm and gcd of Two Polynomials

The preceding example also illustrates a method for calculating least common multiples and greatest common divisors of two polynomials in several variables.

Suppose that f,g are non-zero polynomials in $K[X_1,\dots,X_k]$, a unique factorization domain. Let

$$f = p_1^{d_1}p_2^{d_2}\cdots p_r^{d_r} \text{ and } g = p_1^{e_1}p_2^{e_2}\cdots p_r^{e_r},$$

with the exponents $d_i, e_i \ge 0$, be the unique prime factorizations of f and g. The polynomial

$$\ell = p_1^{\max(d_1,e_1)}p_2^{\max(d_2,e_2)}\cdots p_r^{\max(d_r,e_r)}$$

is clearly divisible by both f and g. This puts ℓ in the ideal intersection $\langle f\rangle \cap \langle g\rangle$. Since ℓ also divides every polynomial that is divisible by f and g, we see that ℓ is the generator of the intersection $\langle f\rangle \cap \langle g\rangle$. This shows that the intersection of two principal ideals in a polynomial ring remains a principal ideal.

The generator ℓ of the intersection (unique up to a scalar multiple) is of course the least common multiple of f and g, denoted by $\mathrm{lcm}(f,g)$. We see that in the case when two ideals $\langle f\rangle, \langle g\rangle$ are principal, the algorithm for finding a Gröbner basis for their intersection will calculate $\mathrm{lcm}(f,g)$.

In the preceding example in the ring $\mathbb{Q}[X, Y]$ we learned, in a purely mechanical way, that $\mathrm{lcm}(X + Y, XY) = X^2Y + XY^2$. Of course this could have been seen with a bit of insight. Indeed, the polynomial $X + Y$ is irreducible. One way to see this is to take a monomial order where $X > Y$ and suppose $X + Y = uv$ for some polynomials u, v. Then $X = \text{lm}(uv) = \text{lm}(u)\,\text{lm}(v)$. This forces one of $\text{lm}(u)$ or $\text{lm}(v)$ to be 1, whence u or v must be a scalar. Clearly XY is the product of irreducibles X, Y. Hence, by its very definition, $\text{lcm}(X + Y, XY) = (X + Y)XY = X^2Y + XY^2$, just as the algorithm had obtained. While it can be satisfying to find this least common multiple by such ad hoc reasoning, we should appreciate that the algorithm, despite its plodding computations, is reliable when unique factorizations are not so readily available.

Finally, the greatest common divisor of f, g from above is given by

$$\gcd(f,g) = p_1^{\min(d_1,e_1)} p_2^{\min(d_2,e_2)} \cdots p_r^{\min(d_r,e_r)}.$$

This is a common factor of f, g that is divisible by all common factors of f and g. The formulas for the least common multiple and greatest common divisor reveal that

$$\gcd(f,g)\,\text{lcm}(f,g) = fg.$$

Hence the discovery of $\text{lcm}(f,g)$ by means of the algorithm that gets a Gröbner basis of $\langle f\rangle \cap \langle g\rangle$, also discovers $\gcd(f,g)$. Just divide $\text{lcm}(f,g)$ into the product fg using the division algorithm.

9.4.2 Units and Zero Divisors in Finitely Generated Algebras

A commutative ring R, containing a field K as a subring, is commonly known as a commutative ***K-algebra*** and also as a commutative algebra over K. Such R is said to be ***finitely generated*** over K when there exist elements $\alpha_1, \ldots, \alpha_k$ in R such that

$$A = K[\alpha_1, \ldots, \alpha_k].$$

This means that there is the surjective substitution map

$$\varphi : K[X_1, \ldots, X_k] \to A, \text{ given by } f(X_1, \ldots, X_k) \mapsto f(\alpha_j, \ldots, \alpha_k).$$

For instance, every field extension of finite degree over K is a finitely generated K-algebra. Also, if T is a linear operator on a vector space, the ring $K[T]$ of all operators that are polynomials in T is finitely generated (in fact singly generated) over K. Clearly all homomorphic images of polynomial rings in finitely many variables are finitely generated algebras. Finitely generated commutative algebras are pervasive in algebraic geometry. It is valuable to have tools for computing within them.

The first isomorphism theorem reveals that A is isomorphic to the quotient ring R/J where $J = \ker\varphi$. So, we might as well work with quotients of polynomial rings modulo their ideals. This opens up a way for Gröbner bases to play a role.

Detecting Units in Quotient Rings

Gröbner bases afford an algorithm for deciding if an element in R/J has an inverse. In general, if R is any commutative ring with an ideal J, an element $f + J$ of the quotient ring R/J has an inverse if and only if there exists a $g + J$ in R/J such that $fg + J = 1 + J$. This is the same as saying that there is a g in R and an h in J such that $fg + h = 1$. In turn this is equivalent to saying that the enlarged ideal $\langle J,f\rangle$, generated by J and f, is all of R.

In the special case where R is a polynomial ring with a selected monomial order, the equality $\langle J,f\rangle = R$ can be determined in practice by calculating a Gröbner basis of the ideal $\langle J,f\rangle$. From the definition of Gröbner bases, an ideal is all of R if and only if its Gröbner basis contains a constant polynomial, which can always be rescaled to be 1.

Here is an illustration of the above principle. Suppose R is the polynomial ring $\mathbb{Q}[X, Y]$ with lexicographic order on monomials, taking $X > Y$. Let J be the ideal generated by

$$g_1 = X - 2Y^2,\ g_2 = Y^3.$$

If

$$f = Y + 1,$$

let us see if $f + J$ has an inverse in R/J. As explained, this comes down to finding a Gröbner basis of the larger ideal generated by g_1, g_2, f, and seeing whether or not 1 lies in such a basis.

The relevant S-polynomials and their standard expressions with complete reductions modulo g_1, g_2, f, found by the division algorithm if need be, are shown below:

$$\begin{aligned} S(g_1,g_2) &= -2Y^5 = 0g_1 - 2Y^2g_2 + 0f + 0 \\ S(g_1,f) &= -X - 2Y^3 = -1g_1 - 2g_2 + (-2Y + 2)f + 2 \\ S(g_2,f) &= -Y^2 = 0g_1 + 0g_2 + 0f - Y^2. \end{aligned}$$

From the standard expression for $S(g_1,f)$ it is apparent that 2 lies in the ideal $\langle g_1, g_2, f\rangle$, and thereby so does 1 lie in this ideal. In other words, this ideal is all of R with minimal Gröbner basis consisting of 1. Furthermore, rearrangement of the standard expression for $S(g_1,f)$ reveals that

$$(Y + 1)g_1 + 2g_2 + (-X + 2Y - 2)f = 2.$$

In the quotient ring R/J we obtain

$$(-X + 2Y - 2 + J)(f + J) = 2 + J.$$

With $\alpha = X + J$ and $\beta = Y + J$, the above translates into saying that

$$(-\alpha + 2\beta - 2)(\beta + 1) = 2,$$

which reveals the inverse of $\beta + 1$ in R/J to be $-\frac{1}{2}\alpha + \beta - 1$.

With more complicated examples a computer algebra system will expedite calculations such as the preceding.

Detecting Zero Divisors in Quotient Rings

If J is a proper ideal in a polynomial ring R, a polynomial f causes the element $f + J$ in the quotient ring R/J to be a zero divisor if and only if $f \notin J$ and there exists a polynomial g also not in J such that $fg \in J$. Let

$$J_f = \{g \in R : fg \in J\}.$$

A moment's reflection shows that J_f is an ideal, and that $J \subseteq J_f$. Also $f \notin J$ if and only if $J_f \neq R$. Thus $f + J$ is a zero divisor in R/J if and only if $J \subsetneq J_f \subsetneq R$.

If $f \neq 0$, every element of the intersection $J \cap \langle f \rangle$ is divisible by f and thereby the set

$$\frac{1}{f}(J \cap \langle f \rangle) = \left\{ \frac{h}{f} : h \in J \cap \langle f \rangle \right\}$$

is inside R. In fact, it is an ideal of R.

Proposition 9.33. *If J is an ideal in a polynomial ring R and f is a non-zero polynomial, then*

$$J_f = \frac{1}{f}(J \cap \langle f \rangle).$$

Proof. This is the same as showing that $fJ_f = J \cap \langle f \rangle$. Well, if $g \in J_f$, then $fg \in J$, and since obviously $fg \in \langle f \rangle$, it follows that $fg \in J \cap \langle f \rangle$. Thus $fJ_f \subseteq J \cap \langle f \rangle$. For the reverse inclusion let $h \in J \cap \langle f \rangle$. Thus $h = fg$ for some g in R, and $fg \in J$. This puts g inside J_f, whereby $h = fg \in fJ_f$. □

It follows from Proposition 9.33 that a non-zero f in R causes $f + J$ to be a zero divisor in R/J if and only if $f \notin J$ and the ideal J is properly inside $\frac{1}{f}(J \cap \langle f \rangle)$. If B is a basis for the ideal $J \cap \langle f \rangle$, the ideal J is properly inside $\frac{1}{f}(J \cap \langle f \rangle)$ if and only if $f \notin J$ and some g from the basis B is such that $\frac{g}{f} \notin J$.

This points to an algorithm for detecting zero divisors in R/J.

- Select a monomial order on R and obtain a Gröbner basis C of J, possibly using Buchberger's algorithm.
- Given a non-zero f in R, determine whether or not $f \in J$ by finding its complete reduction modulo C. If the complete reduction is 0, then $f \in J$ which makes $f + J = 0$, leaving nothing left to check. Continue as follows if $f \notin J$.
- Starting with the generating set C of J and the generator f of $\langle f \rangle$, obtain a basis B of $J \cap \langle f \rangle$ by means of the ideal intersection algorithm. That basis will be a Gröbner basis, but all we shall need is that it generates the intersection.
- To see whether J is properly inside $\frac{1}{f}(J \cap \langle f \rangle)$, check if $\frac{g}{f} \notin J$ for some g in B. This checking can be done for each g in B by finding the complete reduction of $\frac{g}{f}$ modulo the Gröbner basis C of J.
- The element $f + J$ is a zero divisor if and only if at least one of the above complete reductions is non-zero.

Here is an illustration of the above procedure for detecting zero divisors. Take the ideal J inside $\mathbb{Q}[X, Y]$ generated by

$$g_1 = X^2 + 1 \text{ and } g_2 = Y^2 + 1.$$

Let us see if the polynomial $YX + 1$ reduces to a zero divisor in the quotient ring $\mathbb{Q}[X, Y]/J$.

According to the algorithm above we need a monomial order. Select the lexicographic order on the monomials of $\mathbb{Q}[X,Y]$ taking $X < Y$. Since the leading monomial X^2 of g_1 is coprime with the leading monomial Y^2 of g_2, these polynomials form a Gröbner basis of J, according to Proposition 9.29. The complete reduction of f modulo g_1, g_2 is f itself, because the leading monomials of g_1, g_2 do not divide the leading monomial YX of f. Thus $f \notin J$.

Next we seek a basis of $J \cap \langle YX + 1\rangle$ using the ideal intersection algorithm. Accordingly, take the larger polynomial ring $\mathbb{Q}[X, Y, Z]$ with lexicographic order on monomials where $X < Y < Z$. This extends the lexicographic order that was imposed on $\mathbb{Q}[X, Y]$ in a way that the monomials in Z are dominant. Let

$$
\begin{aligned}
f_1 &= Zg_1 = ZX^2 + Z \\
f_2 &= Zg_2 = ZY^2 + Z \\
f_3 &= (1 - Z)f = (1 - Z)(YX + 1) = -ZYX - Z + YX + 1.
\end{aligned}
$$

According to the ideal intersection algorithm, a Gröbner basis for the ideal of $\mathbb{Q}[X, Y, Z]$ generated by f_1, f_2, f_3 is needed. Buchberger's algorithm will give such a basis after a suitable number of steps.

The first three S-polynomials are as follows:

$$
\begin{aligned}
S(f_1, f_2) &= ZY^2 - ZX^2 \\
S(f_1, f_3) &= ZY - ZX + YX^2 + X \\
S(f_2, f_3) &= -ZY + ZX + Y^2X + Y.
\end{aligned}
$$

The division algorithm reveals that 0 is a complete reduction of $S(f_1, f_2)$ modulo f_1, f_2, f_3. Since the leading monomials of f_1, f_2, f_3 do not divide the leading monomial of $S(f_1, f_3)$, this S-polynomial is its own complete reduction modulo f_1, f_2, f_3. According to Buchberger's algorithm we can put

$$f_4 = ZY - ZX + YX^2 + X.$$

Now the complete reduction of $S(f_2, f_3)$ modulo f_1, f_2, f_3, f_4 obtained by the division algorithm is $Y^2X + YX^2 + Y + X$, which is non-zero. So, we can put

$$f_5 = Y^2X + YX^2 + Y + X.$$

Next we can work out the complete reductions of the $S(f_i, f_j)$ modulo $f_1, \ldots, f_5$ to see if any reduction is non-zero. For instance, $S(f_2, f_4) = ZYX + Z - YX - 1$, and its complete reduction modulo the current batch of five f_j is $-Y^2X^2 + 1$. As this is non-zero we can put

$$f_6 = -Y^2X^2 + 1.$$

In the process of calculating more complete reductions of the $S(f_i, f_j)$ modulo our expanded list of six f_j we see that $S(f_3, f_4) = ZX^2 + Z - YX^3 - YX - X^2 - 1$. The complete reduction of $S(f_3, f_4)$ modulo the f_j is $-YX^3 - YX - X^2 - 1$. Since this reduction is non-zero we can put

$$f_7 = -YX^3 - YX - X^2 - 1.$$

Now it can be checked that the complete reductions of all $S(f_i,f_j)$, modulo the current set of seven f_j, are zero. We omit the tedious details.

Our implementation of Buchberger's algorithm has revealed that a Gröbner basis of the ideal generated by f_1,f_2,f_3 consists of these three polynomials as well as f_4,f_5,f_6,f_7. Since $\mathrm{lm}(f_4)$ divides $\mathrm{lm}(f_2)$ as well as $\mathrm{lm}(f_3)$, and $\mathrm{lm}(f_5)$ divides $\mathrm{lm}(f_6)$, the polynomials f_2,f_3,f_6 can be removed to retain the polynomials f_1,f_4,f_5,f_7 as a minimal Gröbner basis of the ideal generated by f_1,f_2,f_3.

According to the ideal intersection algorithm, a Gröbner basis for the ideal intersection $J \cap \langle YX+1\rangle$ is made up of those polynomials among f_1,f_4,f_5,f_7 that are in $\mathbb{Q}[X,Y]$. That basis consists of

$$f_5 = Y^2X + YX^2 + Y + X = (YX+1)(Y+X)$$

and

$$f_7 = -YX^3 - YX - X^2 - 1 = (YX+1)(X^2+1).$$

These polynomials have been factored to reveal, as expected, that they are divisible by $YX+1$

Finally, we notice that $f_5/(YX+1) = Y+X$, and $Y+X$ is not in the ideal J generated by X^2+1, Y^2+1. Indeed, $Y+X$ is its own complete reduction modulo the Gröbner basis X^2+1, Y^2+1 of J. This proves that $YX+1$ projects down to a zero divisor in the quotient ring $\mathbb{Q}[X,Y]/J$.

Of course, with hindsight, we could let $x = X+J, y = Y+J$ in the quotient ring $\mathbb{Q}[X,Y]/J$. Since we have $x^2 = -1$ and $y^2 = -1$, we can see readily that

$$(yx+1)(y+x) = y^2x + yx^2 + y + x = -x - y + y + x = 0.$$

The point of our lengthy algorithm, however, is not to offer ad hoc solutions, but rather a method that works for all examples.

9.4.3 Normal Forms

Let us recap a familiar story for polynomials in one variable.

If $K[X]$ is a polynomial ring over a field K in one variable X, and J is a non-zero ideal of $K[X]$, the canonical projection $\varphi : K[X] \to K[X]/J$, which sends a polynomial f to its coset $f+J$, embeds the field K as a subring of $K[X]/J$. This makes $K[X]/J$ a vector space over K, by the obvious scalar multiplication $(c,\beta) \mapsto c\beta$ where $c \in K$ and $\beta \in K[X]/J$. If we let α be the coset $X+J$, then the projection map φ coincides with the substitution map $f(X) \mapsto f(\alpha)$, and the elements of $K[X]/J$ take the form $f(\alpha)$ where $f \in K[X]$.

It is possible to have $f(\alpha) = h(\alpha)$ for different f,h in $K[X]$. Indeed, $f(\alpha) = h(\alpha)$ if and only if $f-h$ is in the kernel of φ, which is J. The ideals of $K[X]$ are principal. Let g be a generator of J, and say $\deg g = n$. As $K[X]$ is Euclidean, for any f in $K[X]$ there is a quotient q and a remainder r such that $f = gq + r$ and $\deg r < \deg g$ or $r = 0$. Since $g(\alpha) = 0$ we see by substitution that

$$f(\alpha) = g(\alpha)q(\alpha) + r(\alpha) = r(\alpha).$$

Thus every element of the quotient ring $K[X]/J$ takes the form $r(\alpha)$ where $\deg r < n$ or $r = 0$. Now different polynomials r of degree less than n represent different elements of $K[X]/J$. Indeed, if $r, s \in K[X]$ and both $\deg r$, $\deg s$ are less than n and $r(\alpha) = s(\alpha)$, then g divides $r - s$ in $K[X]$. Since $\deg g = n$ while $\deg(r - s) < n$, it follows that $r - s = 0$, whence $r = s$.

As every element of $K[X]/J$ takes the form $r(\alpha)$ for a unique r in $K[X]$ of degree less than n, or for $r = 0$, it is proper to say that the representations $r(\alpha)$ give ***normal forms*** for the elements of $K[X]/J$. It is also clear that a basis for $K[X]/J$ as a vector space over K is the list

$$\varphi(1) = 1, \varphi(X) = \alpha, \varphi(X^2) = \alpha^2, \ldots, \varphi(X^{n-1}) = \alpha^{n-1},$$

which reveals that $K[X]/J$ has dimension n over K.

Our goal now is to tell this story when the polynomial ring has more than one variable. As might be expected, the replacement for the single generator of an ideal J in $K[X]$ will be a Gröbner basis. Taking that point of view, we might note that $1, X, X^2, \ldots, X^{n-1}$ are precisely the monomials in one variable which are not divisible by the leading monomial of the Gröbner basis g of the principal ideal J in $K[X]$.

Monomials that Are Not Reducible Modulo a Set of Polynomials

Let R be a polynomial ring in several variables over a field K, with a monomial order imposed on the set of monomials M. Let B be a non-empty set of non-zero polynomials. For instance, we have in mind that B could be a Gröbner basis of an ideal, relative to the leading monomial function into M. Define M_B to be the following set of monomials:

$$M_B = \{h \in M : \operatorname{lm}(g) \nmid h \text{ whenever } g \in B\}.$$

Using other parlance, M_B consists of all monomials that are not reducible modulo any g in B. As a tie-in, note that if B consists of finitely many $g_1, \ldots, g_n$ and r is a non-zero complete reduction of a polynomial f modulo the g_i, this means that $\operatorname{lm}(r) \in M_B$. Let us refer to M_B as the set of ***monomials not reducible modulo B***.

Since they are monomials, the elements of M_B form a basis for the vector subspace of R that they span. Denote that span of M_B by V_B.

For an illustration, take $\mathbb{Q}[X, Y]$ with lexicographic order using $X > Y$, and let B equal the set $\{X - 2Y^2, Y^3\}$. By inspection, the monomials not strictly reducible modulo B are $1, Y, Y^2$. Thus V_B is the 3-dimensional space of polynomials in Y having degree at most 2.

For another example, take $\mathbb{Z}_5[X, Y]$ with lexicographic order using $X > Y$. We saw in Section 9.3.2 that the set

$$B = \{X^2 + Y^2 + 1, XY + 3X + 2Y^3 + 2Y, Y^5 + 2Y^4 + 4Y^2 + 4Y + 2\}$$

is a Gröbner basis for the ideal that it generates. By inspection

$$M_B = \{1, X, Y, Y^2, Y^3, Y^4\},$$

whose span V_B is 6-dimensional.

To see yet one more example, take $\mathbb{Q}[X, Y, Z]$ with lexicographic order taking $X > Y > Z$. Let $B = \{X + Z, Y - Z\}$. Here, M_B consists of infinitely many monomials. Namely, $1, Z, Z^2, \ldots$.

It is straightforward to detect when M_B is finite.

Proposition 9.34. *Let B be a non-empty set of non-zero polynomials in a polynomial ring $K[X_1, \ldots, X_n]$ endowed with a monomial order. The set M_B of monomials, which are not reducible modulo B, is finite if and only if for every variable X_j there is a polynomial g in B such that* $\mathrm{lm}(g)$ *is a power of X_j.*

Proof. Suppose that for every variable X_j there is a g in B whose leading monomial is $X_j^{e_j}$ for some exponent e_j. Evidently the monomials not divisible by one of these $X_j^{e_j}$ are $X_1^{d_1} X_2^{d_2} \cdots X_n^{d_n}$ where $0 \leq d_j < e_j$. The number of such monomials is the product $e_1 e_2 \cdots e_n$. Since other elements of B may prevent some of these monomials from sitting in M_B, the size of M_B is at most this product.

Conversely, suppose that no power of some variable, say X_j, is the leading monomial of any g in B. This tells us that the infinitely many monomials $1, X_j, X_j^2, X_j^3, \ldots$ lie in M_B, which of course is not finite. □

A Vector Space Complement for a Polynomial Ideal

Continue with a non-empty set B of non-zero polynomials in the polynomial ring R, endowed with a monomial order on the set of monomials M. Let J be the ideal generated by B. The canonical map $\varphi : R \to R/J$ is a surjection, but every coset $f + J$ in the quotient ring has multiple pre-images in R. This raises the problem of choosing a useful representative for each coset. In case B is a Gröbner basis of J, we now show that the restriction of φ to the subspace V_B spanned by M_B is a K-linear bijection onto R/J. Thus the space V_B provides unique representatives for all elements of R/J.

Proposition 9.35. *Suppose R is a polynomial ring in several variables over a field K, endowed with a monomial order on the set M of monomials. If B is a non-empty set of non-zero polynomials generating an ideal J, then*

$$J + V_B = R.$$

Furthermore, if B is a Gröbner basis of J, the above sum of K-linear spaces is direct.

Proof. Recall that $\mathrm{lt}(h)$ denotes the leading term of a polynomial h, whereby the leading coefficient is multiplied with $\mathrm{lm}(h)$.

Suppose on the contrary that $J + V_B \subsetneq R$. In that case, among the polynomials of $R \setminus (J + V_B)$, select f such that $\mathrm{lm}(f)$ is minimal under the monomial order.

If $\mathrm{lm}(f) \in M_B$, examine $f - \mathrm{lt}(f)$, which is not 0 because $f \notin V_B$. So $\mathrm{lm}(f - \mathrm{lt}(f)) < \mathrm{lm}(f)$. By the minimality of $\mathrm{lm}(f)$, there exists g in J and r in V_B such that $f - \mathrm{lt}(f) = g + r$. Hence $f = g + r + \mathrm{lt}(f)$, which puts f inside $J + V_B$, in contradiction to the choice of f.

If $\mathrm{lm}(f) \notin M_B$, this means that $\mathrm{lm}(g) \mid \mathrm{lm}(f)$ for some g in B and thereby in J. Since lm is a Gröbner function, there exist q, r in R such that

$$f = gq + r, \text{ where either } r = 0 \text{ or else } \operatorname{lm}(r) < \operatorname{lm}(f).$$

Now $r \neq 0$ because $f = gq + 0$ would then be in J. Thus $\operatorname{lm}(r) < \operatorname{lm}(f)$. By the minimality of $\operatorname{lm}(f)$, there exist h in J and s in V_B such that $r = h + s$. Then $f = gq + r = gq + h + s$, which puts f inside $J + V_B$, because $gq + h$ is in the ideal J and $s \in V_B$. This contradiction forces $J + V_B$ to be all of R.

For the second claim, suppose B is a Gröbner basis of J. We need to check that $J \cap V_B = \langle 0 \rangle$. Well, if $f \in J$ and $f \neq 0$, the definition of Gröbner basis provides a g in B such that $\operatorname{lm}(g) \mid \operatorname{lm}(f)$. The definition of V_B then ensures that $f \notin V_B$. Hence, the intersection is trivial. □

If $B = \{g_1, \dots, g_n\}$ generates the ideal J and $f \in R$, the proof of Proposition 9.35 could be adapted to provide an algorithm which obtains quotients q_i in R and a remainder r in V_B such that

$$f = g_1 q_1 + \cdots + g_n q_n + r.$$

Such r would be a complete reduction of f modulo the g_i. This adaptation would prove Proposition 9.20. In fact it would yield Proposition 9.21, but our original division algorithm is good enough for our purposes.

Normal Forms in R/J and Its Dimension

Proposition 9.36. *If B is a Gröbner basis for an ideal J in a polynomial ring R relative to a monomial order, then the canonical projection $\varphi : R \to R/J$ restricted to the K-linear space V_B is a K-linear bijection onto R/J.*

Proof. According to Proposition 9.35 every f in R takes the form $f = g + r$ where $g \in J$ and $r \in V_B$. If $\varphi : R \to R/J$ is the canonical projection, then $\varphi(f) = \varphi(g) + \varphi(r) = \varphi(r)$. Thus the restriction of φ to V_B is a K-linear mapping onto R/J. Furthermore if $f \in V_B$ and $\varphi(f) = 0$ in R/J, then $f \in \ker \varphi$ which is J. By Proposition 9.35, $f = 0$, and so φ is a bijection from V_B to R/J. □

We have just learned that every coset $f + J$ in the quotient ring R/J has a unique representative r in V_B such that $f + J = r + J$. If $X_1, \dots, X_m$ are the variables for the polynomial ring R over a field K, put $\alpha_i = X_i + J$. The canonical projection $\varphi : R \to R/J$ is the substitution map $f(X_1, \dots, X_m) \mapsto f(\alpha_1, \dots, \alpha_m)$. Thus every element of R/J takes the form $r(\alpha_1, \dots, \alpha_m)$ for a unique r in V_B. This $r(\alpha_1, \dots, \alpha_m)$ is the ***normal form*** of an element in R/J. To get a basis for R/J as a vector space over K take the monomials in M_B and evaluate them at $(\alpha_1, \dots, \alpha_m)$.

From Proposition 9.36 the dimension of R/J immediately reveals itself.

Proposition 9.37. *If B is a Gröbner basis for an ideal J in a polynomial ring R relative to some monomial order, then the dimension of R/J equals the size of the set M_B.*

For an illustration of Proposition 9.37, take $\mathbb{Q}[X, Y]$ with degree lexicographic order and $X > Y$. Let J be the ideal generated by

$$g_1 = X^3Y^2 - Y, \quad g_2 = XY^3 - X^2.$$

To get $\dim R/J$ we need a Gröbner basis of J. Our tool is the Buchberger algorithm. To avoid tedium one might resort to a computer algebra system, but for the sake of practice, here is another run through the algorithm. To keep track of things we write the monomials of each polynomial in descending degree lexicographic order.

- The initial S-polynomial is

$$S(g_1, g_2) = Yg_1 - X^2g_2 = X^4 - Y^2.$$

  Since neither X^3Y^2 nor XY^3 divides $\mathrm{lm}(S(g_1, g_2))$, which is X^4, the complete reduction of $S(g_1, g_2)$ modulo g_1, g_2 is $S(g_1, g_2)$ itself. Thus g_1, g_2 do not give a Gröbner basis of J. So let

$$g_3 = X^4 - Y^2,$$

  and adjoin g_3 to the basis g_1, g_2.
- The three $S(g_i, g_j)$ for $1 \le i < j \le 3$ are $S(g_1, g_2)$ as well as

$$S(g_1, g_3) = Y^4 - XY$$
$$S(g_2, g_3) = -X^5 + Y^5.$$

  Since none of the leading monomials X^3Y^2, XY^3, X^4 divides Y^4, this means that $Y^4 - XY$ is already the complete reduction of $S(g_1, g_3)$ modulo g_1, g_2, g_3. Let

$$g_4 = Y^4 - XY$$

  and adjoin this to the basis g_1, g_2, g_3 of J.
- With this expanded basis the six S-polynomials are the three already found as well as

$$S(g_1, g_4) = X^4 - Y^3$$
$$S(g_2, g_4) = 0$$
$$S(g_3, g_4) = -X^5Y + Y^6.$$

  Standard expressions for the six S-polynomials modulo g_1, g_2, g_3, g_4 can be obtained from the division algorithm, if not by inspection. Here they are for verification:

$$S(g_1, g_2) = X^4 - Y^2 = 0g_1 + 0g_2 + 1g_3 + 0g_4 + 0$$
$$S(g_1, g_3) = Y^4 - XY = 0g_1 + 0g_2 + 0g_3 + 1g_4 + 0$$
$$S(g_2, g_3) = -X^5 + Y^5 = 0g_1 + 0g_2 - Xg_3 + Yg_4 + 0$$
$$S(g_1, g_4) = X^4 - Y^3 = 0g_1 + 0g_2 + Yg_3 + 0g_4 + 0$$
$$S(g_2, g_4) = 0 = 0g_1 + 0g_2 + 0g_3 + 0g_4 + 0$$
$$S(g_3, g_4) = -X^5Y + Y^6 = 0g_1 + 0g_2 - XYg_3 + Y^2g_4 + 0.$$

Since the complete reductions of the S-polynomials modulo g_1, g_2, g_3, g_4 are zero, these four polynomials form a Gröbner basis B of J.

The leading monomials of this Gröbner basis, in descending graded lexicographic order, are

$$X^3Y^2, X^4, XY^3, Y^4.$$

None of these leading monomials divides any of the others, which makes the Gröbner basis B minimal. The set of monomials not divisible by any of the above four is our desired M_B. By inspection,

$$M_B = \{1, X, X^2, X^3, Y, Y^2, Y^3, XY, X^2Y, X^3Y, XY^2, X^2Y^2\}.$$

According to Proposition 9.37 the dimension of $\mathbb{Q}[X, Y]/J$ equals the size of M_B, which is 12.

We might note that with a different monomial order on $\mathbb{Q}[X, Y]$ the Gröbner basis for the above ideal J will change. For instance, taking the lexicographic order on the monomials of $\mathbb{Q}[X, Y]$ with $Y > X$, and after some work with Buchberger's algorithm (which we omit), a minimal Gröbner basis for J becomes

$$C = \{Y - X^7, X^{12} - X^2\}.$$

Now

$$M_C = \{1, X, X^2, \ldots, X^{11}\}.$$

Reassuringly, the size of M_C is 12, which is the dimension of $\mathbb{Q}[X, Y]/J$.

Using Normal Forms to Find Inverses

In case R/J is finite dimensional, the normal form of a coset in R/J can be exploited to decide if an element in a quotient ring has an inverse and if so, to find the normal form of that inverse. Here is an illustration of how this can be done.

Take the graded lexicographic order on $\mathbb{Q}[X, Y]$ with $X > Y$. The polynomials

$$g_1 = XY^2 - X + Y,\ g_2 = X^2 - XY - Y^2,\ g_3 = Y^3 + X - 2Y$$

form a Gröbner basis B for the ideal J that they generate. This can be seen by Buchberger's theorem, Proposition 9.26. For instance, $S(g_1, g_2) = XY^3 + Y^4 - X^2 + XY$. Using the division algorithm we obtain the following standard expression:

$$S(g_1, g_2) = Yg_1 - 1g_2 + Yg_3 + 0,$$

as can be readily verified. Evidently 0 is a complete reduction of $S(g_1, g_2)$ modulo g_1, g_2, g_3. We omit the verification, but the division algorithm also reveals that $S(g_1, g_3)$ and $S(g_2, g_3)$ have 0 as their complete reduction. Thus g_1, g_2, g_3 form a Gröbner basis of J.

By inspection of the Gröbner basis,

$$M_B = \{1, X, Y, Y^2, XY\}.$$

For brevity of notation, let $\alpha = X + J$ and $\beta = Y + J$, and write $1 + J$ as 1. According to Proposition 9.36 a basis for $\mathbb{Q}[X, Y]/J$ consists of

$$1, \alpha, \beta, \beta^2, \alpha\beta.$$

The normal forms of the elements of $\mathbb{Q}[X, Y]/J$ are the $\mathbb{Q}$-linear combinations of this vector space basis over $\mathbb{Q}$.

Let us now determine if $1 - \alpha + \beta$ has an inverse in $\mathbb{Q}[X, Y]/J$, and if so find that inverse in normal form. Despite the tediousness of the upcoming calculations, what makes it interesting is that the process of finding inverses is systematic. We are seeking possible scalars a, b, c, d, e such that

$$(a + b\alpha + c\beta + d\beta^2 + e\alpha\beta)(1 - \alpha + \beta) = 1.$$

By multiplying the left side out and collecting like terms we get the equation:

$$a + (-a + b)\alpha + (a + c)\beta + (c + d)\beta^2 + (b - c + e)\alpha\beta \\ - b\alpha^2 - e\alpha^2\beta + (-d + e)\alpha\beta^2 + d\beta^3 = 1.$$

Now the normal form of the left side is needed. To that end we need the normal forms of $\alpha^2, \alpha^2\beta, \alpha\beta^2, \beta^3$. So we need the unique pre-images of these elements in the span V_B of M_B. Some obvious pre-images are

$$X^2, X^2Y, XY^2, Y^3,$$

but these are not in V_B. Suitable complete reductions of these monomials modulo the Gröbner basis B are what we need. And so the division algorithm is called into play.

For X^2, a standard expression modulo B is:

$$X^2 = 0g_1 + 1g_2 + 0g_2 + XY + Y^2.$$

By inspection the V_B-component of X^2 in the direct sum $\mathbb{Q}[X, Y] = J + V_B$ is $XY + Y^2$. So the normal form of α^2 is $\alpha\beta + \beta^2$.

For X^2Y a standard expression modulo B is:

$$X^2Y = 1g_1 + Yg_2 + 1g_3 + Y.$$

By inspection the V_B-component of X^2Y is Y. The normal form of $\alpha^2\beta$ is β. Since $\alpha^2\beta = \beta$ at this point one might be tempted to cancel β and conclude that $\alpha^2 = 1$, whereas $\alpha^2 = \alpha\beta + \beta^2$. We see instead that our quotient ring possesses zero divisors such as β and $\alpha^2 - 1$.

For XY^2 a standard expression modulo B is:

$$XY^2 = 1g_1 + 0g_2 + 0g_3 + X - Y.$$

By inspection the V_B-component of XY^2 is $X - Y$, and then the normal form of $\alpha\beta^2$ is $\alpha - \beta$.

Finally, for Y^3 a standard expression is:

$$Y^3 = 0g_1 + 0g_2 + 1g_3 - X + 2Y.$$

The V_B-component of Y^3 is $-X + 2Y$, and so the normal form of β^3 is $-\alpha + 2\beta$.

After substituting these normal forms into the left side of the equation we are trying to solve and collecting like terms, we come up with the equation:

$$a + (-a + b - 2d + e)\alpha + (a + c + 3d - 2e)\beta + (-b + c + d)\beta^2 + (-c + e)\alpha\beta = 1.$$

Since $1, \alpha, \beta, \beta^2, \alpha\beta$ form a basis of $\mathbb{Q}[X, Y]/J$ the above equation comes down to solving the following system of five linear equations in five unknowns:

$$\begin{aligned} a &= 1 \\ -a + b - 2d + e &= 0 \\ a + c + 3d - 2e &= 0 \\ -b + c + d &= 0 \\ -c + e &= 0. \end{aligned}$$

By the usual methods the unique solution to this system is given by:

$$a = 1, \quad b = 1/5, \quad c = 2/5, \quad d = -1/5, \quad e = 2/5.$$

Thus we learn that $1 - \alpha + \beta$ is invertible, and that

$$1 + \frac{1}{5}\alpha + \frac{2}{5}\beta - \frac{1}{5}\beta^2 + \frac{2}{5}\alpha\beta,$$

is its inverse.

9.4.4 Finite Varieties

Gröbner bases and the normal forms for the finitely generated algebras which they yield can be useful in the solution of systems of polynomial equations when the algebras are finite dimensional over K.

A Gröbner Test for Finite Varieties

If K is a field and f is a polynomial in m variables over K, a point $(a_1, \ldots, a_m)$ in K^m is called a ***zero*** of f whenever $f(a_1, \ldots, a_m) = 0$. For instance, the zeroes of the polynomial $X^2 + Y^2 + Z^2 - 1$ in $\mathbb{R}[X, Y, Z]$ form the unit sphere of $\mathbb{R}^3$.

If S is a set of polynomials, the set of common zeroes of all f in S is known as the ***variety*** of S, and is denoted by $\text{var}(S)$. A moment's reflection shows that $(a_1, \ldots, a_m)$ is a zero for all f in S if and only if f is a zero for all f in the ideal $\langle S \rangle$ generated by S inside $K[X_1, \ldots, X_m]$. Thus we often speak of the ***variety of an ideal***. By the Hilbert basis theorem, Proposition 5.42, all ideals of $K[X_1, \ldots, X_m]$ are generated by a finite set of polynomials. Thus a variety is also the set of common zeroes of a finite number of polynomials. Algebraic geometry, a field of great importance in modern mathematics, originated to a large extent in the study of varieties.

A Gröbner basis for an ideal is helpful to determine whether its variety is finite, and then in solving for those finitely many points of the variety. The next result makes some connections.

Proposition 9.38. *Suppose R is a polynomial ring $K[X_1, \dots, X_m]$ in m variables over a field K, and that R is endowed with a monomial order. For any ideal J of R with Gröbner basis B, the following statements are equivalent.*

1. *R/J is finite dimensional over K.*
2. *For each variable X_j, there is a non-zero polynomial p in J whose monomials contain only the variable X_j (i.e. $p \in J \cap K[X_j]$, where $K[X_j]$ is the subring of polynomials in the single variable X_j).*
3. *For each variable X_j, there exists a polynomial in the Gröbner basis B whose leading monomial is a power of X_j.*

Proof. We prove $1 \implies 2 \implies 3 \implies 1$.

Assuming item 1 that R/J is finite dimensional, put $x = X_j + J$ inside R/J. The powers $1, x, x^2, \dots, x^k, \dots$ are K-linearly dependent once k exceeds the dimension of R/J. For such a k we have a_i in K that are not all 0 while

$$a_0 + a_1 x + a_2 x^2 + \cdots + a_k x^k = 0.$$

A pre-image in R, under the canonical projection $R \to R/J$, of the above expression is the non-zero polynomial

$$p = a_0 + a_1 X_j + a_2 X_j^2 + \cdots + a_k X_j^k.$$

Evidently, the monomials of p contain only the variable X_j, and the choice of p puts it inside J.

Assuming item 2, take a non-zero polynomial p in $K[X_j] \cap J$. The Gröbner basis B contains a polynomial g such that $\operatorname{lm}(g)$ divides $\operatorname{lm}(p)$. Since $\operatorname{lm}(p)$ is a power of X_j, it follows that $\operatorname{lm}(g)$ is a power of X_j.

Assuming item 3, use Proposition 9.34 to get that the set M_B of monomials which are not reducible modulo B is finite. Hence $\dim R/J$ is finite because, according to Proposition 9.37, this dimension equals the size of M_B. □

Because of Proposition 9.38 a Gröbner basis offers a practical way to show that the dimension of the K-algebra R/J is finite. Furthermore, when this happens, Proposition 9.38 reveals that the variety of the ideal J is finite.

Proposition 9.39. *If J is an ideal in a polynomial ring R in several variables over a field K and R/J is finite dimensional, then* $\operatorname{var}(J)$ *is finite.*

Proof. By item 2 of Proposition 9.38 there is a non-zero polynomial p_j in each intersection $J \cap K[X_j]$. Every point $(a_1, \dots, a_m)$ in $\operatorname{var}(J)$ must also be a zero of these p_j. This means that $p_j(a_j) = 0$. There are only finitely many roots for each such p_j, in one variable. Since each coordinate a_j of $(a_1, \dots, a_m)$ is constrained to lie in a finite set, the point $(a_1, \dots, a_m)$ itself is constrained to lie in a finite set. □

Solving for a Finite Variety Using Normal Forms

Suppose $f_1, \ldots, f_k$ generate the ideal J in the polynomial ring $K[X_1, \ldots, X_m]$. We are asked to decide if var(J) is finite or not. If var(J) is finite, we wish to calculate its points explicitly. On the basis of the preceding discussion, here is a technique for finding the points of a variety should that variety turn out to be finite.

- Select a monomial order for $K[X_1, \ldots, X_m]$.
- Use Buchberger's algorithm to find a Gröbner basis B for the ideal J generated by $f_1, \ldots, f_k$. As might be inferred from previous runs with this algorithm, this step could take a while.
- If for every variable X_j there is a polynomial g in the acquired Gröbner basis B, such that $\text{lm}(g)$ is a power of X_j, then in accordance with Proposition 9.38 the algebra R/J is finite dimensional, and for each variable X_j there is a non-zero polynomial p_j in $J \cap K[X_j]$. Every point $(c_1, c_2, \ldots, c_m)$ in the variety of $f_1, \ldots, f_k$ must be such that $p_j(c_j) = 0$ for every j from 1 to m. This reduces the problem of computing the points of the variety to one of separately calculating the finitely many roots of each polynomial equation $p_j(X_j) = 0$.

  If the Gröbner basis fails test number 3 in Proposition 9.38, then var(J) could be infinite. In this case we abandon our algorithm.
- If R/J is finite dimensional, then for each variable X_j a suitable non-zero polynomial p_j in $J \cap K[X_j]$ can be found by exploiting the normal form of R/J as follows.

  Take a basis $\beta_1, \ldots, \beta_d$ of R/J. By Proposition 9.36 these can be taken as the images, under the canonical projection $R \to R/J$, of the finitely many monomials that are not reducible modulo any of the g in B. If α_j is the coset $X_j + J$, the powers α^k for k from 0 to d are linearly dependent over K. A non-trivial dependency relationship $\sum_{k=0}^{d} a_k \alpha_j^k = 0$ gives the requisite polynomial $p_j(X_j) = \sum_{k=0}^{d} a_k X_j^k$.

  To get the coefficients a_k (not all 0) write each $\alpha^k = \sum_{i=1}^{d} c_{ik}\beta_i$. In other words, get the normal form of α^k. Then solve

$$\sum_{k=0}^{d} a_k \left(\sum_{i=1}^{d} c_{ik}\beta_i \right) = \sum_{i=1}^{d} \left(\sum_{k=0}^{d} c_{ik} a_k \right) \beta_i = 0.$$

  This comes down to finding a non-trivial solution $(a_0, a_1, \ldots, a_d)$ to the homogeneous linear system

$$\sum_{k=0}^{d} c_{ik} a_k = 0 \text{ where } i \text{ runs from 1 to } d.$$

- With each non-zero polynomial p_j in $J \cap K[X_j]$ in hand, solve $p_j(X_j) = 0$ for its finitely many roots in K. This may of course present its own difficulties, but if $K = \mathbb{Q}$, the rational roots theorem will do nicely.
- For each point $(b_1, \ldots, b_m)$ such that b_j is a root of p_j, test if that point satisfies the original equations $f_1 = f_2 = \cdots = f_k = 0$. The ones that satisfy these equations will constitute var(J).

The upshot of the above procedure is that several simultaneous polynomial equations in several unknowns get replaced by several polynomial equations each with one unknown to be solved individually. The calculations are clearly arduous, but at least there is a method.

Here is an example to illustrate the method.

Take $R = \mathbb{C}[X, Y, Z]$ with the degree lexicographic order and $X > Y > Z$. Using the polynomials

$$f_1 = X^2 + Y + Z - 1, f_2 = Y^2 + X + Z - 1, f_3 = Z^2 + X + Y - 1,$$

let us solve the equations $f_1 = f_2 = f_3 = 0$. In other words, find the variety for the ideal J that they generate.

Of course, a quick inspection reveals that $(1,0,0), (0,1,0), (0,0,1)$ are in $\text{var}(J)$. The remaining points will emerge from the preceding algorithm.

With the degree lexicographic order, the polynomials f_1, f_2, f_3 constitute a Gröbner basis already, because their leading monomials are coprime, which allows Proposition 9.29 to be used. Since for each variable X, Y, Z, there is a polynomial in the Gröbner basis whose leading monomial is a power of that variable, the quotient algebra R/J is finite dimensional, and $\text{var}(J)$ will be finite. For the variable X we need a non-zero polynomial $p(X)$ that also lies in J. Likewise for Y and Z.

To get $p(X)$ we exploit the normal form of the elements of R/J. If we let x, y, z be the cosets $X+J, Y+J, Z+J$ in R/J, respectively, we can see by inspecting the Gröbner basis of J that a basis for R/J, as a vector space over K, is given by

$$1,\ x,\ y,\ z,\ xy\ ,xz\ ,yz\ ,xyz.$$

So, R/J is 8-dimensional.

We seek a non-trivial dependency relation

$$a_0 + a_1x + a_2x^2 + a_3x^3 + a_4x^4 + a_5x^5 + a_6x^6 + a_7x^7 + a_8x^8 = 0,$$

for suitable a_j in $\mathbb{C}$. To that end, the powers of x need to be expressed in terms of the above basis of R/J. Inside R/J we certainly have

$$x^0 = 1, x^1 = x \text{ as well as } x^2 = 1 - y - z,\ y^2 = 1 - x - z,\ z^2 = 1 - x - y.$$

Thus

$$x^3 = x - xy - xz,$$

and then

$$\begin{aligned}
x^4 &= x^2 - x^2y - x^2z \\
&= 1 - y - z - (1 - y - z)y - (1 - y - z)z \\
&= 1 - y - z - y + y^2 + yz - z + yz + z^2 \\
&= 1 - y - z - y + 1 - x - z + yz - z + yz + 1 - x - y \\
&= 3 - 2x - 3y - 3z + 2yz.
\end{aligned}$$

By following this pattern of multiplication by x and substitutions for x^2, y^2, z^2 we also obtain

$$x^5 = -2 + 3x + 2y + 2z - 3xy - 3xz + 2xyz$$
$$x^6 = 13 - 12x - 13y - 13z + 4xy + 4xz + 8yz$$
$$x^7 = -20 + 21x + 20y + 20z - 13xy - 13xz - 8yz + 8xyz$$
$$x^8 = 50 - 62x - 50y - 50z + 28xy + 28xz + 34yz - 8xyz.$$

With the x^j expressed in terms of the ordered basis $1, x, y, z, xy, xz, yz, xyz$ the unknown a_j are the non-trivial solutions of the homogeneous system given by the 8×9 matrix

$$\begin{pmatrix} 1 & 0 & 1 & 0 & 3 & -2 & 13 & -20 & 50 \\ 0 & 1 & 0 & 1 & -2 & 3 & -12 & 21 & -62 \\ 0 & 0 & -1 & 0 & -3 & 2 & -13 & 20 & -50 \\ 0 & 0 & -1 & 0 & -3 & 2 & -13 & 20 & -50 \\ 0 & 0 & 0 & -1 & 0 & -3 & 4 & -13 & 28 \\ 0 & 0 & 0 & -1 & 0 & -3 & 4 & -13 & 28 \\ 0 & 0 & 0 & 0 & 2 & 0 & 8 & -8 & 34 \\ 0 & 0 & 0 & 0 & 0 & 2 & 0 & 8 & -8 \end{pmatrix}.$$

By routine Gaussian elimination, a non-trivial solution of this homogeneous system is the 9-tuple

$$(0, 0, -1, 4, -4, 0, 1, 0, 0),$$

which reveals that

$$x^6 - 4x^4 + 4x^3 - x^2 = 0 \text{ inside } R/J.$$

Hence $X^6 - 4X^4 + 4X^3 - X^2$ is a non-zero polynomial in J. By an application of the rational roots test we get the factorization

$$X^6 - 4X^4 + 4X^3 - X^2 = X^2(X-1)^2(X^2 + 2X - 1).$$

The roots of this polynomial are

$$0,\ 1,\ -1 + \sqrt{2},\ -1 - \sqrt{2}.$$

By the symmetry in the variables X, Y, Z among the polynomials f_1, f_2, f_3, we can see that $Y^6 - 4Y^4 + 4Y^3 - Y^2$ and $Z^6 - 4Z^4 + 4Z^3 - Z^2$ are also non-zero polynomials in J, each of course with the above four roots.

It follows that var(J) is to be found among the 64 points (b_1, b_2, b_3) where each b_i is one of our four roots. A routine substitution of these 64 possible points into the original equations $f_1 = f_2 = f_3 = 0$ shows that only five will satisfy them. Those five points are the originally noted $(1, 0, 0), (0, 1, 0), (0, 0, 1)$ as well as

$$(-1 + \sqrt{2}, -1 + \sqrt{2}, -1 + \sqrt{2}),\ (-1 - \sqrt{2}, -1 - \sqrt{2}, -1 - \sqrt{2}).$$

9.4.5 Hilbert's Nullstellensatz and Idempotents

The search for idempotents in finite-dimensional algebras can be approached by means of polynomial equations, which in turn can be solved by Gröbner methods. In pursuing this line of thought we first touch upon David Hilbert's renowned theorem known as the ***Nullstellensatz***, which translates as "the theorem about zero points."

Hilbert's Nullstellensatz

Hilbert's celebrated result lies at the heart of algebraic geometry. One aspect of it says that if a field K is algebraically closed and if $f_1, \dots, f_n$ are polynomials in $K[X_1, \dots, X_m]$ for which their variety in K^m is empty, then the ideal generated by these polynomials is all of $K[X_1, \dots, X_m]$. In the special case of one variable X, this comes down to saying that if a nonzero polynomial f in the principal ideal domain $K[X]$ is such that the equation $f = 0$ has no solution in K, then f is a constant polynomial. Of course, this is another way to say that K is algebraically closed. The proof for several variables, which we now offer, is based on an approach due to Oscar Zariski (1899–1986).

We open the way to the Nullstellensatz with a result that is quite subtle.

Proposition 9.40. *If F is a field extension of a field K and F is also finitely generated as an algebra over K, then F is an algebraic extension of K.*

Proof. Let $\alpha_1, \dots, \alpha_n$ be a finite list of generators for F as an algebra over K. This means that every element of F is a finite sum of the form

$$\sum_{i_1,\dots,i_n} a_{i_1\cdots i_n}\alpha_1^{i_1}\cdots\alpha_n^{i_n},$$

where the exponents i_j run through non-negative integers and the coefficients $a_{i_1\cdots i_n}$ are in K.

We have the tower of fields:

$$K \subseteq K(\alpha_1) \subseteq K(\alpha_1,\alpha_2) \subseteq K(\alpha_1,\alpha_2,\alpha_3) \subseteq \cdots \subseteq K(\alpha_1,\dots,\alpha_n) = F.$$

The last field F coincides with the algebra $K[\alpha_1, \dots, \alpha_n]$.

The goal is to show that each field in the tower is an algebraic extension of the field that precedes it. From this it will follow by Proposition 6.19 that F is algebraic over K, as desired.

We suppose, to the contrary, that some field in the tower is not an algebraic extension of its preceding field, and seek a contradiction.

Let $K(\alpha_1, \dots, \alpha_{\ell-1}, \alpha_\ell)$ be the last field in the tower which is not an algebraic extension of its preceding field $K(\alpha_1, \dots, \alpha_{\ell-1})$. For simplicity of notation, put

$$E = K(\alpha_1, \dots, \alpha_{\ell-1}) \text{ and } x = \alpha_\ell.$$

Now x is not algebraic over E. This means that the ring $E[x]$ is a ring of polynomials with coefficients in E in what we can legitimately call the variable x. Thus we can speak of irreducibles in $E[x]$ as familiar polynomials. And $E(x)$ is the field of rational functions in x. Furthermore since $E(x)$ is the last field in the tower which is not algebraic over its preceding field, it follows from Proposition 6.19 that F is an algebraic extension of $E(x)$. Since F is finitely generated as a K-algebra, F is also finitely generated as an algebra over $E(x)$. (Just use the original generators α_j.) But a field extension that is both algebraic and finitely generated as an algebra over its subfield is finite dimensional over that subfield, due to Proposition 6.15.

In light of the above, let $\beta_1, \ldots, \beta_r$ be a basis of F as a vector space over $E(x)$. Each γ in F can be expressed as a unique $E(x)$-linear combination

$$\gamma = \sum_{j=1}^{r} \frac{f_j(x)}{g_j(x)} \beta_j,$$

where the unique coefficients $\frac{f_j(x)}{g_j(x)}$ are quotients of coprime $f_j(x), g_j(x)$ in the polynomial ring $E[x]$. For the rest of the proof we shall refer to the irreducible factors of all the denominators $g_j(x)$ as the *irreducible poles in* γ. Each γ has only a finite number of irreducible poles.

In particular there are only finitely many irreducible poles in the products $\beta_i\beta_k$ where i, k run from 1 to r. Let S be the set of irreducible poles in the totality of the products $\beta_i\beta_k$. With a bit of reflection on the distributive law, it can be seen that if T is the finite set of irreducible poles in an element γ of F and U is the finite set of irreducible poles in an element δ of F, then the set of irreducible poles in any E-linear combination $a\gamma + b\delta$ and in the product $\gamma\delta$ sits inside the finite set $T \cup U \cup S$.

As noted at the outset, every element of F takes the form

$$\sum_{i_1,\ldots,i_n} a_{i_1\cdots i_n} \alpha_1^{i_1} \cdots \alpha_n^{i_n},$$

where the $a_{i_1\cdots i_n}$ are in K and thus also in E. Let T_j be the finite set of irreducible poles in α_j. Also let T_0 be the set of irreducible poles in 1. From the preceding observations we see that the irreducible poles in *all* of the elements of F lie inside the finite set $T_0 \cup T_1 \cup T_2 \cup \cdots \cup T_n \cup S$.

However, the polynomial ring $E[x]$ has infinitely many irreducible polynomials, by a simple replication of Euclid's argument that there are infinitely many primes in $\mathbb{Z}$. So, take g to be an irreducible in $E[x]$ but outside of $T_0 \cup T_1 \cup T_2 \cup \cdots \cup T_n \cup S$. The element $\frac{1}{g}\beta_1$ lies in F, and has the irreducible pole g which lies outside of $T_0 \cup T_1 \cup T_2 \cup \cdots \cup T_n \cup S$. We have our contradiction. $\square$

Among other things, the Nullstellensatz says something about the maximal ideals of polynomial rings over algebraically closed fields, and Proposition 9.40 provides a way to get there. But first a more routine result.

Proposition 9.41. *If K is a field and $(a_1, \ldots, a_m)$ is a point in K^m, then the set of polynomials f in $K[X_1, \ldots, X_m]$ for which $f(a_1, \ldots, a_m) = 0$ equals the ideal $\langle X_1 - a_1, X_2 - a_2, \ldots, X_m - a_m\rangle$. This ideal is a maximal ideal.*

Proof. The kernel of the substitution map

$$\varphi : K[X_1, \ldots, X_m] \to K \text{ given by } f(X_1, \ldots, X_m) \mapsto f(a_1, \ldots, a_m)$$

is the ideal M consisting of all polynomials f that vanish at $(a_1, \dots, a_m)$. Since φ is onto (because constant polynomials already pick up all of K), the first isomorphism theorem, Proposition 4.43, gives that $K[X_1, \dots, X_m]/M \cong K$, and since K is a field, Proposition 4.51 shows that the ideal M is maximal. Clearly every $X_j - a_j$ lies in M, and therefore M contains the ideal $\langle X_1 - a_1, \dots, X_m - a_m\rangle$ generated by these linear polynomials.

It now suffices to see that the latter ideal picks up all of M. A quick way to get there is from Proposition 9.20 using a suitable monomial order. For f in M a standard expression modulo $X_1 - a_1, \dots, X_m - a_m$ with a complete reduction gives polynomials $q_1, \dots, q_m$ and a polynomial r such that

$$f = (X_1 - a_1)q_1 + \cdots + (X_m - a_m)q_m + r$$

such that either $r = 0$ or $\mathrm{lm}(r)$ is not divisible by any X_j. The second case means that $\mathrm{lm}(r) = 1$, and thereby that r is a constant. Since $f(a_1, \dots, a_m) = 0$ we can see from the above expression for f that $r = 0$, which puts f inside $\langle X - a_1, \dots, X - a_m\rangle$. □

Now we suppose that the field K is algebraically closed. In that case one version of the Nullstellensatz tells us that the ideals $\langle X_1 - a_1, \dots, X_m - a_m\rangle$ provide all the maximal ideals of $K[X_1, \dots, X_m]$.

Proposition 9.42 (Nullstellensatz 1). *If K is an algebraically closed field and M is a maximal ideal of $K[X_1, \dots, X_m]$, then $M = \langle X_1 - a_1, \dots, X_m - a_m\rangle$ for some point $(a_1, \dots, a_m)$ in K^m.*

Proof. Let $F = K[X_1, \dots, X_m]/M$. The canonical projection $\varphi : K[X_1, \dots, X_m] \to F$ embeds the field K of constant polynomials as a subring of F. In this way F is a K-algebra. Clearly F is finitely generated as an algebra over K by the images $\varphi(X_1), \dots, \varphi(X_m)$ under the canonical projection. Since M is a maximal ideal F is field, by Proposition 4.51. Because K is algebraically closed, Proposition 9.40 reveals that $F = K$.

The above conclusion translates into saying that for every f in $K[X_1, \dots, X_m]$ there is an a in K such that $f - a \in M$. In particular, for each X_j there is an a_j in K such that $X_j - a_j \in M$. Thus the maximal ideal M contains the ideal $\langle X_1 - a_1, \dots, X_m - a_m\rangle$. Because the latter ideal is also maximal according to Proposition 9.41, it must be that $M = \langle X_1 - a_1, \dots, X_m - a_m\rangle$. □

Another version tells us that with an algebraically closed field, varieties of proper ideals are never empty.

Proposition 9.43 (Nullstellensatz 2). *If the field K is algebraically closed and J is a proper ideal of $K[X_1, \dots, X_m]$, then* $\mathrm{var}(J)$ *is non-empty.*

Proof. Let M be a maximal ideal containing J. By Proposition 9.42 M is the ideal generated by $X_1 - a_1, \dots, X_m - a_m$ for some point $(a_1, \dots, a_m)$ in K^m. Clearly this point lies in $\mathrm{var}(M)$, and since M contains J, this point also lies in $\mathrm{var}(J)$. So $\mathrm{var}(J)$ is non-empty. □

In more concrete terms the above says that if $f_1, \dots, f_k$ are polynomials in m variables over an algebraically closed field and if the equations $f_j = 0$ have no simultaneous solutions in K^m then 1 lies in the ideal that they generate.

A third version says something about polynomials that vanish on a variety.

Proposition 9.44 (Nullstellensatz 3). *Let K be an algebraically closed field and J an ideal of $K[X_1, \ldots, X_m]$. If f is a polynomial in $K[X_1, \ldots, X_m]$ that vanishes at all points of* var(J)*, then there is an exponent r such that f^r lies in J.*

Proof. The case $f = 0$ requires no further consideration. For $f \neq 0$, the proof relies on a good trick.

Let R be the larger polynomial ring $K[X_1, \ldots, X_m, Y]$ with one more variable called Y. Inside R take the ideal L generated by J along with the polynomial $Yf - 1$.

Suppose a point $(a_1, \ldots, a_m, b)$ is in var(L). Since L contains J we get that $(a_1, \ldots, a_m)$ is also in var(J). Thus $f(a_1, \ldots, a_m) = 0$ by the assumption on f. In addition $bf(a_1, \ldots, a_m) - 1 = 0$, because $Yf - 1$ is also in L. Therefore $-1 = 0$, a contradiction which shows that var(L) is empty.

By Proposition 9.43 the ideal L is all of R. Keeping in mind that the ideal L, now shown to be R, is generated by J and $Yf - 1$, there exist finitely many f_i in J and corresponding g_i in R and h in R such that

$$\sum_i f_i q_i + (Yf - 1)h = 1.$$

Next consider the substitution map $R \to K(X_1, \ldots, X_m)$ given by

$$f(X_1, \ldots, X_m, Y) \mapsto f\left(X_1, \ldots, X_m, \frac{1}{f}\right).$$

Apply this substitution to the preceding identity to obtain

$$\sum_i f_i(X_1, \ldots, X_m) g_i\left(X_1, \ldots, X_m, \frac{1}{f}\right) + \left(\frac{1}{f}f - 1\right) h\left(X_1, \ldots, X_m, \frac{1}{f}\right) = 1,$$

and thereby

$$\sum_i f_i(X_1, \ldots, X_m)\, g_i\left(X_1, \ldots, X_m, \frac{1}{f}\right) = 1.$$

By collecting denominators, the rational functions $g_i\left(X_1, \ldots, X_m, \frac{1}{f}\right)$ can all be expressed in the form

$$\frac{h_i(X_1, \ldots, X_m)}{f^r},$$

for some exponent r and some polynomials h_i in $K[X_1, \ldots, X_m]$. Thus

$$\sum_i f_i(X_1, \ldots, X_m) h_i(X_1, \ldots, X_m) = f^r,$$

which is the desired result. □

Finite Varieties when the Field Is Algebraically Closed

As noted in Proposition 9.39, if an ideal J in a polynomial ring R over a field K causes R/J to be finite dimensional, then var(J) is finite. We cannot expect the converse to hold over all fields. For example, take the principal ideal J generated by $X^2 + Y^2$ in $\mathbb{R}[X, Y]$. Its variety is the singleton $\{(0, 0)\}$, while the quotient ring $\mathbb{R}[X, Y]/J$ is infinite dimensional over $\mathbb{R}$. One quick way to see the latter point is to observe that $X^2 + Y^2$ is a Gröbner basis for J, which does not satisfy item 3 of Proposition 9.38.

When the field is algebraically closed, the converse comes into effect.

Proposition 9.45. *If $K[X_1, \ldots, X_m]$ is a polynomial ring over an algebraically closed field K and J is an ideal in R such that* var(J) *is finite, then the quotient ring $K[X_1, \ldots, X_m]/J$ is finite dimensional.*

Proof. If var(J) is empty, then J is all of R due to Hilbert's Nullstellensatz, Proposition 9.43, and then R/J is the zero-dimensional zero ring. So, take var(J) to be non-empty.

Pick a j from 1 to m, and let S be the finite set of all jth coordinates of all points in var(J). Then put

$$p = \prod_{a \in S} (X_j - a).$$

By inspection, this polynomial p in one variable X_j vanishes on var(J). From Hilbert's Nullstellensatz, Proposition 9.44, a power of p, say p^r, lies in J. Since for every variable X_j, we have a non-zero polynomial in J with only the single variable X_j, item 2 of Proposition 9.38 applies, and R/J is finite dimensional. □

Idempotents

How does one find the idempotents of a commutative ring? Proposition 9.45 will have a bearing on this question when the ring is a finite-dimensional algebra over a field. But first a few general remarks about idempotents.

An ***idempotent*** in a commutative ring R is an element e such that $e^2 = e$.

The zero element 0 and the identity element 1 immediately come to mind as examples. There could be other idempotents. For instance, take two commutative rings A and B and form the product ring $A \times B$. In such a ring the elements $(1, 0)$ and $(0, 1)$ are idempotents. For another example, take the quotient ring $R = \mathbb{Q}[X]/J$, where J is the principal ideal generated by $X^2 - 1$. If x is the coset $X + J$, then $\frac{x+1}{2}$ is an idempotent of R. Indeed, $x^2 = 1$, and so

$$\left(\frac{x+1}{2}\right)^2 = \frac{x^2 + 2x + 1}{4} = \frac{1 + 2x + 1}{4} = \frac{x+1}{2}.$$

Furthermore $\frac{x+1}{2}$ is neither 0 nor 1. Otherwise, $\frac{x+1}{2} = 0$ implies that $x + 1 = 0$ and thereby that $X+1 \in J$, contradicting the fact X^2+1 does not divide $X+1$. An equally trivial argument shows that this element is not 1.

For yet one more example, take a vector space V and subspaces U, W such that $V = U \oplus W$. This means that every element v of V has a unique representation as a sum $v = u + w$ where $u \in U$ and $w \in W$. Then the projection mapping $T : V \to V$ given by $v \mapsto u$ is an idempotent in the algebra $K[T]$ of linear operators on V that are polynomials in T.

If e is an idempotent, then

$$e(1-e) = e - e^2 = e - e = 0 \text{ and } (1-e)^2 = (1-e)1 - (1-e)e = 1 - e.$$

Thus $1 - e$ is an idempotent too. When two idempotents e, f are such that $ef = 0$ we say that they are ***orthogonal***. As noted, e and $1 - e$ are orthogonal.

Let us call an idempotent in a ring ***proper*** when that idempotent is neither 0 nor 1. What can a proper idempotent e tell us about a commutative ring R? Well, it can be readily checked that the principal ideals Re and $R(1-e)$ become rings in their own right with e and $1-e$ as their respective identity elements. (However, they are not subrings of R.) Viewing the ideals $Re, R(1-e)$ as R-modules we get the module sum $R = Re + R(1-e)$. Indeed, every a in R can be decomposed as $a = ae + a(1-e)$. Furthermore this sum is direct. For that it suffices to check that $Re \cap R(1-e)$ is the zero ideal. Well, suppose $a \in Re \cap R(1-e)$. This means that $a = be = c(1-e)$ for some b, c in R. And then

$$a = be = be^2 = bee = c(1-e)e = 0.$$

The preceding verifications in essence demonstrate the following result.

Proposition 9.46. *If e is an idempotent in a commutative ring R, then the ideals $Re, R(1-e)$ are rings in their own right with respective identity elements e and $1-e$, and the mapping $R \to Re \times R(1-e)$ given by $a \mapsto (ae, a(1-e))$ is an isomorphism of rings.*

The Number of Idempotents in Finite-Dimensional Algebras

Since idempotents decompose rings, the problem of finding them acquires some significance. We concentrate on the idempotents of finite-dimensional K-algebras. One question is to ask how many there could be.

We shall say that a non-zero idempotent in a commutative ring A is ***minimal*** when it cannot be written as $f + g$ where f, g are orthogonal non-zero idempotents. For instance, with any idempotent e, the identity 1 is the sum of the orthogonal idempotents e and $1-e$. So, 1 is minimal in A if and only if A has no proper idempotents.

Proposition 9.47. *If A is a finite-dimensional commutative algebra over a field K, then 1 is expressible as a sum of minimal idempotents that are pairwise orthogonal, and A has only a finite number of idempotents.*

Proof. If $e_1, \ldots, e_n$ are pairwise orthogonal, non-zero idempotents, then they are linearly independent over K. Indeed, suppose $t_1, \ldots, t_n \in K$ and that $\sum_{i=1}^n t_i e_i = 0$. Multiply through by each e_j and use their orthogonality to get

$$0 = \sum_{i=1}^{n} t_i e_i e_j = t_j e_j e_j = t_j e_j.$$

Since e_j is a non-zero vector in the K-linear space A, it follows that $t_j = 0$. Thus any list of non-zero, pairwise orthogonal idempotents has no more elements than the dimension of A over K.

In that light, let $e_1, \ldots, e_n$ be a maximal list of non-zero, pairwise orthogonal idempotents such that

$$e_1 + \cdots + e_n = 1.$$

Each e_j must be minimal. To see that, say to the contrary that e_1 is not minimal. In that case $e_1 = f + g$ where f, g are non-zero orthogonal idempotents. For each $j > 1$, the equations

$$fe_j = f^2 e_j + fge_j = f(f + g)e_j = fe_1 e_j = 0$$

reveal that f is orthogonal to the other e_j. Likewise so is g. This leads to a longer list $f, g, e_2, \ldots, e_n$ of non-zero, pairwise orthogonal idempotents adding up to 1, in contradiction to the maximality of the original list.

Now, for every idempotent f in A and every e_j we have $e_j = fe_j + (1 - f)e_j$. These two summands are orthogonal idempotents. By the minimality of e_j, either $fe_j = 0$ or $(1-f)e_j = 0$. In other words, $fe_j = 0$ or $fe_j = e_j$. In the equation $1 = e_1 + \cdots + e_n$ multiply through by f to get $f = fe_1 + \cdots + fe_n$. This reveals that every idempotent in A is the sum of a subset of e_j. There are 2^n such sums, and thus there can be only a finite number of idempotents in A. □

The Idempotence Quadratics

As noted in Section 9.4.2, finite-dimensional K-algebras are isomorphic to quotient rings of $K[X_1, \ldots, X_m]$ modulo some ideal J. Let $R = K[X_1, \ldots, X_m]$, let B be a Gröbner basis of J relative to a monomial order, and let $\varphi : R \to R/J$ be the canonical projection. If M_B is the set of monomials that are not reducible modulo any g in B, Proposition 9.36 says that φ restricted to the K-linear span of M_B is a K-linear bijection onto R/J. A basis of R/J is given by the image $\varphi(M_B)$. Let us call this image the ***B*-induced basis of R/J**.

If $M_B = \{m_1, \ldots, m_d\}$, the products $\varphi(m_i)\varphi(m_j)$ have a unique expansion with respect to the B-induced basis $\varphi(m_1), \ldots, \varphi(m_d)$. That is

$$\varphi(m_i)\varphi(m_j) = \sum_{k=1}^{d} t_{kij}\varphi(m_k), \text{ where } t_{kij} \in K.$$

We shall refer to the coefficients t_{kij} as the ***structure constants for R/J relative to the B-induced basis*** $\varphi(m_1), \ldots, \varphi(m_d)$. These structure constants depend on a listing of the m_i, which we shall take as given.

If an element ϵ of R/J is expanded as $\epsilon = \sum_{i=1}^{d} s_i\varphi(m_i)$ with $s_i \in K$, then the expansion of its square is given by

$$\epsilon^2 = \left(\sum_{i=1}^{d} s_i\varphi(m_i)\right)\left(\sum_{j=1}^{d} s_j\varphi(m_j)\right) = \sum_{i,j=1}^{d} s_is_j\varphi(m_i)\varphi(m_j)$$
$$= \sum_{i,j=1}^{d} s_is_j\left(\sum_{k=1}^{d} t_{kij}\varphi(m_k)\right) = \sum_{k=1}^{d}\left(\sum_{i,j=1}^{d} t_{kij}s_is_j\right)\varphi(m_k).$$

Thus we learn that $\epsilon^2 = \epsilon$ if and only if

$$\sum_{i,j=1}^{d} t_{kij}s_is_j = s_k,$$

for every k from 1 to d.

Let $Y_1, \ldots, Y_d$ be variables for a new polynomial ring $K[Y_1, \ldots, Y_d]$. The polynomials in $K[Y_1, \ldots, Y_d]$ defined by

$$f_k^B = \sum_{i,j=1}^{d} t_{kij}Y_iY_j - Y_k, \text{ with } k \text{ running from 1 to } d,$$

will be called the ***idempotence quadratics arising from the B-induced basis*** of R/J. The ideal I_B of $K[Y_1, \ldots, Y_d]$ generated by these idempotence quadratics will be called their ***idempotence ideal***. The preceding calculations yield the following observation.

Proposition 9.48. *Let R be the polynomial ring $K[X_1, \ldots, X_m]$ over a field K. Suppose J is a proper ideal of R with Gröbner basis B with respect to a monomial order, and that R/J is finite dimensional over K. An element in R/J is an idempotent if and only if its coordinate d-tuple in K^d, relative to the B-induced basis of R/J, is a common zero of the idempotence quadratics arising from that basis.*

If I_B is the idempotence ideal in $K[Y_1, \ldots, Y_d]$ generated by the idempotence quadratics $f_1^B, \ldots, f_d^B$, the problem of finding idempotents in R/J has come down to finding the points of $\mathrm{var}(I_B)$. For instance, the d-tuple $(0, \ldots, 0)$ is a zero of the idempotence quadratics and thereby lies in $\mathrm{var}(I_B)$, which simply corroborates that the element 0 in R/J is an idempotent. Which d-tuple in K^d corresponds to the identity element 1 in R/J? Well, 1 always appears in the B-induced basis $\beta_1, \ldots, \beta_d$ of R/J. Say $1 = \beta_1$. An inspection of the structure constants for the basis $\beta_1 = 1, \beta_2, \ldots, \beta_d$ reveals that

$$t_{k11} = \begin{cases} 1 & \text{if } k = 1 \\ 0 & \text{if } k > 1. \end{cases}$$

The d-tuple $(1,0, \ldots, 0)$ gives the coordinates of 1, and from the preceding observation it solves the equations $f_k^D = 0$, which puts it inside $\mathrm{var}(I_B)$.

If $K[Y_1, \ldots, Y_d]/I_B$ were finite dimensional, then according to Proposition 9.39 var(I_B) would be finite. We would be in a position to find a Gröbner basis for I_B, use the normal form of $K[Y_1, \ldots, Y_d]/I_B$ to solve for var(I_B), and thereby find the idempotents of $K[X_1, \ldots, X_m]/J$.

The goal at this point is to prove that $K[Y_1, \ldots, Y_d]/I_B$ is finite dimensional.

Invariance of the Idempotence Quadratics under Field Extensions

As shown in Proposition 9.47, a finite-dimensional algebra $K[X_1, \ldots, X_m]/J$ has only finitely many idempotents. If d is the dimension of this quotient algebra, and B is a Gröbner basis of J, and I_B is the ideal inside $K[Y_1, \ldots, Y_d]$ generated by the idempotence quadratics arising from the B-induced basis of R/J, then Proposition 9.48 shows that its variety var(I_B) is finite. If K were also algebraically closed it would follow from Proposition 9.45 that the algebra $K[Y_1, \ldots, Y_d]/I_B$ is finite dimensional over K. But K need not be algebraically closed. To remedy this, an intricate excursion though its algebraic closure seems to be needed.

Proposition 9.49. *Suppose that L is a field extension of a field K, and let*

$$R = K[X_1, \ldots, X_m], \quad S = L[X_1, \ldots, X_m]$$

be polynomial rings over K and L with the same variables X_j and the same monomial order. If J is an ideal of R such that R/J is finite dimensional over K and B is a Gröbner basis of J, then

- *B remains a Gröbner basis of the ideal SJ generated by J inside the larger ring S,*
- $\dim_K R/J = \dim_L S/SJ$,
- *the structure constants for S/SJ relative to the B-induced basis of S/SJ coincide with the structure constants for R/J relative to the B-induced basis of R/J,*
- *the idempotence quadratics arising from the B-induced basis of S/SJ are identical to the idempotence quadratics arising from the B-induced basis of R/J.*

Proof. Since B generates J in R, the set B also generates SJ in S. To see that B remains a Gröbner basis of the latter ideal, apply Buchberger's theorem, Proposition 9.3. Because B is a Gröbner basis of J, each pair f, g in B is such that its S-polynomial $S(f, g)$ has a standard expression in R with complete reduction 0 modulo B. This remains a standard expression with complete reduction 0 modulo B in the larger ring S. By Buchberger's theorem, Proposition 9.3, B is a Gröbner basis of SJ.

For the second item, recall that M_B is the set of monomials that are not divisible by $\mathrm{lm}(g)$ for every g in B. By Proposition 9.37, $\dim_K R/J$ equals the number of elements in M_B. But M_B depends only on B, not on the field of coefficients, and B is the same for J as it is for SJ. Thus $\dim_L S/SJ$ also equals the number of elements in M_B.

For the third item, let

$$\varphi : R \to R/J \text{ and } \psi : S \to S/SJ$$

be the respective canonical projections. Let $M_B = \{m_1, \ldots, m_d\}$, and for k, i, j running from 1 to d, let t_{kij} in K be the structure constants for R/J relative to the basis $\varphi(m_1), \ldots, \varphi(m_d)$. Thus

$$\varphi(m_i)\varphi(m_j) = \sum_{k=1}^{d} t_{kij}\varphi(m_k).$$

Since φ is a ring homomorphism that fixes the elements of K, the above says that

$$m_i m_j - \sum_{k=1}^{d} t_{kij} m_k \in J.$$

But $J \subseteq SJ$. Use the projection ψ to deduce that

$$\psi(m_i)\psi(m_j) = \sum_{k=1}^{d} t_{kij}\psi(m_k).$$

Evidently the structure constants of S/SJ relative to the basis $\psi(m_1), \ldots, \psi(m_d)$, now viewed as elements of the larger field L, are t_{kij} once more.

The fourth item follows immediately from the third because the idempotence quadratics $f_1^B, \ldots, f_d^B$ are completely defined in terms of the structure constants. □

Finding Idempotents by Normal Forms

We are in a position to show that in a finite-dimensional algebra, idempotents can be found algorithmically.

Proposition 9.50. *Let R be the polynomial ring $K[X_1, \ldots, X_m]$ over a field K, and let J be an ideal of R with Gröbner basis B relative to some monomial order. If the quotient K-algebra R/J has finite dimension d and I_B is the ideal inside $K[Y_1, \ldots, Y_d]$ generated by the idempotence quadratics arising from the B-induced basis of R/J, then* $\dim_K K[Y_1, \ldots, Y_d]/I_B$ *is finite.*

Proof. Let L be an algebraically closed field extension of K, for instance the algebraic closure of K. (That such extensions exist is shown in Appendix A as Proposition A.22. If $K = \mathbb{Q}$, a suitable choice for L would be $\mathbb{C}$.) With $S = L[X_1, \ldots, X_m]$, the first item of Proposition 9.49 says that the ideal SJ generated by J inside S retains B as its Gröbner basis. The second item reveals that the quotient ring S/SJ has finite dimension over L. By Proposition 9.47, S/SJ has only finitely many idempotents.

Put

$$T = K[Y_1, \ldots, Y_d] \text{ and } U = L[Y_1, \ldots, Y_d].$$

Clearly $T \subseteq U$. By Proposition 9.49, now applied to these two quotient rings,

$$\dim_K T/I_B = \dim_L U/UI_B.$$

So, to see that T/I_B is finite dimensional over K it suffices to show that U/UI_B is finite dimensional over L. Since L is algebraically closed, Proposition 9.45 will yield the latter provided $\mathrm{var}(UI_B)$ inside L^d is finite.

Well, let $f_1^B, \ldots, f_d^B$ in T be the idempotence quadratics arising from the B-induced basis of R/J. Since they generate I_B in T, they also generate UI_B in the larger ring U. By the fourth item of Proposition 9.49, these $f_1^B, \ldots, f_d^B$ as polynomials in U are the idempotence quadratics arising from the B-induced basis of S/SJ. Since S/SJ has only finitely many idempotents, these f_i^B have only finitely many common zeroes in L^d, due to Proposition 9.48. Because these polynomials generate UI_B, we get that $\mathrm{var}(UI_B)$ is finite, as was desired. □

An Algorithm for Finding Idempotents

The result just obtained points to the following algorithm for finding idempotents in a finite-dimensional K-algebra. This algorithm assumes that we can find roots of polynomials in one variable over K.

- Starting with an ideal J inside the polynomial ring R over a field K, use Buchberger's algorithm to get a Gröbner basis B of J using an appropriate monomial order.
- According to Proposition 9.38 such a Gröbner basis will determine whether or not R/J is finite dimensional.
- If R/J is finite dimensional, calculate the structure constants for R/J relative to the B-induced basis of R/J, and thereby find the idempotence quadratics $f_1^B, \ldots, f_d^B$ inside the polynomial ring $K[Y_1, \ldots, Y_d]$ arising from this B-induced basis.
- Solve for the common zeroes inside K^d of the f_i^B. If need be, apply the algorithm following Proposition 9.39 inside the ring $K[Y_1, \ldots, Y_d]$.
- The above zeroes give the coordinates of the desired idempotents relative to the B-induced basis of R/J.

An Example of Hunting for Idempotents

Let J be the ideal inside $\mathbb{Q}[X, Y]$ generated by

$$f_1 = X^2 - Y - 1, f_2 = XY, \text{ and let } A = \mathbb{Q}[X, Y]/J.$$

We ask if A is a finite-dimensional algebra over $\mathbb{Q}$, and look for its idempotents.

- Take the degree lexicographic order on the monomials of $\mathbb{Q}[X, Y]$, with $X > Y$.
- Use Buchberger's algorithm to find a Gröbner basis of J.

  By routine calculation

$$S(f_1,f_2) = Y(X^2 - Y - 1) - X(XY) = -Y^2 - Y.$$

  Since this S-polynomial is not reducible modulo f_1,f_2, we let

$$f_3 = Y^2 + Y$$

and adjoin this to get a larger basis f_1,f_2,f_3 of J. (Here $-Y^3 - Y$ is replaced by $Y^2 + Y$ without harm, and merely for visual convenience.) Then

$$S(f_1,f_3) = Y^2(X^2 - Y - 1) - X^2(Y^2 + Y) = -X^2Y - Y^3 - Y^2.$$

Also

$$S(f_2,f_3) = Y(XY) - X(Y^2 + Y) = -XY.$$

By the division algorithm, as well as by inspection, we get

$$\begin{aligned} S(f_1,f_2) &= -f_3 = 0f_1 + 0f_2 - 1f_3 + 0 \\ S(f_1,f_3) &= 0f_1 - Xf_2 - Yf_3 + 0 \\ S(f_2,f_3) &= -XY = 0f_1 - 1f_2 + 0f_3 + 0. \end{aligned}$$

As these S-polynomials have 0 complete reductions modulo f_1,f_2,f_3, we can take

$$B = \{f_1,f_2,f_3\} = \{X^2 - Y - 1, XY, Y^2 + Y\}$$

as a Gröbner basis of J.

- There are three monomials not reducible modulo B, namely $1, X, Y$. Let $1, x, y$ be their images respectively, under the canonical projection $\mathbb{Q}[X, Y] \to \mathbb{Q}[X, Y]/J = A$. According to Proposition 9.36 $1, x, y$ constitute a basis of A as a vector space over $\mathbb{Q}$. We had called this the B-induced basis of the algebra A. By Proposition 9.47, A has only finitely many idempotents. To find these idempotents we need the structure constants of A relative to this basis. From the structure constants we can then obtain the idempotence quadratics arising from this basis of A. The structure constants worth noting are given by:

$$x^2 = y + 1, \quad xy = 0, \quad y^2 = -y,$$

by inspection of the Gröbner basis B of J.

- To get the idempotence quadratics let $\epsilon = a + bx + cy$, and ask which triples (a, b, c) in $\mathbb{Q}^3$ cause $\epsilon^2 = \epsilon$ to happen. Well,

$$\begin{aligned} \epsilon^2 &= a^2 + b^2x^2 + c^2y^2 + 2abx + 2acy + 2bcxy \\ &= a^2 + b^2(y + 1) + c^2(-y) + 2abx + 2acy + 2bc(0) \\ &= a^2 + b^2 + 2abx + (b^2 - c^2 + 2ac)y. \end{aligned}$$

So, $\epsilon^2 = \epsilon$ if and only if

$$a^2 + b^2 = a \text{ and } 2ab = b \text{ and } b^2 - c^2 + 2ac = c.$$

At this point these quadratic equations could be solved with an ad hoc inspection. For instance, the middle equation gives that either $b = 0$ or $a = 1/2$. Substitute these possibilities into the first equation to get corresponding a, b values, and then put these into the third equation to get the c. However, another situation might not be so forgiving. With that thought in mind let us pursue the relevant algorithm following Proposition 9.39.

Let X, Y, Z be the variables corresponding to the unknowns a, b, c, respectively. Take the lexicographic order in $\mathbb{Q}[X, Y, Z]$ with $Z > X > Y$. The idempotence quadratics in $\mathbb{Q}[X, Y, Z]$ for the above equations are

$$g_1 = X^2 - X + Y^2,\ g_2 = 2XY - Y,\ g_3 = Z^2 - 2ZX + Z - Y^2.$$

If I is the ideal in $\mathbb{Q}[X, Y, Z]$ generated by these three polynomials, Proposition 9.50 says that $\mathbb{Q}[X, Y, Z]/I$ is finite dimensional. Then Proposition 9.38 ensures that I contains non-zero polynomials in each subring $\mathbb{Q}[X], \mathbb{Q}[Y], \mathbb{Q}[Z]$. We could get the solutions to $g_1 = 0$, $g_2 = 0, g_3 = 0$ by first finding those polynomials purely in X, in Y and in Z, then solving for their roots, say a, b, c, respectively. The solutions to $g_1 = 0, g_2 = 0, g_3 = 0$ will be among the resulting triples (a, b, c).

We need a Gröbner basis for the ideal I.

Using the lexicographic order with $Z > X > Y$, we calculate one of our S-polynomials, and using the division algorithm we calculate its standard expression modulo g_1, g_2, g_3. The calculations are suppressed.

$$\begin{aligned} S(g_1, g_2) &= -\frac{1}{2}XY + Y^3 \\ &= (2XY - Y)\left(-\frac{1}{4}\right) + \left(Y^3 - \frac{1}{4}Y\right) \\ &= g_1 \cdot 0 + g_2\left(-\frac{1}{4}\right) + g_3 \cdot 0 + \left(Y^3 - \frac{1}{4}Y\right). \end{aligned}$$

Since the leading monomial Y^3 of the complete reduction $Y^3 - \frac{1}{4}Y$ modulo g_1, g_2, g_3 is not divisible by the leading monomials X^2, XY, Z^2 of g_1, g_2, g_3, Buchberger's algorithm tells us to let

$$g_4 = Y^3 - \frac{1}{4}Y,$$

and adjoin this to the basis g_1, g_2, g_3 of I.

By Proposition 9.28 each of the S-polynomials

$$S(g_1, g_3), S(g_1, g_4), S(g_2, g_3), S(g_3, g_4)$$

either is 0 or has a standard expression modulo g_1, g_2, g_3, g_4 with 0 complete reduction. This is because the leading monomials in each of the preceding pairs g_i, g_j are coprime. Since

$$S(g_1, g_2) = g_1 \cdot 0 + g_2\left(-\frac{1}{4}\right) + g_3 \cdot 0 + g_4 \cdot 1 + 0,$$

we have a standard expression for $S(g_1, g_2)$ with remainder 0 as well. By Buchberger's theorem, g_1, g_2, g_3, g_4 will form a Gröbner basis of I if $S(g_2, g_4)$ has a standard expression with zero complete reduction modulo g_1, g_2, g_3, g_4. Well, with the calculations suppressed:

$$
\begin{aligned}
S(g_2, g_4) &= S\left(2XY - Y, Y^3 - \frac{1}{4}Y\right) \\
&= \frac{1}{4}XY - \frac{1}{2}Y^3 \\
&= (2XY - Y)\left(\frac{1}{8}\right) + \left(Y^3 - \frac{1}{4}Y\right)\left(-\frac{1}{2}\right) + 0.
\end{aligned}
$$

This confirms that g_1, g_2, g_3, g_4 form a Gröbner basis of I.

At this point we might initiate the laborious process of finding non-zero polynomials in each $\mathbb{Q}[X] \cap I, \mathbb{Q}[Y] \cap I, \mathbb{Q}[Z] \cap I$. We already have one in $\mathbb{Q}[Y] \cap I$, namely $g_4 = Y^3 - \frac{1}{4}Y$. The roots of g_4 are $0, \pm\frac{1}{2}$. This means that the triplet (a, b, c) of solutions to the idempotence quadratics is such that $b = 0$ or $b = \pm 1/2$.

Rather than fish for non-zero polynomials in $\mathbb{Q}[X] \cap I$ and $\mathbb{Q}[Z] \cap I$ we might notice that both g_1 and g_2 do not have a Z variable, i.e. $g_1 = g_1(X, Y)$ and $g_2 = g_2(X, Y)$. For that reason substitute $Y = 0$ and $Y = \pm 1/2$ into both of these polynomials to get

$$
g_1(X, 0) = X^2 - X, \;\; g_1(X, 1/2) = X^2 - X + 1/4, \;\; g_1(X, -1/2) = X^2 - X + 1/4
$$

and

$$
g_2(X, 0) = 0, \;\; g_2(X, 1/2) = X - 1/2, \;\; g_2(X, -1/2) = -X + 1/2.
$$

We learn from these calculations that the solutions of the simultaneous equations $g_1 = g_2 = g_4 = 0$ are of the form

$$
(0, 0, c), \;\; (1, 0, c), \;\; (1/2, 1/2, c), \;\; (1/2, -1/2, c)
$$

where c remains to be determined. Now

$$
\begin{aligned}
g_3(0, 0, Z) = Z^2 + Z, &\quad g_3(1, 0, Z) = Z^2 - Z, \\
g_3(1/2, 1/2, Z) = Z^2 - 1/4, &\quad g_3(1/2, -1/2, Z) = Z^2 - 1/4.
\end{aligned}
$$

By equating these polynomials in Z to zero, we learn that the solutions to the simultaneous equations $g_1 = g_2 = g_3 = g_4 = 0$ are

$$
\begin{aligned}
&(0, 0, 0), (0, 0, -1), (1, 0, 0), (1, 0, 1), \\
&\qquad (1/2, 1/2, 1/2), (1/2, 1/2, -1/2), (1/2, -1/2, 1/2), (1/2, -1/2, -1/2).
\end{aligned}
$$

This is the variety of the ideal I determined by the idempotence quadratics.

These solutions give the idempotents of the algebra $A = \mathbb{Q}[X, Y]/J$. Recalling that A is a 3-dimensional $\mathbb{Q}$-algebra with basis $1, x, y$, the idempotents of A, obtained from the preceding coordinate triples, are:

$$
0, \; -y, \; 1, \; 1 + y, \;\; \frac{1}{2}(1 + x + y), \;\; \frac{1}{2}(1 + x - y), \;\; \frac{1}{2}(1 - x + y), \;\; \frac{1}{2}(1 - x - y).
$$

Using the relations $x^2 = y + 1, xy = 0, y^2 = -y$ on this vector space basis of A, we can in hindsight readily verify that these are idempotents. Clearly $0^2 = 0$ and $1^2 = 1$, but also

$$(-y)^2 = y^2 = -y, \quad (1+y)^2 = 1 + 2y + y^2 = 1 + 2y - y = 1 + y,$$

as well as

$$\begin{aligned}\left(\frac{1}{2}(1 \pm x \pm y)\right)^2 &= \frac{1}{4}(1 + x^2 + y^2 \pm 2x \pm 2y \pm 2xy) \\ &= \frac{1}{4}(1 + y + 1 - y \pm 2x \pm 2y) = \frac{1}{2}(1 \pm x \pm y).\end{aligned}$$

Now consider the following three of the eight idempotents:

$$e_1 = -y, \quad e_2 = \frac{1}{2}(1 + x + y), \quad e_3 = \frac{1}{2}(1 - x + y).$$

By a verification they are pairwise orthogonal (i.e. $e_1e_2 = e_1e_3 = e_2e_3 = 0$), a well as linearly independent over $\mathbb{Q}$. In fact, the independence also follows from the orthogonality. Since A is 3-dimensional over $\mathbb{Q}$, the mapping

$$\varphi : \mathbb{Q} \times \mathbb{Q} \times \mathbb{Q} \to A \text{ defined by } (r_1, r_2, r_3) \mapsto r_1e_1 + r_2e_2 + r_3e_3$$

is an isomorphism of vector spaces over $\mathbb{Q}$. The pairwise orthogonality of e_1, e_2, e_3, also causes φ to be a ring isomorphism, as can be seen by a routine check.

Our search for idempotents in A has revealed its complete structure.

EXERCISES

1. Find a basis for the algebra $\mathbb{R}[X, Y]/\langle X^2 + Y^2\rangle$ as a vector space over $\mathbb{R}$.
2. Suppose that A is an algebra which is finite dimensional over a field K. Show that a non-zero element α in A is a unit if and only if α is not a zero divisor.
3. Let J be the ideal in $\mathbb{Q}[X, Y]$ generated by X^2Y and XY^2. Using an appropriate monomial order and Buchberger's algorithm decide whether the element $X + Y + 1 + J$ in the quotient ring R/J is a unit, and if it is a unit, find its inverse.
4. (a) An element a in a commutative ring R is ***nilpotent*** when $a^n = 0$ for some positive integer n. Prove that a is nilpotent if and only if the element $1 - aX$ is a unit in the polynomial ring $R[X]$.

   (b) Let $F[X_1, X_2, \ldots, X_n]$ be the polynomial ring over a field F in the variables X_j, and let $q_1, \ldots, q_k$ be polynomials generating an ideal J. Let R be the quotient algebra $F[X_1, X_2, \ldots, X_n]/J$. Given p in $F[X_1, X_2, \ldots, X_n]$, describe an algorithm to decide if its image $p + J$ in R is nilpotent.

   Hint. If t is a new variable, explain why the problem comes down to deciding whether 1 is in the ideal generated by $q_1, \ldots, q_k, 1 - pt$.
5. In the polynomial ring $R = \mathbb{Q}[X, Y]$ let J be the ideal generated by $f = XY + X + Y$ and $g = Y^2 + X^2$. Show that the ideal intersection $XR \cap J$ is generated by the polynomials $Xf, Xg, X(Y^2 + Y - X)$.

6. Using the lexicographic order on the monomials of $\mathbb{C}[X, Y]$ with $X > Y$ find a Gröbner basis for the ideal J generated by

$$X^3 - 2XY + Y^3 \text{ and } X^5 - 2X^2Y^2 + Y^5.$$

Show that the variety of J is finite, and find the points of this variety.

7. Use the lexicographic order on the monomials in X, Y with $X > Y$ to find a Gröbner basis of the ideal in $\mathbb{R}[X, Y]$ generated by $X^2 + XY + Y^2 - 1$ and $X^2 + 4Y^2 - 4$. Then find all points in the variety of this ideal. The points are the intersection points of two ellipses.

8. Use Gröbner bases to find the solutions in $\mathbb{C}$ of the simultaneous equations

$$2X^3 + 2X^2Y^2 + 3Y^3 = 0, \quad 3X^5 + 2X^3Y^3 + 2Y^5 = 0.$$

9. Use the ideal intersection algorithm to find the intersection of

$$\langle X^3Y - XY^2 + 1, X^2Y^2 - Y^3 - 1\rangle \text{ with } \langle X^2 - Y^2, X^3 + Y^3\rangle,$$

inside $\mathbb{Q}[X, Y]$.

10. Is the algebra $\mathbb{Q}[X, Y]/\langle X^3 - Y^2, XY - 1\rangle$ finite dimensional over $\mathbb{Q}$?
Is the algebra $\mathbb{Z}_2[X, Y]/\langle X^3 - Y^2, XY - 1\rangle$ finite dimensional over $\mathbb{Z}_2$?
If so, what are their dimensions?

11. On the monomials of $\mathbb{Q}[X, Y]$ take the lexicographic order with $Y > X$. Use Buchberger's algorithm to find a Gröbner basis for the ideal J generated by $Y^2X^3 - Y$ and $Y^3X - X^2$.
Find a basis for the quotient space $\mathbb{Q}[X, Y]/J$ as a vector space over $\mathbb{Q}$.
If $\mathbb{Q}[X, Y]/J$ is finite dimensional, find its idempotents.

12.* This exercise illustrates a fairly general approach for showing that some finitely generated algebras are integral domains and for detecting irreducibles in these domains. The algebra will resemble a polynomial ring in a number of aspects.

Let F be a field, and let m, n be positive, coprime integers. Show that $X^m - Y^n$ is an irreducible polynomial in $F[X, Y]$ by following the steps indicated. Familiarity with the notion of a vector space basis, in the case where the dimension is not finite, will be required.

(a) Put $R = F[X, Y]/\langle X^m - Y^n\rangle$, and let x, y be the images of X, Y, respectively, under the canonical projection $F[X, Y] \to R$. Using the lexicographic ordering on the monomials of $F[X, Y]$, with $X > Y$, explain why a basis for R as a vector space over F is

$$M = \{x^iy^j \text{ where } 0 \leq j \text{ and } 0 \leq i < m\}.$$

(b) Show that the function

$$D : M \to \mathbb{N}, \text{ defined by } x^iy^j \mapsto ni + mj,$$

is injective.

(c) Show that M is closed under multiplication. That is, prove that $ab \in M$ whenever $a, b \in M$.

Hint. Write $a = x^i y^j, b = x^k y^\ell$, where i, k run from 0 to $m - 1$ and j, ℓ are non-negative. Euclidean division gives $i + k = mq + r$ with $0 \leq q$ and $0 \leq r < m$. Also, in R the identity $x^m = y^n$ prevails.

(d) If $a, b \in M$, verify that $D(ab) = D(a) + D(b)$.

(e) If r is an integer in the range $D(M)$, let m_r be its unique pre-image in M. Thus $D(m_r) = r$. If $r, s \in D(M)$, explain why $m_r m_s = m_{r+s}$.

(f) Every non-zero element p of the ring R takes the form

$$p = \lambda_t m_t + \sum_{\substack{s \in D(M) \\ s < t}} \lambda_s m_s,$$

where the λ_t, λ_s are unique coefficients in F, $\lambda_t \neq 0$, and the summation only involves a finite number of indices s from $D(M)$. Define $D(p)$ to be $D(m_t) = t$, and call this integer the *degree* of p.

If p, q in R are non-zero, verify that $pq \neq 0$ and that

$$D(pq) = D(p) + D(q).$$

(g) Conclude that R is an integral domain, and that $X^m - Y^n$ is irreducible.

(h) Show that p in R is a unit if and only if $D(p) = 0$. Then show that the units of R are the elements of F.

(i) Show that x is irreducible in R, but not prime.

13.* This exercise picks up on the thread of the preceding exercise.

Let $F[U, V, X, Y]$ be a polynomial ring in four variables over a field F, and J the ideal generated by $XY - UV$. Let u, v, x, y be the images of U, V, X, Y, respectively, in the quotient algebra $R = F[U, V, X, Y]/J$.

(a) Take the lexicographic monomial ordering with $Y > X > U > V$, and use that to explain why a basis for R as a vector space over F is given by

$$M = \{x^i u^j v^k, y^\ell u^m v^n : i, j, k, \ell, m, n \in \mathbb{N} \text{ but also } \ell \geq 1\}.$$

The condition $\ell \geq 1$ avoids duplication in the presentation of M.

(b) Show that M is closed under multiplication.

Hint. In R the identity $xy = uv$ prevails.

(c) Define the function $D : M \to \mathbb{R}$ by

$$D(x^i u^j v^k) = i + \sqrt{2}j + \sqrt{3}k$$

and

$$D(y^\ell u^m v^n) = (\sqrt{2} + \sqrt{3} - 1)\ell + \sqrt{2}m + \sqrt{3}n.$$

Use the fact that $1, \sqrt{2}, \sqrt{3}$ are linearly independent over $\mathbb{Q}$ to show that D is injective.

(d) Verify that $D(ab) = D(a) + D(b)$ for all a, b in B.

(e) Since M is a basis for R as a vector space over F, every non-zero f in R comes with unique elements $a_1, a_2, \dots, a_r$ in M and unique non-zero coefficients $\lambda_1, \lambda_2, \dots, \lambda_r$ in F such that

$$f = \lambda_1 a_1 + \lambda_2 a_2 + \cdots + \lambda_r a_r.$$

Explain why the a_j can be arranged such that $D(a_1) > D(a_i)$ for all the other $i > 1$.

Call $\lambda_1 a_1$ the *leading component* of f, and define $D(f)$ to equal $D(a_1)$.

If f, g are non-zero elements of R with leading components $\lambda_1 a_1, \mu_1 b_1$, respectively, explain why fg is non-zero and that $\lambda_1 \mu_1 a_1 b_1$ is its leading component. Thereby, R is an integral domain.

Also deduce that

$$D(fg) = D(f) + D(g).$$

(f) Prove that f is a unit of R if and only if $D(f) = 0$.

Deduce that the units of R are precisely the elements of F.

(g) Show that x, y, u, v are irreducible in R.

(h) Show that x is not a prime in R.

One might be tempted to speculate on what kind of F-algebras R will tolerate a useful degree function D such as in the preceding two exercises.

APPENDIX A
Infinite Sets

From time to time in the development of the abstractions of algebra, it becomes convenient to invoke some aspects of set theory. The notions of cardinality and the widely used Zorn's lemma are particularly useful. We will now go over some of these ideas.

Leaving the intricacies of formal set theory aside, we *pretend* that we know what a set is. Very crudely, A is a set when every object x satisfies the statement "x belongs to A" (written as $x \in A$) or its negation (written as $x \notin A$), but not both. Thus the entity $B = \{A : A \text{ is a set and } A \notin A\}$, whatever it might be, is not itself a set. Indeed $B \in B$ if and only if $B \notin B$ by the definition of B. This is Bertrand Russell's famous paradox.

A.1 Zorn's Lemma

On a few occasions, for example, to see that every proper ideal in a ring is contained in a maximal ideal, Zorn's lemma is invoked. As this axiom does not seem so readily acceptable, it may be worthwhile to explore its equivalence to the somewhat more intuitive axiom of choice. For non-believers in the axiom of choice the material to follow could be difficult to swallow.

A.1.1 Choice Functions

Here is one version of the **axiom of choice**.

If X is a non-empty set and $\mathcal{P}$ is the family of all *non-empty* subsets of X, then there is a function $f : \mathcal{P} \to X$ such that $f(A) \in A$ for each A in $\mathcal{P}$.

A function such as f is called a ***choice function*** for the set X. Most mathematicians accept the axiom of choice. After all, each non-empty subset has elements in it to choose from. Surely we ought to be able to pick one element from each non-empty subset A and declare $f(A)$ to be the element so chosen. The upcoming results culminating in Zorn's lemma, Proposition A.3, are predicated on the assumption – albeit one that is not universally accepted – that every non-empty set has a choice function.

A.1.2 Chains of Subsets

We need to plough through a result about families of subsets of a given set. Not only is the result intricate, but the motivations behind it remain obscure until it is put to use. The approach we take is a modification of what is found in *Naive Set Theory* by Paul Halmos (1916–2006), a gem of a book on the theory of sets, published in 1960. In turn, Halmos attributes this approach to the set theorist Ernst Zermelo (1871–1953). First, there is some terminology that needs getting used to.

Given a set X and a family $\mathcal{F}$ of subsets of X, we shall refer to the elements of $\mathcal{F}$ as *sets in* $\mathcal{F}$, not to be confused with the subsets of $\mathcal{F}$, which will be called *sub-families of* $\mathcal{F}$.

By a ***chain inside*** $\mathcal{F}$ we mean a sub-family $\mathcal{K}$ of $\mathcal{F}$ with the property that for all pairs of sets A, B in $\mathcal{K}$, either $A \subseteq B$ or $B \subseteq A$. Chains are also known as ***totally ordered*** families of sets.

For example, X could be a ring and $\mathcal{F}$ could be the family of all proper ideals of X. Using $X = \mathbb{Z}$, an example of a chain inside the family $\mathcal{F}$ of proper ideals of $\mathbb{Z}$ would be the family of ideals of the form $\mathbb{Z}n!$, where $n = 2, 3, \ldots$. For another example of a chain, take X to be the set of real numbers $\mathbb{R}$, and $\mathcal{F}$ the family of all closed intervals $[a, b]$ inside $\mathbb{R}$. The sub-family of intervals of type $[0, b]$, where b runs through all positive reals, is a chain inside $\mathcal{F}$. The foregoing example should dispel any prejudice that a chain is merely a sequence of sets that are going up or down in discrete steps.

Proposition A.1. *Let X be a non-empty set, and $\mathcal{F}$ a family of subsets of X such that the empty set $\emptyset \in \mathcal{F}$. If every chain $\mathcal{K}$ inside $\mathcal{F}$ has the union of its sets $\bigcup_{A \in \mathcal{K}} A$ also in $\mathcal{F}$, then there is a set M in $\mathcal{F}$ with the maximality property that for any x in the complement $X \setminus M$, the augmented set $M \cup \{x\}$ is not in $\mathcal{F}$.*

Proof. In search of a contradiction, let us suppose that every set A in $\mathcal{F}$ can be augmented by some x in $X \setminus A$ while still maintaining $A \cup \{x\}$ as a set in $\mathcal{F}$. Using the axiom of choice, let f be a choice function for the set X. By our assumption, which we expect to contradict, each set A in $\mathcal{F}$ comes with the non-empty set $A^* = \{x \in X \setminus A : A \cup \{x\} \in \mathcal{F}\}$. The choice function f gives $f(A^*) \in A^*$, and thus $A \cup \{f(A^*)\} \in \mathcal{F}$. This permits the construction of the related function $\alpha : \mathcal{F} \to \mathcal{F}$, according to $\alpha(A) = A \cup \{f(A^*)\}$. Note that $A \subsetneq \alpha(A)$ for every set A in $\mathcal{F}$, and that the only sets B in $\mathcal{F}$ (or anywhere else for that matter) such that $A \subseteq B \subseteq \alpha(A)$ are A and $\alpha(A)$.

Following language adopted by Halmos, we say that a sub-family $\mathcal{T}$ of $\mathcal{F}$ is a *tower*[1] whenever:

- $\emptyset \in \mathcal{T}$,
- $\mathcal{T}$ is α-invariant, that is $\alpha(A) \in \mathcal{T}$ for all A in $\mathcal{T}$,
- for every chain $\mathcal{K}$ inside $\mathcal{T}$, the union $\bigcup_{A \in \mathcal{K}} A$ of all sets in $\mathcal{K}$ is a set in $\mathcal{T}$.

Towers exist, for instance, $\mathcal{F}$ is a tower. A bit of reflection reveals that the intersection of all possible towers is another tower. For instance, in case of the third tower requirement, if $\mathcal{K}$ is a chain inside all towers, and B is the union of the sets in $\mathcal{K}$, then B is a set in all towers. Thus we obtain a tower, hereby named $\mathcal{P}$, that is a sub-family of all other towers.

Our objective is to show that this smallest tower $\mathcal{P}$ is, in addition, a chain inside $\mathcal{F}$. Once we do that, the union B of all sets in this chain $\mathcal{P}$ will, by the nature of towers, be a set in $\mathcal{P}$.

[1] Maybe not the best choice of words, but its use is restricted to this proof.

A contradiction will then emerge, because $\alpha(B)$ becomes a set in $\mathcal{P}$ that properly contains the biggest possible set B in $\mathcal{P}$.

A set C in $\mathcal{P}$ will be called *comparable* in $\mathcal{P}$ provided that for any set A in $\mathcal{P}$, one of $A \subseteq C$ or $C \subseteq A$ holds. To be sure, $\emptyset$ is comparable in $\mathcal{P}$, and our objective is to show that all C in $\mathcal{P}$ are comparable in $\mathcal{P}$, for that is what makes a chain.

For a fixed comparable set C in $\mathcal{P}$, any set A in $\mathcal{P}$ fits into one of four disjoint categories as follows:

- $A \subsetneq C$,
- $A = C$,
- $\alpha(C) \subseteq A$,
- none of the above.

Let $\mathcal{R}$ be the sub-family of $\mathcal{P}$ consisting of those sets A in $\mathcal{P}$ that fall into one of the first three categories.

We verify that $\mathcal{R}$, which depends on the fixed comparable set C, is a tower. Well, $\emptyset \in \mathcal{R}$ because $\emptyset \subseteq C$. To check the α-invariance of $\mathcal{R}$, we test it separately for A in each of the first three categories. If $A \subsetneq C$, then, because C is comparable in $\mathcal{P}$, we get that $\alpha(A) \subseteq C$ or $C \subsetneq \alpha(A)$. Now if the latter option occurred, we would get that $A \subsetneq C \subsetneq \alpha(A)$, contrary to the construction of the mapping α. Thus $\alpha(A) \subseteq C$ when $A \subsetneq C$, which puts A into one of the first two categories. If $A = C$, then $\alpha(A) = \alpha(C)$, putting $\alpha(A)$ into the third category. If $\alpha(C) \subseteq A$, then $\alpha(C) \subsetneq \alpha(A)$, using the fact $A \subsetneq \alpha(A)$; and thus $\alpha(A)$ stays in category three in this last case. We also need to see that $\mathcal{R}$ is closed under unions of chains inside $\mathcal{R}$. Let $\mathcal{K}$ be a chain inside $\mathcal{R}$ with union $B = \bigcup_{A \in \mathcal{K}} A$. Either $A \subseteq C$ for all A in $\mathcal{K}$, or $\alpha(C) \subseteq A$ for some A in $\mathcal{K}$. The first case gives that $B \subseteq C$. The second case gives $\alpha(C) \subseteq B$. In either case $B \in \mathcal{R}$.

Having checked that $\mathcal{R}$ is a tower contained within the smallest tower $\mathcal{P}$, we conclude that $\mathcal{R} = \mathcal{P}$. This tells us that for every comparable set C in $\mathcal{P}$ and all A in $\mathcal{P}$, either

$$A \subseteq C \text{ or } \alpha(C) \subseteq A.$$

This revelation, along with the fact $C \subseteq \alpha(C)$, makes it clear that $\alpha(C)$ is comparable in $\mathcal{P}$ whenever C is comparable in $\mathcal{P}$. Thus the sub-family $\mathcal{Q}$ of sets in $\mathcal{P}$ that are comparable is α-invariant, and we saw that $\emptyset \in \mathcal{Q}$. We confirm that $\mathcal{Q}$ is a tower with a routine check of its invariance under unions of chains inside it. Take a chain $\mathcal{K}$ inside $\mathcal{Q}$ with union $B = \bigcup_{C \in \mathcal{K}} C$. If A is any set in $\mathcal{Q}$, then either $C \subseteq A$ for all sets C in $\mathcal{K}$, or $A \subseteq C$ for some set C in $\mathcal{K}$. In the first instance $B \subseteq A$, while in the second instance $A \subseteq B$. In either case B remains comparable in $\mathcal{P}$. Once more, using the minimal nature of the tower $\mathcal{P}$, we conclude that $\mathcal{Q} = \mathcal{P}$, which says that all C in $\mathcal{P}$ are comparable, as was desired to obtain a contradiction. □

A.1.3 Chain Maximality in Partially Ordered Sets

We need to enhance Proposition A.1 to a more useable form. For that, a general consideration of partial orders should make the discussion clearer.

Recall that a ***partial order*** on a non-empty set F is a relation on pairs from F, generally denoted by the standard inequality symbol $\leq$, with the following familiar properties.

- *Reflexivity*: $a \leq a$ for all a in F.
- *Transitivity*: if $a \leq b$ and $b \leq c$, then $a \leq c$.
- *Anti-symmetry*: if $a \leq b$ and $b \leq a$, then $a = b$.

The set F is then called a ***partially ordered set***. In a partially ordered set write

$$a < b \text{ to mean that } a \leq b \text{ but } a \neq b.$$

For example, take the set of integers $\mathbb{Z}$ with the usual ordering. For a wider class of examples, take any set X and let $\mathcal{F}$ be any family of subsets of X. On $\mathcal{F}$ take the relation $\leq$ to be set inclusion. That is, for subsets A, B of X let $A \leq B$ mean that $A \subseteq B$. Note that not all pairs a, b in a partially ordered set have to be related. Just look at the family $\mathcal{F}$ of all possible subsets of $\mathbb{Z}$, wherein the sets $A = \{1, 2\}$ and $B = \{2, 3\}$ are not related by inclusion.

We have seen the notion of chains involving families of subsets of a given set. This notion has an obvious extension to any partially ordered set.

Chains in Partially Ordered Sets

A subset K of a partially ordered set F is called a ***chain*** in F, alternatively also a ***totally ordered set***, provided that for any two elements a, b in K either $a \leq b$ or $b \leq a$. A chain K in a partially ordered set F is called ***maximal*** provided that for every element x in $F \setminus K$, the enlarged set $K \cup \{x\}$ is no longer a chain.

For a trivial example, take F to be the set of integers $\mathbb{Z}$ with their usual ordering. The full set $\mathbb{Z}$ itself is a maximal chain in $\mathbb{Z}$.

Another example comes from the family $\mathcal{F}$ of all ideals in the ring $\mathbb{Z}$, partially ordered by inclusion. Here take the sub-family $\mathcal{K}$ of ideals of the form $\mathbb{Z}2^n$, where $n = 0, 1, 2, \ldots$, as well as the zero ideal. This $\mathcal{K}$ is clearly a chain. To see that $\mathcal{K}$ is maximal, let J be a non-zero ideal of $\mathbb{Z}$ such that for every n either $\mathbb{Z}2^n \subseteq J$ or $J \subseteq \mathbb{Z}2^n$. The ideal J has to be principal; say $J = \mathbb{Z}m$ for some positive integer m. For every exponent $n \geq 0$, either $m \mid 2^n$ or $2^n \mid m$. Since $m > 0$ it can not be that $2^n \mid m$ for all n. Thus $m \mid 2^n$ for some n, in which case $m = 2^k$ for some k, making $J = \mathbb{Z}2^k$, a member of $\mathcal{K}$.

For an example of a non-maximal chain, consider the set $\mathcal{F}$ of all subsets of the real interval $[0, \infty)$, partially ordered by inclusion. The sub-family $\mathcal{K}$ of all closed intervals $[0, a]$, where $a \geq 0$, is clearly a chain. Now enlarge the family $\mathcal{K}$ by throwing in the interval $[0, 1)$, where 1 is excluded. Every closed interval $[0, a]$ contains $[0, 1)$ when $a \geq 1$, or is contained in $[0, 1)$ when $a < 1$. Having thus seen that $\mathcal{K} \cup \{[0, 1)\}$ remains a chain, we conclude that the chain $\mathcal{K}$ is not maximal.

Proposition A.2 (Hausdorff maximality). *Every partially ordered set contains a maximal chain.*

Proof. If X is our partially ordered set, let $\mathcal{F}$ be the family of all chains in X. This $\mathcal{F}$ is non-empty since the empty set $\emptyset$ is a chain in X, on the trivial grounds that there are no elements in $\emptyset$ to dispute the claim. The conditions of Proposition A.1 apply to this family $\mathcal{F}$. Indeed, let

$\mathcal{K}$ be a chain in $\mathcal{F}$. Caution! The chain $\mathcal{K}$ is a chain inside $\mathcal{F}$ partially ordered by set inclusion, not a chain inside X. To repeat, a subset A of X is an element of $\mathcal{K}$ if and only if A is a chain in X. We need to see that the union $U = \bigcup_{A \in \mathcal{K}} A$ is a chain in X. This is routine, if not outright tedious. Let $a, b \in U$. Thus $a \in A$ and $b \in B$ for some A, B in $\mathcal{K}$. Since $\mathcal{K}$ is a chain inside $\mathcal{F}$, we have either $A \subseteq B$ or $B \subseteq A$. Say $B \subseteq A$. Then both a and b belong to the chain A. Therefore $a \le b$ or $b \le a$, and that means $U \in \mathcal{F}$. By Proposition A.1, there is a set M belonging to $\mathcal{F}$ such that for any x in $X \setminus M$ the augmented set $M \cup \{x\}$ is no longer in $\mathcal{F}$. In other words, there is a chain M in X that is not the proper subset of another chain. Such M is our maximal chain. □

A.1.4 Choice Implies Zorn

A subset K of a partially ordered set F is ***bounded above*** provided there is an element b in F such that $a \le b$ for all a in K. An element m in F is called ***maximal*** provided the only a in F that satisfies $m \le a$ is m itself.

The next result was formulated originally in 1922 by Kazimierz Kuratowski (1896–1980), and later independently in 1935 by Max Zorn (1906–1992), not to mention the related work of Ernst Zermelo back in 1904. For some reason, the name "Zorn's lemma" has caught on, in glaring neglect of others' work. It is a very valuable result, and not only in algebra.

Proposition A.3 (Zorn's lemma). *If F is a partially ordered set in which every chain is bounded above, then F contains a maximal element.*

Proof. Using Proposition A.2 take a maximal chain K in F. By assumption, this chain has an upper bound m, and $m \in F$. Such m has to be maximal in F, for otherwise there would be some a in F such that $m < a$. Then we would be able to lengthen the chain K to the bigger chain $K \cup \{a\}$, in contradiction to the maximality of K. □

A.1.5 The Well-Ordering Principle

Having seen that the axiom of choice implies Zorn's lemma, it would be gratifying to confirm that the converse holds. In moving towards that proof we can incorporate the notion of a well-ordered set.

A partially ordered set F is said to be ***well-ordered*** provided *every* non-empty subset A of F has a first element, i.e. an element a in A such that $a \le x$ for all x in A. The partial order relation itself is then called a ***well-ordering*** of F.

Note that a well-ordered set F is already a chain, for if we take any a, b in F, the set $\{a, b\}$ has a first element. The classic example of a well-ordered set is the set $\mathbb{P}$ of positive integers $1, 2, 3, \ldots$ with the usual ordering. On the other hand, the set $\mathbb{Q} \cap (0,1)$ of rational numbers strictly between 0 and 1, although a chain with its usual ordering, is not well-ordered. However, a well-ordering can be placed on $\mathbb{Q} \cap (0,1)$ according to the following display of $\mathbb{Q}^+$ as a sequence:

$$\frac{1}{2} \prec \frac{1}{3} \prec \frac{3}{3} \prec \frac{1}{4} \prec \frac{3}{4} \prec \frac{1}{5} \prec \frac{2}{5} \prec \frac{3}{5} \prec \frac{4}{5} \prec \frac{1}{6} \prec \frac{5}{6} \prec \frac{1}{7} \prec \frac{2}{7} \prec \frac{3}{7} \prec \frac{4}{7} \cdots .$$

The pattern shown writes the rational numbers in lowest terms, orders them first according to increasing denominators, and among equal denominators orders them according to increasing numerators. Of course, the inequalities shown are not those of the usual ordering on $\mathbb{Q}^+$.

For another example, take the Cartesian product $\mathbb{P} \times \mathbb{P}$, and order it lexicographically as shown:

$$\begin{aligned} &(1,1) < (1,2) < (1,3) < \cdots < (1,n) < \cdots \\ &< (2,1) < (2,2) < (2,3) < \cdots < (2,n) < \cdots \\ &< (3,1) < (3,2) < (3,3) < \cdots < (3,n) < \cdots \\ &< \cdots . \end{aligned}$$

More precisely, we declare that $(a,b) \leq (c,d)$ when $a < c$, or $a = c$ together with $b < d$, or $(a,c) = (b,d)$. To see formally that $\mathbb{Z} \times \mathbb{Z}$ is well-ordered in this way, take a non-empty subset A of $\mathbb{Z} \times \mathbb{Z}$. Let a be the first positive integer such that $(a,b) \in A$ for some positive integer b. Then, for this a, let b be the first positive integer such that $(a,b) \in A$. The resulting (a,b) is the first element of A.

Zermelo originally used the axiom of choice to show that every set can be well-ordered. Let us now demonstrate how such well-orderings emerge from Zorn's lemma. The proof of the next result is intricate because it involves an entanglement of several kinds of partially ordered sets.

Proposition A.4. *Assuming Zorn's lemma holds, every non-empty set possesses a well-ordering.*

Proof. If F is our non-empty set, a well-ordering $\leq_A$, defined on a non-empty subset A of F, will be called an *initial well-ordering*. For instance, every two-element subset A has the obvious initial well-ordering of putting one element after the other one. Let us refer to A as the *domain* of its well-ordering $\leq_A$. Possibly the same subset A of F can be the domain of different well-orderings. While the notation $\leq_A$ does not reflect those possibilities, we live with that to avoid making notations more cumbersome. The idea is to apply Zorn's lemma to the family $\mathcal{F}$ of all initial well-orderings, and for that a partial order on $\mathcal{F}$ is needed.

Once a partial order on $\mathcal{F}$ is defined, we are confronted with the simply ghastly notation "$\leq_A \leq \leq_B$." In order to avoid such an eyesore, we shall take refuge in a more verbal notation and say that an initial well-ordering $\leq_A$ *is continued by the initial well-ordering* $\leq_B$ provided:

- $A \subseteq B$,
- the relation $x \leq_A y$ for any x, y in A is the same as the relation $x \leq_B y$, and
- for every x in A and every y in $B \setminus A$ we have $x \leq_B y$.

The second of the three items tells us that the well-ordering on the domain A is the restriction of the well-ordering on B. The third item tells us that all additional elements coming from B are ordered above those of A. Another way of stating the third condition is that A consists of all elements less than the first element of $B \setminus A$. Sometimes these conditions are summarized in the statement that A is an *initial segment* of B, and that $\mathcal{F}$ is *partially ordered by continuation.*

With the intention of applying Zorn's lemma, Proposition A.3, to this partial order on $\mathcal{F}$, take a chain $\mathcal{K}$ in $\mathcal{F}$. The chain is now totally ordered by continuation. The well-ordered sets A arising as domains of the well-orderings $\leq_A$ in $\mathcal{K}$ form a chain themselves under the ordering of set inclusion. Let C be their union, namely

$$C = \bigcup\{A : \text{ the initial well-ordering } \leq_A \text{ is in the chain } \mathcal{K}\}.$$

Now any x, y in C belong to a common domain A in this union, since the family of these domains arising from $\mathcal{K}$ is totally ordered by set inclusion. We can then declare that

$$x \leq_C y \text{ if and only if } x \leq_A y.$$

If B is another domain forming the union C and both x and y belong to B, the fact $\mathcal{K}$ is a chain means that $\leq_A$ is continued by $\leq_B$, or vice versa. Say $\leq_A$ is continued by $\leq_B$. By the nature of continuation we have $x \leq_A y$ if and only if $x \leq_B y$, and thus the definition of $\leq_C$ does not depend on the choice of domain A to which x and y belong. In other words, $\leq_C$ is defined without ambiguity, and it is easy to check that $\leq_C$ is a partial order on C.

Let us confirm that $\leq_C$ is a well-ordering of C and thus a member of $\mathcal{F}$. Take a non-empty subset D of C. The union C yields a domain A, coming from some initial well-ordering $\leq_A$ in the chain $\mathcal{K}$, such that $D \cap A$ is non-empty. The subset $D \cap A$ of A has a first element a using the initial well-ordering $\leq_A$. It turns out that this a is the first element of D using $\leq_C$. Indeed, any other element x of D belongs to some domain B such that $\leq_B$ is in the chain $\mathcal{K}$. Now either $x \in D \cap A$, or $x \in B \setminus A$. In the first case $a \leq_A x$, which, by definition of $\leq_C$, gives $a \leq_C x$. In the second case, the total ordering of the domains forces $A \subsetneq B$. By the third requirement in the definition of continuation, we automatically get $a \leq_B x$, and then, by the definition of $\leq_C$, we deduce once more that $a \leq_C x$. Hence $\leq_C$ is in $\mathcal{F}$, that is, an initial well-ordering.

Next, the definition of $\leq_C$ is such that every initial well-ordering $\leq_A$ in the chain $\mathcal{K}$ is continued by $\leq_C$. The only point here worthy of checking might be the last item in the definition of continuation. To that end take x in A and y in $C \setminus A$. Then there exists an initial well-ordering $\leq_B$ in the chain $\mathcal{K}$ such that $\leq_A$ is continued by $\leq_B$ with y in $B \setminus A$. Then $x \leq_B y$, and from the definition of $\leq_C$ we get $x \leq_C y$, as desired.

We have checked that $\leq_C$ is an upper bound in $\mathcal{F}$ for the chain $\mathcal{K}$. This makes the partially ordered set $\mathcal{F}$ susceptible to Zorn's lemma, Proposition A.3. Accordingly, there is an initial well-ordering $\leq_G$ on subset G of F such that $\leq_G$ is maximal with respect to continuation, i.e. maximal in $\mathcal{F}$. If G is a proper subset of F, take an element z from $F \setminus G$, and put that z above every element of F. More formally, let $H = G \cup \{z\}$. For x, y in G declare that $x \leq_H y$ exactly when $x \leq_G y$, and declare that $x \leq_H z$ for all x in G. We can see that $\leq_G$ is properly continued by the initial well-ordering $\leq_H$, in violation of the maximality of $\leq_G$. Thus it had to be that $G = F$, which results in the original set F being well-ordered. $\square$

A.1.6 Zorn Implies Choice

Proposition A.5. *Assuming Zorn's lemma, every non-empty set has a choice function.*

Proof. For a given non-empty set X, apply Proposition A.4, and well-order X. Now for every non-empty subset A of X define $f(A)$ to be the first element of A using the well-ordering. The resulting function f defined on the family of non-empty subsets of X is a choice function. □

We have demonstrated that the axiom of choice, Zorn's lemma and the well-ordering principle must be accepted or rejected together. We accept them (with a touch of trepidation).

A.2 The Size of Infinite Sets

There is no problem in deciding which of two finite sets is bigger. It is the one with more elements in it. However, with infinite sets such a simple answer evaporates. For example, is the set of integers $\mathbb{Z}$ bigger than the set of positive integers $\mathbb{P}$? After all, both sets are infinite. To get a meaningful answer, sets need to be compared by means of functions, which also go by their various synonyms of mappings, maps, and correspondences.

For starters we need a notion of infinite set. Let us agree that a set A is ***infinite*** whenever there exists is an injection $A \to A$ which is not a surjection. Indeed, the well accepted ***pigeon-hole principle*** says that a finite set will not tolerate such a function into itself.

A.2.1 Infinite Sets and the Positive Integers

If $\mathbb{P}$ denotes the set of positive integers, the mapping $\mathbb{P} \to \mathbb{P}$ given by $n \mapsto n+1$ is an injection of $\mathbb{P}$ into itself that is not a surjection, and thereby $\mathbb{P}$ is infinite.

We can use the axiom of choice to look at infinite sets by means of $\mathbb{P}$.

Proposition A.6. *A set A is infinite if and only if there is an injection $f : \mathbb{P} \to A$.*

Proof. In the presence of such an injection f, define the mapping $g : A \to A$ by

$$g(x) = \begin{cases} f(n+1) & \text{if } x = f(n) \text{ for some } n \text{ in } \mathbb{P} \\ x & \text{if } x \text{ is not in the range of } f. \end{cases}$$

By its very definition g is an injection which never picks up $f(1)$ in its range.

Conversely, suppose that A is a set with an injection $g : A \to A$ that is not a surjection. The following chain of proper inclusions ensues:

$$A \supset g(A) \supset g^2(A) \supset \cdots \supset g^n(A) \supset g^{n+1}(A) \supset \cdots .$$

Here, g^n stands for the n-fold iterate of the function g. Consequently, as n varies through $\mathbb{P}$, the sets $g^{n-1}(A) \setminus g^n(A)$ are non-empty and mutually disjoint.

If $\mathcal{P}$ is the family of non-empty subsets of A, the axiom of choice yields a function $\varphi : \mathcal{P} \to A$ such that $\varphi(X) \in X$ for all X in $\mathcal{P}$. In particular

$$\varphi(g^{n-1}(A) \setminus g^n(A)) \in g^{n-1}(A) \setminus g^n(A) \subseteq A.$$

From that it follows that the function

$$f : \mathbb{P} \to A \text{ given by } n \mapsto \varphi(g^{n-1}(A) \setminus g^n(A))$$

is a suitable injection. □

A.2.2 Comparing Sets

If A, B are sets, we say that the ***cardinality of A is less than or equal to the cardinality of B*** provided there exists an injection $f : A \to B$. While such a definition is accurate, the terminology is somewhat distracting because the meaning of the word "cardinality" is not given. In a full discussion of set theory the term "cardinality" has a sound meaning, but it is not necessary for our purposes to take such a large digression. As long as the term "cardinality" is not taken as a stand-alone concept, and is only used as a language for comparing sets, no issues should arise. Since injections are also called ***embeddings***, we can also say that A ***embeds*** into B when there exists an injection from A to B. When that happens we write

$$A \preceq B.$$

When $A \preceq B$ it is also common to say that B ***dominates*** A. If there is a bijection $f : A \to B$, we say that A and B have ***equal cardinality***, or that they are ***equivalent***, and write

$$A \sim B.$$

If A embeds into B but there is no bijection between A and B, we say that the cardinality of A is strictly ***less than the cardinality*** of B, or that A is ***strictly dominated*** by B. In that case we write

$$A \prec B.$$

According to these notations Proposition A.6 says that a set A is infinite if and only if $\mathbb{P} \preceq A$.

Comparing Sets by Surjections

An alternative way to establish domination is by means of surjections.

Proposition A.7. *If A, B are sets, there is a surjection $B \to A$ if and only if there is an injection $A \to B$, i.e. if and only if $A \preceq B$.*

Proof. Let $\mathcal{P}$ be the family of non-empty subsets of B and let $\varphi : \mathcal{P} \to B$ be a choice function for B. If $f : B \to A$ is a surjection, the inverse image $f^{-1}(x)$ of every x in A is a non-empty subset of B. Then the mapping

$$g : A \to B \text{ given by } x \mapsto \varphi(f^{-1}(x))$$

is an injection of A into B. For every x in A the mapping g chooses an element $g(x)$ in $f^{-1}(x)$.

For the converse, say there is an injection $g: A \to B$. Define the requisite surjection $f: B \to A$ by

$$f(y) = \begin{cases} x \text{ if } y = g(x) \text{ for some (unique) } x \text{ in } A \\ \text{any fixed value } a \text{ in } A \text{ if } y \text{ is not in the range of } g. \end{cases}$$

□

A.2.3 Countable Sets

Leaving finite sets aside, a set A is ***countable*** if there is a bijection $\mathbb{P} \to A$, i.e. $\mathbb{P} \sim A$. In other words there is a sequence of non-repeating terms that lists all elements of A.

For example, the set A of rational numbers strictly between 0 and 1 is countable, for here is a sequence which enumerates this set:

$$\frac{1}{2}, \frac{1}{3}, \frac{2}{3}, \frac{1}{4}, \frac{3}{4}, \frac{1}{5}, \frac{2}{5}, \frac{3}{5}, \frac{4}{5}, \frac{1}{6}, \frac{5}{6}, \frac{1}{7}, \frac{2}{7}, \frac{3}{7}, \frac{4}{7}, \frac{5}{7}, \frac{6}{7}, \frac{1}{8}, \frac{3}{8}, \frac{5}{8}, \frac{7}{8}, \frac{1}{9}, \frac{2}{9}, \frac{4}{9}, \frac{5}{9}, \frac{7}{9}, \frac{8}{9}, \ldots.$$

The pattern is to list all fractions between 0 and 1 written in lowest terms, in order of increasing denominators, and in order of increasing numerators when denominators are equal.

Not all infinite sets are countable. In particular, the set $\mathbb{R}$ of real numbers is uncountable. The common proof of this is the famous Cantor diagonalization argument, which shows that no function $\mathbb{P} \to \mathbb{R}$ can be surjective. In fact there is no surjection into the set $\mathbb{R}^+$ of positive reals. Indeed, any such function could be represented as a sequence of real numbers. Keeping in mind that every positive real number has a decimal expansion, tabulate the function as a sequence of decimal expansions like so:

$$\begin{aligned} x_1 &= a_1.d_{11}d_{12}d_{13}d_{14}d_{15}d_{16}\ldots \\ x_2 &= a_2.d_{21}d_{22}d_{23}d_{24}d_{25}d_{26}\ldots \\ x_3 &= a_3.d_{31}d_{32}d_{33}d_{34}d_{35}d_{36}\ldots \\ x_4 &= a_4.d_{41}d_{42}d_{43}d_{44}d_{45}d_{46}\ldots \\ x_5 &= a_5.d_{51}d_{52}d_{53}d_{54}d_{55}d_{56}\ldots \\ x_6 &= a_6.d_{61}d_{62}d_{63}d_{64}d_{65}d_{66}\ldots \\ &\vdots \end{aligned}$$

Here each a_n represents the integer part of the nth real number x_n in the sequence (written from the top on down), and d_{nm} represents the mth decimal digit of the fractional part that remains when a_n is subtracted from x_n.

Regardless of what this sequence of real numbers is, we can build a real number that never appears in the sequence. If $d_{nn} = 1$, put $b_n = 2$ and if $d_{nn} \neq 1$ put $b_n = 1$. Then take the real number x whose decimal expansion is

$$x = 0.b_1b_2b_3b_4b_5b_6\ldots.$$

This x never appears in the given sequence because it was built to disagree with each x_n at the nth decimal position. Furthermore (for those who are aware of the ambiguities of decimal expansions), the fact the expansion of x never involves 0s or 9s ensures there is no alternative expansion for x which could put x in the given sequence of x_n. We have just seen that

$$\mathbb{P} \prec \mathbb{R}.$$

The uncountability of $\mathbb{R}$ puts to rest the naive belief that nothing is bigger than infinity. In mathematics at least, there exist higher orders of infinity.

A.2.4 The Cantor–Schröder–Bernstein Theorem

The identity map on any set A reveals that $A \preceq A$, in fact $A \sim A$. Thus $\preceq$ is a reflexive relation. By composing the appropriate injections it becomes also obvious that if A embeds into B and B embeds into C, then A embeds into C. That is $A \preceq B$ and $B \preceq C$ implies $A \preceq C$. Thus $\preceq$ is transitive.

Here is something which is not so evident.

Proposition A.8 (Cantor–Schröder–Bernstein theorem). *If A, B are sets such that A embeds into B and B embeds into A, then A and B are equivalent.*

Proof. If $f : A \to B$ and $g : B \to A$ are the assumed injections, the subtlety is to come up with a bijection between these two sets.

For every subset X of A let

$$\varphi(X) = A \setminus g\,(B \setminus f(X)),$$

which is a convoluted way to build another subset of A. If X, Y are subsets of A, it is routine to see that

$$\begin{aligned} X \subseteq Y &\implies f(X) \subseteq f(Y) \\ &\implies B \setminus f(X) \supseteq B \setminus f(Y) \\ &\implies g(B \setminus f(X)) \supseteq g(B \setminus f(Y)) \\ &\implies A \setminus g(B \setminus f(X)) \subseteq A \setminus g(B \setminus f(Y)) \\ &\implies \varphi(X) \subseteq \varphi(Y). \end{aligned}$$

The set function φ preserves inclusion. This will reveal that there is a subset E of A such that $\varphi(E) = E$.

Indeed, let $\mathcal{E}$ be the family of all subsets X of A such that $X \subseteq \varphi(X)$. There is at least one such subset of A in $\mathcal{E}$, the empty set $\emptyset$. If $X \in \mathcal{E}$, then $X \subseteq \varphi(X)$, and since φ preserves inclusion we obtain $\varphi(X) \subseteq \varphi(\varphi(X))$. So $\varphi(X) \in \mathcal{E}$ whenever $X \in \mathcal{E}$. Now let

$$E = \bigcup_{X \in \mathcal{E}} X,$$

the union of all X in $\mathcal{E}$.

For every X in $\mathcal{E}$ we have $X \subseteq \varphi(X) \in \mathcal{E}$. Hence

$$E = \bigcup_{X \in \mathcal{E}} X \subseteq \bigcup_{X \in \mathcal{E}} \varphi(X) \subseteq E.$$

And so $\bigcup_{X \in \mathcal{E}} \varphi(X) = E$, which leads to

$$E = \bigcup_{X \in \mathcal{E}} \varphi(X) = \varphi(\bigcup_{X \in \mathcal{E}} X) = \varphi(E).$$

We have found a subset E of A such that

$$E = A \setminus g\,(B \setminus f(E))\,.$$

By taking complements, $A \setminus E = g\,(B \setminus f(E))$. Since g is injective, there is a bijection

$$h : A \setminus E \to B \setminus f(E),$$

given by the inverse of g acting on $A \setminus E$. And there is the obvious bijection

$$f : E \to f(E).$$

The subsets E and $A \setminus E$ partition A, while the subsets $f(E)$ and $B \setminus f(E)$ partition B. Furthermore $E \sim f(E)$ and $A \setminus E \sim B \setminus f(E)$. Glue these two bijections together to obtain the needed bijection between A and B. □

In symbolic notation we have seen that

$$A \preceq B \text{ and } B \preceq C \text{ implies } A \sim B.$$

Our relation $\preceq$ is anti-symmetric, as we might expect of something that mimics an inequality.

For a simple application of Proposition A.8 let us verify that every infinite subset B of a countable set A is again countable. Since B is infinite we have $\mathbb{P} \preceq B$, and using the inclusion mapping from B to A we see that $B \preceq A \sim \mathbb{P}$. By Cantor–Schröder–Bernstein, $\mathbb{P} \sim B$.

A.2.5 Total Ordering by Cardinality

The Cantor–Schröder–Bernstein theorem reveals that on any collection of sets, the relation $\preceq$ of domination is a partial ordering up to set equivalence. But $\preceq$ is also a total ordering.

Proposition A.9. *If A, B are non-empty sets, then either $A \preceq B$ or $B \preceq A$.*

Proof. Zorn's lemma will be applied to the following partially ordered set. Despite its abstract flavor the proof is quite routine.

If X is a non-empty subset of A, an embedding $f : X \to B$ will be called a ***partial embedding***. Let $\mathcal{P}$ be the family of all partial embeddings. If $f : X \to B$ and $g : Y \to B$ are partial embeddings declare that

$$f \leq g \text{ if and only if } X \subseteq Y \text{ and } f(x) = g(x) \text{ for all } x \text{ in } X.$$

In other words, $f \leq g$ if and only if f is an extension of g to a larger domain. That this is a partial ordering on $\mathcal{P}$ is quite evident.

In order to apply Zorn's lemma to $\mathcal{P}$, let $\mathcal{C}$ be a chain in $\mathcal{P}$, i.e. a totally ordered subset of $\mathcal{P}$. We need a partial embedding $f : D \to B$ that extends every partial embedding in $\mathcal{C}$. The obvious candidate for D is the union of all domains of all partial embeddings in the chain $\mathcal{C}$. If $x \in D$, then x lies in at least one domain X for an embedding $f : X \to B$ from the chain $\mathcal{C}$. Put $g(x) = f(x)$. The choice of partial embedding $f : X \to B$ from the chain used to define $g(x)$ does not matter because the ordering in $\mathcal{C}$ is by extension. This well-defined g extends all $f : X \to B$ from the chain $\mathcal{C}$.

Since every chain in $\mathcal{P}$ has an upper bound, Zorn's lemma yields a partial embedding $f : X \to B$ which is maximal with respect to extension. The possibility that $X \subsetneq A$ and $f(X) \subsetneq B$ does not occur. Indeed, if that happened, we could pick $a \in A \backslash X$ and $b \in B \backslash f(X)$, and make a proper extension of f by sending a to b, contrary to the maximality of f. So $X = A$ or $f(X) = B$. The first case gives $A \preceq B$. In the second case, the inverse embedding $f^{-1} : B \to X \subseteq A$ gives $B \preceq A$. □

A.2.6 The Unboundedness of Cardinality

Having seen that for every pair of sets the trichotomy

$$A \prec B, \ B \prec A \ \text{ or } \ A \sim B$$

must prevail, the question arises as to whether there is a biggest set, i.e. a set A such that $B \preceq A$ for every other set B.

If A is a set, let $\mathcal{P}(A)$ be the set of all subsets of A. This is known as the ***power set*** of A.

Proposition A.10. *If A is a set, then $A \prec \mathcal{P}(A)$.*

Proof. The mapping that sends every x in A to the singleton set $\{x\}$ in $\mathcal{P}(A)$ shows that $A \preceq \mathcal{P}(A)$. For this comparison to be strict we show that no function $f : A \to \mathcal{P}(A)$ can be surjective. If f is such a function and $x \in A$, its value $f(x)$ is a subset of A, which may contain x or not. Let

$$B = \{x \in A : x \notin f(x)\}.$$

It turns out that this set B is not in the range of f. Indeed, if $B = f(y)$ for some y in A, then the very definition of B says that $y \in B$ if and only if $y \notin B$, a contradictory situation. □

It might be worth observing that the argument in Proposition A.10 is a variation of Cantor's diagonal argument that $\mathbb{P} \prec \mathbb{R}$. To keep matters simple, let us reprove that $\mathbb{P} \prec \mathcal{P}(\mathbb{P})$. Every subset X of $\mathbb{P}$ is determined by its characteristic function. This is the sequence (x_n) which takes value 1 if $n \in X$ and value 0 if $n \notin X$. Thus $\mathcal{P}(\mathbb{P})$ is equivalent to the set $\mathcal{B}$ of all sequences of 0s and 1s, frequently called ***binary strings***.

Now, suppose that $f : \mathbb{P} \to \mathcal{B}$ is any function. We wish to show f cannot be surjective. Such f can be described by a sequence of binary strings as follows:

$$
\begin{aligned}
f(1) &= d_{11}d_{12}d_{13}d_{14}d_{15}d_{16}\ldots\\
f(2) &= d_{21}d_{22}d_{23}d_{24}d_{25}d_{26}\ldots\\
f(3) &= d_{31}d_{32}d_{33}d_{34}d_{35}d_{36}\ldots\\
f(4) &= d_{41}d_{42}d_{43}d_{44}d_{45}d_{46}\ldots\\
f(5) &= d_{51}d_{52}d_{53}d_{54}d_{55}d_{56}\ldots\\
f(6) &= d_{61}d_{62}d_{63}d_{64}d_{65}d_{66}\ldots\\
&\vdots
\end{aligned}
$$

Here the d_{nj} consist of 0s and 1s. Take the string $b = b_1b_2b_3b_4\ldots$ given by $b_j = 1$ if $d_{jj} = 0$ and $b_j = 0$ if $d_{jj} = 1$. Clearly b is not equal to any $f(j)$ because b was built to disagree with $f(j)$ in the jth term. Thus $\mathbb{P} \prec \mathcal{B} \sim \mathcal{P}(\mathbb{P})$.

If A_n is the subset of $\mathbb{P}$ corresponding to the string $f(n)$ and B is the set corresponding to the string b, we can see that $x \in B$ if and only if $x \notin A_n$. So the preceding diagonalization method is nothing but the repetition of the key step in the proof of Proposition A.10.

A curiosity emerges from Proposition A.10. Namely, there is no such thing as "the set of all sets." For if such a set, say Ω, existed, then $\mathcal{P}(\Omega)$ the family of all subsets of Ω would be a subset of Ω. Hence we would have $\Omega \prec \mathcal{P}(\Omega) \preceq \Omega$, an obvious contradiction.

A.2.7 A Bit of Cardinal Arithmetic

Having seen from Proposition A.10 that every set comes with a set whose cardinality is higher, a number of facts regarding the comparison of sets remain to be examined. The required arguments are quite straightforward if somewhat fussy.

Proposition A.11. *The set of positive integers is equivalent to its Cartesian product with itself, i.e.* $\mathbb{P} \sim \mathbb{P} \times \mathbb{P}$.

Proof. It suffices to tabulate $\mathbb{P} \times \mathbb{P}$ as one sequence. The infinite table on display picks up all of $\mathbb{P} \times \mathbb{P}$:

$$
\begin{array}{l}
(1,1)\ (1,2)\ (1,3)\ (1,4)\ (1,5)\ (1,6)\ldots\\
(2,1)\ (2,2)\ (2,3)\ (2,4)\ (2,5)\ (2,6)\ldots\\
(3,1)\ (3,2)\ (3,3)\ (3,4)\ (3,5)\ (3,6)\ldots\\
(4,1)\ (4,2)\ (4,3)\ (4,4)\ (4,5)\ (4,6)\ldots\\
(5,1)\ (5,2)\ (5,3)\ (5,4)\ (5,5)\ (5,6)\ldots\\
(6,1)\ (6,2)\ (6,3)\ (6,4)\ (6,5)\ (6,6)\ldots\\
\vdots
\end{array}
$$

Now travel along the cross diagonals to build one sequence that picks up all of this table as follows:

$$(1,1)\ (2,1)\ (1,2)\ (3,1)\ (2,2)\ (1,3)\ (4,1)\ (3,2)\ (2,3)\ (1,4)$$
$$(5,1)\ (4,2)\ (3,3)\ (2,4)\ (1,5)\ (6,1)\ (5,2)\ (4,3)\ (3,4)\ (2,5)\ (1,6)\ldots.$$

In case the pattern is not clear, we first put down the pair that adds up to 2, then the two pairs that add up to 3, then the three pairs that add up to 4, then the four pairs that add up to 5, then the five pairs that add up to 6, and so on. □

A variation of the above result is that a Cartesian product of a countable set with a countable set is a countable set. By iteration of the argument, any Cartesian product $A_1 \times A_2 \times \cdots \times A_k$ of countable sets is countable. The evident details are left to the reader.

Proposition A.12. *If A_n is a sequence of countable infinite sets, then its union $\bigcup_{n=1}^{\infty} A_n$ remains countable.*

Proof. Each A_n can be enumerated as a single sequence:

$$x_{n1}, x_{n2}, x_{n3}, \ldots, x_{nm}, \ldots.$$

The function $f : \mathbb{P} \times \mathbb{P} \to \bigcup_{n=1}^{\infty} A_n$ given by $(n,m) \mapsto x_{nm}$ is a surjection. Since the union of the A_n is infinite, and using Proposition A.11, we obtain

$$\mathbb{P} \preceq \bigcup_{n=1}^{\infty} A_n \preceq \mathbb{P} \times \mathbb{P} \sim \mathbb{P}.$$

By Cantor–Schröder–Bernstein, $\mathbb{P} \sim \bigcup_{n=1}^{\infty} A_n$. □

Our goal is to extend Proposition A.11 to all infinite sets, but the loss of a sequential approach makes it necessary to take some preliminary steps.

Proposition A.13. *If A is a finite set and B is an infinite set, then $A \cup B \sim B$.*

Proof. No harm results from supposing A is disjoint from B. There is an embedding $f : \mathbb{P} \to B$, and A can be listed as $a_1, a_2, \ldots, a_m$. The requisite bijection $g : B \to A \cup B$ is given by

$$g(x) = \begin{cases} f(j) & \text{if } x \text{ is one of the } a_j \text{ in } A \\ f(k+m) & \text{if } x = f(k) \text{ in the range of } f \\ x & \text{if } x \text{ is not in the range of } f. \end{cases}$$

Figuratively speaking, bump the sequence $f(n)$ over by m steps in order to make room for the a_j. □

Proposition A.14. *If B is an infinite set and $T = \{0,1\}$ is a two element set, then $B \times T \sim B$.*

Proof. The proof is a bit of a long story, but at its heart are the details needed to apply Zorn's lemma.

For each subset X of B, a bijection $f : X \to X \times T$ will be called a ***partial bijection*** with domain X. Since B is infinite such partial bijections do exist. Indeed, $\mathbb{P} \preceq B$, which means that B contains a countable subset X. If

$$x_1, x_2, x_3, \ldots, x_n, \ldots$$

is an enumeration of X, create an enumeration of $X \times T$ according to

$$(x_1, 0), (x_1, 1), (x_2, 0), (x_2, 1), (x_3, 0), (x_3, 1), \ldots, (x_n, 0), (x_n, 1), \ldots.$$

The partial bijection comes from matching up $(x_n, 0)$ with x_{2n-1} and $(x_n, 1)$ with x_{2n}.

Let $\mathcal{P}$ be the non-empty family of partial bijections. If $f : X \to X \times T$ and $g : Y \to Y \times T$ are partial bijections, declare $f \leq g$ provided g extends f, i.e. $X \subseteq Y$ and $f(x) = g(x)$ whenever $x \in X$.

This partial ordering on $\mathcal{P}$ satisfies the condition of Zorn's lemma. This can be readily visualized, but here are the details. Let $\mathcal{C}$ be a chain in $\mathcal{P}$. To get an upper bound for $\mathcal{C}$ let A be the union of the domains X for every partial bijection in the chain $\mathcal{C}$. For each x in A there is a partial bijection $g : X \to X \times T$ in the chain $\mathcal{C}$ such that $x \in X$. Define $f : A \to A \times T$ by $f(x) = g(x)$. Since $\mathcal{C}$ is a chain the choice of partial bijection g in $\mathcal{C}$ has no effect on the definition of $f(x)$.

Check that f is a surjection. Well, if $z \in A \times T$, then $z \in X \times T$ for some X which is the domain of a partial bijection $g : X \to X \times T$ taken from $\mathcal{C}$. Thus there is an x in X such that $g(x) = z$. And by the definition of f we also have $f(x) = z$.

To check that f is an injection, suppose $f(x) = f(y)$ for some x, y in A. Since $\mathcal{C}$ is a chain, there is some $g : X \to X \times T$ in $\mathcal{C}$, such that X contains both x and y. And then $g(x) = g(y)$ by the definition of f. Because g is a bijection, conclude that $x = y$.

Clearly f is an upper bound for the chain $\mathcal{C}$. Using Zorn's lemma let $f : X \to X \times T$ be a maximal partial bijection.

It turns out that the complement $B \setminus X$ is finite. To see that, suppose $B \setminus X$ is infinite. Then there is an infinite countable subset K inside $B \setminus X$. The set $K \times T$ lies in the complement $(B \times T) \setminus (X \times T)$. Also, there is a partial bijection $K \to K \times T$ as argued at the start of this proof. Now, splice the above bijection together with the partial bijection $f : X \to X \times T$ to get a partial bijection $X \cup K \to (X \cup K) \times T$ that properly extends f. Since this contradicts the maximality of f, the complement $B \setminus X$ must be finite.

The complement $(B \times T) \setminus (X \times T)$ is also finite. Then, according to Proposition A.13,

$$B = X \cup (B \setminus X) \sim X \sim X \times T \sim (X \times T) \cup ((B \times T) \setminus (X \times T)) = B \times T,$$

as desired. □

Proposition A.15. *If A, B are sets, B is infinite and $A \preceq B$, then $A \cup B \sim B$.*

Proof. Clearly $B \preceq A \cup B$. For the reverse domination a surjection $B \to A \cup B$ will suffice. If $T = \{0, 1\}$, there is a bijection $f : B \to B \times T$, due to Proposition A.14. And there is a surjection $g : B \to A$ since $A \preceq B$. This leads to the surjection

$$h : B \times T \to A \cup B \text{ given by } (y, 0) \mapsto y \text{ in } B \text{ and } (y, 1) \mapsto g(y) \text{ in } A.$$

A suitable surjection of B onto $A \cup B$ is provided by $h \circ f$. □

We come to our principal result on cardinal arithmetic.

Proposition A.16. *If A is an infinite set, then $A \sim A \times A$.*

Proof. We imitate somewhat the proof of Proposition A.14.

For any subset X of A a bijection $X \to X \times X$ will be called a *partial bijection*. If X is a countable subset of A, then there does exist a partial bijection $X \to X \times X$ because of Proposition A.11. Thereby the set $\mathcal{P}$ of all partial bijections on A is non-empty.

If $f : X \to X \times X$ and $g : Y \to Y \times Y$ are partial bijections declare that $f \leq g$ provided g is an extension of f. This gives a partial ordering on $\mathcal{P}$ for which every chain $\mathcal{C}$ in $\mathcal{P}$ is bounded. The way to get an upper bound for $\mathcal{C}$ is identical to the method used in Proposition A.14, so it might be safe to omit its repetition.

According to Zorn's lemma, let $f : X \to X \times X$ be a maximal partial bijection. Now, either $X \sim A$ or $X \prec A$. We need to rule out the latter possibility.

To that end suppose $X \prec A$. Since $A = X \cup (A \setminus X)$, it follows from Proposition A.15 that $(A \setminus X) \sim A$. And then $X \prec A \setminus X$. Thus there is a set Y contained in $A \setminus X$ such that $Y \sim X$. With this Y we will build a bijection $X \cup Y \to (X \cup Y) \times (X \cup Y)$ that will extend f.

The product $(X \cup Y) \times (X \cup Y)$ is the disjoint union of the four sets

$$X \times X,\ X \times Y,\ Y \times X,\ Y \times Y.$$

Because of the partial bijection f we have

$$Y \sim X \sim X \times X \sim X \times Y \sim Y \times X \sim Y \times Y.$$

By Proposition A.15 there exists a bijection $g : Y \to (X \times Y) \cup (Y \times X) \cup (Y \times Y)$. Glue this to the bijection $f : X \to X \times X$ in the obvious way, to get a bijection

$$X \cup Y \to (X \times X) \cup (X \times Y) \cup (Y \times X) \cup (Y \times Y) = (X \cup Y) \times (X \cup Y).$$

Since the maximality of f does not tolerate such an extension, $X \sim A$. And then

$$A \sim X \sim X \times X \sim A \times A,$$

as desired. □

Proposition A.17. *If $A \preceq B$ and B is infinite, then $A \times B \sim B$.*

Proof. Clearly $A \times B \preceq B \times B \sim B$, and the comparison $B \preceq A \times B$ should be evident. Cantor–Schröder–Bernstein then gives the desired equivalence. □

Proposition A.18. *If A is an infinite set, then any product $A \times A \times \cdots \times A$, of finitely many copies of A, is equivalent to A.*

Proof. By Proposition A.17

$$A \sim A \times A \sim A \times (A \times A) = A \times A \times A \sim A \times (A \times A \times A) = A \times A \times A \times A,$$

and so on. □

Here is the last of our results on cardinal arithmetic.

Proposition A.19. *If A is an infinite set and A^n denotes the n-fold Cartesian product of A with itself, then $\bigcup_{n=1}^{\infty} A^n \sim A$.*

Proof. Since A is infinite, $\mathbb{P} \preceq A$. Then by Proposition A.17

$$A \sim \mathbb{P} \times A = \bigcup_{n=1}^{\infty} (\{n\} \times A).$$

From Proposition A.18 each of the sets $\{n\} \times A$ in the preceding disjoint union is equivalent to A^n. And so $A \sim \bigcup_{n=1}^{\infty} A^n$. □

The Cardinality of Polynomial Rings

By way of an application, let us observe that if A is an infinite ring and $A[X]$ is the ring of polynomials in one variable over A, then $A[X] \sim A$. To see this, note that for each positive integer n, there is an embedding from the set of polynomials of degree $n-1$ into A^n. Just send the polynomial of degree $n-1$ to its n-tuple of coefficients. (The zero polynomial goes into A^1 along with the other constant polynomials.) Glue these embeddings together to get the obvious embedding $A[X] \to \bigcup_{n=1}^{\infty} A^n$. And so we get

$$A[X] \preceq \bigcup_{n=1}^{\infty} A^n \sim A.$$

Obviously $A \preceq A[X]$. By Cantor–Schröder–Bernstein it follows that $A[X] \sim A$.

While we are at it, we might as well record the cardinality of $A[X]$ when A is finite. In this case $A \preceq \mathbb{Z}$. Lift this embedding to polynomials in the obvious way to get $A[X] \preceq \mathbb{Z}[X]$. As we just saw $\mathbb{Z}[X] \sim \mathbb{Z}$. Thus $A[X] \preceq \mathbb{Z} \sim \mathbb{P}$. Since $A[X]$ is infinite we also have $\mathbb{P} \preceq A[X]$. By Cantor–Schröder–Bernstein we conclude that $A[X] \sim \mathbb{P}$. In other words, $A[X]$ is countable when A is finite.

EXERCISES

1. Show that $\mathbb{R} \sim \mathcal{P}(\mathbb{P})$.

   Hint. Every real number has a binary expansion, and Cantor–Schröder–Bernstein can be useful.

A.3 The Algebraic Closure of a Field

It might be fitting to close our story with an application of Zorn's lemma and the idea of cardinality to a standard construction in field theory.

Definition A.20. A field K is called ***algebraically closed*** provided every non-constant polynomial f in $K[X]$ splits in $K[X]$. If a field K is both algebraically closed and an algebraic extension of a field F, we say that K is an ***algebraic closure*** of F.

To say that K is algebraically closed is equivalent to saying that every non-constant f in $K[X]$ has a root in K. To see this, say every non-constant polynomial in $K[X]$ has root in K, and let f in $K[X]$ be such a non-constant polynomial. Now f has a linear factor in $K[X]$ because it has a root in K. Then the other factor of f has a root in K, which leads to another linear factor in $K[X]$ of f, and so on until f splits into linear factors.

Another point of view is that a field K is algebraically closed if and only if the only algebraic field extension of K is K itself. Indeed, if K has a proper algebraic extension L, then the minimal polynomial of any element in $L \setminus K$ will fail to split over K. Conversely, if there is a polynomial in $K[X]$ that fails to split over K, then the splitting field of such a polynomial will be a proper algebraic extension of K.

An Algebraic Closure of $\mathbb{Q}$

By the fundamental theorem of algebra, Proposition 7.3, the field $\mathbb{C}$ of complex numbers is an algebraically closed extension of $\mathbb{Q}$. Due to Proposition A.22 the set $\mathbb{A}$ of all elements in $\mathbb{C}$ that are algebraic over $\mathbb{Q}$ is an algebraic field extension of $\mathbb{Q}$. This $\mathbb{A}$ is called the field of ***algebraic numbers***. Though it is not immediate to see, $\mathbb{C}$ contains numbers that are not algebraic over $\mathbb{Q}$. Such numbers are called ***transcendental***. For instance, π and the natural base e of the exponential function are transcendental. Thus $\mathbb{A}$ is a proper subfield of $\mathbb{C}$. We will also see shortly that $\mathbb{A}$ is proper inside $\mathbb{C}$ because of their differing cardinalities.

Yet, $\mathbb{A}$ remains algebraically closed. Indeed, every non-constant f in $\mathbb{A}[X]$ splits over $\mathbb{C}$, meaning that the splitting field K of f can be taken inside $\mathbb{C}$. Since K is an algebraic extension of $\mathbb{A}$ and $\mathbb{A}$ is an algebraic extension of $\mathbb{Q}$, it follows from Proposition 6.19 that K is an algebraic extension of $\mathbb{Q}$. Therefore $K = \mathbb{A}$, over which f splits.

The Cardinality of Algebraic Extensions

There is a hard bound on the cardinality of an algebraic extension of a field.

Proposition A.21. *If a field K is an algebraic extension of a field F, then $K \sim F$ when F is infinite, and $K \preceq \mathbb{P}$ when F is finite.*

Proof. Use the cardinality of the ring of polynomials $F[X]$ as an intermediary. As noted after Proposition A.19, $F[X] \sim F$ when F is infinite, and $F[X] \sim \mathbb{P}$ when F is finite.

Let $N = F[X] \setminus F$, the set of non-constant polynomials. It follows readily from Cantor–Schröder–Bernstein that $N \sim F$ when F is infinite and $N \sim \mathbb{P}$ when F is finite. For each f in N, let R_f be the finite set of roots of f that lie in K.

Since K is algebraic over F this means that $K = \bigcup_{f \in N} R_f$. Clearly $\mathbb{P} \times N$ equals the disjoint union $\bigcup_{f \in N}(\mathbb{P} \times \{f\})$. Since each R_f is finite, there is a surjection $\mathbb{P} \times \{f\} \to R_f$ (many in fact). Glue these surjections together to make a surjection

$$\mathbb{P} \times N = \bigcup_{f \in N} (\mathbb{P} \times \{f\}) \to \bigcup_{f \in N} R_f = K.$$

Using Proposition A.7 deduce that $K \preceq \mathbb{P} \times N$. By Proposition A.17, $\mathbb{P} \times N \sim N$. And so $K \preceq N$.

If F is infinite, $N \sim F$ yields $K \preceq F$, and since obviously $F \preceq K$, it follows that $K \sim F$.

If F is finite, $N \sim \mathbb{P}$ yields $K \preceq \mathbb{P}$. □

Because of the above, the field $\mathbb{A}$ of complex numbers that are algebraic over $\mathbb{Q}$ is not all of $\mathbb{C}$. This is because $\mathbb{A} \sim \mathbb{Q}$. Since $\mathbb{Q} \sim \mathbb{P}$, this implies that $\mathbb{A}$ is countable, while $\mathbb{C}$ is not countable. Indeed, $\mathbb{C}$ contains the set of positive reals, which the Cantor diagonal argument shows to be uncountable.[2]

A Synthesis of Algebraic Closures

The field $\mathbb{A}$ of complex numbers that are algebraic over $\mathbb{Q}$ is an algebraic closure of $\mathbb{Q}$. The existence of an algebraic closure for $\mathbb{Q}$ is assured by the fundamental theorem of algebra. To obtain an algebraic closure for an arbitrary field demands a synthesis out of the blue.

Beneath everything else, a field K is a set. The ring operations $+$ and $\cdot$ on K that define the field are not specified in its notation. Yet different addition and multiplication operations on the same set K will make different fields, which are not revealed by the notation K. Fortunately, no confusion results most of the time. In the theorem that follows we shall continue to denote a field K using only the name K of its underlying set, and continue to hope no confusion results.

Proposition A.22. *Every field has an algebraic closure.*

Proof. Let F be the field in question.

Choose a set S that contains the underlying set F, and so big that S is uncountable and that $F \prec S$. The set S has enough room to contain a subset E which can be endowed with ring operations such that E becomes an algebraic field extension of F as well as an algebraically closed field.

Let $\mathcal{F}$ be the family of fields L such that

- the underlying set L sits between F and S, i.e. $F \subseteq L \subseteq S$,

[2] Given that there exists a preponderance of transcendental numbers, the general difficulty of showing that a specific number, such as π for instance, is actually transcendental seems puzzling. Some facetiously refer to this as the problem of "finding hay in a haystack."

- L is a field extension of F, i.e. the ring operations on F are those of L restricted to F, and
- the extension L is algebraic over F.

Partially order $\mathcal{F}$ by extension. That is, for K, L in $\mathcal{F}$, declare that $K \leq L$ when L is a field extension of K. That is, $K \subseteq L$ and the ring operations on K are the restrictions of the ring operations on L. At the very least, $F \in \mathcal{F}$.

This partial ordering on $\mathcal{F}$ by extension satisfies the requirements of Zorn's lemma. Indeed, the union of any chain of fields in $\mathcal{F}$ is a field in $\mathcal{F}$ that extends all fields in the chain. The routine verifications are omitted.

Thus we have a maximal E in $\mathcal{F}$. This means that $F \subseteq E \subseteq S$ as underlying sets, and as a field E is an algebraic extension of F. Furthermore, the only algebraic field extension L of E such that $E \subseteq L \subseteq S$ is E. Since E has no proper algebraic extensions that sit inside S, it seems that E ought to be algebraically closed. But we require that E shall have no proper algebraic extension, regardless of where the underlying set sits. There might conceivably be a proper algebraic extension L of E that sits outside of S.

So, let L be a proper algebraic extension of E. By Proposition 6.19, L is algebraic over F as well. Proposition A.21 reveals that $F \sim E \sim L$ when F is infinite and $E \preceq L \preceq \mathbb{P}$ when F is finite. In either case the set S was taken so big that $E \prec S$. From the fact $S = E \cup (S \setminus E)$, it follows from Proposition A.15 that $(S \setminus E) \sim S$. Since

$$L \setminus E \preceq L \sim F \prec S \sim S \setminus E \text{ when } F \text{ is infinite and}$$

$$L \setminus E \preceq L \preceq \mathbb{P} \prec S \sim S \setminus E \text{ when } F \text{ is finite,}$$

an injection $\varphi : L \setminus E \to S \setminus E$ is available. We have the disjoint union $L = E \cup (L \setminus E)$ as well as the disjoint union $E \cup \varphi(L \setminus E)$, a set which we call K. By extending φ to be the identity on E, we get a bijection $\varphi : L \to K$. Here L is a field extension of E, while K is a subset of S containing the field E. The diagram below captures the situation.

$$\begin{array}{ccc} & & S \\ & & | \\ L & \xrightarrow{\varphi} & K \\ | & & | \\ E & = & E \\ | & & | \\ F & = & F \end{array} \qquad \text{and} \qquad \varphi(L \setminus E) \subseteq S \setminus E,\ \varphi_{|E} = \mathrm{id},\ \varphi(L) = K.$$

Transport the ring operations on L over to K by means of the bijection φ. More explicitly, if $\varphi(x), \varphi(y) \in K$, define $\varphi(x) + \varphi(y)$ to be $\varphi(x + y)$, and do the same for the multiplication. Obviously K is now a field custom built to be isomorphic to L. Since φ is the identity on E, the ring operations transported from L over to K by φ do not alter the original ring operations on E. So, the new field K inside the set S is a proper algebraic extension of E. This K contradicts the maximality of E. And so E had to be algebraically closed. $\square$

An Issue with Sets

One may rightly wonder why the fuss in the proof above with the set S. Why not just let $\mathcal{F}$ be the family of *all* algebraic field extensions of F regardless of where the underlying set sits? The answer is that there is no such family. If there were such a family $\mathcal{F}$, here is the kind of anomaly that could result.

Pretend $\mathcal{F}$ is the family of all algebraic field extensions of $\mathbb{Z}_2$. Take the four element set $A = \{0, 1, \emptyset, \mathcal{F}\}$, which is an unusual assemblage to be sure. On this four element set impose the addition and multiplication tables corresponding to the field with four elements, putting $0, 1$ into their usual roles. Now A is an algebraic extension of the field $\mathbb{Z}_2$. Then we deduce something bizarre:

$$\cdots \mathcal{F} \in A \in \mathcal{F} \in A \in \mathcal{F} \in A \in \mathcal{F} \in A \in \cdots$$

This should lead us to question what is meant by a set. But that is for another day.

The Uniqueness of Algebraic Closures

As might be expected, two algebraic closures of the same field are isomorphic. For this reason it is common to speak of *the* algebraic closure of a field.

Proposition A.23. *If L, M are algebraic closures of a common subfield F, then there is an isomorphism $\varphi : L \to M$ which fixes F.*

Proof. We need to use Zorn's lemma on a suitable partially ordered set.

For each intermediate subfield E between F and L we shall refer to an F-map $\sigma : E \to M$ as a *partial F-embedding of L into M*. The goal is to show there is a partial F-embedding for which $E = L$. (In which case the embedding would no longer be partial.)

Let $\mathcal{P}$ be the family of all partial F-embeddings into M. Certainly $\mathcal{P}$ is non-empty because the inclusion map $F \to M$ is in $\mathcal{P}$. Given partial F-embeddings $\sigma : E \to M, \tau : K \to M$ in $\mathcal{P}$ declare that

$$(\sigma : E \to M) \preceq (\tau : K \to M)$$

provided

$$E \text{ is a subfield of } K \text{ and } \sigma(a) = \tau(a) \text{ for all } a \text{ in } E.$$

More informally, partial F-embeddings are partially ordered by extension. It is routine to see that this puts a partial order on $\mathcal{P}$.

Zorn's lemma can be applied to this partial order provided every chain in $\mathcal{P}$ has an upper bound in $\mathcal{P}$. Suppose, as j runs through some indexing set J, that $\{\sigma_j : E_j \to M\}_{j \in J}$ is a chain in $\mathcal{P}$ (i.e. a totally ordered subset of partial F-embeddings). The E_j form a chain of subfields of L, all of them containing F. Let $E = \cup_{j \in J} E_j$. Because the E_j form a chain, their union E is a subfield of L. Indeed, if $x, y \in E$, there must be an E_j to which both x and y belong. Hence $x \pm y$ and xy are in E_j, and thereby in E. If $x \neq 0$, so is x^{-1} in E_j, and thus in E.

A partial F-embedding $\sigma : E \to M$ can be defined like so. If $a \in E$ take an E_j such that $a \in E_j$, and define $\sigma(a)$ to be $\sigma_j(a)$. This definition does not depend on which E_j was chosen to contain a. To see that, suppose a is also in E_i for some i in J. Since we have a chain in $\mathcal{P}$, one of E_j or E_i is a subfield of the other. Say $E_i \subseteq E_j$. By the definition of the partial order $\preceq$, we must have $\sigma_i(a) = \sigma_j(a)$, as claimed. The final verifications that $\sigma : E \to M$ is a partial F-embedding is a routine matter, which we can omit. By the very definition of σ it is also clear that $(\sigma_j : E_j \to M) \preceq (\sigma : E \to M)$ for all j in J. We have found an upper bound in $\mathcal{P}$ for the given chain.

Using Zorn's lemma let $\varphi : E \to M$ be a maximal partial F-embedding. It now turns out that $E = L$. To see that suppose, on the contrary, that there exists an element a in $L \setminus E$. Let g in $E[X]$ be the minimal polynomial of a over the field E. Such g is available because a is algebraic over F and thus over E.

The isomorphism $\varphi : E \to \varphi(E)$ has a natural extension to a ring isomorphism $\varphi : E[X] \to \varphi(E)[X]$ defined by $\sum a_j X^j \mapsto \sum \varphi(a_j) X^j$. This is described in Definition 6.30. The name of the natural extension can stay φ without harm. If $f \in E[X]$, we call its value $\varphi(f)$ in $\varphi(E)[X]$ the φ-transfer of f.

Since g is irreducible over E, its φ-transfer $\varphi(g)$ is irreducible over $\varphi(E)$. Given that M is algebraically closed, take a root b of $\varphi(g)$ inside M. By Proposition 6.31 the isomorphism $\varphi : E \to \varphi(E)$ has an extension to an isomorphism $\psi : E[a] \to \varphi(E)[b]$. The following diagram keeps track of the situation.

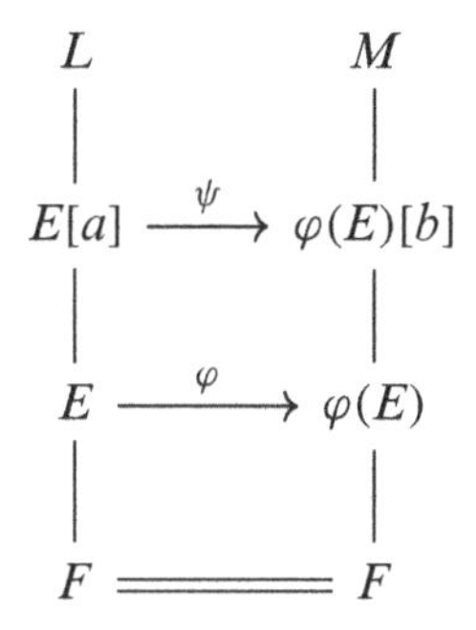

From the assumption that $L \setminus E$ is non-empty, we built a partial F-embedding ψ into M which properly extends the maximal F-embedding $\varphi : E \to M$. To evade this contradiction the domain E of the maximal φ must be all of L.

To conclude we check that $\varphi : L \to M$ is onto M. It suffices to see that the field $\varphi(L)$ is algebraically closed. For then M, being algebraic over $\varphi(L)$, will equal $\varphi(L)$, as desired. So, let f be a non-constant polynomial in $\varphi(L)[X]$. Its φ^{-1}-transfer $\varphi^{-1}(f)$ in $L[X]$ splits over the algebraically closed L. The φ-transfer of $\varphi^{-1}(f)$ is the original f, which must split over $\varphi(E)$ because the transfer mapping $\varphi : L[X] \to \varphi(L)[X]$ is a ring isomorphism that extends the field isomorphism $\varphi : L \to \varphi(L)$. $\square$

EXERCISES

1. Show that the algebraic closure of a finite field is infinite and countable.
2. Show that the algebraic closure of $\mathbb{Q}$ is countable.
3. If K is an algebraic extension of a field K and A is an algebraic closure of K, show that A is an algebraic closure of F.
4. For each power p^n of a prime p there is exactly one field $\mathbb{F}_{p^n}$ of size p^n, up to isomorphism. Furthermore we take F_{p^d} as a subfield of F_{p^n} whenever $d \mid n$.

   Show that for any two finite fields of characteristic p there is a finite field that contains them both.

   Show that the algebraic closure of $\mathbb{Z}_p$ is the union $\bigcup_{n=1}^{\infty} \mathbb{F}_{p^n}$ taken with a suitable addition and multiplication operation.

Index

Printed in the United States
by Baker & Taylor Publisher Services